" Etraterrestrial Life "

Edited by

Paul F. Kisak

Virginia, USA

Visit our website at: https://www.createspace.com/6039320

Printed in The United States of America
First Trade Edition: 2016

10 9 8 7 6 5 4 3 2 1

Black & White on White paper
532 pages

ISBN-13: 978-1523809394
ISBN-10: 1523809396

Library of Congress Control Number :

Virginia, USA

Extraterrestrial Life

" An Overview of Theories & Possibilities "

Edited by Paul F. Kisak

Contents

Chapter 1

Extraterrestrial life

For other uses, see Astrobiology.

Extraterrestrial life[n 1] is life that does not originate from Earth. It is also called **alien life**, or, if it is a sentient and/or relatively complex individual, an "extraterrestrial" or "alien" (or, to avoid confusion with the legal sense of "alien", a "**space alien**"). These as-yet-hypothetical life forms range from simple bacteria-like organisms to beings with civilizations far more advanced than humanity. Although many scientists expect extraterrestrial life to exist, so far no unambiguous evidence for its existence exists.[1][2]

The science of extraterrestrial life is known as exobiology. The science of astrobiology also considers life on Earth as well, and in the broader astronomical context. Meteorites that have fallen to Earth have sometimes been examined for signs of microscopic extraterrestrial life. In 2015, "remains of biotic life" were found in 4.1 billion-year-old rocks in Western Australia, when the young Earth was about 400 million years old.[3][4] According to one of the researchers, "If life arose relatively quickly on Earth ... then it could be common in the universe."[3]

Since the mid-20th century, there has been an ongoing search for signs of extraterrestrial intelligence, from radios used to detect possible extraterrestrial signals, to telescopes used to search for potentially habitable extrasolar planets.[5] It has also played a major role in works of science fiction. Over the years, science fiction works, especially Hollywood's involvement, has increased the public's interest in the possibility of extraterrestrial life. Some encourage aggressive methods to try to get in contact with life in outer space, whereas others argue that it might be dangerous to actively call attention to Earth.[6][7]

1.1 Background

Alien life, such as microorganisms, has been hypothesized to exist in the Solar System and throughout the universe. This hypothesis relies on the vast size and consistent physical laws of the observable universe. According to this argument, made by scientists such as Carl Sagan and Stephen Hawking,[8] it would be improbable for life *not* to exist somewhere other than Earth.[9][10] This argument is embodied in the Copernican principle, which states that Earth does not occupy a unique position in the Universe, and the mediocrity principle, which states that there is nothing special about life on Earth.[11] The chemistry of life may have begun shortly after the Big Bang, 13.8 billion years ago, during a habitable epoch when the universe was only 10–17 million years old.[12][13][14] Life may have emerged independently at many places throughout the universe. Alternatively, life may have formed less frequently, then spread—by meteoroids, for example—between habitable planets in a process called panspermia.[15][16] In any case, complex organic molecules may have formed in the protoplanetary disk of dust grains surrounding the Sun before the formation of Earth.[17] According to these studies, this process may occur outside Earth on several planets and moons of the Solar System and on planets of other stars.[17]

Since the 1950s, scientists have argued the idea that "habitable zones" around stars are the most likely places to find life. Numerous discoveries in these zones since 2007 have generated estimations of frequencies of Earth-like planets —in terms of composition— numbering in the many billions[18] though as of 2013, only a small number of planets have been discovered in these zones.[19] Nonetheless, on November 4, 2013, astronomers reported, based on *Kepler* space mission data, that there could be as many as 40 billion Earth-sized planets orbiting in the habitable zones of Sun-like stars and red dwarfs in the Milky Way.[20][21] 11 billion of which may be orbiting Sun-like stars.[22] The nearest such planet may be 12 light-years away, according to the scientists.[20][21] Astrobiologists have also considered a "follow the energy" view of potential habitats.[23][24]

1.2 Possible basis

1.2.1 Biochemistry

Main articles: Biochemistry, Hypothetical types of biochemistry and Water and life

It is often hypothesized that life forms elsewhere in the universe would, like life on Earth, be based on carbon chemistry and rely on liquid water. Life forms based on ammonia (rather than water) have been suggested, though this solvent appears less suitable than water. It is also conceivable that there are forms of life whose solvent is a liquid hydrocarbon, such as methane, ethane or propane.[25]

About 29 chemical elements are thought to play an active positive role in living organisms on Earth.[26] About 95% of this living matter is built upon only six elements: carbon, hydrogen, nitrogen, oxygen, phosphorus and sulfur. These six elements form the basic building blocks of virtually all life on Earth, whereas most of the remaining elements are found only in trace amounts.[27] The unique characteristics of carbon made it unlikely that any other element could replace carbon, even on another planet, to generate the biochemistry necessary for life. The carbon atom has the unique ability to make four strong chemical bonds with other atoms, including other carbon atoms. These covalent bonds have a direction in space, so that carbon atoms can form the skeletons of complex 3-dimensional structures with definite architectures such as nucleic acids and proteins. Carbon forms more compounds than all other elements combined. The great versatility of the carbon atom makes it the element most likely to provide the bases—even exotic ones—to the chemical composition of life on other planets.[28]

Life on Earth requires water as its solvent in which biochemical reactions take place. Sufficient quantities of carbon and the other elements along with water, may enable the formation of living organisms on other planets with a chemical make-up and temperature range similar to that of Earth.[29] Terrestrial planets such as Earth are formed in a process that allows for the possibility of having compositions similar to Earth's.[30] The combination of carbon, hydrogen and oxygen in the chemical form of carbohydrates (e.g. sugar) can be a source of chemical energy on which life depends, and can provide structural elements for life. Plants derive energy through the conversion of light energy into chemical energy via photosynthesis. Life, as currently recognized, requires carbon in both reduced (methane derivatives) and partially oxidized (carbon oxides) states. Nitrogen is needed as a reduced ammonia derivative in all proteins, sulfur as a derivative of hydrogen sulfide in some necessary proteins, and phosphorus oxidized to phosphates in genetic material and in energy transfer.

1.3 Planetary habitability in the Solar System

See also: Planetary habitability and Natural satellite habitability

Some bodies in the Solar System have been suggested as having the potential for an environment that could host extraterrestrial life, particularly those with possible subsurface oceans.[31] Should life be discovered elsewhere in the Solar System, astrobiologists suggest that it will more likely be in the form of extremophile microorganisms.

Some speculate that Mars may possess niche subsurface environments where microbial life might exist.[32][33][34] A subsurface marine environment on Jupiter's moon Europa might be the most likely habitat in the Solar System, outside Earth, for extremophile microorganisms.[35][36][37]

The panspermia hypothesis proposes that life elsewhere in the Solar System may have a common origin. If extraterrestrial life was found on another body in the Solar System, it could have originated from Earth just as life on Earth may have been seeded from elsewhere (exogenesis). The first known mention of the term 'panspermia' was in the writings of the 5th century BC Greek philosopher Anaxagoras.[38] In the nineteenth century it was again revived in modern form by several scientists, including Jöns Jacob Berzelius (1834),[39] Kelvin (1871),[40] Hermann von Helmholtz (1879)[41] and, somewhat later, by Svante Arrhenius (1903).[42] Sir Fred Hoyle (1915–2001) and Chandra Wickramasinghe (born 1939) are important proponents of the hypothesis who further contended that lifeforms continue to enter Earth's atmosphere, and may be responsible for epidemic outbreaks, new diseases, and the genetic novelty necessary for macroevolution.[43]

Directed panspermia concerns the deliberate transport of microorganisms in space, sent to Earth to start life here, or sent from Earth to seed new stellar systems with life. The Nobel prize winner Francis Crick, along with Leslie Orgel proposed that seeds of life may have been purposely spread by an advanced extraterrestrial civilization,[44] but considering an early "RNA world" Crick noted later that life may have originated on Earth.[45]

1.3.1 Mars

Main article: Life on Mars

Life on Mars has been long speculated. Liquid water is widely thought to have existed on Mars in the past, and now can occasionally be found as low-volume liquid brines in shallow Martian soil.[46] The origin of the potential biosignature of methane observed in Mars atmosphere is unexplained, although abiotic hypotheses have also been proposed.[47] By July 2008, laboratory tests aboard NASA's *Phoenix* Mars lander identified water in a surface soil sample. Photographs from the Mars Global Surveyor from 2006 showed evidence of recent (i.e. within 10 years) flows of a liquid on Mars's frigid surface.[48] There is evidence that Mars had a warmer and wetter past: dried-up river beds, polar ice caps, volcanos, and minerals that form in the presence of water have all been found. Nevertheless, present conditions on Mars subsurface may support life.[49][50] Evidence obtained by the *Curiosity* rover studying Aeolis Palus, Gale Crater in 2013, strongly suggest an ancient freshwater lake that could have been a hospitable environment for microbial life.[51][52]

Current studies on Mars by the *Curiosity* and *Opportunity* rovers are now searching for evidence of ancient life, including a biosphere based on autotrophic, chemotrophic and/or chemolithoautotrophic microorganisms, as well as ancient water, including fluvio-lacustrine environments (plains related to ancient rivers or lakes) that may have been habitable.[53][54][55][56] The search for evidence of habitability, taphonomy (related to fossils), and organic carbon on Mars is now a primary NASA objective.[53]

1.3.2 Ceres

Ceres, the only dwarf planet in the asteroid belt, was confirmed by the Herschel Space Observatory to have a thin water vapor atmosphere.[57][58] Frost on the surface may also have been detected in the form of bright spots.[59][60][61] The presence of water on Ceres has led to speculation that life may be possible there.[62][63][64]

1.3.3 Jupiter system

Jupiter

Carl Sagan and others in the 1960s and 1970s computed conditions for hypothetical microorganisms living in the atmosphere of Jupiter,[65] however, the intense radiation and other conditions do not appear to permit encapsulation and molecular biochemistry, so life there is thought unlikely.[66] In contrast, some of Jupiter's moons may have habitats capable of sustaining life. Scientists have indications that heated subsurface oceans of liquid water may exist deep under the crusts of the three outer Galilean moons —Europa,[35][67][68] Ganymede,[69][70][71][72][73] and Callisto.[74][75][76] The EJSM/Laplace mission is planned to determine the habitability of these environments.

Europa

Main article: Life on Europa

Jupiter's moon Europa has been subject to speculation about the existence of life due to the strong possibility of a liquid water ocean beneath its ice surface.[35][37] Hydrothermal vents on the bottom of the ocean, if they exist, may warm the ice and could be capable of supporting multicellular microorganisms.[78] It is also possible that Europa could support aerobic macrofauna using oxygen created by cosmic rays impacting its surface ice.[79]

The case for life on Europa was greatly enhanced in 2011 when it was discovered that vast lakes exist within Europa's thick, icy shell. Scientists found that ice shelves surrounding the lakes appear to be collapsing into them, thereby providing a mechanism through which life-forming chemicals created in sunlit areas on Europa's surface could be transferred to its interior.[80][81]

On December 11, 2013, NASA reported the detection of "clay-like minerals" (specifically, phyllosilicates), often associated with organic materials, on the icy crust of Europa.[82] The presence of the minerals may have been the result of a collision with an asteroid or comet according to the scientists.[82] The *Europa Multiple-Flyby Mission*, which would assess the habitability of Europa, is planned for launch in 2025.[83][84] Europa's subsurface ocean is considered the best target for the discovery of life.[35][37]

1.3.4 Saturn system

Titan and Enceladus have been speculated to host possible habitats supportive of life.

Enceladus

Enceladus, a moon of Saturn, has some of the conditions for life, including geothermal activity and water vapor, as well as possible under-ice oceans heated by tidal effects.[85][86] The *Cassini–Huygens* probe detected carbon, hydrogen, nitrogen and oxygen—all key elements for supporting life—during its 2005 flyby through one of Enceladus's geysers spewing ice and gas. The temperature and density of the plumes indicate a warmer, watery source beneath the surface.[47]

Titan

Main article: Life on Titan

Titan, the largest moon of Saturn, is the only known moon in the Solar System with a significant atmosphere. Data from the *Cassini–Huygens* mission refuted the hypothesis of a global hydrocarbon ocean, but later demonstrated the existence of liquid hydrocarbon lakes in the polar regions—the first stable bodies of surface liquid discovered outside Earth.[87][88][89] Analysis of data from the mission has uncovered aspects of atmospheric chemistry near the surface that are consistent with—but do not prove—the hypothesis that organisms there if present, could be consuming hydrogen, acetylene and ethane, and producing methane.[90][91][92]

1.3.5 Small Solar System bodies

Small Solar System bodies have also been speculated to host habitats for extremophiles. Fred Hoyle and Chandra Wickramasinghe have proposed that microbial life might exist on comets and asteroids.[93][94][95][96]

1.4 Scientific search

The scientific search for extraterrestrial life is being carried out both directly and indirectly.

1.4.1 Direct search

Scientists search for biosignatures within the Solar System by studying planetary surfaces and examining meteorites.[12][13] Some claim to have identified evidence that microbial life has existed on Mars.[97][98][99][100][101][102] An experiment on the two Viking Mars landers reported gas emissions from heated Martian soil samples that some scientists argue are consistent with the presence of living microorganisms.[103] Lack of corroborating evidence from other experiments on the same samples, indicates that a non-biological reaction is a more likely hypothesis.[103][104][105][106] In 1996, a controversial report stated that structures resembling nanobacteria were discovered in a meteorite, ALH84001, formed of rock ejected from Mars.[97][98]

In February 2005, NASA scientists reported that they may have found some evidence of present life on Mars.[107] The two scientists, Carol Stoker and Larry Lemke of NASA's Ames Research Center, based their claim on methane signatures found in Mars's atmosphere resembling the methane production of some forms of primitive life on Earth, as well as on their own study of primitive life near the Rio Tinto river in Spain. NASA officials soon distanced NASA from the scientists' claims, and Stoker herself backed off from her initial assertions.[108] Though such methane findings are still debated, support among some scientists for the existence of life on Mars seems to be growing.[109]

In November 2011, NASA launched the Mars Science Laboratory that landed the *Curiosity* rover on Mars. It is designed to assess the past and present habitability on Mars using a variety of scientific instruments. The rover landed on Mars at Gale Crater in August 2012.[110][111]

The Gaia hypothesis stipulates that any planet with a robust population of life will have an atmosphere in chemical disequilibrium, which is relatively easy to determine from a distance by spectroscopy. However, significant advances in the ability to find and resolve light from smaller rocky worlds near their star are necessary before such spectroscopic methods can be used to analyze extrasolar planets. To that effect, the Carl Sagan Institute was founded in 2014 dedicated to the atmospheric characterization of exoplanets in circumstellar habitable zones.[112][113] Planetary spectroscopic data will be obtained from telescopes like WFIRST and E-ELT.[114]

In August 2011, findings by NASA, based on studies of meteorites found on Earth, suggests DNA and RNA components (adenine, guanine and related organic molecules), building blocks for life as we know it, may be formed extraterrestrially in outer space.[115][116][117] In October 2011, scientists reported that cosmic dust contains complex organic matter ("amorphous organic solids with a mixed aromatic-aliphatic structure") that could be created naturally, and rapidly, by stars.[118][119][120] One of the scientists suggested that these compounds may have been related to the development of life on Earth and said that, "If this is the case, life on Earth may have had an easier time getting started as these organics can serve as basic ingredients for life."[118]

In August 2012, and in a world first, astronomers at Copenhagen University reported the detection of a specific sugar molecule, glycolaldehyde, in a distant star system. The molecule was found around the protostellar binary *IRAS 16293-2422*, which is located 400 light years from Earth.[121][122] Glycolaldehyde is needed to form ribonucleic acid, or RNA, which is similar in function to DNA. This finding suggests that complex organic molecules may form in stellar systems prior to the formation of planets, eventually arriving on young planets early in their formation.[123]

1.4.2 Indirect search

Projects such as SETI are monitoring the galaxy for electromagnetic interstellar communications from civilizations on other worlds.[124][125] If there is an advanced extraterrestrial civilization, there is no guarantee that it is transmitting radio communications in the direction of Earth or that this information could be interpreted as such by humans. The length of time required for a signal to travel across the vastness of space means that any signal detected, would come from the distant past.[126]

Another biosignature of intelligence, would be to find an abundance of anomalies of heavy elements in the light spectrum of a star, which would take place if the star was used as a repository for nuclear waste material.[127]

1.4.3   Extrasolar planets

Main article: Extrasolar planets
See also: List of planetary systems
Some astronomers search for extrasolar planets that may be conducive to life, narrowing the search to terrestrial planets within the habitable zone of their star.[128][129] Since 1992,over 2000 exoplanets have been discovered (2030 planets in 1288 planetary systems including 502 multiple planetary systems as of 11 December 2015).[130] The extrasolar planets so far discovered range in size from that of terrestrial planets similar to Earth's size to that of gas giants larger than Jupiter.[130] The number of observed exoplanets is expected to increase greatly in the coming years.[131]

The *Kepler space telescope* has also detected a few thousand[132][133] candidate planets,[134][135] of which about 11% may be false positives.[136] There is at least one planet on average per star.[137]

About 1 in 5 Sun-like stars[lower-alpha 1] have an "Earth-sized"[lower-alpha 2] planet in the habitable zone,[lower-alpha 3] with the nearest expected to be within 12 light-years distance from Earth.[138][139] Assuming 200 billion stars in the Milky Way,[lower-alpha 4] that would be 11 billion potentially habitable Earth-sized planets in the Milky Way, rising to 40 billion if red dwarfs are included.[22] The rogue planets in the Milky Way possibly number in the trillions.[140]

The nearest known exoplanet, if confirmed, would be Alpha Centauri Bb, located 4.37 light-years from Earth in the southern constellation of Centaurus.[141] As of March 2014, the least massive planet known is PSR B1257+12 A, which is about twice the mass of the Moon. The most massive planet listed on the NASA Exoplanet Archive is DENIS-P J082303.1-491201 b,[142][143] about 29 times the mass of Jupiter, although according to most definitions of a planet, it is too massive to be a planet and may be a brown dwarf instead. Almost all of the planets detected so far are within the Milky Way, but there have also been a few possible detections of extragalactic planets. The study of planetary habitability also considers a wide range of other factors in determining the suitability of a planet for hosting life.[5]

1.5   The Drake equation

Main article: Drake equation

In 1961, University of California, Santa Cruz, astronomer and astrophysicist Frank Drake devised the Drake equation as a way to stimulate scientific dialogue at a meeting on the search for extraterrestrial intelligence (SETI).[144] The Drake equation is a probabilistic argument used to estimate the number of active, communicative extraterrestrial civilizations in the Milky Way galaxy. The equation is best understood not as an equation in the strictly mathematical sense, but to summarize all the various concepts which scientists must contemplate when considering the question of life elsewhere.[145] The Drake equation is:

$$N = R_* \cdot f_p \cdot n_e \cdot f_l \cdot f_i \cdot f_c \cdot L$$

where:

> N = the number of civilizations in our galaxy with which radio-communication might be possible (i.e. which are on our current past light cone);

and

> R^* = the average rate of star formation in our galaxy
>
> f_p = the fraction of those stars that have planets
>
> n_e = the average number of planets that can potentially support life per star that has planets
>
> f_l = the fraction of planets that could support life that actually develop life at some point
>
> f_i = the fraction of planets with life that actually go on to develop intelligent life (civilizations)

fc = the fraction of civilizations that develop a technology that releases detectable signs of their existence into space

L = the length of time for which such civilizations release detectable signals into space[146]

The Drake equation has proved controversial since several of its factors are extremely uncertain and are entirely based on conjecture. Thus the equation cannot be used to draw firm conclusions of any kind.[147] This has led critics to label the equation a guesstimate, or even meaningless.

Based on observations from the Hubble Space Telescope, there are at least 125 billion galaxies in the observable universe. It is estimated that at least ten percent of all Sun-like stars have a system of planets,[148] i.e. there are 6.25×10^{18} stars with planets orbiting them in the observable universe. Even if we assume that only one out of a billion of these stars have planets supporting life, there would be some 6.25×10^{9} (billion) life-supporting planetary systems in the observable universe.

The apparent contradiction between high estimates of the probability of the existence of extraterrestrial civilizations and the lack of evidence for such civilizations, is known as the Fermi paradox.[149]

1.6 Cultural impact

1.6.1 Cosmic pluralism

Main article: Cosmic pluralism

Cosmic pluralism, the plurality of worlds, or simply pluralism, describes the philosophical belief in numerous "worlds" in addition to Earth, which might harbor extraterrestrial life. Before the development of the heliocentric theory and a recognition that our Sun is just one of many stars,[150] the notion of pluralism was largely mythological and philosophical.[152][153] With the scientific and Copernican revolutions, and later, during the later, during the Enlightenment, cosmic pluralism became a mainstream notion, supported by the likes of Bernard le Bovier de Fontenelle in his 1686 work *Entretiens sur la pluralité des mondes*.[154] Pluralism was also championed by philosophers such as John Locke, Giordano Bruno and astronomers such as William Herschel. The astronomer Camille Flammarion promoted the notion of cosmic pluralism in his 1862 book *La pluralité des mondes habités*.[155] None of these notions of pluralism were based on any specific observation or scientific information.

1.6.2 Early modern period

There was a dramatic shift in thinking initiated by the invention of the telescope and the Copernican assault on geocentric cosmology. Once it became clear that Earth was merely one planet amongst countless bodies in the universe, the theory of extraterrestrial life started to become a topic in the scientific community. The best known early-modern proponent of such ideas was the Italian philosopher Giordano Bruno, who argued in the 16th century for an infinite universe in which every star is surrounded by its own planetary system. Bruno wrote that other worlds "have no less virtue nor a nature different to that of our earth" and, like Earth, "contain animals and inhabitants".[156]

In the early 17th century, the Czech astronomer Anton Maria Schyrleus of Rheita mused that "if Jupiter has (...) inhabitants (...) they must be larger and more beautiful than the inhabitants of Earth, in proportion to the [characteristics] of the two spheres".[157]

In Baroque literature such as *The Other World: The Societies and Governments of the Moon* by Cyrano de Bergerac, extraterrestrial societies are presented as humoristic or ironic parodies of earthly society. The didactic poet Henry More took up the classical theme of the Greek Democritus in "Democritus Platonissans, or an Essay Upon the Infinity of Worlds" (1647). In "The Creation: a Philosophical Poem in Seven Books" (1712), Sir Richard Blackmore observed: "We may pronounce each orb sustains a race / Of living things adapted to the place". With the new relative viewpoint that the Copernican revolution had wrought, he suggested "our world's sunne / Becomes a starre elsewhere". Fontanelle's

"Conversations on the Plurality of Worlds" (translated into English in 1686) offered similar excursions on the possibility of extraterrestrial life, expanding, rather than denying, the creative sphere of a Maker.

The possibility of extraterrestrials remained a widespread speculation as scientific discovery accelerated. William Herschel, the discoverer of Uranus, was one of many 18th–19th-century astronomers who believed that the Solar System is populated by alien life. Other luminaries of the period who championed "cosmic pluralism" included Immanuel Kant and Benjamin Franklin. At the height of the Enlightenment, even the Sun and Moon were considered candidates for extraterrestrial inhabitants.

1.6.3   19th century

Speculation about life on Mars increased in the late 19th century, following telescopic observation of apparent Martian canal—which soon, however, turned out to be optical illusions.[158] Despite this, in 1895, American astronomer Percival Lowell published his book *Mars*, followed by *Mars and its Canals* in 1906, proposing that the canals were the work of a long-gone civilization.[159] This idea led British writer H. G. Wells to write the novel *The War of the Worlds* in 1897, telling of an invasion by aliens from Mars who were fleeing the planet's desiccation.

Spectroscopic analysis of Mars's atmosphere began in earnest in 1894, when U.S. astronomer William Wallace Campbell showed that neither water nor oxygen was present in the Martian atmosphere.[160] By 1909 better telescopes and the best perihelic opposition of Mars since 1877 conclusively put an end to the canal hypothesis.

The science fiction genre, although not so named during the time, develops during the late 19th century. Jules Verne's *Around the Moon* (1870) features a discussion of the possibility of life on the Moon, but with the conclusion that it is barren. Stories involving extraterrestrials are found in e.g. Garrett P. Serviss's *Edison's Conquest of Mars* (1898), an unauthorized sequel to *The War of the Worlds* by H. G. Wells was published in 1897 which stands at the beginning of the popular idea of the "Martian invasion" of Earth prominent in 20th-century pop culture.

1.6.4   20th century

See also: Space exploration
Most unidentified flying objects or UFO sightings[161] can be readily explained as sightings of Earth-based aircraft, known astronomical objects, or as hoaxes.[162] Nonetheless, a certain fraction of the public believe that UFOs might actually be of extraterrestrial origin, and, indeed, the notion has had influence on popular culture.

The possibility of extraterrestrial life on the Moon was ruled out in the 1960s, and during the 1970s it became clear that most of the other bodies of the Solar System do not harbor highly developed life, although the question of primitive life on bodies in the Solar System remains an open question.

1.6.5   Recent history

The failure so far of the SETI program to detect an intelligent radio signal after decades of effort has at least partially dimmed the prevailing optimism of the beginning of the space age. Notwithstanding, belief in extraterrestrial beings continues to be voiced in pseudoscience, conspiracy theories, and in popular folklore, notably "Area 51" and legends. It has become a pop culture trope given less-than-serious treatment in popular entertainment.

In the words of SETI's Frank Drake, "All we know for sure is that the sky is not littered with powerful microwave transmitters".[163] Drake noted that it is entirely possible that advanced technology results in communication being carried out in some way other than conventional radio transmission. At the same time, the data returned by space probes, and giant strides in detection methods, have allowed science to begin delineating habitability criteria on other worlds, and to confirm that at least other planets are plentiful, though aliens remain a question mark. The Wow! signal, detected in 1977 by a SETI project, remains a subject of speculative debate.

In 2000, geologist and paleontologist Peter Ward and astrobiologist Donald Brownlee published a book entitled *Rare Earth: Why Complex Life is Uncommon in the Universe*.[164] In it, they discussed the Rare Earth hypothesis, in which they claim that Earth-like life is rare in the universe, whereas microbial life is common. Ward and Brownlee are open to the idea of evolution on other planets that is not based on essential Earth-like characteristics (such as DNA and carbon).

Theoretical physicist Stephen Hawking in 2010 warned that humans should not try to contact alien life forms. He warned that aliens might pillage Earth for resources. "If aliens visit us, the outcome would be much as when Columbus landed in America, which didn't turn out well for the Native Americans", he said.[165] Jared Diamond has expressed similar concerns.[166]

In November 2011, the White House released an official response to two petitions asking the U.S. government to acknowledge formally that aliens have visited Earth and to disclose any intentional withholding of government interactions with extraterrestrial beings. According to the response, "The U.S. government has no evidence that any life exists outside our planet, or that an extraterrestrial presence has contacted or engaged any member of the human race."[167][168] Also, according to the response, there is "no credible information to suggest that any evidence is being hidden from the public's eye."[167][168] The response noted "odds are pretty high" that there may be life on other planets but "the odds of us making contact with any of them—especially any intelligent ones—are extremely small, given the distances involved."[167][168]

In 2013, the exoplanet Kepler-62f was discovered, along with Kepler-62e and Kepler-62c. A related special issue of the journal Science, published earlier, described the discovery of the exoplanets.[169]

On 17 April 2014, the discovery of the Earth-size exoplanet Kepler-186f, 500 light-years from Earth, was publicly announced;[170] it is the first Earth-size planet to be discovered in the habitable zone and it has been hypothesized that there may be liquid water on its surface.

On 13 February 2015, scientists (including Geoffrey Marcy, Seth Shostak, Frank Drake and David Brin) at a convention of the American Association for the Advancement of Science, discussed Active SETI and whether transmitting a message to possible intelligent extraterrestrials in the Cosmos was a good idea;[171][172] one result was a statement, signed by many, that a "worldwide scientific, political and humanitarian discussion must occur before any message is sent".[173]

On 20 July 2015, Stephen Hawking, British physicist, and Yuri Milner, Russian billionaire, along with the SETI Institute, announced a well-funded effort, called the Breakthrough Initiatives, to expand efforts to search for extraterrestrial life. The group contracted the services of the 100-meter Robert C. Byrd Green Bank Telescope in West Virginia in the United States and the 64-meter Parkes Telescope in New South Wales, Australia.[174]

1.7 See also

Searches for extraterrestrial life

- Allen Telescope Array
- Astrobiology
- Communication with extraterrestrial intelligence
- Nexus for Exoplanet System Science
- SETI

Subjects

- Extraterrestrial liquid water
- Planetary protection
- Potential cultural impact of extraterrestrial contact

Hypotheses

- Aurelia and Blue Moon

- Fermi paradox

- Kardashev scale

- Metalaw

- Rare Earth hypothesis

- Sentience quotient

- Zoo hypothesis

1.8   Notes

[1] Where "extraterrestrial" is derived from the Latin *extra* ("beyond", "not of") and *terrestris* ("of Earth", "belonging to Earth").

[1] For the purpose of this 1 in 5 statistic, "Sun-like" means G-type star. Data for Sun-like stars wasn't available so this statistic is an extrapolation from data about K-type stars

[2] For the purpose of this 1 in 5 statistic, Earth-sized means 1–2 Earth radii

[3] For the purpose of this 1 in 5 statistic, "habitable zone" means the region with 0.25 to 4 times Earth's stellar flux (corresponding to 0.5–2 AU for the Sun).

[4] About 1/4 of stars are GK Sun-like stars. The number of stars in the galaxy is not accurately known, but assuming 200 billion stars in total, the Milky Way would have about 50 billion Sun-like (GK) stars, of which about 1 in 5 (22%) or 11 billion would be Earth-sized in the habitable zone. Including red dwarfs would increase this to 40 billion.

1.9   References

[1] Davies, Paul (18 November 2013). "Are We Alone in the Universe?". *New York Times*. Retrieved 20 November 2013.

[2] Pickrell, John (4 September 2006). "Top 10: Controversial pieces of evidence for extraterrestrial life". *New Scientist*. Retrieved 2011-02-18.

[3] Borenstein, Seth (19 October 2015). "Hints of life on what was thought to be desolate early Earth". *Excite* (Yonkers, NY: Mindspark Interactive Network). Associated Press. Retrieved 2015-10-20.

[4] Bell, Elizabeth A.; Boehnike, Patrick; Harrison, T. Mark; et al. (19 October 2015). "Potentially biogenic carbon preserved in a 4.1 billion-year-old zircon" (PDF). *Proc. Natl. Acad. Sci. U.S.A.* (Washington, D.C.: National Academy of Sciences). doi:10.1073/pnas.1517557112. ISSN 1091-6490. Retrieved 2015-10-20. Early edition, published online before print.

[5] Overbye, Dennis (January 6, 2015). "As Ranks of Goldilocks Planets Grow, Astronomers Consider What's Next". *New York Times*. Retrieved January 6, 2015.

[6] BBC News – Scientists in US are urged to seek contact with aliens

[7] Baum, Seth; Haqq-Misra, Jacob; Domagal-Goldman, Shawn. "Would Contact with Extraterrestrials Benefit or Harm Humanity? A Scenario Analysis". *Acta Astronautica, 2011, 68 (11–12):2014–2129*, April 22, 2011, accessed August 18, 2011.

[8] Weaver, Rheyanne. "Ruminations on other worlds". *State Press*. Retrieved 10 March 2014.

[9] Brad Steiger, John White, eds. (1986). *Other Worlds, Other Universes*. Health Research Books. p. 3. ISBN 0-7873-1291-6.

[10] Filkin, David; Hawking, Stephen W. (1998). *Stephen Hawking's universe: the cosmos explained*. Art of Mentoring Series. Basic Books. p. 194. ISBN 0-465-08198-3.

[11] Rauchfuss, Horst (2008). *Chemical Evolution and the Origin of Life*. T. N. Mitchell. Springer. ISBN 3-540-78822-0

[12] Loeb, Abraham (October 2014). "The Habitable Epoch of the Early Universe". *International Journal of Astrobiology* **13** (04): 337–339. arXiv:1312.0613. Bibcode:2014IJAsB..13..337L. doi:10.1017/S1473550414000196. Retrieved 15 December 2014.

[13] Loeb, Abraham (2 December 2013). "The Habitable Epoch of the Early Universe" (PDF). *Arxiv*. arXiv:1312.0613v3. Retrieved 15 December 2014.

[14] Dreifus, Claudia (2 December 2014). "Much-Discussed Views That Go Way Back – Avi Loeb Ponders the Early Universe, Nature and Life". *New York Times*. Retrieved 3 December 2014.

[15] Rampelotto, P.H. (2010). "Panspermia: A Promising Field Of Research" (PDF). Astrobiology Science Conference. Retrieved 3 December 2014. External link in |publisher= (help)

[16] Gonzalez, Guillermo; Richards, Jay Wesley (2004). *The privileged planet: how our place in the cosmos is designed for discovery*. Regnery Publishing. pp. 343–345. ISBN 0-89526-065-4.

[17] Moskowitz, Clara (29 March 2012). "Life's Building Blocks May Have Formed in Dust Around Young Sun". Space.com. Retrieved 30 March 2012.

[18] Choi, Charles Q. (21 March 2011). "New Estimate for Alien Earths: 2 Billion in Our Galaxy Alone". Space.com. Retrieved 2011-04-24.

[19] Torres, Abel Mendez (April 26, 2013). "Ten potentially habitable exoplanets now". *Habitable Exoplanets Catalog*. University of Puerto Rico. Retrieved April 29, 2013.

[20] Overbye, Dennis (November 4, 2013). "Far-Off Planets Like the Earth Dot the Galaxy". *New York Times*. Retrieved November 5, 2013.

[21] Petigura, Eric A.; Howard, Andrew W.; Marcy, Geoffrey W. (October 31, 2013). "Prevalence of Earth-size planets orbiting Sun-like stars". *Proceedings of the National Academy of Sciences of the United States of America*. arXiv:1311.6806. Bibcode:2013PNAS..11019273P. doi:10.1073/pnas.1319909110. Retrieved November 5, 2013.

[22] Khan, Amina (November 4, 2013). "Milky Way may host billions of Earth-size planets". *Los Angeles Times*. Retrieved November 5, 2013.

[23] Hoehler, Tori M.; Amend, Jan P.; Shock, Everett L. (2007). "A "Follow the Energy" Approach for Astrobiology". *Astrobiology* **7** (6): 819–823. Bibcode:2007AsBio...7..819H. doi:10.1089/ast.2007.0207. ISSN 1531-1074.

[24] Jones, Eriita G.; Lineweaver, Charles H. (2010). "To What Extent Does Terrestrial Life "Follow The Water"?". *Astrobiology* **10** (3): 349–361. Bibcode:2010AsBio..10..349J. doi:10.1089/ast.2009.0428. ISSN 1531-1074.

[25] Committee on the Limits of Organic Life in Planetary Systems, Committee on the Origins and Evolution of Life, National Research Council; The Limits of Organic Life in Planetary Systems; The National Academies Press. 2007; p 74

[26] Ultratrace minerals. Authors: Nielsen, Forrest H. USDA, ARS Source: Modern nutrition in health and disease / editors, Maurice E. Shils ... et al.. Baltimore : Williams & Wilkins, c1999., p. 283-303. Issue Date: 1999 URI:

[27] Mix, Lucas John (2009). *Life in space: astrobiology for everyone*. Harvard University Press. p. 76. ISBN 0-674-03321-3. Retrieved 2011-08-08.

[28] Norman H. Horowitz, To Utopia and Back: The Search for Life in the Solar System 1986 W.H. Freeman & Co, NY ISBN 0-7167-1765-4 ISBN 0-7167-1766-2

[29] Pace, Norman R. (January 20, 2001). "The universal nature of biochemistry". *Proceedings of the National Academy of Sciences of the United States of America* **98** (3): 805–808. Bibcode:2001PNAS...98..805P. doi:10.1073/pnas.98.3.805. PMC 33372. PMID 11158550.

[30] Bond, Jade C.; O'Brien, David P.; Lauretta, Dante S. (June 2010). "The Compositional Diversity of Extrasolar Terrestrial Planets. I. In Situ Simulations". *The Astrophysical Journal* **715** (2): 1050–1070. arXiv:1004.0971. Bibcode:2010ApJ...715.1050B. doi:10.1088/0004-637X/715/2/1050.

[31] Dyches, Preston; Chou, Felcia (7 April 2015). "The Solar System and Beyond is Awash in Water". *NASA*. Retrieved 8 April 2015.

12   CHAPTER 1. EXTRATERRESTRIAL LIFE

[32] Summons, Roger E.; Amend, Jan P.; Bish, David; Buick, Roger; Cody, George D.; Des Marais, David J.; Dromart, Gilles; Eigenbrode, Jennifer L.; et al. (2011). "Preservation of Martian Organic and Environmental Records: Final Report of the Mars Biosignature Working Group". *Astrobiology* **11** (2): 157–81. Bibcode:2011AsBio..11..157S. doi:10.1089/ast.2010.0506. PMID 21417945. There is general consensus that extant microbial life on Mars would probably exist (if at all) in the subsurface and at low abundance.

[33] Michalski, Joseph R.; Cuadros, Javier; Niles, Paul B.; Parnell, John; Deanne Rogers, A.; Wright, Shawn P. (2013). "Groundwater activity on Mars and implications for a deep biosphere". *Nature Geoscience* **6** (2): 133–8. Bibcode:2013NatGe...6..133M. doi:10.1038/ngeo1706.

[34] "Habitability and Biology: What are the Properties of Life?". *Phoenix Mars Mission*. The University of Arizona. Retrieved 2013-06-06. If any life exists on Mars today, scientists believe it is most likely to be in pockets of liquid water beneath the Martian surface.

[35] Tritt, Charles S. (2002). "Possibility of Life on Europa". Milwaukee School of Engineering. Archived from the original on 9 June 2007. Retrieved 10 August 2007.

[36] Jeffrey S. Kargel, Jonathan Z. Kaye, James W. Head, III; et al. (2000). "Europa's Crust and Ocean: Origin, Composition, and the Prospects for Life" (PDF). *Icarus* (Planetary Sciences Group, Brown University) **148** (1): 226–265. Bibcode:2000Icar..148..226K. doi:10.1006/icar.2000.6471.

[37] Schulze-Makuch, Dirk; and Irwin, Louis N. (2001). "Alternative Energy Sources Could Support Life on Europa" (PDF). *Departments of Geological and Biological Sciences, University of Texas at El Paso*. Archived from the original (PDF) on 3 July 2006. Retrieved 21 December 2007.

[38] Margaret O'Leary (2008) Anaxagoras and the Origin of Panspermia Theory, iUniverse publishing Group, # ISBN 978-0-595-49596-2

[39] Berzelius (1799–1848), J. J. "Analysis of the Alais meteorite and implications about life in other worlds".

[40] Thomson (Lord Kelvin), W. (1871). "Inaugural Address to the British Association Edinburgh. "We must regard it as probably to the highest degree that there are countless seed-bearing meteoritic stones moving through space."". *Nature* **4** (92): 261–278 [262]. Bibcode:1871Natur...4..261.. doi:10.1038/004261a0.

[41] Darwin's contribution to the development of the Panspermia theory. By Demets R. *Astrobiology*. 2012 October: 12(10) pages: 946–50. doi: 10.1089/ast.2011.0790

[42] Arrhenius, S., *Worlds in the Making: The Evolution of the Universe*. New York, Harper & Row, 1908.

[43] Fred Hoyle, Chandra Wickramasinghe and John Watson, *Viruses from Space and Related Matters*, University College Cardiff Press, 1986.

[44] Crick, F. H.; Orgel, L. E. (1973). "Directed Panspermia". *Icarus* **19**: 341–348. Bibcode:1979JBIS...32..419M. doi:10.1016/0019-1035(73)90110-3+.

[45] "Anticipating an RNA world. Some past speculations on the origin of life: where are they today?" by L. E. Orgel and F. H. C. Crick in *FASEB J.* (1993) Volume 7 pages 238–239.

[46] Ojha, L.; Wilhelm, M. B.; Murchie, S. L.; McEwen, A. S.; Wray, J. J.; Hanley, J.; Massé, M.; Chojnacki, M. (2015). "Spectral evidence for hydrated salts in recurring slope lineae on Mars". *Nature Geoscience*. doi:10.1038/ngeo2546.

[47] "Top 10 Places To Find Alien Life : Discovery News". News.discovery.com. 2010-06-08. Retrieved 2012-06-13.

[48] "Water 'flowed recently' on Mars". *BBC News*. 2006-12-06. Retrieved 2010-05-02.

[49] Baldwin, Emily (26 April 2012). "Lichen survives harsh Mars environment". Skymania News. Retrieved 27 April 2012.

[50] de Vera, J.-P.; Kohler, Ulrich (26 April 2012). "The adaptation potential of extremophiles to Martian surface conditions and its implication for the habitability of Mars" (PDF). European Geosciences Union. Retrieved 27 April 2012.

[51] Chang, Kenneth (December 9, 2013). "On Mars, an Ancient Lake and Perhaps Life". *New York Times*. Retrieved December 9, 2013.

[52] Various (December 9, 2013). "Science – Special Collection – Curiosity Rover on Mars". *Science*. Retrieved December 9, 2013.

[53] Grotzinger, John P. (January 24, 2014). "Introduction to Special Issue – Habitability, Taphonomy, and the Search for Organic Carbon on Mars". *Science* **343** (6169): 386–387. Bibcode:2014Sci...343..386G. doi:10.1126/science.1249944. Retrieved January 24, 2014.

[54] Various (January 24, 2014). "Special Issue – Table of Contents – Exploring Martian Habitability". *Science* **343** (6169): 345–452. Retrieved 24 January 2014.

[55] Various (January 24, 2014). "Special Collection – Curiosity – Exploring Martian Habitability". *Science*. Retrieved January 24, 2014.

[56] Grotzinger, J.P.; et al. (January 24, 2014). "A Habitable Fluvio-Lacustrine Environment at Yellowknife Bay, Gale Crater, Mars". *Science* **343** (6169). Bibcode:2014Sci...343A.386G. doi:10.1126/science.1242777. Retrieved January 24, 2014.

[57] Küppers, M.; O'Rourke, L.; Bockelée-Morvan, D.; Zakharov, V.; Lee, S.; Von Allmen, P.; Carry, B.; Teyssier, D.; Marston, A.; Müller, T.; Crovisier, J.; Barucci, M. A.; Moreno, R. (2014-01-23). "Localized sources of water vapour on the dwarf planet (1) Ceres". *Nature* **505** (7484): 525–527. Bibcode:2014Natur.505..525K. doi:10.1038/nature12918. ISSN 0028-0836. PMID 24451541.

[58] Campins, H.; Comfort, C. M. (23 January 2014). "Solar system: Evaporating asteroid". *Nature* **505** (7484): 487–488. Bibcode:2014Natur.505..487C. doi:10.1038/505487a. PMID 24451536.

[59] A'Hearn, Michael F.; Feldman, Paul D. (1992). "Water vaporization on Ceres". *Icarus* **98**(1): 54–60. Bibcode:1992Icar...98....5 doi:10.1016/0019-1035(92)90206-M.

[60] A. Duffy – Cosmos – What on Ceres are those bright spots?

[61] Rivkin, Andrew (21 July 2015). "Dawn at Ceres: A haze in Occator crater?". *The Planetary Society*. Retrieved 2015-07-24.

[62] O'Neill, Ian (5 March 2009). "Life on Ceres: Could the Dwarf Planet be the Root of Panspermia". *Universe Today*. Retrieved 30 January 2012.

[63] Catling, David C. (2013). *Astrobiology: A Very Short Introduction*. Oxford: Oxford University Press. p. 99. ISBN 0-19-958645-4.

[64] Boyle, Alan (22 January 2014). "Is There Life on Ceres? Dwarf Planet Spews Water Vapor". NBC. Retrieved 10 February 2015.

[65] Ponnamperuma, Cyril; Molton, Peter (January 1973). "The prospect of life on Jupiter". *Space Life Sciences* **4** (1): 32–44. Bibcode:1973SLSci...4....32P. doi:10.1007/BF02626340.

[66] Irwin, Louis Neal; Schulze-Makuch, Dirk (June 2001). "Assessing the Plausibility of Life on Other Worlds". *Astrobiology* **1** (2): 143–160. Bibcode:2001AsBio...1..143I. doi:10.1089/153110701753198918. PMID 12467118.

[67] Dyches, Preston; Brown, Dwayne (12 May 2015). "NASA Research Reveals Europa's Mystery Dark Material Could Be Sea Salt". *NASA*. Retrieved 12 May 2015.

[68] Jeffrey S. Kargel, Jonathan Z. Kaye, James W. Head, III; et al. (2000). "Europa's Crust and Ocean: Origin, Composition, and the Prospects for Life" (PDF). *Icarus* (Planetary Sciences Group, Brown University) **148** (1): 226–265. Bibcode:2000Icar..148..226K. doi:10.1006/icar.2000.6471.

[69] Staff (March 12, 2015). "NASA's Hubble Observations Suggest Underground Ocean on Jupiter's Largest Moon". *NASA News*. Retrieved 2015-03-15.

[70] "Jupiter moon Ganymede could have ocean with more water than Earth – NASA". *Russia Today (RT)*. 13 March 2015. Retrieved 2015-03-13.

[71] Clavin, Whitney (1 May 2014). "Ganymede May Harbor 'Club Sandwich' of Oceans and Ice". *NASA* (Jet Propulsion Laboratory). Retrieved 2014-05-01.

[72] Vance, Steve; Bouffard, Mathieu; Choukroun, Mathieu; Sotina, Christophe (12 April 2014). "Ganymede's internal structure including thermodynamics of magnesium sulfate oceans in contact with ice". *Planetary and Space Science*. Bibcode:2014P&SS...96...62V. doi:10.1016/j.pss.2014.03.011. Retrieved 2014-05-02.

[73] Staff (1 May 2014). "Video (00:51) – Jupiter's 'Club Sandwich' Moon". *NASA*. Retrieved 2014-05-02.

[74] Chang, Kenneth (March 12, 2015). "Suddenly, It Seems, Water Is Everywhere in Solar System". *New York Times*. Retrieved March 12, 2015.

[75] Kuskov, O.L.; Kronrod, V.A. (2005). "Internal structure of Europa and Callisto".*Icarus*177(2): 550–369. Bibcode:2005Icar..1 doi:10.1016/j.icarus.2005.04.014.

[76] Showman, Adam P.; Malhotra, Renu (1999). "The Galilean Satellites"(PDF).*Science*286(5437): 77–84. doi:10.1126/science PMID 10506564.

[77] Possibility of Life on Europa. University of Victoria. Department of Physics And Astronomy. Canada

[78] Friedman, Louis (December 14, 2005). "Projects: Europa Mission Campaign". The Planetary Society. Retrieved 2011-08-08

[79] Nancy Atkinson (2009). "Europa Capable of Supporting Life. Scientist Says". Universe Today. Retrieved 2011-08-18.

[80] Phil Plait, "Huge lakes of water may exist under Europa's ice", "Bad Astronomy Blog"

[81] "SCIENTISTS FIND EVIDENCE FOR "GREAT LAKE" ON EUROPA AND POTENTIAL NEW HABITAT FOR LIFE"

[82] Cook, Jia-Rui c. (December 11, 2013). "Clay-Like Minerals Found on Icy Crust of Europa". *NASA*. Retrieved December 11, 2013.

[83] Wall, Mike (5 March 2014). "NASA hopes to launch ambitious mission to icy Jupiter moon". *Space.com*. Retrieved 2014-04-15.

[84] Clark, Stephen (14 March 2014). "Economics, water plumes to drive Europa mission study". *Spaceflight Now*. Retrieved 2014-04-15.

[85] Coustenis, A.; et al. (March 2009). "TandEM: Titan and Enceladus mission". *Experimental Astronomy* 23 (3): 893–946. Bibcode:2009ExA....23..893C. doi:10.1007/s10686-008-9103-z.

[86] Lovett, Richard A. (31 May 2011). "Enceladus named sweetest spot for alien life". *Nature* (Nature). doi:10.1038/news.2011.337. Retrieved 2011-06-03.

[87] SPACE.com – Scientists Reconsider Habitability of Saturn's Moon

[88] SPACE.com – Lakes Found on Saturn's Moon Titan

[89] "Lakes on Titan, Full-Res: PIA08630". 2006-07-24.

[90] "What is Consuming Hydrogen and Acetylene on Titan?". NASA/JPL. 2010. Archived from the original on 6 June 2010. Retrieved 2010-06-06.

[91] Darrell F. Strobel (2010). "Molecular hydrogen in Titan's atmosphere: Implications of the measured tropospheric and thermospheric mole fractions". *Icarus* 208 (2): 878. Bibcode:2010Icar..208..878S. doi:10.1016/j.icarus.2010.03.003.

[92] McKay, C. P.; Smith, H. D. (2005). "Possibilities for methanogenic life in liquid methane on the surface of Titan". *Icarus* 178 (1): 274–276. Bibcode:2005Icar..178..274M. doi:10.1016/j.icarus.2005.05.018.

[93] Hoyle, Fred. *Evolution from Space*, Omni Lecture, Royal Institution, London, 12 January 1982; *Evolution from Space* (1982) pp. 27–28 ISBN 0-89490-083-8; *Evolution from Space: A Theory of Cosmic Creationism* (1984) ISBN 0-671-49263-2

[94] Hoyle, Fred (1985). *Living Comets*. Cardiff: University College, Cardiff Press.

[95] Wickramasinghe, Chandra (June 2011). "Viva Panspermia". *The Observatory*.

[96] Wesson, P (2010). "Panspermia, Past and Present: Astrophysical and Biophysical Conditions for the Dissemination of Life in Space". *Sp. Sci.Rev.* 1–4 156: 239–252. arXiv:1011.0101. Bibcode:2010SSRv..156..239W. doi:10.1007/s11214-010-9671-x.

[97] Crenson, Matt (6 August 2006). "Experts: Little Evidence of Life on Mars". Associated Press (on discovery.com). Archived from the original on 16 April 2011. Retrieved 8 March 2011. External link in |publisher= (help)

[98] McKay DS, Gibson EK, ThomasKeprta KL, Vali H, Romanek CS, Clemett SJ, Chillier XDF, Maechling CR, Zare RN (1996). "Search for past life on Mars: Possible relic biogenic activity in Martian meteorite ALH84001". *Science* 273 (5277): 924–930. Bibcode:1996Sci...273..924M. doi:10.1126/science.273.5277.924. PMID 8688069.

[99] McKay DS, Thomas-Keprta KL, Clemett, SJ, Gibson, EK Jr, Spencer L, Wentworth SJ (2009). Hoover, Richard B; Levin, Gilbert V; Rozanov, Alexei Y; Retherford, Kurt D, eds. "Life on Mars: new evidence from martian meteorites". *Proc. SPIE*. Proceedings of SPIE **7441** (1): 744102. doi:10.1117/12.832317. Retrieved 8 March 2011.

[100] Webster, Guy (27 February 2014). "NASA Scientists Find Evidence of Water in Meteorite, Reviving Debate Over Life on Mars". *NASA*. Retrieved 27 February 2014.

[101] White, Lauren M.; Gibson, Everett K.; Thomnas-Keprta, Kathie L.; Clemett, Simon J.; McKay, David (19 February 2014). "Putative Indigenous Carbon-Bearing Alteration Features in Martian Meteorite Yamato 000593". *Astrobiology* **14** (2): 170–181. Bibcode:2014AsBio..14..170W. doi:10.1089/ast.2011.0733. Retrieved 27 February 2014.

[102] Gannon, Megan (28 February 2014). "Mars Meteorite with Odd 'Tunnels' & 'Spheres' Revives Debate Over Ancient Martian Life". *Space.com*. Retrieved 28 February 2014.

[103] Chambers, Paul (1999). *Life on Mars; The Complete Story*. London: Blandford. ISBN 0-7137-2747-0.

[104] Klein, Harold P.; Levin, Gilbert V.; Levin, Gilbert V.; Oyama, Vance I.; Lederberg, Joshua; Rich, Alexander; Hubbard, Jerry S.; Hobby, George L.; Straat, Patricia A.; Berdahl, Bonnie J.; Carle, Glenn C.; Brown, Frederick S.; Johnson, Richard D. (1976-10-01). "The Viking Biological Investigation: Preliminary Results". *Science* **194** (4260): 99–105. Bibcode:1976Sci...194...99K. doi:10.1126/science.194.4260.99. PMID 17793090. Retrieved 2008-08-15.

[105] Beegle, Luther W.; Wilson, Michael G.; Abilleira, Fernando; Jordan, James F.; Wilson, Gregory R. (August 2007). "A Concept for NASA's Mars 2016 Astrobiology Field Laboratory". *Astrobiology* **7** (4): 545–577. Bibcode:2007AsBio...7..545B. doi:10.1089/ast.2007.0153. PMID 17723090. Retrieved 2009-07-20.

[106] "ExoMars rover". ESA. Retrieved 2014-04-14.

[107] Berger, Brian (2005). "Exclusive: NASA Researchers Claim Evidence of Present Life on Mars".

[108] "NASA denies Mars life reports". spacetoday.net. 2005.

[109] Spotts, Peter N. (2005-02-28). "Sea boosts hope of finding signs of life on Mars". The Christian Science Monitor. Retrieved 2006-12-18.

[110] Chow, Dennis (22 July 2011). "NASA's Next Mars Rover to Land at Huge Gale Crater". Space.com. Retrieved 2011-07-22.

[111] Amos, Jonathan (22 July 2011). "Mars rover aims for deep crater". *BBC News*. Retrieved 2011-07-22.

[112] Glaser, Linda (January 27, 2015). "Introducing: The Carl Sagan Institute". Retrieved 2015-05-11.

[113] "Carl Sagan Institute - Research". May 2015. Retrieved 2015-05-11.

[114] Cofield, Calla (30 March 2015). "Catalog of Earth Microbes Could Help Find Alien Life". *Space.com*. Retrieved 2015-05-11.

[115] Callahan, M.P.; Smith, K.E.; Cleaves, H.J.; Ruzica, J.; Stern, J.C.; Glavin, D.P.; House, C.H.; Dworkin, J.P. (11 August 2011). "Carbonaceous meteorites contain a wide range of extraterrestrial nucleobases". PNAS. doi:10.1073/pnas.1106493108. Retrieved 2011-08-15.

[116] Steigerwald, John (8 August 2011). "NASA Researchers: DNA Building Blocks Can Be Made in Space". NASA. Retrieved 2011-08-10.

[117] ScienceDaily Staff (9 August 2011). "DNA Building Blocks Can Be Made in Space, NASA Evidence Suggests". ScienceDaily. Retrieved 2011-08-09.

[118] Chow, Denise (26 October 2011). "Discovery: Cosmic Dust Contains Organic Matter from Stars". Space.com. Retrieved 2011-10-26.

[119] ScienceDaily Staff (26 October 2011). "Astronomers Discover Complex Organic Matter Exists Throughout the Universe". ScienceDaily. Retrieved 2011-10-27.

[120] Kwok, Sun; Zhang, Yong (26 October 2011). "Mixed aromatic–aliphatic organic nanoparticles as carriers of unidentified infrared emission features". *Nature* **479** (7371): 80–3. Bibcode:2011Natur.479...80K. doi:10.1038/nature10542. PMID 22031328.

[121] Than, Ker (August 29, 2012). "Sugar Found In Space". *National Geographic*. Retrieved August 31, 2012.

[122] Staff (August 29, 2012). "Sweet! Astronomers spot sugar molecule near star". Associated Press. Retrieved August 31, 2012.

[123] Jørgensen, J. K.; Favre, C.; Bisschop, S.; Bourke, T.; Dishoeck, E.; Schmalzl, M. (2012). "Detection of the simplest sugar, glycolaldehyde, in a solar-type protostar with ALMA" (PDF). eprint.

[124] Schenkel, Peter (May 2006). "SETI Requires a Skeptical Reappraisal". Skeptical Inquirer. Retrieved June 28, 2009.

[125] Moldwin, Mark (November 2004). "Why SETI is science and UFOlogy is not". Skeptical Inquirer.

[126] "The Search for Extraterrestrial Intelligence (SETI) in the Optical Spectrum". The Columbus Optical SETI Observatory.

[127] Whitmire, Daniel P.; Wright, David P. (April 1980). "Nuclear waste spectrum as evidence of technological extraterrestrial civilizations". Icarus 42 (1): 149–156. Bibcode:1980Icar...42..149W. doi:10.1016/0019-1035(80)90253-5.

[128] "Discovery of OGLE 2005-BLG-390Lb, the first cool rocky/icy exoplanet". IAP.fr. 25 January 2006.

[129] SPACE.com – Major Discovery: New Planet Could Harbor Water and Life

[130] Schneider, Jean (10 September 2011). "Interactive Extra-solar Planets Catalog". The Extrasolar Planets Encyclopaedia. Retrieved 2012-01-30.

[131] Mike Wall. "NASA's Kepler Observatory to continue hunt for strange new worlds". "The Christian Science Monitor"

[132] "NASA – Kepler". Retrieved 4 November 2013.

[133] Harrington, J. D.; Johnson, M. (4 November 2013). "NASA Kepler Results Usher in a New Era of Astronomy".

[134] Tenenbaum, P.; Jenkins, J. M.; Seader, S.; Burke, C. J.; Christiansen, J. L.; Rowe, J. F.; Caldwell, D. A.; Clarke, B. D.; Li, J.; Quintana, E. V.; Smith, J. C.; Thompson, S. E.; Twicken, J. D.; Borucki, W. J.; Batalha, N. M.; Cote, M. T.; Haas, M. R.; Hunter, R. C.; Sanderfer, D. T.; Girouard, F. R.; Hall, J. R.; Ibrahim, K.; Klaus, T. C.; McCauliff, S. D.; Middour, C. K.; Sabale, A.; Uddin, A. K.; Wohler, B.; Barclay, T.; Still, M. (2013). "Detection of Potential Transit Signals in the First 12 Quarters of Kepler Mission Data". The Astrophysical Journal Supplement Series 206: 5. arXiv:1212.2915. Bibcode:2013ApJS..206....5T. doi:10.1088/0067-0049/206/1/5.

[135] "My God, it's full of planets! They should have sent a poet." (Press release). Planetary Habitability Laboratory, University of Puerto Rico at Arecibo. 3 January 2012.

[136] Santerne, A.; Diaz, R. F.; Almenara, J.-M.; Lethuillier, A.; Deleuil, M.; Moutou, C. (2013). "Astrophysical false positives in exoplanet transit surveys: Why do we need bright stars?". arXiv:1310.2133 [astro-ph.EP].

[137] Cassan, A.; et al. (January 11, 2012). "One or more bound planets per Milky Way star from microlensing observations". Nature 481 (7380): 167–169. arXiv:1202.0903. Bibcode:2012Natur.481..167C. doi:10.1038/nature10684. PMID 22237108.

[138] Sanders, R. (4 November 2013). "Astronomers answer key question: How common are habitable planets?". newscenter.berkeley.edu.

[139] Petigura, E. A.; Howard, A. W.; Marcy, G. W. (2013). "Prevalence of Earth-size planets orbiting Sun-like stars". Proceedings of the National Academy of Sciences 110(48): 19273. arXiv:1311.6806. Bibcode:2013PNAS..11019273P.doi:10.1073/pnas.1319

[140] Strigari, L. E.; Barnabè, M.; Marshall, P. J.; Blandford, R. D. (2012). "Nomads of the Galaxy". Monthly Notices of the Royal Astronomical Society 423 (2): 1856–1865. arXiv:1201.2687. Bibcode:2012MNRAS.423.1856S. doi:10.1111/j.1365-2966.2012.21009.x. estimates 700 objects >10^{-8} solar masses (roughly the mass of Mars) per main-sequence star between 0.08 and 1 Solar mass, of which there are billions in the Milky Way.

[141] Dumusque, X.; Pepe, F.; Lovis, C.; Ségransan, D.; Sahlmann, J.; Benz, W.; Bouchy, F.; Mayor, M.; et al. (17 October 2012). "An Earth mass planet orbiting Alpha Centauri B" (PDF). Nature 490 (7423): 207–11. Bibcode:2012Natur.491..207D. doi:10.1038/nature11572. PMID 23075844. Retrieved 17 October 2012.

[142] "DENIS-P J082303.1-491201 b". Caltech. Retrieved 8 March 2014.

[143] Sahlmann, J.; Lazorenko, P. F.; Ségransan, D.; Martin, E. L.; Queloz, D.; Mayor, M.; Udry, S. (August 2013). "Astrometric orbit of a low-mass companion to an ultracool dwarf". Harvard University 556: 133. arXiv:1306.3225. Bibcode:2013A&A...556 A.133S.doi:10.1051/0004-6361/201321871.

[144] "Chapter 3 — Philosophy: "Solving the Drake Equation". SETI League. December 2002. Retrieved July 24, 2015.

[145] Burchell, M.J. (2006). "W(h)ither the Drake equation?". *International Journal of Astrobiology* **5**(3): 243–250. Bibcode:2006IJ doi:10.1017/S1473550406003107.

[146] Aguirre, L. (1 July 2008). "The Drake Equation". *Nova ScienceNow*. PBS. Retrieved 2010-03-07.

[147] Jack Cohen and Ian Stewart (2002). *Evolving the Alien*. John Wiley and Sons, Inc., Hoboken, NJ. Chapter 6, *What does a Martian look like?*

[148] Marcy, G.; Butler, R.; Fischer, D.; et al. (2005). "Observed Properties of Exoplanets: Masses, Orbits and Metallicities". *Progress of Theoretical Physics Supplement* **158**: 24–42. arXiv:astro-ph/0505003. Bibcode:2005PThPS.158...24M.doi:10.114

[149] Overbye, Dennis (August 3, 2015). "The Flip Side of Optimism About Life on Other Planets". *New York Times*. Retrieved October 29, 2015.

[150] *Who discovered that the Sun was a Star?* Stanford Solar Center.

[151] Michael J. Crowe (1999). *The Extraterrestrial Life Debate, 1750–1900*. Courier Dover Publications. ISBN 0-486-40675-X.

[152] Wiker, Benjamin D. (November 4, 2002). "Alien Ideas: Christianity and the Search for Extraterrestrial Life". *Crisis Magazine*. Archived from the original on February 10, 2003.

[153] Irwin, Robert (2003). *The Arabian Nights: A Companion*. Tauris Parke Paperbacks. p. 204 & 209. ISBN 1-86064-983-1.

[154] Conversations on the Plurality of Worlds— Bernard le Bovier de Fontenelle

[155] Flammarion, (Nicolas) Camille (1842–1925)— The Internet Encyclopedia of Science

[156] "Giordano Bruno: On the Infinite Universe and Worlds (De l'Infinito Universo et Mondi) Introductory Epistle: Argument of the Third Dialogue". Retrieved 4 October 2014.

[157] "Rheita.htm". cosmovisions.com.

[158] Evans, J. E. and Maunder, E. W. (1903) "Experiments as to the Actuality of the 'Canals' observed on Mars", MNRAS, **63** (1903) 488

[159] *Is Mars habitable? A critical examination of Professor Percival Lowell's book "Mars and its canals.", an alternative explanation*, by Alfred Russel Wallace, F.R.S., etc. *London, Macmillan and co., 1907.*

[160] Chambers, Paul (1999). *Life on Mars; The Complete Story*. London: Blandford. ISBN 0-7137-2747-0.

[161] Cross, Anne (2004). "The Flexibility of Scientific Rhetoric: A Case Study of UFO Researchers". *Qualitative Sociology* **27** (1): 3–34. doi:10.1023/B:QUAS.0000015542.28438.41.

[162] Ailleris, Philippe (January–February 2011). "The lure of local SETI: Fifty years of field experiments". *Acta Astronautica* **68** (1–2): 2–15. Bibcode:2011AcAau..68....2A. doi:10.1016/j.actaastro.2009.12.011.

[163] "LECTURE 4: MODERN THOUGHTS ON EXTRATERRESTRIAL LIFE". *The University of Antarctica*. Retrieved 2015-07-25.

[164] Amazon.com: Rare Earth: Why Complex Life is Uncommon in the Universe: Books: Peter Ward, Donald Brownlee

[165] "Hawking warns over alien beings". *BBC News*. 2010-04-25. Retrieved 2010-05-02.

[166] Diamond, Jared. "The Third Chimpanzee", Harper Perennial, 2006, Chapter 12.

[167] Larson, Phil (5 November 2011). "Searching for ET, But No Evidence Yet". White House. Retrieved 2011-11-06.

[168] Atkinson, Nancy (5 November 2011). "No Alien Visits or UFO Coverups, White House Says". UniverseToday. Retrieved 2011-11-06.

[169] Staff (May 3, 2013). "Special Issue: Exoplanets". *Science*. Retrieved May 18, 2013.

[170] Chang, Kenneth (17 April 2014). "Scientists Find an 'Earth Twin', or Maybe a Cousin". *New York Times*.

[171] Borenstein, Seth (of AP News) (13 February 2015). "Should We Call the Cosmos Seeking ET? Or Is That Risky?". *New York Times*. Retrieved 14 February 2015.

[172] Ghosh, Pallab (12 February 2015). "Scientist: 'Try to contact aliens'". *BBC News*. Retrieved 12 February 2015.

[173] Various (13 February 2015). "Statement – Regarding Messaging To Extraterrestrial Intelligence (METI) / Active Searches For Extraterrestrial Intelligence (Active SETI)". *University of California, Berkeley*. Retrieved 14 February 2015.

[174] Katz, Gregory (20 July 2015). "Searching for ET: Hawking to look for extraterrestrial life". *AP News*. Retrieved 20 July 2015.

1.10   Further reading

- Baird, John C. (1987). *The Inner Limits of Outer Space: A Psychologist Critiques Our Efforts to Communicate With Extraterrestrial Beings*. Hanover: University Press of New England. ISBN 0-87451-406-1.

- Cohen, Jack; Stewart, Ian (2002). *Evolving the Alien: The Science of Extraterrestrial Life*. Ebury Press. ISBN 0-09-187927-2.

- Crowe, Michael J. (1986). *The Extraterrestrial Life Debate, 1750–1900*. Cambridge. ISBN 0-521-26305-0.

- Crowe, Michael J. (2008). *The extraterrestrial life debate Antiquity to 1915: A Source Book*. University of Notre Dame Press. ISBN 0-268-02368-9.

- Dick, Steven J. (1984). *Plurality of Worlds: The Extraterrestrial Life Debate from Democritis to Kant*. Cambridge.

- Dick, Steven J. (1996). *The Biological Universe: The Twentieth Century Extraterrestrial Life Debate and the Limits of Science*. Cambridge. ISBN 0-521-34326-7.

- Dick, Steven J. (2001). *Life on Other Worlds: The 20th Century Extraterrestrial Life Debate*. Cambridge. ISBN 0-521-79912-0.

- Dick, Steven J.; Strick, James E. (2004). *The Living Universe: NASA And the Development of Astrobiology*. Rutgers. ISBN 0-8135-3447-X.

- Fasan, Ernst (1970). *Relations with alien intelligences – the scientific basis of metalaw*. Berlin: Berlin Verlag.

- Goldsmith, Donald (1997). *The Hunt for Life on Mars*. New York: A Dutton Book. ISBN 0-525-94336-6.

- Grinspoon, David (2003). *Lonely Planets: The Natural Philosophy of Alien Life*. HarperCollins. ISBN 0-06-018540-6.

- Lemnick, Michael T. (1998). *Other Worlds: The Search for Life in the Universe*. New York: A Touchstone Book.

- Michaud, Michael (2006). *Contact with Alien Civilizations – Our Hopes and Fears about Encountering Extraterrestrials*. Berlin: Springer. ISBN 0-387-28598-9.

- Pickover, Cliff (2003). *The Science of Aliens*. New York: Basic Books. ISBN 0-465-07315-8.

- Roth, Christopher F. (2005). Debbora Battaglia, ed. *Ufology as Anthropology: Race, Extraterrestrials, and the Occult. E.T. Culture: Anthropology in Outerspaces* (Durham, NC: Duke University Press).

- Sagan, Carl; Shklovskii, I. S. (1966). *Intelligent Life in the Universe*. Random House.

- Sagan, Carl (1973). *Communication with Extraterrestrial Intelligence*. MIT Press. ISBN 0-262-19106-7.

- Ward, Peter D. (2005). *Life as we do not know it-the NASA search for (and synthesis of) alien life*. New York: Viking. ISBN 0-670-03458-4.

- Tumminia, Diana G. (2007). *Alien Worlds – Social and Religious Dimensions of Extraterrestrial Contact*. Syracuse: Syracuse University Press. ISBN 978-0-8156-0858-5.

1.11 External links

- "Is it true that there could be intelligent life out there?". physics.org. Retrieved 2 November 2012.

- Minerals and the Origins of Life (Robert Hazen, NASA) (video, 60m, April 2014).

- "Search for Life in the Universe" (NASA) (video, 87m, July 14, 2014).

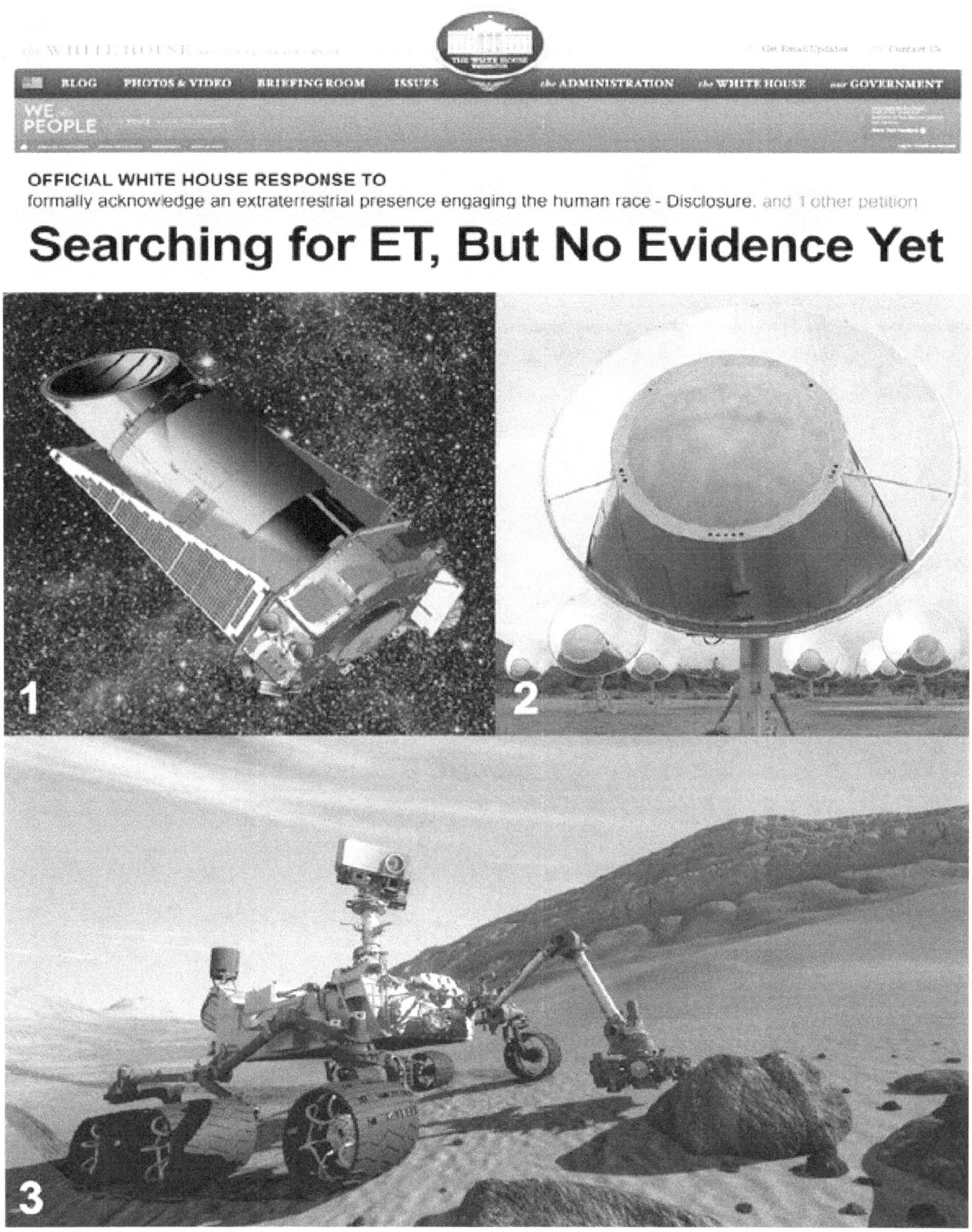

OFFICIAL WHITE HOUSE RESPONSE TO
formally acknowledge an extraterrestrial presence engaging the human race - Disclosure. and 1 other petition

Searching for ET, But No Evidence Yet

Some major international efforts to search for extraterrestrial life. Clockwise from top left:
1 The search for extrasolar planets
(image: Kepler *telescope)*
2 Listening for extraterrestrial signals
indicating intelligence (image: Allen array*)*
3 Robotic exploration of the Solar System
(image: Curiosity *rover on Mars)*

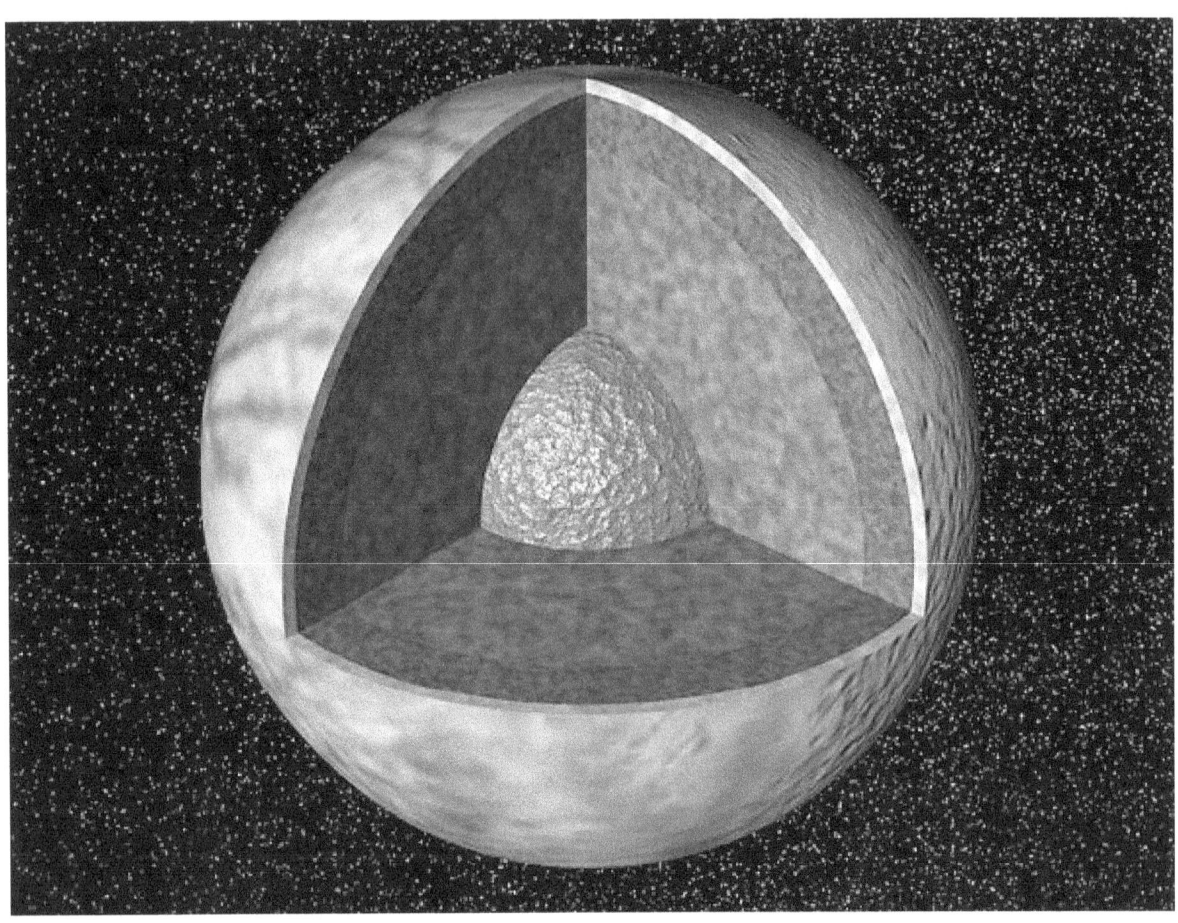

Subsurface oceans such as the one pictured of Europa could possibly harbor life.[77]

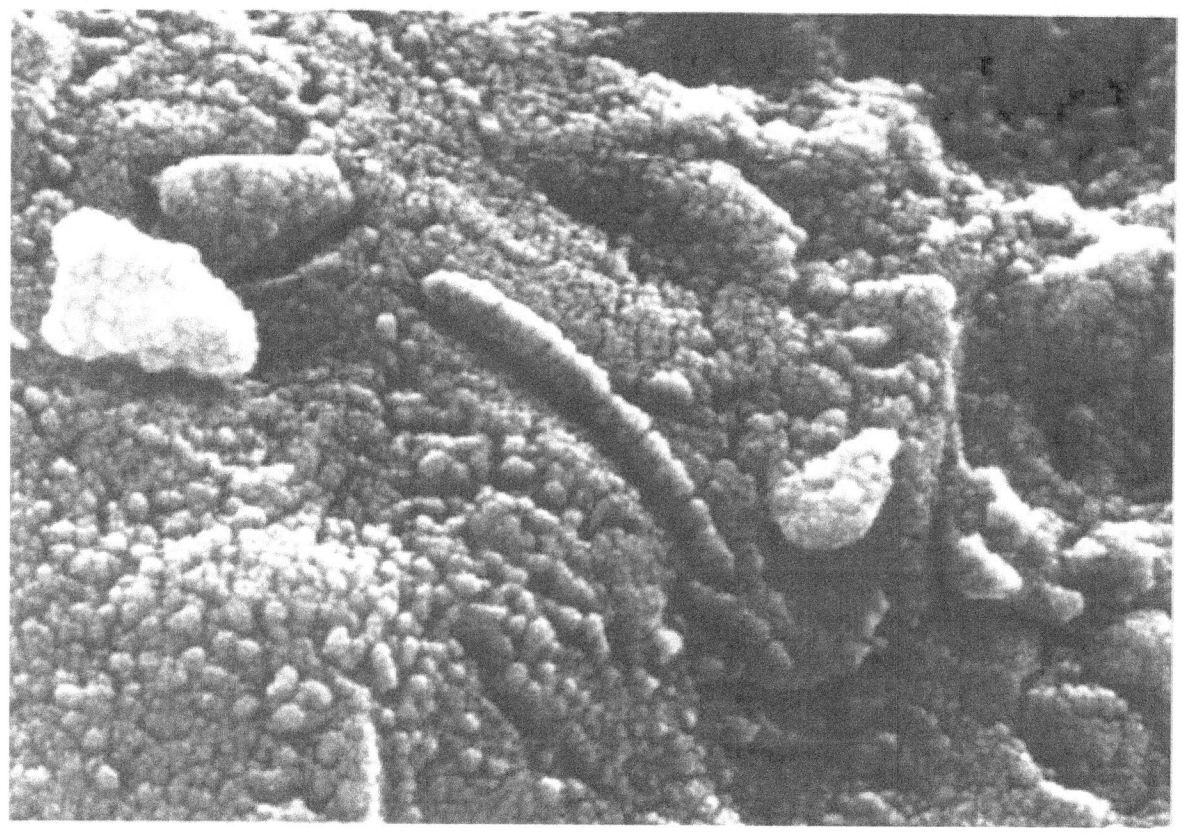

Electron micrograph of martian meteorite ALH84001 showing structures that some scientists think could be fossilized bacteria-like life forms.

Artist's Impression of Gliese 581 c, the first terrestrial extrasolar planet discovered within its star's habitable zone.

Artist's impression of the Kepler telescope in space.

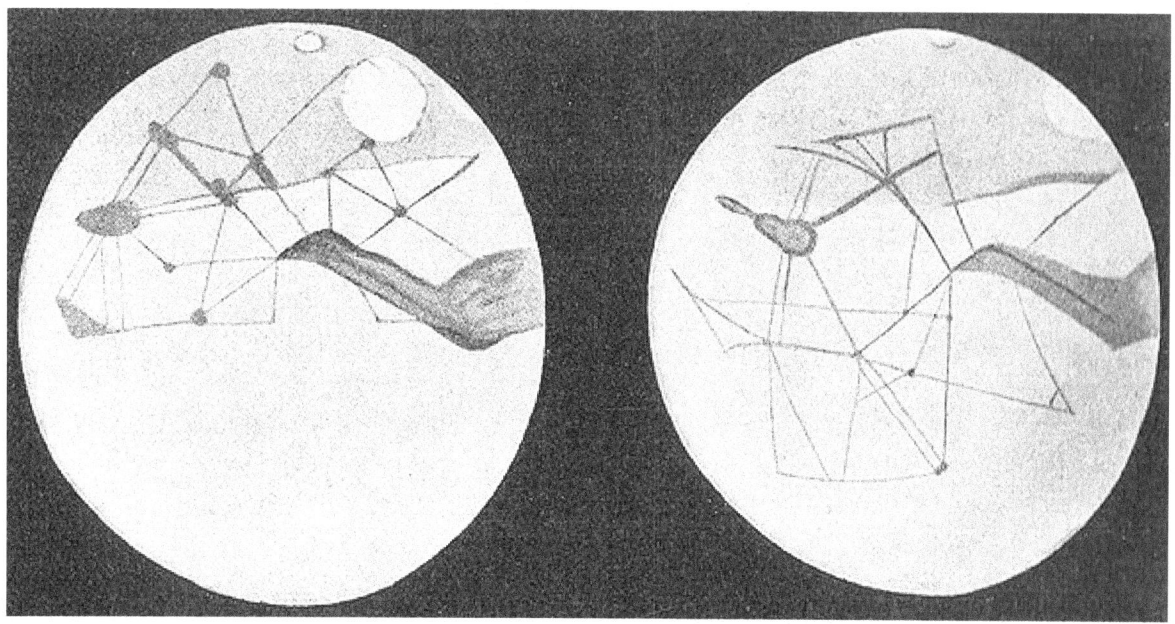

Artificial Martian channels, depicted by Percival Lowell

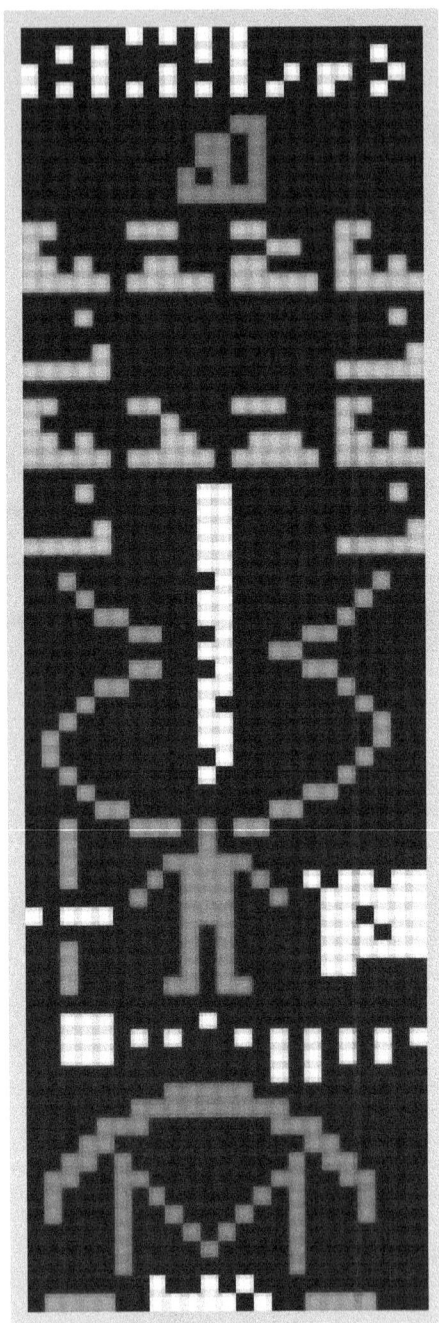

The Arecibo message is a digital message sent to globular star cluster M13, and is a well-known symbol of human attempts to contact extraterrestrials.

Chapter 2

Astrobiology

For the journal, see Astrobiology (journal).

Astrobiology is the study of the origin, evolution, distribution, and future of life in the universe: extraterrestrial life and life on Earth. This interdisciplinary field encompasses the search for habitable environments in our Solar System and habitable planets outside our Solar System, the search for evidence of prebiotic chemistry, laboratory and field research into the origins and early evolution of life on Earth, and studies of the potential for life to adapt to challenges on Earth and in outer space.[2] Astrobiology addresses the question of whether life exists beyond Earth, and how humans can detect it if it does.[3] (The term **exobiology** is similar but more specific—it covers the search for life beyond Earth, and the effects of extraterrestrial environments on living things.)[4]

Astrobiology makes use of physics, chemistry, astronomy, biology, molecular biology, ecology, planetary science, geography, and geology to investigate the possibility of life on other worlds and help recognize biospheres that might be different from the biosphere on Earth.[5] The origin and early evolution of life is an inseparable part of the discipline of astrobiology.[6] Astrobiology concerns itself with interpretation of existing scientific data; given more detailed and reliable data from other parts of the universe, the roots of astrobiology itself—physics, chemistry and biology—may have their theoretical bases challenged. Although speculation is entertained to give context, astrobiology concerns itself primarily with hypotheses that fit firmly into existing scientific theories.

The chemistry of life may have begun shortly after the Big Bang, 13.8 billion years ago, during a habitable epoch when the Universe was only 10–17 million years old.[7][8][9] According to the panspermia hypothesis, microscopic life—distributed by meteoroids, asteroids and other small Solar System bodies—may exist throughout the universe.[10] According to research published in August 2015, very large galaxies may be more favorable to the creation and development of habitable planets than smaller galaxies, like the Milky Way galaxy.[11] Nonetheless, Earth is the only place in the universe known to harbor life.[12][13] Estimates of habitable zones around other stars,[14][15] along with the discovery of hundreds of extrasolar planets and new insights into the extreme habitats here on Earth, suggest that there may be many more habitable places in the universe than considered possible until very recently.[16][17][18]

Current studies on the planet Mars by the *Curiosity* and *Opportunity* rovers are now searching for evidence of ancient life as well as plains related to ancient rivers or lakes that may have been habitable.[19][20][21][22] The search for evidence of habitability, taphonomy (related to fossils), and organic molecules on the planet Mars is now a primary NASA objective on Mars.[19]

2.1 Overview

Astrobiology is etymologically derived from the Greek ἄστρον, *astron*, "constellation, star"; βίος, *bios*, "life"; and -λογία, -logia, *study*. The synonyms of astrobiology are diverse; however, the synonyms were structured in relation to the most important sciences implied in its development: astronomy and biology. A close synonym is *exobiology* from the Greek Ἔξω, "external"; Βίος, *bios*, "life"; and λογία, -logia, *study*. The term exobiology was coined by molecular biologist Joshua Lederberg.[24] Exobiology is considered to have a narrow scope limited to search of life external to Earth, whereas subject area of astrobiology is wider and investigates the link between life and the universe, which includes the search for extraterrestrial life, but also includes the study of life on Earth, its origin, evolution and limits. Exobiology as a term tends to be replaced by astrobiology.

Another term used in the past is **xenobiology**, ("biology of the foreigners") a word used in 1954 by science fiction writer Robert Heinlein in his work The Star Beast.[25] The term *xenobiology* is now used in a more specialized sense, to mean "biology based on foreign chemistry", whether of extraterrestrial or terrestrial (possibly synthetic) origin. Since alternate chemistry analogs to some life-processes have been created in the laboratory, xenobiology is now considered as an extant subject.[26]

While it is an emerging and developing field, the question of whether life exists elsewhere in the universe is a verifiable hypothesis and thus a valid line of scientific inquiry. Though once considered outside the mainstream of scientific inquiry, astrobiology has become a formalized field of study. Planetary scientist David Grinspoon calls astrobiology a field of natural philosophy, grounding speculation on the unknown, in known scientific theory.[27] NASA's interest in exobiology first began with the development of the U.S. Space Program. In 1959, NASA funded its first exobiology project, and in 1960, NASA founded an Exobiology Program, which is now one of four main elements of NASA's current Astrobiology Program.[3][28] In 1971, NASA funded the Search for Extra-Terrestrial Intelligence (SETI) to search radio frequencies of the electromagnetic spectrum for interstellar communications transmitted by extraterrestrial life outside the Solar System. NASA's Viking missions to Mars, launched in 1976, included three biology experiments designed to look for metabolism of present life on Mars.

Advancements in the fields of astrobiology, observational astronomy and discovery of large varieties of extremophiles with extraordinary capability to thrive in the harshest environments on Earth, have led to speculation that life may possibly be thriving on many of the extraterrestrial bodies in the universe. A particular focus of current astrobiology research is the search for life on Mars due to its proximity to Earth and geological history. There is a growing body of evidence to suggest that Mars has previously had a considerable amount of water on its surface, water being considered an essential precursor to the development of carbon-based life.[29]

Missions specifically designed to search for current life on Mars were the Viking program and Beagle 2 probes. The Viking results were inconclusive,[30] and Beagle 2 failed minutes after landing.[31] A future mission with a strong astrobiology role would have been the Jupiter Icy Moons Orbiter, designed to study the frozen moons of Jupiter—some of which may have liquid water—had it not been cancelled. In late 2008, the Phoenix lander probed the environment for past and present planetary habitability of microbial life on Mars, and to research the history of water there.

In November 2011, NASA launched the Mars Science Laboratory mission carrying the Curiosity *rover, which landed on Mars at Gale Crater in August 2012.*[32][33][34] *The* Curiosity *rover is currently probing the environment for past and present planetary habitability of microbial life on Mars. On 9 December 2013, NASA reported that, based on evidence from* Curiosity *studying Aeolis Palus, Gale Crater contained an ancient freshwater lake which could have been a hospitable environment for microbial life.*[35][36]

The European Space Agency is currently collaborating with the Russian Federal Space Agency (Roscosmos) and developing the ExoMars astrobiology rover, which is to be launched in 2018.[37] While NASA is developing the Mars 2020 astrobiology rover and sample cacher for a later return to Earth.

2.2   Methodology

2.2.1   Planetary habitability

Main article: Planetary habitability

When looking for life on other planets like Earth, some simplifying assumptions are useful to reduce the size of the task of the astrobiologist. One is the informed assumption that the vast majority of life forms in our galaxy are based on carbon chemistries, as are all life forms on Earth.[38] Carbon is well known for the unusually wide variety of molecules that can be formed around it. Carbon is the fourth most abundant element in the universe and the energy required to make or break a bond is just at an appropriate level for building molecules which are not only stable, but also reactive. The fact that carbon atoms bond readily to other carbon atoms allows for the building of arbitrarily long and complex molecules.

The presence of liquid water is an assumed requirement, as it is a common molecule and provides an excellent environment for the formation of complicated carbon-based molecules that could eventually lead to the emergence of life.[39] Some researchers posit environments of ammonia, or more likely, water-ammonia mixtures as possible solvents for hypothetical types of biochemistry.[40]

A third assumption is to focus on planets orbiting Sun-like stars for increased probabilities of planetary habitability.[41] Very large stars have relatively short lifetimes, meaning that life might not have time to emerge on planets orbiting them. Very small stars provide so little heat and warmth that only planets in very close orbits around them would not be frozen solid, and in such close orbits these planets would be tidally "locked" to the star.[42] The long lifetimes of red dwarfs could allow the development of habitable environments on planets with thick atmospheres. This is significant, as red dwarfs are extremely common. (See Habitability of red dwarf systems).

Since Earth is the only planet known to harbor life, there is no evident way to know if any of the simplifying assumptions are correct.

2.2.2 Communication attempts

Main article: Communication with extraterrestrial intelligence
 Research on communication with extraterrestrial intelligence (CETI) focuses on composing and deciphering messages that could theoretically be understood by another technological civilization. Communication attempts by humans have included broadcasting mathematical languages, pictorial systems such as the Arecibo message and computational approaches to detecting and deciphering 'natural' language communication. The SETI program, for example, uses both radio telescopes and optical telescopes to search for deliberate signals from an extraterrestrial intelligence.

While some high-profile scientists, such as Carl Sagan, have advocated the transmission of messages,[43][44] scientist Stephen Hawking has warned against it, suggesting that aliens might simply raid Earth for its resources and then move on.

2.2.3 Elements of astrobiology

Astronomy

Main article: Astronomy
 Most astronomy-related astrobiology research falls into the category of extrasolar planet (exoplanet) detection, the hypothesis being that if life arose on Earth, then it could also arise on other planets with similar characteristics. To that end, a number of instruments designed to detect Earth-sized exoplanets have been considered, most notably NASA's Terrestrial Planet Finder (TPF) and ESA's Darwin programs, both of which have been cancelled. NASA launched the Kepler mission in March 2009, and the French Space Agency launched the COROT space mission in 2006.[45][46] There are also several less ambitious ground-based efforts underway.

The goal of these missions is not only to detect Earth-sized planets, but also to directly detect light from the planet so that it may be studied spectroscopically. By examining planetary spectra, it would be possible to determine the basic composition of an extrasolar planet's atmosphere and/or surface. Given this knowledge, it may be possible to assess the likelihood of life being found on that planet. A NASA research group, the Virtual Planet Laboratory,[47] is using computer modeling to generate a wide variety of virtual planets to see what they would look like if viewed by TPF or Darwin. It is hoped that once these missions come online, their spectra can be cross-checked with these virtual planetary spectra for features that might indicate the presence of life.

An estimate for the number of planets with intelligent *communicative* extraterrestrial life can be gleaned from the Drake equation, essentially an equation expressing the probability of intelligent life as the product of factors such as the fraction of planets that might be habitable and the fraction of planets on which life might arise:[48]

$$N = R^* \times f_p \times n_e \times f_l \times f_i \times f_c \times L$$

where:

- N = The number of communicative civilizations

- R^* = The rate of formation of suitable stars (stars such as our Sun)

- fp = The fraction of those stars with planets (current evidence indicates that planetary systems may be common for stars like the Sun)

- ne = The number of Earth-sized worlds per planetary system

- fl = The fraction of those Earth-sized planets where life actually develops

- fi = The fraction of life sites where intelligence develops

- fc = The fraction of communicative planets (those on which electromagnetic communications technology develops)

- L = The "lifetime" of communicating civilizations

However, whilst the rationale behind the equation is sound, it is unlikely that the equation will be constrained to reasonable error limits any time soon. The first term, N, number of stars, is generally constrained within a few orders of magnitude. The second and third terms, fp, stars with planets and fe, planets with habitable conditions, are being evaluated for the star's neighborhood. The problem with the formula is that it is not usable to generate or support hypotheses because it contains factors that can never be verified. Drake originally formulated the equation merely as an agenda for discussion at the Green Bank conference,[49] but some applications of the formula had been taken literally and related to simplistic or pseudoscientific arguments.[50] Another associated topic is the Fermi paradox, which suggests that if intelligent life is common in the universe, then there should be obvious signs of it.

Another active research area in astrobiology is planetary system formation. It has been suggested that the peculiarities of our Solar System (for example, the presence of Jupiter as a protective shield)[51] may have greatly increased the probability of intelligent life arising on our planet.[52][53]

Biology

See also: Abiogenesis, Biology and Extremophile
Biology cannot state that a process or phenomenon, by being mathematically possible, has to exist forcibly in an extraterrestrial body. Biologists specify what is speculative and what is not.[50]

Until the 1970s, life was thought to be entirely dependent on energy from the Sun. Plants on Earth's surface capture energy from sunlight to photosynthesize sugars from carbon dioxide and water, releasing oxygen in the process that is then consumed by oxygen-respiring organisms, passing their energy up the food chain. Even life in the ocean depths, where sunlight cannot reach, was thought to obtain its nourishment either from consuming organic detritus rained down from the surface waters or from eating animals that did.[54] The world's ability to support life was thought to depend on its access to sunlight. However, in 1977, during an exploratory dive to the Galapagos Rift in the deep-sea exploration submersible *Alvin*, scientists discovered colonies of giant tube worms, clams, crustaceans, mussels, and other assorted creatures clustered around undersea volcanic features known as black smokers.[54] These creatures thrive despite having no access to sunlight, and it was soon discovered that they comprise an entirely independent ecosystem. Instead of plants, the basis for this food chain is a form of bacterium that derives its energy from oxidization of reactive chemicals, such as hydrogen or hydrogen sulfide, that bubble up from the Earth's interior. This chemosynthesis revolutionized the study of biology and astrobiology by revealing that life need not be sun-dependent; it only requires water and an energy gradient in order to exist.

Extremophiles, organisms able to survive in extreme environments, are a core research element for astrobiologists. Such organisms include biota which are able to survive several kilometers below the ocean's surface near hydrothermal vents and microbes that thrive in highly acidic environments.[55] It is now known that extremophiles thrive in ice, boiling water, acid, alkali, the water core of nuclear reactors, salt crystals, toxic waste and in a range of other extreme habitats that were previously thought to be inhospitable for life.[56] It opened up a new avenue in astrobiology by massively expanding the number of possible extraterrestrial habitats. Characterization of these organisms, their environments and their evolutionary pathways, is considered a crucial component to understanding how life might evolve elsewhere in the universe. For example, some organisms able to withstand exposure to the vacuum and radiation of outer space include the lichen fungi

Rhizocarpon geographicum and *Xanthoria elegans*,[57] the bacterium *Bacillus safensis*,[58] *Deinococcus radiodurans*,[58] *Bacillus subtilis*,[58] yeast *Saccharomyces cerevisiae*,[58] seeds from *Arabidopsis thaliana* ('mouse-ear cress'),[58] as well as the invertebrate animal Tardigrade.[58]

Jupiter's moon, Europa,[56][59][60][61][62][63] and Saturn's moon, Enceladus,[64][65] are now considered the most likely locations for extant extraterrestrial life in the Solar System.

The origin of life, known as abiogenesis, distinct from the evolution of life, is another ongoing field of research. Oparin and Haldane postulated that the conditions on the early Earth were conducive to the formation of organic compounds from inorganic elements and thus to the formation of many of the chemicals common to all forms of life we see today. The study of this process, known as prebiotic chemistry, has made some progress, but it is still unclear whether or not life could have formed in such a manner on Earth. The alternative hypothesis of panspermia is that the first elements of life may have formed on another planet with even more favorable conditions (or even in interstellar space, asteroids, etc.) and then have been carried over to Earth — the panspermia hypothesis.

The cosmic dust permeating the universe contains complex organic matter ("amorphous organic solids with a mixed aromatic-aliphatic structure") that could be created naturally, and rapidly, by stars.[66][67][68] Further, a scientist suggested that these compounds may have been related to the development of life on Earth and said that, "If this is the case, life on Earth may have had an easier time getting started as these organics can serve as basic ingredients for life."[66] In September 2012, NASA scientists reported that polycyclic aromatic hydrocarbons (PAHs), subjected to interstellar medium conditions, are transformed through hydrogenation, oxygenation and hydroxylation, to more complex organics - "a step along the path toward amino acids and nucleotides, the raw materials of proteins and DNA, respectively".[69][70]

More than 20% of the carbon in the universe may be associated with PAHs, possible starting materials for the formation of life. PAHs seem to have been formed shortly after the Big Bang, are widespread throughout the universe, and are associated with new stars and exoplanets.[71]

Astroecology

Main article: Astroecology

Astroecology concerns the interactions of life with space environments and resources, in planets, asteroids and comets. On a larger scale, astroecology concerns resources for life about stars in the galaxy through the cosmological future. Astroecology attempts to quantify future life in space, addressing this area of astrobiology.

Experimental astroecology investigates resources in planetary soils, using actual space materials in meteorites.[72] The results suggest that Martian and carbonaceous chondrite materials can support bacteria, algae and plant (asparagus, potato) cultures, with high soil fertilities. The results support that life could have survived in early aqueous asteroids and on similar materials imported to Earth by dust, comets and meteorites, and that such asteroid materials can be used as soil for future space colonies.[72][73]

On the largest scale, cosmoecology concerns life in the universe over cosmological times. The main sources of energy may be red giant stars and white and red dwarf stars, sustaining life for 10^{20} years.[72][72][74] Astroecologists suggest that their mathematical models may quantify the potential amounts of future life in space, allowing a comparable expansion in biodiversity, potentially leading to diverse intelligent life forms.[75]

Astrogeology

Main article: Geology of solar terrestrial planets

Astrogeology is a planetary science discipline concerned with the geology of the celestial bodies such as the planets and their moons, asteroids, comets, and meteorites. The information gathered by this discipline allows the measure of a planet's or a natural satellite's potential to develop and sustain life, or planetary habitability.

An additional discipline of astrogeology is geochemistry, which involves study of the chemical composition of the Earth and other planets, chemical processes and reactions that govern the composition of rocks and soils, the cycles of matter and energy and their interaction with the hydrosphere and the atmosphere of the planet. Specializations include cosmochemistry, biochemistry and organic geochemistry.

The fossil record provides the oldest known evidence for life on Earth.[76] By examining the fossil evidence, paleontologists are able to better understand the types of organisms that arose on the early Earth. Some regions on Earth, such as the Pilbara in Western Australia and the McMurdo Dry Valleys of Antarctica, are also considered to be geological analogs to regions of Mars, and as such, might be able to provide clues on how to search for past life on Mars.

The various organic functional groups, composed of hydrogen, oxygen, nitrogen, phosphorus, sulfur, and a host of metals, such as iron, magnesium, and zinc, provide the enormous diversity of chemical reactions necessarily catalyzed by a living organism. Silicon, in contrast, interacts with only a few other atoms, and the large silicon molecules are monotonous compared with the combinatorial universe of organic macromolecules.[50][77] Indeed, it seems likely that the basic building blocks of life anywhere will be similar those on Earth, in the generality if not in the detail.[77] Although terrestrial life and life that might arise independently of Earth are expected to use many similar, if not identical, building blocks, they also are expected to have some biochemical qualities that are unique. If life has had a comparable impact elsewhere in the Solar System, the relative abundances of chemicals key for its survival - whatever they may be - could betray its presence. Whatever extraterrestrial life may be, its tendency to chemically alter its environment might just give it away.[78]

2.3 Life in the Solar System

See also: Abiogenesis, Life on Mars, Life on Europa, Life on Titan and Hypothetical types of biochemistry

People have long speculated about the possibility of life in settings other than Earth, however, speculation on the nature of life elsewhere often has paid little heed to constraints imposed by the nature of biochemistry.[77] The likelihood that life throughout the universe is probably carbon-based, is suggested by the fact that carbon is one of the most abundant of the higher elements. Only two of the natural atoms, carbon and silicon, are known to serve as the backbones of molecules sufficiently large to carry biological information. As the structural basis for life, one of carbon's important features is that unlike silicon, it can readily engage in the formation of chemical bonds with many other atoms, thereby allowing for the chemical versatility required to conduct the reactions of biological metabolism and propagation.

Thought on where in the Solar System life might occur, was limited historically by the understanding that life relies ultimately on light and warmth from the Sun and, therefore, is restricted to the surfaces of planets.[77] The three most likely candidates for life in the Solar System are the planet Mars, the Jovian moon Europa, and Saturn's moon Titan. [79][80][81][82][83] More recently, Saturn's moon Enceladus may be considered a likely candidate as well.[65][84]

Mars, Enceladus and Europa are considered likely candidates in the search for life primarily because they may have liquid water, a molecule essential for life as we know it for its use as a solvent in cells.[29] Water on Mars is found in its polar ice caps, and newly carved gullies recently observed on Mars suggest that liquid water may exist, at least transiently, on the planet's surface.[85][86] At the Martian low temperatures and low pressure, liquid water is likely to be highly saline.[87] As for Europa, liquid water likely exists beneath the moon's icy outer crust.[60][79][80] This water may be warmed to a liquid state by volcanic vents on the ocean floor, but the primary source of heat is probably tidal heating.[88] On 11 December 2013, NASA reported the detection of "clay-like minerals" (specifically, phyllosilicates), often associated with organic materials, on the icy crust of Europa.[89] The presence of the minerals may have been the result of a collision with an asteroid or comet according to the scientists.[89]

Another planetary body that could potentially sustain extraterrestrial life is Saturn's largest moon, Titan.[83] Titan has been described as having conditions similar to those of early Earth.[90] On its surface, scientists have discovered the first liquid lakes outside Earth, but they seem to be composed of ethane and/or methane, not water.[91] Some scientists think it possible that these liquid hydrocarbons might take the place of water in living cells different from those on Earth.[92][93] After Cassini data was studied, it was reported on March 2008 that Titan may also have an underground ocean composed of liquid water and ammonia.[94] Additionally, Saturn's moon Enceladus may have an ocean below its icy surface[95] and, according to NASA scientists in May 2011, "is emerging as the most habitable spot beyond Earth in the Solar System for life as we know it".[65][84]

Measuring the ratio of hydrogen and methane levels on Mars may help determine the likelihood of life on Mars.[96][97] According to the scientists, "...low H_2/CH_4 ratios (less than approximately 40) indicate that life is likely present and active."[96] Other scientists have recently reported methods of detecting hydrogen and methane in extraterrestrial atmospheres.[98][99]

Complex organic compounds of life, including uracil, cytosine and thymine, have been formed in a laboratory under outer space conditions, using starting chemicals such as pyrimidine, found in meteorites. Pyrimidine, like polycyclic aromatic hydrocarbons (PAHs), the most carbon-rich chemical found in the universe.[100]

2.4 Rare Earth hypothesis

Main article: Rare Earth hypothesis

The Rare Earth hypothesis postulates that multicellular life forms found on Earth may actually be more of a rarity than scientists assume. It provides a possible answer to the Fermi paradox which suggests, "If extraterrestrial aliens are common, why aren't they obvious?" It is apparently in opposition to the principle of mediocrity, assumed by famed astronomers Frank Drake, Carl Sagan, and others. The Principle of Mediocrity suggests that life on Earth is not exceptional, but rather that life is more than likely to be found on innumerable other worlds.

The anthropic principle states that fundamental laws of the universe work specifically in a way that life would be possible. The anthropic principle supports the Rare Earth Hypothesis by arguing the overall elements that are needed to support life on Earth are so fine-tuned that it is nearly impossible for another just like it to exist by random chance (note that these terms are used by scientists in a different way from the vernacular conception of them).

2.5 Research

See also: Extraterrestrial life

The systematic search for possible life outside Earth is a valid multidisciplinary scientific endeavor.[101] However, hypotheses and predictions as to its existence and origin vary widely, and at the present, the development of hypotheses firmly grounded on science may be considered astrobiology's most concrete practical application. It has been proposed that viruses are likely to be encountered on other life-bearing planets.[102]

2.5.1 Research outcomes

As of 2015, no evidence of extraterrestrial life has been identified. Examination of the Allan Hills 84001 meteorite, which was recovered in Antarctica in 1984 and originated from Mars, is thought by David McKay, as well as few other scientists, to contain microfossils of extraterrestrial origin; this interpretation is controversial.[103][104][105]

Yamato 000593 is the second largest meteorite from Mars, and was found on Earth in 2000. At a microscopic level, spheres are found in the meteorite that are rich in carbon compared to surrounding areas that lack such spheres. The carbon-rich spheres may have been formed by biotic activity according to some NASA scientists.[106][107][108]

On 5 March 2011, Richard B. Hoover, a scientist with the Marshall Space Flight Center, speculated on the finding of alleged microfossils similar to cyanobacteria in CI1 carbonaceous meteorites.[109][110] However, NASA formally distanced itself from Hoover's claim.[111] According to American astrophysicist Neil deGrasse Tyson: "At the moment, life on Earth is the only known life in the universe, but there are compelling arguments to suggest we are not alone."[112]

Extreme environments on Earth

On 17 March 2013, researchers reported that microbial life forms thrive in the Mariana Trench, the deepest spot on the Earth.[113][114] Other researchers reported related studies that microbes thrive inside rocks up to 1900 feet below the sea floor under 8500 feet of ocean off the coast of the northwestern United States.[113][115] According to one of the researchers, "You can find microbes everywhere — they're extremely adaptable to conditions, and survive wherever they are."[113] These finds expand the potential habitability of certain niches of other planets.

Methane

In 2004, the spectral signature of methane (CH
4) was detected in the Martian atmosphere by both Earth-based telescopes as well as by the Mars Express orbiter. Because of solar radiation and cosmic radiation, methane is predicted to disappear from the Martian atmosphere within several years, so the gas must be actively replenished in order to maintain the present concentration.[116][117] The *Curiosity* rover will perform precision measurements of oxygen and carbon isotope ratios in carbon dioxide (CO_2) and methane (CH_4) in the atmosphere of Mars in order to distinguish between a geochemical and a biological origin.[118][119][120]

Planetary systems

It is possible that some exoplanets may have moons with solid surfaces or liquid oceans that are hospitable. Most of the planets so far discovered outside our Solar System are hot gas giants thought to be inhospitable to life, so it is not yet known whether our Solar System, with a warm, rocky, metal-rich inner planet such as Earth, is of an aberrant composition. Improved detection methods and increased observing time will undoubtedly discover more planetary systems, and possibly some more like ours. For example, NASA's Kepler Mission seeks to discover Earth-sized planets around other stars by measuring minute changes in the star's light curve as the planet passes between the star and the spacecraft. Progress in infrared astronomy and submillimeter astronomy has revealed the constituents of other star systems.

Planetary habitability

Main article: Planetary habitability

Efforts to answer questions such as the abundance of potentially habitable planets in habitable zones and chemical precursors have had much success. Numerous extrasolar planets have been detected using the wobble method and transit method, showing that planets around other stars are more numerous than previously postulated. The first Earth-sized extrasolar planet to be discovered within its star's habitable zone is Gliese 581 c.[121]

2.6 Missions

Research into the environmental limits of life and the workings of extreme ecosystems is ongoing, enabling researchers to better predict what planetary environments might be most likely to harbor life. Missions such as the Phoenix lander, Mars Science Laboratory, ExoMars, Mars 2020 rover to Mars, and the *Cassini* probe to Saturn's moons aim to further explore the possibilities of life on other planets in our Solar System.

2.6.1 Viking program

Main article: Viking biological experiments
 The two Viking landers each carried four types of biological experiments to the surface of Mars in the late 1970s. These were the only Mars landers to carry out experiments to look specifically for metabolism by current microbial life on Mars. The landers used a robotic arm to collect soil samples into sealed test containers on the craft. The two landers were identical, so the same tests were carried out at two places on Mars' surface: Viking 1 near the equator and Viking 2 further north.[122] The result was inconclusive,[123] and is still disputed by some scientists.[124][125][126][127]

2.6.2 Beagle 2

Main article: Beagle 2
 Beagle 2 was an unsuccessful British Mars lander that formed part of the European Space Agency's 2003 Mars Express mission. Its primary purpose was to search for signs of life on Mars, past or present. Although it landed safely, it was unable to correctly deploy its solar panels and telecom antenna.[128]

2.6.3 EXPOSE

Main article: EXPOSE

EXPOSE is a multi-user facility mounted in 2008 outside the International Space Station dedicated to astrobiology.[129][130] EXPOSE was developed by the European Space Agency (ESA) for long-term spaceflights that allows to expose organic chemicals and biological samples to outer space in low Earth orbit.[131]

2.6.4 Mars Science Laboratory

Main article: Mars Science Laboratory

The Mars Science Laboratory (MSL) mission landed a rover that is currently in operation on Mars.[132] It was launched 26 November 2011, and landed at Gale Crater on 6 August 2012.[34] Mission objectives are to help assess Mars' habitability and in doing so, determine whether Mars is or has ever been able to support life,[133] collect data for a future human mission, study Martian geology, its climate, and further assess the role that water, an essential ingredient for life as we know it, played in forming minerals on Mars.

2.6.5 ExoMars

Main article: ExoMars

ExoMars is a robotic mission to Mars to search for possible biosignatures of Martian life, past or present. This astrobiological mission is currently under development by the European Space Agency (ESA) in partnership with the Russian Federal Space Agency (Roscosmos); it is planned for a 2018 launch.[134][135][136]

2.6.6 Mars 2020

Main article: Mars 2020

The 'Mars 2020' rover mission is a concept under development by NASA with a possible launch in 2020. It is intended to investigate environments on Mars relevant to astrobiology, investigate its surface geological processes and history, including the assessment of its past habitability and potential for preservation of biosignatures and biomolecules within accessible geological materials.[137] The Science Definition Team is proposing the rover collect and package at least 31 samples of rock cores and soil for a later mission to bring back for more definitive analysis in laboratories on Earth. The rover could make measurements and technology demonstrations to help designers of a human expedition understand any hazards posed by Martian dust and demonstrate how to collect carbon dioxide (CO_2), which could be a resource for making molecular oxygen (O_2) and rocket fuel.[138][139]

2.6.7 Proposed concepts

Red Dragon

Main article: Red Dragon (spacecraft)

Red Dragon is a proposed concept for a low-cost Mars lander mission that would utilize the SpaceX Falcon Heavy launch vehicle, and a modified Dragon capsule to enter the Martian atmosphere. The lander's primary mission would be a technology demonstration, and to search for evidence of life on Mars (biosignatures), past or present. The concept had been scheduled to be propose for funding on 2012/2013 as a NASA Discovery mission, for launch in 2018.[140][141]

Icebreaker Life

Main article: Icebreaker Life

Icebreaker Life is a lander mission that is being proposed for NASA's Discovery Program for the 2018 launch opportunity.[142] If selected and funded, the stationary lander would be a near copy of the successful 2008 *Phoenix* and it would carry an upgraded astrobiology scientific payload, including a 1-meter-long core drill to sample ice-cemented ground in the northern plains to conduct a search for organic molecules and evidence of current or past life on Mars.[143][144] One of the key goals of the *Icebreaker Life* mission is to test the hypothesis that the ice-rich ground in the polar regions has significant concentrations of organics due to protection by the ice from oxidants and radiation.

Journey to Enceladus and Titan

Journey to Enceladus and Titan (JET) is an orbiter astrobiology mission concept to assess the habitability potential of Saturn's moons Enceladus and Titan.[145][146][147]

Enceladus Life Finder

Enceladus Life Finder (ELF) is a proposed astrobiology mission concept for a space probe intended to assess the habitability of the internal aquatic ocean of Enceladus, Saturn's sixth-largest moon.[148][149]

Life Investigation For Enceladus

Life Investigation For Enceladus (LIFE) is a proposed astrobiology sample-return mission concept for Enceladus. The spacecraft would enter into Saturn orbit and enable multiple flybys through Enceladus' icy plumes to collect icy plume particles and volatiles and return them to Earth on a capsule. The spacecraft may sample Enceladus' plumes, the E ring of Saturn, and the Titan upper atmosphere.[150][151][152]

Europa Multiple-Flyby Mission

Main article: Europa Multiple-Flyby Mission

Europa Multiple-Flyby Mission is a mission planned by NASA for a 2025 launch that will conduct detailed reconnaissance of Jupiter's moon Europa and will investigate whether the icy moon could harbor conditions suitable for life.[153][154] It will also aid in the selection of future landing sites.[155][156]

2.7 See also

- Abiogenesis
- Active SETI
- Astrochemistry/Cosmochemistry
- Astrosciences
- Cosmic dust
- Exoplanetology
- Extraterrestrial life

- Forward-contamination

- Hypothetical types of biochemistry

- List of microorganisms tested in outer space

- Nexus for Exoplanet System Science

- Planetary habitability

- Planetary protection

2.8   References

[1] "Launching the Alien Debates (part 1 of 7)". *Astrobiology Magazine*. NASA. 8 December 2006. Retrieved 5 May 2014.

[2] Cockell, Charles S. (4 October 2012). "How the search for aliens can help sustain life on Earth". *CNN News*. Retrieved 8 October 2012.

[3] "About Astrobiology". *NASA Astrobiology Institute*. NASA. 21 January 2008. Archived from the original on 11 October 2008. Retrieved 20 October 2008.

[4] Mirriam Webster Dictionary entry "Exobiology" (accessed 11 April 2013)

[5] Ward, P. D.; Brownlee, D. (2004). *The life and death of planet Earth*. New York: Owl Books. ISBN 0-8050-7512-7.

[6] "Origins of Life and Evolution of Biospheres". *Journal: Origins of Life and Evolution of Biospheres*. Retrieved 2015-04-06.

[7] Loeb, Abraham (October 2014). "The Habitable Epoch of the Early Universe". *International Journal of Astrobiology* **13** (04): 337–339. arXiv:1312.0613. Bibcode:2014IJAsB..13..337L. doi:10.1017/S1473550414000196. Retrieved 15 December 2014.

[8] Loeb, Abraham (2 December 2013). "The Habitable Epoch of the Early Universe" (PDF). *Arxiv*. arXiv:1312.0613v3. Retrieved 15 December 2014.

[9] Dreifus, Claudia (2 December 2014). "Much-Discussed Views That Go Way Back - Avi Loeb Ponders the Early Universe, Nature and Life". *New York Times*. Retrieved 3 December 2014.

[10] Rampelotto, P.H. (2010). "Panspermia: A Promising Field Of Research" (PDF). *Astrobiology Science Conference*. Retrieved 3 December 2014.

[11] Choi, Charles Q. (21 August 2015). "Giant Galaxies May Be Better Cradles for Habitable Planets". *Space.com*. Retrieved 24 August 2015.

[12] Graham, Robert W. (February 1990). "NASA Technical Memorandum 102363 - Extraterrestrial Life in the Universe" (PDF). *NASA* (Lewis Research Center, Ohio). Retrieved July 7, 2014.

[13] Altermann, Wladyslaw (2008). "From Fossils to Astrobiology - A Roadmap to Fata Morgana?". In Seckbach, Joseph; Walsh, Maud. *From Fossils to Astrobiology: Records of Life on Earth and the Search for Extraterrestrial Biosignatures* **12**. p. xvii. ISBN 1-4020-8836-1.

[14] Horneck, Gerda; Petra Rettberg (2007). *Complete Course in Astrobiology*. Wiley-VCH. ISBN 3-527-40660-3.

[15] Davies, Paul (18 November 2013). "Are We Alone in the Universe?". *New York Times*. Retrieved 20 November 2013.

[16] Overbye, Dennis (4 November 2013). "Far-Off Planets Like the Earth Dot the Galaxy". *New York Times*. Retrieved 5 November 2013.

[17] Petigura, Eric A.; Howard, Andrew W.; Marcy, Geoffrey W. (31 October 2013). "Prevalence of Earth-size planets orbiting Sun-like stars". *Proceedings of the National Academy of Sciences of the United States of America* **110**: 19273–19278. arXiv:1311.6806. Bibcode:2013PNAS..11019273P. doi:10.1073/pnas.1319909110. Retrieved 5 November 2013.

[18] Khan, Amina (4 November 2013). "Milky Way may host billions of Earth-size planets". *Los Angeles Times*. Retrieved 5 November 2013.

[19] Grotzinger, John P. (24 January 2014). "Introduction to Special Issue - Habitability, Taphonomy, and the Search for Organic Carbon on Mars". *Science* **343** (6169): 386–387. Bibcode:2014Sci...343..386G. doi:10.1126/science.1249944. Retrieved 24 January 2014.

[20] Various (24 January 2014). "Special Issue - Table of Contents - Exploring Martian Habitability". *Science* **343** (6169): 345–452. Retrieved 24 January 2014.

[21] Various (24 January 2014). "Special Collection - Curiosity - Exploring Martian Habitability". *Science*. Retrieved 24 January 2014.

[22] Grotzinger, J.P.; et al. (24 January 2014). "A Habitable Fluvio-Lacustrine Environment at Yellowknife Bay, Gale Crater, Mars". *Science* **343** (6169): 1242777. Bibcode:2014Sci...343A.386G. doi:10.1126/science.1242777. Retrieved 24 January 2014.

[23] Gutro, Robert (4 November 2007). "NASA Predicts Non-Green Plants on Other Planets". Goddard Space Flight Center. Archived from the original on 6 October 2008. Retrieved 20 October 2008.

[24] Launching a New Science: Exobiology and the Exploration of Space *The National Library of Medicine*.

[25] Heinlein R and Harold W (21 July 1961). "Xenobiology". *Science* **134** (3473): 223, 225. Bibcode:1961Sci...134..223H. doi:10.1126/science.134.3473.223. JSTOR 1708323.

[26] Markus Schmidt (9 March 2010). "Xenobiology: A new form of life as the ultimate biosafety tool". *BioEssays* **32** (4): 322–331. doi:10.1002/bies.200900147. PMC 2909387. PMID 20217844.

[27] Grinspoon 2004

[28] Steven J. Dick and James E. Strick (2004). *The Living Universe: NASA and the Development of Astrobiology*. New Brunswick, NJ: Rutgers University Press.

[29] NOVA | Mars | Life's Little Essential | PBS

[30] Klein HP and Levin GV (1 October 1976). "The Viking Biological Investigation: Preliminary Results". *Science* **194** (4260): 99–105. Bibcode:1976Sci...194...99K. doi:10.1126/science.194.4260.99. PMID 17793090. Retrieved 15 August 2008.

[31] Amos, Jonathan (16 January 2015). "Lost Beagle2 probe found 'intact' on Mars". *BBC*. Retrieved 16 January 2015.

[32] Webster, Guy; Brown, Dwayne (22 July 2011). "NASA's Next Mars Rover To Land At Gale Crater". NASA JPL. Retrieved 22 July 2011.

[33] Chow, Dennis (22 July 2011). "NASA's Next Mars Rover to Land at Huge Gale Crater". Space.com. Retrieved 22 July 2011.

[34] Amos, Jonathan (22 July 2011). "Mars rover aims for deep crater". *BBC News*. Archived from the original on 22 July 2011. Retrieved 22 July 2011.

[35] Chang, Kenneth (9 December 2013). "On Mars, an Ancient Lake and Perhaps Life". *New York Times*. Retrieved 9 December 2013.

[36] Various (9 December 2013). "Science - Special Collection - Curiosity Rover on Mars". *Science*. Retrieved 9 December 2013.

[37] "ExoMars: ESA and Roscosmos set for Mars missions". *European Space Agency (ESA)*. 14 March 2013. Retrieved 14 March 2013.

[38] "Polycyclic Aromatic Hydrocarbons: An Interview With Dr. Farid Salama". *Astrobiology magazine*. 2000. Retrieved 20 October 2008.

[39] "Astrobiology". Macmillan Science Library: Space Sciences. 2006. Retrieved 20 October 2008.

[40] Penn State (19 August 2006). "The Ammonia-Oxidizing Gene". Astrobiology Magazine. Retrieved 20 October 2008.

[41] "Stars and Habitable Planets". Sol Company. 2007. Archived from the original on 1 October 2008. Retrieved 20 October 2008.

[42] "M Dwarfs: The Search for Life is On". Red Orbit & Astrobiology Magazine. 29 August 2005. Retrieved 20 October 2008.

[43] Sagan, Carl. Communication with Extraterrestrial Intelligence. MIT Press. 1973, 428 pgs.

[44] "You Never Get a Seventh Chance to Make a First Impression: An Awkward History of Our Space Transmissions". *Lightspeed Magazine*. Retrieved 13 March 2015.

[45] "Kepler Mission". NASA. 2008. Archived from the original on 31 October 2008. Retrieved 20 October 2008.

[46] "The COROT space telescope". CNES. 17 October 2008. Archived from the original on 8 November 2008. Retrieved 20 October 2008.

[47] "The Virtual Planet Laboratory". NASA. 2008. Retrieved 20 October 2008.

[48] Ford, Steve (August 1995). "What is the Drake Equation?". SETI League. Archived from the original on 29 October 2008. Retrieved 20 October 2008.

[49] Amir Alexander. "The Search for Extraterrestrial Intelligence: A Short History - Part 7: The Birth of the Drake Equation".

[50] "Astrobiology". Biology Cabinet. 26 September 2006. Archived from the original on 12 December 2010. Retrieved 17 January 2011.

[51] Horner, Jonathan; Barrie Jones (24 August 2007). "Jupiter: Friend or foe?". Europlanet. Retrieved 20 October 2008.

[52] Jakosky, Bruce; David Des Marais; et al. (14 September 2001). "The Role Of Astrobiology in Solar System Exploration". NASA. SpaceRef.com. Retrieved 20 October 2008.

[53] Bortman, Henry (29 September 2004). "Coming Soon: "Good" Jupiters". *Astrobiology Magazine*. Retrieved 20 October 2008.

[54] Chamberlin, Sean (1999). "Black Smokers and Giant Worms". *Fullerton College*. Retrieved 11 February 2011.

[55] Carey, Bjorn (7 February 2005). "Wild Things: The Most Extreme Creatures". *Live Science*. Retrieved 20 October 2008.

[56] Cavicchioli, R. (Fall 2002). "Extremophiles and the search for extraterrestrial life". *Astrobiology* 2(3): 281–92. Bibcode:2002 doi:10.1089/153110702762027862. PMID 12530238.

[57] "Lichens survive in harsh environment of outer space". Retrieved 13 March 2015.

[58] *The Planetary Report*, Volume XXIX, number 2, March/April 2009. "We make it happen! Who will survive? Ten hardy organisms selected for the LIFE project, by Amir Alexander

[59] "Jupiter's Moon Europa Suspected Of Fostering Life" (PDF). *Daily University Science News*. 2002. Retrieved 8 August 2009.

[60] Weinstock, Maia (24 August 2000). "Galileo Uncovers Compelling Evidence of Ocean On Jupiter's Moon Europa". *Space.com*. Retrieved 20 October 2008.

[61] Cavicchioli, R. (Fall 2002). "Extremophiles and the search for extraterrestrial life". *Astrobiology* 2(3): 281–92. Bibcode:2002 doi:10.1089/153110702762027862. PMID 12530238.

[62] David, Leonard (7 February 2006). "Europa Mission: Lost In NASA Budget". Space.com. Retrieved 8 August 2009.

[63] "Clues to possible life on Europa may lie buried in Antarctic ice". *Marshal Space Flight Center* (NASA). 5 March 1998. Archived from the original on 31 July 2009. Retrieved 8 August 2009.

[64] Lovett, Richard A. (31 May 2011). "Enceladus named sweetest spot for alien life". *Nature* (Nature). doi:10.1038/news.2011.337. Retrieved 3 June 2011.

[65] Kazan, Casey (2 June 2011). "Saturn's Enceladus Moves to Top of "Most-Likely-to-Have-Life" List". The Daily Galaxy. Retrieved 3 June 2011.

[66] Chow, Denise (26 October 2011). "Discovery: Cosmic Dust Contains Organic Matter from Stars". Space.com. Retrieved 26 October 2011.

[67] ScienceDaily Staff (26 October 2011). "Astronomers Discover Complex Organic Matter Exists Throughout the Universe". ScienceDaily. Retrieved 27 October 2011.

[68] Kwok, Sun; Zhang, Yong (26 October 2011). "Mixed aromatic–aliphatic organic nanoparticles as carriers of unidentified infrared emission features". *Nature* **479** (7371): 80–3. Bibcode:2011Natur.479...80K. doi:10.1038/nature10542. PMID 22031328.

[69] Staff (20 September 2012). "NASA Cooks Up Icy Organics to Mimic Life's Origins". Space.com. Retrieved 22 September 2012.

[70] Gudipati, Murthy S.; Yang, Rui (1 September 2012). "In-Situ Probing Of Radiation-Induced Processing Of Organics In Astrophysical Ice Analogs—Novel Laser Desorption Laser Ionization Time-Of-Flight Mass Spectroscopic Studies". *The Astrophysical Journal Letters* **756** (1): L24. Bibcode:2012ApJ...756L..24G. doi:10.1088/2041-8205/756/1/L24. Retrieved 22 September 2012.

[71] Hoover, Rachel (21 February 2014). "Need to Track Organic Nano-Particles Across the Universe? NASA's Got an App for That". *NASA*. Retrieved 22 February 2014.

[72] Mautner, Michael N. (2002). "Planetary bioresources and astroecology. 1. Planetary microcosm bioessays of Martian and meteorite materials: soluble electrolytes, nutrients, and algal and plant responses" (PDF). *Icarus* **158** (1): 72–86. Bibcode:2002Icar..158...72M.doi:10.1006/icar.2002.6841. PMID 12449855.

[73] Mautner, Michael N. (2002). "Planetary resources and astroecology. Planetary microcosm models of asteroid and meteorite interiors: electrolyte solutions and microbial growth. Implications for space populations and panspermia" (PDF). *Astrobiology* **2** (1): 59–76. Bibcode:2002Icar..158...72M. doi:10.1006/icar.2002.6841. PMID 12449855.

[74] Mautner, Michael N. (2005). "Life in the cosmological future: Resources, biomass and populations" (PDF). *Journal of the British Interplanetary Society* **58**: 167–180. Bibcode:2005JBIS...58..167M.

[75] Mautner, Michael N. (2000). *Seeding the Universe with Life: Securing Our Cosmological Future* (PDF). Washington D. C.: Legacy Books (www.amazon.com). ISBN 0-476-00330-X.

[76] "Fossil Succession". U.S. Geological Survey. 14 August 1997. Archived from the original on 14 October 2008. Retrieved 20 October 2008.

[77] Pace, Norman R. (30 January 2001). "The universal nature of biochemistry". *Proceedings of the National Academy of Sciences of the USA* **98** (3): 805–808. Bibcode:2001PNAS...98..805P. doi:10.1073/pnas.98.3.805. PMC 33372. PMID 11158550. Retrieved 20 March 2010.

[78] Marshall, Michael (21 January 2011). "Telltale chemistry could betray ET". *New Scientists*. Retrieved 22 January 2011.

[79] Tritt, Charles S. (2002). "Possibility of Life on Europa". Milwaukee School of Engineering. Retrieved 20 October 2008.

[80] Friedman, Louis (14 December 2005). "Projects: Europa Mission Campaign". The Planetary Society. Archived from the original on 20 September 2008. Retrieved 20 October 2008.

[81] David, Leonard (10 November 1999). "Move Over Mars – Europa Needs Equal Billing". Space.com. Retrieved 20 October 2008.

[82] Than, Ker (28 February 2007). "New Instrument Designed to Sift for Life on Mars". Space.com. Retrieved 20 October 2008.

[83] Than, Ker (13 September 2005). "Scientists Reconsider Habitability of Saturn's Moon". *Science.com*. Retrieved 11 February 2011.

[84] Lovett, Richard A. (31 May 2011). "Enceladus named sweetest spot for alien life". *Nature* (Nature). doi:10.1038/news.2011.337. Retrieved 3 June 2011.

[85] "NASA Images Suggest Water Still Flows in Brief Spurts on Mars". NASA. 2006. Archived from the original on 16 October 2008. Retrieved 20 October 2008.

[86] "Water ice in crater at Martian north pole". European Space Agency. 28 July 2005. Archived from the original on 23 September 2008. Retrieved 20 October 2008.

[87] Landis, Geoffrey A. (1 June 2001). "Martian Water: Are There Extant Halobacteria on Mars?". *Astrobiology* **1** (2): 161–164. Bibcode:2001AsBio...1..161L. doi:10.1089/153110701753198927. PMID 12467119. Retrieved 20 October 2008.

[88] Kruszelnicki, Karl (5 November 2001). "Life on Europa, Part 1". ABC Science. Retrieved 20 October 2008.

[89] Cook, Jia-Rui c. (11 December 2013). "Clay-Like Minerals Found on Icy Crust of Europa". *NASA*. Retrieved 11 December 2013.

[90] "Titan: Life in the Solar System?". *BBC - Science & Nature*. Retrieved 20 October 2008.

[91] Britt, Robert Roy (28 July 2006). "Lakes Found on Saturn's Moon Titan". *Space.com*. Archived from the original on 4 October 2008. Retrieved 20 October 2008.

[92] Committee on the Limits of Organic Life in Planetary Systems, Committee on the Origins and Evolution of Life, National Research Council; The Limits of Organic Life in Planetary Systems; The National Academies Press, 2007; p 74

[93] McKay, C. P.; Smith, H. D. (2005). "Possibilities for methanogenic life in liquid methane on the surface of Titan". *Icarus* **178** (1): 274–276. Bibcode:2005Icar..178..274M. doi:10.1016/j.icarus.2005.05.018.

[94] Lovett, Richard A. (20 March 2008). "Saturn Moon Titan May Have Underground Ocean". *National Geographic News*. Archived from the original on 24 September 2008. Retrieved 20 October 2008.

[95] "Saturn moon 'may have an ocean'". *BBC News*. 10 March 2006. Retrieved 5 August 2008.

[96] Oze, Christopher; Jones, Camille; Goldsmith, Jonas I.; Rosenbauer, Robert J. (7 June 2012). "Differentiating biotic from abiotic methane genesis in hydrothermally active planetary surfaces". *PNAS* **109** (25): 9750–9754. Bibcode:2012PNAS..109.9750O. doi:10.1073/pnas.1205223109. PMC 3382529. PMID 22679287. Retrieved 27 June 2012.

[97] Staff (25 June 2012). "Mars Life Could Leave Traces in Red Planet's Air: Study". Space.com. Retrieved 27 June 2012.

[98] Brogi, Matteo; Snellen, Ignas A. G.; de Krok, Remco J.; Albrecht, Simon; Birkby, Jayne; de Mooij, Ernest J. W. (28 June 2012). "The signature of orbital motion from the dayside of the planet t Boötis b". *Nature* **486** (7404): 502–504. arXiv:1206.6109. Bibcode:2012Natur.486..502B. doi:10.1038/nature11161. Retrieved 28 June 2012.

[99] Mann, Adam (27 June 2012). "New View of Exoplanets Will Aid Search for E.T.". Wired (magazine). Retrieved 28 June 2012.

[100] Marlaire, Ruth (3 March 2015). "NASA Ames Reproduces the Building Blocks of Life in Laboratory". *NASA*. Retrieved 5 March 2015.

[101] "NASA Astrobiology: Life in the Universe". Retrieved 13 March 2015.

[102] Griffin, Dale Warren (14 August 2013). "The Quest for Extraterrestrial Life: What About the Viruses?". *Astrobiology (journal)* **13** (8): 774–783. Bibcode:2013AsBio..13..774G. doi:10.1089/ast.2012.0959. Retrieved 6 September 2013.

[103] Crenson, Matt (6 August 2006). "Experts: Little Evidence of Life on Mars". Associated Press (on discovery.com). Archived from the original on 16 April 2011. Retrieved 8 March 2011. External link in |publisher= (help)

[104] McKay DS, Gibson E. K., Thomas-Keprta K. L., Vali H., Romanek C. S., Clemett S. J., Chillier X. D. F., Maechling C. R., Zare R. N. (1996). "Search for past life on Mars: Possible relic biogenic activity in Martian meteorite ALH84001". *Science* **273** (5277): 924–930. Bibcode:1996Sci...273..924M. doi:10.1126/science.273.5277.924. PMID 8688069.

[105] McKay David S., Thomas-Keprta K. L., Clemett, S. J., Gibson, E. K. Jr, Spencer L., Wentworth S. J. (2009). Hoover, Richard B.; Levin, Gilbert V.; Rozanov, Alexei Y.; Retherford, Kurt D., eds. "Life on Mars: new evidence from martian meteorites". *Proc. SPIE*. Proceedings of SPIE **7441** (1): 744102. doi:10.1117/12.832317. Retrieved 8 March 2011.

[106] Webster, Guy (27 February 2014). "NASA Scientists Find Evidence of Water in Meteorite, Reviving Debate Over Life on Mars". *NASA*. Retrieved 27 February 2014.

[107] White, Lauren M.; Gibson, Everett K.; Thomnas-Keprta, Kathie L.; Clemett, Simon J.; McKay, David (19 February 2014). "Putative Indigenous Carbon-Bearing Alteration Features in Martian Meteorite Yamato 000593". *Astrobiology* **14** (2): 170–181. Bibcode:2014AsBio..14..170W. doi:10.1089/ast.2011.0733. Retrieved 27 February 2014.

[108] Gannon, Megan (28 February 2014). "Mars Meteorite with Odd 'Tunnels' & 'Spheres' Revives Debate Over Ancient Martian Life". *Space.com*. Retrieved 28 February 2014.

[109] Tenney, Garrett (5 March 2011). "Exclusive: NASA Scientist Claims Evidence of Alien Life on Meteorite". Fox News. Archived from the original on 6 March 2011. Retrieved 6 March 2011.

[110] Hoover, Richard B. (2011). "Fossils of Cyanobacteria in CI1 Carbonaceous Meteorites: Implications to Life on Comets, Europa, and Enceladus". *Journal of Cosmology* **13**: xxx. Retrieved 6 March 2011.

[111] Sheridan, Kerry (7 March 2011). "NASA shoots down alien fossil claims". *ABC News*. Retrieved 7 March 2011.

[112] Tyson, Neil deGrasse (23 July 2001). "The Search for Life in the Universe". *Department of Astrophysics and Hayden Planetarium*. NASA. Retrieved 7 March 2011.

[113] Choi, Charles Q. (17 March 2013). "Microbes Thrive in Deepest Spot on Earth". LiveScience. Retrieved 17 March 2013.

[114] Glud, Ronnie; Wenzhöfer, Frank; Middleboe, Mathias; Oguri, Kazumasa; Turnewitsch, Robert; Canfield, Donald E.; Kitazato, Hiroshi (17 March 2013). "High rates of microbial carbon turnover in sediments in the deepest oceanic trench on Earth". *Nature Geoscience* **6**: 284–288. Bibcode:2013NatGe...6..284G. doi:10.1038/ngeo1773. Retrieved 17 March 2013.

[115] Oskin, Becky (14 March 2013). "Intraterrestrials: Life Thrives in Ocean Floor". LiveScience. Retrieved 17 March 2013.

[116] Vladimir A. Krasnopolsky (February 2005). "Some problems related to the origin of methane on Mars". *Icarus* **180** (2): 359–367. Bibcode:2006Icar..180..359K. doi:10.1016/j.icarus.2005.10.015.

[117] Planetary Fourier Spectrometer website (ESA, Mars Express)

[118] "Sample Analysis at Mars (SAM) Instrument Suite". NASA. October 2008. Archived from the original on 7 October 2008. Retrieved 9 October 2008.

[119] Tenenbaum, David (9 June 2008). "Making Sense of Mars Methane". *Astrobiology Magazine*. Archived from the original on 23 September 2008. Retrieved 8 October 2008.

[120] Tarsitano CG and Webster CR (2007). "Multilaser Herriott cell for planetary tunable laser spectrometers". *Applied Optics*, **46** (28): 6923–6935. Bibcode:2007ApOpt..46.6923T. doi:10.1364/AO.46.006923. PMID 17906720.

[121] Than, Ker (24 April 2007). "Major Discovery: New Planet Could Harbor Water and Life". Space.com. Archived from the original on 15 October 2008. Retrieved 20 October 2008.

[122] Chambers, Paul (1999). *Life on Mars; The Complete Story*. London: Blandford. ISBN 0-7137-2747-0.

[123] Levin, G and P. Straat. 1976. Viking Labeled Release Biology Experiment: Interim Results. Science: 194. 1322-1329.

[124] Bianciardi, Giorgio; Miller, Joseph D.; Straat, Patricia Ann; Levin, Gilbert V. (March 2012). "Complexity Analysis of the Viking Labeled Release Experiments". *IJASS* **13** (1): 14–26. Bibcode:2012IJASS..13...14B. doi:10.5139/IJASS.2012.13.1.14. Retrieved 15 April 2012.

[125] Klotz, Irene (12 April 2012). "Mars Viking Robots 'Found Life'". Discovery News. Retrieved 16 April 2012.

[126] Navarro-González, R.; et al. (2006). "The limitations on organic detection in Mars-like soils by thermal volatilization–gas chromatography–MS and their implications for the Viking results" (PDF). *PNAS* **103** (44): 16089–16094. Bibcode:2006PNAS..10316089N.doi:10.1073/pnas.0604210103. PMC1621051. PMID17060639. Retrieved2April2012.

[127] Paepe, Ronald (2007). "The Red Soil on Mars as a proof for water and vegetation" (PDP). *Geophysical Research Abstracts* **9** (1794). Retrieved 2 May 2012.

[128] "Beagle 2 : the British led exploration of Mars". Retrieved 13 March 2015.

[129] Gerda Horneck, Petra Rettberg, Jobst-Ulrich Schott, Corinna Panitz, Andrea L'Afflitto, Ralf von Heise-Rotenburg, Reiner Willnecker, Pietro Baglioni, Jason Hatton, Jan Dettmann, René Demets and Günther Reitz., Elke Rabbow (9 July 2009). "EXPOSE, an Astrobiological Exposure Facility on the International Space Station - from Proposal to Flight" (PDF). *Orig Life Evol Biosph* **39** (6): 581–98. Bibcode:2009OLEB...39..581R. doi:10.1007/s11084-009-9173-6. PMID 19629743. Retrieved 8 July 2013.

[130] Karen Olsson-Francis; Charles S. Cockell (23 October 2009). "Experimental methods for studying microbial survival in extraterrestrial environments" (PDF). *Journal of Microbiological Methods* **80** (1): 1–13. doi:10.1016/j.mimet.2009.10.004. PMID 19854226. Retrieved 31 July 2013.

[131] Centre national d'études spatiales (CNES). "EXPOSE - home page". Retrieved 8 July 2013.

[132] "Name NASA's Next Mars Rover". NASA/JPL. 27 May 2009. Archived from the original on 22 May 2009. Retrieved 27 May 2009.

[133] "Mars Science Laboratory: Mission". NASA/JPL. Retrieved 12 March 2010.

[134] Amos, Jonathan (15 March 2012). "Europe still keen on Mars missions". *BBC News*. Retrieved 16 March 2012.

[135] Svitak, Amy (16 March 2012). "Europe Joins Russia on Robotic ExoMars". *Aviation Week*. Retrieved 16 March 2012.

[136] Selding, Peter B. de (15 March 2012). "ESA Ruling Council OKs ExoMars Funding". *Space News*. Retrieved 16 March 2012.

[137] Cowing, Keith (21 December 2012). "Science Definition Team for the 2020 Mars Rover". *NASA*. Science Ref. Retrieved 21 December 2012.

[138] "Science Team Outlines Goals for NASA's 2020 Mars Rover". *Jet Propulsion Laboratory* (NASA). 9 July 2013. Retrieved 10 July 2013.

[139] "Mars 2020 Science Definition Team Report - Frequently Asked Questions" (PDF). *NASA*. 9 July 2013. Retrieved 10 July 2013.

[140] Wall, Mike (31 July 2011). "'Red Dragon' Mission Mulled as Cheap Search for Mars Life". *SPACE.com*. Retrieved 1 May 2012.

[141] "NASA ADVISORY COUNCIL (NAC) - Science Committee Report" (PDF). *Ames Research Center, NASA*. 1 November 2011. Retrieved 1 May 2012.

[142] McKay, Christopher P.; Carol R. Stoker, Brian J. Glass, Arwen I. Davé, Alfonso F. Davila, Jennifer L. Heldmann, Margarita M. Marinova, Alberto G. Fairen, Richard C. Quinn, Kris A. Zacny, Gale Paulsen, Peter H. Smith, Victor Parro, Dale T. Andersen, Michael H. Hecht, Denis Lacelle, and Wayne H. Pollard. (5 April 2013). "The *Icebreaker Life* Mission to Mars: A Search for Biomolecular Evidence for Life". *Astrobiology* 13 (4): 334–353. Bibcode:2013AsBio..13..334M. doi:10.1089/ast.2012.0878. PMID 23560417. Retrieved 30 June 2013.

[143] Choi, Charles Q. (16 May 2013). "Icebreaker Life Mission". *Astrobiology Magazine*. Retrieved 1 July 2013.

[144] McKay, C. P.; Carol R. Stoker, Brian J. Glass, Arwen I. Davé, Alfonso F. Davila, Jennifer L. Heldmann, Margarita M. Marinova, Alberto G. Fairen, Richard C. Quinn, Kris A. Zacny, Gale Paulsen, Peter H. Smith, Victor Parro, Dale T. Andersen, Michael H. Hecht, Denis Lacelle, and Wayne H. Pollard. (2012). "THE ICEBREAKER LIFE MISSION TO MARS: A SEARCH FOR BIOCHEMICAL EVIDENCE FOR LIFE". *Concepts and Approaches for Mars Exploration* (PDF). Lunar and Planetary Institute. retrieved 1 July 2013

[145] Sotin, C.; Altwegg, K.; Brown, R.H.; et al. (2011). *JET: Journey to Enceladus and Titan* (PDF). 42nd Lunar and Planetary Science Conference. Lunar and Planetary Institute.

[146] Kane, Van (3 April 2014). "Discovery Missions for an Icy Moon with Active Plumes". *The Planetary Society*. Retrieved 2015-04-09.

[147] Matousek, Steve; Sotin, Christophe; Goebel, Dan; Lang, Jared (June 18–21, 2013). *JET: Journey to Enceladus and Titan* (PDF). Low Cost Planetary Missions Conference. California Institute of Technology.

[148] Lunine, J.I.; Waite, J.H.; Postberg, F.; Spilker, L. (2015). *Enceladus Life Finder: The search for life in a habitable moon* (PDF). 46th Lunar and Planetary Science Conference (2015). Houston, Texas.: Lunar and Planetary Institute.

[149] Clark, Stephen (April 6, 2015). "Diverse destinations considered for new interplanetary probe". *Space Flight Now*. Retrieved 2015-04-07.

[150] Tsou, Peter; Brownlee, D.E.; McKay, Christopher; Anbar, A.D.; Yano, H. (August 2012). "LIFE: Life Investigation For Enceladus A Sample Return Mission Concept in Search for Evidence of Life.". *Astrobiology Journal* 12 (8): 730–742. Bibcode:2012AsBio..12..730T. doi:10.1089/ast.2011.0813. Retrieved 2015-04-10.

[151] Tsou, Peter; Anbar, Ariel; Atwegg, Kathrin; Porco, Carolyn; Baross, John; McKay, Christopher (2014). "LIFE - Enceladus Plume Sample Return via Discovery" (PDF). *45th Lunar and Planetary Science Conference*. Retrieved 2015-04-10.

[152] Tsou, Peter (2013). "LIFE: Life Investigation For Enceladus - A Sample Return Mission Concept in Search for Evidence of Life." (.doc). *Jet Propulsion Laboratory*. Retrieved 2015-04-10.

[153] "Europa Clipper". *Jet Propulsion Laboratory* (NASA). November 2013. Retrieved 13 December 2013.

[154] Kane, Van (26 May 2013). "Europa Clipper Update". *Future Planetary Exploration*. Retrieved 13 December 2013.

[155] Pappalardo, Robert T.; S. Vance, F. Bagenal, B.G. Bills, D.L. Blaney, D.D. Blankenship, W.B. Brinckerhoff; et al. (2013). "Science Potential from a Europa Lander". *Astrobiology* **13** (8): 740–773. Bibcode:2013AsBio..13..740P. doi:10.1089/ast.2013 .1003.PMID23924246. Retrieved14 December2013.

[156] Senske, D. (2 October 2012). "Europa Mission Concept Study Update", *Presentation to Planetary Science Subcommittee* (PDF), retrieved 14 December 2013

2.9 Bibliography

- The *International Journal of Astrobiology*, published by Cambridge University Press, is the forum for practitioners in this interdisciplinary field.

- *Astrobiology*, published by Mary Ann Liebert, Inc., is a peer-reviewed journal that explores the origins of life, evolution, distribution, and destiny in the universe.

- Catling, David C. (2013). *Astrobiology: A Very Short Introduction*. Oxford: Oxford University Press. ISBN 0-19-958645-4.

- Cockell, Charles S. (2015). *Astrobiology: Understanding Life in the Universe*. NJ: Wiley-Blackwell. ISBN 978-1-118-91332-1.

- Kolb, Vera M. (Ed) (2015). *Astrobiology: An Evolutionary Approach*. Boca Raton: CRC Press. ISBN 978-1-4665-8461-7.

- Dick, Steven J.; James Strick (2005). *The Living Universe: NASA and the Development of Astrobiology*. Piscataway, NJ: Rutgers University Press. ISBN 0-8135-3733-9.

- Grinspoon, David (2004) [2003]. *Lonely planets. The natural philosophy of alien life*. New York: ECCO. ISBN 0-06-018540-6.

- Mautner, Michael N. (2000). *Seeding the Universe with Life: Securing Our Cosmological Future* (PDF). Washington D. C.: Legacy Books (www.amazon.com). ISBN 0-476-00330-X.

- Jakosky, Bruce M. (2006). *Science, Society, and the Search for Life in the Universe*. Tucson: University of Arizona Press. ISBN 0-8165-2613-3.

- Lunine, Jonathan I. (2005). *Astrobiology. A Multidisciplinary Approach*. San Francisco: Pearson Addison-Wesley. ISBN 0-8053-8042-6.

- Gilmour, Iain; Mark A. Sephton (2004). *An introduction to astrobiology*. Cambridge: Cambridge Univ. Press. ISBN 0-521-83736-7.

- Ward, Peter; Brownlee, Donald (2000). *Rare Earth: Why Complex Life is Uncommon in the Universe*. New York: Copernicus. ISBN 0-387-98701-0.

- Chyba, C. F.; Hand, K. P. (2005). "ASTROBIOLOGY: The Study of the Living Universe". *Annual Review of Astronomy and Astrophysics* **43**: 31. Bibcode:2005ARA&A..43...31C. doi:10.1146/annurev.astro.43.051804.102202.

2.10 External links

- Astrobiology.nasa.gov
- Spanish Centro de Astrobiología
- UK Centre for Astrobiology
- Astrobiology Web
- Astrobiology Research at The Library of Congress

2.11   Further reading

- D. Goldsmith, T. Owen, *The Search For Life In The Universe*, Addison-Wesley Publishing Company, 2001 (3rd edition). ISBN 978-1891389160

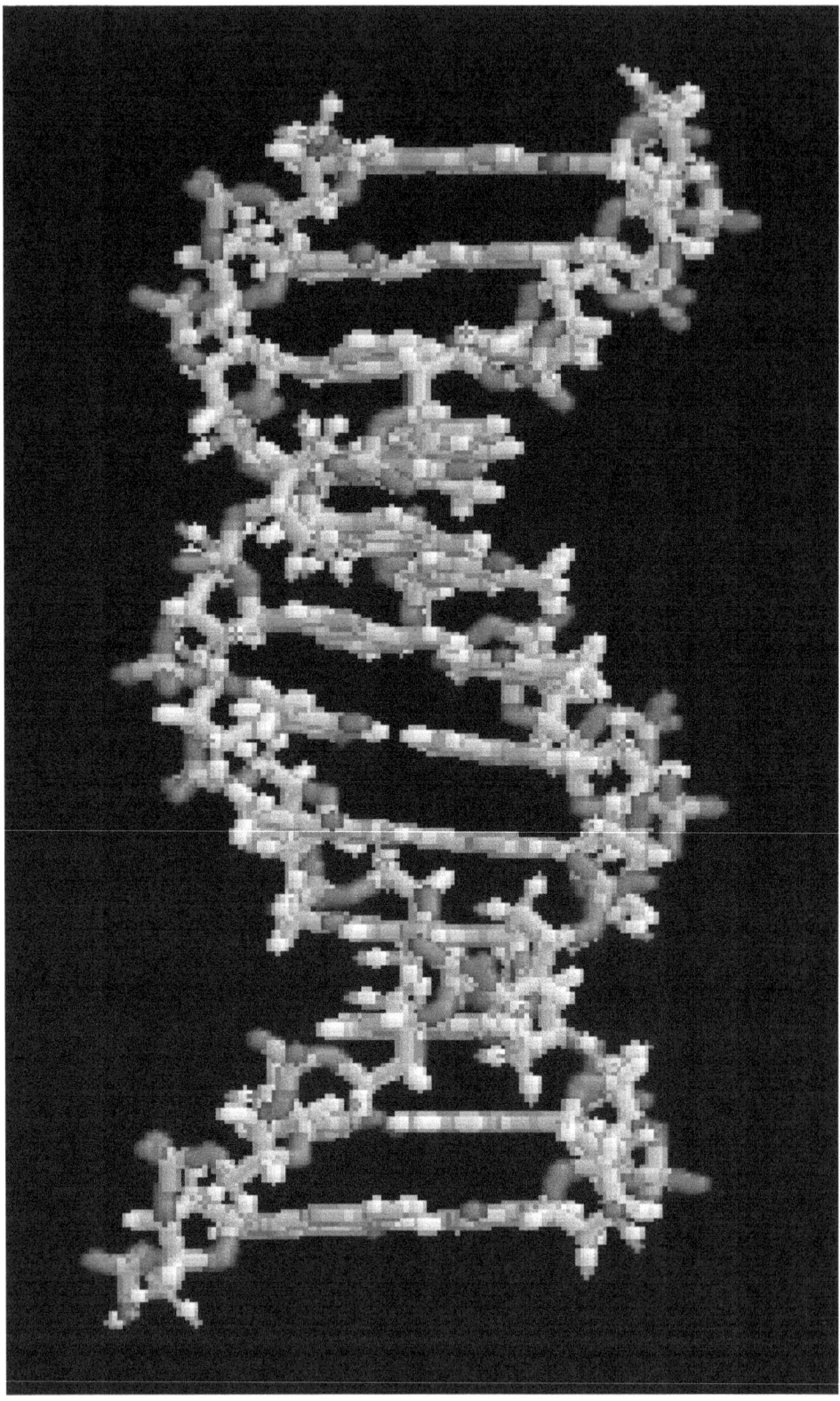

Nucleic acids may not be the only biomolecules in the Universe capable of coding for life processes.[1]

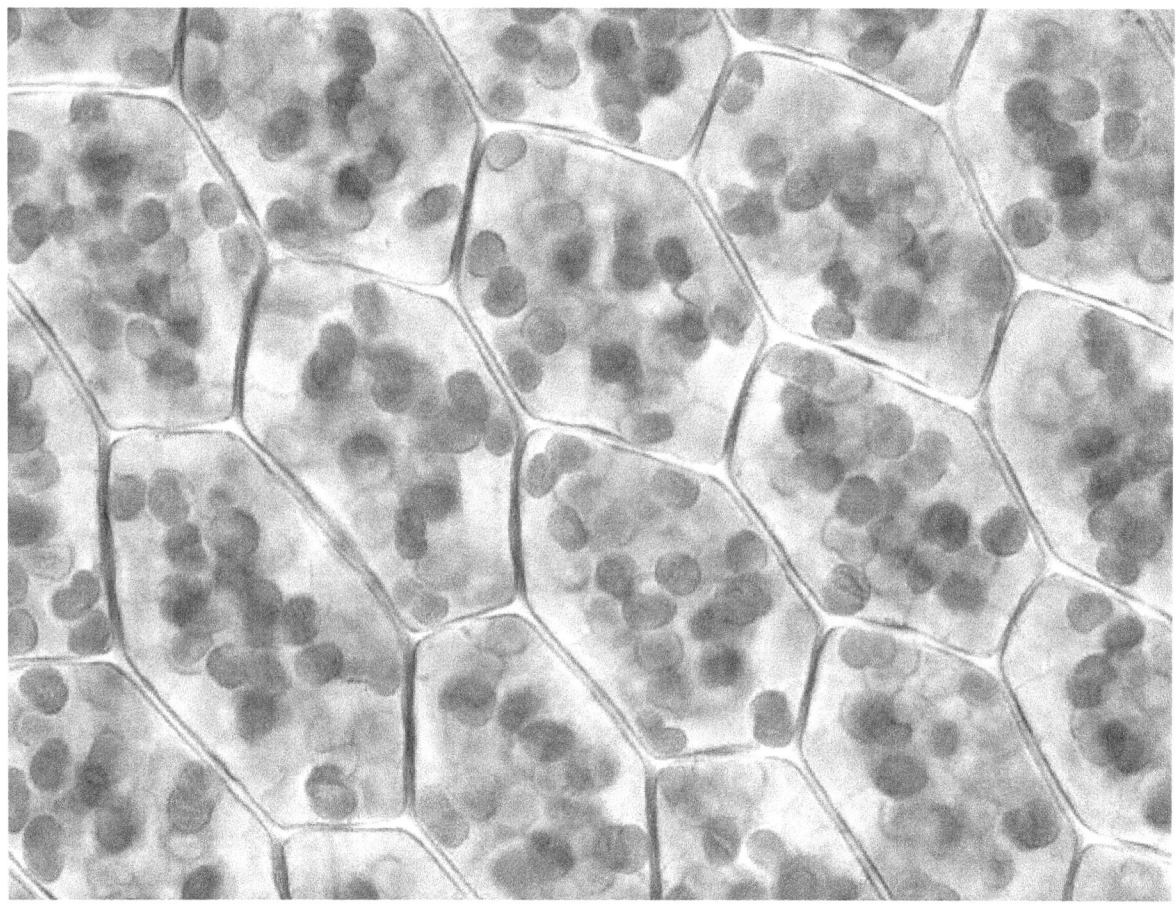

It is not known whether life elsewhere in the universe would utilize cell structures like those found on Earth. (Chloroplasts within plant cells shown here.)[23]

In June 2014, the John W. Kluge Center of the Library of Congress held a seminar focusing on astrobiology. Panel members (L to R) Robin Lovin, Derek Malone-France, and Steven J. Dick

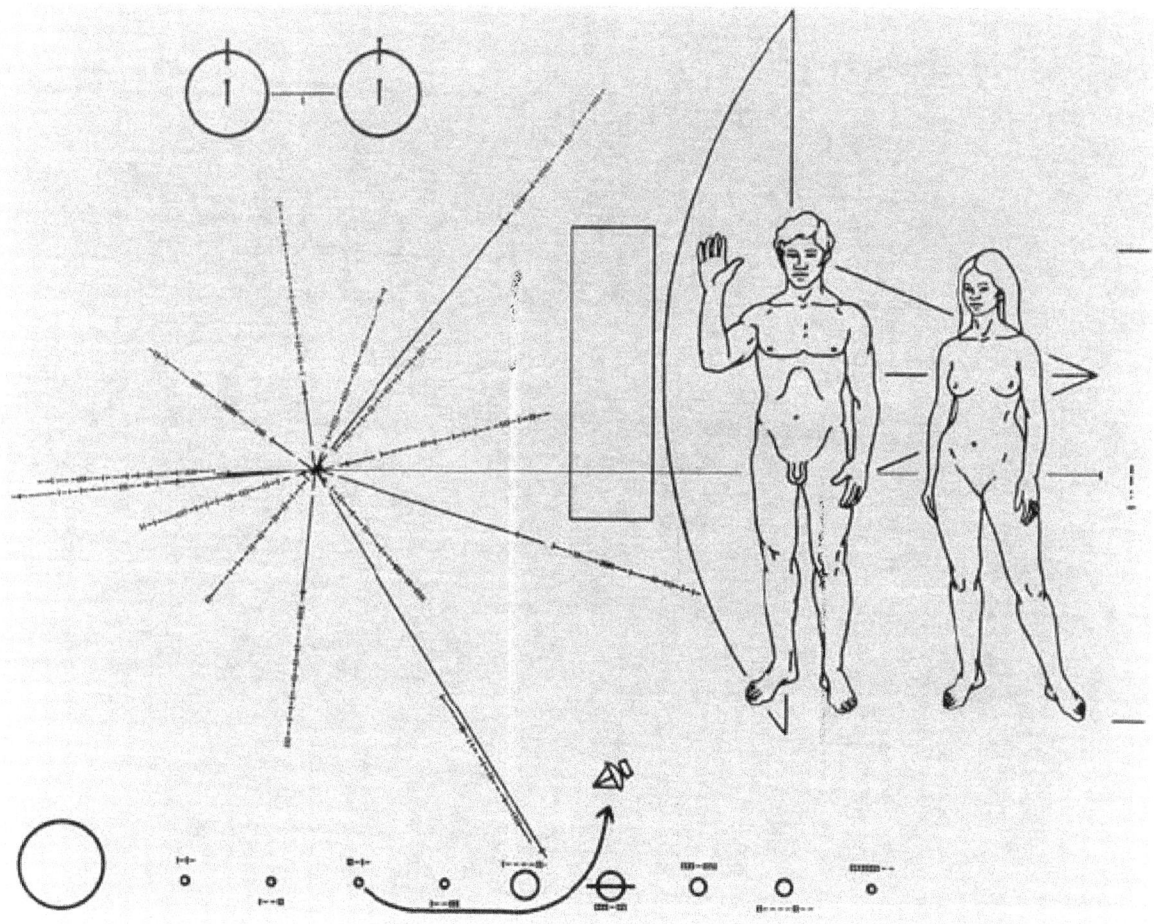

The illustration on the Pioneer plaque

Artist's impression of the extrasolar planet OGLE-2005-BLG-390Lb orbiting its star 20,000 light-years from Earth; this planet was discovered with gravitational microlensing.

The NASA Kepler mission, launched in March 2009, searches for extrasolar planets.

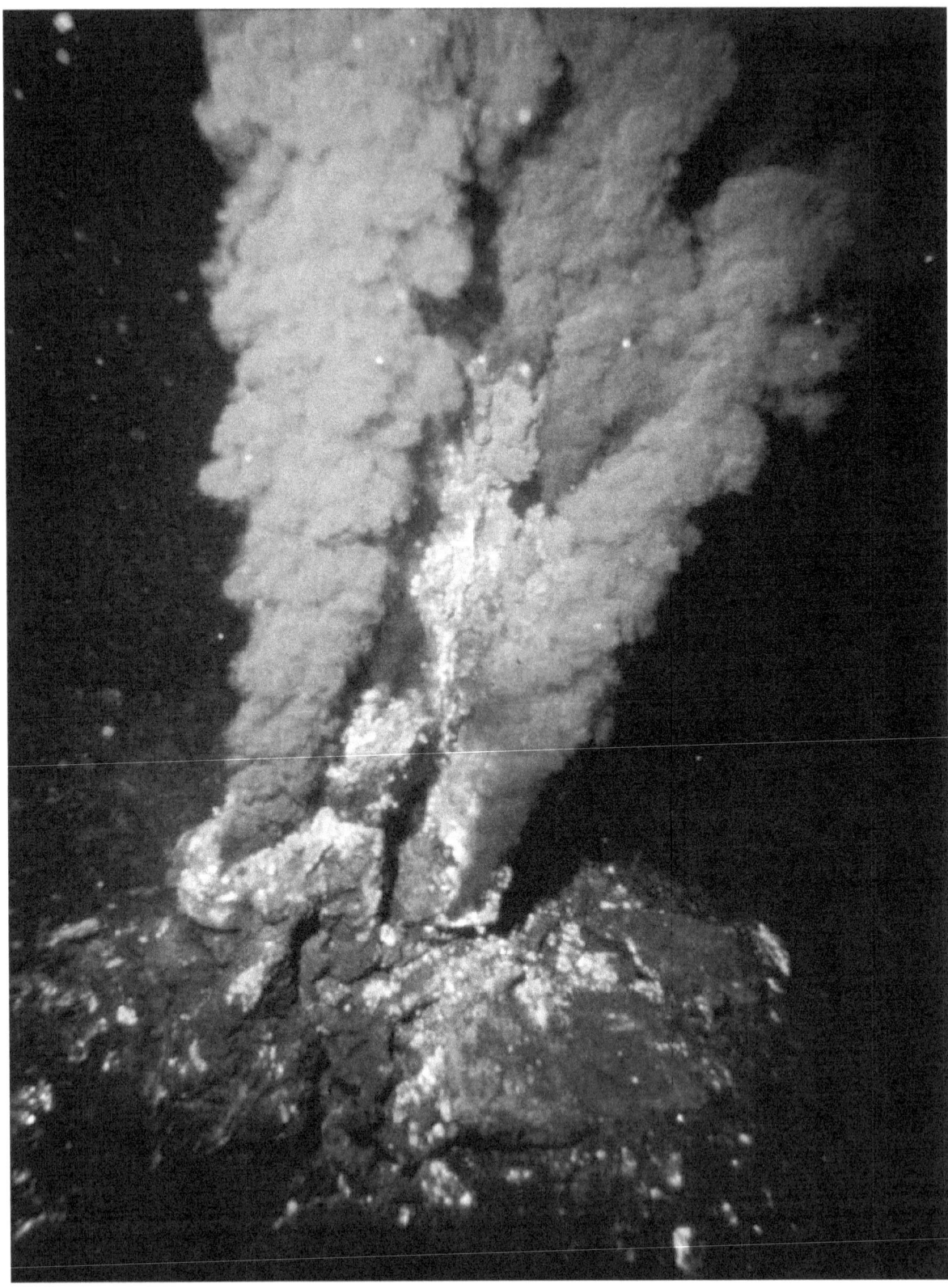

Hydrothermal vents are able to support extremophile bacteria on Earth and may also support life in other parts of the cosmos.

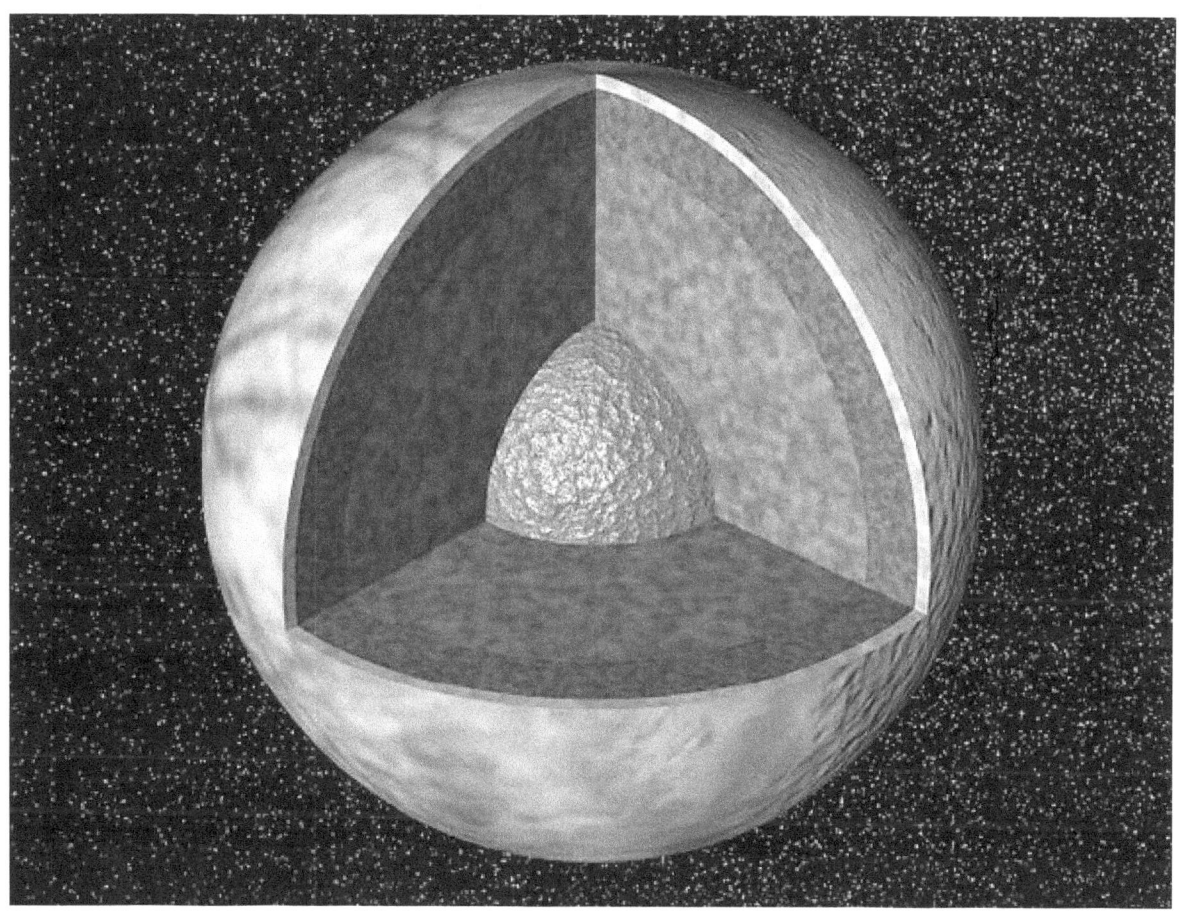

Europa, due to the ocean that exists under its icy surface, might host some form of microbial life.

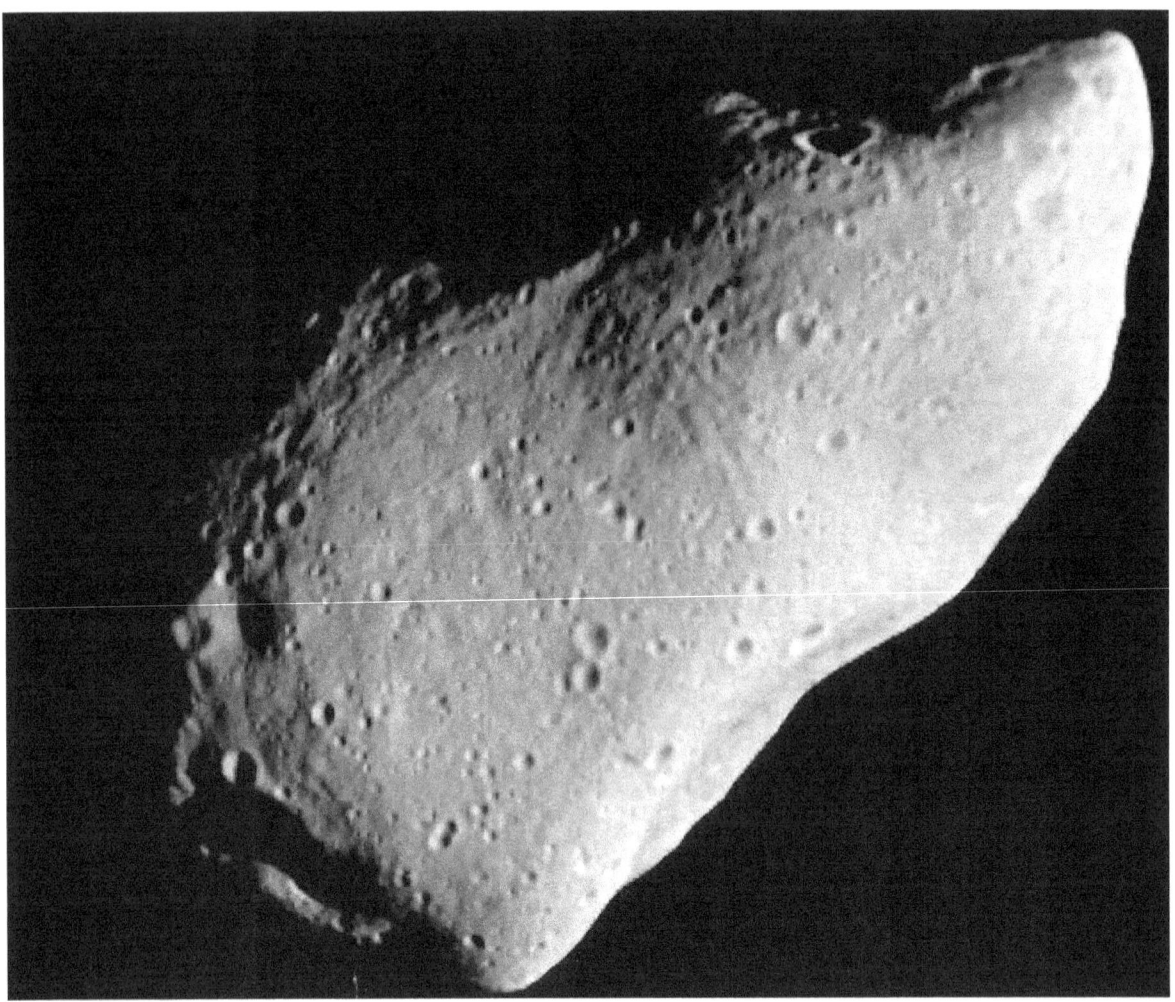

Asteroid(s) may have transported life to Earth.

Carl Sagan posing with a model of the Viking Lander.

Replica of the 33.2 kg Beagle-2 lander

Mars Science Laboratory rover concept artwork

ExoMars rover model

Chapter 3

Biotic material

Biotic material or **biological derived material** is any material that originates from living organisms. Most such materials contain carbon and are capable of decay.

The earliest life on Earth arose at least 3.5 billion years ago.[1][2][3] Earlier physical evidences of life include graphite, a biogenic substance, in 3.7 billion-year-old metasedimentary rocks discovered in southwestern Greenland,[4] as well as, "remains of biotic life" found in 4.1 billion-year-old rocks in Western Australia.[5][6] Earth's biodiversity has expanded continually except when interrupted by mass extinctions.[7] Although scholars estimate that over 99 percent of all species of life (over five billion)[8] that ever lived on Earth are extinct,[9][10] there are still an estimated 10–14 million extant species,[11][12] of which about 1.2 million have been documented and over 86% have not yet been described.[13]

Examples of biotic materials are wood, linoleum, straw, humus, manure, bark, crude oil, cotton, spider silk, chitin, fibrin, and bone.

The use of biotic materials, and processed biotic materials (bio-based material) as alternative natural materials, over synthetics is popular with those who are environmentally conscious because such materials are usually biodegradable, renewable, and the processing is commonly understood and has minimal environmental impact. However, not all biotic materials are used in an environmentally friendly way, such as those that require high levels of processing, are harvested unsustainably, or are used to produce carbon emissions.

When the source of the recently living material has little importance to the product produced, such as in the production of biofuels, biotic material is simply called biomass. Many fuel sources may have biological sources, and may be divided roughly into fossil fuels, and biofuel.

In soil science, biotic material is often referred to as *organic matter*. Biotic materials in soil include glomalin, Dopplerite and humic acid. Some biotic material may not be considered to be organic matter if it is low in organic compounds, such as a clam's shell, which is an essential component of the living organism, but contains little organic carbon.

Examples of the use of biotic materials include:

- Alternative natural materials

- building material, for a stylistic reasons, or to reduce allergic reactions.

- clothing

- energy production

- food

- medicine

- ink

- composting and mulch

3.1 References

[1] Schopf, JW, Kudryavtsev, AB, Czaja, AD, and Tripathi, AB. (2007). *Evidence of Archean life: Stromatolites and microfossils.* Precambrian Research 158:141–155.

[2] Schopf, JW (2006). *Fossil evidence of Archaean life.* Philos Trans R Soc Lond B Biol Sci 29:361(1470) 869-85.

[3] Hamilton Raven, Peter; Brooks Johnson, George (2002). *Biology.* McGraw-Hill Education. p. 68. ISBN 978-0-07-112261-0. Retrieved 7 July 2013.

[4] Ohtomo, Yoko; Kakegawa, Takeshi; Ishida, Akizumi; et al. (January 2014). "Evidence for biogenic graphite in early Archaean Isua metasedimentary rocks". *Nature Geoscience* (London: Nature Publishing Group) 7 (1): 25–28. Bibcode:2014NatGe...7...250.doi:10.1038/ngeo2025. ISSN 1752-0894.

[5] Borenstein, Seth (19 October 2015). "Hints of life on what was thought to be desolate early Earth". *Excite* (Yonkers, NY: Mindspark Interactive Network). Associated Press. Retrieved 2015-10-20.

[6] Bell, Elizabeth A.; Boehnike, Patrick; Harrison, T. Mark; et al. (19 October 2015). "Potentially biogenic carbon preserved in a 4.1 billion-year-old zircon" (PDF). *Proc. Natl. Acad. Sci. U.S.A.* (Washington, D.C.: National Academy of Sciences). doi:10.1073/pnas.1517557112. ISSN 1091-6490. Retrieved 2015-10-20. Early edition, published online before print.

[7] Sahney, S., Benton, M.J. and Ferry, P.A. (27 January 2010). "Links between global taxonomic diversity, ecological diversity and the expansion of vertebrates on land" (PDF). *Biology Letters* 6 (4): 544–47. doi:10.1098/rsbl.2009.1024. PMC 2936204. PMID 20106856.

[8] Kunin, W.E.; Gaston, Kevin, eds. (31 December 1996). *The Biology of Rarity: Causes and consequences of rare—common differences.* ISBN 978-0412633805. Retrieved 26 May 2015.

[9] Stearns, Beverly Peterson; Stearns, S. C.; Stearns, Stephen C. (1 August 2000). *Watching, from the Edge of Extinction.* Yale University Press. p. 1921. ISBN 978-0-300-08469-6. Retrieved 27 December 2014.

[10] Novacek, Michael J. (8 November 2014). "Prehistory's Brilliant Future". *New York Times.* Retrieved 25 December 2014.

[11] May, Robert M. (1988). "How many species are there on earth?". *Science* 241 (4872): 1441–1449. Bibcode:1988Sci...241.1441M. doi:10.1126/science.241.4872.1441. PMID 17790039.

[12] Miller, G.; Spoolman, Scott (1 January 2012). "Biodiversity and Evolution". *Environmental Science.* Cengage Learning. p. 62. ISBN 1-133-70787-4. Retrieved 27 December 2014.

[13] Mora, C.; Tittensor, D.P.; Adl, S.; Simpson, A.G.; Worm, B. (23 August 2011). "How many species are there on Earth and in the ocean?". *PLOS Biology.* doi:10.1371/journal.pbio.1001127. Retrieved 26 May 2015.

Chapter 4

Exoplanet

Artist's view gives an impression of how commonly planets orbit the stars in the Milky Way.[1]

An **exoplanet** or **extrasolar planet** is a planet that orbits a star other than the Sun. Over 2000 exoplanets have been discovered since 1988 (2030 planets in 1288 planetary systems including 502 multiple planetary systems as of 11 December 2015).[3]

The *Kepler space telescope* has also detected a few thousand[4][5] candidate planets,[6][7] of which about 11% may be false positives.[8] There is at least one planet on average per star.[9] About 1 in 5 Sun-like stars[lower-alpha 1] have an "Earth-sized"[lower-alpha 2] planet in the habitable zone,[lower-alpha 3] with the nearest expected to be within 12 light-years distance from Earth.[10][11] Assuming 200 billion stars in the Milky Way,[lower-alpha 4] that would be 11 billion potentially habitable Earth-sized planets in the Milky Way, rising to 40 billion if planets orbiting the numerous red dwarfs are included.[12]

The least massive planet known is PSR B1257+12 A, which is about twice the mass of the Moon. The most massive

Size comparison of Jupiter and the exoplanet TrES-3b. TrES-3b has an orbital period of only 31 hours[21] and is classified as a Hot Jupiter for being large and close to its star, making it one of the easiest planets to detect by the transit method.

planet listed on the NASA Exoplanet Archive is DENIS-P J082303.1-491201 b,[13][14] about 29 times the mass of Jupiter, although according to most definitions of a planet, it is too massive to be a planet and may be a brown dwarf instead. There are planets that are so near to their star that they take only a few hours to orbit and there are others so far away that they take thousands of years to orbit. Some are so far out that it is difficult to tell whether they are gravitationally bound to the star. Almost all of the planets detected so far are within the Milky Way, but there have also been a few possible detections of extragalactic planets.

The discovery of exoplanets has intensified interest in the search for extraterrestrial life. There is special interest in planets that orbit in a star's habitable zone, where it is possible for liquid water (and therefore life) to exist on the surface. The study of planetary habitability also considers a wide range of other factors in determining the suitability of a planet for hosting life.[15]

Besides exoplanets, there are also rogue planets, which do not orbit any star and which tend to be considered separately, especially if they are gas giants, in which case they are often counted, like WISE 0855−0714, as sub-brown dwarfs.[16] The rogue planets in the Milky Way possibly number in the trillions.[17]

4.1   Definition

4.1.1   IAU

The official definition of "planet" used by the International Astronomical Union (IAU) only covers the Solar System and thus does not apply to exoplanets.[18][19] As of April 2011, the only defining statement issued by the IAU that pertains to exoplanets is a working definition issued in 2001 and modified in 2003.[20] That definition contains the following criteria:

- Objects with true masses below the limiting mass for thermonuclear fusion of deuterium (currently calculated to be 13 Jupiter masses for objects of solar metallicity) that orbit stars or stellar remnants are "planets" (no matter how they formed). The minimum mass/size required for an extrasolar object to be considered a planet should be the same as that used in the Solar System.

- Substellar objects with true masses above the limiting mass for thermonuclear fusion of deuterium are "brown dwarfs", no matter how they formed or where they are located.

- Free-floating objects in young star clusters with masses below the limiting mass for thermonuclear fusion of deuterium are not "planets", but are "sub-brown dwarfs" (or whatever name is most appropriate).

4.1.2 Alternatives

However, the IAU's working definition is not universally accepted. One alternate suggestion is that planets should be distinguished from brown dwarfs on the basis of formation. It is widely believed that giant planets form through core accretion, and that process may sometimes produce planets with masses above the deuterium fusion threshold;[21][22][23] massive planets of that sort may have already been observed.[24] Brown dwarfs form like stars from the direct collapse of clouds of gas and this formation mechanism also produces objects that are below the 13 MJ_{up} limit and can be as low as 1 MJ_{up}.[25] Objects in this mass range that orbit their stars with wide separations of hundreds or thousands of AU and have large star/object mass ratios likely formed as brown dwarfs; their atmospheres would likely have a composition more similar to their host star than accretion-formed planets which would contain increased abundances of heavier elements. Most directly imaged planets as of April 2014 are massive and have wide orbits so probably represent the low-mass end of brown dwarf formation.[26]

Also, the 13-Jupiter-mass cutoff does not have precise physical significance. Deuterium fusion can occur in some objects with a mass below that cutoff.[23] The amount of deuterium fused depends to some extent on the composition of the object.[27] The Extrasolar Planets Encyclopaedia includes objects up to 25 Jupiter masses, saying, "The fact that there is no special feature around 13 MJ_{up} in the observed mass spectrum reinforces the choice to forget this mass limit".[28] The Exoplanet Data Explorer includes objects up to 24 Jupiter masses with the advisory: "The 13 Jupiter-mass distinction by the IAU Working Group is physically unmotivated for planets with rocky cores, and observationally problematic due to the sin i ambiguity."[29] The NASA Exoplanet Archive includes objects with a mass (or minimum mass) equal to or less than 30 Jupiter masses.[30] Another criterion for separating planets and brown dwarfs, rather than deuterium fusion, formation process or location, is whether the core pressure is dominated by coulomb pressure or electron degeneracy pressure with the dividing line at around 5 Jupiter masses.[31][32] Another suggestion, based on mass-density relationships, is that the dividing line should be at 60 Jupiter masses.[33]

4.2 History of detection

For centuries philosophers and scientists supposed that extrasolar planets existed, but there was no way of detecting them or of knowing their frequency or how similar they might be to the planets of the Solar System. Various detection claims made in the nineteenth century were rejected by astronomers. The first confirmed detection came in 1992, with the discovery of several terrestrial-mass planets orbiting the pulsar PSR B1257+12.[34] The first confirmation of an exoplanet orbiting a main-sequence star was made in 1995, when a giant planet was found in a four-day orbit around the nearby star 51 Pegasi. Some exoplanets have been imaged directly by telescopes, but the vast majority have been detected through indirect methods such as the transit method and the radial-velocity method.

4.2.1 Early speculations

This space we declare to be infinite... In it are an infinity of worlds of the same kind as our own.

Giordano Bruno (1584)[35]

In the sixteenth century the Italian philosopher Giordano Bruno, an early supporter of the Copernican theory that Earth and other planets orbit the Sun (heliocentrism), put forward the view that the fixed stars are similar to the Sun and are likewise accompanied by planets.

In the eighteenth century the same possibility was mentioned by Isaac Newton in the "General Scholium" that concludes his *Principia*. Making a comparison to the Sun's planets, he wrote "And if the fixed stars are the centers of similar systems, they will all be constructed according to a similar design and subject to the dominion of *One*."[36]

In 1952, more than 40 years before the first hot Jupiter was discovered, Otto Struve wrote that there is no compelling reason why planets could not be much closer to their parent star than is the case in the Solar System, and proposed that Doppler spectroscopy and the transit method could detect super-Jupiters in short orbits.[37]

4.2.2 Discredited claims

Claims of exoplanet detections have been made since the nineteenth century. Some of the earliest involve the binary star 70 Ophiuchi. In 1855 Capt. W. S. Jacob at the East India Company's Madras Observatory reported that orbital anomalies made it "highly probable" that there was a "planetary body" in this system.[38] In the 1890s, Thomas J. J. See of the University of Chicago and the United States Naval Observatory stated that the orbital anomalies proved the existence of a dark body in the 70 Ophiuchi system with a 36-year period around one of the stars.[39] However, Forest Ray Moulton published a paper proving that a three-body system with those orbital parameters would be highly unstable.[40] During the 1950s and 1960s, Peter van de Kamp of Swarthmore College made another prominent series of detection claims, this time for planets orbiting Barnard's Star.[41] Astronomers now generally regard all the early reports of detection as erroneous.[42]

In 1991 Andrew Lyne, M. Bailes and S. L. Shemar claimed to have discovered a pulsar planet in orbit around PSR 1829-10, using pulsar timing variations.[43] The claim briefly received intense attention, but Lyne and his team soon retracted it.[44]

4.2.3 Confirmed discoveries

Main article: Discoveries of exoplanets
See also: List of exoplanet firsts

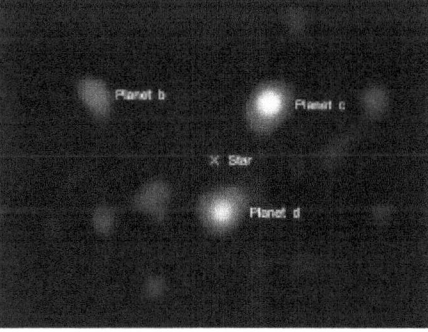

The three known planets of the star HR8799, as imaged by the Hale Telescope. The light from the central star was blanked out by a vector vortex coronagraph.

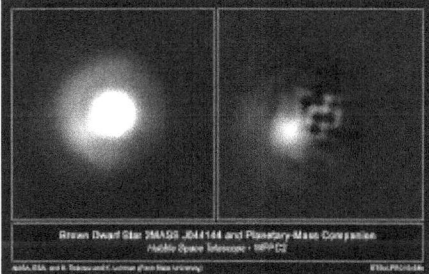

2MASS J044144 is a brown dwarf with a companion about 5–10 times the mass of Jupiter. It is not clear whether this companion object is a sub-brown dwarf or a planet.

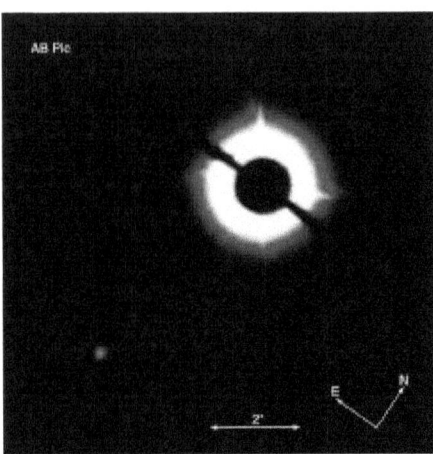

Coronagraphic image of AB Pictoris showing a companion (bottom left), which is either a brown dwarf or a massive planet. The data was obtained on 16 March 2003 with NACO on the VLT, using a 1.4 arcsec occulting mask on top of AB Pictoris.

As of 11 December 2015, a total of 2030 confirmed exoplanets are listed in the Extrasolar Planets Encyclopaedia, including a few that were confirmations of controversial claims from the late 1980s.[3] The first published discovery to receive subsequent confirmation was made in 1988 by the Canadian astronomers Bruce Campbell, G. A. H. Walker, and Stephenson Yang of the University of Victoria and the University of British Columbia.[45] Although they were cautious about claiming a planetary detection, their radial-velocity observations suggested that a planet orbits the star Gamma Cephei. Partly because the observations were at the very limits of instrumental capabilities at the time, astronomers remained skeptical for several years about this and other similar observations. It was thought some of the apparent planets might instead have been brown dwarfs, objects intermediate in mass between planets and stars. In 1990 additional observations were published that supported the existence of the planet orbiting Gamma Cephei,[46] but subsequent work in 1992 again raised serious doubts.[47] Finally, in 2003, improved techniques allowed the planet's existence to be confirmed.[48]

On 9 January 1992, radio astronomers Aleksander Wolszczan and Dale Frail announced the discovery of two planets orbiting the pulsar PSR 1257+12.[34] This discovery was confirmed, and is generally considered to be the first definitive detection of exoplanets. Follow-up observations solidified these results, and confirmation of a third planet in 1994 revived the topic in the popular press.[49] These pulsar planets are believed to have formed from the unusual remnants of the supernova that produced the pulsar, in a second round of planet formation, or else to be the remaining rocky cores of gas giants that somehow survived the supernova and then decayed into their current orbits.

On 6 October 1995, Michel Mayor and Didier Queloz of the University of Geneva announced the first definitive detection of an exoplanet orbiting a main-sequence star, namely the nearby G-type star 51 Pegasi.[50][51] This discovery, made at the Observatoire de Haute-Provence, ushered in the modern era of exoplanetary discovery. Technological advances, most notably in high-resolution spectroscopy, led to the rapid detection of many new exoplanets: astronomers could detect exoplanets indirectly by measuring their gravitational influence on the motion of their host stars. More extrasolar planets were later detected by observing the variation in a star's apparent luminosity as an orbiting planet passed in front of it.

Initially, most known exoplanets were massive planets that orbited very close to their parent stars. Astronomers were surprised by these "hot Jupiters", because theories of planetary formation had indicated that giant planets should only form at large distances from stars. But eventually more planets of other sorts were found, and it is now clear that hot Jupiters are a minority of exoplanets. In 1999, Upsilon Andromedae became the first main-sequence star known to have multiple planets.[52] Kepler-16 contains the first discovered planet that orbits around a binary main-sequence star system.[53]

On 26 February 2014, NASA announced the discovery of 715 newly verified exoplanets around 305 stars by the Kepler Space Telescope. These exoplanets were checked using a statistical technique called "verification by multiplicity".[54][55][56] Prior to these results, most confirmed planets were gas giants comparable in size to Jupiter or larger as they are more easily detected, but the Kepler planets are mostly between the size of Neptune and the size of Earth.[54]

On 23 July 2015, NASA announced Kepler-452b, a near-Earth-size planet orbiting the habitable zone of a G2-type star.[57]

4.2.4 Candidate discoveries

As of March 2014, NASA's Kepler mission had identified more than 2,900 planetary candidates, several of them being nearly Earth-sized and located in the habitable zone, some around Sun-like stars.[4][5][58]

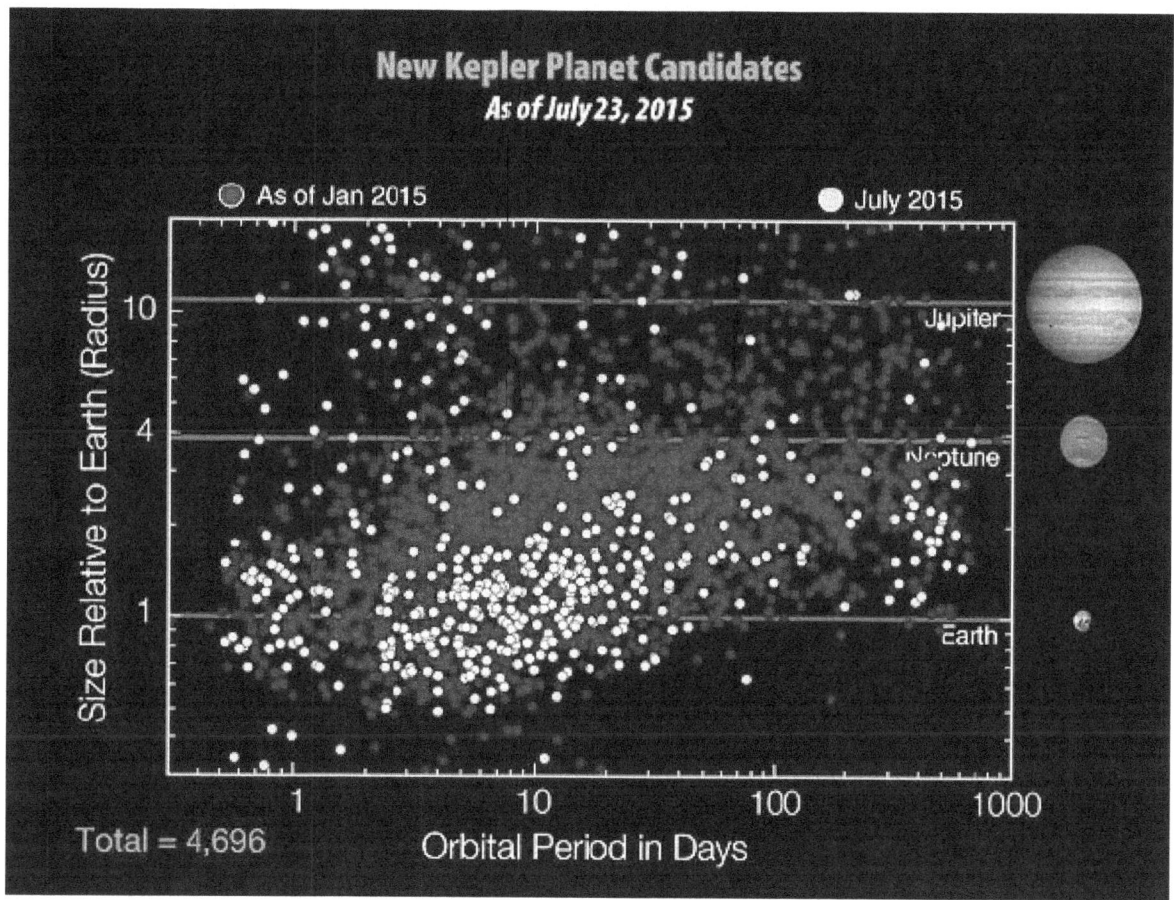

Kepler mission – new exoplanet candidates – as of 23 July 2015.[59]

4.3 Detection methods

Main article: Methods of detecting extrasolar planets

4.3.1 Direct imaging

Main article: Direct imaging

Planets are extremely faint compared to their parent stars. At visible wavelengths, they usually have less than a millionth of their host star's brightness. It is difficult to detect such a faint light source, and furthermore the parent star causes a glare that tends to wash it out. It is necessary to block the light from the parent star in order to reduce the glare while leaving the light from the planet detectable; doing so is a major technical challenge which requires extreme optothermal stability.[60]

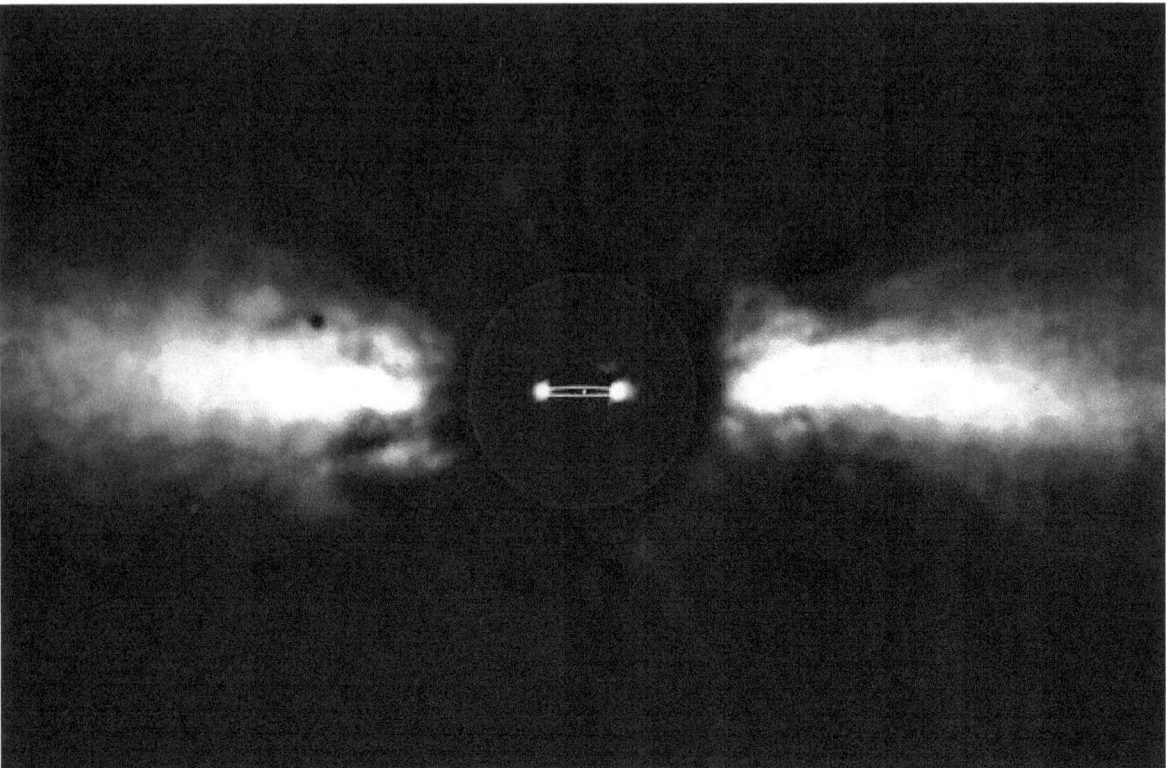

Directly imaged planet, Beta Pictoris b

All exoplanets that have been directly imaged are both large (more massive than Jupiter) and widely separated from their parent star. Most of them are also very hot, so that they emit intense infrared radiation; the images have then been made at infrared where the planet is brighter than it is at visible wavelengths. During the gas-accretion phase of giant-planet formation the star–planet contrast may be even better in H alpha than it is in infrared—an H alpha survey is currently underway.[61]

Specially designed direct-imaging instruments such as Gemini Planet Imager, VLT-SPHERE, and SCExAO will image dozens of gas giants, however the vast majority of known extrasolar planets have only been detected through indirect methods. The following are the indirect methods that have proven useful:

4.3.2   Indirect methods

- Transit method

  If a planet crosses (or transits) in front of its parent star's disk, then the observed brightness of the star drops by a small amount. The amount by which the star dims depends on its size and on the size of the planet, among other factors. This method suffers from a substantial rate of false positives and confirmation from another method is usually considered necessary. The transit method reveals the radius of a planet, and it has the benefit that it sometimes allows a planet's atmosphere to be investigated through spectroscopy. Because the transit method requires that part of the planet's orbit intersect a line-of-sight between the host star and Earth, the probability that an exoplanet in a randomly oriented orbit will be observed to transit the star is somewhat small. The Kepler telescope uses this method.

- Radial velocity or Doppler method

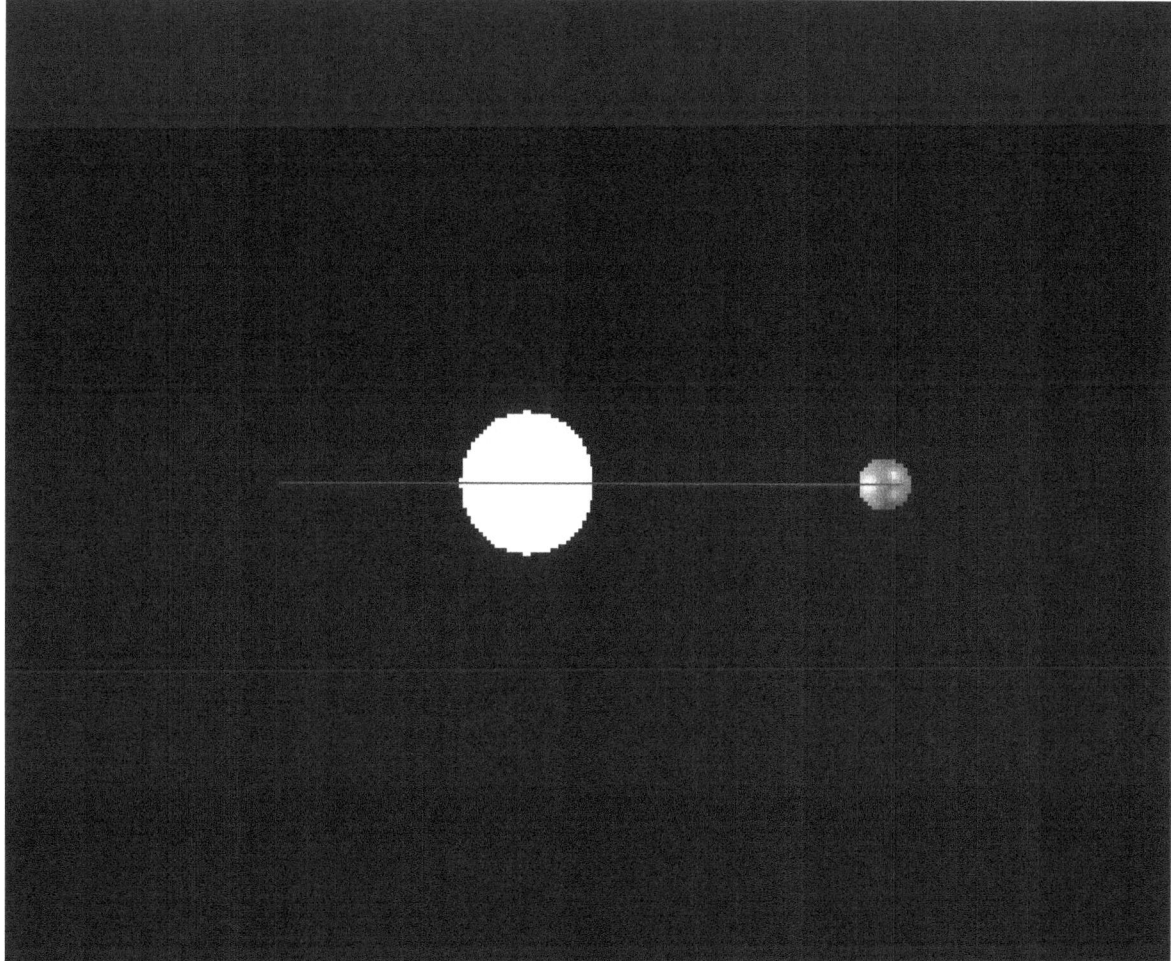

When the star is behind a planet, its brightness will seem to dim

As a planet orbits a star, the star also moves in its own small orbit around the system's center of mass. Variations in the star's radial velocity—that is, the speed with which it moves towards or away from Earth—can be detected from displacements in the star's spectral lines due to the Doppler effect. Extremely small radial-velocity variations can be observed, of 1 m/s or even somewhat less.[62] This method has the advantage of being applicable to stars with a wide range of characteristics. One of its disadvantages is that it cannot determine a planet's true mass, but can only set a lower limit on that mass. However, if the radial velocity of the planet itself can be distinguished from the radial velocity of the star, then the true mass can be determined.[63]

- Transit timing variation (TTV)

When multiple planets are present, each one slightly perturbs the others' orbits. Small variations in the times of transit for one planet can thus indicate the presence of another planet, which itself may or may not transit. For example, variations in the transits of the planet Kepler-19b suggest the existence of a second planet in the system, the non-transiting Kepler-19c.[64][65] If multiple transiting planets exist in one system, then this method can be used to confirm their existence.[66] In another form of the method, timing the eclipses in an eclipsing binary star can reveal an outer planet that orbits both stars; as of August 2013, a few planets have been found in that way with numerous planets confirmed with this method.

- Transit duration variation (TDV)

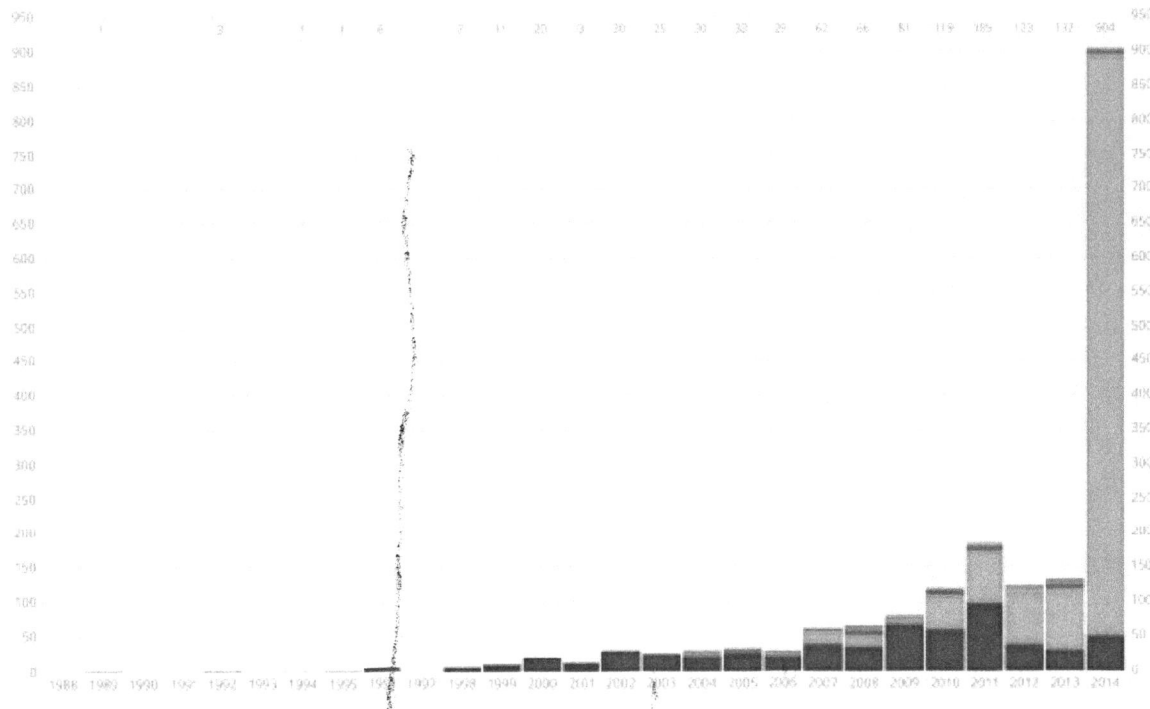

Discovered extrasolar planets per year and by detection method (as of September 2014):

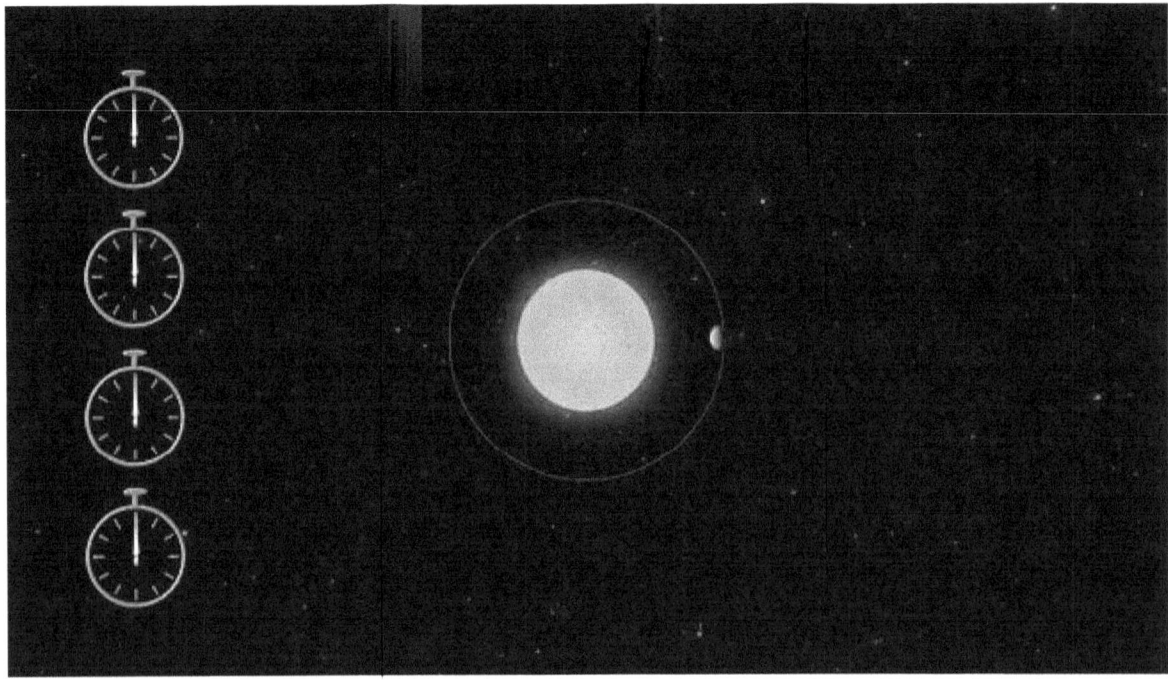

Animation showing difference between planet transit timing of 1-planet and 2-planet systems

When a planet orbits multiple stars or if the planet has moons, its transit time can significantly vary per transit. Although no new planets or moons have been discovered with this method, it is used to successfully confirm many transiting circumbinary planets.[67]

- Gravitational microlensing

Microlensing occurs when the gravitational field of a star acts like a lens, magnifying the light of a distant background star. Planets orbiting the lensing star can cause detectable anomalies in the magnification as it varies over time. Unlike most other methods which have detection bias towards planets with small (or for resolved imaging, large) orbits, microlensing method is most sensitive to detecting planets around 1–10 AU away from Sun-like stars.

- Astrometry

Astrometry consists of precisely measuring a star's position in the sky and observing the changes in that position over time. The motion of a star due to the gravitational influence of a planet may be observable. Because the motion is so small, however, this method has not yet been very productive. It has produced only a few disputed detections, though it has been successfully used to investigate the properties of planets found in other ways.

- Pulsar timing

A pulsar (the small, ultradense remnant of a star that has exploded as a supernova) emits radio waves extremely regularly as it rotates. If planets orbit the pulsar, they will cause slight anomalies in the timing of its observed radio pulses. The first confirmed discovery of an extrasolar planet was made using this method. But as of 2011, it has not been very productive; five planets have been detected in this way, around three different pulsars.

- Variable star timing (pulsation frequency)

Like pulsars, there are some other types of stars which exhibit periodic activity. Deviations from the periodicity can sometimes be caused by a planet orbiting it. As of 2013, a few planets have been discovered with this method.[68]

- Reflection/emission modulations

When a planet orbits very close to the star, it catches a considerable amount of starlight. As the planet orbits around the star, the amount of light changes due to planets having phases from Earth's viewpoint or planet glowing more from one side than the other due to temperature differences.[69]

- Relativistic beaming

Relativistic beaming measures the observed flux from the star due to its motion. The brightness of the star changes as the planet moves closer or further away from its host star.[70]

- Ellipsoidal variations

Massive planets close to their host stars can slightly deform the shape of the star. This causes the brightness of the star to slightly deviate depending how it is rotated relative to Earth.[71]

- Polarimetry

With polarimetry method, a polarized light reflected off the planet is separated from unpolarized light emitted from the star. No new planets have been discovered with this method although a few already discovered planets have been detected with this method.[72][73]

- Circumstellar disks

Disks of space dust surround many stars, believed to originate from collisions among asteroids and comets. The dust can be detected because it absorbs starlight and re-emits it as infrared radiation. Features in the disks may suggest the presence of planets, though this is not considered a definitive detection method.

4.4 Nomenclature

4.4.1 Proper names

Most exoplanets have catalog names which are explained in the following sections, but in July 2014 the IAU launched a process for giving proper names to exoplanets.[74][75] The process involves public nomination and voting for the new names, and the IAU plans to announce new names for 32 planets in 20 systems in mid-December 2015.[76] The decision to give the planets new names followed the private company Uwingu's exoplanet naming contest, which the IAU harshly criticized.[77] Previously a few planets had received unofficial names: notably Osiris (HD 209458 b), Bellerophon (51 Pegasi b), and Methuselah (PSR B1620-26 b).

4.4.2 Multiple-star standard

The convention for naming exoplanets is an extension of the one used by the Washington Multiplicity Catalog (WMC) for multiple-star systems, and adopted by the International Astronomical Union.[78] The brightest member of a star system receives the letter "A". Distinct components not contained within "A" are labeled "B", "C", etc. Subcomponents are designated by one or more suffixes with the primary label, starting with lowercase letters for the 2nd hierarchical level and then numbers for the 3rd.[79] For example, if there is a triple star system in which two stars orbit each other closely with a third star in a more distant orbit, the two closely orbiting stars would be named Aa and Ab, whereas the distant star would named B. For historical reasons, this standard is not always followed: for example Alpha Centauri A, B and C are *not* labelled Alpha Centauri Aa, Ab and B.

4.4.3 Extrasolar planet standard

Following an extension of the above standard, an exoplanet's name is normally formed by taking the name of its parent star and adding a lowercase letter. The first planet discovered in a system is given the designation "b" and later planets are given subsequent letters. If several planets in the same system are discovered at the same time, the closest one to the star gets the next letter, followed by the other planets in order of orbital size.

For instance, in the 55 Cancri system the first planet – 55 Cancri b – was discovered in 1996; two additional farther planets were simultaneously discovered in 2002 with the nearest to the star being named 55 Cancri c and the other 55 Cancri d; a fourth planet was claimed (its existence was later disputed) in 2004 and named 55 Cancri e despite lying closer to the star than 55 Cancri b; and the most recently discovered planet, in 2007, was named 55 Cancri f despite lying between 55 Cancri c and 55 Cancri d.[80] As of April 2012 the highest letter in use is "j", for the unconfirmed planet HD 10180 j, and with "h" being the highest letter for a confirmed planet, belonging to the same host star).[3]

If a planet orbits one member of a binary star system, then an uppercase letter for the star will be followed by a lowercase letter for the planet. Examples are 16 Cygni Bb[81] and HD 178911 Bb.[82] Planets orbiting the primary or "A" star should have 'Ab' after the name of the system, as in HD 41004 Ab.[83] However, the "A" is sometimes omitted; for example the first planet discovered around the primary star of the Tau Boötis binary system is usually called simply Tau Boötis b.[84] The star designation is necessary when more than one star in the system has its own planetary system such as in case of WASP-94 A and WASP-94 B.[85]

If the parent star is a single star, then it may still be regarded as having an "A" designation, though the "A" is not normally written. The first exoplanet found to be orbiting such a star could then be regarded as a secondary subcomponent that should be given the suffix "Ab". For example, 51 Peg Aa is the host star in the system 51 Peg; and the first exoplanet is then 51 Peg Ab. Because most exoplanets are in single-star systems, the implicit "A" designation was simply dropped, leaving the exoplanet name with the lower-case letter only: 51 Peg b.

A few exoplanets have been given names that do not conform to the above standard. For example, the planets that orbit the pulsar PSR 1257 are often referred to with capital rather than lowercase letters. Also, the underlying name of the star system itself can follow several different systems. In fact, some stars (such as Kepler-11) have only received their names due to their inclusion in planet-search programs, previously only being referred to by their celestial coordinates.

4.4.4 Circumbinary planets and 2010 proposal

Hessman et al. state that the implicit system for exoplanet names utterly failed with the discovery of circumbinary planets.[178] They note that the discoverers of the two planets around HW Virginis tried to circumvent the naming problem by calling them "HW Vir 3" and "HW Vir 4", i.e. the latter is the 4th object – stellar or planetary – discovered in the system. They also note that the discoverers of the two planets around NN Serpentis were confronted with multiple suggestions from various official sources and finally chose to use the designations "NN Ser c" and "NN Ser d".

The proposal of Hessman et al. starts with the following two rules:

> **Rule 1**. The formal name of an exoplanet is obtained by appending the appropriate suffixes to the formal name of the host star or stellar system. The upper hierarchy is defined by upper-case letters, followed by lower-case letters, followed by numbers, etc. The naming order within a hierarchical level is for the order of discovery only. (This rule corresponds to the present provisional WMC naming convention.)
>
> **Rule 2**. Whenever the leading capital letter designation is missing, this is interpreted as being an informal form with an implicit "A" unless otherwise explicitly stated. (This rule corresponds to the present exoplanet community usage for planets around single stars.)

They note that under these two proposed rules all of the present names for 99% of the planets around single stars are preserved as informal forms of the IAU sanctioned provisional standard. They would rename Tau Boötis b formally as Tau Boötis Ab, retaining the prior form as an informal usage (using Rule 2, above).

To deal with the difficulties relating to circumbinary planets, the proposal contains two further rules:

> **Rule 3**. As an alternative to the nomenclature standard in Rule 1, a hierarchical relationship can be expressed by concatenating the names of the higher order system and placing them in parentheses, after which the suffix for a lower order system is added.
>
> **Rule 4**. When in doubt (i.e. if a different name has not been clearly set in the literature), the hierarchy expressed by the nomenclature should correspond to dynamically distinct (sub)systems in order of their dynamical relevance. The choice of hierarchical levels should be made to emphasize dynamical relationships, if known.

They submit that the new form using parentheses is the best for known circumbinary planets and has the desirable effect of giving these planets identical sublevel hierarchical labels and stellar component names that conform to the usage for binary stars. They say that it requires the complete renaming of only two exoplanetary systems: The planets around HW Virginis would be renamed HW Vir (AB) b & (AB) c, whereas those around NN Serpentis would be renamed NN Ser (AB) b & (AB) c. In addition the previously known single circumbinary planets around PSR B1620-26 and DP Leonis) can almost retain their names (PSR B1620-26 b and DP Leonis b) as unofficial informal forms of the "(AB)b" designation where the "(AB)" is left out.

The discoverers of the circumbinary planet around Kepler-16 followed the naming scheme proposed by Hessman et al. when naming the body Kepler-16 (AB)-b, or simply Kepler-16b when there is no ambiguity.[153]

4.4.5 Other naming systems

Another nomenclature, often seen in science fiction, uses Roman numerals in the order of planets' positions from the star. (This was inspired by an old system for naming moons of the outer planets, such as "Jupiter IV" for Callisto.) But such a system is impractical for scientific use, because new planets may be found closer to the star, changing all numerals.

4.5 Formation and evolution

See also: Nebular hypothesis, Planetary migration, Formation and evolution of the Solar System and Future of the Earth

Planets form within a few tens of millions of years of their star forming,[86][87][88] and there are stars that are forming today and other stars that are ten billion years old, so unlike the planets of the Solar System, which can only be observed as they are today, studying exoplanets allows the observation of exoplanets at different stages of evolution. When planets form they have hydrogen envelopes that cool and contract over time and, depending on the mass of the planet, some or all of the hydrogen is eventually lost to space. This means that even terrestrial planets can start off with large radii.[89][90][91] An example is Kepler-51b which has only about twice the mass of Earth but is almost the size of Saturn which is a hundred times the mass of Earth. Kepler-51b is quite young at a few hundred million years old.[92]

4.6   Planet-hosting stars

Main article: Planetary system § Planet-hosting stars

There is at least one planet on average per star.[9] About 1 in 5 Sun-like stars[lower alpha 1] have an "Earth-sized"[lower alpha 2] planet in the habitable zone.[11]

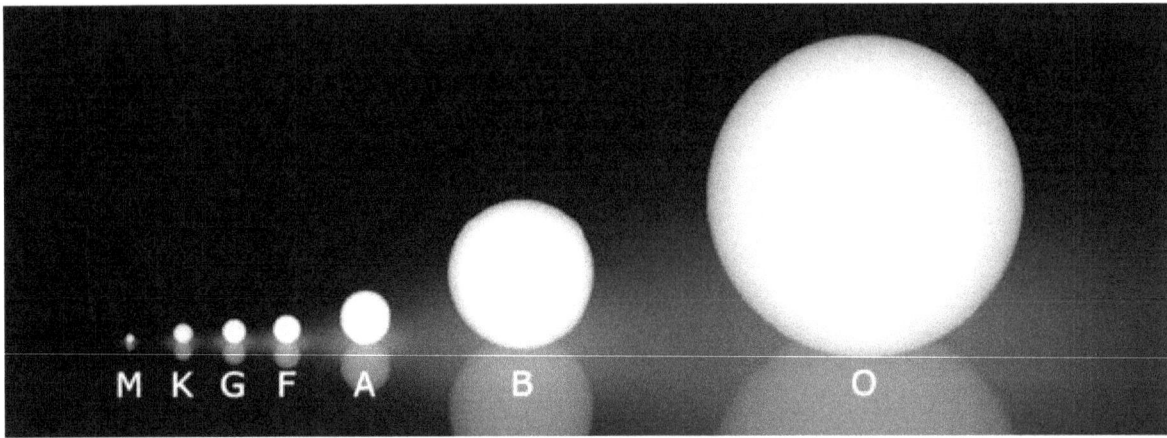

The Morgan-Keenan spectral classification

Most known exoplanets orbit stars roughly similar to the Sun, that is, main-sequence stars of spectral categories F, G, or K. Lower-mass stars (red dwarfs, of spectral category M) are less likely to have planets massive enough to detect by the radial-velocity method.[93][94] Although several tens of planets around red dwarfs have been discovered by the Kepler spacecraft which uses the transit method which can detect smaller planets.

Stars with a higher metallicity than the Sun are more likely to have planets, especially giant planets, than stars with lower metallicity.[95]

Some planets orbit one member of a binary star system,[96] and several circumbinary planets have been discovered which orbit around both members of binary star. A few planets in triple star systems are known[97] and one in the quadruple system Kepler 64.

4.7   Orbital parameters

Most known extrasolar planet candidates have been discovered using indirect methods and therefore only some of their physical and orbital parameters can be determined. For example, out of the six independent parameters that define an orbit, the radial-velocity method can determine four: semi-major axis, eccentricity, longitude of periastron, and time of periastron. Two parameters remain unknown: inclination and longitude of the ascending node.

4.7.1   Distance from star, semi-major axis and orbital period

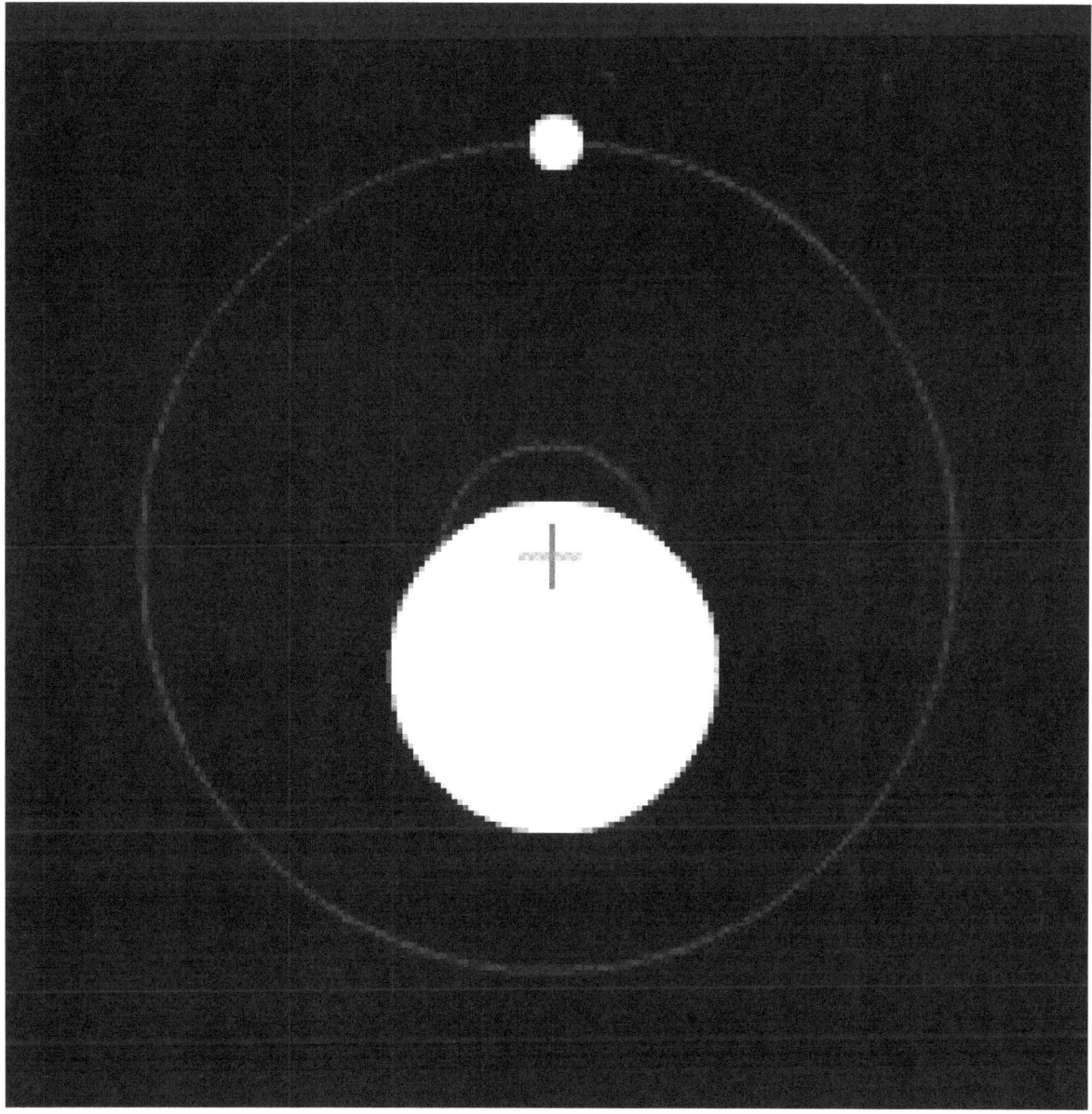

Diagram showing how a planet and a star orbit their common center of mass (red cross).

There are exoplanets that are much closer to their parent star than any planet in the Solar System is to the Sun, and there are also exoplanets that are much further from their star. Mercury, the closest planet to the Sun at 0.4 astronomical units (AU), takes 88 days for an orbit, but the smallest known orbits of exoplanets have orbital periods of only a few hours, e.g. Kepler-70b. The Kepler-11 system has five of its planets in smaller orbits than Mercury's. Neptune is 30 AU from the Sun and takes 165 years to orbit it, but there are exoplanets that are thousands of AU from their star and take tens of thousands of years to orbit, e.g. GU Piscium b.[98]

The orbit of a planet is not centered on the star but on their common center of mass (see diagram on right). For circular orbits, the semi-major axis is the distance between the planet and the center of mass of the system. For elliptical orbits, the planet–star distance varies over the course of the orbit, in which case the semi-major axis is the average of the largest and smallest distances between the planet and the center of mass of the system. If the sizes of the star and planet are relatively small compared to the size of the orbit and the orbit is nearly circular and the center of mass is not too far from

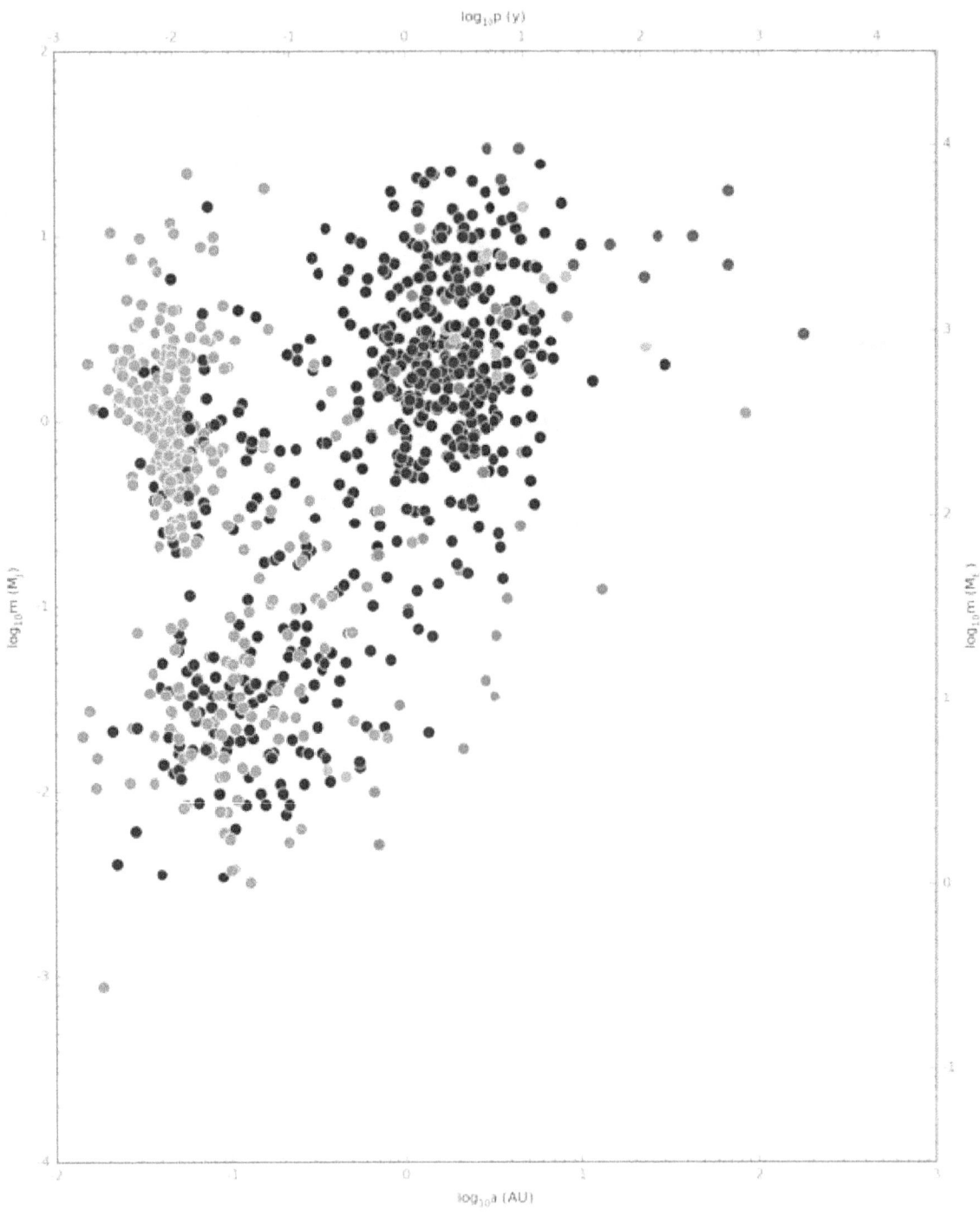

Scatterplot showing masses and orbital radii (and period) of all extrasolar planets discovered through September 2014, with colors indicating method of detection: For reference, Solar System planets are marked as gray circles. The horizontal axis plots the log of the semi-major axis, and the vertical axis plots the log of the mass.

the star's center, such as in the Earth–Sun system, then the distance from any point on the star to any point on the planet is approximately the same as the semi-major axis. However, when a star's radius expands when it turns into a red giant, then the distance between the planet and the star's surface can become close to zero, or even less than zero if the planet has been engulfed by the expanding red giant, whereas the center of mass from which the semi-major axis is measured will still be near the center of the red giant.

Orbital period is the time taken to complete one orbit. For any given star, the shorter the semi-major axis of a planet, the shorter the orbital period. Also comparing planets around different stars but with the same semi-major axis, the more massive the star, the shorter the orbital period.

Over the lifetime of a star, the semi-major axes of its planets changes. This planetary migration happens especially during the formation of the planetary system when planets interact with the protoplanetary disk and each other until a relatively stable position is reached, and later in the red-giant and asymptotic-giant-branch phases when the star expands and engulfs the nearest planets that can cause them to move inwards, and when the red giant loses mass as the outer layers dissipate causing planets to move outwards as a result of the red giant's reduced gravitational field.

The radial-velocity and transit methods are most sensitive to planets with small orbits. The earliest discoveries such as 51 Peg b were gas giants with orbits of a few days.[93] These "hot Jupiters" likely formed further out and migrated inwards. The Kepler spacecraft has found planets with even shorter orbits of only a few hours, which places them within the star's upper atmosphere or corona, and these planets are Earth-sized or smaller and are probably the left-over solid cores of giant planets that have evaporated due to being so close to the star,[99] or even being engulfed by the star in its red-giant phase in the case of Kepler-70b. As well as evaporation, other reasons why larger planets are unlikely to survive orbits only a few hours long include orbital decay caused by tidal force, tidal-inflation instability, and Roche-lobe overflow.[100] The Roche limit implies that small planets with orbits of a few hours are likely made mostly of iron.[100]

The direct imaging method is most sensitive to planets with large orbits, and has discovered some planets that have planet–star separations of hundreds of AU. However, protoplanetary disks are usually only around 100 AU in radius, and core accretion models predict giant planet formation to be within 10 AU, where the planets can coalesce quickly enough before the disk evaporates. Very-long-period giant planets may have been rogue planets that were captured,[101] or formed close-in and gravitationally scattered outwards, or the planet and star could be a mass-imbalanced wide binary system with the planet being the primary object of its own separate protoplanetary disk. Gravitational instability models might produce planets at multi-hundred AU separations but this would require unusually large disks.[102][103] For planets with very wide orbits up to several hundred thousand AU it may be difficult to observationally determine whether the planet is gravitationally bound to the star.

Most planets that have been discovered are within a couple of AU from their host star because the most used methods (radial-velocity and transit) require observation of several orbits to confirm that the planet exists and there has only been enough time since these methods were first used to cover small separations. Some planets with larger orbits have been discovered by direct imaging but there is a middle range of distances, roughly equivalent to the Solar System's gas giant region, which is largely unexplored. Direct imaging equipment for exploring that region is being installed on the world's largest telescopes and should begin operation in 2014, e.g. Gemini Planet Imager and VLT-SPHERE. The microlensing method has detected a few planets in the 1–10 AU range.[104] It appears plausible that in most exoplanetary systems, there are one or two giant planets with orbits comparable in size to those of Jupiter and Saturn in the Solar System. Giant planets with substantially larger orbits are now known to be rare, at least around Sun-like stars.[105]

The distance of the habitable zone from a star depends on the type of star and this distance changes during the star's lifetime as the size and temperature of the star changes.

- The Fate of Scattered Planets, Benjamin C. Bromley, Scott J. Kenyon, 10 October 2014

- Planetary Populations in the Mass-Period Diagram: A Statistical Treatment of Exoplanet Formation and the Role of Planet Traps, Yasuhiro Hasegawa, Ralph E. Pudritz, 8 October 2013

4.7.2   Eccentricity

The eccentricity of an orbit is a measure of how elliptical (elongated) it is. All the planets of the Solar System except for Mercury have near-circular orbits (e<0.1).[106] Most exoplanets with orbital periods of 20 days or less have near-circular orbits, i.e. very low eccentricity. That is believed to be due to tidal circularization: reduction of eccentricity over time due to gravitational interaction between two bodies. The mostly sub-Neptune-sized planets found by the Kepler spacecraft with short orbital periods have very circular orbits.[54] By contrast, the giant planets with longer orbital periods discovered by radial-velocity methods have quite eccentric orbits. (As of July 2010, 55% of such exoplanets have eccentricities greater than 0.2, whereas 17% have eccentricities greater than 0.5.[3]) Moderate to high eccentricities (e>0.2) of giant planets are *not* an observational selection effect, because a planet can be detected about equally well regardless of the eccentricity of its orbit. The prevalence of elliptical orbits for giant planets is a major puzzle, because current theories of planetary formation strongly suggest planets should form with circular (that is, non-eccentric) orbits.[107]

However, for weak Doppler signals near the limits of the current detection ability the eccentricity becomes poorly constrained and biased towards higher values. It is suggested that some of the high eccentricities reported for low-mass exoplanets may be overestimates, because simulations show that many observations are also consistent with two planets on circular orbits. Reported observations of single planets in moderately eccentric orbits have about a 15% chance of being a pair of planets.[108] This misinterpretation is especially likely if the two planets orbit with a 2:1 resonance. With the exoplanet sample known in 2009, a group of astronomers has concluded that "(1) around 35% of the published eccentric one-planet solutions are statistically indistinguishable from planetary systems in 2:1 orbital resonance, (2) another 40% cannot be statistically distinguished from a circular orbital solution" and "(3) planets with masses comparable to Earth could be hidden in known orbital solutions of eccentric super-Earths and Neptune mass planets".[109]

Radial velocity surveys found exoplanet orbits beyond 0.1 AU to be eccentric, particularly for large planets. Kepler spacecraft transit data is consistent with the RV surveys and also revealed that smaller planets tend to have less eccentric orbits.[110]

4.7.3   Inclination vs. spin–orbit angle

Orbital inclination is the angle between a planet's orbital plane and another plane of reference. For exoplanets the inclination is usually stated with respect to an observer on Earth: the angle used is that between the normal to the planet's orbital plane and the line of sight from Earth to the star. Therefore, most planets observed by the transit method are close to 90 degrees.[111] Because the word 'inclination' is used in exoplanet studies for this line-of-sight inclination then the angle between the planet's orbit and the star's rotation must use a different word and is termed the spin–orbit angle or spin–orbit alignment. In most cases the orientation of the star's rotational axis is unknown. The Kepler spacecraft has found a few hundred multi-planet systems and in most of these systems the planets all orbit in nearly the same plane, much like the Solar System.[54] However, a combination of astrometric and radial-velocity measurements has shown that some planetary systems contain planets whose orbital planes are significantly tilted relative to each other.[112] More than half of hot Jupiters have orbital planes substantially misaligned with their parent star's rotation. A substantial fraction of hot-Jupiters even have retrograde orbits, meaning that they orbit in the opposite direction from the star's rotation.[113] Rather than a planet's orbit having been disturbed, it may be that the star itself flipped early in their system's formation due to interactions between the star's magnetic field and the planet-forming disc.[114]

4.7.4   Periastron precession

Periastron precession is the rotation of a planet's orbit within the orbital plane, i.e. the axes of the ellipse change direction. Various factors cause the precession. In the Solar System perturbations from other planets are the main cause, but for close-in exoplanets the largest factor can be tidal forces between the star and planet. For close-in exoplanets, the general relativistic contribution to the precession is also significant and can be orders of magnitude larger than the same effect for Mercury. Some exoplanets have significantly eccentric orbits, which makes it easier to detect the precession. The effect of general relativity can be detectable in timescales of roughly 10 years or less.[115]

4.7.5   Nodal precession

Nodal precession is rotation of a planet's orbital plane. This differs from periastron precession, which is rotation of a planet's orbit within that plane. Nodal precession is more easily seen as distinct from periastron precession when the orbital plane is inclined to the star's rotation, the extreme case being a polar orbit.

WASP-33 is a fast-rotating star that hosts a hot Jupiter in an almost polar orbit. The quadrupole mass moment and the proper angular momentum of the star are 1900 and 400 times, respectively, larger than those of the Sun. This causes significant classical and relativistic deviations from Kepler's laws. In particular, the fast rotation causes large nodal precession because of the star's oblateness and the Lense–Thirring effect.[116]

4.8   Rotation and axial tilt

Plot of equatorial spin velocity vs. mass for planets comparing Beta Pictoris b to the Solar System planets. (ESO/I. Snellen (Leiden University)

In April 2014 the first measurement of a planet's rotation period was announced: the length of day for the super-Jupiter gas giant Beta Pictoris b is 8 hours (based on the assumption that the axial tilt of the planet is small.)[117][118][119] With an equatorial rotational velocity of 25 km per second, this is faster than for the giant planets of the Solar System, in line with the expectation that the more massive a giant planet, the faster it spins. Beta Pictoris b's distance from its star is 9AU. At such distances the rotation of Jovian planets is not slowed by tidal effects.[120] Beta Pictoris b is still warm and young and over the next hundreds of millions of years, it will cool down and shrink to about the size of Jupiter, and if its angular momentum is preserved then as it shrinks the length of its day will decrease to about 3 hours and its equatorial rotation velocity will speed up to about 40 km per second.[118] The images of Beta Pictoris b do not have high enough resolution to directly see details but doppler spectroscopy techniques were used to show that different parts of the planet were moving at different speeds and in opposite directions from which it was inferred that the planet is rotating.[117] With the next generation of large ground-based telescopes it will be possible to use doppler imaging techniques to make a global map of the planet, like the recent mapping of the brown dwarf Luhman 16B.[121][122]

4.8.1   Origin of spin and tilt of terrestrial planets

Giant impacts have a large effect on the spin of terrestrial planets. The last few giant impacts during planetary formation tend to be the main determiner of a terrestrial planet's rotation rate. On average the spin angular velocity will be about 70% of the velocity that would cause the planet to break up and fly apart: the natural outcome of planetary embryo impacts at speeds slightly larger than escape velocity. In later stages terrestrial planet spin is also affected by impacts with planetesimals. During the giant impact stage, the thickness of a protoplanetary disk is far larger than the size of planetary embryos so collisions are equally likely to come from any direction in three-dimensions. This results in the axial tilt of accreted planets ranging from 0 to 180 degrees with any direction as likely as any other with both prograde and retrograde spins equally probable. Therefore, prograde spin with a small axial tilt, common for the Solar System's terrestrial planets except Venus, is not common in general for terrestrial planets built by giant impacts. The initial axial tilt of a planet determined by giant impacts can be substantially changed by stellar tides if the planet is close to its star and by satellite tides if the planet has a large satellite.[123]

4.8.2 Tidal effects

For most planets the rotation period and axial tilt (also called obliquity) are not known, but a large number of planets have been detected with very short orbits (where tidal effects are greater) and will probably have reached an equilibrium rotation that can be predicted.

Tidal effects are the result of forces acting on a body differing from one part of the body to another.[120] For example, the gravitational effect of a star varies with distance from one side of a planet to another. Also heat from a star creates a temperature gradient between the day and nightsides which is another source of tides. For example, on Earth, air pressure variations on the ground are affected more by temperature differences than gravitational ones.

Tides modify the rotation and orbit of planets until an equilibrium is reached. Whenever the rotation rate is slowed, there is an increase of the orbit semi-major axis due to the conservation of angular momentum. Most of the large moons in the Solar System, including the Moon, are tidally locked to their host planet: the same side of the moon is always facing the planet. This means the moons' rotation periods are synchronous with their orbital period. However, when an orbit is eccentric, as is the case with many exoplanets' orbits of their host stars, there are equilibrium states such as spin–orbit resonances that are far more likely than synchronous rotation. A spin–orbit resonance is when the rotation period and the orbital period are in an integer ratio – this is called a commensurability. Non-resonant equilibriums such as the retrograde rotation of Venus can also occur when both gravitational and thermal atmospheric tides are both significant.

A synchronous tidal lock isn't necessarily particularly slow – there are planets with orbits that take only a few hours.

Gravitational tides tend to reduce the axial tilt to zero but over a longer time-scale than the rotation rate reaches equilibrium. However, the presence of multiple planets in a system can cause axial tilt to be captured in a resonance called a Cassini state. There are small oscillations around this state and in the case of Mars these axial tilt variations are chaotic.

Hot Jupiters' close proximity to their host star means that their spin–orbit evolution is mostly due to the star's gravity and not the other effects. Hot Jupiters rotation rate is not thought to be captured into spin–orbit resonance due to way fluid-body reacts to tides, and therefore slows down to synchronous rotation if it is on a circular orbit or slows to a non-synchronous rotation if on an eccentric orbit. Hot Jupiters are likely to evolve towards zero axial tilt even if they had been in a Cassini state during planetary migration when they were further from their star. Hot Jupiters' orbits will become more circular over time, however the presence of other planets in the system on eccentric orbits, even ones as small as Earth and as far away as the habitable zone, can continue to maintain the eccentricity of the Hot Jupiter so that the length of time for tidal circularization can be billions instead of millions of years.

The rotation rate of planet HD 80606 b is predicted to be about 1.9 days. HD 80606 b avoids spin–orbit resonance because it is a gas giant. The eccentricity of its orbit means that it avoids becoming tidally locked.

4.9 Physical parameters

See also: Planet § Physical characteristics

4.9.1 Mass

When a planet is found by the radial-velocity method, its orbital inclination i is unknown and can range from 0 to 90 degrees. The method is unable to determine the true mass (M) of the planet, but rather gives a lower limit for its mass, M

sin*i*. In a few cases an apparent exoplanet may be a more massive object such as a brown dwarf or red dwarf. However, the probability of a small value of i (say less than 30 degrees, which would give a true mass at least double the observed lower limit) is relatively low (1−(√3)/2 ≈ 13%) and hence most planets will have true masses fairly close to the observed lower limit.[93]

If a planet's orbit is nearly perpendicular to the line of vision (i.e. *i* close to 90°), a planet can be detected through the transit method. The inclination will then be known, and the inclination combined with *M* sin*i* from radial-velocity observations will give the planet's true mass.

Also, astrometric observations and dynamical considerations in multiple-planet systems can sometimes provide an upper limit to the planet's true mass.

The mass of a transiting exoplanet can also be determined from the transmission spectrum of its atmosphere, as it can be used to constrain independently the atmospheric composition, temperature, pressure, and scale height.[124]

Transit-timing variation can also be used to find planets' masses.[125]

4.9.2   Radius, density and bulk composition

Prior to recent results from the Kepler spacecraft most confirmed planets were gas giants comparable in size to Jupiter or larger because they are most easily detected. However, the planets detected by Kepler are mostly between the size of Neptune and the size of Earth.[54]

If a planet is detectable by both the radial-velocity and the transit methods, then both its true mass and its radius can be found. The planet's density can then be calculated. Planets with low density are inferred to be composed mainly of hydrogen and helium, whereas planets of intermediate density are inferred to have water as a major constituent. A planet of high density is inferred to be rocky, like Earth and the other terrestrial planets of the Solar System.

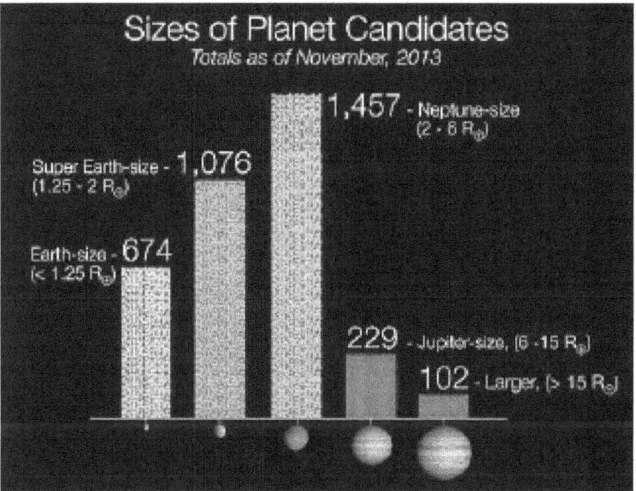

Sizes of *Kepler* Planet Candidates – based on 2,740 candidates orbiting 2,036 stars as of 4 November 2013 (NASA).

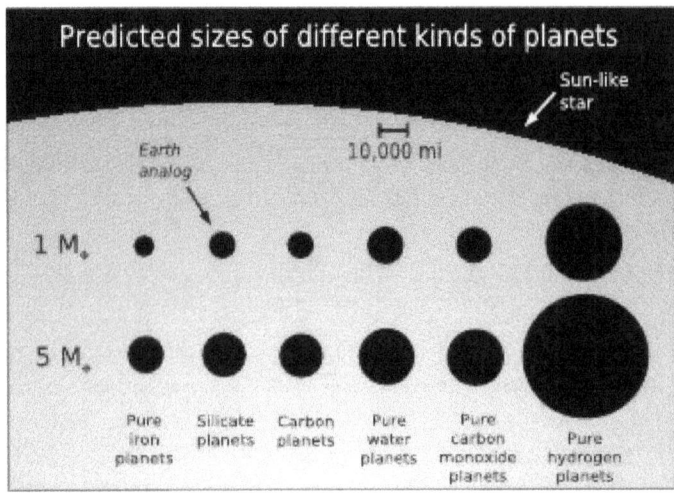

Comparison of sizes of planets with different compositions.

Gas giants, puffy planets, and super-Jupiters

Size comparison of WASP-17b (right) with Jupiter (left).

Main articles: Gas giant, Puffy planet and Super-Jupiter

Gaseous planets that are hot because they are close to their star or because they are still hot from their formation are expanded by the heat. For colder gas planets there is a maximum radius which is slightly larger than Jupiter which occurs when the mass reaches a few Jupiter-masses. Adding mass beyond this point causes the radius to shrink.[31][126][127]

Even when taking heating from the star into account, many transiting exoplanets are much larger than expected given their mass, meaning that they have surprisingly low density.[128] See the magnetic field section for one possible explanation.

Besides those inflated hot Jupiters there is another type of low-density planet: occurring at around 0.6 times the size of Jupiter where there are very few planets. The planets around Kepler-51[92] are far less dense (far more diffuse) than the inflated hot Jupiters as can be seen in the plots on the right where the three Kepler-51 planets stand out in the diffusity vs. radius plot. A more detailed study taking into account star spots may modify these results to produce less extreme

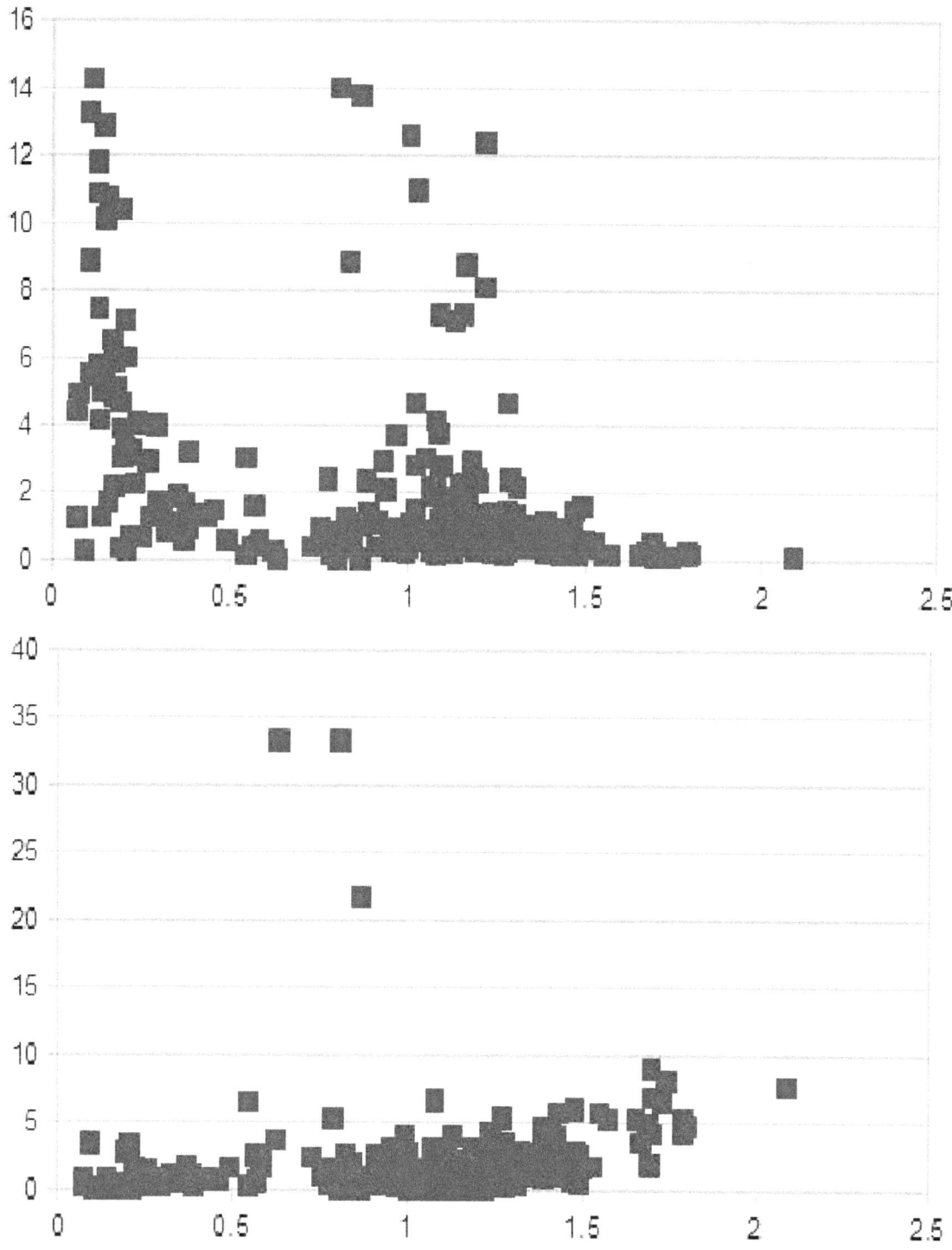

Plots of exoplanet density and radius.[lower-alpha 5] Top: Density vs. Radius. Bottom: Diffusity=1/Density vs. Radius. Units: Radius in RJup. Density in g/cm³. Diffusity in cm³/g. These plots show that there are a wide range of densities for planets between Earth and Neptune size, then the planets of 0.6RJup size are very low-density and there are very few of them, then the gas giants have a large range of densities.

values.[92]

Ice giants and super-Neptunes

Main article: Ice giant

Kepler-101b is the first super-Neptune planet. It has three times Neptune's mass but a Neptune-like composition with more than 60% heavy elements unlike hydrogen/helium-dominated gas giants.[129]

Super-Earths, mini-Neptunes, and gas dwarfs

Main articles: Super-Earth, Mini-Neptune, Helium planet and Gas dwarf

If a planet has a radius and/or mass between that of Earth and Neptune then there is a question about whether the planet is rocky like Earth, a mixture of volatiles and gas like Neptune, a small planet with a hydrogen/helium envelope (mini-Jupiter), or of some other composition.

Some of the Kepler transiting planets with radii in the range 1–4 Earth radii have had their masses measured by radial-velocity or transit-timing methods. The calculated densities show that up to 1.5 Earth radii, these planets are rocky and that density increases with increasing radius due to gravitational compression. However, between 1.5 and 4 Earth radii the density decreases with increasing radius. This indicates that above 1.5 Earth radii planets tend to have increasing amounts of volatiles and gas. Despite this general trend there is a wide range of masses at a given radius, which could be because gas planets can have rocky cores of different masses or compositions[130] and could also be due to photoevaporation of volatiles.[131] Thermal evolutionary atmosphere models suggest a radius of 1.75 times that of Earth as a dividing line between rocky and gaseous planets.[132] Excluding close-in planets that have lost their gas envelope due to stellar irradiation, studies of the metallicity of stars suggest a dividing line of 1.7 Earth radii between rocky planets and gas dwarfs; then another dividing line at 3.9 Earth radii between gas dwarfs and gas giants. These dividing lines are statistical trends and do not necessarily apply to specific planets because there are many other factors besides metallicity that affect planet formation, including distance from star – there may be larger rocky planets formed at larger distances.[133] An independent reanalysis of the data suggests that there are no such dividing lines and that there is a continuum of planet formation between 1 and 4 Earth radii and no reason to expect that the amount of solid material in a protoplanetary disk determines whether super-Earths or mini-Neptunes are formed.[134]

The discovery of the low-density Earth-mass planet Kepler-138d shows that there is an overlapping range of *masses* in which both rocky planets and low-density planets occur.[135] Low-mass low-density planets could be ocean planets or super-Earths with a remnant hydrogen atmosphere, or hot planets with a steam atmosphere, or mini-Neptunes with a hydrogen-helium atmosphere.[136] Other possibilities for low-mass low-density planets are large atmospheres of carbon monoxide, carbon dioxide, methane, or nitrogen.[137]

Massive solid planets and giant planets with massive cores

In 2014, new measurements of Kepler-10c found that it is a Neptune-mass planet (17 Earth masses) with a density higher than Earth's, indicating that Kepler-10c is made mostly of rock with possibly up to 20% high-pressure water ice but without a hydrogen-dominated envelope. Because this is well above the 10-Earth-mass upper limit that is commonly used for the term 'super-Earth', the term **mega-Earth** has been coined.[138][139] A similarly massive and dense planet could be Kepler-131b, although its density is not as well measured as that of Kepler 10c. The next most massive known solid planets are half this mass: 55 Cancri e and Kepler-20b.[140]

Gas planets can also have large solid cores: the Saturn-mass planet HD 149026 b has only two-thirds of Saturn's radius, so it may have a rock–ice core of 60 Earth masses or more.[31] Corot-20b has 4.24 times Jupiter's mass but a radius of only 0.84 that of Jupiter—it may have a metal core of 800 Earth masses if the heavy elements are concentrated in the core or a core of 300 Earth masses if the heavy elements are more distributed throughout the planet.[141][142]

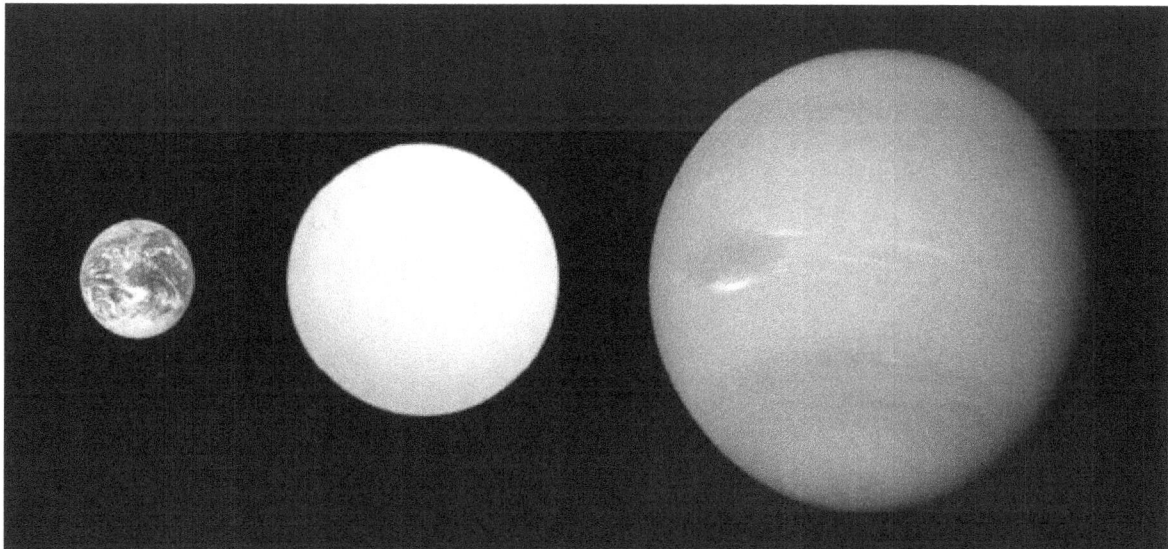

Size comparison of Kepler-10c with Earth and Neptune

Transit-timing variation measurements indicate that Kepler-52b, Kepler-52c and Kepler-57b have maximum-masses between 30 and 100 times the mass of Earth, although the actual masses could be much lower. With radii about 2 Earth radii in size, they might have densities larger than that of an iron planet of the same size. They orbit very close to their stars, so they could be the remnant cores (chthonian planets) of evaporated gas giants or brown dwarfs. If cores are massive enough they could remain compressed for billions of years despite losing the atmospheric mass.[143][144]

Solid planets up to thousands of Earth masses may be able to form around massive stars (B-type and O-type stars; 5–120 solar masses), where the protoplanetary disk would contain enough heavy elements. Also, these stars have high UV radiation and winds that could photoevaporate the gas in the disk, leaving just the heavy elements.[145] For comparison, Neptune's mass equals 17 Earth masses, Jupiter has 318 Earth masses, and the 13 Jupiter-mass limit used in the IAU's working definition of an exoplanet equals approximately 4000 Earth masses.[145]

Another way of forming massive solid planets is when a white dwarf in a close binary system loses material to a companion neutron star. The white dwarf can be reduced to planetary-mass, leaving just its crystallised carbon–oxygen core. A likely example of this is PSR J1719-1438 b.

Cold planets have a maximum radius because adding more mass at that point causes the planet to compress under the weight instead of increasing the radius. The maximum radius for solid planets is smaller than the maximum radius for gas planets.[145]

4.9.3   Shape

When the size of a planet is described using its radius this is approximating the shape by a sphere. However, the rotation of a planet causes it to be flattened at the poles so that the equatorial radius is larger than the polar radius, making it closer to an oblate spheroid. The oblateness of transiting exoplanets will affect the transit light curves. At the limits of current technology it has been possible to show that HD 189733b is less oblate than Saturn.[146] If the planet is close to its star, then gravitational tides will elongate the planet in the direction of the star, so that the planet will be closer to a triaxial ellipsoid.[147] Because tidal deformation is along a line between the planet and the star, it is difficult to detect from transit photometry—it will have an order of magnitude less effect on the transit light curves than that caused by rotational deformation even in cases where the tidal deformation is larger than rotational deformation (such as is the case for tidally locked hot Jupiters).[146] Material rigidity of rocky planets and rocky cores of gas planets will cause further deviations from the aforementioned shapes.[146] Thermal tides caused by unevenly irradiated surfaces are another factor.[148]

4.10 Atmosphere

Sunset studies on Titan by Cassini help understand exoplanet atmospheres (artist's concept).

As of February 2014, more than fifty transiting and five directly imaged exoplanet atmospheres have been observed,[149] resulting in detection of molecular spectral features; observation of day–night temperature gradients; and constraints on vertical atmospheric structure.[150] Also, an atmosphere has been detected on the non-transiting hot Jupiter Tau Boötis b.[151][152]

Spectroscopic measurements can be used to study a transiting planet's atmospheric composition,[153] temperature, pressure, and scale height, and hence can be used to determine its mass.[124]

Stellar light is polarized by atmospheric molecules; this could be detected with a polarimeter. HD 189733 b has been studied by polarimetry.

Extrasolar planets have phases similar to the phases of the Moon. By observing the exact variation of brightness with phase, astronomers can calculate atmospheric-particle sizes.

4.10.1   Atmospheric composition

In 2001 sodium was detected in the atmosphere of HD 209458 b.[154]

In 2008 water, carbon monoxide, carbon dioxide[155] and methane[156] were detected in the atmosphere of HD 189733 b.

In 2013 water was detected in the atmospheres of HD 209458 b, XO-1b, WASP-12b, WASP-17b, and WASP-19b.[157][158]

In July 2014, NASA announced finding very dry atmospheres on three exoplanets (HD 189733b, HD 209458b, WASP-12b) orbiting Sun-like stars.[160]

In September 2014, NASA reported that HAT-P-11b is the first Neptune-sized exoplanet known to have a relatively cloud-free atmosphere and, as well, the first time molecules of any kind have been found, specifically water vapor, on such a relatively small exoplanet.[161]

The presence of oxygen may be detectable by ground-based telescopes,[162] which, if discovered, would suggest the presence of life on an exoplanet.

In June 2015, NASA reported that WASP-33b has a stratosphere. Ozone and hydrocarbons absorb large amounts of ultraviolet radiation, heating the upper parts of atmosphere's that contain them, creating a temperature inversion and a stratosphere. However, these molecules are destroyed at the temperatures of hot exoplanets, creating doubt if the hot exoplanets could have a stratosphere. A temperature inversion, and stratosphere was identified on WASP-33b caused by titanium oxide, which is a strong absorber of visible and ultraviolet radiation, and can only exist as a gas in a hot atmosphere. WASP-33b is the hottest exoplanet known, with a temperature of 3,200 °C (5,790 °F)[163] and is approximately four and a half times the mass of Jupiter.[164][165]

4.10.2   Atmospheric circulation

The atmospheric circulation of planets that rotate more slowly or have a thicker atmosphere allows more heat to flow to the poles which reduces the temperature differences between the poles and the equator.[166]

4.10.3   Clouds

In October 2013, the detection of clouds in the atmosphere of Kepler-7b was announced.[167][168] and, in December 2013, also in the atmospheres of GJ 436 b and GJ 1214 b.[169][170][171][172]

4.10.4   Precipitation

Precipitation in the form of liquid (rain) or solid (snow) varies in composition depending on atmospheric temperature, pressure, composition, and altitude. Hot atmospheres could have iron rain,[173] molten-glass rain,[174] and rain made from rocky minerals such as enstatite, corundum, spinel, and wollastonite.[175] Deep in the atmospheres of gas giants it could rain diamonds[176] and helium containing dissolved neon.[177]

4.10.5   Abiotic oxygen

There are non-biological processes that produce oxygen, so the detection of oxygen is not necessarily an indication of life.

The processes of life result in a mixture of chemicals that are not in chemical equilibrium but there are also abiotic disequilibrium processes that need to be considered. The most robust atmospheric biosignature is often considered to be molecular oxygen O_2 and its photochemical byproduct ozone O_3. The photolysis of water H_2O by UV rays followed by hydrodynamic escape of hydrogen can lead to a build-up of oxygen in planets close to their star undergoing runaway greenhouse effect. For planets in the habitable zone it was believed that water photolysis would be strongly limited by cold-trapping of water vapour in the lower atmosphere. However, the extent of H_2O cold-trapping depends strongly on the amount of non-condensible gases in the atmosphere such as nitrogen N_2 and argon. In the absence of such gases the likelihood of build-up of oxygen also depends in complex ways on the planet's accretion history, internal chemistry, atmospheric dynamics and orbital state. Therefore, oxygen on its own cannot be considered a robust biosignature.[178] The ratio of nitrogen and argon to oxygen could be detected by studying thermal phase curves[179] or by transit transmission spectroscopy measurement of the spectral Rayleigh scattering slope in a clear-sky (i.e. aerosol-free) atmosphere.[180]

4.11 Surface

4.11.1 Surface composition

Surface features can be distinguished from atmospheric features by comparing emission and reflection spectroscopy with transmission spectroscopy. Mid-infrared spectroscopy of exoplanets may detect rocky surfaces, and near-infrared may identify magma oceans or high-temperature lavas, hydrated silicate surfaces and water ice, giving an unambiguous method to distinguish between rocky and gaseous exoplanets.[181]

4.11.2 Surface temperature

See also: § Habitability

One can estimate the temperature of an exoplanet based on the intensity of the light it receives from its parent star.

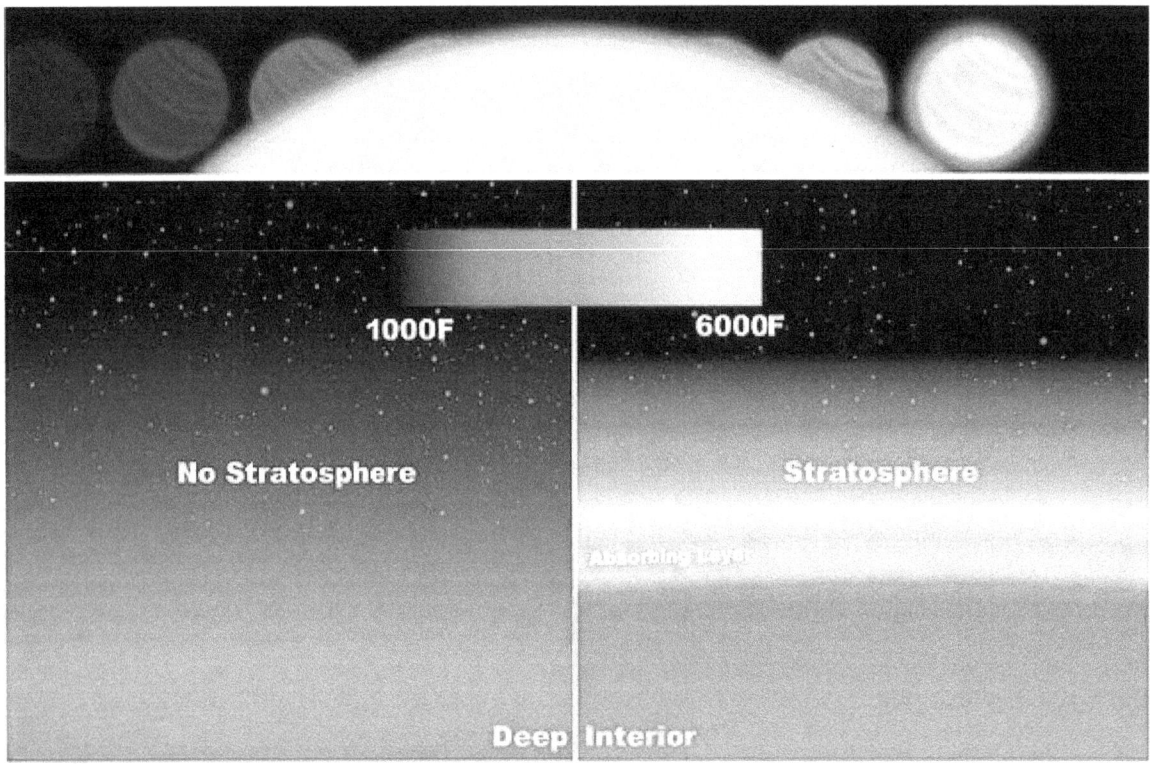

Artist's illustration of temperature inversion in exoplanet's atmosphere.[182]

For example, the planet OGLE-2005-BLG-390Lb is estimated to have a surface temperature of roughly −220 °C (50 K). However, such estimates may be substantially in error because they depend on the planet's usually unknown albedo, and because factors such as the greenhouse effect may introduce unknown complications. A few planets have had their temperature measured by observing the variation in infrared radiation as the planet moves around in its orbit and is eclipsed by its parent star. For example, the planet HD 189733b has been found to have an average temperature of 1205±9 K (932±9 °C) on its dayside and 973±33 K (700±33 °C) on its nightside.[183]

4.12   General features

4.12.1   Color and brightness

See also: Sudarsky's gas giant classification
In 2013 the color of an exoplanet was found for the first time. The best-fit albedo measurements of HD 189733b suggest

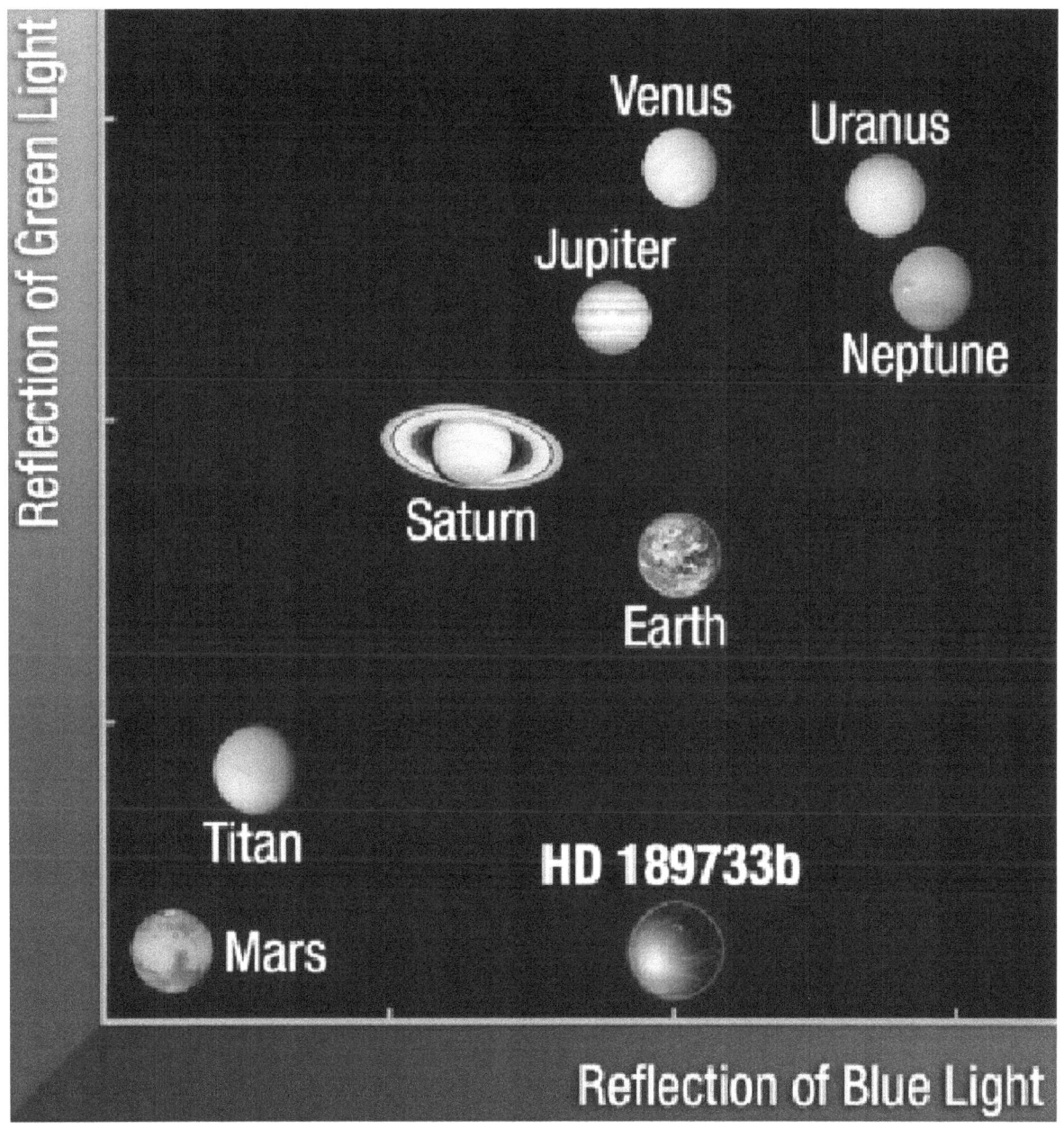

This color–color diagram compares the colors of planets in the Solar System to exoplanet HD 189733b. The exoplanet's deep blue color is produced by silicate droplets, which scatter blue light in its atmosphere.

that it is deep dark blue.[184][185]

Visually, GJ 504 b would have a magenta color.[186]

Kappa Andromedae b if seen up close, would appear reddish in color.[187]

The apparent brightness (apparent magnitude) of a planet depends on how far away the observer is, how reflective the planet is (albedo), and how much light the planet receives from its star, which depends on how far the planet is from the star and how bright the star is. So, a planet with a low albedo that is close to its star can appear brighter than a planet with high albedo that is far from the star.[188]

The darkest known planet in terms of geometric albedo is TrES-2b, a hot Jupiter that reflects less than 1% of the light from its star, making it less reflective than coal or black acrylic paint. Hot Jupiters are expected to be quite dark due to sodium and potassium in their atmospheres but it is not known why TrES-2b is so dark—it could be due to an unknown chemical.[189][190][191]

For gas giants, geometric albedo generally decreases with increasing metallicity or atmospheric temperature unless there are clouds to modify this effect. Increased cloud-column depth increases the albedo at optical wavelengths, but decreases it at some infrared wavelengths. Optical albedo increases with age, because older planets have higher cloud-column depths. Optical albedo decreases with increasing mass, because higher-mass giant planets have higher surface gravities, which produces lower cloud-column depths. Also, elliptical orbits can cause major fluctuations in atmospheric composition, which can have a significant effect.[192]

There is more thermal emission than reflection at some near-infrared wavelengths for massive and/or young gas giants. So, although optical brightness is fully phase-dependent, this is not always the case in the near infrared.[192]

Temperatures of gas giants reduce over time and with distance from their star. Lowering the temperature increases optical albedo even without clouds. At a sufficiently low temperature, water clouds form, which further increase optical albedo. At even lower temperatures ammonia clouds form, resulting in the highest albedos at most optical and near-infrared wavelengths.[192]

4.12.2 Magnetic field

In 2014, a magnetic field around HD 209458 b was inferred from the way hydrogen was evaporating from the planet. It is the first (indirect) detection of a magnetic field on an exoplanet. The magnetic field is estimated to be about one tenth as strong as Jupiter's.[193][194]

Interaction between a close-in planet's magnetic field and a star can produce spots on the star in a similar way to how the Galilean moons produce aurorae on Jupiter.[195] Auroral radio emissions could be detected with radio telescopes such as LOFAR.[196][197] The radio emissions could enable determination of the rotation rate of a planet which is difficult to detect otherwise.[198]

Earth's magnetic field results from its flowing liquid metallic core, but in super-Earths the mass can produce high pressures with large viscosities and high melting temperatures which could prevent the interiors from separating into different layers and so result in undifferentiated coreless mantles. Magnesium oxide, which is rocky on Earth, can be a liquid metal at the pressures and temperatures found in super-Earths and could generate a magnetic field in the mantles of super-Earths.[199]

Hot Jupiters have been observed to have a larger radius than expected. This could be caused by the interaction between the stellar wind and the planet's magnetosphere creating an electric current through the planet that heats it up causing it to expand. The more magnetically active a star is the greater the stellar wind and the larger the electric current leading to more heating and expansion of the planet. This theory matches the observation that stellar activity is correlated with inflated planetary radii.[200]

4.12.3 Plate tectonics

In 2007 two independent teams of researchers came to opposing conclusions about the likelihood of plate tectonics on larger super-earths[201][202] with one team saying that plate tectonics would be episodic or stagnant[203] and the other team saying that plate tectonics is very likely on super-earths even if the planet is dry.[204]

If super-earths have more than 80 times as much water as Earth then they become ocean planets with all land completely submerged. However, if there is less water than this limit, then the deep water cycle will move enough water between the oceans and mantle to allow continents to exist.[205][206]

4.12.4 Volcanism

Large surface temperature variations on 55 Cancri e have been attributed to possible volcanic activity releasing large clouds of dust which blanket the planet and block thermal emissions.[207][208]

4.12.5 Rings

The star 1SWASP J140747.93-394542.6 is orbited by an object that is circled by a ring system much larger than Saturn's rings. However, the mass of the object is not known; it could be a brown dwarf or low-mass star instead of a planet.[209][210]

The brightness of optical images of Fomalhaut b could be due to starlight reflecting off a circumplanetary ring system with a radius between 20 to 40 times that of Jupiter's radius, about the size of the orbits of the Galilean moons.[211]

The rings of the Solar System's gas giants are aligned with their planet's equator. However, for exoplanets that orbit close to their star, tidal forces from the star would lead to the outermost rings of a planet being aligned with the planet's orbital plane around the star. A planet's innermost rings would still be aligned with the planet's equator so that if the planet has a tilted rotational axis, then the different alignments between the inner and outer rings would create a warped ring system.[212]

4.12.6 Moons

In December 2013 a candidate exomoon of a rogue planet was announced.[213] No exomoons have been confirmed so far.

4.12.7 Comet-like tails

KIC 12557548 b is a small rocky planet, very close to its star, that is evaporating and leaving a trailing tail of cloud and dust like a comet.[214] The dust could be ash erupting from volcanos and escaping due to the small planet's low surface-gravity, or it could be from metals that are vaporized by the high temperatures of being so close to the star with the metal vapor then condensing into dust.[215]

In June 2015, scientists reported that the atmosphere of GJ 436 b was evaporating, resulting in a giant cloud around the planet and, due to radiation from the host star, a long trailing tail 14×10^6 km (9×10^6 mi) long.[216]

4.12.8 Insolation pattern

Tidally locked planets in a 1:1 spin-orbit resonance would have their star always shining directly overhead on one spot which would be hot with the opposite hemisphere receiving no light and being freezing cold. Such a planet could resemble an eyeball with the hotspot being the pupil.[217] Planets with an eccentric orbit could be locked in other resonances. 3:2 and 5:2 resonances would result in a double-eyeball pattern with hotspots in both eastern and western hemispheres.[218] Planets with both an eccentric orbit and a tilted axis of rotation would have more complicated insolation patterns.[219]

4.13 Habitability

Main articles: Planetary habitability, Astrobiology and Drake equation

4.13.1 Habitable zone

The habitable zone around a star is the region where the temperature is just right to allow liquid water to exist on a planet: that is, not too close to the star for the water to evaporate and not too far away from the star for the water to freeze.

The heat produced by stars varies depending on the size and age of the star so that the habitable zone can be at different distances. Also, the atmospheric conditions on the planet influence the planet's ability to retain heat so that the location of the habitable zone is also specific to each type of planet: desert planets (also known as dry planets), with very little water, will have less water vapor in the atmosphere than Earth and so have a reduced greenhouse effect, meaning that a desert planet could maintain oases of water closer to its star than Earth is to the Sun. The lack of water also means there is less ice to reflect heat into space, so the outer edge of desert-planet habitable zones is further out.[220][221] Rocky planets with a thick hydrogen atmosphere could maintain surface water much further out than the Earth–Sun distance.[222] Planets with larger mass have wider habitable zones because the gravity reduces the water cloud column depth which reduces the greenhouse effect of water vapor thus moving the inner edge of the HZ closer to the star.[223]

Planetary rotation rate is one of the major factors determining the circulation of the atmosphere and hence the pattern of clouds: slowly rotating planets create thick clouds that reflect more and so can be habitable much closer to their star. Earth with its current atmosphere would be habitable in Venus's orbit, if it had Venus's slow rotation, so Venus must have had a higher rotation rate in the past if it lost its water ocean as a result of going through a runaway greenhouse effect, but if Venus never had an ocean because water vapor was lost to space during its formation before it could cool to form an ocean,[224] Venus could have had its slow rotation throughout its history.[225]

Tidally locked planets (a.k.a. "eyeball" planets[226]) can be habitable closer to their star than previously thought due to the effect of clouds: at high stellar flux, strong convection produces thick water clouds near the substellar point that greatly increase the planetary albedo and reduce surface temperatures.[227]

Habitable zones have usually been defined in terms of surface temperature, however over half of Earth's biomass is from subsurface microbes,[228] and the temperature increases as you go deeper underground, so the subsurface can be conducive for life when the surface is frozen and if this is considered, the habitable zone extends much further from the star,[229] even rogue planets could have liquid water at sufficient depths underground.[230] In an earlier era of the universe the temperature of the cosmic microwave background would have allowed any rocky planets that existed to have liquid water on their surface regardless of their distance from a star.[231] Jupiter-like planets might not be habitable, but they could have habitable moons.[232]

4.13.2 Ice ages and snowball states

See also: Ice age and Snowball Earth

The outer edge of the habitable zone is where planets will be completely frozen but even planets well inside the habitable zone can periodically become frozen. If orbital fluctuations or other causes produce cooling then this creates more ice but ice reflects sunlight causing even more cooling creating a feedback loop until the planet is completely or nearly completely frozen. When the surface is frozen this stops carbon dioxide weathering resulting in a build-up of carbon dioxide in the atmosphere from volcanic emissions. This creates a greenhouse effect which unfreezes the planet again. Planets with a large axial tilt[233] are less likely to enter snowball states and can retain liquid water further from their star. Large fluctuations of axial tilt can have even more of a warming effect than a fixed large tilt.[234][235] Paradoxically planets around cooler stars, such as red dwarfs, are less likely to enter snowball states because the infrared radiation emitted by cooler stars is mostly at wavelengths that are absorbed by ice which heats it up.[236][237]

4.13.3 Tidal heating

If a planet has an eccentric orbit then tidal heating can provide another source of energy besides stellar irradiation. This means that eccentric planets in the radiative habitable zone can be too hot for liquid water (Tidal Venus). Tides also circularize orbits over time so there could be planets in the habitable zone with circular orbits that have no water because they used to have eccentric orbits.[238] Eccentric planets further out than the radiative habitable zone would still have frozen surfaces but the tidal heating could create a subsurface ocean similar to Europa's.[239] In some planetary systems, such as in the Upsilon Andromedae system, the eccentricity of orbits is maintained or even periodically varied by perturbations from other planets in the system. Tidal heating can cause outgassing from the mantle, contributing to the formation and replenishment of an atmosphere.[240]

4.13.4 Potentially habitable planets

See also: List of potentially habitable exoplanets and List of nearest terrestrial exoplanet candidates

A review in 2015 came to the conclusion that the exoplanets Kepler-62f, Kepler-186f and Kepler-442b were likely the best candidates for being potentially habitable.[241] These are at a distance of 1200, 490 and 1,120 light-years away, respectively. Of these, Kepler-186f is in similar size to Earth with its 1.2-Earth-radius measure, and it is located towards the outer edge of the habitable zone around its red dwarf sun.

When looking at the nearest terrestrial exoplanet candidates, Tau Ceti e is merely 11.9 light-years away. Its average surface temperature is estimated to be 68 °C (154 °F).[242]

Earth-size planets

See also: Earth analog

- In November 2013 it was announced that 22±8% of Sun-like[lower alpha 1] stars have an Earth-sized[lower alpha 2] planet in the habitable[lower alpha 3] zone.[10][11] Assuming 200 billion stars in the Milky Way,[lower alpha 4] that would be 11 billion potentially habitable Earths, rising to 40 billion if red dwarfs are included.[12]

- Kepler-186f, a 1.2-Earth-radius planet in the habitable zone of a red dwarf, announced in April 2014.

- In February 2013, researchers calculated that up to 6% of small red dwarfs may have planets with Earth-like properties. This suggests that the closest "alien Earth" to the Solar System could be 13 light-years away. The estimated distance increases to 21 light-years when a 95 percent confidence interval is used.[243] In March 2013 a revised estimate based on a more accurate consideration of the size of the habitable zone around red dwarfs gave an occurrence rate of 50% for Earth-size planets in the habitable zone of red dwarfs.[244]

- At 1.63 times Earth's radius Kepler-452b is the first discovered near-Earth-size planet in the "habitable zone" around a G2-type Sun-like star (July 23, 2015).[245]

4.14 Cultural impact

On 9 May 2013, a congressional hearing by two United States House of Representatives subcommittees discussed "Exoplanet Discoveries: Have We Found Other Earths?", prompted by the discovery of exoplanet Kepler-62f, along with Kepler-62e and Kepler-62c. A related special issue of the journal *Science*, published earlier, described the discovery of the exoplanets.[246]

4.15 See also

- List of exoplanets

- Exocomet

- Exomoon

- Extragalactic planet

- List of exoplanet extremes

- List of exoplanet research projects

- Nexus for Exoplanet System Science

- Planetary system
- List of planets

4.16 Notes

[1] For the purpose of this 1 in 5 statistic, "Sun-like" means G-type star. Data for Sun-like stars wasn't available so this statistic is an extrapolation from data about K-type stars

[2] For the purpose of this 1 in 5 statistic, Earth-sized means 1–2 Earth radii

[3] For the purpose of this 1 in 5 statistic, "habitable zone" means the region with 0.25 to 4 times Earth's stellar flux (corresponding to 0.5–2 AU for the Sun).

[4] About 1/4 of stars are GK Sun-like stars. The number of stars in the galaxy is not accurately known, but assuming 200 billion stars in total, the Milky Way would have about 50 billion Sun-like (GK) stars, of which about 1 in 5 (22%) or 11 billion would be Earth-sized in the habitable zone. Including red dwarfs would increase this to 40 billion.

[5] Data from NASA catalog July 2014, excluding objects described as having unphysically high density

4.17 References

[1] "Planet Population is Plentiful", ESO, 11 January 2012. Retrieved 13 January 2012.

[2] "Exoplanet Transit Database: TrES-3b", *astro.cz*, Czech Astronomical Society. Retrieved 7 July 2015.

[3] Schneider, J. "Interactive Extra-solar Planets Catalog". *The Extrasolar Planets Encyclopedia*.

[4] "NASA – Kepler". Retrieved 4 November 2013.

[5] Harrington, J. D.; Johnson, M. (4 November 2013). "NASA Kepler Results Usher in a New Era of Astronomy".

[6] Tenenbaum, P.; Jenkins, J. M.; Seader, S.; Burke, C. J.; Christiansen, J. L.; Rowe, J. F.; Caldwell, D. A.; Clarke, B. D.; Li, J.; Quintana, E. V.; Smith, J. C.; Thompson, S. E.; Twicken, J. D.; Borucki, W. J.; Batalha, N. M.; Cote, M. T.; Haas, M. R.; Hunter, R. C.; Sanderfer, D. T.; Girouard, F. R.; Hall, J. R.; Ibrahim, K.; Klaus, T. C.; McCauliff, S. D.; Middour, C. K.; Sabale, A.; Uddin, A. K.; Wohler, B.; Barclay, T.; Still, M. (2013). "Detection of Potential Transit Signals in the First 12 Quarters of *Kepler* Mission Data". *The Astrophysical Journal Supplement Series* **206**: 5. arXiv:1212.2915. Bibcode:2013ApJS..206....5T. doi:10.1088/0067-0049/206/1/5.

[7] "My God, it's full of planets! They should have sent a poet." (Press release). Planetary Habitability Laboratory, University of Puerto Rico at Arecibo. 3 January 2012.

[8] Santerne, A.; Díaz, R. F.; Almenara, J.-M.; Lethuillier, A.; Deleuil, M.; Moutou, C. (2013). "Astrophysical false positives in exoplanet transit surveys: Why do we need bright stars?". arXiv:1310.2133 [astro-ph.EP].

[9] Cassan, A.; et al. (January 11, 2012). "One or more bound planets per Milky Way star from microlensing observations". *Nature* **481** (7380): 167–169. arXiv:1202.0903. Bibcode:2012Natur.481..167C. doi:10.1038/nature10684. PMID 22237108.

[10] Sanders, R. (4 November 2013). "Astronomers answer key question: How common are habitable planets?". *newscenter.berkeley.edu*.

[11] Petigura, E. A.; Howard, A. W.; Marcy, G. W. (2013). "Prevalence of Earth-size planets orbiting Sun-like stars". *Proceedings of the National Academy of Sciences* **110**(48): 19273–19278. arXiv:1311.6806. Bibcode:2013PNAS..11019273P.doi:10.1073/pna

[12] Khan, Amina (4 November 2013). "Milky Way may host billions of Earth-size planets". *Los Angeles Times*. Retrieved 5 November 2013.

[13] "DENIS-P J082303.1-491201 b". *Caltech*. Retrieved 8 March 2014.

[14] Sahlmann, J.; Lazorenko, P. F.; Ségransan, D.; Martín, E. L.; Queloz, D.; Mayor, M.; Udry, S. (August 2013). "Astrometric orbit of a low-mass companion to an ultracool dwarf". *Harvard University* **556**: 133. arXiv:1306.3225. Bibcode:2013A&A...556A.133S.doi:10.1051/0004-6361/201321871.

[15] Overbye, Dennis (6 January 2015). "As Ranks of Goldilocks Planets Grow, Astronomers Consider What's Next". *New York Times*.

[16] Beichman, C.; Gelino, Christopher R.; Kirkpatrick, J. Davy; Cushing, Michael C.; Dodson-Robinson, Sally; Marley, Mark S.; Morley, Caroline V.; Wright, E. L. (2014). "WISE Y Dwarfs As Probes of the Brown Dwarf-Exoplanet Connection". *The Astrophysical Journal* **783** (2): 68. arXiv:1401.1194v2. Bibcode:2014ApJ...783...68B. doi:10.1088/0004-637X/783/2/68.

[17] Strigari, L. E.; Barnabè, M.; Marshall, P. J.; Blandford, R. D. (2012). "Nomads of the Galaxy". *Monthly Notices of the Royal Astronomical Society* **423** (2): 1856–1865. arXiv:1201.2687. Bibcode:2012MNRAS.423.1856S. doi:10.1111/j.1365-2966.2012.21009.x. estimates 700 objects >10^{-8} solar masses (roughly the mass of Mars) per main-sequence star between 0.08 and 1 Solar mass, of which there are billions in the Milky Way.

[18] "IAU 2006 General Assembly: Result of the IAU Resolution votes". 2006. Retrieved 25 April 2010.

[19] Brit, R. R. (2006). "Why Planets Will Never Be Defined". *Space.com*. Retrieved 13 February 2008.

[20] "Working Group on Extrasolar Planets: Definition of a "Planet"". *IAU position statement*. 28 February 2003. Retrieved 23 November 2014.

[21] Mordasini, C.; Alibert; Benz; Naef; et al. (2007). "Giant Planet Formation by Core Accretion". arXiv:0710.5667v1 [astro-ph].

[22] Baraffe, I.; Chabrier, G.; Barman, T. (2008). "Structure and evolution of super-Earth to super-Jupiter exoplanets. I. Heavy element enrichment in the interior". *Astronomy and Astrophysics* **482** (1): 315–332. arXiv:0802.1810. Bibcode:2008A&A...482..315B.doi:10.1051/0004-6361:20079321.

[23] Bodenheimer, P.; D'Angelo, G.; Lissauer, J. J.; Fortney, J. J.; Saumon, D. (2013). "Deuterium Burning in Massive Giant Planets and Low-mass Brown Dwarfs Formed by Core-nucleated Accretion". *The Astrophysical Journal* **770** (2): 120. arXiv:1305.0980. Bibcode:2013ApJ...770..120B. doi:10.1088/0004-637X/770/2/120.

[24] Bouchy, F.; Hébrard, G.; Udry, S.; Delfosse, X.; Boisse, I.; Desort, M.; Bonfils, X.; Eggenberger, A.; Ehrenreich, D.; Forveille, T.; Lagrange, A. M.; Le Coroller, H.; Lovis, C.; Moutou, C.; Pepe, F.; Perrier, C.; Pont, F.; Queloz, D.; Santos, N. C.; Ségransan, D.; Vidal-Madjar, A. (2009). "TheSOPHIEsearch for northern extrasolar planets". *Astronomy and Astrophysics* **505** (2): 853–858. Bibcode:2009A&A...505..853B. doi:10.1051/0004-6361/200912427.

[25] Kumar, Shiv S. (2003). "Nomenclature: Brown Dwarfs, Gas Giant Planets, and ?".*Brown Dwarfs***211**: 532. Bibcode:2003IAUS

[26] Brandt, T. D.; McElwain, M. W.; Turner, E. L.; Mede, K.; Spiegel, D. S.; Kuzuhara, M.; Schlieder, J. E.; Wisniewski, J. P.; Abe, L.; Biller, B.; Brandner, W.; Carson, J.; Currie, T.; Egner, S.; Feldt, M.; Golota, T.; Goto, M.; Grady, C. A.; Guyon, O.; Hashimoto, J.; Hayano, Y.; Hayashi, M.; Hayashi, S.; Henning, T.; Hodapp, K. W.; Inutsuka, S.; Ishii, M.; Iye, M.; Janson, M.; et al. (2014). "A Statistical Analysis of Seeds and Other High-Contrast Exoplanet Surveys: Massive Planets or Low-Mass Brown Dwarfs?". *The Astrophysical Journal* **794** (2): 159. arXiv:1404.5335. Bibcode:2014ApJ...794..159B. doi:10.1088/0004-637X/794/2/159.

[27] Spiegel, D. S.; Burrows, A.; Milsom, J. A. (2011). "The Deuterium-Burning Mass Limit for Brown Dwarfs and Giant Planets". *The Astrophysical Journal* **727**: 57. arXiv:1008.5150. Bibcode:2011ApJ...727...57S. doi:10.1088/0004-637X/727/1/57.

[28] Schneider, J.; Dedieu, C.; Le Sidaner, P.; Savalle, R.; Zolotukhin, I. (2011). "Defining and cataloging exoplanets: The exoplanet.eu database". *Astronomy & Astrophysics* **532** (79): A79. arXiv:1106.0586. Bibcode:2011A&A...532A..79S. doi:10.1051/0004-6361/201116713.

[29] Wright, J. T.; et al. (2010). "The Exoplanet Orbit Database". arXiv:1012.5676v1 [astro-ph.SR].

[30] Exoplanet Criteria for Inclusion in the Archive, NASA Exoplanet Archive

[31] Basri, Gibor; Brown, Michael E. (2006). "Planetesimals To Brown Dwarfs: What is a Planet?". *Ann. Rev. Earth Planet. Sci.* **34**: 193–216. arXiv:astro-ph/0608417. Bibcode:2006AREPS..34..193B. doi:10.1146/annurev.earth.34.031405.125058.

[32] Liebert, James (2003). "Nomenclature: Brown Dwarfs, Gas Giant Planets, and ?". *Brown Dwarfs* **211**: 533. Bibcode:2003IAUS..211..529B.

[33] A Definition for Giant Planets Based on the Mass-Density Relationship, Artie P. Hatzes, Heike Rauer, (Submitted on 16 Jun 2015)

[34] Wolszczan, A.; Frail, D. A. (1992). "A planetary system around the millisecond pulsar PSR1257 + 12". *Nature* **355** (6356): 145–147. Bibcode:1992Natur.355..145W. doi:10.1038/355145a0.

[35] Bruno, Giordano (1584). *On the Infinite Universe and Worlds.*

[36] Newton, Isaac; I. Bernard Cohen and Anne Whitman (1999) [1713]. *The Principia: A New Translation and Guide.* University of California Press. p. 940. ISBN 0-520-20217-1.

[37] Struve, Otto (1952). "Proposal for a project of high-precision stellar radial velocity work". *The Observatory* **72**: 199–200. Bibcode:1952Obs....72..199S.

[38] Jacob, W. S. (1855). "On Certain Anomalies presented by the Binary Star 70 Ophiuchi". *Monthly Notices of the Royal Astronomical Society* **15**: 228–230. Bibcode:1855MNRAS..15..228J. doi:10.1093/mnras/15.9.228.

[39] See, T. J. J. (1896). "Researches on the orbit of 70 Ophiuchi, and on a periodic perturbation in the motion of the system arising from the action of an unseen body". *The Astronomical Journal* **16**: 17–23. Bibcode:1896AJ.....16...17S. doi:10.1086/102368.

[40] Sherrill, T. J. (1999). "A Career of Controversy: The Anomaly of T. J. J. See" (PDF). *Journal for the History of Astronomy* **30** (98): 25–50. Bibcode:1999JHA....30...25S. doi:10.1177/002182869903000102.

[41] van de Kamp, P. (1969). "Alternate dynamical analysis of Barnard's star". *Astronomical Journal* **74**: 757–759. Bibcode:1969AJ. doi:10.1086/110852.

[42] Boss, Alan (2009). *The Crowded Universe: The Search for Living Planets.* Basic Books. pp. 31–32. ISBN 978-0-465-00936-7.

[43] Bailes, M.; Lyne, A. G.; Shemar, S. L. (1991). "A planet orbiting the neutron star PSR1829–10". *Nature* **352** (6333): 311–313. Bibcode:1991Natur.352..311B. doi:10.1038/352311a0.

[44] Lyne, A. G.; Bailes, M. (1992). "No planet orbiting PS R1829–10". *Nature* **355** (6357): 213. Bibcode:1992Natur.355..213L. doi:10.1038/355213b0.

[45] Campbell, B.; Walker, G. A. H.; Yang, S. (1988). "A search for substellar companions to solar-type stars". *The Astrophysical Journal* **331**: 902. Bibcode:1988ApJ...331..902C. doi:10.1086/166608.

[46] Lawton, A. T.; Wright, P. (1989). "A planetary system for Gamma Cephei?". *Journal of the British Interplanetary Society* **42**: 335–336. Bibcode:1989JBIS...42..335L.

[47] Walker, G. A. H; Bohlender, D. A.; Walker, A. R.; Irwin, A. W.; Yang, S. L. S.; Larson, A. (1992). "Gamma Cephei – Rotation or planetary companion?". *Astrophysical Journal Letters* **396** (2): L91–L94. Bibcode:1992ApJ...396L..91W. doi:10.1086/186524.

[48] Hatzes, A. P.; Cochran, William D.; Endl, Michael; McArthur, Barbara; Paulson, Diane B.; Walker, Gordon A. H.; Campbell, Bruce; Yang, Stephenson (2003). "A Planetary Companion to Gamma Cephei A". *Astrophysical Journal* **599** (2): 1383–1394. arXiv:astro-ph/0305110. Bibcode:2003ApJ...599.1383H. doi:10.1086/379281.

[49] Holtz, Robert (22 April 1994). "Scientists Uncover Evidence of New Planets Orbiting Star". *Los Angeles Times via The Tech Online.*

[50] Mayor, M.; Queloz, D. (1995). "A Jupiter-mass companion to a solar-type star". *Nature* **378** (6555): 355–359. Bibcode:1995Na doi:10.1038/378355a0.

[51] Gibney, Elizabeth (18 December 2013). "In search of sister earths". *Nature* **504**: 361. Bibcode:2013Natur.504..357.. doi:10.10

[52] Lissauer, J. J. (1999). "Three planets for Upsilon Andromedae". *Nature* **398** (6729): 659. Bibcode:1999Natur.398..659L. doi:10.1038/19409.

[53] Doyle, L. R.; Carter, J. A.; Fabrycky, D. C.; Slawson, R. W.; Howell, S. B.; Winn, J. N.; Orosz, J. A.; Pr Sa, A.; Welsh, W. F.; Quinn, S. N.; Latham, D.; Torres, G.; Buchhave, L. A.; Marcy, G. W.; Fortney, J. J.; Shporer, A.; Ford, E. B.; Lissauer, J. J.; Ragozzine, D.; Rucker, M.; Batalha, N.; Jenkins, J. M.; Borucki, W. J.; Koch, D.; Middour, C. K.; Hall, J. R.; McCauliff, S.; Fanelli, M. N.; Quintana, E. V.; et al. (2011). "Kepler-16: A Transiting Circumbinary Planet". *Science* **333** (6049): 1602–6. arXiv:1109.3432. Bibcode:2011Sci...333.1602D. doi:10.1126/science.1210923. PMID 21921192.

[54] Johnson, Michele; Harrington, J.D. (26 February 2014). "NASA's Kepler Mission Announces a Planet Bonanza, 715 New Worlds". *NASA.* Retrieved 26 February 2014.

[55] Wall, Mike (26 February 2014). "Population of Known Alien Planets Nearly Doubles as NASA Discovers 715 New Worlds".

[56] "Kepler telescope bags huge haul of planets". Retrieved 27 February 2014.

[57] Johnson, Michelle; Chou, Felicia (23 July 2015). "NASA's Kepler Mission Discovers Bigger, Older Cousin to Earth". *NASA*.

[58] "NASA's Exoplanet Archive KOI table". NASA. Retrieved 28 February 2014.

[59] Johnson, Michele (23 July 2015). "Kepler Planet Candidates, July 2015". *NASA*. Retrieved 24 July 2015.

[60] Perryman, Michael (2011). *The Exoplanet Handbook*. Cambridge University Press. p. 149. ISBN 978-0-521-76559-6.

[61] Close, L. M.; Follette, K. B.; Males, J. R.; Puglisi, A.; Xompero, M.; Apai, D.; Najita, J.; Weinberger, A. J.; Morzinski, K.; Rodigas, T. J.; Hinz, P.; Bailey, V.; Briguglio, R. (2014). "Discovery of Hα Emission from the Close Companion Inside the Gap of Transitional Disk HD142527". *The Astrophysical Journal* **781** (2): L30. arXiv:1401.1273. Bibcode:2014ApJ...781L..30C. doi:10.1088/2041-8205/781/2/L30.

[62] Pepe, F.; Lovis, C.; Ségransan, D.; Benz, W.; Bouchy, F.; Dumusque, X.; Mayor, M.; Queloz, D.; Santos, N. C.; Udry, S. (2011). "The HARPS search for Earth-like planets in the habitable zone". *Astronomy & Astrophysics* **534**: A58. arXiv:1108.3447. Bibcode:2011A&A...534A..58P. doi:10.1051/0004-6361/201117055.

[63] Rodler, F.; Lopez-Morales, M.; Ribas, I. (2012). "Weighing the Non-Transiting Hot Jupiter Tau BOO b". *The Astrophysical Journal Letters* **753** (25): L25. arXiv:1206.6197. Bibcode:2012ApJ...753L..25R. doi:10.1088/2041-8205/753/1/L25.

[64] Planet Hunting: Finding Earth-like Planets. Scientific Computing. 19 July 2010

[65] Ballard, S.; Fabrycky, D.; Fressin, F.; Charbonneau, D.; Desert, J. M.; Torres, G.; Marcy, G.; Burke, C. J.; Isaacson, H.; Henze, C.; Steffen, J. H.; Ciardi, D. R.; Howell, S. B.; Cochran, W. D.; Endl, M.; Bryson, S. T.; Rowe, J. F.; Holman, M. J.; Lissauer, J. J.; Jenkins, J. M.; Still, M.; Ford, E. B.; Christiansen, J. L.; Middour, C. K.; Haas, M. R.; Li, J.; Hall, J. R.; McCauliff, S.; Batalha, N. M.; et al. (2011). "The Kepler-19 System: A Transiting 2.2 R⊕ Planet and a Second Planet Detected Via Transit Timing Variations". *The Astrophysical Journal* **743** (2): 200. arXiv:1109.1561. Bibcode:2011ApJ...743..200B. doi:10.1088/0004-637X/743/2/200.

[66] Lissauer, J. J.; Fabrycky, D. C.; Ford, E. B.; Borucki, W. J.; Fressin, F.; Marcy, G. W.; Orosz, J. A.; Rowe, J. F.; Torres, G.; Welsh, W. F.; Batalha, N. M.; Bryson, S. T.; Buchhave, L. A.; Caldwell, D. A.; Carter, J. A.; Charbonneau, D.; Christiansen, J. L.; Cochran, W. D.; Desert, J. M.; Dunham, E. W.; Fanelli, M. N.; Fortney, J. J.; Gautier III, T. N.; Geary, J. C.; Gilliland, R. L.; Haas, M. R.; Hall, J. R.; Holman, M. J.; Koch, D. G.; et al. (2011). "A closely packed system of low-mass, low-density planets transiting Kepler-11". *Nature* **470** (7332): 53–58. arXiv:1102.0291. Bibcode:2011Natur.470...53L. doi:10.1038/nature09760.

[67] Pál, A.; Kocsis, B. (2008). "Periastron Precession Measurements in Transiting Extrasolar Planetary Systems at the Level of General Relativity". *Monthly Notices of the Royal Astronomical Society* **389**: 191–198. arXiv:0806.0629. Bibcode:2008MNRAS.389..191P. doi:10.1111/j.1365-2966.2008.13512.x.

[68] Silvotti, R.; Schuh, S.; Janulis, R.; Solheim, J. -E.; Bernabei, S.; Østensen, R.; Oswalt, T. D.; Bruni, I.; Gualandi, R.; Bonanno, A.; Vauclair, G.; Reed, M.; Chen, C. -W.; Leibowitz, E.; Paparo, M.; Baran, A.; Charpinet, S.; Dolez, N.; Kawaler, S.; Kurtz, D.; Moskalik, P.; Riddle, R.; Zola, S. (2007). "A giant planet orbiting the 'extreme horizontal branch' star V 391 Pegasi" (PDF). *Nature* **449** (7159): 189–91. Bibcode:2007Natur.449..189S. doi:10.1038/nature06143. PMID 17851517.

[69] Jenkins, J.M.; Laurance R. Doyle (20 September 2003). "Detecting reflected light from close-in giant planets using space-based photometers". *Astrophysical Journal* 1(595): 429–445. arXiv:astro-ph/0305473. Bibcode:2003ApJ...595..429J.doi:10.1086/

[70] Loeb, A.; Gaudi, B. S. (2003). "Periodic Flux Variability of Stars due to the Reflex Doppler Effect Induced by Planetary Companions". *The Astrophysical Journal Letters* **588** (2): L117. arXiv:astro-ph/0303212. Bibcode:2003ApJ...588L.117L. doi:10.1086/375551.

[71] Atkinson, Nancy (13 May 2013) Using the Theory of Relativity and BEER to Find Exoplanets. *Universe Today*.

[72] Schmid, H. M.; Beuzit, J. -L.; Feldt, M.; Gisler, D.; Gratton, R.; Henning, T.; Joos, F.; Kasper, M.; Lenzen, R.; Mouillet, D.; Moutou, C.; Quirrenbach, A.; Stam, D. M.; Thalmann, C.; Tinbergen, J.; Verinaud, C.; Waters, R.; Wolstencroft, R. (2006). "Search and investigation of extra-solar planets with polarimetry". *Proceedings of the International Astronomical Union* **1**: 165. doi:10.1017/S1743921306009252.

[73] Berdyugina, S. V.; Berdyugin, A. V.; Fluri, D. M.; Piirola, V. (2008). "First Detection of Polarized Scattered Light from an Exoplanetary Atmosphere". *The Astrophysical Journal* **673**: L83. arXiv:0712.0193. Bibcode:2008ApJ...673L..83B.doi:10.1086/

[74] NameExoWorlds: An IAU Worldwide Contest to Name Exoplanets and their Host Stars. IAU.org. 9 July 2014

[75] NameExoWorlds.

[76] NameExoWorlds.

[77] Stromberg, Joseph (10 July 2014). "We've found hundreds of new planets. And now they're going to get cool names". Vox. Retrieved 10 July 2014.

[78] Hessman, F. V.; Dhillon, V. S.; Winget, D. E.; Schreiber, M. R.; Horne, K.; Marsh, T. R.; Guenther, E.; Schwope, A.; Heber, U. (2010). "On the naming convention used for multiple star systems and extrasolar planets". arXiv:1012.0707 [astro-ph.SR].

[79] Hartkopf, William I. and Mason, Brian D. "Addressing confusion in double star nomenclature: The Washington Multiplicity Catalog". United States Naval Observatory. Retrieved 12 September 2008.

[80] Schneider, J. (2011). "Notes for star 55 Cnc". Extrasolar Planets Encyclopaedia. Retrieved 26 September 2011.

[81] Schneider, J. (2011). "Notes for Planet 16 Cyg B b". Extrasolar Planets Encyclopaedia. Retrieved 26 September 2011.

[82] Schneider, J. (2011). "Notes for Planet HD 178911 B b". Extrasolar Planets Encyclopaedia. Retrieved 26 September 2011.

[83] Schneider, J. (2011). "Notes for Planet HD 41004 A b". Extrasolar Planets Encyclopaedia. Retrieved 26 September 2011.

[84] Schneider, J. (2011). "Notes for Planet Tau Boo b". Extrasolar Planets Encyclopaedia. Retrieved 26 September 2011.

[85] Neveu-Vanmalle, M.; Queloz, D.; Anderson, D. R.; Charbonnel, C.; Collier Cameron, A.; Delrez, L.; Gillon, M.; Hellier, C.; Jehin, E.; Lendl, M.; Maxted, P. F. L.; Pepe, F.; Pollacco, D.; Ségransan, D.; Smalley, B.; Smith, A. M. S.; Southworth, J.; Triaud, A. H. M. J.; Udry, S.; West, R. G. (2014). "WASP-94 a and B planets: Hot-Jupiter cousins in a twin-star system". *Astronomy & Astrophysics* **572**: A49. arXiv:1409.7566. Bibcode:2014A&A...572A..49N. doi:10.1051/0004-6361/201424744.

[86] Mamajek, Eric E. (26 June 2009) Initial Conditions of Planet Formation: Lifetimes of Primordial Disks. arxiv.org

[87] Rice, W. K. M.; Armitage, P. J. (2003). "On the Formation Timescale and Core Masses of Gas Giant Planets". *The Astrophysical Journal* **598**: L55. arXiv:astro-ph/0310191. Bibcode:2003ApJ...598L..55R. doi:10.1086/380390.

[88] Yin, Q.; Jacobsen, S. B.; Yamashita, K.; Blichert-Toft, J.; Télouk, P.; Albarède, F. (2002). "A short timescale for terrestrial planet formation from Hf–W chronometry of meteorites". *Nature* **418** (6901): 949–952. Bibcode:2002Natur.418..949Y. doi:10.1038/nature00995. line feed character in |title= at position 61 (help)

[89] Lammer, H.; Stokl, A.; Erkaev, N. V.; Dorfi, E. A.; Odert, P.; Gudel, M.; Kulikov, Y. N.; Kislyakova, K. G.; Leitzinger, M. (2014). "Origin and loss of nebula-captured hydrogen envelopes from 'sub'- to 'super-Earths' in the habitable zone of Sun-like stars". *Monthly Notices of the Royal Astronomical Society* **439** (4): 3225–3238. arXiv:1401.2765. Bibcode:2014MNRAS.439.3 225L. doi:10.1093/mnras/stu085.

[90] Johnson, R.E. (17 February 2010) Thermally-Driven Atmospheric Escape. arxiv.org

[91] Zendejas, Jesus; Segura, Antigona and Raga, Alejandro (31 May 2010) Atmospheric mass loss by stellar wind from planets around main sequence M stars. arxiv.org

[92] Masuda, K. (2014). "Very Low Density Planets Around Kepler-51 Revealed with Transit Timing Variations and an Anomaly Similar to a Planet-Planet Eclipse Event". *The Astrophysical Journal* **783**: 53. arXiv:1401.2885. Bibcode:2014ApJ...783...53M. doi:10.1088/0004-637X/783/1/53.

[93] Cumming, Andrew; Butler, R. Paul; Marcy, Geoffrey W.; Vogt, Steven S.; Wright, Jason T.; et al. (2008). "The Keck Planet Search: Detectability and the Minimum Mass and Orbital Period Distribution of Extrasolar Planets". *Publications of the Astronomical Society of the Pacific* **120** (867): 531–554. arXiv:0803.3357. Bibcode:2008PASP..120..531C. doi:10.1086/588487.

[94] Bonfils, X.; Forveille, T.; Delfosse, X.; Udry, S.; Mayor, M.; Perrier, C.; Bouchy, F.; Pepe, F.; Queloz, D.; Bertaux, J. -L. (2005). "The HARPS search for southern extra-solar planets". *Astronomy and Astrophysics* **443** (3): L15–L18. arXiv:astro-ph/0509211. Bibcode:2005A&A...443L..15B. doi:10.1051/0004-6361:200500193.

[95] Wang, J.; Fischer, D. A. (2014). "Revealing a Universal Planet–Metallicity Correlation for Planets of Different Solar-Type Stars". *The Astronomical Journal* **149**: 14. Bibcode:2015AJ....149...14W. doi:10.1088/0004-6256/149/1/14.

[96] Schwarz, Richard. BINARY CATALOGUE OF EXOPLANETS. Universität Wien

[97] Schwarz, Richard. STAR-DATA. Universität Wien

[98] Odd planet, so far from its star... Université de Montréal. 13 May 2014

[99]   Klotz, Irene (15 August 2013) "Time Really Flies on These Kepler Planets". News.discovery.com

[100] Rappaport, S.; Sanchis-Ojeda, R.; Rogers, L. A.; Levine, A.; Winn, J. N. (2013). "The Roche limit for close-orbiting planets: Minimum density, composition constraints, and application to the 4.2 hr planet KOI 1843.03". *The Astrophysical Journal* **773**: L15. arXiv:1307.4080. Bibcode:2013ApJ...773L..15R. doi:10.1088/2041-8205/773/1/L15.

[101] Perets, H. B.; Kouwenhoven, M. B. N. (2012). "On the Origin of Planets at Very Wide Orbits from the Recapture of Free Floating Planets". *The Astrophysical Journal* **750**: 83. arXiv:1202.2362. Bibcode:2012ApJ...750...83P. doi:10.1088/0004-637X/750/1/83.

[102] Scharf, Caleb; Menou, Kristen (2009). "Long-Period Exoplanets from Dynamical Relaxation". *The Astrophysical Journal* **693** (2): L113. arXiv:0811.1981. Bibcode:2009ApJ...693L.113S. doi:10.1088/0004-637X/693/2/L113.

[103] D'Angelo, G.; Durisen, R. H.; Lissauer, J. J. (2011). "Giant Planet Formation". In Seager, S. *Exoplanets*. University of Arizona Press, Tucson, AZ. pp. 319–346. arXiv:1006.5486. Bibcode:2010exop.book..319D.

[104] Catalog Listing. Extrasolar Planet Encyclopaedia

[105] Nielsen, E. L.; Close, L. M. (2010). "A Uniform Analysis of 118 Stars with High-Contrast Imaging: Long-Period Extrasolar Giant Planets Are Rare Around Sun-Like Stars". *The Astrophysical Journal* **717** (2): 878–896. arXiv:0909.4531. Bibcode:2010ApJ...717..878N. doi:10.1088/0004-637X/717/2/878.

[106] Marcy, G.; et al. (2005). "Observed Properties of Exoplanets: Masses, Orbits and Metallicities". *Progress of Theoretical Physics Supplement* **158**: 24–42. arXiv:astro-ph/0505003. Bibcode:2005PThPS.158...24M. doi:10.1143/PTPS.158.24.

[107] Boss, Alan (2009). *The Crowded Universe: The Search for Living Planets*. Basic Books. p. 26. ISBN 978-0-465-00936-7.

[108] Rodigas, T. J.; Hinz, P. M. (2009). "Which Radial Velocity Exoplanets Have Undetected Outer Companions?". *The Astrophysical Journal* **702**: 716–723. arXiv:0907.0020. Bibcode:2009ApJ...702..716R. doi:10.1088/0004-637X/702/1/716.

[109] Anglada-Escudé, G.; López-Morales, M.; Chambers, J. E. (2010). "How Eccentric Orbital Solutions Can Hide Planetary Systems in 2:1 Resonant Orbits". *The Astrophysical Journal* **709**: 168–178. arXiv:0809.1275. Bibcode:2010ApJ...709..168A. doi:10.1088/0004-637X/709/1/168.

[110] Kane, Stephen R. *et al.* (7 March 2012) The Exoplanet Eccentricity Distribution from Kepler Planet Candidates. arxiv.org

[111] Mason, John (2008) *Exoplanets: Detection, Formation, Properties, Habitability*. Springer. ISBN 3540740074. p. 2

[112] Out of Flatland: Orbits Are Askew in a Nearby Planetary System. *Scientific American*. 24 May 2010.

[113] "Turning planetary theory upside down". Astro.gla.ac.uk. 13 April 2010.

[114] "Tilting stars may explain backwards planets". *New Scientist*. 1 September 2010, Vol. 2776.

[115] Observability of the General Relativistic Precession of Periastra in Exoplanets. Andres Jordan, Gaspar A. Bakos, 3 June 2008

[116] Classical and relativistic node precessional effects in WASP-33b and perspectives for detecting them. Lorenzo Iorio, 25 August 2010

[117] Length of Exoplanet Day Measured for First Time. Eso.org. 30 April 2014

[118] Snellen, I. A. G.; Brandl, B. R.; De Kok, R. J.; Brogi, M.; Birkby, J.; Schwarz, H. (2014). "Fast spin of the young extrasolar planet β Pictoris b". *Nature* **509** (7498): 63–65. arXiv:1404.7506. Bibcode:2014Natur.509...63S. doi:10.1038/nature13253. PMID 24784216.

[119] Klotz, Irene (30 April 2014) Newly Clocked Exoplanet Spins a Whole Day in 8 Hours. Discovery.com.

[120] Correia, Alexandre C. M. and Laskar, Jacques (2010) "Tidal Evolution of Exoplanets", in *Exoplanets*, ed. S. Seager, University of Arizona Press. ISBN 978-0-8165-2945-2

[121] Cowen, Ron (30 April 2014) Exoplanet Rotation Detected for the First Time. Scientific American

[122] Crossfield, I. J. M. (2014). "Doppler imaging of exoplanets and brown dwarfs". *Astronomy & Astrophysics* **566**: A130. arXiv:1404.7853. Bibcode:2014A&A...566A.130C. doi:10.1051/0004-6361/201423750.

[123] Raymond, Sean N. *et al.* (5 December 2013) Terrestrial Planet Formation at Home and Abroad. arxiv.org

[124] de Wit, Julien; Seager, S. (19 December 2013). "Constraining Exoplanet Mass from Transmission Spectroscopy". *Science* **342** (6165): 1473–1477. arXiv:1401.6181. Bibcode:2013Sci...342.1473D. doi:10.1126/science.1245450. PMID 24357312.

[125] Nesvorný, D.; Morbidelli, A. (2008). "Mass and Orbit Determination from Transit Timing Variations of Exoplanets". *The Astrophysical Journal* **688**: 636–646. Bibcode:2008ApJ...688..636N. doi:10.1086/592230.

[126] Seager, S. and Lissauer, J. J. (2010) "Introduction to Exoplanets", pp. 3–13 in *Exoplanets*, Sara Seager (ed.), University of Arizona Press. ISBN 0816529450

[127] Lissauer, J. J. and de Pater, I. (2013) *Fundamental Planetary Science: Physics, Chemistry and Habitability*. Cambridge University Press. ISBN 052161855X. p. 74

[128] Baraffe, I.; Chabrier, G.; Barman, T. (2010). "The physical properties of extra-solar planets". *Reports on Progress in Physics* **73**: 016901. arXiv:1001.3577. Bibcode:2010RPPh...73a6901B. doi:10.1088/0034-4885/73/1/016901.

[129] Bonomo, A. S.; Sozzetti, A.; Lovis, C.; Malavolta, L.; Rice, K.; Buchhave, L. A.; Sasselov, D.; Cameron, A. C.; Latham, D. W.; Molinari, E.; Pepe, F.; Udry, S.; Affer, L.; Charbonneau, D.; Cosentino, R.; Dressing, C. D.; Dumusque, X.; Figueira, P.; Fiorenzano, A. F. M.; Gettel, S.; Harutyunyan, A.; Haywood, R. D.; Horne, K.; Lopez-Morales, M.; Mayor, M.; Micela, G.; Motalebi, F.; Nascimbeni, V.; Phillips, D. F.; et al. (2014). "Characterization of the planetary system Kepler-101 with HARPS-N". *Astronomy & Astrophysics* **572**: A2. Bibcode:2014A&A...572A...2B. doi:10.1051/0004-6361/201424617.

[130] Weiss, L. M.; Marcy, G. W. (2014). "The Mass-Radius Relation for 65 Exoplanets Smaller Than 4 Earth Radii". *The Astrophysical Journal* **783**: L6. arXiv:1312.0936. Bibcode:2014ApJ...783L...6W. doi:10.1088/2041-8205/783/1/L6.

[131] Marcy, G. W.; Weiss, L. M.; Petigura, E. A.; Isaacson, H.; Howard, A. W.; Buchhave, L. A. (2014). "Occurrence and core-envelope structure of 1–4× Earth-size planets around Sun-like stars". *Proceedings of the National Academy of Sciences* **111** (35): 12655–12660. arXiv:1404.2960. Bibcode:2014PNAS..11112655M. doi:10.1073/pnas.1304197111.

[132] Lopez, E. D.; Fortney, J. J. (2014). "Understanding the Mass-Radius Relation for Sub-Neptunes: Radius As a Proxy for Composition". *The Astrophysical Journal* **792**: 1. arXiv:1311.0329. Bibcode:2014ApJ...792....1L. doi:10.1088/0004-637X/792/1/1.

[133] Buchhave, L. A.; Bizzarro, M.; Latham, D. W.; Sasselov, D.; Cochran, W. D.; Endl, M.; Isaacson, H.; Juncher, D.; Marcy, G. W. (2014). "Three regimes of extrasolar planet radius inferred from host star metallicities". *Nature* **509** (7502): 593–595. arXiv:1405.7695. Bibcode:2014Natur.509..593B. doi:10.1038/nature13254.

[134] A Continuum of Planet Formation Between 1 and 4 Earth Radii, Kevin C. Schlaufman, (Submitted on 23 Jan 2015)

[135] Cowen, Ron (6 January 2014) "Earth-mass exoplanet is no Earth twin". *Nature News*, doi:10.1038/nature.2014.14477

[136] Cabrera, Juan; Grenfell, John Lee and Nettelmann, Nadine (2014) PS6.3. Observations and Modeling of Low Mass Low Density (LMLD) Exoplanets. *European Geosciences Union General Assembly 2014*

[137] Benneke, Björn and Seager, Sara (26 June 2013) How to Distinguish between Cloudy Mini-Neptunes and Water/Volatile-Dominated Super-Earths.

[138] Sasselov, Dimitar (2 June 2014) Exoplanets: From Exhilarating to Exasperating. 22:59, Kepler-10c: The "Mega-Earth", YouTube

[139] Astronomers Find a New Type of Planet: The "Mega-Earth"

[140] Dumusque, X.; Bonomo, A. S.; Haywood, R. L. D.; Malavolta, L.; Ségransan, D.; Buchhave, L. A.; Cameron, A. C.; Latham, D. W.; Molinari, E.; Pepe, F.; Udry, S. P.; Charbonneau, D.; Cosentino, R.; Dressing, C. D.; Figueira, P.; Fiorenzano, A. F. M.; Gettel, S.; Harutyunyan, A.; Horne, K.; Lopez-Morales, M.; Lovis, C.; Mayor, M.; Micela, G.; Motalebi, F.; Nascimbeni, V.; Phillips, D. F.; Piotto, G.; Pollacco, D.; Queloz, D.; et al. (2014). "The Kepler-10 Planetary System Revisited by Harps-N: A Hot Rocky World and a Solid Neptune-Mass Planet". *The Astrophysical Journal* **789** (2): 154. arXiv:1405.7881. Bibcode:2014ApJ...789..154D. doi:10.1088/0004-637X/789/2/154.

[141] Tidal Downsizing Model. IV. Destructive feedback in planets. Sergei Nayakshin, (Submitted on 6 Oct 2015)

[142] XX. CoRoT-20b: A very high density, high eccentricity transiting giant planet, M. Deleuil, A.S. Bonomo, S. Ferraz-Mello, A. Erikson, F. Bouchy, M. Havel, S. Aigrain, J.-M. Almenara, R. Alonso, M. Auvergne, A. Baglin, P. Barge, P. Bordé, H. Bruntt, J. Cabrera, S. Carpano, C. Cavarroc, Sz. Csizmadia, C. Damiani, H.J. Deeg, R. Dvorak, M. Fridlund, G. Hébrard, D. Gandolfi, M. Gillon, E. Guenther, T. Guillot, A. Hatzes, L. Jorda, A. Léger, H. Lammer, T. Mazeh, C. Moutou, A. Ollivier, A. Ofir, H. Parviainen, D. Queloz, H. Rauer, A. Rodriguez, D. Rouan, A. Santerne, J. Schneider, L. Tal-Or, B. Tingley, J. Weingrill, G. Wuchterl (Submitted on 14 Sep 2011)

[143] Mocquet, A.; Grasset, O. and Sotin, C. (2013) Super-dense remnants of gas giant exoplanets, EPSC Abstracts, Vol. 8, EPSC2013-986-1, European Planetary Science Congress 2013

[144] Mocquet, A.; Grasset, O.; Sotin, C. (2014). "Very high-density planets: a possible remnant of gas giants". *Phil. Trans. R. Soc. A* **372** (2014): 20130164. Bibcode:2014RSPTA.37230164M. doi:10.1098/rsta.2013.0164.

[145] Seager, S.; Kuchner, M.; Hier-Majumder, C. A.; Militzer, B. (2007). "Mass-Radius Relationships for Solid Exoplanets". *The Astrophysical Journal* **669** (2): 1279–1297. arXiv:0707.2895. Bibcode:2007ApJ...669.1279S. doi:10.1086/521346.

[146] Carter, J. A.; Winn, J. N. (2010). "Empirical Constraints on the Oblateness of an Exoplanet". *The Astrophysical Journal* **709** (2): 1219–1229. arXiv:0912.1594. Bibcode:2010ApJ...709.1219C. doi:10.1088/0004-637X/709/2/1219.

[147] Leconte, J.; Lai, D.; Chabrier, G. (2011). "Distorted, nonspherical transiting planets: Impact on the transit depth and on the radius determination". *Astronomy & Astrophysics* **528**: A41. arXiv:1101.2813. Bibcode:2011A&A...528A..41L. doi:10.1051/0004-6361/201015811.

[148] Arras, Phil (7 January 2009) Thermal Tides in Short Period Exoplanets. arxiv.org

[149] Madhusudhan, Nikku (5 February 2014) Exoplanetary Atmospheres. arxiv.org

[150] Seager, S.; Deming, D. (2010). "Exoplanet Atmospheres". arXiv:1005.4037 [astro-ph.EP].

[151] Brogi, M.; Snellen, I. A. G.; De Kok, R. J.; Albrecht, S.; Birkby, J.; De Mooij, E. J. W. (2012). "The signature of orbital motion from the dayside of the planet τ Boötis b". *Nature* **486** (7404): 502–504. arXiv:1206.6109. Bibcode:2012Natur.486..502B. doi:10.1038/nature11161.

[152] Rodler, F.; Lopez-Morales, M.; Riba, I. (2012). "Weighing the Non-transiting Hot Jupiter τ Boo b". *The Astrophysical Journal Letters* **753** (1): L25. arXiv:1206.6197. Bibcode:2012ApJ...753L..25R. doi:10.1088/2041-8205/753/1/L25.

[153] Charbonneau, D.; Brown, T.; Burrows, A.; Laughlin, G. (2006). "When Extrasolar Planets Transit Their Parent Stars". *Protostars and Planets V*. University of Arizona Press. arXiv:astro-ph/0603376.

[154] Charbonneau, D.; Brown, T. M.; Noyes, R. W.; Gilliland, R. L. (2002). "Detection of an Extrasolar Planet Atmosphere". *The Astrophysical Journal* **568**: 377–384. arXiv:astro-ph/0111544. Bibcode:2002ApJ...568..377C. doi:10.1086/338770.

[155] Swain, M. R.; Vasisht, G.; Tinetti, G.; Bouwman, J.; Chen, P.; Yung, Y.; Deming, D.; Deroo, P. (2009). "Molecular Signatures in the Near Infrared Dayside Spectrum of HD 189733b". *The Astrophysical Journal* **690** (2): L114. arXiv:0812.1844. Bibcode:2009ApJ...690L.114S. doi:10.1088/0004-637X/690/2/L114.

[156] NASA – Hubble Finds First Organic Molecule on an Exoplanet. NASA. 19 March 2008

[157] "Hubble Traces Subtle Signals of Water on Hazy Worlds". NASA. 3 December 2013. Retrieved 4 December 2013.

[158] Deming, D.; Wilkins, A.; McCullough, P.; Burrows, A.; Fortney, J. J.; Agol, E.; Dobbs-Dixon, I.; Madhusudhan, N.; Crouzet, N.; Desert, J. M.; Gilliland, R. L.; Haynes, K.; Knutson, H. A.; Line, M.; Magic, Z.; Mandell, A. M.; Ranjan, S.; Charbonneau, D.; Clampin, M.; Seager, S.; Showman, A. P. (2013). "Infrared Transmission Spectroscopy of the Exoplanets HD 209458b and XO-1b Using the Wide Field Camera-3 on the Hubble Space Telescope". *The Astrophysical Journal* **774** (2): 95. arXiv:1302.1141. Bibcode:2013ApJ...774...95D. doi:10.1088/0004-637X/774/2/95.

[159] Mandell, A. M.; Haynes, K.; Sinukoff, E.; Madhusudhan, N.; Burrows, A.; Deming, D. (2013). "Exoplanet Transit Spectroscopy Using WFC3: WASP-12 b, WASP-17 b, and WASP-19 b". *The Astrophysical Journal* **779** (2): 128. arXiv:1310.2949. Bibcode:2013ApJ...779..128M. doi:10.1088/0004-637X/779/2/128.

[160] Harrington, J.D.; Villard, Ray (24 July 2014). "RELEASE 14–197 – Hubble Finds Three Surprisingly Dry Exoplanets". *NASA*. Retrieved 25 July 2014.

[161] Clavin, Whitney; Chou, Felicia; Weaver, Donna; Villard; Johnson, Michele (24 September 2014). "NASA Telescopes Find Clear Skies and Water Vapor on Exoplanet". *NASA*. Retrieved 24 September 2014.

[162] Kawahara, H.; Matsuo, T.; Takami, M.; Fujii, Y.; Kotani, T.; Murakami, N.; Tamura, M.; Guyon, O. (2012). "Can Ground-based Telescopes Detect the Oxygen 1.27 μm Absorption Feature as a Biomarker in Exoplanets?". *The Astrophysical Journal* **758**: 13. arXiv:1206.0558. Bibcode:2012ApJ...758...13K. doi:10.1088/0004-637X/758/1/13.

[163] "Hottest planet is hotter than some stars". Retrieved 2015-06-12.

[164] "NASA's Hubble Telescope Detects 'Sunscreen' Layer on Distant Planet". Retrieved 2015-06-11.

[165] Haynes, Korey; Mandell, Avi M.; Madhusudhan, Nikku; Deming, Drake; Knutson, Heather (2015-05-06). "Spectroscopic Evidence for a Temperature Inversion in the Dayside Atmosphere of the Hot Jupiter WASP-33b". *arXiv:1505.01490 [astro-ph]*. External link in |journal= (help)

[166] Showman, Adam P.; Wordsworth, Robin D.; Merlis, Timothy M. and Kaspi, Yohai (11 June 2013) Atmospheric Circulation of Terrestrial Exoplanets, arxiv.org

[167] Chu, Jennifer (2 October 2013). "Scientists generate first map of clouds on an exoplanet". *MIT*. Retrieved 2 January 2014.

[168] Demory, B. O.; De Wit, J.; Lewis, N.; Fortney, J.; Zsom, A.; Seager, S.; Knutson, H.; Heng, K.; Madhusudhan, N.; Gillon, M.; Barclay, T.; Desert, J. M.; Parmentier, V.; Cowan, N. B. (2013). "Inference of Inhomogeneous Clouds in an Exoplanet Atmosphere". *The Astrophysical Journal* **776** (2): L25. arXiv:1309.7894. Bibcode:2013ApJ...776L..25D. doi:10.1088/2041-8205/776/2/L25.

[169] Harrington, J.D.; Weaver, Donna; Villard, Ray (31 December 2013). "Release 13-383 – NASA's Hubble Sees Cloudy Super-Worlds With Chance for More Clouds". *NASA*.

[170] Moses, Julianne (2014). "Extrasolar planets: Cloudy with a chance of dustballs". *Nature* **505**(7481): 31–32. Bibcode:2014Natu doi:10.1038/505031a.

[171] Knutson, Heather; Benecke, Björn; Deming, Drake; Homeier, Derek (2014). "A featureless transmission spectrum for the Neptune-mass exoplanet GJ 436b". *Nature* **505**(7481): 66–68. arXiv:1401.3350. Bibcode:2014Natur.505...66K.doi:10.1038/

[172] Kreidberg, Laura; Bean, Jacob L.; Desert, Jean-Michel; Benecke, Björn; Deming, Drake; Stevenson, Kevin B.; Seager, Sara; Berta-Thompson, Zachory; Seifahrt, Andreas; Homeier, Derek (2014). "Clouds in the atmosphere of the super-Earth exoplanet GJ 1214b". *Nature* **505** (7481): 69–72. arXiv:1401.0022. Bibcode:2014Natur.505...69K. doi:10.1038/nature12888.

[173] New World of Iron Rain. *Astrobiology Magazine*. 8 January 2003

[174] Howell, Elizabeth (30 August 2013) On Giant Blue Alien Planet, It Rains Molten Glass. SPACE.com

[175] Raining Pebbles: Rocky Exoplanet Has Bizarre Atmosphere, Simulation Suggests. Science Daily. 1 October 2009

[176] Morgan, James (14 October 2013) 'Diamond rain' falls on Saturn and Jupiter. BBC.

[177] Sanders, Robert (22 March 2010) Helium rain on Jupiter explains lack of neon in atmosphere. newscenter.berkeley.edu

[178] Wordsworth, R.; Pierrehumbert, R. (2014). "Abiotic Oxygen-Dominated Atmospheres on Terrestrial Habitable Zone Planets". *The Astrophysical Journal* **785** (2): L20. arXiv:1403.2713. Bibcode:2014ApJ...785L..20W. doi:10.1088/2041-8205/785/2/L20.

[179] Selsis, F.; Wordsworth, R. D.; Forget, F. (2011). "Thermal phase curves of nontransiting terrestrial exoplanets". *Astronomy & Astrophysics* **532**: A1. arXiv:1104.4763. Bibcode:2011A&A...532A...1S. doi:10.1051/0004-6361/201116654.

[180] Benneke, B.; Seager, S. (2012). "Atmospheric Retrieval for Super-Earths: Uniquely Constraining the Atmospheric Composition with Transmission Spectroscopy". *The Astrophysical Journal* **753** (2): 100. arXiv:1203.4018. Bibcode:2012ApJ...753..100B. doi:10.1088/0004-637X/753/2/100.

[181] Hu, Renyu (6 April 2012) Theoretical Spectra of Terrestrial Exoplanet Surfaces. arxiv.org

[182] "NASA, ESA, and K. Haynes and A. Mandell (Goddard Space Flight Center)". Retrieved 15 June 2015.

[183] Knutson, H. A.; Charbonneau, D.; Allen, L. E.; Fortney, J. J.; Agol, E.; Cowan, N. B.; Showman, A. P.; Cooper, C. S.; Megeath, S. T. (2007). "A map of the day–night contrast of the extrasolar planet HD 189733b" (PDF). *Nature* **447** (7141): 183–6. arXiv:0705.0993. Bibcode:2007Natur.447..183K. doi:10.1038/nature05782. PMID 17495920.

[184] NASA Hubble Finds a True Blue Planet. NASA. 11 July 2013

[185] Evans, T. M.; Pont, F. D. R.; Sing, D. K.; Aigrain, S.; Barstow, J. K.; Désert, J. M.; Gibson, N.; Heng, K.; Knutson, H. A.; Lecavelier Des Etangs, A. (2013). "The Deep Blue Color of HD189733b: Albedo Measurements with Hubble Space Telescope/Space Telescope Imaging Spectrograph at Visible Wavelengths". *The Astrophysical Journal* **772** (2): L16. arXiv:1307.3239. Bibcode:2013ApJ...772L..16E. doi:10.1088/2041-8205/772/2/L16.

[186] Kuzuhara, M.; Tamura, M.; Kudo, T.; Janson, M.; Kandori, R.; Brandt, T. D.; Thalmann, C.; Spiegel, D.; Biller, B.; et al. (2013). "Direct Imaging of a Cold Jovian Exoplanet in Orbit around the Sun-like Star GJ 504". *The Astrophysical Journal* **774** (11): 11. arXiv:1307.2886. Bibcode:2013ApJ...774...11K. doi:10.1088/0004-637X/774/1/11.

[187] Carson; Thalmann; Janson; Kozakis; Bonnefoy; Biller; Schlieder; Currie; McElwain (November 15, 2012). "Direct Imaging Discovery of a 'Super-Jupiter' Around the late B-Type Star Kappa And". arXiv:1211.3744 [astro-ph.SR].

[188] The Apparent Brightness and Size of Exoplanets and their Stars, Abel Mendez, updated 30 June 2012, 12:10 PM

[189] "Coal-Black Alien Planet Is Darkest Ever Seen". Space.com. Retrieved 12 August 2011.

[190] Kipping, David M. and Spiegel, David S. (10 August 2011) Detection of visible light from the darkest world. arxiv.org

[191] Barclay, T.; Huber, D.; Rowe, J. F.; Fortney, J. J.; Morley, C. V.; Quintana, E. V.; Fabrycky, D. C.; Barentsen, G.; Bloemen, S.; Christiansen, J. L.; Demory, B. O.; Fulton, B. J.; Jenkins, J. M.; Mullally, F.; Ragozzine, D.; Seader, S. E.; Shporer, A.; Tenenbaum, P.; Thompson, S. E. (2012). "Photometrically derived masses and radii of the planet and star in the TrES-2 system". *The Astrophysical Journal* **761**: 53. arXiv:1210.4592. Bibcode:2012ApJ...761...53B. doi:10.1088/0004-637X/761/1/53.

[192] Burrows, Adam (18 December 2014) Scientific Return of Coronagraphic Exoplanet Imaging and Spectroscopy Using WFIRST. arxiv.org

[193] Unlocking the Secrets of an Alien World's Magnetic Field. Space.com. by Charles Q. Choi, November 20, 2014

[194] Kislyakova, K. G.; Holmstrom, M.; Lammer, H.; Odert, P.; Khodachenko, M. L. (2014). "Magnetic moment and plasma environment of HD 209458b as determined from Ly observations". *Science* **346** (6212): 981. arXiv:1411.6875. Bibcode:2014Sci...346..981K. doi:10.1126/science.1257829. PMID 25414310.

[195] Footprint of a Magnetic Exoplanet, www.skyandtelescope.com, 9 January 2004, Robert Naeye

[196] Nichols, J. D. (2011). "Magnetosphere-ionosphere coupling at Jupiter-like exoplanets with internal plasma sources: Implications for detectability of auroral radio emissions". *Monthly Notices of the Royal Astronomical Society* **414** (3): 2125–2138. arXiv:1102.2737. Bibcode:2011MNRAS.414.2125N. doi:10.1111/j.1365-2966.2011.18528.x.

[197] Radio Telescopes Could Help Find Exoplanets. RedOrbit. 18 April 2011

[198] "Radio Detection of Extrasolar Planets: Present and Future Prospects" (PDF). *NRL, NASA/GSFC, NRAO, Observatoire de Paris*. Retrieved 15 October 2008.

[199] Super-Earths Get Magnetic 'Shield' from Liquid Metal. Charles Q. Choi, SPACE.com. 22 November 2012 02:01pm ET.

[200] Buzasi, D. (2013). "Stellar Magnetic Fields As a Heating Source for Extrasolar Giant Planets". *The Astrophysical Journal* **765** (2): L25. arXiv:1302.1466. Bibcode:2013ApJ...765L..25B. doi:10.1088/2041-8205/765/2/L25.

[201] Valencia, Diana; O'Connell, Richard J. (2009). "Convection scaling and subduction on Earth and super-Earths". *Earth and Planetary Science Letters* **286** (3–4): 492–502. Bibcode:2009E&PSL.286..492V. doi:10.1016/j.epsl.2009.07.015.

[202] Van Heck, H.J.; Tackley, P.J. (2011). "Plate tectonics on super-Earths: Equally or more likely than on Earth". *Earth and Planetary Science Letters* **310** (3–4): 252–261. Bibcode:2011E&PSL.310..252V. doi:10.1016/j.epsl.2011.07.029.

[203] O'Neill, C.; Lenardic, A. (2007). "Geological consequences of super-sized Earths". *Geophysical Research Letters* **34** (19). Bibcode:2007GeoRL..3419204O. doi:10.1029/2007GL030598.

[204] Valencia, Diana; O'Connell, Richard J.; Sasselov, Dimitar D (November 2007). "Inevitability of Plate Tectonics on Super-Earths". *Astrophysical Journal Letters* **670**(1): L45–L48. arXiv:0710.0699. Bibcode:2007ApJ...670L..45V. doi:10.1086/5240

[205] Super Earths Likely To Have Both Oceans and Continents. astrobiology.com. 7 January 2014

[206] Cowan, N. B.; Abbot, D. S. (2014). "Water Cycling Between Ocean and Mantle: Super-Earths Need Not Be Waterworlds". *The Astrophysical Journal* **781**: 27. arXiv:1401.0720. Bibcode:2014ApJ...781...27C. doi:10.1088/0004-637X/781/1/27.

[207] Astronomers May Have Found Volcanoes 40 Light-Years From Earth

[208] Variability in the super-Earth 55 Cnc e, Brice-Olivier Demory, Michael Gillon, Nikku Madhusudhan, Didier Queloz, (Submitted on 1 May 2015)

[209] Scientists Discover a Saturn-like Ring System Eclipsing a Sun-like Star. Space Daily. 13 January 2012

[210] Mamajek, E. E.; Quillen, A. C.; Pecaut, M. J.; Moolekamp, F.; Scott, E. L.; Kenworthy, M. A.; Cameron, A. C.; Parley, N. R. (2012). "Planetary Construction Zones in Occultation: Discovery of an Extrasolar Ring System Transiting a Young Sun-Like Star and Future Prospects for Detecting Eclipses by Circumsecondary and Circumplanetary Disks". *The Astronomical Journal* **143** (3): 72. arXiv:1108.4070. Bibcode:2012AJ....143...72M. doi:10.1088/0004-6256/143/3/72.

[211] Kalas, P.; Graham, J. R.; Chiang, E.; Fitzgerald, M. P.; Clampin, M.; Kite, E. S.; Stapelfeldt, K.; Marois, C.; Krist, J. (2008). "Optical Images of an Exosolar Planet 25 Light-Years from Earth". *Science* **322** (5906): 1345–8. arXiv:0811.1994. Bibcode:2008Sci...322.1345K. doi:10.1126/science.1166609. PMID 19008414.

[212] Schlichting, Hilke E. and Chang, Philip (19 Apr 2011) Warm Saturns: On the Nature of Rings around Extrasolar Planets that Reside Inside the Ice Line. arxiv.org

[213] Bennett, D. P.; Batista, V.; Bond, I. A.; Bennett, C. S.; Suzuki, D.; Beaulieu, J. -P.; Udalski, A.; Donatowicz, J.; Bozza, V.; Abe, F.; Botzler, C. S.; Freeman, M.; Fukunaga, D.; Fukui, A.; Itow, Y.; Koshimoto, N.; Ling, C. H.; Masuda, K.; Matsubara, Y.; Muraki, Y.; Namba, S.; Ohnishi, K.; Rattenbury, N. J.; Saito, T.; Sullivan, D. J.; Sumi, T.; Sweatman, W. L.; Tristram, P. J.; Tsurumi, N.; et al. (2014). "MOA-2011-BLG-262Lb: A sub-Earth-mass moon orbiting a gas giant or a high-velocity planetary system in the galactic bulge". *The Astrophysical Journal* **785** (2): 155. arXiv:1312.3951. Bibcode:2014ApJ...785..155B. doi:10.1088/0004-637X/785/2/155.

[214] Evaporating exoplanet stirs up dust. Phys.org. 28 August 2012

[215] Woollacott, Emma (18 May 2012) New-found exoplanet is evaporating away. *TG Daily*

[216] Bhanoo, Sindya N. (25 June 2015). "A Planet with a Tail Nine Million Miles Long". *New York Times*. Retrieved 25 June 2015.

[217] Forget "Earth-Like"—We'll First Find Aliens on Eyeball Planets. Nautilus. POSTED BY SEAN RAYMOND ON FEB 20, 2015

[218] Insolation patterns on eccentric exoplanets. Anthony R. Dobrovolskis, Icarus, Volume 250, April 2015, Pages 395–399

[219] Patterns of Sunlight on Extra-Solar Planets, Tony Dobrovolskis, 18 March 2014

[220] Choi, Charles Q. (1 September 2011) Alien Life More Likely on 'Dune' Planets. *Astrobiology Magazine*

[221] Abe, Y.; Abe-Ouchi, A.; Sleep, N. H.; Zahnle, K. J. (2011). "Habitable Zone Limits for Dry Planets". *Astrobiology* **11** (5): 443–460. Bibcode:2011AsBio..11..443A. doi:10.1089/ast.2010.0545. PMID 21707386.

[222] Seager, S. (2013). "Exoplanet Habitability". *Science* **340** (6132): 577–81. Bibcode:2013Sci...340..577S. doi:10.1126/science. PMID 23641111.

[223] Habitable Zones Around Main-Sequence Stars: Dependence on Planetary Mass, Ravi kumar Kopparapu, Ramses M. Ramirez, James SchottelKotte, James F. Kasting, Shawn Domagal-Goldman, Vincent Eymet, 21 Apr 2014

[224] Hamano, K.; Abe, Y.; Genda, H. (2013). "Emergence of two types of terrestrial planet on solidification of magma ocean". *Nature* **497** (7451): 607–10. Bibcode:2013Natur.497..607H. doi:10.1038/nature12163. PMID 23719462.

[225] Yang, J.; Boué, G. L.; Fabrycky, D. C.; Abbot, D. S. (2014). "Strong Dependence of the Inner Edge of the Habitable Zone on Planetary Rotation Rate" (PDF). *The Astrophysical Journal* **787**: L2. arXiv:1404.4992. Bibcode:2014ApJ...787L...2Y. doi:10.1088/2041-8205/787/1/L2.

[226] Real-life Sci-Fi World #2: the Hot Eyeball planet

[227] Stabilizing Cloud Feedback Dramatically Expands the Habitable Zone of Tidally Locked Planets, Jun Yang, Nicolas B. Cowan, Dorian S. Abbot, 1 Jul 2013

[228] Amend, J. P.; Teske, A. (2005). "Expanding frontiers in deep subsurface microbiology". *Palaeogeography, Palaeoclimatology, Palaeoecology* **219**: 131–155. doi:10.1016/j.palaeo.2004.10.018.

[229] Further away planets 'can support life' say researchers. BBC. 7 January 2014

[230] Abbot, D. S.; Switzer, E. R. (2011). "The Steppenwolf: A Proposal for a Habitable Planet in Interstellar Space". *The Astrophysical Journal* **735** (2): L27. arXiv:1102.1108. Bibcode:2011ApJ...735L..27A. doi:10.1088/2041-8205/735/2/L27.

[231] Loeb, A. (2014). "The habitable epoch of the early Universe". *International Journal of Astrobiology* **13** (4): 337–339. arXiv:1312.0613. Bibcode:2014IJAsB..13..337L. doi:10.1017/S1473550414000196.

[232] Home, sweet exomoon: The new frontier in the search for ET. New Scientist, 29 July 2015

[233] Linsenmeier, Manuel *et al.* (21 January 2014) Habitability of Earth-like planets with high obliquity and eccentric orbits: results from a general circulation model. arxiv.org

[234] Kelley, Peter (15 April 2014) Astronomers: 'Tilt-a-worlds' could harbor life. www.washington.edu

[235] Armstrong, J. C.; Barnes, R.; Domagal-Goldman, S.; Breiner, J.; Quinn, T. R.; Meadows, V. S. (2014). "Effects of Extreme Obliquity Variations on the Habitability of Exoplanets". *Astrobiology* **14** (4): 277–291. arXiv:1404.3686. Bibcode:2014AsBio..14..277A.doi:10.1089/ast.2013.1129.

[236] Kelley, Peter (18 July 2013) A warmer planetary haven around cool stars, as ice warms rather than cools. www.washington.edu

[237] Shields, A. L.; Bitz, C. M.; Meadows, V. S.; Joshi, M. M.; Robinson, T. D. (2014). "Spectrum-Driven Planetary Deglaciation Due to Increases in Stellar Luminosity". *The Astrophysical Journal* **785**: L9. arXiv:1403.3695. Bibcode:2014ApJ...785L...9S. doi:10.1088/2041-8205/785/1/L9.

[238] Barnes, R.; Mullins, K.; Goldblatt, C.; Meadows, V. S.; Kasting, J. F.; Heller, R. (2013). "Tidal Venuses: Triggering a Climate Catastrophe via Tidal Heating". *Astrobiology* **13** (3): 225–250. arXiv:1203.5104. Bibcode:2013AsBio..13..225B. doi:10.1089/ast.2012.0851.

[239] Heller, R.; Armstrong, J. (2014). "Superhabitable Worlds".*Astrobiology***14**: 50–66. arXiv:1401.2392. Bibcode:2014AsBio..14 doi:10.1089/ast.2013.1088.

[240] Jackson, B.; Barnes, R.; Greenberg, R. (2008). "Tidal heating of terrestrial extrasolar planets and implications for their habitability". *Monthly Notices of the Royal Astronomical Society* **391**: 237–245. arXiv:0808.2770. Bibcode:2008MNRAS.391..237J. doi:10.1111/j.1365-2966.2008.13868.x.

[241] Paul Gilster, Andrew LePage (2015-01-30). "A Review of the Best Habitable Planet Candidates". Centauri Dreams, Tau Zero Foundation. Retrieved 2015-07-24.

[242] Giovanni F. Bignami (2015). *The Mystery of the Seven Spheres: How Homo sapiens will Conquer Space.* Springer. ISBN 9783319170046., Page 110

[243] Howell, Elizabeth (6 February 2013). "Closest 'Alien Earth' May Be 13 Light-Years Away". *Space.com*. TechMediaNetwork. Retrieved 7 February 2013.

[244] Kopparapu, Ravi Kumar (March 2013). "A revised estimate of the occurrence rate of terrestrial planets in the habitable zones around kepler m-dwarfs". *The Astrophysical Journal Letters* **767**: L8. arXiv:1303.2649. Bibcode:2013ApJ...767L...8K. doi:10.1088/2041-8205/767/1/L8.

[245] "NASA's Kepler Mission Discovers Bigger, Older Cousin to Earth". Retrieved 2015-07-23.

[246] "Special Issue: Exoplanets". *Science*, 3 May 2013.

4.18   Further reading

- Boss, Alan (2009). *The Crowded Universe: The Search for Living Planets*. Basic Books. ISBN 978-0-465-00936-7 (Hardback); ISBN 978-0-465-02039-3 (Paperback).

- Dorminey, Bruce (2001). *Distant Wanderers*. Springer-Verlag. ISBN 978-0-387-95074-7 (Hardback); ISBN 978-1-4419-2872-6 (Paperback).

- Jayawardhana, Ray (2011). *Strange New Worlds: The Search for Alien Planets and Life beyond Our Solar System*. Princeton, NJ: Princeton University Press. ISBN 978-0-691-14254-8 (Hardcover).

- Perryman, Michael (2011). *The Exoplanet Handbook*. Cambridge University Press. ISBN 978-0-521-76559-6.

- Seager, Sara (2010). *Exoplanet Atmospheres: Physical Processes*. Princeton University Press. ISBN 978-0-691-11914-4 (Hardback); ISBN 978-0-691-14645-4 (Paperback).

- Seager, Sara, ed. (2011). *Exoplanets*. University of Arizona Press. ISBN 978-0-8165-2945-2.

- Villard, Ray; Cook, Lynette R. (2005). *Infinite Worlds: An Illustrated Voyage to Planets Beyond Our Sun*. University of California Press. ISBN 978-0-520-23710-0.

- Yaqoob, Tahir (2011). *Exoplanets and Alien Solar Systems*. New Earth Labs (Education and Outreach). ISBN 978-0-974-16892-0 (Paperback).

4.18.1 Formation and evolution

- Co-evolution of atmospheres, life, and climate, John Lee Grenfell et al., 20 May 2010

- Phase Separation in Giant Planets: Inhomogeneous Evolution of Saturn, Jonathan J. Fortney, William B. Hubbard, 1 May 2003

- Magnetodynamo Lifetimes for Rocky, Earth-Mass Exoplanets with Contrasting Mantle Convection Regimes, Joost van Summeren, Eric Gaidos, Clinton P. Conrad, 9 Apr 2013

- The effect of evaporation on the evolution of close-in giant planets, I. Baraffe, F. Selsis, G. Chabrier, T. S. Barman, F. Allard, P.H. Hauschildt, H. Lammer, 5 Apr 2004

- Observational Evidence for Tidal Destruction of Exoplanets, Brian Jackson, Rory Barnes, Richard Greenberg, 7 Apr 2009

- A Dying Universe: The Long Term Fate and Evolution of Astrophysical Objects, Fred C. Adams, Gregory Laughlin, 18 Jan 1997

4.18.2 Volcanism

- Detecting Volcanism on Extrasolar Planets, L. Kaltenegger, W. G. Henning, D. D. Sasselov, 7 September 2010

- Detecting planetary geochemical cycles on exoplanets: Atmospheric signatures and the case of SO2, L. Kaltenegger, D. Sasselov, 17 November 2009

- Geodynamics and Rate of Volcanism on Massive Earth-like Planets, Edwin S. Kite, Michael Manga, Eric Gaidos, 31 May 2009

- Tidal Heating of Terrestrial Extra-Solar Planets and Implications for their Habitability, Brian Jackson, Rory Barnes, Richard Greenberg, 20 August 2008

4.18.3 Interior structure

- Planetary internal structures, I. Baraffe, G. Chabrier, J. Fortney, C. Sotin, 19 January 2014

4.18.4 Surface mapping

- Global Mapping of Earth-like Exoplanets from Scattered Light Curves, Hajime Kawahara, Yuka Fujii, 16 July 2010

- A Two-Dimensional Infrared Map of the Extrasolar Planet HD 189733b, C. Majeau, E. Agol, N. Cowan, 19 September 2012

4.18.5   Climate and weather

- Patterns of Sunlight on Extra-Solar Planets, Tony Dobrovolskis, 18 March 2014

- Possible climates on terrestrial exoplanets, Francois Forget, Jeremy Leconte, 18 November 2013

- Indication of insensitivity of planetary weathering behavior and habitable zone to surface land fraction, Dorian S. Abbot, Nicolas B. Cowan, Fred J. Ciesla, 8 August 2012

- Clouds and Hazes in Exoplanet Atmospheres, Mark S. Marley, Andrew S. Ackerman, Jeffrey N. Cuzzi, Daniel Kitzmann, 23 January 2013

- Atmospheric Circulation of Exoplanets, Adam P. Showman, James Y-K. Cho, Kristen Menou, 16 November 2009

- New Technique Could Measure Exoplanet Atmospheric Pressure, an Indicator of Habitability, Shannon Hall on 6 March 2014, www.universetoday.com

4.18.6   Water

See also: § Habitability, § Plate tectonics, Ocean planet, Desert planet and Extraterrestrial liquid water

After hydrogen and helium, oxygen is the most common element in many planetary systems (in some systems carbon is more common than oxygen), and water H_2O one of the most common compounds. Gas giants are composed mostly of hydrogen and helium, but most planets are between the size of Earth and Neptune, where many planets will have deep water oceans covering the entire surface in addition to a H–He envelope.

- Water: from clouds to planets, Ewine F. van Dishoeck, Edwin A. Bergin, Dariusz C. Lis, Jonathan I. Lunine, 25 February 2014

- Are Exoplanets Orbiting Red Dwarf Stars too Dry for Life?, Michael Schirber, Astrobiology Magazine, 27 August 2013

- Carbon-Rich Exoplanets May Lack Surface Water, 26 October 2013

- 'Water-Trapped' Worlds, Adam Hadhazy, Astrobiology Magazine, 18 July 2013

- Lobster-Shaped Extrasolar Oceans, 10 March 2014, Charles Q. Choi, Astrobiology Magazine

- Alien Moons Could Bake Dry from Young Gas Giants' Hot Glow, Adam Hadhazy, Astrobiology Magazine 25 March 2014

- The Longevity of Oceans on Terrestrial Exoplanets, Bullock, Mark Alan; Grinspoon, D. H.

- False Positive For Ocean Glint on Exoplanets: the Latitude-Albedo Effect, Nicolas B. Cowan, Dorian S. Abbot, Aiko Voigt, 4 May 2012

4.18.7   Orbital dynamics

Eccentricity dynamics

See also: Tidal circularization and Kozai mechanism

- High Orbital Eccentricities of Extrasolar Planets Induced by the Kozai Mechanism, G. Takeda, F.A. Rasio, last revised 9 June 2005

- Extreme Climate Variations from Milankovitch-like Eccentricity Oscillations in Extrasolar Planetary Systems, David S. Spiegel, 11 October 2010

- Orbital Dynamics of Multi-Planet Systems with Eccentricity Diversity, Stephen R. Kane, Sean N. Raymond, 8 February 2014

- Type II migration of planets on eccentric orbits, Althea V. Moorhead, Eric B. Ford, 21 April 2009

- Evolution of Giant Planets in Eccentric Disks, Gennaro D'Angelo, Stephen H. Lubow, Matthew R. Bate, 1 December 2006

Inclination dynamics

- A Class of Warm Jupiters with Mutually Inclined, Apsidally Misaligned, Close Friends, Rebekah Dawson, Eugene Chiang, 9 October 2014

4.19 External links

- The Extrasolar Planets Encyclopaedia (Paris Observatory)

- NASA Exoplanet Archive

- Open Exoplanet Catalogue

- The Habitable Exoplanets Catalog (PHL/UPR Arecibo)

- The Habitable Zone Gallery

- Exoplanets: Interactive Visual of XKCD 1071

- NASA's PlanetQuest

- A Zoo of Extra-Solar Planets (audio and transcript) —Astronomy Cast on 9 February 2009 with Pamela Gay and Chris Lintott

- Transiting Exoplanet Light Curves Using Differential Photometry

- Extrasolar Planets – D. Montes, UCM

- Exoplanets at Paris Observatory

- "Exoplanets in relation to host star's current habitable zone". *planetarybiology.com*.

- Doyle, Laurence R. (19 March 2009). "Naming New Extrasolar Planets". *SETI institute*. SPACE.com. Retrieved 2 June 2010.

- Exomol Project Spectroscopic database of molecules of importance for the characterization of exoplanets.

- Characterizing bulk composition of Solid Planets

- Graphical Comparison of Extrasolar Planets

- Video (86:49) – "Search for Life in the Universe" – NASA (14 July 2014).

- Kepler's Tally of Planets

4.19.1   News

- Arxiv: Earth and Planetary Astrophysics

- Extrasolar News and Discoveries

- astrobites the astro-ph reader's digest

- Virtual Planetary Laboratory

Chapter 5

Copernican principle

In physical cosmology, the **Copernican principle**, named after Nicolaus Copernicus, is a working assumption that arises from a modified cosmological extension of Copernicus' heliocentric universe. Under the Copernican principle neither the Sun nor the Earth are in a central, specially favored position in the universe.[1] In some sense, it is equivalent to the mediocrity principle. More recently, the principle has been generalized to the relativistic concept that humans are not privileged observers of the universe.[2]

5.1   Origin and implications

Michael Rowan-Robinson emphasizes the Copernican principle as the threshold test for modern thought, asserting that: "It is evident that in the post-Copernican era of human history, no well-informed and rational person can imagine that Earth occupies a unique position in the universe.".[3]

Hermann Bondi named the principle after Copernicus in the mid-20th century, although the principle itself dates back to the 16th-17th century paradigm shift away from the Ptolemaic system, which placed Earth at the center of the universe. Copernicus proposed that the motion of the planets can be explained by reference to an assumption that the Sun and not Earth is centrally located and stationary. He argued that the apparent retrograde motion of the planets is an illusion caused by Earth's movement around the Sun, which the Copernican model placed at the centre of the universe. Copernicus himself was mainly motivated by technical dissatisfaction with the earlier system and not by support for any mediocrity principle.[4] In fact, although the Copernican heliocentric model is often described as "demoting" Earth from its central role it had in the Ptolemaic geocentric model, neither Copernicus nor other 15th- and 16th-century scientists and philosophers viewed it as such.[5][6] In the late 20th Century, Carl Sagan asked, "Who are we? We find that we live on an insignificant planet of a humdrum star lost in a galaxy tucked away in some forgotten corner of a universe in which there are far more galaxies than people.".[7]

In cosmology, if one assumes the Copernican principle and observes that the universe appears isotropic or the same in all directions from our vantage-point on Earth, then one can infer that the universe is generally homogeneous or the same everywhere (at any given time) and is also isotropic about any given point. These two conditions make up the cosmological principle.[3] In practice, astronomers observe that the universe has heterogeneous or non-uniform structures up to the scale of galactic superclusters, filaments and great voids. It becomes more and more homogeneous and isotropic when observed on larger and larger scales, with little detectable structure on scales of more than about 200 million parsecs. However, on scales comparable to the radius of the observable universe, we see systematic changes with distance from Earth. For instance, galaxies contain more young stars and are less clustered, and quasars appear more numerous. While this might suggest that Earth is at the center of the universe, the Copernican principle requires us to interpret it as evidence for the evolution of the universe with time: this distant light has taken most of the age of the universe to reach and shows us the universe when it was young. The most distant light of all, cosmic microwave background radiation, is isotropic to at least one part in a thousand.

Modern mathematical cosmology is based on the assumption that the Cosmological principle is almost, but not exactly, true on the largest scales. The Copernican principle represents the irreducible philosophical assumption needed to justify this, when combined with the observations.

Bondi and Thomas Gold used the Copernican principle to argue for the perfect cosmological principle which maintains that the universe is also homogeneous in time, and is the basis for the steady-state cosmology.[8] However, this strongly conflicts with the evidence for cosmological evolution mentioned earlier: the universe has progressed from extremely different conditions at the Big Bang, and will continue to progress toward extremely different conditions, particularly under the rising influence of dark energy, apparently toward the Big Freeze or Big Rip.

Since the 1990s the term has been used (interchangeably with "the Copernicus method") for J. Richard Gott's Bayesian-inference-based prediction of duration of ongoing events, a generalized version of the Doomsday argument.

5.2   Tests of the principle

The Copernican principle has never been proven, and in the most general sense cannot be proven, but it is implicit in many modern theories of physics. Cosmological models are often derived with reference to the Cosmological principle, slightly more general than the Copernican principle, and many tests of these models can be considered tests of the Copernican principle.[9]

5.2.1   Historical

Before the term Copernican principle was even coined, Earth was repeatedly shown not to have any special location in the universe. The Copernican Revolution dethroned Earth to just one of many planets orbiting the Sun. William Herschel found that the Solar System is moving through space within our disk-shaped Milky Way galaxy. Edwin Hubble showed that our galaxy is just one of many galaxies in the universe. Examination of our galaxy's position and motion in the universe led to the Big Bang theory and the whole of modern cosmology.

5.2.2   Modern tests

Recent and planned tests relevant to the cosmological and Copernican principles include:

- time drift of cosmological redshifts;[10]

- modelling the local gravitational potential using reflection of cosmic microwave background (CMB) photons;[11]

- the redshift dependence of the luminosity of supernovae;[12]

- the kinetic Sunyaev-Zel'dovich effect in relation to dark energy;[13]

- cosmic neutrino background;[14]

- the integrated Sachs-Wolfe effect[15]

- testing the isotropy and homogeneity of the CMB;[16][17][18][19][20]

5.3   Physics without the principle

The standard model of cosmology, the Lambda-CDM model, assumes the Copernican principle and the more general Cosmological principle and observations are largely consistent but there are always unsolved problems. Some cosmologists and theoretical physicists design models lacking the Cosmological or Copernican principles, to constrain the valid values of observational results, to address specific known issues, and to propose tests to distinguish between current models and other possible models.

A prominent example in this context is the observed accelerating universe and the cosmological constant issue. An alternative proposal to dark energy is that the universe is much more inhomogeneous than currently assumed, and specifically that we are in an extremely large low-density void.[21] To match observations we would have to be very close to the centre of this void, immediately contradicting the Copernican principle.

5.4 See also

- Anthropic principle
- Copernican heliocentrism
- Doomsday argument
- Geocentric model
- Hubble Bubble (astronomy)
- List of unsolved problems in physics
- Mediocrity principle
- Modified Newtonian dynamics
- Quantum gravity
- Rare Earth hypothesis

5.5 References

[1] Bondi, Hermann (1952). *Cosmology*. Cambridge University Press. p. 13.

[2] Peacock, John A. (1998). *Cosmological Physics*. Cambridge University Press. p. 66. ISBN 0-521-42270-1.

[3] Rowan-Robinson, Michael (1996). *Cosmology* (3rd ed.). Oxford University Press. pp. 62–63. ISBN 978-0-19-851884-6.

[4] Kuhn, Thomas S. (1957). *The Copernican Revolution: Planetary Astronomy in the Development of Western Thought*. Harvard University Press. ISBN 978-0-674-17103-9.

[5] Musser, George (2001). "Copernican Counterrevolution". *Scientific American* **284** (3): 24. doi:10.1038/scientificamerican0301-24a.

[6] Danielson, Dennis (2009). "The Bones of Copernicus". *American Scientist* **97** (1): 50–57. doi:10.1511/2009.76.50.

[7] Sagan C. Cosmos (1980) p.193

[8] Bondi, H.; Gold, T. (1948). "The Steady-State Theory of the Expanding Universe". *Monthly Notices of the Royal Astronomical Society* **108** (3): 252–270. Bibcode:1948MNRAS.108..252B. doi:10.1093/mnras/108.3.252.

[9] Clarkson, C.; Bassett, B.; Lu, T. (2008). "A General Test of the Copernican Principle". *Physical Review Letters* **101**. arXiv:0712 Bibcode:2008PhRvL.101a1301C. doi:10.1103/PhysRevLett.101.011301.

[10] Uzan, J. P.; Clarkson, C.; Ellis, G. (2008). "Time Drift of Cosmological Redshifts as a Test of the Copernican Principle". *Physical Review Letters* **100** (19). arXiv:0801.0068. Bibcode:2008PhRvL.100s1303U. doi:10.1103/PhysRevLett.100.191303.

[11] Caldwell, R.; Stebbins, A. (2008). "A Test of the Copernican Principle". *Physical Review Letters* **100** (19). arXiv:0711.3459. Bibcode:2008PhRvL.100s1302C. doi:10.1103/PhysRevLett.100.191302.

[12] Clifton, T.; Ferreira, P.; Land, K. (2008). "Living in a Void: Testing the Copernican Principle with Distant Supernovae". *Physical Review Letters* **101** (13). arXiv:0807.1443. Bibcode:2008PhRvL.101m1302C. doi:10.1103/PhysRevLett.101.131302.

[13] Zhang, P.; Stebbins, A. (2011). "Confirmation of the Copernican principle through the anisotropic kinetic Sunyaev Zel'dovich effect". *Philosophical Transactions of the Royal Society A: Mathematical, Physical and Engineering Sciences* **369** (1957): 5138. Bibcode:2011RSPTA.369.5138Z. doi:10.1098/rsta.2011.0294.

[14] Jia, J.; Zhang, H. (2008). "Can the Copernican principle be tested using the cosmic neutrino background?". *Journal of Cosmology and Astroparticle Physics* **2008** (12): 002. arXiv:0809.2597. Bibcode:2008JCAP...12..002J. doi:10.1088/1475-7516/2008/12/002.

[15] Tomita, K.; Inoue, K. (2009). "Probing violation of the Copernican principle via the integrated Sachs-Wolfe effect". *Physical Review D* **79** (10). arXiv:0903.1541. Bibcode:2009PhRvD..79j3505T. doi:10.1103/PhysRevD.79.103505.

[16] Clifton, T.; Clarkson, C.; Bull, P. (2012). "Isotropic Blackbody Cosmic Microwave Background Radiation as Evidence for a Homogeneous Universe".*Physical Review Letters***109**(5). arXiv:1111.3794. Bibcode:2012PhRvL.109e1303C.doi:10.1103/Phys

[17] Kim, J.; Naselsky, P. (2011). "Lack of Angular Correlation and Odd-Parity Preference in Cosmic Microwave Background Data". *The Astrophysical Journal* **739** (2): 79. arXiv:1011.0377. Bibcode:2011ApJ...739...79K. doi:10.1088/0004-637X/739/2/79.

[18] Copi, C. J.; Huterer, D.; Schwarz, D. J.; Starkman, G. D. (2010). "Large-Angle Anomalies in the CMB". *Advances in Astronomy* **2010**: 1. arXiv:1004.5602. Bibcode:2010AdAst2010E..92C. doi:10.1155/2010/847541.

[19] Planck Collaboration; Ade; Aghanim; Armitage-Caplan; Arnaud; Ashdown; Atrio-Barandela; Aumont; Baccigalupi (2013). "Planck 2013 results. XXIII. Isotropy and Statistics of the CMB". arXiv:1303.5083 [astro-ph.CO].

[20] Longo (2007). "Does the Universe Have a Handedness?". arXiv:astro-ph/0703325 [astro-ph].

[21] February, S.; Larena, J.; Smith, M.; Clarkson, C. (2010). "Rendering dark energy void". *Monthly Notices of the Royal Astronomical Society*: no. arXiv:0909.1479. Bibcode:2010MNRAS.405.2231F. doi:10.1111/j.1365-2966.2010.16627.x.

5.6   External links

- Spiked-online Article

- Slate: How will the Universe End?

Chapter 6

Mediocrity principle

For other uses of "mediocrity", see mediocrity (disambiguation).

The **mediocrity principle** is the philosophical notion that "if an item is drawn at random from one of several sets or categories, it's likelier to come from the most numerous category than from any one of the less numerous categories".[1] The principle has been taken to suggest that there is nothing very unusual about the evolution of the Solar System, Earth's history, the evolution of biological complexity, human evolution, or any one nation. It is a heuristic in the vein of the Copernican principle, and is sometimes used as a philosophical statement about the place of humanity. The idea is to assume mediocrity, rather than starting with the assumption that a phenomenon is special, privileged, exceptional, or even superior.[2][3]

Consistent with the notion, astronomers reported, on 4 November 2013, that there could be as many as 40 billion Earth-sized planets orbiting in the habitable zones of sun-like stars and red dwarf stars within the Milky Way Galaxy alone, based on *Kepler* space mission data.[4][5] 11 billion of these estimated planets may be orbiting sun-like stars.[6] The nearest such planet may be 12 light-years away, according to the scientists.[4][5]

6.1 Extraterrestrial life

The mediocrity principle suggests, given the existence of life on Earth, that life typically exists on Earth-like planets throughout the universe.[7]

6.2 Other uses of the heuristic

David Deutsch argues that the mediocrity principle is not actually correct from a physical point of view, either in reference to our part of the universe or to our species. Deutsch refers to Stephen Hawking's quote that "The human race is just a chemical scum on a moderate-sized planet, orbiting around a very average star in the outer suburb of one among a hundred billion galaxies", writing that our neighborhood in the universe is not typical (80% of the universe's matter is dark matter) and that a concentration of mass such as our solar system is an "isolated, uncommon phenomenon". He also argues with Richard Dawkins's opinion that humans, as result of natural evolution, are limited to the capabilities of our species — Deutsch responds that even though evolution did not give humans the ability to detect neutrinos, scientists can currently detect them, significantly expanding their capabilities beyond what is available as a result of evolution.[8]

6.3 See also

- Abiogenesis

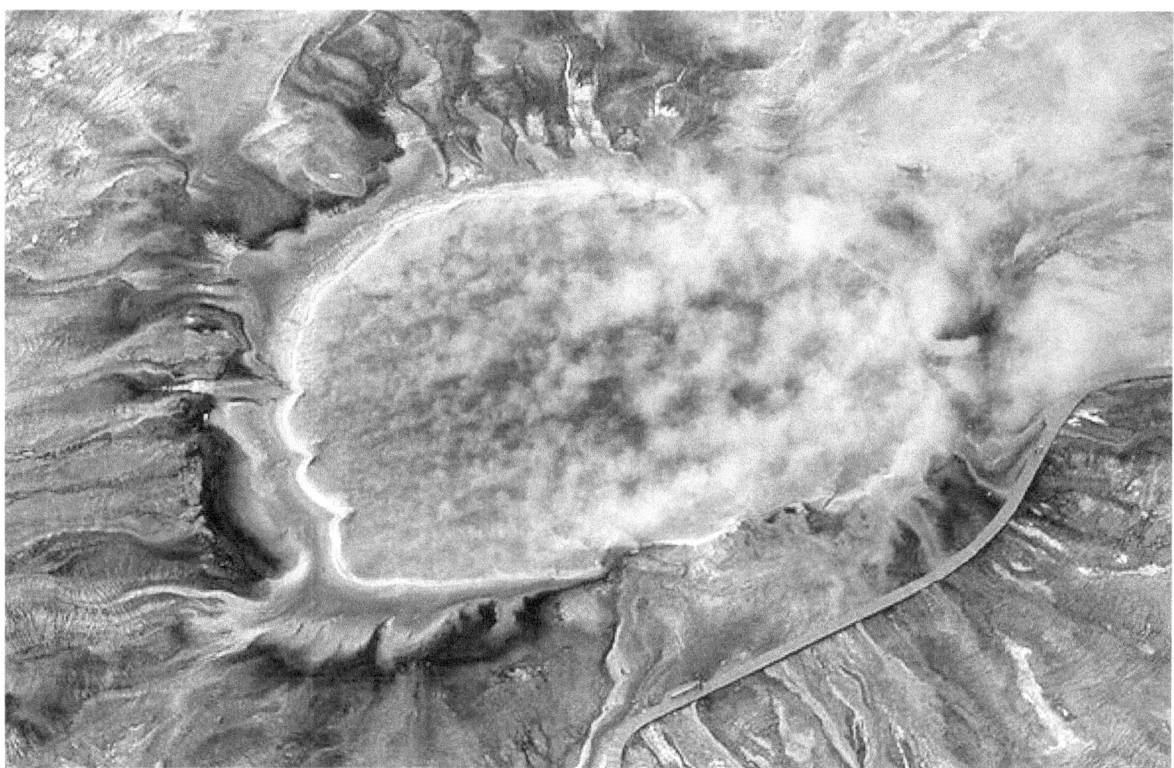

Life on Earth is ubiquitous, but does it exist elsewhere?

- Copernican principle

- Cosmicism

- Cosmological principle

- Cosmic pluralism

- Deep ecology

- Drake equation

- Exceptionalism

- Neocatastrophism

- Plenitude principle

- Rare Earth hypothesis

- Uniformitarianism

- Anthropic principle

6.4   Notes

[1] Kukla, A. (2009). *Extraterrestrials: A Philosophical Perspective*. Lexington Books. p. 20. ISBN 9780739142455. LCCN 2009032272.

[2] Encyclopædia Britannica

[3] PZ Myers explains the Mediocrity principle at edge.org

[4] Overbye, Dennis (4 November 2013). "Far-Off Planets Like the Earth Dot the Galaxy". *New York Times*. Retrieved 5 November 2013.

[5] Petigura, Eric A.; Howard, Andrew W.; Marcy, Geoffrey W. (31 October 2013). "Prevalence of Earth-size planets orbiting Sun-like stars". *Proceedings of the National Academy of Sciences of the United States of America*. Bibcode:2013PNAS..11019273P. doi:10.1073/pnas.1319909110. Retrieved 5 November 2013.

[6] Khan, Amina (4 November 2013). "Milky Way may host billions of Earth-size planets". *Los Angeles Times*. Retrieved 5 November 2013.

[7] Chaisson, Eric, and Steve McMillan. *Astronomy: A Beginner's Guide to the Universe*. Ed. Nancy Whilton. San Francisco: Pearson, 2010.

[8] David Deutsch (2011). *The Beginning of Infinity*. ISBN 978-0-14-196969-5.

6.5 References

- Gonzalez, Richards, *The Privileged Planet: How Our Place in the Cosmos is Designed for Discovery*, Regnery Publishing, 2004, ISBN 0-89526-065-4

- Peter Ward and Donald Brownlee, *Rare Earth: Why Complex Life is Uncommon in the Universe*, Copernicus Books, January 2000, ISBN 0-387-98701-0

6.6 External links

- Goodwin, Gribbin, and Hendry's 1997 Hubble Parameter measurement relying on the mediocrity principle The authors call this the 'Principle of Terrestrial Mediocrity' even though the assumption they make is that the Milky Way *Galaxy* is typical (rather than Earth). This term was coined by Alexander Vilenkin (1995).

Chapter 7

Biochemistry

For the journal, see Biochemistry (journal).
"Biological Chemistry" redirects here. For the journal formerly named Biological Chemistry Hoppe-Seyler, see Biological Chemistry (journal).

Biochemistry, sometimes called **biological chemistry**, is the study of chemical processes within and relating to living organisms.[1] By controlling information flow through biochemical signaling and the flow of chemical energy through metabolism, biochemical processes give rise to the complexity of life. Over the last decades of the 20th century, biochemistry has become so successful at explaining living processes that now almost all areas of the life sciences from botany to medicine to genetics are engaged in biochemical research.[2] Today, the main focus of pure biochemistry is in understanding how biological molecules give rise to the processes that occur within living cells, which in turn relates greatly to the study and understanding of whole organisms.

Biochemistry is closely related to molecular biology, the study of the molecular mechanisms by which genetic information encoded in DNA is able to result in the processes of life. Depending on the exact definition of the terms used, molecular biology can be thought of as a branch of biochemistry, or biochemistry as a tool with which to investigate and study molecular biology.

Much of biochemistry deals with the structures, functions and interactions of biological macromolecules, such as proteins, nucleic acids, carbohydrates and lipids, which provide the structure of cells and perform many of the functions associated with life. The chemistry of the cell also depends on the reactions of smaller molecules and ions. These can be inorganic, for example water and metal ions, or organic, for example the amino acids which are used to synthesize proteins. The mechanisms by which cells harness energy from their environment via chemical reactions are known as metabolism. The findings of biochemistry are applied primarily in medicine, nutrition, and agriculture. In medicine, biochemists investigate the causes and cures of disease. In nutrition, they study how to maintain health and study the effects of nutritional deficiencies. In agriculture, biochemists investigate soil and fertilizers, and try to discover ways to improve crop cultivation, crop storage and pest control.

7.1 History

Main article: History of biochemistry

At its broadest definition, biochemistry can be seen as a study of the components and composition of living things and how they come together to become life, and the history of biochemistry may therefore go back as far as the ancient Greeks.[3] However, biochemistry as a specific scientific discipline has its beginning some time in the 19th century, or a little earlier, depending on which aspect of biochemistry is being focused on. Some argued that the beginning of biochemistry may have been the discovery of the first enzyme, diastase (today called amylase), in 1833 by Anselme Payen,[4] while others considered Eduard Buchner's first demonstration of a complex biochemical process alcoholic fermentation in cell-free extracts in 1897 to be the birth of biochemistry.[5][6] Some might also point as its beginning to the influential 1842 work by

Gerty Cori and Carl Cori jointly won the Nobel Prize in 1947 for their discovery of the Cori cycle at RPMI.

Justus von Liebig, *Animal chemistry, or, Organic chemistry in its applications to physiology and pathology*, which presented a chemical theory of metabolism,[3] or even earlier to the 18th century studies on fermentation and respiration by Antoine Lavoisier.[7][8] Many other pioneers in the field who helped to uncover the layers of complexity of biochemistry have been proclaimed founders of modern biochemistry, for example Emil Fischer for his work on the chemistry of proteins,[9] and F. Gowland Hopkins on enzymes and the dynamic nature of biochemistry.[10]

The term "biochemistry" itself is derived from a combination of biology and chemistry. In 1877, Felix Hoppe-Seyler used the term (*biochemie* in German) as a synonym for physiological chemistry in the foreword to the first issue of *Zeitschrift für Physiologische Chemie* (Journal of Physiological Chemistry) where he argued for the setting up of institutes dedicated to this field of study.[11][12] The German chemist Carl Neuberg however is often cited to have been coined the word in 1903,[13][14][15] while some credited it to Franz Hofmeister.[16]

It was once generally believed that life and its materials had some essential property or substance (often referred to as the "vital principle") distinct from any found in non-living matter, and it was thought that only living beings could produce the molecules of life.[17] Then, in 1828, Friedrich Wöhler published a paper on the synthesis of urea, proving that organic compounds can be created artificially.[18] Since then, biochemistry has advanced, especially since the mid-20th century, with the development of new techniques such as chromatography, X-ray diffraction, dual polarisation interferometry, NMR spectroscopy, radioisotopic labeling, electron microscopy, and molecular dynamics simulations. These techniques allowed for the discovery and detailed analysis of many molecules and metabolic pathways of the cell, such as glycolysis and the Krebs cycle (citric acid cycle).

Another significant historic event in biochemistry is the discovery of the gene and its role in the transfer of information in the cell. This part of biochemistry is often called molecular biology.[19] In the 1950s, James D. Watson, Francis Crick, Rosalind Franklin, and Maurice Wilkins were instrumental in solving DNA structure and suggesting its relationship with genetic transfer of information.[20] In 1958, George Beadle and Edward Tatum received the Nobel Prize for work in fungi showing that one gene produces one enzyme.[21] In 1988, Colin Pitchfork was the first person convicted of murder with DNA evidence, which led to growth of forensic science.[22] More recently, Andrew Z. Fire and Craig C. Mello received the 2006 Nobel Prize for discovering the role of RNA interference (RNAi), in the silencing of gene expression.[23]

7.2   Starting materials: the chemical elements of life

Main articles: Composition of the human body and Dietary mineral

Around two dozen of the 92 naturally occurring chemical elements are essential to various kinds of biological life. Most rare elements on Earth are not needed by life (exceptions being selenium and iodine), while a few common ones (aluminum and titanium) are not used. Most organisms share element needs, but there are a few differences between plants and animals. For example, ocean algae use bromine but land plants and animals seem to need none. All animals require sodium, but some plants do not. Plants need boron and silicon, but animals may not (or may need ultra-small amounts).

Just six elements—carbon, hydrogen, nitrogen, oxygen, calcium, and phosphorus—make up almost 99% of the mass of a human body (see composition of the human body for a complete list). In addition to the six major elements that compose most of the human body, humans require smaller amounts of possibly 18 more.[24]

7.3   Biomolecules

The four main classes of molecules in biochemistry (often called biomolecules) are carbohydrates, lipids, proteins, and nucleic acids. Many biological molecules are polymers: in this terminology, *monomers* are relatively small micromolecules that are linked together to create large macromolecules known as *polymers*. When monomers are linked together to synthesize a biological polymer, they undergo a process called dehydration synthesis. Different macromolecules can assemble in larger complexes, often needed for biological activity.

7.3.1   Carbohydrates

Main articles: Carbohydrate, Monosaccharide, Disaccharide and Polysaccharide
Carbohydrates

Glucose, a monosaccharide

A molecule of sucrose (glucose + fructose), a disaccharide

Amylose, a polysaccharide made up of several thousand glucose units

The function of carbohydrates includes energy storage and providing structure. Sugars are carbohydrates, but not all carbohydrates are sugars. There are more carbohydrates on Earth than any other known type of biomolecule; they are used to store energy and genetic information, as well as play important roles in cell to cell interactions and communications.

The simplest type of carbohydrate is a monosaccharide, which between other properties contains carbon, hydrogen, and oxygen, mostly in a ratio of 1:2:1 (generalized formula $C_nH_{2n}O_n$, where n is at least 3). Glucose ($C_6H_{12}O_6$) is one of the most important carbohydrates, others include fructose ($C_6H_{12}O_6$), the sugar commonly associated with the sweet taste of fruits,[25][a] and deoxyribose ($C_5H_{10}O_4$).

When two monosaccharides undergo dehydration synthesis whereby a molecule of water is released, as two hydrogen atoms and one oxygen atom are lost from the two monosaccharides. The new molecule, consisting of two monosaccharides, is called a *disaccharide* and is conjoined together by a glycosidic or ether bond. The reverse reaction can also occur, using a molecule of water to split up a disaccharide and break the glycosidic bond; this is termed *hydrolysis*. The most well-known disaccharide is sucrose, ordinary sugar (in scientific contexts, called *table sugar* or *cane sugar* to differentiate it from other sugars). Sucrose consists of a glucose molecule and a fructose molecule joined together. Another important disaccharide is lactose, consisting of a glucose molecule and a galactose molecule. As most humans age, the production of lactase, the enzyme that hydrolyzes lactose back into glucose and galactose, typically decreases. This results in lactase deficiency, also called *lactose intolerance*.

When a few (around three to six) monosaccharides are joined, it is called an *oligosaccharide* (oligo- meaning "few"). These molecules tend to be used as markers and signals, as well as having some other uses. Many monosaccharides joined together make a polysaccharide. They can be joined together in one long linear chain, or they may be branched. Two of the most common polysaccharides are cellulose and glycogen, both consisting of repeating glucose monomers. Examples are *Cellulose* which is an important structural component of plant's cell walls, and *glycogen*, used as a form of energy storage in animals.

Sugar can be characterized by having reducing or non-reducing ends. A reducing end of a carbohydrate is a carbon atom that can be in equilibrium with the open-chain aldehyde (aldose) or keto form (ketose). If the joining of monomers takes place at such a carbon atom, the free hydroxy group of the pyranose or furanose form is exchanged with an OH-side-chain of another sugar, yielding a full acetal. This prevents opening of the chain to the aldehyde or keto form and renders the modified residue non-reducing. Lactose contains a reducing end at its glucose moiety, whereas the galactose moiety form a full acetal with the C4-OH group of glucose. Saccharose does not have a reducing end because of full acetal formation between the aldehyde carbon of glucose (C1) and the keto carbon of fructose (C2).

7.3.2   Lipids

Main articles: Lipid, Glycerol and Fatty acid
Lipids comprises a diverse range of molecules and to some extent is a catchall for relatively water-insoluble or nonpolar

A triglyceride with a glycerol molecule on the left and three fatty acids coming off it.

compounds of biological origin, including waxes, fatty acids, fatty-acid derived phospholipids, sphingolipids, glycolipids, and terpenoids (e.g., retinoids and steroids). Some lipids are linear aliphatic molecules, while others have ring structures. Some are aromatic, while others are not. Some are flexible, while others are rigid.[26]Lipids are usually made from one molecule of glycerol combined with other molecules. In triglycerides, the main group of bulk lipids, there is one molecule of glycerol and three fatty acids. Fatty acids are considered the monomer in that case, and may be saturated (no double bonds in the carbon chain) or unsaturated (one or more double bonds in the carbon chain).

Most lipids have some polar character in addition to being largely nonpolar. In general, the bulk of their structure is nonpolar or hydrophobic ("water-fearing"), meaning that it does not interact well with polar solvents like water. Another part of their structure is polar or hydrophilic ("water-loving") and will tend to associate with polar solvents like water. This makes them amphiphilic molecules (having both hydrophobic and hydrophilic portions). In the case of cholesterol, the polar group is a mere -OH (hydroxyl or alcohol). In the case of phospholipids, the polar groups are considerably larger and more polar, as described below.

Lipids are an integral part of our daily diet. Most oils and milk products that we use for cooking and eating like butter, cheese, ghee etc., are composed of fats. Vegetable oils are rich in various polyunsaturated fatty acids (PUFA). Lipid-containing foods undergo digestion within the body and are broken into fatty acids and glycerol, which are the final degradation products of fats and lipids. Lipids, especially phospholipids, are also used in various pharmaceutical products, either as co-solubilisers (e.g., in parenteral infusions) or else as drug carrier components (e.g., in a liposome or transfersome).

7.3.3   Proteins

Main articles: Protein and Amino acid
 Proteins are very large molecules – macro-biopolymers – made from monomers called amino acids. An amino acid consists of a carbon atom bound to four groups. One is an amino group, —NH_2, and one is a carboxylic acid group,

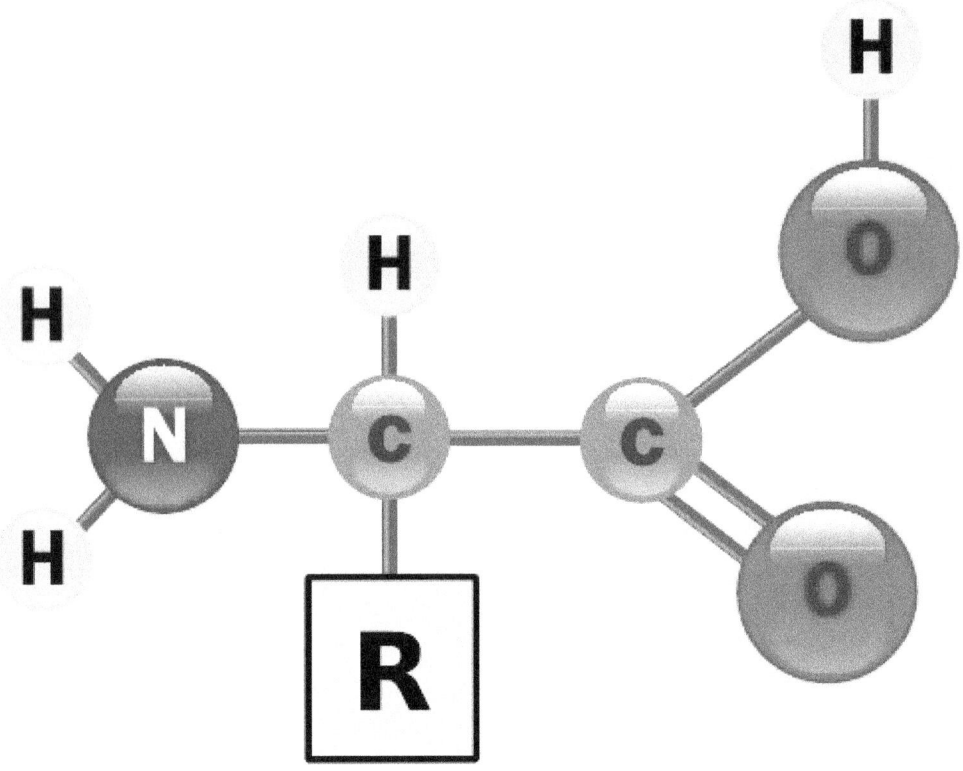

The general structure of an α-amino acid, with the amino group on the left and the carboxyl group on the right.

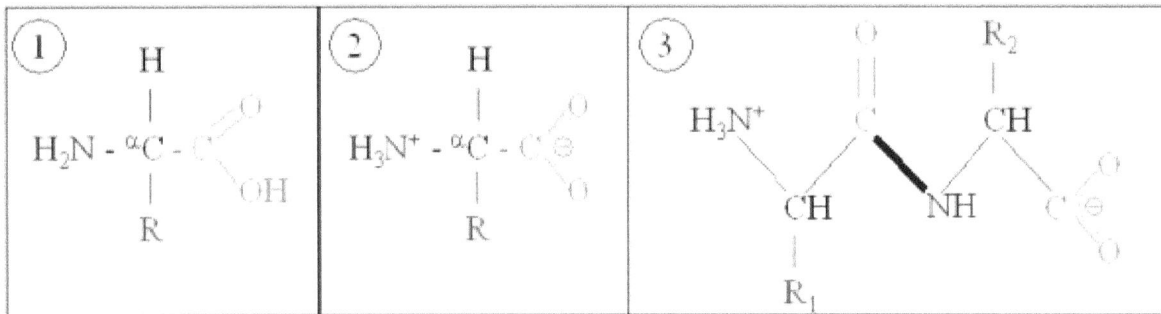

Generic amino acids (1) in neutral form, (2) as they exist physiologically, and (3) joined together as a dipeptide.

—COOH (although these exist as —NH$_3^+$ and —COO$^-$ under physiologic conditions). The third is a simple hydrogen atom. The fourth is commonly denoted "—R" and is different for each amino acid. There are 20 standard amino acids, each containing a carboxyl group, an amino group, and a side-chain (known as an "R" group). The "R" group is what makes each amino acid different, and the properties of the side-chains greatly influence the overall three-dimensional conformation of a protein. Some amino acids have functions by themselves or in a modified form; for instance, glutamate functions as an important neurotransmitter. Amino acids can be joined via a peptide bond. In this dehydration synthesis, a water molecule is removed and the peptide bond connects the nitrogen of one amino acid's amino group to the carbon of the other's carboxylic acid group. The resulting molecule is called a *dipeptide*, and short stretches of amino acids (usually, fewer than thirty) are called *peptides* or polypeptides. Longer stretches merit the title *proteins*. As an example, the important blood serum protein albumin contains 585 amino acid residues.[27]

Some proteins perform largely structural roles. For instance, movements of the proteins actin and myosin ultimately are

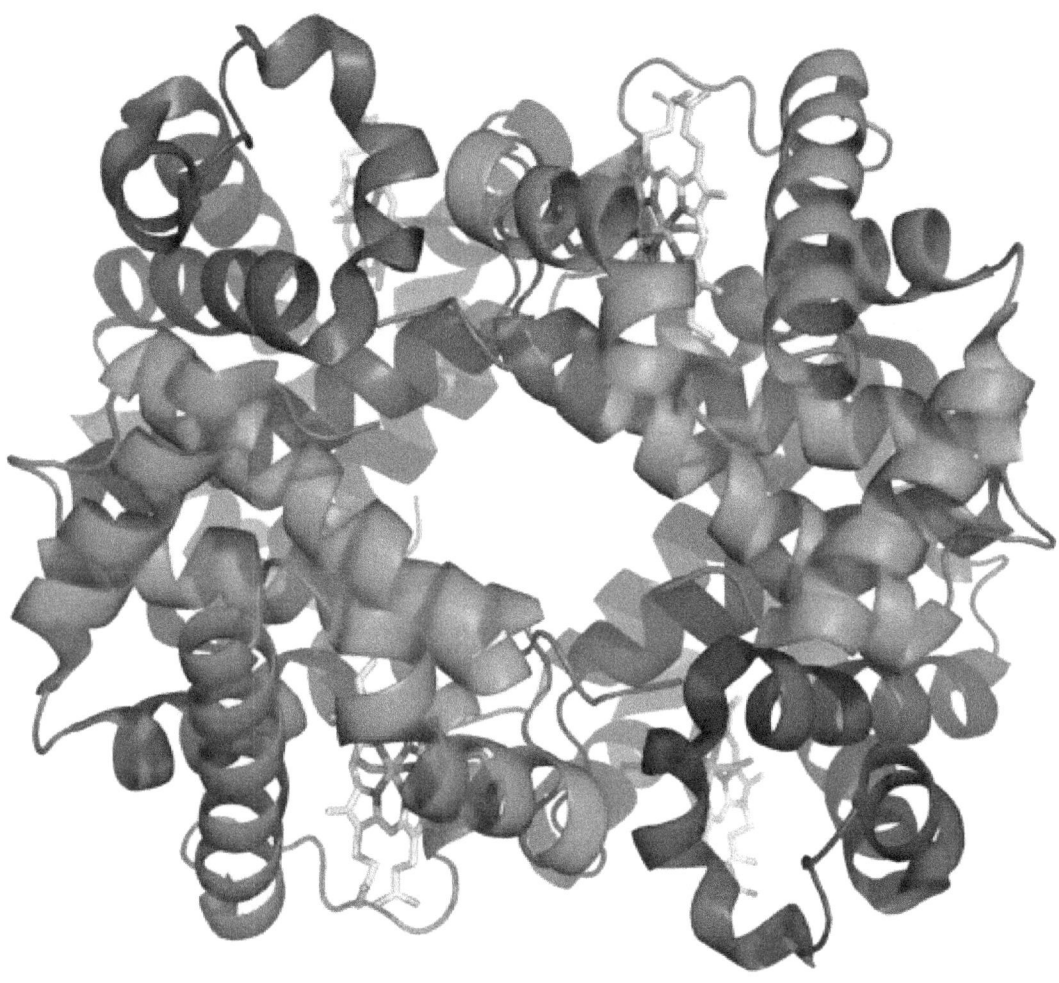

A schematic of hemoglobin. The red and blue ribbons represent the protein globin; the green structures are the heme groups.

responsible for the contraction of skeletal muscle. One property many proteins have is that they specifically bind to a certain molecule or class of molecules—they may be *extremely* selective in what they bind. Antibodies are an example of proteins that attach to one specific type of molecule. In fact, the enzyme-linked immunosorbent assay (ELISA), which uses antibodies, is one of the most sensitive tests modern medicine uses to detect various biomolecules. Probably the most important proteins, however, are the enzymes. These molecules recognize specific reactant molecules called *substrates*; they then catalyze the reaction between them. By lowering the activation energy, the enzyme speeds up that reaction by a rate of 10^{11} or more: a reaction that would normally take over 3,000 years to complete spontaneously might take less than a second with an enzyme. The enzyme itself is not used up in the process, and is free to catalyze the same reaction with a new set of substrates. Using various modifiers, the activity of the enzyme can be regulated, enabling control of the biochemistry of the cell as a whole.

The structure of proteins is traditionally described in a hierarchy of four levels. The primary structure of a protein simply consists of its linear sequence of amino acids; for instance, "alanine-glycine-tryptophan-serine-glutamate-asparagine-glycine-lysine-...". Secondary structure is concerned with local morphology (morphology being the study of structure).

Some combinations of amino acids will tend to curl up in a coil called an α-helix or into a sheet called a β-sheet; some α-helixes can be seen in the hemoglobin schematic above. Tertiary structure is the entire three-dimensional shape of the protein. This shape is determined by the sequence of amino acids. In fact, a single change can change the entire structure. The alpha chain of hemoglobin contains 146 amino acid residues; substitution of the glutamate residue at position 6 with a valine residue changes the behavior of hemoglobin so much that it results in sickle-cell disease. Finally, quaternary structure is concerned with the structure of a protein with multiple peptide subunits, like hemoglobin with its four subunits. Not all proteins have more than one subunit.[28]

Ingested proteins are usually broken up into single amino acids or dipeptides in the small intestine, and then absorbed. They can then be joined to make new proteins. Intermediate products of glycolysis, the citric acid cycle, and the pentose phosphate pathway can be used to make all twenty amino acids, and most bacteria and plants possess all the necessary enzymes to synthesize them. Humans and other mammals, however, can synthesize only half of them. They cannot synthesize isoleucine, leucine, lysine, methionine, phenylalanine, threonine, tryptophan, and valine. These are the essential amino acids, since it is essential to ingest them. Mammals do possess the enzymes to synthesize alanine, asparagine, aspartate, cysteine, glutamate, glutamine, glycine, proline, serine, and tyrosine, the nonessential amino acids. While they can synthesize arginine and histidine, they cannot produce it in sufficient amounts for young, growing animals, and so these are often considered essential amino acids.

If the amino group is removed from an amino acid, it leaves behind a carbon skeleton called an α-keto acid. Enzymes called transaminases can easily transfer the amino group from one amino acid (making it an α-keto acid) to another α-keto acid (making it an amino acid). This is important in the biosynthesis of amino acids, as for many of the pathways, intermediates from other biochemical pathways are converted to the α-keto acid skeleton, and then an amino group is added, often via transamination. The amino acids may then be linked together to make a protein.[29]

A similar process is used to break down proteins. It is first hydrolyzed into its component amino acids. Free ammonia (NH_3), existing as the ammonium ion (NH_4^+) in blood, is toxic to life forms. A suitable method for excreting it must therefore exist. Different tactics have evolved in different animals, depending on the animals' needs. Unicellular organisms, of course, simply release the ammonia into the environment. Likewise, bony fish can release the ammonia into the water where it is quickly diluted. In general, mammals convert the ammonia into urea, via the urea cycle.[30]

In order to determine whether two proteins are related, or in other words to decide whether they are homologous or not, scientists use sequence-comparison methods. Methods like Sequence Alignments and Structural Alignments are powerful tools that help scientists identify homologies between related molecules.[31] The relevance of finding homologies among proteins goes beyond forming an evolutionary pattern of protein families. By finding how similar two protein sequences are, we acquire knowledge about their structure and therefore their function.

7.3.4   Nucleic acids

Main articles: Nucleic acid, DNA, RNA and Nucleotides

Nucleic acids, so called because of its prevalence in cellular nuclei, is the generic name of the family of biopolymers. They are complex, high-molecular-weight biochemical macromolecules that can convey genetic information in all living cells and viruses.[32] The monomers are called nucleotides, and each consists of three components: a nitrogenous heterocyclic base (either a purine or a pyrimidine), a pentose sugar, and a phosphate group. The most common nucleic acids are deoxyribonucleic acid (DNA) and ribonucleic acid (RNA).[33] The phosphate group and the sugar of each nucleotide bond with each other to form the backbone of the nucleic acid, while the sequence of nitrogenous bases stores the information. The most common nitrogenous bases are adenine, cytosine, guanine, thymine, and uracil. The nitrogenous bases of each strand of a nucleic acid will form hydrogen bonds with certain other nitrogenous bases in a complementary strand of nucleic acid (similar to a zipper). Adenine binds with thymine and uracil; Thymine binds only with adenine; and cytosine and guanine can bind only with one another.

Aside from the genetic material of the cell, nucleic acids often play a role as second messengers, as well as forming the base molecule for adenosine triphosphate (ATP), the primary energy-carrier molecule found in all living organisms. Also, the nitrogenous bases possible in the two nucleic acids are different: adenine, cytosine, and guanine occur in both RNA and DNA, while thymine occurs only in DNA and uracil occurs only in RNA.

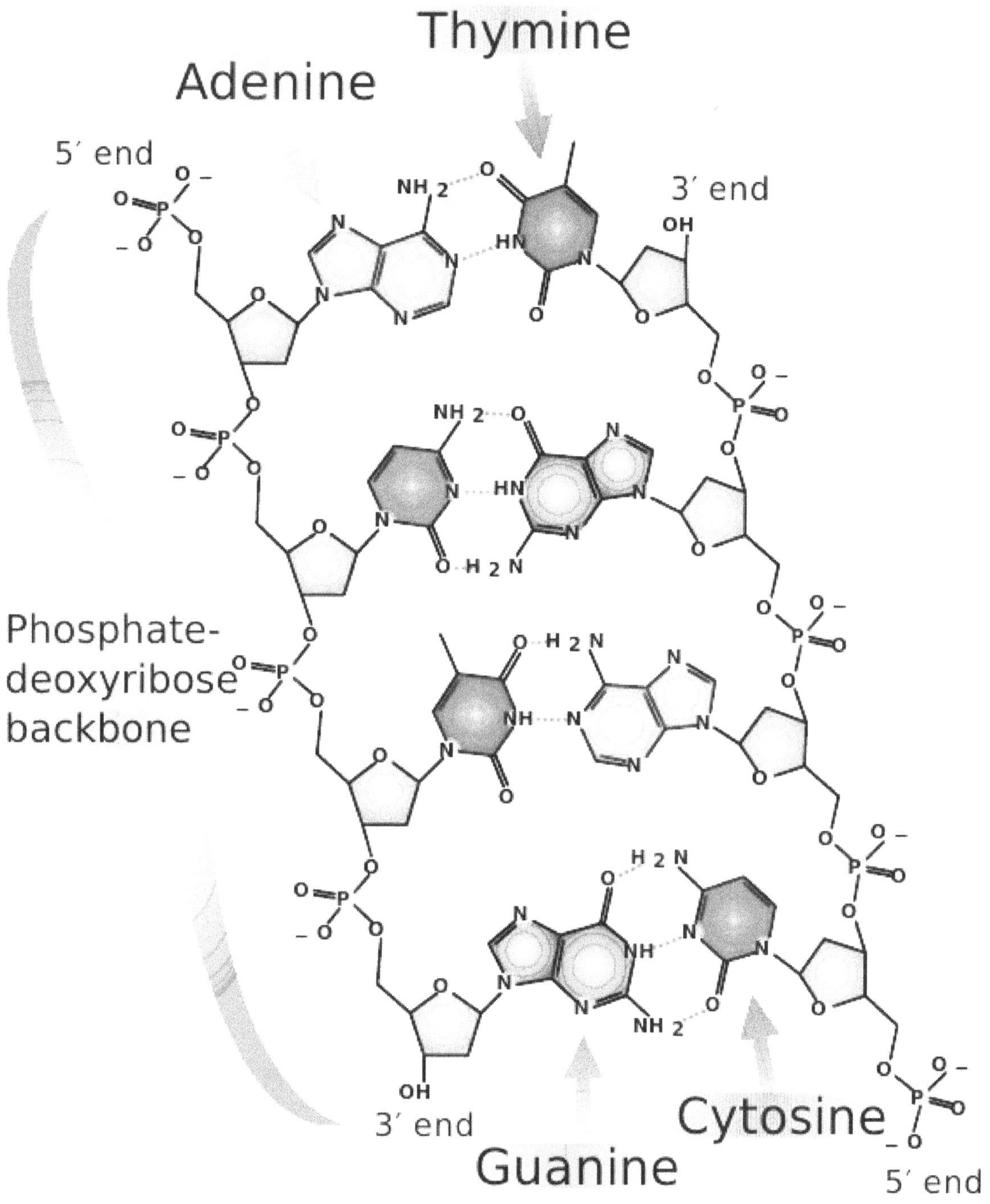

The structure of deoxyribonucleic acid (DNA), the picture shows the monomers being put together.

7.4   Metabolism

7.4.1   Carbohydrates as energy source

Main article: Carbohydrate metabolism

Glucose is the major energy source in most life forms. For instance, polysaccharides are broken down into their monomers (glycogen phosphorylase removes glucose residues from glycogen). Disaccharides like lactose or sucrose are cleaved into their two component monosaccharides.

Glycolysis (anaerobic)

Glucose is mainly metabolized by a very important ten-step pathway called glycolysis, the net result of which is to break down one molecule of glucose into two molecules of pyruvate; this also produces a net two molecules of ATP, the energy currency of cells, along with two reducing equivalents as converting NAD^+ to NADH. This does not require oxygen; if no oxygen is available (or the cell cannot use oxygen), the NAD is restored by converting the pyruvate to lactate (lactic acid) (e.g., in humans) or to ethanol plus carbon dioxide (e.g., in yeast). Other monosaccharides like galactose and fructose can be converted into intermediates of the glycolytic pathway.[34]

Aerobic

In aerobic cells with sufficient oxygen, as in most human cells, the pyruvate is further metabolized. It is irreversibly converted to acetyl-CoA, giving off one carbon atom as the waste product carbon dioxide, generating another reducing equivalent as NADH. The two molecules acetyl-CoA (from one molecule of glucose) then enter the citric acid cycle, producing two more molecules of ATP, six more NADH molecules and two reduced (ubi)quinones (via $FADH_2$ as enzyme-bound cofactor), and releasing the remaining carbon atoms as carbon dioxide. The produced NADH and quinol molecules then feed into the enzyme complexes of the respiratory chain, an electron transport system transferring the electrons ultimately to oxygen and conserving the released energy in the form of a proton gradient over a membrane (inner mitochondrial membrane in eukaryotes). Thus, oxygen is reduced to water and the original electron acceptors NAD^+ and quinone are regenerated. This is why humans breathe in oxygen and breathe out carbon dioxide. The energy released from transferring the electrons from high-energy states in NADH and quinol is conserved first as proton gradient and converted to ATP via ATP synthase. This generates an additional 28 molecules of ATP (24 from the 8 NADH + 4 from the 2 quinols), totaling to 32 molecules of ATP conserved per degraded glucose (two from glycolysis + two from the citrate cycle). It is clear that using oxygen to completely oxidize glucose provides an organism with far more energy than any oxygen-independent metabolic feature, and this is thought to be the reason why complex life appeared only after Earth's atmosphere accumulated large amounts of oxygen.

Gluconeogenesis

Main article: Gluconeogenesis

In vertebrates, vigorously contracting skeletal muscles (during weightlifting or sprinting, for example) do not receive enough oxygen to meet the energy demand, and so they shift to anaerobic metabolism, converting glucose to lactate. The liver regenerates the glucose, using a process called gluconeogenesis. This process is not quite the opposite of glycolysis, and actually requires three times the amount of energy gained from glycolysis (six molecules of ATP are used, compared to the two gained in glycolysis). Analogous to the above reactions, the glucose produced can then undergo glycolysis in tissues that need energy, be stored as glycogen (or starch in plants), or be converted to other monosaccharides or joined into di- or oligosaccharides. The combined pathways of glycolysis during exercise, lactate's crossing via the bloodstream to the liver, subsequent gluconeogenesis and release of glucose into the bloodstream is called the Cori cycle.[35]

7.5 Relationship to other "molecular-scale" biological sciences

Researchers in biochemistry use specific techniques native to biochemistry, but increasingly combine these with techniques and ideas developed in the fields of genetics, molecular biology and biophysics. There has never been a hard-line among these disciplines in terms of content and technique. Today, the terms *molecular biology* and *biochemistry* are nearly interchangeable. The following figure is a schematic that depicts one possible view of the relationship between the fields:

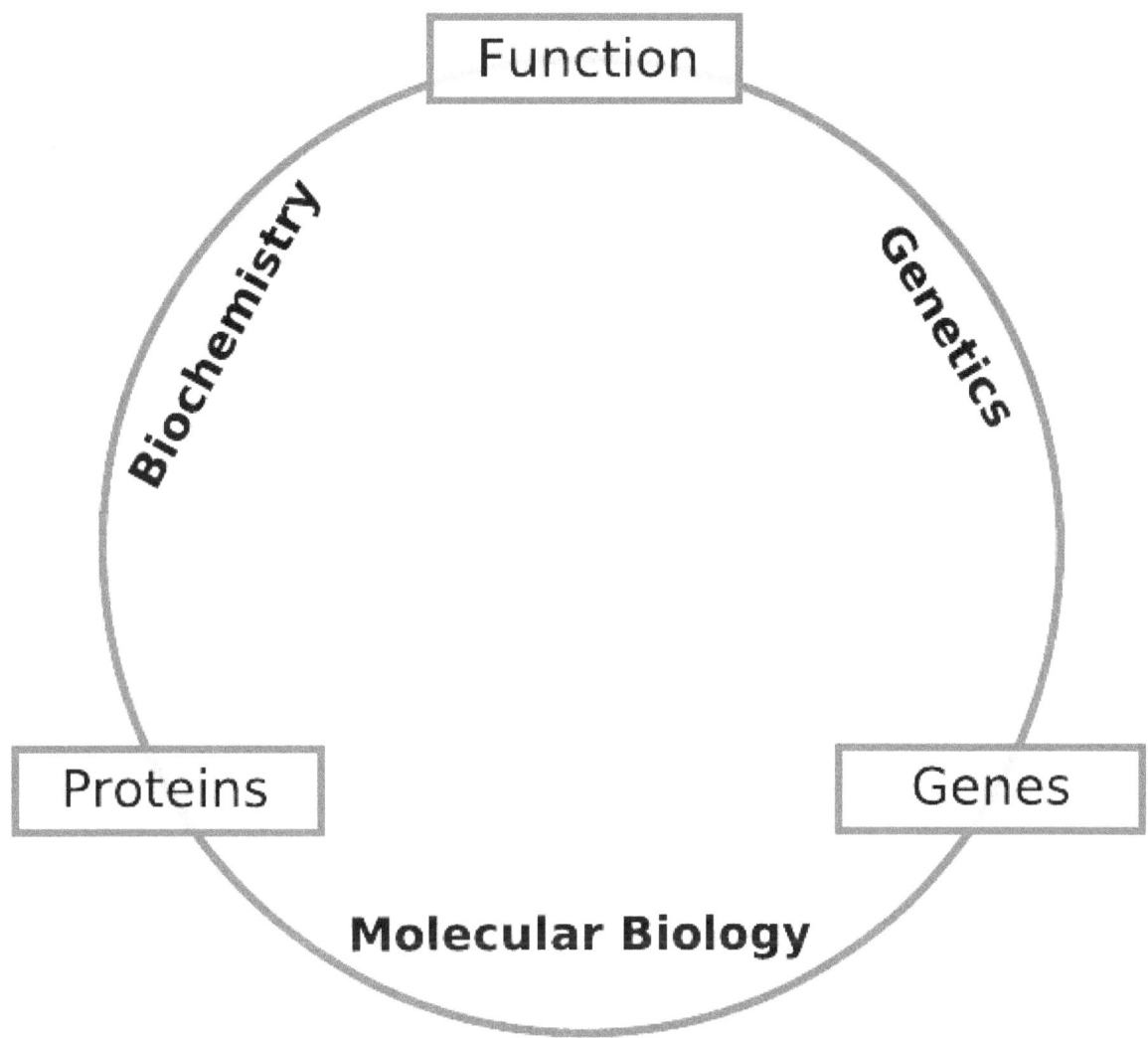

Schematic relationship between biochemistry, genetics, and molecular biology

- *Biochemistry* is the study of the chemical substances and vital processes occurring in living organisms. Biochemists focus heavily on the role, function, and structure of biomolecules. The study of the chemistry behind biological processes and the synthesis of biologically active molecules are examples of biochemistry.

- *Genetics* is the study of the effect of genetic differences on organisms. Often this can be inferred by the absence of a normal component (e.g., one gene). The study of "mutants" – organisms with a changed gene that leads to the organism being different with respect to the so-called "wild type" or normal phenotype. Genetic interactions (epistasis) can often confound simple interpretations of such "knock-out" or "knock-in" studies.

- *Molecular biology* is the study of molecular underpinnings of the process of replication, transcription and translation of the genetic material. The central dogma of molecular biology where genetic material is transcribed into RNA and then translated into protein, despite being an oversimplified picture of molecular biology, still provides a good starting point for understanding the field. This picture, however, is undergoing revision in light of emerging novel roles for RNA.[36]

- *Chemical biology* seeks to develop new tools based on small molecules that allow minimal perturbation of biological systems while providing detailed information about their function. Further, chemical biology employs biological systems to create non-natural hybrids between biomolecules and synthetic devices (for example emptied viral capsids that can deliver gene therapy or drug molecules).

7.6 See also

Main article: Outline of biochemistry

7.6.1 Lists

- Important publications in biochemistry (chemistry)
- List of biochemistry topics
- List of biochemists
- List of biomolecules

7.6.2 See also

- Chemical ecology
- Computational biomodeling
- EC number
- Hypothetical types of biochemistry
- International Union of Biochemistry and Molecular Biology
- Metabolome
- Metabolomics
- Molecular medicine
- Plant biochemistry
- Proteolysis
- Small molecule
- Structural biology
- TCA cycle

7.7 Notes

a. ^ Fructose is not the only sugar found in fruits. Glucose and sucrose are also found in varying quantities in various fruits, and indeed sometimes exceed the fructose present. For example, 32% of the edible portion of date is glucose, compared with 23.70% fructose and 8.20% sucrose. However, peaches contain more sucrose (6.66%) than they do fructose (0.93%) or glucose (1.47%).[37]

7.8   References

[1] "Biochemistry". *acs.org*.

[2] "scientific term 'biochemistry'".

[3] Ton van Helvoort (2000). Arne Hessenbruch, ed. *Reader's Guide to the History of Science*. Fitzroy Dearborn Publishing. p. 81.

[4] Hunter (2000), p. 75.

[5] Jacob Darwin Hamblin. *Science in the Early Twentieth Century: An Encyclopedia*. ABC-CLIO. p. 26. ISBN 978-1-85109-665-7.

[6] Hunter (2000), pp. 96–98.

[7] Clarence Peter Berg (1980). "The University of Iowa and Biochemistry from Their Beginnings": 1–2. ISBN 9780874140149.

[8] Frederic Lawrence Holmes (1987). *Lavoisier and the Chemistry of Life: An Exploration of Scientific Creativity*. University of Wisconsin Press. p. xv. ISBN 978-0299099848.

[9] Burton Feldman (2001). *The Nobel Prize: A History of Genius, Controversy, and Prestige*. Arcade Publishing. p. 206. ISBN 978-1559705929.

[10] Marelene F. Rayner-Canham, Marelene Rayner-Canham, Geoffrey Rayner-Canham (2005). *Women in Chemistry: Their Changing Roles from Alchemical Times to the Mid-Twentieth Century*. Chemical Heritage Foundation. p. 136. ISBN 978-0941901277.

[11] Anne-Katrin Ziesak, Hans-Robert Cram (18 October 1999). *Walter de Gruyter Publishers, 1749-1999*. Walter de Gruyter & Co. p. 169. ISBN 978-3110167412.

[12] Horst Kleinkauf, Hans von Döhren, Lothar Jaenicke (1988). *The Roots of Modern Biochemistry: Fritz Lippmann's Squiggle and its Consequences*. Walter de Gruyter & Co. p. 116. ISBN 9783110852455.

[13] Ben-Menahem, Ari (2009). *Historical Encyclopedia of Natural and Mathematical Sciences*. Springer. p. 2982. ISBN 978-3-540-68831-0.

[14] Mark Amsler (1986). *The Languages of Creativity: Models, Problem-solving, Discourse*. University of Delaware Press. p. 55. ISBN 978-0874132809.

[15] *Advances in Carbohydrate Chemistry and Biochemistry, Volume 70*. Academic Press. 28 November 2013. p. 36. ASIN B00H7E78BG.

[16] Koscak Maruyama (1988). Horst Kleinkauf, Hans von Döhren, Lothar Jaenickem, eds. *The Roots of Modern Biochemistry: Fritz Lippmann's Squiggle and its Consequences*. Walter de Gruyter & Co. p. 43. ISBN 9783110852455.

[17] Fiske, John (1890). *Outlines of Cosmic Philosophy Based on the Doctrines of Evolution, with Criticisms on the Positive Philosophy, Volume 1*. Boston and New York: Houghton, Mifflin. pp. 419–20. Retrieved 16 February 2015.

[18] Kauffman, G.B.; Chooljian, S.H. (2001). "Friedrich Wöhler (1800–1882), on the bicentennial of his birth". *The Chemical Educator* **6** (2): 121–133. doi:10.1007/s00897010444a.

[19] Tropp (2012), p. 2.

[20] Tropp (2012), pp. 19–20.

[21] Krebs, Jocelyn E.; Goldstein, Elliott S.; Lewin, Benjamin; Kilpatrick, Stephen T. (2012). *Essential Genes*. Jones & Bartlett Publishers. p. 32. ISBN 978-1-4496-1265-8.

[22] Butler, John M. (2009). *Fundamentals of Forensic DNA Typing*. Academic Press. p. 5. ISBN 978-0-08-096176-7.

[23] Sen, Chandan K.; Roy, Sashwati (2007). "miRNA: Licensed to kill the messenger". *DNA Cell Biology* **26** (4): 193–194. doi:10.1089/dna.2006.0567. PMID 17465885.

[24] Ultratrace minerals. Authors: Nielsen, Forrest H. USDA, ARS Source: Modern nutrition in health and disease / editors, Maurice E. Shils ... et al.. Baltimore : Williams & Wilkins, c1999., p. 283-303. Issue Date: 1999 URI:

[25] Whiting, G.C (1970). "Sugars". In A.C. Hulme. *The Biochemistry of Fruits and their Products.* Volume 1. London & New York: Academic Press. pp. 1–31.

[26] Fromm and Hargrove (2012). pp. 22–27.

[27] Metzler, David Everett; Metzler, Carol M. (2001). *Biochemistry: The Chemical Reactions of Living Cells* **1**. Academic Press. p. 58. ISBN 978-0-12-492540-3.

[28] Fromm and Hargrove (2012). pp. 35–51.

[29] Fromm and Hargrove (2012). pp. 279–292.

[30] Sherwood, Lauralee; Klandorf, Hillar; Yancey, Paul H. (2012). *Animal Physiology: From Genes to Organisms.* Cengage Learning. p. 558. ISBN 978-0-8400-6865-1.

[31] Fariselli, Piero; Rossi, Ivan; Capriotti, Emidio; Casadio, Rita (2007). "The WWWH of remote homolog detection: the state of the art". *Briefings in Bioinformatics* **8** (2): 78–87. doi:10.1093/bib/bbl032. PMID 17003074.

[32] Voet, D. and Voet, J. G. (2011). Biochemistry (4th ed.). John Wiley & Sons Inc.: Hoboken, NJ

[33] Tropp (2012). pp. 5–9.

[34] Fromm and Hargrove (2012). pp. 163–180.

[35] Fromm and Hargrove (2012). pp. 183–194.

[36] Ulveling, Damien; Francastel, Claire; Hubé, Florent (2011). "When one is better than two: RNA with dual functions". *Biochimie* **93** (4): 633–644. doi:10.1016/j.biochi.2010.11.004. PMID 21111023.

[37] Whiting, G.C. (1970). p.5

7.8.1 Cited literature

- Fromm, Herbert J.; Hargrove, Mark (2012). *Essentials of Biochemistry.* Springer. ISBN 978-3-642-19623-2.

- Hunter, Graeme K. (2000). *Vital Forces: The Discovery of the Molecular Basis of Life.* Academic Press. ISBN 978-0-12-361811-5.

- Tropp, Burton E. (2012). *Molecular Biology* (4th ed.). Jones & Bartlett Learning. ISBN 978-1-4496-0091-4.

7.9 External links

- The Virtual Library of Biochemistry and Cell Biology

- Biochemistry, 5th ed. Full text of Berg, Tymoczko, and Stryer, courtesy of NCBI.

- Biochemistry, 2nd ed. Full text of Garrett and Grisham.

- Biochemistry Animation (Narrated Flash animations.)

- SystemsX.ch - The Swiss Initiative in Systems Biology

- Biochemistry Online Resources – Lists of Biochemistry departments, websites, journals, books and reviews, employment opportunities and events.

- Full text of Biochemistry by Kevin and Indira, an introductory biochemistry textbook.

Chapter 8

Panspermia

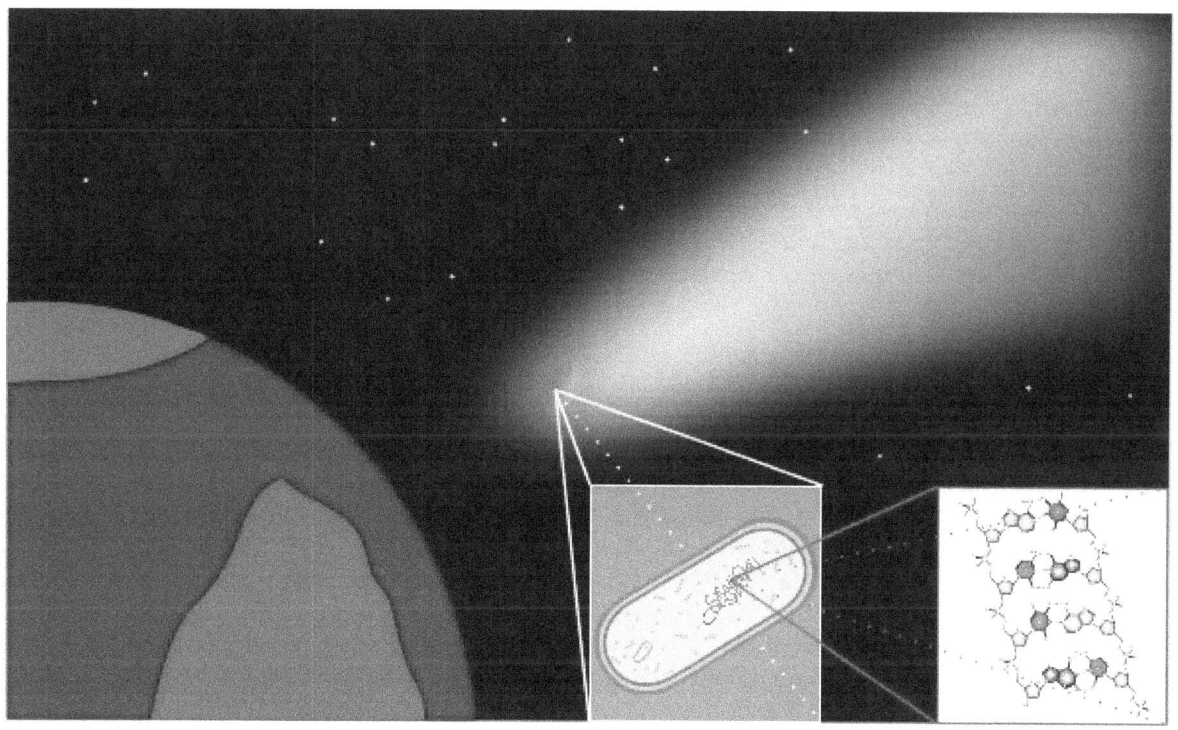

Illustration of a comet (center) transporting a bacterial life form (inset) through space to the Earth (left)

Panspermia (from Greek πᾶν (*pan*), meaning "all", and σπέρμα (*sperma*), meaning "seed") is the hypothesis that life exists throughout the Universe, distributed by meteoroids, asteroids, comets,[1][2] planetoids[3] and, also, by spacecraft in the form of unintended contamination by microorganisms.[4][5]

Panspermia is a hypothesis proposing that microscopic life forms that can survive the effects of space, such as extremophiles, become trapped in debris that is ejected into space after collisions between planets and small Solar System bodies that harbor life. Some organisms may travel dormant for an extended amount of time before colliding randomly with other planets or intermingling with protoplanetary disks. If met with ideal conditions on a new planet's surfaces, the organisms become active and the process of evolution begins. Panspermia is not meant to address how life began, just the method that may cause its distribution in the Universe.[6][7][8]

Pseudo-panspermia (sometimes called *"soft panspermia"* or *"molecular panspermia"*) argues that the pre-biotic organic building blocks of life originated in space and were incorporated in the solar nebula from which the planets condensed and were further —and continuously— distributed to planetary surfaces where life then emerged (abiogenesis).[9][10] From the early 1970s it was becoming evident that interstellar dust consisted of a large component of organic molecules. Interstellar molecules are formed by chemical reactions within very sparse interstellar or circumstellar clouds of dust and gas.[11] The dust plays a critical role of shielding the molecules from the ionizing effect of ultraviolet radiation emitted by stars.[12]

Several simulations in laboratories and in low Earth orbit suggest that ejection, entry and impact is survivable for some simple organisms. In 2015, "remains of biotic life" were found in 4.1 billion-year-old rocks in Western Australia, when the young Earth was about 400 million years old.[13][14] According to one of the researchers, "If life arose relatively quickly on Earth ... then it could be common in the universe."[13]

8.1 History

The first known mention of the term was in the writings of the 5th century BC Greek philosopher Anaxagoras.[15] Panspermia began to assume a more scientific form through the proposals of Jöns Jacob Berzelius (1834),[16] Hermann E. Richter (1865),[17] Kelvin (1871),[18] Hermann von Helmholtz (1879)[19][20] and finally reaching the level of a detailed hypothesis through the efforts of the Swedish chemist Svante Arrhenius (1903).[21]

Sir Fred Hoyle (1915–2001) and Chandra Wickramasinghe (born 1939) were influential proponents of panspermia.[22][23] In 1974 they proposed the hypothesis that some dust in interstellar space was largely organic (containing carbon), which Wickramasinghe later proved to be correct.[24][25][26] Hoyle and Wickramasinghe further contended that life forms continue to enter the Earth's atmosphere, and may be responsible for epidemic outbreaks, new diseases, and the genetic novelty necessary for macroevolution.[27]

In an Origins Symposium presentation on April 7, 2009, physicist Stephen Hawking stated his opinion about what humans may find when venturing into space, such as the possibility of alien life through the theory of panspermia: "Life could spread from planet to planet or from stellar system to stellar system, carried on meteors."[28]

8.2 Proposed mechanisms

Panspermia can be said to be either interstellar (between star systems) or interplanetary (between planets in the same star system);[29][30] its transport mechanisms may include comets,[31][32] radiation pressure and lithopanspermia (microorganisms embedded in rocks).[33][34][35] Interplanetary transfer of nonliving material is well documented, as evidenced by meteorites of Martian origin found on Earth.[35] Space probes may also be a viable transport mechanism for interplanetary cross-pollination in our Solar System or even beyond. However, space agencies have implemented planetary protection procedures to reduce the risk of planetary contamination,[36][37] although, as recently discovered, some microorganisms, such as Tersicoccus phoenicis, may be resistant to procedures used in spacecraft assembly clean room facilities.[4][5] In 2012, mathematician Edward Belbruno and astronomers Amaya Moro-Martín and Renu Malhotra proposed that gravitational low energy transfer of rocks among the young planets of stars in their birth cluster is commonplace, and not rare in the general galactic stellar population.[38][39] Deliberate directed panspermia from space to seed Earth[40] or sent from Earth to seed other solar systems have also been proposed.[41][42][43][44] One twist to the hypothesis by engineer Thomas Dehel (2006), proposes that plasmoid magnetic fields ejected from the magnetosphere may move the few spores lifted from the Earth's atmosphere with sufficient speed to cross interstellar space to other systems before the spores can be destroyed.[45][46]

8.2.1 Radiopanspermia

In 1903, Svante Arrhenius published in his article *The Distribution of Life in Space*,[47] the hypothesis now called radiopanspermia, that microscopic forms of life can be propagated in space, driven by the radiation pressure from stars.[48] Arrhenius argued that particles at a critical size below 1.5 μm would be propagated at high speed by radiation pressure of the Sun. However, because its effectiveness decreases with increasing size of the particle, this mechanism holds for very tiny particles only, such as single bacterial spores.[49] The main criticism of radiopanspermia hypothesis came from Shklovskii and Sagan, who pointed out the proofs of the lethal action of space radiations (UV and X-rays) in the cosmos.[50] Regardless of the evidence, Wallis and Wickramasinghe argued in 2004 that the transport of individual bacteria or clumps of bacteria, is overwhelmingly more important than lithopanspermia in terms of numbers of microbes transferred, even accounting for the death rate of unprotected bacteria in transit.[51]

Then, data gathered by the orbital experiments ERA, BIOPAN, EXOSTACK and EXPOSE, determined that isolated spores, including those of *B. subtilis*, were killed by several orders of magnitude if exposed to the full space environment for a mere few seconds, but if shielded against solar UV, the spores were capable of surviving in space for up to 6 years while embedded in clay or meteorite powder (artificial meteorites).[49][52] Though minimal protection is required to shelter a spore against UV radiation, exposure to solar UV and cosmic ionizing radiation of unprotected DNA, break it up into its bases.[53][54][55] Also, exposing DNA to the ultrahigh vacuum of space alone is sufficient to cause DNA damage, so the transport of unprotected DNA or RNA during interplanetary flights powered solely by light pressure is extremely unlikely.[55] The feasibility of other means of transport for the more massive shielded spores into the outer solar system – for example, thru gravitational capture by comets – is at this time unknown.

Based on experimental data on radiation effects and DNA stability, it has been concluded that for such long travel times, boulder sized rocks which are greater than or equal to 1 meter in diameter are required to effectively shield resistant microorganisms, such as bacterial spores against galactic cosmic radiation.[56][57] These results clearly negate the radiopanspermia hypothesis, which requires single spores accelerated by the radiation pressure of the Sun, requiring many years to travel between the planets, and support the likelihood of interplanetary transfer of microorganisms within asteroids or comets, the so-called **lithopanspermia** hypothesis.[49][52]

8.2.2   Lithopanspermia

Lithopanspermia, the transfer of organisms in rocks from one planet to another either through interplanetary or interstellar space, remains speculative. Although there is no evidence that lithopanspermia has occurred in our own Solar System, the various stages have become amenable to experimental testing.[58]

- **Planetary ejection** — For lithopanspermia to occur, microorganisms must survive ejection from a planetary surface which involves extreme forces of acceleration and shock with associated temperature excursions. Hypothetical values of shock pressures experienced by ejected rocks are obtained with Martian meteorites, which suggest the shock pressures of approximately 5 to 55 GPa, acceleration of 3×10^6 m/s^2 and jerk of 6×10^9 m/s^3 and post-shock temperature increases of about 1 K to 1000 K.[59][60] To determine the effect of acceleration during ejection on microorganisms, rifle and ultracentrifuge methods were successfully used under simulated outer space conditions.[58]

- **Survival in transit** — The survival of microorganisms has been studied extensively using both simulated facilities and in low Earth orbit. A large number of microorganisms have been selected for exposure experiments. It is possible to separate these microorganisms into two groups, the human-borne, and the extremophiles. Studying the human-borne microorganisms is significant for human welfare and future manned missions; whilst the extremophiles are vital for studying the physiological requirements of survival in space.[58]

- **Atmospheric entry** — An important aspect of the lithopanspermia hypothesis to test is that microbes situated on or within rocks could survive hypervelocity entry from space through Earth's atmosphere (Cockell, 2008). As with planetary ejection, this is experimentally tractable, with sounding rockets and orbital vehicles being used for microbiological experiments.[58][59] *B. subtilis* spores inoculated onto granite domes were subjected to hypervelocity atmospheric transit (twice) by launch to a ~120 km altitude on an Orion two-stage rocket. The spores were shown to have survived on the sides of the rock, but they did not survive on the forward-facing surface that was subjected to a maximum temperature of 145 °C.[61] In separate experiments, as part of the ESA STONE experiment, numerous organisms were embedded in different types or rocks and were mounted in the heat shield of six Foton re-entry capsules. During reentry, the rock samples were subjected to temperatures and pressure loads comparable to those experienced in meteorites.[62] The exogenous arrival of photosynthetic microorganisms could have quite profound consequences for the course of biological evolution on the inoculated planet. As photosynthetic organisms must be close to the surface of a rock to obtain sufficient light energy, atmospheric transit might act as a filter against them by ablating the surface layers of the rock. Although cyanobacteria have been shown to survive the desiccating, freezing conditions of space in orbital experiments, this would be of no benefit as the STONE experiment showed that they cannot survive atmospheric entry.[63] Thus, non-photosynthetic organisms deep within rocks have a chance to survive the exit and entry process. (See also: Impact survival.) Research presented at the European Planetary Science Congress in 2015 suggests that ejection, entry and impact is survivable for some simple organisms.[64]

8.2.3   Accidental panspermia

Thomas Gold, a professor of astronomy, suggested in 1960 the hypothesis of "Cosmic Garbage", that life on Earth might have originated accidentally from a pile of waste products dumped on Earth long ago by extraterrestrial beings.[65]

8.2.4   Directed panspermia

Main article: Directed panspermia

Directed panspermia concerns the deliberate transport of microorganisms in space, sent to Earth to start life here, or sent from Earth to seed new solar systems with life by introduced species of microorganisms on lifeless planets. The Nobel prize winner Francis Crick, along with Leslie Orgel proposed that life may have been purposely spread by an advanced extraterrestrial civilization,[40] but considering an early "RNA world" Crick noted later that life may have originated on Earth.[66] It has been suggested that 'directed' panspermia was proposed in order to counteract various objections, including the argument that microbes would be inactivated by the space environment and cosmic radiation before they could make a chance encounter with Earth.[67]

Conversely, active directed panspermia has been proposed to secure and expand life in space.[43] This may be motivated by biotic ethics that values, and seeks to propagate, the basic patterns of our organic gene/protein life-form.[68] The panbiotic program would seed new solar systems nearby, and clusters of new stars in interstellar clouds. These young targets, where local life would not have formed yet, avoid any interference with local life.

For example, microbial payloads launched by solar sails at speeds up to 0.0001 c (30,000 m/s) would reach targets at 10 to 100 light-years in 0.1 million to 1 million years. Fleets of microbial capsules can be aimed at clusters of new stars in star-forming clouds, where they may land on planets or captured by asteroids and comets and later delivered to planets. Payloads may contain extremophiles for diverse environments and cyanobacteria similar to early microorganisms. Hardy multicellular organisms (rotifer cysts) may be included to induce higher evolution.[69]

The probability of hitting the target zone can be calculated from $P(target) = \frac{A(target)}{\pi(dg)^2} = \frac{ar(target)^2 v^2}{(tp)^2 d^4}$ where A(target) is the cross-section of the target area, dg is the positional uncertainty at arrival; a – constant (depending on units), r(target) is the radius of the target area; v the velocity of the probe; (tp) the targeting precision (arcsec/yr); and d the distance to the target, guided by high-resolution astrometry of 1×10^{-5} arcsec/yr (all units in SIU). These calculations show that relatively near target stars(Alpha PsA, Beta Pictoris) can be seeded by milligrams of launched microbes; while seeding the Rho Ophiochus star-forming cloud requires hundreds of kilograms of dispersed capsules.[43]

Directed panspermia to secure and expand life in space is becoming possible because of developments in solar sails, precise astrometry, extrasolar planets, extremophiles and microbial genetic engineering. After determining the composition of chosen meteorites, astroecologists performed laboratory experiments that suggest that many colonizing microorganisms and some plants could obtain many of their chemical nutrients from asteroid and cometary materials.[70] However, the scientists noted that phosphate (PO_4) and nitrate (NO_3–N) critically limit nutrition to many terrestrial lifeforms.[70] With such materials, and energy from long-lived stars, microscopic life planted by directed panspermia could find an immense future in the galaxy.[71]

A number of publications since 1979 have proposed the idea that directed panspermia could be demonstrated to be the origin of all life on Earth if a distinctive 'signature' message were found, deliberately implanted into either the genome or the genetic code of the first microorganisms by our hypothetical progenitor.[72][73][74][75] In 2013 a team of physicists claimed that they had found mathematical and semiotic patterns in the genetic code which, they believe, is evidence for such a signature.[76][77][78] Further investigations are needed.

A microscopic ball made of titanium and vanadium was found in Earth's upper atmosphere in early 2015. Milton Wainwright, a UK researcher and astrobiologist at the University of Buckingham claimed in a tabloid that the metal ball "could contain DNA." He speculates that it could be an alien device sent to Earth by extraterrestrials in order to continue seeding the planet with life.[79]

8.2.5   Pseudo-panspermia

Further information: List of interstellar and circumstellar molecules and Abiogenesis § Extraterrestrial organic molecules

Pseudo-panspermia (sometimes called soft panspermia, molecular panspermia or quasi-panspermia) proposes that the organic molecules used for life originated in space and were incorporated in the solar nebula, from which the planets condensed and were further —and continuously— distributed to planetary surfaces where life then emerged (abiogenesis).[9][10] From the early 1970s it was becoming evident that interstellar dust consisted of a large component of organic molecules. The first suggestion came from Chandra Wickramasinghe, who proposed a polymeric composition based on the molecule formaldehyde (CH_2O).[80] Interstellar molecules are formed by chemical reactions within very sparse interstellar or circumstellar clouds of dust and gas. Usually this occurs when a molecule becomes ionized, often as the result of an interaction with cosmic rays. This positively charged molecule then draws in a nearby reactant by electrostatic attraction of the neutral molecule's electrons. Molecules can also be generated by reactions between neutral atoms and molecules, although this process is generally slower.[11] The dust plays a critical role of shielding the molecules from the ionizing effect of ultraviolet radiation emitted by stars.[12]

A 2008 analysis of $^{12}C/^{13}C$ isotopic ratios of organic compounds found in the Murchison meteorite indicates a non-terrestrial origin for these molecules rather than terrestrial contamination. Biologically relevant molecules identified so far include uracil, an RNA nucleobase, and xanthine.[81][82] These results demonstrate that many organic compounds which are components of life on Earth were already present in the early Solar System and may have played a key role in life's origin.[83]

In August 2009, NASA scientists identified one of the fundamental chemical building-blocks of life (the amino acid glycine) in a comet for the first time.[84]

On August 2011, a report, based on NASA studies with meteorites found on Earth, was published suggesting building blocks of DNA (adenine, guanine and related organic molecules) may have been formed extraterrestrially in outer space.[85][86][87] In October 2011, scientists reported that cosmic dust contains complex organic matter ("amorphous organic solids with a mixed aromatic-aliphatic structure") that could be created naturally, and rapidly, by stars.[88][89][90] One of the scientists suggested that these complex organic compounds may have been related to the development of life on Earth and said that, "If this is the case, life on Earth may have had an easier time getting started as these organics can serve as basic ingredients for life."[88]

On August 2012, and in a world first, astronomers at Copenhagen University reported the detection of a specific sugar molecule, glycolaldehyde, in a distant star system. The molecule was found around the protostellar binary *IRAS 16293-2422*, which is located 400 light years from Earth.[91][92] Glycolaldehyde is needed to form ribonucleic acid, or RNA, which is similar in function to DNA. This finding suggests that complex organic molecules may form in stellar systems prior to the formation of planets, eventually arriving on young planets early in their formation.[93]

In September 2012, NASA scientists reported that polycyclic aromatic hydrocarbons (PAHs), subjected to interstellar medium (ISM) conditions, are transformed, through hydrogenation, oxygenation and hydroxylation, to more complex organics - "a step along the path toward amino acids and nucleotides, the raw materials of proteins and DNA, respectively". Further, as a result of these transformations, the PAHs lose their spectroscopic signature which could be one of the reasons" for the lack of PAH detection in interstellar ice grains, particularly the outer regions of cold, dense clouds or the upper molecular layers of protoplanetary disks."[94][95]

In 2013, the Atacama Large Millimeter Array (ALMA Project) confirmed that researchers have discovered an important pair of prebiotic molecules in the icy particles in interstellar space (ISM). The chemicals, found in a giant cloud of gas about 25,000 light-years from Earth in ISM, may be a precursor to a key component of DNA and the other may have a role in the formation of an important amino acid. Researchers found a molecule called cyanomethanimine, which produces adenine, one of the four nucleobases that form the "rungs" in the ladder-like structure of DNA. The other molecule, called ethanamine, is thought to play a role in forming alanine, one of the twenty amino acids in the genetic code. Previously, scientists thought such processes took place in the very tenuous gas between the stars. The new discoveries, however, suggest that the chemical formation sequences for these molecules occurred not in gas, but on the surfaces of ice grains in interstellar space.[96] NASA ALMA scientist Anthony Remijan stated that finding these molecules in an interstellar gas cloud means that important building blocks for DNA and amino acids can 'seed' newly formed planets with the chemical precursors for life.[97]

In March 2013, a simulation experiment indicate that dipeptides (pairs of amino acids) that can be building blocks of proteins, can be created in interstellar dust.[98]

In February 2014, NASA announced a greatly upgraded database for tracking polycyclic aromatic hydrocarbons (PAHs) in the universe. According to scientists, more than 20% of the carbon in the universe may be associated with PAHs, possible starting materials for the formation of life. PAHs seem to have been formed shortly after the Big Bang, are widespread throughout the universe, and are associated with new stars and exoplanets.[99]

In March 2015, NASA scientists reported that, for the first time, complex DNA and RNA organic compounds of life, including uracil, cytosine and thymine, have been formed in the laboratory under outer space conditions, using starting chemicals, such as pyrimidine, found in meteorites. Pyrimidine, like polycyclic aromatic hydrocarbons (PAHs), the most carbon-rich chemical found in the Universe, may have been formed in red giants or in interstellar dust and gas clouds, according to the scientists.[100]

8.3 Extraterrestrial life

Main article: Extraterrestrial life

The chemistry of life may have begun shortly after the Big Bang, 13.8 billion years ago, during a habitable epoch when the Universe was only 10–17 million years old.[101][102][103] According to the panspermia hypothesis, microscopic life—distributed by meteoroids, asteroids and other small Solar System bodies—may exist throughout the universe.[104] Nonetheless, Earth is the only place in the universe known to harbor life.[105][106] The sheer number of planets in the Milky Way galaxy, however, may make it probable that life has arisen somewhere else in the galaxy and the universe. It is generally agreed that the conditions required for the evolution of intelligent life as we know it are probably exceedingly rare in the universe, while simultaneously noting that simple single-celled microorganisms may be more likely.[107]

The extrasolar planet results from the Kepler mission estimate 100–400 billion exoplanets, with over 3,500 as candidates or confirmed exoplanets.[108] On 4 November 2013, astronomers reported, based on Kepler space mission data, that there could be as many as 40 billion Earth-sized planets orbiting in the habitable zones of sun-like stars and red dwarf stars within the Milky Way Galaxy.[109][110] 11 billion of these estimated planets may be orbiting sun-like stars.[111] The nearest such planet may be 12 light-years away, according to the scientists.[109][110]

It is estimated that space travel over cosmic distances would take an incredibly long time to an outside observer, and with vast amounts of energy required. However, there are reasons to hypothesize that faster-than-light interstellar space travel might be feasible. This has been explored by NASA scientists since at least 1995.[112]

8.3.1 Hypotheses on extraterrestrial sources of illnesses

Hoyle and Wickramasinghe have speculated that several outbreaks of illnesses on Earth are of extraterrestrial origins, including the 1918 flu pandemic, and certain outbreaks of polio and mad cow disease. For the 1918 flu pandemic they hypothesized that cometary dust brought the virus to Earth simultaneously at multiple locations—a view almost universally dismissed by experts on this pandemic. Hoyle also speculated that HIV came from outer space.[113] After Hoyle's death, The Lancet published a letter to the editor from Wickramasinghe and two of his colleagues,[114] in which they hypothesized that the virus that causes severe acute respiratory syndrome (SARS) could be extraterrestrial in origin and not originated from chickens. The Lancet subsequently published three responses to this letter, showing that the hypothesis was not evidence-based, and casting doubts on the quality of the experiments referenced by Wickramasinghe in his letter.[115][116][117] A 2008 encyclopedia notes that "Like other claims linking terrestrial disease to extraterrestrial pathogens, this proposal was rejected by the greater research community."[113]

8.3.2   Case studies

- A meteorite originating from Mars known as ALH84001 was shown in 1996 to contain microscopic structures resembling small terrestrial nanobacteria. When the discovery was announced, many immediately conjectured that these were fossils and were the first evidence of extraterrestrial life — making headlines around the world. Public interest soon started to dwindle as most experts started to agree that these structures were not indicative of life, but could instead be formed abiotically from organic molecules. However, in November 2009, a team of scientists at Johnson Space Center, including David McKay, reasserted that there was "strong evidence that life may have existed on ancient Mars", after having reexamined the meteorite and finding magnetite crystals.[118][119]

- On May 11, 2001, two researchers from the University of Naples claimed to have found live extraterrestrial bacteria inside a meteorite. Geologist Bruno D'Argenio and molecular biologist Giuseppe Geraci claim the bacteria were wedged inside the crystal structure of minerals, but were resurrected when a sample of the rock was placed in a culture medium. They believe that the bacteria were not terrestrial because they survived when the sample was sterilized at very high temperature and washed with alcohol. They also claim that the bacteria's DNA is unlike any on Earth.[120][121] They presented a report on May 11, 2001, concluding that this is the first evidence of extraterrestrial life, documented in its genetic and morphological properties. Some of the bacteria they discovered were found inside meteorites that have been estimated to be over 4.5 billion years old, and were determined to be related to modern day *Bacillus subtilis* and *Bacillus pumilis* bacteria on Earth but appears to be a different strain.[122]

- An Indian and British team of researchers led by Chandra Wickramasinghe reported on 2001 that air samples over Hyderabad, India, gathered from the stratosphere by the Indian Space Research Organization, contained clumps of living cells. Wickramasinghe calls this "unambiguous evidence for the presence of clumps of living cells in air samples from as high as 41 km, above which no air from lower down would normally be transported".[123][124] Two bacterial and one fungal species were later independently isolated from these filters which were identified as *Bacillus simplex*, *Staphylococcus pasteuri* and *Engyodontium album* respectively.[125][126] The experimental procedure suggested that these were not the result of laboratory contamination, although similar isolation experiments at separate laboratories were unsuccessful.

  A reaction report at NASA Ames indicated skepticism towards the premise that Earth life cannot travel to and reside at such altitudes.[127]

  Pushkar Ganesh Vaidya from the Indian Astrobiology Research Centre reported in 2009 that "the three microorganisms captured during the balloon experiment do not exhibit any distinct adaptations expected to be seen in microorganisms occupying a cometary niche".[128][129]

- In 2005 an improved experiment was conducted by ISRO. On April 10, 2005 air samples were collected from the upper atmosphere at altitudes ranging from 20 km to more than 40 km. The samples were tested at two labs in India. The labs found 12 bacterial and 6 different fungal species in these samples. The fungi were *Penicillium decumbens*, *Cladosporium cladosporioides*, *Alternaria sp.* and *Tilletiopsis albescens*. Out of the 12 bacterial samples, three were identified as new species and named *Janibacter hoyeli.sp.nov* (after Fred Hoyle), *Bacillus isronensis.sp.nov* (named after ISRO) and *Bacillus aryabhati* (named after the ancient Indian mathematician, Aryabhata). These three new species showed that they were more resistant to UV radiation than similar bacteria.[130][131]

  Atmospheric sampling by NASA in 2010 before and after hurricanes, collected 314 different types of bacteria; the study suggests that large-scale convection during tropical storms and hurricanes can then carry this material from the surface higher up into the atmosphere.[132][133]

- On January 10, 2013, Chandra Wickramasinghe found fossil diatom frustules in what he thinks is a new kind of carbonaceous meteorite called Polonnaruwa that landed in the North Central Province of Sri Lanka on 29 December 2012.[134] Early on, there was criticism that that Wickramasinghe's report was not an examination of an actual meteorite but of some terrestrial rock passed off as a meteorite.[135]

  Wickramasinghe's team remark that they are aware that a large number of unrelated stones have been submitted for analysis, and have no knowledge regarding the nature, source or origin of the stones their critics have examined, so Wickramasinghe clarifies that he is using the stones submitted by the Medical Research Institute in Sri Lanka.[136] In response to the criticism from other scientists, Wickramasinghe performed X-ray diffraction[137] and isotope[136] analyses to verify its meteoritic origin. His analysis revealed a 95% silica and 3% quartz content,[137] and interpreted this result as a "carbonaceous meteorite of unknown type".[137] In addition, Wickramasinghe's team remarked that the temperature at which sand must be heated by lightning to melt and form a fulgurite (1770 °C) would have vaporized and burned all carbon-rich organisms and melted and thus destroyed the delicately marked silica frustules of the diatoms,[136] and that the oxygen isotope data confirms its meteoric origin.[136] Wickramasinghe's team also argues that since living diatoms require nitrogen fixation to synthetize amino acids, proteins, DNA, RNA and other life-critical biomolecules, a population of extraterrestrial cyanobacteria must have been a required component of the comet (Polonnaruwa meteorite) "ecosystem".[136]

- In 2013, Dale Warren Griffin, a microbiologist working at the United States Geological Survey noted that viruses are the most numerous entities on Earth. Griffin speculates that viruses evolved in comets and on other planets and moons may be pathogenic to humans, so he proposed to also look for viruses on moons and planets of the Solar System.[138]

8.3.3 Hoaxes

A separate fragment of the Orgueil meteorite (kept in a sealed glass jar since its discovery) was found in 1965 to have a seed capsule embedded in it, whilst the original glassy layer on the outside remained undisturbed. Despite great initial excitement, the seed was found to be that of a European Juncaceae or Rush plant that had been glued into the fragment and camouflaged using coal dust. The outer "fusion layer" was in fact glue. Whilst the perpetrator of this hoax is unknown, it is thought that they sought to influence the 19th century debate on spontaneous generation — rather than panspermia — by demonstrating the transformation of inorganic to biological matter.[139]

8.4 Extremophiles

See also: Extremophile

Until the 1970s, life was believed to depend on its access to sunlight. Even life in the ocean depths, where sunlight cannot reach, was believed to obtain its nourishment either from consuming organic detritus rained down from the surface waters or from eating animals that did.[140] However, in 1977, during an exploratory dive to the Galapagos Rift in the deep-sea exploration submersible *Alvin*, scientists discovered colonies of assorted creatures clustered around undersea volcanic features known as black smokers.[140] It was soon determined that the basis for this food chain is a form of bacterium that derives its energy from oxidation of reactive chemicals, such as hydrogen or hydrogen sulfide, that bubble up from the Earth's interior. This chemosynthesis revolutionized the study of biology by revealing that terrestrial life need not be Sun-dependent; it only requires water and an energy gradient in order to exist.

It is now known that extremophiles, microorganisms with extraordinary capability to thrive in the harshest environments on Earth, can specialize to thrive in the deep-sea,[141][142][143] ice, boiling water, acid, the water core of nuclear reactors, salt crystals, toxic waste and in a range of other extreme habitats that were previously thought to be inhospitable for life.[144][145][146][147] Living bacteria found in ice core samples retrieved from 3,700 metres (12,100 ft) deep at Lake Vostok in Antarctica, have provided data for extrapolations to the likelihood of microorganisms surviving frozen in extraterrestrial habitats or during interplanetary transport.[148] Also, bacteria have been discovered living within warm rock deep in the Earth's crust.[149]

In order to test some these organism's potential resilience in outer space, plant seeds and spores of bacteria, fungi and ferns have been exposed to the harsh space environment.[146][147][150] Spores are produced as part of the normal life cycle of many plants, algae, fungi and some protozoans, and some bacteria produce endospores or cysts during times of stress. These structures may be highly resilient to ultraviolet and gamma radiation, desiccation, lysozyme, temperature, starvation and chemical disinfectants, while metabolically inactive. Spores germinate when favourable conditions are restored after exposure to conditions fatal to the parent organism.

Although computer models suggest that a captured meteoroid would typically take some tens of millions of years before collision with a neighboring solar system planet,[38] there are documented viable Earthly bacterial spores that are 40 million years old that are very resistant to radiation,[38][44] and others able to resume life after being dormant for 25 million years,[151] suggesting that lithopanspermia life-transfers are possible via meteorites exceeding 1 m in size.[38]

The discovery of deep-sea ecosystems, along with advancements in the fields of astrobiology, observational astronomy and discovery of large varieties of extremophiles, opened up a new avenue in astrobiology by massively expanding the number of possible extraterrestrial habitats and possible transport of hardy microbial life through vast distances.[58]

8.4.1   Research in outer space

See also: List of microorganisms tested in outer space

The question of whether certain microorganisms can survive in the harsh environment of outer space has intrigued biologists since the beginning of spaceflight, and opportunities were provided to expose samples to space. The first American tests were made in 1966, during the Gemini IX and XII missions, when samples of bacteriophage T1 and spores of *Penicillium roqueforti* were exposed to outer space for 16.8 h and 6.5 h, respectively.[49][58] Other basic life sciences research in low Earth orbit started in 1966 with the Soviet biosatellite program Bion and the U.S. Biosatellite program. Thus, the plausibility of panspermia can be evaluated by examining life forms on Earth for their capacity to survive in space.[152] The following experiments carried on low Earth orbit specifically tested some aspects of panspermia or lithopanspermia:

ERA

The Exobiology Radiation Assembly (ERA) was a 1992 experiment on board the European Retrievable Carrier (EURECA) on the biological effects of space radiation. EURECA was an unmanned 4.5 tonne satellite with a payload of 15 experiments.[153] It was an astrobiology mission developed by the European Space Agency (ESA). Spores of different strains of *Bacillus subtilis* and the *Escherichia coli* plasmid pUC19 were exposed to selected conditions of space (space vacuum and/or defined wavebands and intensities of solar ultraviolet radiation). After the approximately 11-month mission, their responses were studied in terms of survival, mutagenesis in the *his* (*B. subtilis*) or *lac* locus (pUC19), induction of DNA strand breaks, efficiency of DNA repair systems, and the role of external protective agents. The data were compared with those of a simultaneously running ground control experiment:[154][155]

- The survival of spores treated with the vacuum of space, however shielded against solar radiation, is substantially increased, if they are exposed in multilayers and/or in the presence of glucose as protective.

- All spores in "artificial meteorites", i.e. embedded in clays or simulated Martian soil, are killed.

- Vacuum treatment leads to an increase of mutation frequency in spores, but not in plasmid DNA.

- Extraterrestrial solar ultraviolet radiation is mutagenic, induces strand breaks in the DNA and reduces survival substantially.

- Action spectroscopy confirms results of previous space experiments of a synergistic action of space vacuum and solar UV radiation with DNA being the critical target.

- The decrease in viability of the microorganisms could be correlated with the increase in DNA damage.

- The purple membranes, amino acids and urea were not measurably affected by the dehydrating condition of open space, if sheltered from solar radiation. Plasmid DNA, however, suffered a significant amount of strand breaks under these conditions.[154]

BIOPAN

BIOPAN is a multi-user experimental facility installed on the external surface of the Russian Foton descent capsule. Experiments developed for BIOPAN are designed to investigate the effect of the space environment on biological material after exposure between 13 to 17 days.[156] The experiments in BIOPAN are exposed to solar and cosmic radiation, the space vacuum and weightlessness, or a selection thereof. Of the 6 missions flown so far on BIOPAN between 1992 and 2007, dozens of experiments were conducted, and some analyzed the likelihood of panspermia. Some bacteria, lichens (*Xanthoria elegans*, *Rhizocarpon geographicum* and their mycobiont cultures, the black Antarctic microfungi *Cryomyces minteri* and *Cryomyces antarcticus*), spores, and even one animal (tardigrades) were found to have survived the harsh outer space environment and cosmic radiation.[157][158][159][160]

EXOSTACK

The German EXOSTACK experiment was deployed in 7 April 1984 on board the Long Duration Exposure Facility statellite. 30% of *Bacillus subtilis* spores survived the nearly 6 years exposure when embedded in salt crystals, whereas 80% survived in the presence of glucose, which stabilize the structure of the cellular macromolecules, especially during vacuum-induced dehydration.[49][161]

If shielded against solar UV, spores of *B. subtilis* were capable of surviving in space for up to 6 years, especially if embedded in clay or meteorite powder (artificial meteorites). The data support the likelihood of interplanetary transfer of microorganisms within meteorites, the so-called lithopanspermia hypothesis.[49]

EXPOSE

EXPOSE is a multi-user facility mounted outside the International Space Station dedicated to astrobiology experiments.[150] Results from the orbital mission, especially the experiments *SEEDS*[162] and *LiFE*,[163] concluded that after an 18-month exposure, some seeds and lichens (*Stichococcus sp.* and *Acarospora sp.*, a lichenized fungal genus) may be capable to survive interplanetary travel if sheltered inside comets or rocks from cosmic radiation and UV radiation.[150][164] The survival of some lichen species in space has also been characterized in simulated laboratory experiments.[165][166]

A separate experiment on EXPOSE called Beer was designed to find microbes that could be used in life-support recycling equipment and future "bio-mining" projects on Mars. It carried group of microbes called OU-20 resembling cyanobacteria genus *Gloeocapsa*, and it survived 553 days exposure outside the ISS.[167]

Rosetta

In 2014, the *Rosetta* spacecraft arrived at COMET 67P/Churyumov–Gerasimenko. A few months after arriving at the comet, *Rosetta* released a small lander, named *Philae*, onto its surface. The plan was to investigate Churyumov–Gerasimenko up close for two years. *Philae's* battery has since died; however scientists hope that as the comet travels toward the sun greater solar energy will recharge *Philae* (via its solar panels) and *Philae* will resume operation. Rosetta's Project Scientist, Gerhard Schwehm, stated that sterilization is generally not crucial since comets are usually regarded as objects where prebiotic molecules can be found, but not living microorganisms.[168] Notwithstanding, other scientists think it will be an opportunity to gather evidence for one of panspermia's hypotheses: the possibility of both active and dormant microbes inside comets.[7][18]

In July 2015, scientists reported that upon the first touchdown of the *Philae* lander on comet 67/P 's surface, measurements by the COSAC and Ptolemy instruments revealed sixteen organic compounds, four of which were seen for the first time on a comet, including acetamide, acetone, methyl isocyanate and propionaldehyde.[169][170][171]

Phobos LIFE

The *Phobos LIFE* or *Living Interplanetary Flight Experiment*, was developed by the Planetary Society and intended to send selected microorganisms on a three-year interplanetary round-trip in a small capsule aboard the Russian Fobos-Grunt spacecraft in 2011. Unfortunately, the spacecraft suffered technical difficulties soon after launch and fell back to Earth, so the experiment was never carried out. The experiment would have tested one aspect of panspermia: lithopanspermia, the hypothesis that life could survive space travel, if protected inside rocks blasted by impact off one planet to land on another.[172][173][174][175]

8.5   Criticism

Panspermia is criticized because it does not answer the question of the origin of life but merely places it on another celestial body. It was also criticized because it could not be tested experimentally. Furthermore, it was suggested that single spores will not survive the physical forces and environment of outer space.[176]

The concept of panspermia was revived when technology provided the opportunity to study the survival of bacterial spores in the harsh environment of space.[58] Wallis and Wickramasinghe argued in 2004 that the transport of individual bacteria or clumps of bacteria, is overwhelmingly more important than lithopanspermia in terms of numbers of microbes transferred, even accounting for the death rate of unprotected bacteria in transit.[177] Then it was found that isolated spores of *B. subtilis* were killed by several orders of magnitude if exposed to the full space environment for a mere few seconds. These results clearly negate the original panspermia hypothesis, which requires single spores as space travelers accelerated by the radiation pressure of the Sun, requiring many years to travel between the planets. However, if shielded against solar UV, spores of *Bacillus subtilis* were capable of surviving in space for up to 6 years, especially if embedded in clay or meteorite powder (artificial meteorites). The data support the likelihood of interplanetary transfer of microorganisms within meteorites, the so-called **lithopanspermia** hypothesis.[49]

8.6   Science fiction

- Jack Finney's novel *The Body Snatchers* (1955) and the subsequent film adaptations describe spores drifting through space to arrive on the surface of Earth, though the premise is most fully discussed in the second version *Invasion of the Body Snatchers* (1978 film).

- In Ursula K. Le Guin's series the Hainish Cycle (1964–2014), Earth and other planets are seeded by the Hain using genetic engineering.

- Michael Crichton's 1969 novel, *The Andromeda Strain*, is based on the panspermiatic premise of a meteor bringing an crystalline alien bacterium to Earth. The phrase "Andromeda Strain" has become a shorthand for mysterious infectious diseases.

- Stephen King's short story "Weeds" (1976), later adapted into the Creepshow vignette "The Lonesome Death of Jordy Verrill" (1982; starring King,) involves a meteor crashing to Earth which carries with it a virulent plant/fungus which spreads rapidly.

- In the *Star Trek: The Next Generation* episode, "The Chase" (season 6, episode 20, April 26, 1993), the common humanoid form and genetic compatibility of alien species throughout the Alpha Quadrant is revealed to have resulted from directed panspermia by an earlier species of intelligent humanoid progenitors who seeded the many planets with their own DNA.

- In the 1990s TV series *Space: Above and Beyond*, about a war between humanity and an alien species known as the Chigs, it is eventually revealed that life on the Chig's planet and the Chigs themselves evolved from Earth bacteria carried there by an asteroid billions of years ago.

- Tess Gerritsen's novel, *Gravity* (1999), involves the exposure of astronauts aboard the Space Shuttle and International Space Station, to a chimera based on Archaeons, that were recovered from the Galapagos Rift.

- The plot of 2001 American science fiction comedy *Evolution* follows college professor Ira Kane (David Duchovny) and geologist Harry Block (Orlando Jones) who investigate a meteor crash in Arizona. They discover that the meteor is harboring extraterrestrial life which is evolving very quickly into large, diverse and outlandish creatures.

- The plot of the 2001 short film *Horses on Mars* centers on microbes as characters spreading into the inner solar system from Mars four billion years ago, with the main character making it to Venus while his friends land on Earth. His friends on Earth successfully evolve and send him a message via the Venera 13 lander, and later eventually make the trip back home to Mars as space-faring creatures, but without the main character, whose unsophisticated attempt to make it back to Mars ends in failure.

- In the reimagined *Battlestar Galactica*, season 3, episodes 6 and 7 ("Torn", November 3, 2006; "A Measure of Salvation", November 10, 2006), a Cylon basestar discovers an ancient beacon and takes it on board, whereupon a deadly virus from the beacon infects the Cylons. Doctor Cottle determines the Cylon infection to be a three-thousand-year-old strain of Lymphocytic choriomeningitis. Admiral Adama and President Roslin speculate that the beacon was accidentally infected prior to placement by ancient human colonists on their way from Kobol to Earth. Adama remarks, "An entire race almost wiped out because someone forgot to wipe their nose."

- The premise of Gareth Edwards's 2010 film *Monsters* is that a NASA deep space probe crashes, bringing back with it an alien species requiring the U.S. and Mexican military to quarantine a large district of the border region.

- The opening sequence of Ridley Scott's 2012 *Alien* prequel, *Prometheus* depicts a humanoid species, referred to as 'the Engineers', seeding what is presumably the early Earth by disintegrating the body of one of their members and spilling his DNA into the water of the planet. At the climax of the film it is revealed that for unknown reasons the Engineers deemed their experiment to have been a failure and intended to end it by eradicating all life on Earth.

- The novels, "The Ice Limit" (2000) and "The Lost Island" (2014), by Douglas Preston and Lincoln Child, make references to panspermia.

- The novel *Titan* by Stephen Baxter ends with human astronauts seeding the moon Titan with bacteria. The bacteria eventually evolve into creatures that intentionally spread primitive lifeforms to other star systems.

8.7 See also

- Abiogenesis

- Anthropic principle

- Astrobiology

- Cryptobiosis

- Drake equation

- Fermi paradox

- Fine-tuned Universe

- Interplanetary contamination

- Last universal ancestor

- List of microorganisms tested in outer space

- Planetary protection

- Rare Earth hypothesis

- Red rain in Kerala

8.8   References

[1] Wickramasinghe, Chandra (2011). "Bacterial morphologies supporting cometary panspermia: a reappraisal". *International Journal of Astrobiology* **10** (1): 25–30. Bibcode:2011IJAsB..10...25W. doi:10.1017/S1473550410000157.

[2] Napier, William (October 2011). "Exchange of Biomaterial Between Planetary Systems" (PDF) **16**: 6616–6642.

[3] Rampelotto, P. H. (2010). Panspermia: A promising field of research. In: Astrobiology Science Conference. Abs 5224.

[4] Forward planetary contamination like *Tersicoccus phoenicis*, that has shown resistance to methods usually used in spacecraft assembly clean rooms: Madhusoodanan, Jyoti (May 19, 2014). "Microbial stowaways to Mars identified". *Nature (journal)*. doi:10.1038/nature.2014.15249. Retrieved May 23, 2014.

[5] Webster, Guy (November 6, 2013). "Rare New Microbe Found in Two Distant Clean Rooms". *NASA.gov*. Retrieved November 6, 2013.

[6] A variation of the panspermia hypothesis is **necropanspermia** which is described by astronomer Paul Wesson as follows: "The vast majority of organisms reach a new home in the Milky Way in a technically dead state ... Resurrection may, however, be possible." Grossman, Lisa (2010-11-10). "All Life on Earth Could Have Come From Alien Zombies". *Wired*. Retrieved 10 November 2010.

[7] Hoyle, F. and Wickramasinghe, N.C., 1981. Evolution from Space (Simon & Schuster Inc., NY, 1981 and J.M. Dent and Son, Lond, 1981), ch3 pp. 35-49.

[8] Wickramasinghe, J., Wickramasinghe, C. and Napier, W., 2010. Comets and the Origin of Life (World Scientific, Singapore, 1981), ch6 pp. 137-154.

[9] Klyce, Brig (2001). "Panspermia Asks New Questions". Retrieved 25 July 2013.

[10] Klyce, Brig (2001). Kingsley, Stuart A; Bhathal, Ragbir, eds. "The Search for Extraterrestrial Intelligence (SETI) in the Optical Spectrum III". *Proc. SPIE Vol. 4273*. The Search for Extraterrestrial Intelligence (SETI) in the Optical Spectrum III **4273**: 11. Bibcode:2001SPIE.4273...11K. doi:10.1117/12.435366. |chapter= ignored (help)

[11] Dalgarno, A. (2006). "The galactic cosmic ray ionization rate". *Proceedings of the National Academy of Sciences* **103** (33): 12269–73. Bibcode:2006PNAS..10312269D. doi:10.1073/pnas.0602117103. PMC 1567869. PMID 16894166.

[12] Brown, Laurie M.; Pais, Abraham; Pippard, A. B. (1995). "The physics of the interstellar medium". *Twentieth Century Physics* (2nd ed.). CRC Press. p. 1765. ISBN 0-7503-0310-7.

[13] Borenstein, Seth (19 October 2015). "Hints of life on what was thought to be desolate early Earth". *Excite* (Yonkers, NY: Mindspark Interactive Network). Associated Press. Retrieved 2015-10-20.

[14] Bell, Elizabeth A.; Boehnke, Patrick; Harrison, T. Mark; et al. (19 October 2015). "Potentially biogenic carbon preserved in a 4.1 billion-year-old zircon" (PDF). *Proc. Natl. Acad. Sci. U.S.A.* (Washington, D.C.: National Academy of Sciences). doi:10.1073/pnas.1517557112. ISSN 1091-6490. Retrieved 2015-10-20. Early edition, published online before print.

[15] Margaret O'Leary (2008) Anaxagoras and the Origin of Panspermia Theory, iUniverse publishing Group, # ISBN 978-0-595-49596-2

[16] Berzelius (1799–1848), J. J. "Analysis of the Alais meteorite and implications about life in other worlds".

[17] Lynn J. Rothschild; Adrian M. Lister (June 2003). *Evolution on Planet Earth – The Impact of the Physical Environment*. Academic Press. pp. 109–127. ISBN 978-0-12-598655-7.

[18] Thomson (Lord Kelvin), W. (1871). "Inaugural Address to the British Association Edinburgh. "We must regard it as probably to the highest degree that there are countless seed-bearing meteoritic stones moving through space."". *Nature* **4** (92): 261–278 [262]. Bibcode:1871Natur...4..261.. doi:10.1038/004261a0.

[19] "The word: Panspermia". *New Scientist* (2541). 7 March 2006. Retrieved 25 July 2013.

[20] "History of Panspermia". Retrieved 25 July 2013.

[21] Arrhenius, S., *Worlds in the Making: The Evolution of the Universe*. New York, Harper & Row, 1908.

[22] Napier, W.M. (2007). "Pollination of exoplanets by nebulae". *Int.J.Astrobiol* **6** (3): 223–228. Bibcode:2007IJAsB...6..223N. doi:10.1017/S1473550407003710.

[23] Line, M.A. (2007). "Panspermia in the context of the timing of the origin of life and microbial phylogeny". *Int. J. Astrobiol.* 3 **6** (3): 249–254. Bibcode:2007IJAsB...6..249L. doi:10.1017/S1473550407003813.

[24] Wickramasinghe, D. T.; Allen, D. A. (1980). "The 3.4-μm interstellar absorption feature". *Nature* **287** (5782): 518–519. Bibcode:1980Natur.287..518W. doi:10.1038/287518a0.

[25] Allen, D. A.; Wickramasinghe, D. T. (1981). "Diffuse interstellar absorption bands between 2.9 and 4.0 μm". *Nature* **294** (5838): 239–240. Bibcode:1981Natur.294..239A. doi:10.1038/294239a0.

[26] Wickramasinghe, D. T.; Allen, D. A. (1983). "Three components of 324 ?m absorption bands". *Astrophysics and Space Science* **97** (2): 369–378. Bibcode:1983Ap&SS..97..369W. doi:10.1007/BF00653492.

[27] Fred Hoyle; Chandra Wickramasinghe & John Watson (1986). *Viruses from Space and Related Matters*. University College Cardiff Press.

[28] Weaver, Rheyanne (April 7, 2009). "Ruminations on other worlds". *statepress.com*. Retrieved 25 July 2013.

[29] Khan, Amina (7 March 2014). "Did two planets around nearby star collide? Toxic gas holds hints". *LA Times*. Retrieved 9 March 2014.

[30] Dent, W. R. F.; Wyatt, M. C.; Roberge, A.; Augereau, J.- C.; et al. (6 March 2014). "Molecular Gas Clumps from the Destruction of Icy Bodies in the β Pictoris Debris Disk". *Science* **343** (6178): 1490–1492. Bibcode:2014Sci...343.1490D. doi:10.1126/science.1248726. Retrieved 9 March 2014.

[31] Wickramasinghe, Chandra; Wickramasinghe, Chandra; Napier, William (2009). *Comets and the Origin of Life*. World Scientific Press. doi:10.1142/6008. ISBN 978-981-256-635-5.

[32] Wall, Mike. "Comet Impacts May Have Jump-Started Life on Earth". space.com. Retrieved 1 August 2013.

[33] Weber, P; Greenberg, J. M. (1985). "Can spores survive in interstellar space?".*Nature***316**(6027): 403–407. Bibcode:1985Nat doi:10.1038/316403a0.

[34] Melosh, H. J. (1988). "The rocky road to panspermia". *Nature* **332** (6166): 687–688. Bibcode:1988Natur.332..687M. doi:10.1038/332687a0. PMID 11536601.

[35] C. Mileikowsky; F. A. Cucinotta; J. W. Wilson; B. Gladman; et al. (2000). "Risks threatening viable transfer of microbes between bodies in our solar system". *Planetary and Space Science* **48** (11): 1107–1115. Bibcode:2000P&SS...48.1107M. doi:10.1016/S0032-0633(00)00085-4.

[36] Studies Focus On Spacecraft Sterilization

[37] European Space Agency: Dry heat sterilisation process to high temperatures

[38] Edward Belbruno; Amaya Moro-Martín; Malhotra, Renu & Savransky, Dmitry (2012). "Chaotic Exchange of Solid Material between Planetary". *Astrobiology* **12** (8): 754–74. arXiv:1205.1059. Bibcode:2012AsBio..12..754B. doi:10.1089/ast.2012.0825. PMC 3440031. PMID 22897115.

[39] Slow-moving rocks better odds that life crashed to Earth from space News at Princeton, September 24, 2012.

[40] Crick, F. H.; Orgel, L. E. (1973). "Directed Panspermia".*Icarus***19**(3): 341–348. Bibcode:1979JBIS...32..419M.doi:10.1016/1035(73)90110-3.

[41] Mautner, Michael N. (2000). *Seeding the Universe with Life: Securing Our Cosmological Future* (PDF). Washington D. C.: Legacy Books (www.amazon.com). ISBN 0-476-00330-X.

[42] Mautner, M; Matloff, G. (1979). "Directed panspermia: A technical evaluation of seeding nearby solar systems" (PDF). *J. British Interplanetary Soc.* **32**: 419.

[43] Mautner, M. N. (1997). "Directed panspermia. 3. Strategies and motivation for seeding star-forming clouds" (PDF). *J. British Interplanetary Soc.* **50**: 93–102. Bibcode:1997JBIS...50...93M.

[44] BBC Staff (23 August 2011). "Impacts 'more likely' to have spread life from Earth". BBC. Retrieved 24 August 2011.

[45] "Electromagnetic space travel for bugs? - space – 21 July 2006 – New Scientist Space". Space.newscientist.com. Archived from the original on January 11, 2009. Retrieved December 8, 2014.

[46] Dehel, T. (2006-07-23). "Uplift and Outflow of Bacterial Spores via Electric Field". *36th COSPAR Scientific Assembly. Held 16–23 July 2006* (Adsabs.harvard.edu) **36**: 1. arXiv:hep-ph/0612311. Bibcode:2006cosp....36....1D.

[47] "Die Verbreitung des Lebens im Weltenraum" (the "Distribution of Life in Space"). Published in Die Umschau. 1903.

[48] *Ancient micronauts: interplanetary transport of microbes by cosmic impacts.* Wayne L. Nicholson. Trends in Microbiology. Vol. 17, No. 6. (June 2009), pp. 243-250. doi:10.1016/j.tim.2009.03.004

[49] Horneck, G.; Klaus, D. M.; Mancinelli, R. L. (2010). "Space Microbiology". *Microbiology and Molecular Biology Reviews* **74** (1): 121–56. doi:10.1128/MMBR.00016-09. PMC 2832349. PMID 20197502.

[50] I. S. Shklovskii; Carl Sagan (1966). *Intelligent Life in the Universe.* Emerson-Adams Press, Incorporated. ISBN 978-1-892803-02-3.

[51] Wickramasinghe, M.K.; Wickramasinghe, C. (2004). "Interstellar transfer of planetary microbiota". *Mon. Not.R. Astr. Soc.* **348**: 52–57. Bibcode:2004MNRAS.348...52W. doi:10.1111/j.1365-2966.2004.07355.x.

[52] *Protection of Bacterial Spores in Space, a Contribution to the Discussion on Panspermia.* Gerda Horneck, Petra Rettberg, Günther Reitz, Jörg Wehner, Ute Eschweiler, Karsten Strauch, Corinna Panitz, Verena Starke, Christa Baumstark-Khan. Origins of life and evolution of the biosphere. December 2001. Volume 31, Issue 6, pp. 527-547.

[53] R.O. Rahn, J.L. Hosszu, Influence of relative humidity on the photochemistry of DNA films. Biochim. Biophys Acta 190 (1969) 126–131.

[54] M.H. Patrick, D.M. Gray, Independence of photproduct formation on DNA conformation. Photochem. Photobiol. 24 (1976) 507–513.

[55] Wayne L. Nicholson; Andrew C. Schuerger; Peter Setlow (21 January 2005). "The solar UV environment and bacterial spore UV resistance: considerations for Earth-to-Mars transport by natural processes and human spaceflight" (PDF). *Mutation Research* **571** (1–2): 249–264. doi:10.1016/j.mrfmmm.2004.10.012. PMID 15748651. Retrieved 2 August 2013.

[56] Clark BC., Planetary interchange of bioactive material: probability factors and implications Origins Life Evol Biosphere 2001; 31: 185-97

[57] Mileikowsky C. et al. Natural Transfer of Microbes in space. part I: from Mars to Earth and Earth to Mars Icarus 2000; 145; 391-427

[58] Olsson-Francis, Karen; Cockell, Charles S. (2010). "Experimental methods for studying microbial survival in extraterrestrial environments". *Journal of Microbiological Methods* **80** (1): 1–13. doi:10.1016/j.mimet.2009.10.004. PMID 19854226.

[59] Cockell, Charles S. (2007). "The Interplanetary Exchange of Photosynthesis". *Origins of Life and Evolution of Biospheres* **38**: 87–104. Bibcode:2008OLEB...38...87C. doi:10.1007/s11084-007-9112-3.

[60] Horneck, Gerda; Stöffler, Dieter; Ott, Sieglinde; Hornemann, Ulrich; et al. (2008). "Microbial Rock Inhabitants Survive Hypervelocity Impacts on Mars-Like Host Planets: First Phase of Lithopanspermia Experimentally Tested". *Astrobiology* **8** (1): 17–44. Bibcode:2008AsBio...8...17H. doi:10.1089/ast.2007.0134. PMID 18237257.

[61] Fajardo-Cavazos, Patricia; Link, Lindsey; Melosh, H. Jay; Nicholson, Wayne L. (2005). "Bacillus subtilisSpores on Artificial Meteorites Survive Hypervelocity Atmospheric Entry: Implications for Lithopanspermia". *Astrobiology* **5** (6): 726–36. Bibcode:2005AsBio....5..726F. doi:10.1089/ast.2005.5.726. PMID 16379527.

[62] Brack, A.; Baglioni, P.; Borruat, G.; Brandstätter, F.; et al. (2002). "Do meteoroids of sedimentary origin survive terrestrial atmospheric entry? The ESA artificial meteorite experiment STONE". *Planetary and Space Science* **50** (7–8): 763–772. Bibcode:2002P&SS...50..763B. doi:10.1016/S0032-0633(02)00018-1.

[63] Cockell, Charles S.; Brack, André; Wynn-Williams, David D.; Baglioni, Pietro; et al. (2007). "Interplanetary Transfer of Photosynthesis: An Experimental Demonstration of a Selective Dispersal Filter in Planetary Island Biogeography". *Astrobiology* **7** (1): 1–9. Bibcode:2007AsBio...7....1C. doi:10.1089/ast.2006.0038. PMID 17407400.

[64] "Could Life Have Survived a Fall to Earth?". *EPSC.* 12 September 2013. Retrieved 2015-04-21.

[65] Gold, T. "Cosmic Garbage", Air Force and Space Digest, 65 (May 1960).

[66] "Anticipating an RNA world. Some past speculations on the origin of life: where are they today?" by L. E. Orgel and F. H. C. Crick in *FASEB J.* (1993) Volume 7 pages 238-239.

[67] Clark, Benton C. Clark (February 2001). "Planetary Interchange of Bioactive Material: Probability Factors and Implications". *Origins of life and evolution of the biosphere* **31** (1–2): 185–197. Bibcode:2001OLEB...31..185C. doi:10.1023/A:1006757011007. PMID 11296521.

[68] Mautner, Michael N. (2009). "Life-centered ethics, and the human future in space" (PDF). *Bioethics* **23** (8): 433–440. doi:10.1111/j.1467-8519.2008.00688.x. PMID 19077128.

[69] Mautner, Michael Noah Ph.D. (2000). *Seeding the Universe with Life: Securing our Cosmological Future* (PDF). Legacy Books (www.amazon.com). ISBN 0-476-00330-X.

[70] Mautner, Michael N. (2002). "Planetary bioresources and astroecology. 1. Planetary microcosm bioessays of Martian and meteorite materials: soluble electrolytes, nutrients, and algal and plant responses" (PDF). *Icarus* **158** (1): 72–86. Bibcode:2002Icar..158...72M.doi:10.1006/icar.2002.6841. PMID 12449855.

[71] Mautner, Michael N. (2005). "Life in the cosmological future: Resources, biomass and populations" (PDF). *Journal of the British Interplanetary Society* **58**: 167–180. Bibcode:2005JBIS...58..167M.

[72] G. Marx (1979). "Message through time".*Acta Astronautica***6**(1–2): 221–225. Bibcode:1979AcAau...6..221M.doi:10.1016/0 5765(79)90158-9.

[73] H. Yokoo; T. Oshima (1979). "Is bacteriophage φX174 DNA a message from an extraterrestrial intelligence?". *Icarus* **38** (1): 148–153. Bibcode:1979Icar...38..148Y. doi:10.1016/0019-1035(79)90094-0.

[74] Overbye, Dennis (26 June 2007). "Human DNA, the Ultimate Spot for Secret Messages (Are Some There Now?)". Retrieved 2014-10-09.

[75] Davies, Paul C.W. (2010). *The Eerie Silence: Renewing Our Search for Alien Intelligence*. Boston, Massachusetts: Houghton Mifflin Harcourt. ISBN 978-0-547-13324-9.

[76] V. I. shCherbak; M. A. Makukov (2013). "The "Wow! signal" of the terrestrial genetic code". *Icarus* **224** (1): 228–242. Bibcode:2013Icar..224..228S. doi:10.1016/j.icarus.2013.02.017.

[77] Makukov, Maxim (4 October 2014). "Claim to have identified extraterrestrial signal in the universal genetic code thereby confirming directed panspermia.". *Maxim Makukov*. The New Reddit Journal of Science. Retrieved 2014-10-09.

[78] M. A. Makukov; V. I. shCherbak (2014). "Space ethics to test directed panspermia". *Life Sciences in Space Research* **3**: 10–17. doi:10.1016/j.lssr.2014.07.003.

[79] "'Seed of Life' From Outer Space Suggests Aliens Created Life On Earth, U.K. Scientists Say". *The Inquisitr*. February 13, 2015. Retrieved 2015-03-11.

[80] N.C. Wickramasinghe, Formaldehyde Polymers in Interstellar Space, Nature, 252, 462, 1974.

[81] Martins, Zita; Botta, Oliver; Fogel, Marilyn L.; Sephton, Mark A.; Glavin, Daniel P.; Watson, Jonathan S.; Dworkin, Jason P.; Schwartz, Alan W.; Ehrenfreund, Pascale (2008). "Extraterrestrial nucleobases in the Murchison meteorite". *Earth and Planetary Science Letters* **270**: 130–136. Bibcode:2008E&PSL.270..130M. doi:10.1016/j.epsl.2008.03.026.

[82] AFP Staff (20 August 2009). "We may all be space aliens: study". AFP. Archived from the original on June 17, 2008. Retrieved 8 November 2014.

[83] Martins, Zita; Botta, Oliver; Fogel, Marilyn L.; Sephton, Mark A.; Glavin, Daniel P.; Watson, Jonathan S.; Dworkin, Jason P.; Schwartz, Alan W.; Ehrenfreund, Pascale (2008). "Extraterrestrial nucleobases in the Murchison meteorite". *Earth and Planetary Science Letters* **270**: 130–136. Bibcode:2008E&PSL.270..130M. doi:10.1016/j.epsl.2008.03.026.

[84] "'Life chemical' detected in comet". *NASA* (BBC News). 18 August 2009. Retrieved 6 March 2010.

[85] Callahan, M. P.; Smith, K. E.; Cleaves, H. J.; Ruzicka, J.; et al. (2011). "Carbonaceous meteorites contain a wide range of extraterrestrial nucleobases". *Proceedings of the National Academy of Sciences* **108** (34): 13995–8. Bibcode:2011PNAS..10813 995C.doi:10.1073/pnas.1106493108. PMC3161613. PMID21836052.

[86] Steigerwald, John (8 August 2011). "NASA Researchers: DNA Building Blocks Can Be Made in Space". NASA. Retrieved 10 August 2011.

[87] ScienceDaily Staff (9 August 2011). "DNA Building Blocks Can Be Made in Space, NASA Evidence Suggests". ScienceDaily. Retrieved 9 August 2011.

[88] Chow, Denise (26 October 2011). "Discovery: Cosmic Dust Contains Organic Matter from Stars". Space.com. Retrieved 26 October 2011.

[89] ScienceDaily Staff (26 October 2011). "Astronomers Discover Complex Organic Matter Exists Throughout the Universe". ScienceDaily. Retrieved 27 October 2011.

[90] Kwok, Sun; Zhang, Yong (2011). "Mixed aromatic–aliphatic organic nanoparticles as carriers of unidentified infrared emission features". *Nature* **479** (7371): 80–3. Bibcode:2011Natur.479...80K. doi:10.1038/nature10542. PMID 22031328.

[91] Than, Ker (August 29, 2012). "Sugar Found In Space". *National Geographic*. Retrieved August 31, 2012.

[92] Staff (August 29, 2012). "Sweet! Astronomers spot sugar molecule near star". AP News. Retrieved August 31, 2012.

[93] Jørgensen, Jes K.; Favre, Cécile; Bisschop, Suzanne E.; Bourke, Tyler L.; et al. (2012). "Detection of the Simplest Sugar, Glycolaldehyde, in a Solar-Type Protostar with Alma". *The Astrophysical Journal* **757**: L4. Bibcode:2012ApJ...757L...4J. doi:10.1088/2041-8205/757/1/L4.

[94] Staff (September 20, 2012). "NASA Cooks Up Icy Organics to Mimic Life's Origins". Space.com. Retrieved September 22, 2012.

[95] Gudipati, Murthy S.; Yang, Rui (2012). "In-Situ Probing of Radiation-Induced Processing of Organics in Astrophysical Ice Analogs—Novel Laser Desorption Laser Ionization Time-Of-Flight Mass Spectroscopic Studies". *The Astrophysical Journal* **756**: L24. Bibcode:2012ApJ...756L..24G. doi:10.1088/2041-8205/756/1/L24.

[96] Loomis, Ryan A.; Zaleski, Daniel P.; Steber, Amanda L.; Neill, Justin L.; et al. (2013). "The Detection of Interstellar Ethanimine (Ch3Chnh) from Observations Taken During the Gbt Primos Survey". *The Astrophysical Journal* **765**: L9. Bibcode:2013ApJ...765L...9L. doi:10.1088/2041-8205/765/1/L9.

[97] Finley, Dave.*Discoveries Suggest Icy Cosmic Start for Amino Acids and DNA Ingredients*, The National Radio Astronomy Observatory, Feb. 28, 2013

[98] Kaiser, R. I.; Stockton, A. M.; Kim, Y. S.; Jensen, E. C.; et al. (March 5, 2013). "On the Formation of Dipeptides in Interstellar Model Ices". *The Astrophysical Journal* **765** (2): 111. Bibcode:2013ApJ...765..111K. doi:10.1088/0004-637X/765/2/111. Lay summary – *Phys.org*.

[99] Hoover, Rachel (February 21, 2014). "Need to Track Organic Nano-Particles Across the Universe? NASA's Got an App for That". *NASA*. Retrieved 22 February 2014.

[100] Marlaire, Ruth (3 March 2015). "NASA Ames Reproduces the Building Blocks of Life in Laboratory". *NASA*. Retrieved 5 March 2015.

[101] Loeb, Abraham (October 2014). "The Habitable Epoch of the Early Universe". *International Journal of Astrobiology* (Cambridge) **13** (4): 337–39. Bibcode:2014IJAsB..13..337L. doi:10.1017/S1473550414000196. Retrieved 15 December 2014.

[102] Loeb, Abraham(2 December 2013). "The Habitable Epoch of the Early Universe"(PDF).*ArXiv***13**(4): 337–339. arXiv:1312.0 Bibcode:2014IJAsB..13..337L. doi:10.1017/S1473550414000196. Retrieved 15 December 2014.

[103] Dreifus, Claudia (2 December 2014). "Much-Discussed Views That Go Way Back – Avi Loeb Ponders the Early Universe, Nature and Life". *The New York Times*. Retrieved 3 December 2014.

[104] Rampelotto, P.H. (2010). "Panspermia: A Promising Field Of Research" (PDF). *Astrobiology Science Conference*. Harvard: USRA. Retrieved 3 December 2014. External link in |work= (help)

[105] Graham, Robert W (February 1990). "Extraterrestrial Life in the Universe" (PDF). *Technical Memorandum* (Lewis Research Center, OH: NASA). 102363. Retrieved 7 July 2014.

[106] Altermann, Wladyslaw (2008). "From Fossils to Astrobiology – A Roadmap to Fata Morgana?". In Seckbach, Joseph; Walsh, Maud. *From Fossils to Astrobiology: Records of Life on Earth and the Search for Extraterrestrial Biosignatures* **12**. p. xvii. ISBN 1-4020-8836-1.

[107] Webb, Stephen (2002). *If the universe is teeming with aliens, where is everybody? Fifty solutions to the Fermi paradox and the problem of extraterrestrial life*. Copernicus, Springer.

[108] Steffen, Jason H.; Batalha, Natalie M.; Borucki, William J; Buchhave, Lars A.; et al. (9 November 2010). "Five Kepler target stars that show multiple transiting exoplanet candidates". *Astrophysical Journal* **725**: 1226–41. arXiv:1006.2763. Bibcode:2010ApJ...725.1226S. doi:10.1088/0004-637X/725/1/1226.

[109] Overbye, Dennis (November 4, 2013). "Far-Off Planets Like the Earth Dot the Galaxy". *The New York Times*. Retrieved 5 November 2013.

[110] Petigura, Eric A.; Howard, Andrew W.; Marcy, Geoffrey W (October 31, 2013). "Prevalence of Earth-size planets orbiting Sun-like stars". *Proceedings of the National Academy of Sciences of the United States of America* **110** (48): 19273–78. Bibcode:2013PNAS..11019273P. doi:10.1073/pnas.1319909110. Retrieved 5 November 2013.

[111] Khan, Amina (November 4, 2013). "Milky Way may host billions of Earth-size planets". *The Los Angeles Times*. Retrieved 5 November 2013.

[112] Crawford, I.A. (Sep 1995). "Some Thoughts on the Implications of Faster-Than-Light Interstellar Space Travel". *Quarterly Journal of the Royal Astronomical Society* **36** (3): 205. Bibcode:1995QJRAS..36..205C.

[113] Byrne, Joseph Patrick (2008). "Panspermia". *Encyclopedia of Pestilence, Pandemics, and Plagues* (entry). ABC-CLIO. pp. 454–55. ISBN 978-0-313-34102-1.

[114] Wickramasinghe, C; Wainwright, M; Narlikar, J (May 24, 2003). "SARS—a clue to its origins?". *Lancet* **361** (9371): 1832. doi:10.1016/S0140-6736(03)13440-X. PMID 12781581.

[115] Willerslev, E; Hansen, AJ; Rønn, R; Nielsen, OJ (Aug 2, 2003). "Panspermia – true or false?". *Lancet* **362** (9381): 406; author reply 407–8. doi:10.1016/S0140-6736(03)14039-1. PMID 12907025.

[116] Bhargava, PM (Aug 2, 2003). "Panspermia – true or false?". *Lancet* **362** (9381): 407; author reply 407–8. doi:10.1016/S0140-6736(03)14041-X. PMID 12907028.

[117] Ponce de Leon, S; Lazcano, A (Aug 2, 2003). "Panspermia – true or false?". *Lancet* **362** (9381): 406–7; author reply 407–8. doi:10.1016/s0140-6736(03)14040-8. PMID 12907026.

[118] "New Study Adds to Finding of Ancient Life Signs in Mars Meteorite". NASA. 2009-11-30. Retrieved 1 December 2009.

[119] Thomas-Keprta, K.; Clement, S; McKay, D; Gibson, E & Wentworth, S (2009). "Origin of Magnetite Nanocrystals in Martian Meteorite ALH84001".*Geochimica et Cosmochimica Acta***73**(73): 6631–6677. Bibcode:2009GeCoA..73.6631T.doi:10.1016/

[120] "Alien visitors". *New Scientist Space*. 11 May 2001. Retrieved 20 August 2009.

[121] D'Argenio, Bruno; Geraci, Giuseppe & del Gaudio, Rosanna (March 2001). "Microbes in rocks and meteorites: a new form of life unaffected by time, temperature, pressure". *Rendiconti Lincei* **12** (1): 51–68. doi:10.1007/BF02904521. Retrieved 13 October 2009.

[122] Geraci, Giuseppe; del Gaudio, Rosanna; D'Argenio, Bruno (2001). "Microbes in rocks and meteorites: a new form of life unaffected by time, temperature, pressure" (PDF). *Rend. Fis. Acc. Lincei* **12**: 51–68.

[123] "Scientists Say They Have Found Extraterrestrial Life in the Stratosphere But Peers Are Skeptical: Scientific American". Sciam. 2001-07-31. Retrieved 20 August 2009.

[124] Narlikar, JV; Lloyd, D; Wickramasinghe, NC; Turner; Al-Mufti; Wallis; Wainwright; Rajaratnam; Shivaji; Reddy; Ramadurai; Hoyle (2003). "Balloon experiment to detect micro-organisms in the outer space". *Astrophys Space Science* **285** (2): 555–62. Bibcode:2003Ap&SS.285..555N. doi:10.1023/A:1025442021619. Missing llast4= in Authors list (help)

[125] Wainwright, M; Wickramasinghe, N.C; Narlikar, J.V; Rajaratnam, P. "Microorganisms cultured from stratospheric air samples obtained at 41 km". Retrieved 11 May 2007.

[126] Wainwright, M (2003). "A microbiologist looks at panspermia".*Astrophys Space Science***285**(2): 563–70. Bibcode:2003Ap&S doi:10.1023/A:1025494005689.

[127] Stenger, Richard (2000-11-24). "Space: Scientists discover possible microbe from space". *CNN*. Retrieved 20 August 2009.

[128] Vaidya, Pushkar Ganesh (July 2009). "Critique on Vindication of Panspermia" (PDF). *Apeiron* **16** (3). Retrieved 28 November 2009.

[129] *Mumbai scientist challenges theory that bacteria came from space*. IN: AOL.

[130] *Janibacter hoylei sp. nov., Bacillus isronensis sp. nov.* and *Bacillus aryabhattai sp. nov.*, isolated from cryotubes used for collecting air from upper atmosphere. *International Journal of Systematic and Evolutionary Microbiology* 2009. http://ijs.sgmjournals.org/cgi/content/abstract/ijs.0.002527-0v1

[131] Discovery of New Microorganisms in the Stratosphere.

[132] Timothy Oleson (May 5, 2013). "Lofted by hurricanes, bacteria live the high life". *NASA* (Earth Magazine). Retrieved 21 September 2013.

[133] Helen Shen (28 January 2013). "High-flying bacteria spark interest in possible climate effects".*Nature News*. doi:10.1038/natu Retrieved 21 September 2013.

[134] Wickramasinghe, N. C.; Wallis, J.; Wallis, D. H.; Samaranayake, Anil (January 10, 2013). "Fossil Diatoms in a New Carbonaceous Meteorite" (PDF). *Journal of Cosmology* (Journal of Cosmology) **21** (37): 1–14. arXiv:1303.2398. Bibcode:2013JCos...21.9560W.Retrieved January14, 2013.

[135] Phil Plait (15 January 2013). "No, Diatoms Have Not Been Found in a Meteorite". *Slate.com – Astronomy*. Retrieved 16 January 2013.

[136] Wallis, Jamie; Miyake, Nori; Hoover, Richard B.; Oldroyd, Andrew; et al. (5 March 2013). "The Polonnaruwa meteorite: oxygen isotope, crystalline and biological composition" (PDF). *Journal of Cosmology* **22** (2): 1845. arXiv:1303.1845. Bibcode:2013JCos...2210004W. Retrieved 7 March 2013.

[137] Wickramasinghe, N.C.; J. Wallis; N. Miyake; Anthony Oldroyd; et al. (4 February 2013). "Authenticity of the life-bearing Polonnaruwa meteorite" (PDF). *Journal of Cosmology*. Retrieved 4 February 2013.

[138] Griffin, Dale Warren (14 August 2013). "The Quest for Extraterrestrial Life: What About the Viruses?". *Astrobiology* **13** (8): 774–783. Bibcode:2013AsBio..13..774G. doi:10.1089/ast.2012.0959.

[139] Edward Anders, Eugene R. DuFresne,Ryoichi Hayatsu, Albert Cavaille, Ann DuFresne, and Frank W. Fitch. "Contaminated Meteorite", *Science, New Series*. Volume 146, Issue 3648 (Nov.27, 1964), 1157–1161.

[140] Chamberlin, Sean (1999). "Black Smokers and Giant Worms". *Fullerton College*. Retrieved 11 February 2011.

[141] Choi, Charles Q. (17 March 2013). "Microbes Thrive in Deepest Spot on Earth". LiveScience. Retrieved 17 March 2013.

[142] Oskin, Becky (14 March 2013). "Intraterrestrials: Life Thrives in Ocean Floor". LiveScience. Retrieved 17 March 2013.

[143] Glud, Ronnie; Wenzhöfer, Frank; Middelboe, Mathias; Oguri, Kazumasa; et al. (17 March 2013). "High rates of microbial carbon turnover in sediments in the deepest oceanic trench on Earth". *Nature Geoscience* **6** (4): 284–288. Bibcode:2013NatGe...6..284G.doi:10.1038/ngeo1773. Retrieved17March2013.

[144] Carey, Bjorn (7 February 2005). "Wild Things: The Most Extreme Creatures". *Live Science*. Retrieved 20 October 2008.

[145] Cavicchioli, R. (Fall 2002). "Extremophiles and the search for extraterrestrial life". *Astrobiology* **2** (3): 281–92. Bibcode:2002As doi:10.1089/153110702762027862. PMID 12530238. Bio...2..281C.

[146] The BIOPAN experiment MARSTOX II of the FOTON M-3 mission July 2008.

[147] Surviving the Final Frontier. 25 November 2002.

[148] Christner, Brent C. (2002). "Detection, recovery, isolation, and characterization of bacteria in glacial ice and Lake Vostok accretion ice". *Ohio State University*. Retrieved 4 February 2011.

[149] Nanjundiah, V. (2000). "The smallest form of life yet?" (PDF). *Journal of Biosciences* **25** (1): 9–10. doi:10.1007/BF02985175. PMID 10824192.

[150] Rabbow, Elke Rabbow; Gerda Horneck; Petra Rettberg; Jobst-Ulrich Schott; et al. (9 July 2009). "EXPOSE, an Astrobiological Exposure Facility on the International Space Station – from Proposal to Flight" (PDF). *Orig Life Evol Biosph* **39** (6): 581–98. doi:10.1007/s11084-009-9173-6. PMID 19629743. Retrieved 8 July 2013.

[151] Bacterium revived from 25 million year sleep Digital Center for Microbial Ecology

[152] Tepfer, David Tepfer (December 2008). "The origin of life, panspermia and a proposal to seed the Universe". *Plant Science* **175** (6): 756–760. doi:10.1016/j.plantsci.2008.08.007.

[153] "Exobiology and Radiation Assembly (ERA)". *ESA*. NASA. 1992. Retrieved 22 July 2013.

[154] Zhang, K. Dose; A. Bieger-Dose; R. Dillmann; M. Gill; et al. (1995). "ERA-experiment "space biochemistry"". *Advances in Space Research* **16** (8): 119–129. doi:10.1016/0273-1177(95)00280-R. PMID 11542696.

[155] Vaisberg, Horneck G; Eschweiler U; Reitz G; Wehner J; et al. (1995). "Biological responses to space: results of the experiment "Exobiological Unit" of ERA on EURECA I". *Adv Space Res.* **16** (8): 105–18. Bibcode:1995AdSpR..16..105V. doi:10.1016/0273-1177(95)00279-N. PMID 11542695.

[156] "BIOPAN Pan for exposure to space environment". *Kayser Italia*. 2013. Retrieved 17 July 2013.

[157] De La Torre Noetzel, Rosa (2008). "Experiment lithopanspermia: Test of interplanetary transfer and re-entry process of epi- and endolithic microbial communities in the FOTON-M3 Mission". *37th COSPAR Scientific Assembly. Held 13–20 July 2008* **37**: 660. Bibcode:2008cosp...37..660D.

[158] "Life in Space for Life ion Earth – Biosatelite Foton M3". June 26, 2008. Retrieved 13 October 2009.

[159] Jönsson, K. Ingemar Jönsson; Elke Rabbow; Ralph O. Schill; Mats Harms-Ringdahl; et al. (9 September 2008). "Tardigrades survive exposure to space in low Earth orbit". *Current Biology* **18** (17): R729–R731. doi:10.1016/j.cub.2008.06.048. PMID 18786368.

[160] de Vera; J.P.P.; et al. (2010). "COSPAR 2010 Conference". Research Gate. Retrieved 17 July 2013

[161] Paul Clancy (Jun 23, 2005). *Looking for Life, Searching the Solar System*. Cambridge University Press. Retrieved 26 March 2014.

[162] Tepfer, David Tepfer; Andreja Zalar & Sydney Leach. (May 2012). "Survival of Plant Seeds, Their UV Screens, and nptII DNA for 18 Months Outside the International Space Station". *Astrobiology* **12** (5): 517–528. Bibcode:2012AsBio..12..517T. doi:10.1089/ast.2011.0744. PMID 22680697.

[163] Scalzi, Giuliano Scalzi; Laura Selbmann; Laura Zucconi; Elke Rabbow; et al. (1 June 2012). "LIFE Experiment: Isolation of Cryptoendolithic Organisms from Antarctic Colonized Sandstone Exposed to Space and Simulated Mars Conditions on the International Space Station". *Origins of Life and Evolution of Biospheres* **42** (2 – 3): 253–262. doi:10.1007/s11084-012-9282-5.

[164] Onofri, Silvano Onofri; Rosa de la Torre; Jean-Pierre de Vera; Sieglinde Ott; et al. (May 2012). "Survival of Rock-Colonizing Organisms After 1.5 Years in Outer Space". *Astrobiology* **12** (5): 508–516. Bibcode:2012AsBio..12..508O. doi:10.1089/ast.2011.0736.PMID22680696.

[165] Baldwin, Emily (26 April 2012). "Lichen survives harsh Mars environment". Skymania News. Retrieved 27 April 2012.

[166] de Vera, J.-P.; Kohler, Ulrich (26 April 2012). "The adaptation potential of extremophiles to Martian surface conditions and its implication for the habitability of Mars" (PDF). European Geosciences Union. Retrieved 27 April 2012.

[167] Amos, Jonathan (23 August 2010). "Beer microbes live 553 days outside ISS". *BBC News – Science and Technology*. Retrieved 31 July 2013.

[168] "No! bugs please, this is a clean planet!". European Space Agency (ESA). 30 July 2002. Retrieved 16 July 2013.

[169] Jordans, Frank (30 July 2015). "Philae probe finds evidence that comets can be cosmic labs". *The Washington Post*. Associated Press. Retrieved 30 July 2015.

[170] "Science on the Surface of a Comet". European Space Agency. 30 July 2015. Retrieved 30 July 2015.

[171] Bibring, J.-P.; Taylor, M.G.G.T.; Alexander, C.; Auster, U.; Biele, J.; Finzi, A. Ercoli; Goesmann, F.; Klingehoefer, G.; Kofman, W.; Mottola, S.; Seidenstiker, K.J.; Spohn, T.; Wright, I. (31 July 2015). "Philae's First Days on the Comet – Introduction to Special Issue". *Science* **349** (6247): 493. doi:10.1126/science.aac5116. Retrieved 30 July 2015.

[172] "LIFE Experiment". Planetary.org. Retrieved 20 August 2009.

[173] "Living interplanetary flight experiment: an experiment on survivability of microorganisms during interplanetary transfer" (PDF). Retrieved 20 August 2009.

[174] "Projects: LIFE Experiment: Phobos". The Planetary Society. Retrieved 2 April 2011.

[175] Zak, Anatoly (1 September 2008). "Mission Possible". *Air & Space Magazine*. Smithsonian Institution. Retrieved 26 May 2009.

[176] Nussinov, M. D; Lysenko, S. V (1983). "Cosmic vacuum prevents radiopanspermia", *Orig. Life* **13**: 153–64.

[177] Wickramasinghe, M.K.; Wickramasinghe, C. (2004). "Interstellar transfer of planetary microbiota.". *Mon. Not.R. Astr. Soc.* **348**: 52–57. Bibcode:2004MNRAS.348...52W. doi:10.1111/j.1365-2966.2004.07355.x.

8.9   Further reading

- Crick, F (1981). *Life, Its Origin and Nature*, Simon & Schuster, ISBN 0-7088-2235-5.

- Hoyle, F (1983). *The Intelligent Universe*, London: Michael Joseph, ISBN 0-7181-2298-4.

8.10   External links

- A.E. Zlobin, 2013, Tunguska similar impacts and origin of life (mathematical theory of origin of life; incoming of pattern recognition algorithm due to comets)

- Francis Crick's notes for a lecture on directed panspermia, dated 5 November 1976.

- "Earth sows its seeds in space". *Nature News*. 23 February 2004. doi:10.1038/news040216-20 (inactive 2015-02-01).

- Warmflash, D.; Weiss, B. (24 October 2005). "Did Life Come from Another World?". *Scientific American* **293**: 64–71. doi:10.1038/scientificamerican1105-64.

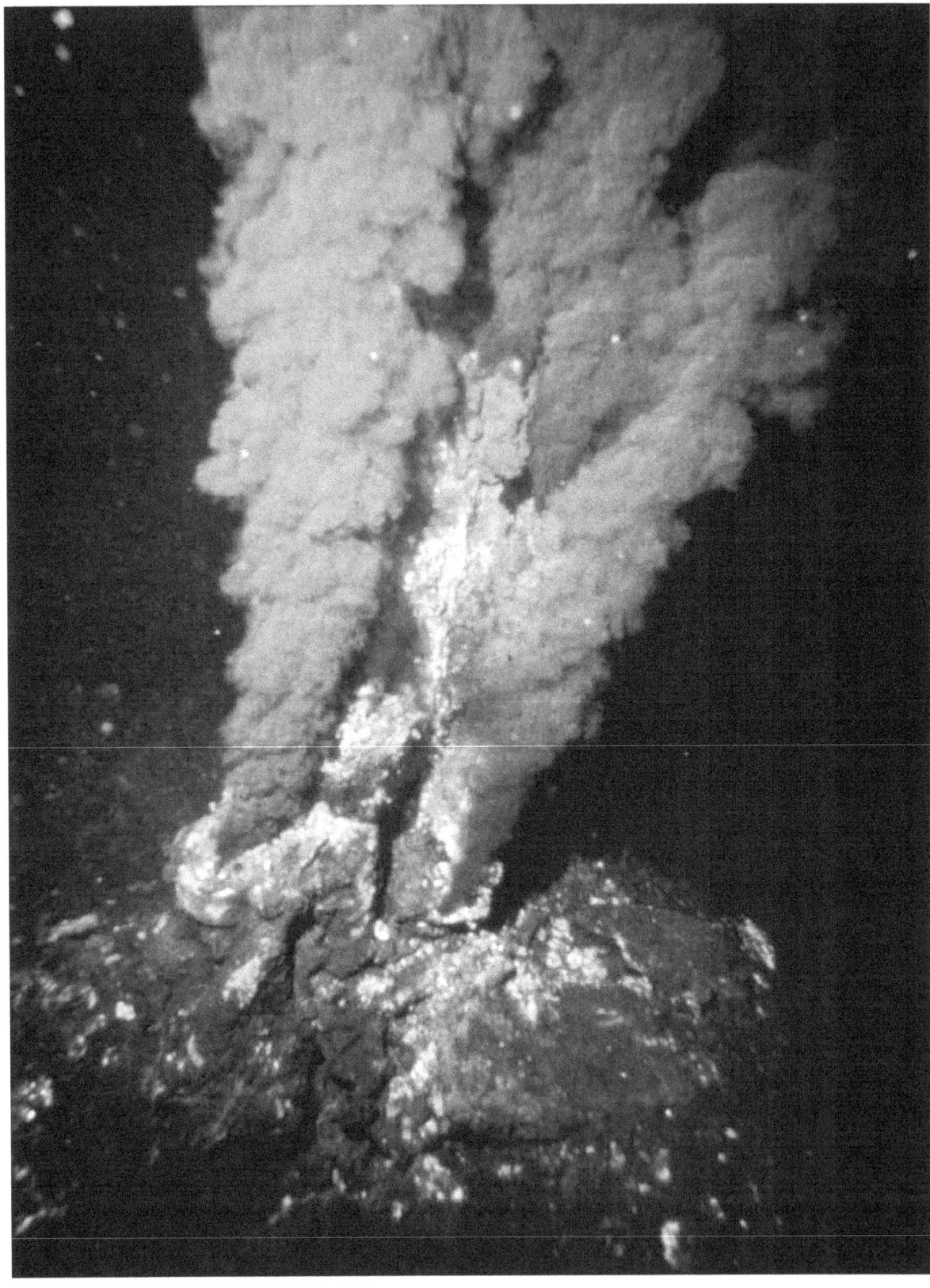

Hydrothermal vents are able to support extremophile bacteria on Earth and may also support life in other parts of the cosmos.

EURECA facility deployment in 1992

EXOSTACK on the Long Duration Exposure Facility satellite.

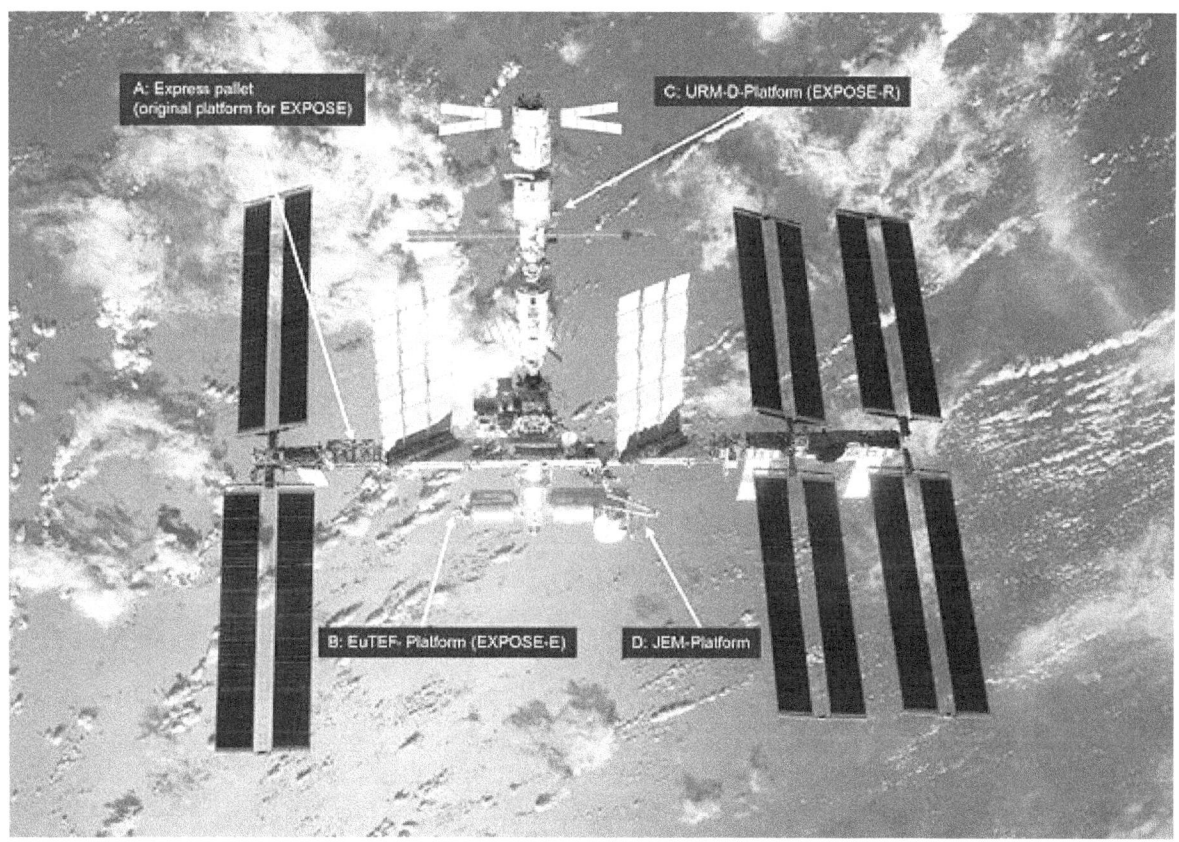

Location of the astrobiology EXPOSE-E and EXPOSE-R facilities on the International Space Station

Chapter 9

Abiogenesis

"Origin of life" redirects here. For non-scientific views on the origins of life, see Creation myth.

Abiogenesis (Brit.: /ˌeɪbaɪoˈdʒɛnɪsɪs/ *AY-by-oh-JEN-ə-siss*[1] U.S. English pronunciation: /ˌeɪˌbaɪoʊˈdʒɛnɪsɪs/),[2] or **biopoiesis**.[3]

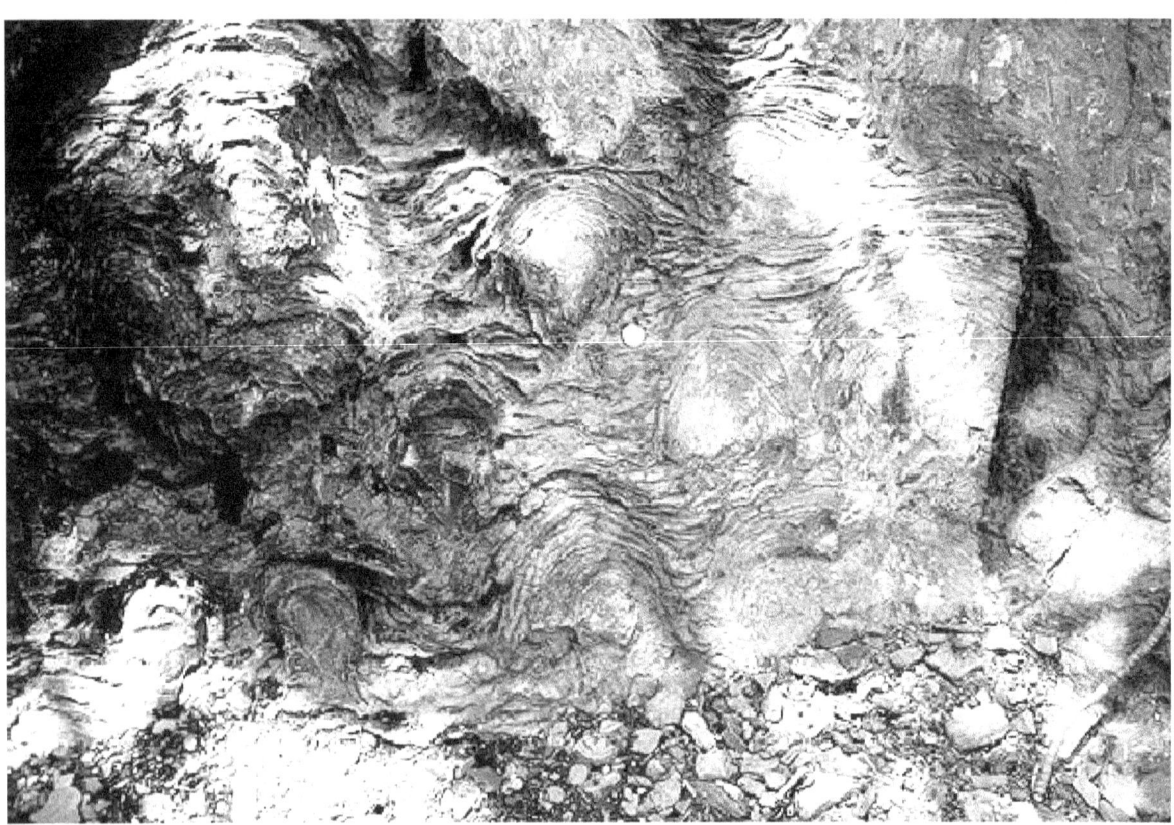

Precambrian stromatolites in the Siyeh Formation, Glacier National Park. In 2002, a paper in the scientific journal Nature *suggested that these 3.5 Ga (billion years) old geological formations contain fossilized cyanobacteria microbes. This suggests they are evidence of one of the earliest known life forms on Earth.*

is the natural process of life arising from non-living matter, such as simple organic compounds.[4][5][6][7] It is thought to have occurred on Earth between 3.8 and 4.1[8] billion years ago, and is studied through a combination of laboratory experiments and extrapolation from the genetic information of modern organisms in order to make reasonable conjectures about what pre-life chemical reactions may have given rise to a living system.[9]

The study of abiogenesis involves three main types of considerations: the geophysical, the chemical, and the biological.[10]

with more recent approaches attempting a synthesis of all three. Many approaches investigate how self-replicating molecules, or their components, came into existence. It is generally accepted that current life on Earth descended from an RNA world,[11] although RNA-based life may not have been the first life to have existed.[12][13] The Miller–Urey experiment and similar experiments demonstrated that most amino acids, basic chemicals of life, can be synthesized from inorganic compounds in conditions intended to be similar to early Earth. Several mechanisms of organic molecule synthesis have been investigated, including lightning and radiation. Other approaches ("metabolism first" hypotheses) focus on understanding how catalysis in chemical systems on the early Earth might have provided the precursor molecules necessary for self-replication.[14] Complex organic molecules have been found in the Solar System and in interstellar space, and these molecules may have provided starting material for the development of life on Earth.[15][16][17][18]

The panspermia hypothesis suggests that microscopic life was distributed by meteoroids, asteroids and other small Solar System bodies and that life may exist throughout the Universe.[19] It is speculated that the biochemistry of life may have begun shortly after the Big Bang, 13.8 billion years ago, during a habitable epoch when the age of the universe was only 10–17 million years.[20][21] Panspermia hypothesis answers the question of whence life, not how life came to be; it only postulates the origin of life to a locale outside the Earth.

Nonetheless, Earth is the only place in the Universe known to harbor life.[22][23] The age of the Earth is about 4.54 billion years.[24][25][26] The earliest undisputed evidence of life on Earth dates at least from 3.5 billion years ago,[27][28][29] during the Eoarchean Era after a geological crust started to solidify following the earlier molten Hadean Eon. There are microbial mat fossils found in 3.48 billion-year-old sandstone discovered in Western Australia.[30][31][32] Other early physical evidence of a biogenic substance is graphite in 3.7 billion-year-old metasedimentary rocks discovered in southwestern Greenland[33] as well as "remains of biotic life" found in 4.1 billion-year-old rocks in Western Australia.[34][35] According to one of the researchers, "If life arose relatively quickly on Earth ... then it could be common in the universe."[34]

9.1   Early geophysical conditions

Main article: Timeline of the evolutionary history of life

Based on recent computer model studies, the complex organic molecules necessary for life may have formed in the protoplanetary disk of dust grains surrounding the Sun before the formation of the Earth.[36] According to the computer studies, this same process may also occur around other stars that acquire planets.[36] (Also see Extraterrestrial organic molecules).

The Hadean Earth is thought to have had a secondary atmosphere, formed through degassing of the rocks that accumulated from planetesimal impactors. At first, it was thought that the Earth's atmosphere consisted of hydrides—methane, ammonia and water vapour—and that life began under such reducing conditions, which are conducive to the formation of organic molecules. During its formation, the Earth lost a significant part of its initial mass, with a nucleus of the heavier rocky elements of the protoplanetary disk remaining.[37] According to later models, suggested by study of ancient minerals, the atmosphere in the late Hadean period consisted largely of nitrogen and carbon dioxide, with smaller amounts of carbon monoxide, hydrogen, and sulfur compounds.[38] As Earth lacked the gravity to hold any molecular hydrogen, this component of the atmosphere would have been rapidly lost during the Hadean period, along with the bulk of the original inert gases. The solution of carbon dioxide in water is thought to have made the seas slightly acidic, giving it a pH of about 5.5.[39] The atmosphere at the time has been characterized as a "gigantic, productive outdoor chemical laboratory."[40] It may have been similar to the mixture of gases released today by volcanoes, which still support some abiotic chemistry.[40]

Oceans may have appeared first in the Hadean Eon, as soon as two hundred million years (200 Ma) after the Earth was formed, in a hot 100 °C (212 °F) reducing environment, and the pH of about 5.8 rose rapidly towards neutral.[41] This has been supported by the dating of 4.404 Ga-old zircon crystals from metamorphosed quartzite of Mount Narryer in Western Australia, which are evidence that oceans and continental crust existed within 150 Ma of Earth's formation.[42] Despite the likely increased vulcanism and existence of many smaller tectonic "platelets," it has been suggested that between 4.4 and 4.3 Ga (billion year), the Earth was a water world, with little if any continental crust, an extremely turbulent atmosphere and a hydrosphere subject to intense ultraviolet (UV) light, from a T Tauri stage Sun, cosmic radiation and continued bolide impacts.[43]

The Hadean environment would have been highly hazardous to modern life. Frequent collisions with large objects, up to

500 kilometres (310 mi) in diameter, would have been sufficient to sterilise the planet and vaporise the ocean within a few months of impact, with hot steam mixed with rock vapour becoming high altitude clouds that would completely cover the planet. After a few months, the height of these clouds would have begun to decrease but the cloud base would still have been elevated for about the next thousand years. After that, it would have begun to rain at low altitude. For another two thousand years, rains would slowly have drawn down the height of the clouds, returning the oceans to their original depth only 3,000 years after the impact event.[44]

9.1.1 The earliest biological evidence for life on Earth

The earliest life on Earth existed before 3.5 billion years ago,[27][28][29] during the Eoarchean Era when sufficient crust had solidified following the molten Hadean Eon. Physical evidence has been found in biogenic graphite in 3.7 billion-year-old metasedimentary rocks from southwestern Greenland[33] and microbial mat fossils found in 3.48 billion-year-old sandstone from Western Australia.[30][32] Evidence of early life in rocks from Akilia Island, near the Isua supracrustal belt in southwestern Greenland, dating to 3.7 billion years ago have shown biogenic carbon isotopes.[45] At Strelley Pool, in the Pilbarra region of Western Australia, compelling evidence of early life has been found in pyrite-bearing sandstone in a fossilized beach, that showed rounded tubular cells that oxidised sulfur by photosynthesis in the absence of oxygen.[46] More recently, geochemists have found evidence that life likely existed on Earth at least 4.1 billion years ago — 300 million years earlier than previous research suggested.[34][35][47]

In the earlier period between 3.8 and 4.1 Ga, changes in the orbits of the giant planets may have caused a heavy bombardment by asteroids and comets[48] that pockmarked the Moon and the other inner planets (Mercury, Mars, and presumably Earth and Venus). This would likely have repeatedly sterilized the planet, had life appeared before that time.[40] Geologically, the Hadean Earth would have been far more active than at any other time in its history. Studies of meteorites suggests that radioactive isotopes such as aluminium-26 with a half-life of 7.17×10^5 years, and potassium-40 with a half-life of 1.250×10^9 years, isotopes mainly produced in supernovae, were much more common.[49] Coupled with internal heating as a result of gravitational sorting between the core and the mantle, there would have been a great deal of mantle convection, with the probable result of many more smaller and much more active tectonic plates than now exist.

The time periods between such devastating environmental events give time windows for the possible origin of life in the early environments. A study by Kevin A. Maher and David J. Stevenson shows that if the deep marine hydrothermal setting provides a suitable site for the origin of life, then abiogenesis could have happened as early as 4.0 to 4.2 Ga, whereas if it occurred at the surface of the Earth, abiogenesis could only have occurred between 3.7 and 4.0 Ga.[50]

9.2 Conceptual history

9.2.1 Spontaneous generation

Main article: Spontaneous generation

Belief in spontaneous generation of certain forms of life from non-living matter goes back to Aristotle and ancient Greek philosophy and continued to have support in Western scholarship until the 19th century.[51] This belief was paired with a belief in heterogenesis, i.e., that one form of life derived from a different form (e.g., bees from flowers).[52] Classical notions of spontaneous generation held that certain complex, living organisms are generated by decaying organic substances. According to Aristotle, it was a readily observable truth that aphids arise from the dew that falls on plants, flies from putrid matter, mice from dirty hay, crocodiles from rotting logs at the bottom of bodies of water, and so on.[53] In the 17th century, people began to question such assumptions. In 1646, Sir Thomas Browne published his *Pseudodoxia Epidemica* (subtitled *Enquiries into Very many Received Tenets, and commonly Presumed Truths*), which was an attack on false beliefs and "vulgar errors." His contemporary, Alexander Ross, erroneously refuted him, stating: "To question this [Ed.: i.e., spontaneous generation], is to question Reason, Sense, and Experience: If he doubts of this, let him go to *Ægypt*, and there he will finde the fields swarming with mice begot of the mud of *Nylus*, to the great calamity of the Inhabitants."[54][55]

In 1665, Robert Hooke published the first drawings of a microorganism. Hooke was followed in 1676 by Antonie van

Leeuwenhoek, who drew and described microorganisms that are now thought to have been protozoa and bacteria.[56] Many felt the existence of microorganisms was evidence in support of spontaneous generation, since microorganisms seemed too simplistic for sexual reproduction, and asexual reproduction through cell division had not yet been observed. Van Leeuwenhoek took issue with the ideas common at the time that fleas and lice could spontaneously result from putrefaction, and that frogs could likewise arise from slime. Using a broad range of experiments ranging from sealed and open meat incubation and the close study of insect reproduction he became, by the 1680s, convinced that spontaneous generation was incorrect.[57]

The first experimental evidence against spontaneous generation came in 1668 when Francesco Redi showed that no maggots appeared in meat when flies were prevented from laying eggs. It was gradually shown that, at least in the case of all the higher and readily visible organisms, the previous sentiment regarding spontaneous generation was false. The alternative seemed to be biogenesis: that every living thing came from a pre-existing living thing (*omne vivum ex ovo*, Latin for "every living thing from an egg").

In 1768, Lazzaro Spallanzani demonstrated that microbes were present in the air, and could be killed by boiling. In 1861, Louis Pasteur performed a series of experiments that demonstrated that organisms such as bacteria and fungi do not spontaneously appear in sterile, nutrient-rich media, but could only appear by invasion from without.

The belief that self-ordering by spontaneous generation was impossible begged for an alternative. By the middle of the 19th century, the theory of biogenesis had accumulated so much evidential support, due to the work of Pasteur and others, that the alternative theory of spontaneous generation had been effectively disproven. John Desmond Bernal, a pioneer in X-ray crystallography, suggested that earlier theories such as spontaneous generation were based upon an explanation that life was continuously created as a result of chance events.[58]

9.2.2   The origin of the terms *biogenesis* and *abiogenesis*

Main article: Biogenesis

The term biogenesis is usually credited to either Henry Charlton Bastian or to Thomas Henry Huxley.[59] Bastian used the term (around 1869) in an unpublished exchange with John Tyndall to mean *life-origination or commencement*. In 1870, Huxley, as new president of the British Association for the Advancement of Science, delivered an address entitled *Biogenesis and Abiogenesis*.[60] In it he introduced the term *biogenesis* (with an opposite meaning to Bastian) and also introduced the term *abiogenesis*:

> And thus the hypothesis that living matter always arises by the agency of pre-existing living matter, took definite shape; and had, henceforward, a right to be considered and a claim to be refuted, in each particular case, before the production of living matter in any other way could be admitted by careful reasoners. It will be necessary for me to refer to this hypothesis so frequently, that, to save circumlocution, I shall call it the hypothesis of *Biogenesis*; and I shall term the contrary doctrine–that living matter may be produced by not living matter–the hypothesis of *Abiogenesis*.[60]

Subsequently, in the preface to Bastian's 1871 book, *The Modes of Origin of Lowest Organisms*,[61] the author refers to the possible confusion with Huxley's usage and he explicitly renounced his own meaning:

> A word of explanation seems necessary with regard to the introduction of the new term *Archebiosis*. I had originally, in unpublished writings, adopted the word *Biogenesis* to express the same meaning—viz., life-origination or commencement. But in the mean time the word *Biogenesis* has been made use of, quite independently, by a distinguished biologist [Huxley], who wished to make it bear a totally different meaning. He also introduced the word *Abiogenesis*. I have been informed, however, on the best authority, that neither of these words can—with any regard to the language from which they are derived—be supposed to bear the meanings which have of late been publicly assigned to them. Wishing to avoid all needless confusion, I therefore renounced the use of the word *Biogenesis*, and being, for the reason just given, unable to adopt the other term, I was compelled to introduce a new word, in order to designate the process by which living matter is supposed to come into being, independently of pre-existing living matter.[62]

9.2.3 Pasteur and Darwin

Pasteur remarked, about a finding of his in 1864 which he considered definitive, "Never will the doctrine of spontaneous generation recover from the mortal blow struck by this simple experiment."[63][64] One alternative was that life's origins on Earth had come from somewhere else in the Universe. Periodically resurrected (see Panspermia, above) Bernal said that this approach "is equivalent in the last resort to asserting the operation of metaphysical, spiritual entities... it turns on the argument of creation by design by a creator or demiurge."[65] Such a theory, Bernal said was unscientific and a number of scientists defined life as a result of an inner *life force*, which in the late 19th century was championed by Henri Bergson.

The concept of evolution proposed by Charles Darwin put an end to these metaphysical theologies. In a letter to Joseph Dalton Hooker on 1 February 1871,[66] Darwin discussed the suggestion that the original spark of life may have begun in a "warm little pond, with all sorts of ammonia and phosphoric salts, light, heat, electricity, &c., present, that a proteine compound was chemically formed ready to undergo still more complex changes." He went on to explain that "at the present day such matter would be instantly devoured or absorbed, which would not have been the case before living creatures were formed." He had written to Hooker in 1863 stating that "It is mere rubbish, thinking at present of the origin of life; one might as well think of the origin of matter.". In *On the Origin of Species* he had referred to life having been "created", by which he "really meant 'appeared' by some wholly unknown process", but had soon regretted using the old-testament term "creation".[67]

9.2.4 "Primordial soup" hypothesis

Further information: Miller–Urey experiment

No new notable research or theory on the subject appeared until 1924, when Alexander Oparin reasoned that atmospheric oxygen prevents the synthesis of certain organic compounds that are necessary building blocks for the evolution of life. In his book *The Origin of Life*,[68][69] Oparin proposed that the "spontaneous generation of life" that had been attacked by Louis Pasteur did in fact occur once, but was now impossible because the conditions found on the early Earth had changed, and preexisting organisms would immediately consume any spontaneously generated organism. Oparin argued that a "primeval soup" of organic molecules could be created in an oxygenless atmosphere through the action of sunlight. These would combine in ever more complex ways until they formed coacervate droplets. These droplets would "grow" by fusion with other droplets, and "reproduce" through fission into daughter droplets, and so have a primitive metabolism in which factors that promote "cell integrity" survive, and those that do not become extinct. Many modern theories of the origin of life still take Oparin's ideas as a starting point.

Robert Shapiro has summarized the "primordial soup" theory of Oparin and J. B. S. Haldane in its "mature form" as follows:[70]

1. The early Earth had a chemically reducing atmosphere.

2. This atmosphere, exposed to energy in various forms, produced simple organic compounds ("monomers").

3. These compounds accumulated in a "soup" that may have concentrated at various locations (shorelines, oceanic vents etc.).

4. By further transformation, more complex organic polymers—and ultimately life—developed in the soup.

About this time, Haldane suggested that the Earth's prebiotic oceans—different from their modern counterparts—would have formed a "hot dilute soup" in which organic compounds could have formed. Bernal called this idea *biopoiesis* or *biopoesis*, the process of living matter evolving from self-replicating but nonliving molecules,[58][71] and proposed that biopoiesis passes through a number of intermediate stages.

One of the most important pieces of experimental support for the "soup" theory came in 1952. Stanley L. Miller and Harold C. Urey, performed an experiment that demonstrated how organic molecules could have spontaneously formed from inorganic precursors, under conditions like those posited by the Oparin-Haldane Hypothesis. The now-famous

Charles Darwin in 1879

Miller–Urey experiment used a highly reducing mixture of gases—methane, ammonia and hydrogen—to form basic organic monomers, such as amino acids.[72] In the Miller–Urey experiment, a mixture of water, hydrogen, methane, and ammonia was cycled through an apparatus that delivered electrical sparks to the mixture. After one week, it was found that about 10% to 15% of the carbon in the system was now in the form of a racemic mixture of organic compounds, including amino acids, which are the building blocks of proteins. This provided direct experimental support for the second point of the "soup" theory, and it is around the remaining two points of the theory that much of the debate now centers.

Bernal shows that based upon this and subsequent work there is no difficulty in principle in forming most of the molecules we recognise as the basic molecules of life from their inorganic precursors. The underlying hypothesis held by Oparin, Haldane, Bernal, Miller and Urey, for instance, was that multiple conditions on the primeval Earth favored chemical reactions that synthesized the same set of complex organic compounds from such simple precursors. A 2011 reanalysis of the saved vials containing the original extracts that resulted from the Miller and Urey experiments, using current and more advanced analytical equipment and technology, has uncovered more biochemicals than originally discovered in the 1950s. One of the more important findings was 23 amino acids, far more than the five originally found.[73] However, Bernal said that "it is not enough to explain the formation of such molecules, what is necessary," he says, "is a physical-chemical explanation of the origins of these molecules that suggests the presence of suitable sources and sinks for free energy."[74]

9.2.5   Proteinoid microspheres

Main article: Proteinoid

In trying to uncover the intermediate stages of abiogenesis mentioned by Bernal, Sidney W. Fox in the 1950s and 1960s studied the spontaneous formation of peptide structures under conditions that might plausibly have existed early in Earth's history. He demonstrated that amino acids could spontaneously form small chains called peptides. In one of his experiments, he allowed amino acids to dry out as if puddled in a warm, dry spot in prebiotic conditions. He found that, as they dried, the amino acids formed long, often cross-linked, thread-like, submicroscopic polypeptide molecules now named "proteinoid microspheres."[75]

In another experiment using a similar method to set suitable conditions for life to form, Fox collected volcanic material from a cinder cone in Hawaii. He discovered that the temperature was over 100 °C (212 °F) just 4 inches (100 mm) beneath the surface of the cinder cone, and suggested that this might have been the environment in which life was created—molecules could have formed and then been washed through the loose volcanic ash and into the sea. He placed lumps of lava over amino acids derived from methane, ammonia and water, sterilized all materials, and baked the lava over the amino acids for a few hours in a glass oven. A brown, sticky substance formed over the surface and when the lava was drenched in sterilized water a thick, brown liquid leached out. It turned out that the amino acids had combined to form proteinoids, and the proteinoids had combined to form small globules that Fox called "microspheres." His proteinoids were not cells, although they formed clumps and chains reminiscent of cyanobacteria, but they contained no functional nucleic acids or any encoded information. Based upon such experiments, Colin S. Pittendrigh stated in December 1967 that "laboratories will be creating a living cell within ten years," a remark that reflected the typical contemporary levels of innocence of the complexity of cell structures.[76]

9.3   Current models

There is still no "standard model" of the origin of life. Most currently accepted models draw at least some elements from the framework laid out by Alexander Oparin (in 1924) and J. B. S. Haldane (in 1925), who postulated the molecular or chemical evolution theory of life.[77] According to them, the first molecules constituting the earliest cells "were synthesized under natural conditions by a slow process of molecular evolution, and these molecules then organized into the first molecular system with properties with biological order."[77] Oparin and Haldane suggested that the atmosphere of the early Earth may have been chemically reducing in nature, composed primarily of methane (CH_4), ammonia (NH_3), water (H_2O), hydrogen sulfide (H_2S), carbon dioxide (CO_2) or carbon monoxide (CO), and phosphate (PO_4^{3-}), with molecular oxygen (O_2) and ozone (O_3) either rare or absent. According to later models, the atmosphere in the late Hadean period consisted largely of nitrogen (N_2) and carbon dioxide, with smaller amounts of carbon monoxide, hydrogen (H_2), and sulfur compounds;[78] while it did lack molecular oxygen and ozone,[79] it wasn't as chemically reducing as Oparin and Haldane supposed. In the atmosphere proposed by Oparin and Haldane, electrical activity can catalyze the creation of certain basic small molecules (monomers) of life, such as amino acids. This was demonstrated in the Miller–Urey experiment reported in 1953.

Bernal coined the term *biopoiesis* in 1949 to refer to the origin of life.[80] In 1967, he suggested that it occurred in three "stages": 1) the origin of biological monomers; 2) the origin of biological polymers; and 3) the evolution from molecules to cells. He suggested that evolution commenced between stage 1 and 2. The first stage is now fairly well understood, and the discovery of alkaline vents and the similarity with the "proton pump" found as the basis of biological life has begun to provide evidence about the second stage. Bernal considered the third, the discovery of methods by which biological reactions were incorporated behind cell walls, to be the most difficult. Modern work on the self organising capacities by which cell membranes self-assemble, and the work on micropores in various substrates is seen as a halfway house towards the development of independent free-living cells, and research into this is an ongoing effort.[39][81][82]

The chemical processes that took place on the early Earth are called *chemical evolution*. Both Manfred Eigen and Sol Spiegelman demonstrated that evolution, including replication, variation, and natural selection, can occur in populations of molecules as well as in organisms.[40] Spiegelman took advantage of natural selection to synthesize the Spiegelman Monster, which had a genome with just 218 nucleotide bases, having deconstructively evolved from a 4500 base bacterial RNA. Eigen built on Spiegelman's work and produced a similar system further degraded to just 48 or 54 nucleotides, which was the minimum required for the binding of the replication enzyme.[83]

Chemical evolution was followed by the initiation of biological evolution, which led to the first cells.[40] No one has yet synthesized a "protocell" using basic components with the necessary properties of life (the so-called "bottom-up-approach"). Without such a proof-of-principle, explanations have tended to focus on chemosynthesis.[84] However, some researchers are working in this field, notably Steen Rasmussen and Jack W. Szostak. Others have argued that a "top-down approach" is more feasible. One such approach, successfully attempted by Craig Venter and others at The Institute for Genomic Research, involves engineering existing prokaryotic cells with progressively fewer genes, attempting to discern at which point the most minimal requirements for life were reached.[85][86]

9.4   Chemical origin of organic molecules

The elements, except for hydrogen, ultimately derive from stellar nucleosynthesis. Complex molecules, including organic molecules, form naturally both in space and on planets.[15] There are two possible sources of organic molecules on the early Earth:

1. Terrestrial origins – organic molecule synthesis driven by impact shocks or by other energy sources (such as UV light, redox coupling, or electrical discharges) (e.g., Miller's experiments)

2. Extraterrestrial origins – formation of organic molecules in interstellar dust clouds and rain down on planets.[87][88] (See pseudo-panspermia)

Estimates of the production of organics from these sources suggest that the Late Heavy Bombardment before 3.5 Ga within the early atmosphere made available quantities of organics comparable to those produced by terrestrial sources.[89][90]

It has been estimated that the Late Heavy Bombardment may also have effectively sterilised the Earth's surface to a depth of tens of metres. If life evolved deeper than this, it would have also been shielded from the early high levels of ultraviolet radiation from the T Tauri stage of the Sun's evolution. Simulations of geothermically heated oceanic crust yield far more organics than those found in the Miller-Urey experiments (see below). In the deep hydrothermal vents, Everett Shock has found "there is an enormous thermodynamic drive to form organic compounds, as seawater and hydrothermal fluids, which are far from equilibrium, mix and move towards a more stable state."[91] Shock has found that the available energy is maximised at around 100 – 150 degrees Celsius, precisely the temperatures at which the hyperthermophilic bacteria and thermoacidophilic archaea have been found, at the base of the phylogenetic tree of life closest to the Last Universal Common Ancestor (LUCA).[92]

9.4.1   Chemical synthesis

While features of self-organization and self-replication are often considered the hallmark of living systems, there are many instances of abiotic molecules exhibiting such characteristics under proper conditions. Stan Palasek showed that self-assembly of ribonucleic acid (RNA) molecules can occur spontaneously due to physical factors in hydrothermal vents.[93] Virus self-assembly within host cells has implications for the study of the origin of life,[94] as it lends further credence to the hypothesis that life could have started as self-assembling organic molecules.[95][96]

Multiple sources of energy were available for chemical reactions on the early Earth. For example, heat (such as from geothermal processes) is a standard energy source for chemistry. Other examples include sunlight and electrical discharges (lightning), among others.[40] Unfavorable reactions can also be driven by highly favorable ones, as in the case of iron-sulfur chemistry. For example, this was probably important for carbon fixation (the conversion of carbon from its inorganic form to an organic one).[note 1] Carbon fixation via iron-sulfur chemistry is highly favorable, and occurs at neutral pH and 100 °C (212 °F). Iron-sulfur surfaces, which are abundant near hydrothermal vents, are also capable of producing small amounts of amino acids and other biological metabolites.[40]

Formamide produces all four ribonucleotides and other biological molecules when warmed in the presence of various terrestrial minerals. Formamide is ubiquitous in the Universe, produced by the reaction of water and hydrogen cyanide (HCN). It has several advantages as a biotic precursor, including the ability to easily become concentrated through the evaporation of water.[97][98] Although HCN is poisonous, it only affects aerobic organisms (eukaryotes and aerobic bacteria), which did not yet exist. It can play roles in other chemical processes as well, such as the synthesis of the amino acid glycine.[40]

In 1961, it was shown that the nucleic acid purine base adenine can be formed by heating aqueous ammonium cyanide solutions.[99] Other pathways for synthesizing bases from inorganic materials were also reported.[100] Leslie E. Orgel and colleagues have shown that freezing temperatures are advantageous for the synthesis of purines, due to the concentrating effect for key precursors such as hydrogen cyanide.[101] Research by Stanley L. Miller and colleagues suggested that while adenine and guanine require freezing conditions for synthesis, cytosine and uracil may require boiling temperatures.[102] Research by the Miller group notes the formation of seven different amino acids and 11 types of nucleobases in ice when ammonia and cyanide were left in a freezer from 1972 to 1997.[103][104] Other work demonstrated the formation of s-triazines (alternative nucleobases), pyrimidines (including cytosine and uracil), and adenine from urea solutions subjected to freeze-thaw cycles under a reductive atmosphere (with spark discharges as an energy source).[105] The explanation given for the unusual speed of these reactions at such a low temperature is eutectic freezing. As an ice crystal forms, it stays pure: only molecules of water join the growing crystal, while impurities like salt or cyanide are excluded. These impurities become crowded in microscopic pockets of liquid within the ice, and this crowding causes the molecules to collide more often. Mechanistic exploration using quantum chemical methods provide a more detailed understanding of some of the chemical processes involved in chemical evolution, and a partial answer to the fundamental question of molecular biogenesis.[106]

At the time of the Miller–Urey experiment, scientific consensus was that the early Earth had a reducing atmosphere with compounds relatively rich in hydrogen and poor in oxygen (e.g., CH_4 and NH_3 as opposed to CO_2 and nitrogen dioxide (NO_2)). However, current scientific consensus describes the primitive atmosphere as either weakly reducing or neutral[107][108] (see also Oxygen Catastrophe). Such an atmosphere would diminish both the amount and variety of amino acids that could be produced, although studies that include iron and carbonate minerals (thought present in early oceans) in the experimental conditions have again produced a diverse array of amino acids.[107] Other scientific research has focused on two other potential reducing environments: outer space and deep-sea thermal vents.[109][110][111]

The spontaneous formation of complex polymers from abiotically generated monomers under the conditions posited by the "soup" theory is not at all a straightforward process. Besides the necessary basic organic monomers, compounds that would have prohibited the formation of polymers were also formed in high concentration during the Miller–Urey and Joan Oró experiments.[112] The Miller–Urey experiment, for example, produces many substances that would react with the amino acids or terminate their coupling into peptide chains.[113]

A research project completed in March 2015 by John D. Sutherland and others found that a network of reactions beginning with hydrogen cyanide and hydrogen sulfide, in streams of water irradiated by UV light, could produce the chemical components of proteins and lipids, as well as those of RNA,[114][115] while not producing a wide range of other compounds.[116] The researchers used the term "cyanosulfidic" to describe this network of reactions.[115]

9.4.2   Autocatalysis

Main article: Autocatalysis

Autocatalysts are substances that catalyze the production of themselves and therefore are "molecular replicators." The simplest self-replicating chemical systems are autocatalytic, and typically contain three components: a product molecule and two precursor molecules. The product molecule joins together the precursor molecules, which in turn produce more product molecules from more precursor molecules. The product molecule catalyzes the reaction by providing a complementary template that binds to the precursors, thus bringing them together. Such systems have been demonstrated both in biological macromolecules and in small organic molecules.[117][118] Systems that do not proceed by template mechanisms, such as the self-reproduction of micelles and vesicles, have also been observed.[118]

It has been proposed that life initially arose as autocatalytic chemical networks.[119] British ethologist Richard Dawkins wrote about autocatalysis as a potential explanation for the origin of life in his 2004 book *The Ancestor's Tale*.[120] In his book, Dawkins cites experiments performed by Julius Rebek, Jr. and his colleagues in which they combined amino adenosine and pentafluorophenyl esters with the autocatalyst amino adenosine triacid ester (AATE). One product was a variant of AATE, which catalysed the synthesis of themselves. This experiment demonstrated the possibility that autocatalysts could exhibit competition within a population of entities with heredity, which could be interpreted as a rudimentary form of natural selection.[121][122]

In the early 1970s, Manfred Eigen and Peter Schuster examined the transient stages between the molecular chaos and a self-replicating hypercycle in a prebiotic soup.[123] In a hypercycle, the information storing system (possibly RNA) produces an enzyme, which catalyzes the formation of another information system, in sequence until the product of the last aids in the formation of the first information system. Mathematically treated, hypercycles could create quasispecies, which through natural selection entered into a form of Darwinian evolution. A boost to hypercycle theory was the discovery of ribozymes capable of catalyzing their own chemical reactions. The hypercycle theory requires the existence of complex biochemicals, such as nucleotides, which do not form under the conditions proposed by the Miller–Urey experiment.

It has been shown that early error prone translation machinery can be stable against an error catastrophe of the type that had been envisaged as problematical known as "Orgel's paradox" caused caused by catalytic activities that would be disruptive.[124][125][126]

9.4.3   Homochirality

Main article: Homochirality

Homochirality refers to the geometric property of some materials that are composed of chiral units. Chiral refers to nonsuperimposable 3D forms that are mirror images of one another, as are left and right hands. Living organisms use molecules that have the same chirality ("handedness"): with almost no exceptions,[127] amino acids are left-handed while nucleotides and sugars are right-handed. Chiral molecules can be synthesized, but in the absence of a chiral source or a chiral catalyst, they are formed in a 50/50 mixture of both enantiomers (called a racemic mixture). Known mechanisms for the production of non-racemic mixtures from racemic starting materials include: asymmetric physical laws, such as the electroweak interaction; asymmetric environments, such as those caused by circularly polarized light, quartz crystals, or the Earth's rotation; and statistical fluctuations during racemic synthesis.[128]

Once established, chirality would be selected for.[129] A small bias (enantiomeric excess) in the population can be amplified into a large one by asymmetric autocatalysis, such as in the Soai reaction.[130] In asymmetric autocatalysis, the catalyst is a chiral molecule, which means that a chiral molecule is catalysing its own production. An initial enantiomeric excess, such as can be produced by polarized light, then allows the more abundant enantiomer to outcompete the other.[131]

Clark has suggested that homochirality may have started in outer space, as the studies of the amino acids on the Murchison meteorite showed that L-alanine is more than twice as frequent as its D form, and L-glutamic acid was more than three times prevalent than its D counterpart. Various chiral crystal surfaces can also act as sites for possible concentration and assembly of chiral monomer units into macromolecules.[132] Compounds found on meteorites suggest that the chirality of life derives from abiogenic synthesis, since amino acids from meteorites show a left-handed bias, whereas sugars show a predominantly right-handed bias, the same as found in living organisms.[133]

9.5 Self-enclosement, reproduction, duplication and the RNA world

9.5.1 Protocells

Main article: Protocell

A protocell is a self-organized, self-ordered, spherical collection of lipids proposed as a stepping-stone to the origin of life.[134] A central question in evolution is how simple protocells first arose and differed in reproductive contribution to the following generation driving the evolution of life. Although a functional protocell has not yet been achieved in a laboratory setting, there are scientists who think the goal is well within reach.[135][136][137]

Self-assembled vesicles are essential components of primitive cells.[134] The second law of thermodynamics requires that the Universe move in a direction in which disorder (or entropy) increases, yet life is distinguished by its great degree of organization. Therefore, a boundary is needed to separate life processes from non-living matter.[138] Researchers Irene A. Chen and Jack W. Szostak amongst others, suggest that simple physicochemical properties of elementary protocells can give rise to essential cellular behaviors, including primitive forms of differential reproduction competition and energy storage. Such cooperative interactions between the membrane and its encapsulated contents could greatly simplify the transition from simple replicating molecules to true cells.[136] Furthermore, competition for membrane molecules would favor stabilized membranes, suggesting a selective advantage for the evolution of cross-linked fatty acids and even the phospholipids of today.[136] Such micro-encapsulation would allow for metabolism within the membrane, the exchange of small molecules but the prevention of passage of large substances across it.[139] The main advantages of encapsulation include the increased solubility of the contained cargo within the capsule and the storage of energy in the form of a electrochemical gradient.

A 2012 study led by Armen Y. Mulkidjanian of Germany's University of Osnabrück, suggests that inland pools of condensed and cooled geothermal vapour have the ideal characteristics for the origin of life.[140] Scientists confirmed in 2002 that by adding a montmorillonite clay to a solution of fatty acid micelles (lipid spheres), the clay sped up the rate of vesicles formation 100-fold.[137]

Another protocell model is the Jeewanu. First synthesized in 1963 from simple minerals and basic organics while exposed to sunlight, it is still reported to have some metabolic capabilities, the presence of semipermeable membrane, amino acids, phospholipids, carbohydrates and RNA-like molecules.[141][142] However, the nature and properties of the Jeewanu remains to be clarified.

Electrostatic interactions induced by short, positively charged, hydrophobic peptides containing 7 amino acids in length or fewer, can attach RNA to a vesicle membrane, the basic cell membrane.[143]

9.5.2 RNA world

Main article: RNA world

The RNA world hypothesis describes an early Earth with self-replicating and catalytic RNA but no DNA or proteins.[145] It is generally accepted that current life on Earth descends from an RNA world,[11][146] although RNA-based life may not have been the first life to exist.[12][13] This conclusion is drawn from many independent lines of evidence, such as the observations that RNA is central to the translation process and that small RNAs can catalyze all of the chemical groups and information transfers required for life.[13][147] The structure of the ribosome has been called the "smoking gun," as it showed that the ribosome is a ribozyme, with a central core of RNA and no amino acid side chains within 18 angstroms of the active site where peptide bond formation is catalyzed.[12] The concept of the RNA world was first proposed in 1962 by Alexander Rich,[148] and the term was coined by Walter Gilbert in 1986.[13][149]

Possible precursors for the evolution of protein synthesis include a mechanism to synthesize short peptide cofactors or from a mechanism for the duplication of RNA. It is likely that the ancestral ribosome was composed entirely of RNA, although some roles have since been taken over by proteins. Major remaining questions on this topic include identifying the selective force for the evolution of the ribosome and determining how the genetic code arose.[150]

Eugene Koonin said, "Despite considerable experimental and theoretical effort, no compelling scenarios currently exist for the origin of replication and translation, the key processes that together comprise the core of biological systems and the apparent pre-requisite of biological evolution. The RNA World concept might offer the best chance for the resolution of this conundrum but so far cannot adequately account for the emergence of an efficient RNA replicase or the translation system. The MWO [Ed.: "many worlds in one"] version of the cosmological model of eternal inflation could suggest a way out of this conundrum because, in an infinite multiverse with a finite number of distinct macroscopic histories (each repeated an infinite number of times), emergence of even highly complex systems by chance is not just possible but inevitable."[151]

Viral origins and the RNA World

Recent evidence for a "virus first" hypothesis, which may support theories of the RNA world have been suggested in new research.[152] One of the difficulties for the study viral origins and evolution is their high rate of mutation; this is particularly the case in RNA retroviruses like HIV/AIDS.[153] A 2015 study compared protein fold structures across different branches of the tree of life, where researchers can reconstruct the evolutionary histories of the folds and of the organisms whose genomes code for those folds. They argue that protein folds are better markers of ancient events as their three-dimensional structures can be maintained even as the sequences that code for those begin to change.[152] Thus, the viral protein repertoire retain traces of ancient evolutionary history that can be recovered using advanced bioinformatics approaches. Those researchers have concluded that, "the prolonged pressure of genome and particle size reduction eventually reduced virocells into modern viruses (identified by the complete loss of cellular makeup), meanwhile other coexisting cellular lineages diversified into modern cells.[154] The data suggest that viruses originated from ancient cells that co-existed with the ancestors of modern cells.[152] These ancient cells likely contained segmented RNA genomes.[152][155]

9.5.3   RNA synthesis and replication

The RNA world hypothesis has spurred scientists to determine if RNA molecules could have spontaneously formed able to catalyze their own replication.[156][157][158] Evidence suggests that the chemical conditions, including the presence of boron, molybdenum and oxygen needed for the initial production of RNA molecules, may have been better on the planet Mars than on the planet Earth.[156][157] If so, life-suitable molecules originating on Mars, may have later migrated to Earth via meteor ejections.[156][157]

A number of hypotheses of formation of RNA have been put forward. As of 1994, there are difficulties in the explanation of the abiotic synthesis of the nucleotides cytosine and uracil.[159] Subsequent research has shown possible routes of synthesis; for example, formamide produces all four ribonucleotides and other biological molecules when warmed in the presence of various terrestrial minerals.[97][98] Early cell membranes could have formed spontaneously from proteinoids, which are protein-like molecules produced when amino acid solutions are heated while in the correct concentration of aqueous solution. These are seen to form micro-spheres which are observed to behave similarly to membrane-enclosed compartments. Other possible means of producing more complicated organic molecules include chemical reactions that take place on clay substrates or on the surface of the mineral pyrite.

Factors supportive of an important role for RNA in early life include its ability to act both to store information and to catalyze chemical reactions (as a ribozyme); its many important roles as an intermediate in the expression of and maintenance of the genetic information (in the form of DNA) in modern organisms; and the ease of chemical synthesis of at least the components of the RNA molecule under the conditions that approximated the early Earth. Relatively short RNA molecules have been artificially produced in labs, which are capable of replication.[160] Such replicase RNA, which functions as both code and catalyst provides its own template upon which copying can occur. Jack W. Szostak has shown that certain catalytic RNAs can join smaller RNA sequences together, creating the potential for self-replication. If these conditions were present, Darwinian natural selection would favour the proliferation of such autocatalytic sets, to which further functionalities could be added.[161] Such autocatalytic systems of RNA capable of self-sustained replication have been identified.[162] The RNA replication systems, which include two ribozymes that catalyze each other's synthesis, showed a doubling time of the product of about one hour, and were subject to natural selection under the conditions that existed in the experiment.[163] In evolutionary competition experiments, this led to the emergence of new systems which replicated more efficiently.[12] This was the first demonstration of evolutionary adaptation occurring in a molecular genetic system.[163]

Depending on the specific definition used, life can be considered to have emerged when RNA chains began to express the basic conditions necessary for natural selection to operate as conceived by Darwin: heritability, variation of type, and differential reproductive output. The fitness of an RNA replicator (its per capita rate of increase) would likely be a function of its adaptive capacities that are intrinsic (in the sense that they were determined by the nucleotide sequence) and the availability of its resources.[164][165] The three primary adaptive capacities may have been (1) the capacity to replicate with moderate fidelity, giving rise to both heritability while allowing variation of type, (2) the capacity to avoid decay, and (3) the capacity to acquire and process resources.[164][165] These capacities would have been determined initially by the folded configurations of the RNA replicators that, in turn, would be encoded in their individual nucleotide sequences. Relative reproductive success, competition, between different replicators would have depended on the relative values of their adaptive capacities.

9.5.4 Pre-RNA world

It is possible that a different type of nucleic acid, such as PNA, TNA or GNA, was the first to emerge as a self-reproducing molecule, only later replaced by RNA.[166][167] Larralde *et al.*, say that "the generally accepted prebiotic synthesis of ribose, the formose reaction, yields numerous sugars without any selectivity."[168] and they conclude that their "results suggest that the backbone of the first genetic material could not have contained ribose or other sugars because of their instability." The ester linkage of ribose and phosphoric acid in RNA is known to be prone to hydrolysis.[169]

Pyrimidine ribonucleosides and their respective nucleotides have been prebiotically synthesised by a sequence of reactions which by-pass the free sugars, and are assembled in a stepwise fashion by using nitrogenous or oxygenous chemistries. Sutherland has demonstrated high yielding routes to cytidine and uridine ribonucleotides built from small 2 and 3 carbon fragments such as glycolaldehyde, glyceraldehyde or glyceraldehyde-3-phosphate, cyanamide and cyanoacetylene. One of the steps in this sequence allows the isolation of enantiopure ribose aminooxazoline if the enantiomeric excess of glyceraldehyde is 60% or greater.[170] This can be viewed as a prebiotic purification step, where the said compound spontaneously crystallised out from a mixture of the other pentose aminooxazolines. Ribose aminooxazoline can then react with cyanoacetylene in a mild and highly efficient manner to give the alpha cytidine ribonucleotide. Photoanomerization with UV light allows for inversion about the 1' anomeric centre to give the correct beta stereochemistry.[171] In 2009 they showed that the same simple building blocks allow access, via phosphate controlled nucleobase elaboration, to 2',3'-cyclic pyrimidine nucleotides directly, which are known to be able to polymerise into RNA. This paper also highlights the possibility for the photo-sanitization of the pyrimidine-2',3'-cyclic phosphates.[172]

9.6 Origin of biological metabolism

Laboratory research suggests that metabolism-like reactions could have occurred naturally in early oceans, before the first organisms evolved.[14][173] The findings suggests that metabolism predates the origin of life and evolved through the chemical conditions that prevailed in the world's earliest oceans. Reconstructions in laboratories show that some of these reactions can produce RNA, and some others resemble two essential reaction cascades of metabolism: glycolysis and the pentose phosphate pathway, that provide essential precursors for nucleic acids, amino acids and lipids.[173] Following are some observed discoveries and related hypotheses.

9.6.1 Iron–sulfur world

Main article: Iron–sulfur world theory

Proposed in the 1980s by Günter Wächtershäuser, encouraged and supported by Karl R. Popper,[174][175][176] in his iron–sulfur world theory, this hypothesis postulates the evolution of pre-biotic chemical pathways as the start toward the evolution of life. It presents a consistent system of tracing today's biochemistry back to ancestral reactions that provide alternative pathways to the synthesis of organic building blocks from simple gaseous compounds.

In contrast to the classical Miller experiments, which depend on external sources of energy (such as simulated lightning or ultraviolet irradiation), "Wächtershäuser systems" come with a built-in source of energy, sulfides of iron and other minerals (e.g., pyrite). The energy released from redox reactions of these metal sulfides is not only available for the synthesis of organic molecules. It is therefore hypothesized that such systems may be able to evolve into autocatalytic sets of self-replicating, metabolically active entities that would predate the life forms known today.[134][173] The experiment produced a relatively small yield of dipeptides (0.4% to 12.4%) and a smaller yield of tripeptides (0.10%) although under the same conditions, dipeptides were quickly broken down.[177]

Several models reject the idea of the self-replication of a "naked-gene" but postulate the emergence of a primitive metabolism which could provide a safe environment for the later emergence of RNA replication. The centrality of the Krebs cycle (citric acid cycle) to energy production in aerobic organisms, and in drawing in carbon dioxide and hydrogen ions in biosynthesis of complex organic chemicals, suggests that it was one of the first parts of the metabolism to evolve.[178] Somewhat in agreement with these notions, geochemist Michael Russell has proposed that "the purpose of life is to hydrogenate carbon dioxide" (as part of a "metabolism-first," rather than a "genetics-first," scenario).[179][180] Physicist Jeremy England of MIT has proposed that thermodynamically, life was bound to eventually arrive, as based on established physics, he mathematically indicates "...that when a group of atoms is driven by an external source of energy (like the sun or chemical fuel) and surrounded by a heat bath (like the ocean or atmosphere), it will often gradually restructure itself in order to dissipate increasingly more energy. This could mean that under certain conditions, matter inexorably acquires the key physical attribute associated with life."[181][182]

One of the earliest incarnations of this idea was put forward in 1924 with Oparin's notion of primitive self-replicating vesicles which predated the discovery of the structure of DNA. Variants in the 1980s and 1990s include Wächtershäuser's iron–sulfur world theory and models introduced by Christian de Duve based on the chemistry of thioesters. More abstract and theoretical arguments for the plausibility of the emergence of metabolism without the presence of genes include a mathematical model introduced by Freeman Dyson in the early 1980s and Stuart Kauffman's notion of collectively autocatalytic sets, discussed later in that decade.

Orgel summarized his analysis of the proposal by stating, "There is at present no reason to expect that multistep cycles such as the reductive citric acid cycle will self-organize on the surface of FeS/FeS$_2$ or some other mineral."[183] It is possible that another type of metabolic pathway was used at the beginning of life. For example, instead of the reductive citric acid cycle, the "open" acetyl-CoA pathway (another one of the five recognised ways of carbon dioxide fixation in nature today) would be compatible with the idea of self-organisation on a metal sulfide surface. The key enzyme of this pathway, carbon monoxide dehydrogenase/acetyl-CoA synthase harbours mixed nickel-iron-sulfur clusters in its reaction centers and catalyses the formation of acetyl-CoA (which may be regarded as a modern form of acetyl-thiol) in a single step.

9.6.2   Zn-World hypothesis

The Zn-World (zinc world) theory of Armen Y. Mulkidjanian[184] is an extension of Wächtershäuser's pyrite hypothesis. Wächtershäuser based his theory of the initial chemical processes leading to informational molecules (i.e., RNA, peptides) on a regular mesh of electric charges at the surface of pyrite that may have made the primeval polymerization thermodynamically more favourable by attracting reactants and arranging them appropriately relative to each other.[185] The Zn-World theory specifies and differentiates further.[184][186] Hydrothermal fluids rich in H$_2$S interacting with cold primordial ocean (or Darwin's "warm little pond") water leads to the precipitation of metal sulfide particles. Oceanic vent systems and other hydrothermal systems have a zonal structure reflected in ancient volcanogenic massive sulfide deposits (VMS) of hydrothermal origin. They reach many kilometers in diameter and date back to the Archean Eon. Most abundant are pyrite (FeS$_2$), chalcopyrite (CuFeS$_2$), and sphalerite (ZnS), with additions of galena (PbS) and alabandite (MnS). ZnS and MnS have a unique ability to store radiation energy, e.g., provided by UV light. Since during the relevant time window of the origins of replicating molecules the primordial atmospheric pressure was high enough (>100 bar, about 100 atmospheres) to precipitate near the Earth's surface and UV irradiation was 10 to 100 times more intense than now, the unique photosynthetic properties mediated by ZnS provided just the right energy conditions to energize the synthesis of informational and metabolic molecules and the selection of photostable nucleobases.

The Zn-World theory has been further filled out with experimental and theoretical evidence for the ionic constitution of the interior of the first proto-cells before archaea, bacteria and proto-eukaryotes evolved. Archibald Macallum noted the resemblance of organism fluids such as blood, and lymph to seawater;[187] however, the inorganic composition of all cells differ from that of modern seawater, which led Mulkidjanian and colleagues to reconstruct the "hatcheries" of the first cells combining geochemical analysis with phylogenomic scrutiny of the inorganic ion requirements of universal components of modern cells. The authors conclude that ubiquitous, and by inference primordial, proteins and functional systems show affinity to and functional requirement for K^+, Zn^{2+}, Mn^{2+}, and phosphate. Geochemical reconstruction shows that the ionic composition conducive to the origin of cells could not have existed in what we today call marine settings but is compatible with emissions of vapor-dominated zones of what we today call inland geothermal systems. Under the oxygen depleted, CO_2-dominated primordial atmosphere, the chemistry of water condensates and exhalations near geothermal fields would resemble the internal milieu of modern cells. Therefore, the precellular stages of evolution may have taken place in shallow "Darwin ponds" lined with porous silicate minerals mixed with metal sulfides and enriched in K^+, Zn^{2+}, and phosphorus compounds.[188][189]

9.6.3   Deep sea vent hypothesis

The deep sea vent, or alkaline hydrothermal vent, theory for the origin of life on Earth posits that life may have begun at submarine hydrothermal vents.[190] William Martin and Michael Russell have suggested "that life evolved in structured iron monosulphide precipitates in a seepage site hydrothermal mound at a redox, pH and temperature gradient between sulphide-rich hydrothermal fluid and iron(II)-containing waters of the Hadean ocean floor. The naturally arising, three-dimensional compartmentation observed within fossilized seepage-site metal sulphide precipitates indicates that these inorganic compartments were the precursors of cell walls and membranes found in free-living prokaryotes. The known capability of FeS and NiS to catalyse the synthesis of the acetyl-methylsulphide from carbon monoxide and methyl-sulphide, constituents of hydrothermal fluid, indicates that pre-biotic syntheses occurred at the inner surfaces of these metal-sulphide-walled compartments,..."[191] These form where hydrogen-rich fluids emerge from below the sea floor, as a result of serpentinization of ultra-mafic olivine with seawater and a pH interface with carbon dioxide-rich ocean water. The vents form a sustained chemical energy source derived from redox reactions, in which electron donors, such as molecular hydrogen, react with electron acceptors, such as carbon dioxide (see Iron–sulfur world theory). These are highly exothermic reactions.[note 2]

Michael Russell demonstrated that alkaline vents created an abiogenic proton motive force (PMF) chemiosmotic gradient, in which conditions are ideal for an abiogenic hatchery for life. Their microscopic compartments "provide a natural means of concentrating organic molecules," composed of iron-sulfur minerals such as mackinawite, endowed these mineral cells with the catalytic properties envisaged by Wächtershäuser.[178] This movement of ions across the membrane depends on a combination of two factors:

1. Diffusion force caused by concentration gradient—all particles including ions tend to diffuse from higher concentration to lower.

2. Electrostatic force caused by electrical potential gradient—cations like protons H^+ tend to diffuse down the electrical potential, anions in the opposite direction.

These two gradients taken together can be expressed as an electrochemical gradient, providing energy for abiogenic synthesis. The proton motive force can be described as the measure of the potential energy stored as a combination of proton and voltage gradients across a membrane (differences in proton concentration and electrical potential).

Jack W. Szostak suggested that geothermal activity provides greater opportunities for the origination of life in open lakes where there is a buildup of minerals. In 2010, based on spectral analysis of sea and hot mineral water, Ignat Ignatov and Oleg Mosin demonstrated that life may have predominantly originated in hot mineral water. The hot mineral water that contains bicarbonate and calcium ions has the most optimal range.[192] This is similar case as the origin of life in hydrothermal vents, but with bicarbonate and calcium ions in hot water. This water has a pH of 9-11 and is possible to have the reactions in seawater. According to Melvin Calvin, certain reactions of condensation-dehydration of amino acids and nucleotides in individual blocks of peptides and nucleic acids can take place in the primary hydrosphere with pH 9-11 at a later evolutionary stage.[193] Some of these compounds like hydrocyanic acid (HCN) have been proven in the experiments of Miller. This is the environment in which the stromatolites have been created. David Ward of Montana State University described the formation of stromatolites in hot mineral water at the Yellowstone National Park. Stromatolites survive in hot mineral water and in proximity to areas with volcanic activity.[194] Processes have evolved in the sea near geysers of hot mineral water. In 2011, Tadashi Sugawara from the University of Tokyo created a protocell in hot water.[195]

Experimental research and computer modeling suggest that the surfaces of mineral particles inside hydrothermal vents have catalytic properties to enzymes and are able to create simple organic molecules, such as methanol (CH_3OH) and formic, acetic and pyruvic acid out of the dissolved CO_2 in the water.[196][197]

9.6.4   Thermosynthesis

Today's bioenergetic process of fermentation is carried out by either the aforementioned citric acid cycle or the Acetyl-CoA pathway, both of which have been connected to the primordial Iron–sulfur world. In a different approach, the thermosynthesis hypothesis considers the bioenergetic process of chemiosmosis, which plays an essential role in cellular respiration and photosynthesis, more basal than fermentation: the ATP synthase enzyme, which sustains chemiosmosis, is proposed as the currently extant enzyme most closely related to the first metabolic process.[198][199]

First, life needed an energy source to bring about the condensation reaction that yielded the peptide bonds of proteins and the phosphodiester bonds of RNA. In a generalization and thermal variation of the binding change mechanism of today's ATP synthase, the "first protein" would have bound substrates (peptides, phosphate, nucleosides, RNA 'monomers') and condensed them to a reaction product that remained bound until after a temperature change it was released by thermal unfolding.

The energy source under the thermosynthesis hypothesis was thermal cycling, the result of suspension of protocells in a convection current, as is plausible in a volcanic hot spring; the convection accounts for the self-organization and dissipative structure required in any origin of life model. The still ubiquitous role of thermal cycling in germination and cell division is considered a relic of primordial thermosynthesis.

By phosphorylating cell membrane lipids, this "first protein" gave a selective advantage to the lipid protocell that contained the protein. This protein also synthesized a library of many proteins, of which only a minute fraction had thermosynthesis capabilities. As proposed by Dyson,[10] it propagated functionally: it made daughters with similar capabilities, but it did not copy itself. Functioning daughters consisted of different amino acid sequences.

Whereas the Iron–sulfur world identifies a circular pathway as the most simple, the thermosynthesis hypothesis does not even invoke a pathway: ATP synthase's binding change mechanism resembles a physical adsorption process that yields free energy,[200] rather than a regular enzyme's mechanism, which decreases the free energy. It has been claimed that the emergence of cyclic systems of protein catalysts is implausible.[201]

9.7   Other models of abiogenesis

9.7.1   Clay hypothesis

Montmorillonite, an abundant clay, is a catalyst for the polymerization of RNA and for the formation of membranes from lipids.[202] A model for the origin of life using clay was forwarded by Alexander Graham Cairns-Smith in 1985 and explored as a plausible mechanism by several scientists.[203] The clay hypothesis postulates that complex organic molecules arose gradually on a pre-existing, non-organic replication surfaces of silicate crystals in solution.

At the Rensselaer Polytechnic Institute, James P. Ferris' studies have also confirmed that clay minerals of montmorillonite catalyze the formation of RNA in aqueous solution, by joining nucleotides to form longer chains.[204]

In 2007, Bart Kahr from the University of Washington and colleagues reported their experiments that tested the idea that crystals can act as a source of transferable information, using crystals of potassium hydrogen phthalate. "Mother" crystals with imperfections were cleaved and used as seeds to grow "daughter" crystals from solution. They then examined the distribution of imperfections in the new crystals and found that the imperfections in the mother crystals were reproduced in the daughters, but the daughter crystals also had many additional imperfections. For gene-like behavior to be observed, the quantity of inheritance of these imperfections should have exceeded that of the mutations in the successive generations, but it did not. Thus Kahr concluded that the crystals "were not faithful enough to store and transfer information from one generation to the next."[205]

9.7.2 Gold's "deep-hot biosphere" model

In the 1970s, Thomas Gold proposed the theory that life first developed not on the surface of the Earth, but several kilometers below the surface. It is claimed that discovery of microbial life below the surface of another body in our Solar System would lend significant credence to this theory. Thomas Gold also asserted that a trickle of food from a deep, unreachable, source is needed for survival because life arising in a puddle of organic material is likely to consume all of its food and become extinct. Gold's theory is that the flow of such food is due to out-gassing of primordial methane from the Earth's mantle; more conventional explanations of the food supply of deep microbes (away from sedimentary carbon compounds) is that the organisms subsist on hydrogen released by an interaction between water and (reduced) iron compounds in rocks.

9.7.3 Panspermia

Main article: Panspermia

Exogenesis is related to, but not the same as, the notion of panspermia. Neither hypothesis actually answers the question of how life first originated, but merely shifts it to another planet or a comet. However, the advantage of an extraterrestrial origin of primitive life is that life is not required to have evolved on each planet it occurs on, but rather in a single location, and then spread about the galaxy to other star systems via cometary and/or meteorite impact. Evidence to support the hypothesis is scant, but it finds support in studies of Martian meteorites found in Antarctica and in studies of extremophile microbes' survival in outer space.[206][207][208][209][210][211][212]

9.7.4 Extraterrestrial organic molecules

See also: List of interstellar and circumstellar molecules and Panspermia § Pseudo-panspermia

An organic compound is any member of a large class of gaseous, liquid, or solid chemicals whose molecules contain carbon. Carbon is the fourth most abundant element in the Universe by mass after hydrogen, helium, and oxygen.[213] Carbon is abundant in the Sun, stars, comets, and in the atmospheres of most planets.[214] Organic compounds are relatively common in space, formed by "factories of complex molecular synthesis" which occur in molecular clouds and circumstellar envelopes, and chemically evolve after reactions are initiated mostly by ionizing radiation.[15][215][216][217] Based on computer model studies, the complex organic molecules necessary for life may have formed on dust grains in the protoplanetary disk surrounding the Sun before the formation of the Earth.[36] According to the computer studies, this same process may also occur around other stars that acquire planets.[36]

Observations suggest that the majority of organic compounds introduced on Earth by interstellar dust particles are considered principal agents in the formation of complex molecules, thanks to their peculiar surface-catalytic activities.[218][219] Studies reported in 2008, based on $^{12}C/^{13}C$ isotopic ratios of organic compounds found in the Murchison meteorite, suggested that the RNA component uracil and related molecules, including xanthine, were formed extraterrestrially.[220][221] On 8 August 2011, a report based on NASA studies of meteorites found on Earth was published suggesting DNA components (adenine, guanine and related organic molecules) were made in outer space.[218][222][223] Scientists also found that the cosmic dust permeating the Universe contains complex organics ("amorphous organic solids with a mixed aromatic–aliphatic structure") that could be created naturally, and rapidly, by stars.[224][225][226] Sun Kwok of The University of Hong Kong suggested that these compounds may have been related to the development of life on Earth said that "If this is the case, life on Earth may have had an easier time getting started as these organics can serve as basic ingredients for life."[224]

Glycolaldehyde, the first example of an interstellar sugar molecule, was detected in the star-forming region near the center of our galaxy. It was discovered in 2000 by Jes Jørgensen and Jan M. Hollis.[227] In 2012, Jørgensen's team reported the detection of glycolaldehyde in a distant star system. The molecule was found around the protostellar binary IRAS 16293-2422 400 light years from Earth.[228][229][230] Glycolaldehyde is needed to form RNA, which is similar in function to DNA. These findings suggest that complex organic molecules may form in stellar systems prior to the formation of planets, eventually arriving on young planets early in their formation.[231] Because sugars are associated with both metabolism and the genetic code, two of the most basic aspects of life, it is thought the discovery of extraterrestrial sugar increases the likelihood that life may exist elsewhere in our galaxy.[227]

NASA announced in 2009 that scientists had identified another fundamental chemical building block of life in a comet for the first time, glycine, an amino acid, which was detected in material ejected from comet Wild 2 in 2004 and grabbed by NASA's *Stardust* probe. Glycine has been detected in meteorites before. Carl Pilcher, who leads the NASA Astrobiology Institute commented that "The discovery of glycine in a comet supports the idea that the fundamental building blocks of life are prevalent in space, and strengthens the argument that life in the Universe may be common rather than rare."[232] Comets are encrusted with outer layers of dark material, thought to be a tar-like substance composed of complex organic material formed from simple carbon compounds after reactions initiated mostly by ionizing radiation. It is possible that a rain of material from comets could have brought significant quantities of such complex organic molecules to Earth.[233][234][235] Amino acids which were formed extraterrestrially may also have arrived on Earth via comets.[40] It is estimated that during the Late Heavy Bombardment, meteorites may have delivered up to five million tons of organic prebiotic elements to Earth per year.[40]

Polycyclic aromatic hydrocarbons (PAH) are the most common and abundant of the known polyatomic molecules in the observable universe, and are considered a likely constituent of the primordial sea.[236][237][238] In 2010, PAHs, along with fullerenes (or "buckyballs"), have been detected in nebulae.[239][240]

In March 2015, NASA scientists reported that, for the first time, complex DNA and RNA organic compounds of life, including uracil, cytosine and thymine, have been formed in the laboratory under outer space conditions, using starting chemicals, such as pyrimidine, found in meteorites. Pyrimidine, like PAHs, the most carbon-rich chemical found in the Universe, may have been formed in red giant stars or in interstellar dust and gas clouds.[241]

9.7.5 Lipid world

Main article: Gard model

The lipid world theory postulates that the first self-replicating object was lipid-like.[242][243] It is known that phospholipids form lipid bilayers in water while under agitation—the same structure as in cell membranes. These molecules were not present on early Earth, but other amphiphilic long-chain molecules also form membranes. Furthermore, these bodies may expand (by insertion of additional lipids), and under excessive expansion may undergo spontaneous splitting which preserves the same size and composition of lipids in the two progenies. The main idea in this theory is that the molecular composition of the lipid bodies is the preliminary way for information storage, and evolution led to the appearance of polymer entities such as RNA or DNA that may store information favorably. Studies on vesicles from potentially prebiotic amphiphiles have so far been limited to systems containing one or two types of amphiphiles. This in contrast to the output of simulated prebiotic chemical reactions, which typically produce very heterogeneous mixtures of compounds.[134] Within the hypothesis of a lipid bilayer membrane composed of a mixture of various distinct amphiphilic compounds there is the opportunity of a huge number of theoretically possible combinations in the arrangements of these amphiphiles in the membrane. Among all these potential combinations, a specific local arrangement of the membrane would have favored the constitution of an hypercycle,[244][245] actually a positive feedback composed of two mutual catalysts represented by a membrane site and a specific compound trapped in the vesicle. Such site/compound pairs are transmissible to the daughter vesicles leading to the emergence of distinct lineages of vesicles which would have allowed Darwinian natural selection.[246]

9.7.6 Polyphosphates

A problem in most scenarios of abiogenesis is that the thermodynamic equilibrium of amino acid versus peptides is in the direction of separate amino acids. What has been missing is some force that drives polymerization. The resolution of this problem may well be in the properties of polyphosphates.[247][248] Polyphosphates are formed by polymerization of ordinary monophosphate ions PO_4^{-3}. Several mechanisms for such polymerization have been suggested. Polyphosphates cause polymerization of amino acids into peptides. They are also logical precursors in the synthesis of such key biochemical compounds as adenosine triphosphate (ATP). A key issue seems to be that calcium reacts with soluble phosphate to form insoluble calcium phosphate (apatite), so some plausible mechanism must be found to keep calcium ions from causing precipitation of phosphate. There has been much work on this topic over the years, but an interesting new idea is that meteorites may have introduced reactive phosphorus species on the early Earth.[249]

9.7.7 PAH world hypothesis

Main article: PAH world hypothesis

Polycyclic aromatic hydrocarbons (PAH) are known to be abundant in the Universe,[236][237][238] including in the interstellar medium, in comets, and in meteorites, and are some of the most complex molecules so far found in space.[214]

Other sources of complex molecules have been postulated, including extraterrestrial stellar or interstellar origin. For example, from spectral analyses, organic molecules are known to be present in comets and meteorites. In 2004, a team detected traces of PAHs in a nebula.[250] In 2010, another team also detected PAHs, along with fullerenes, in nebulae.[239] The use of PAHs has also been proposed as a precursor to the RNA world in the PAH world hypothesis. The Spitzer Space Telescope has detected a star, HH 46-IR, which is forming by a process similar to that by which the Sun formed. In the disk of material surrounding the star, there is a very large range of molecules, including cyanide compounds, hydrocarbons, and carbon monoxide. In September 2012, NASA scientists reported that PAHs, subjected to interstellar medium conditions, are transformed, through hydrogenation, oxygenation and hydroxylation, to more complex organics—"a step along the path toward amino acids and nucleotides, the raw materials of proteins and DNA, respectively."[251][252] Further, as a result of these transformations, the PAHs lose their spectroscopic signature which could be one of the reasons "for the lack of PAH detection in interstellar ice grains, particularly the outer regions of cold, dense clouds or the upper molecular layers of protoplanetary disks."[251][252]

NASA maintains a database for tracking PAHs in the Universe.[214][253] More than 20% of the carbon in the Universe may be associated with PAHs,[214][214] possible starting materials for the formation of life. PAHs seem to have been formed shortly after the Big Bang, are widespread throughout the Universe,[236][237][238] and are associated with new stars and exoplanets.[214]

9.7.8 Radioactive beach hypothesis

Zachary Adam claims that tidal processes that occurred during a time when the Moon was much closer may have concentrated grains of uranium and other radioactive elements at the high-water mark on primordial beaches, where they may have been responsible for generating life's building blocks.[254] According to computer models reported in *Astrobiology*,[255] a deposit of such radioactive materials could show the same self-sustaining nuclear reaction as that found in the Oklo uranium ore seam in Gabon. Such radioactive beach sand might have provided sufficient energy to generate organic molecules, such as amino acids and sugars from acetonitrile in water. Radioactive monazite material also has released soluble phosphate into the regions between sand-grains, making it biologically "accessible." Thus amino acids, sugars, and soluble phosphates might have been produced simultaneously, according to Adam. Radioactive actinides, left behind in some concentration by the reaction, might have formed part of organometallic complexes. These complexes could have been important early catalysts to living processes.

John Parnell has suggested that such a process could provide part of the "crucible of life" in the early stages of any early wet rocky planet, so long as the planet is large enough to have generated a system of plate tectonics which brings radioactive minerals to the surface. As the early Earth is thought to have had many smaller plates, it might have provided a suitable environment for such processes.[256]

9.7.9 Thermodynamic dissipation

Karo Michaelian from the National Autonomous University of Mexico (UNAM) points out that any model for the origin of life must take into account the fact that life is an irreversible thermodynamic process and, like all irreversible processes, its origin and persistence as a "self-organized" system is due to its dissipation an imposed generalized chemical potential, i.e., the production of entropy. That is, entropy production is not incidental to the process of life, but rather the fundamental reason for its existence. Present day life augments the entropy production of Earth in its solar environment by dissipating ultraviolet and visible photons into heat through organic pigments in water. This heat then catalyzes a host of secondary dissipative processes such as the water cycle, ocean and wind currents, hurricanes, etc.[257][258] Michaelian argues that if the thermodynamic function of life today is to produce entropy through photon dissipation, then this probably was its function at its very beginnings.[259] It turns out that both RNA and DNA when in water solution are very strong absorbers and extremely rapid dissipaters of UV light within the 230–290 nm wavelength region, which is a part of the Sun's spectrum that could have penetrated the prebiotic atmosphere.[260] The amount of ultraviolet (UV-C) light reaching the Earth's surface within this spectral range in the Archean could have been on the order of 4 W/m²,[261] or some 31 orders of magnitude greater than it is today at 260 nm where RNA and DNA absorb most strongly.[260] In fact, not only RNA and DNA, but many fundamental molecules of life (those common to all three domains of life, archea, bacteria, and eucaryote) are also pigments that absorb in the UV-C, and many of these also have a chemical affinity to RNA and DNA.[262] Nucleic acids may thus have acted as acceptor molecules to the UV-C photon excited antenna pigment donor molecules by providing an ultrafast channel for dissipation. Michaelian has shown that there would have existed a non-linear, non-equilibrium thermodynamic imperative to the abiogenic UV-C photochemical synthesis [172] and proliferation of these pigments over the entire Earth surface if they augmented the solar photon dissipation rate.[263]

A simple mechanism to explain enzyme-less replication of RNA and DNA can be given within the same dissipative thermodynamic framework by assuming that life arose when the temperature of the primitive seas had cooled to somewhat below the denaturing temperature of RNA or DNA. The ratio of $^{18}O/^{16}O$ found in cherts of the Barberton greenstone belt of South Africa indicates that the Earth's surface temperature was around 80 °C at 3.8 Ga,[264][265] falling to 70±15 °C about 3.5 to 3.2 Ga,[266] suggestively close to RNA or DNA denaturing (uncoiling and separation) temperatures. During the night, the surface water temperature would drop below the denaturing temperature and single strand RNA/DNA could act as extension template for the formation of double strand RNA/DNA. During the daylight hours, RNA and DNA would absorb UV-C light and convert this directly into heat at the ocean surface, thereby raising the local temperature enough to allow for denaturing of RNA and DNA. Direct experimental evidence for the denaturing of DNA through UV-C light dissipation has now been obtained.[267]

The copying process would have been repeated with each diurnal cycle.[259] Such an ultraviolet and temperature assisted RNA/DNA reproduction (UVTAR) bears similarity to polymerase chain reaction (PCR), a routine laboratory procedure employed to multiply DNA segments. Since denaturation would be most probable in the late afternoon when the Archean sea surface temperature would be highest, and since late afternoon submarine sunlight is somewhat circularly polarized, the homochirality of the organic molecules of life can also be explained within the proposed thermodynamic framework.[259][268]

The fact that the aromatic amino acids have been shown to have chemical affinity to their codons, or anti-codons, and that they also absorb strongly in the UV-C, suggests that they might have originally acted as antenna pigments to increase dissipation and to provide more local heat for UVTAR replication of RNA and DNA as the sea surface temperature cooled. The accumulation of information, e.g., coding for the aromatic amino acids, in RNA or DNA would thus be related to reproductive success under this mechanism. Michaelian suggests that the traditional origin of life research, that expects to describe the emergence of life without overwhelming reference to entropy production through dissipation, is erroneous and that imposed environmental potentials, such as the solar photon flux, and the dissipation of this flux, must be considered to understand the emergence, proliferation, and evolution of life.

9.7.10   Multiple genesis

Different forms of life with variable origin processes may have appeared quasi-simultaneously in the early history of Earth.[269] The other forms may be extinct (having left distinctive fossils through their different biochemistry—e.g., hypothetical types of biochemistry). It has been proposed that:

> The first organisms were self-replicating iron-rich clays which fixed carbon dioxide into oxalic and other dicarboxylic acids. This system of replicating clays and their metabolic phenotype then evolved into the sulfide rich region of the hotspring acquiring the ability to fix nitrogen. Finally phosphate was incorporated into the evolving system which allowed the synthesis of nucleotides and phospholipids. If biosynthesis recapitulates biopoiesis, then the synthesis of amino acids preceded the synthesis of the purine and pyrimidine bases. Furthermore the polymerization of the amino acid thioesters into polypeptides preceded the directed polymerization of amino acid esters by polynucleotides.[270]

9.7.11 Fluctuating hydrothermal pools on volcanic islands

Bruce Damer and David Deamer have come to the conclusion that cell membranes cannot be formed in salty seawater, and must therefore have originated in freshwater. Before the continents formed, the only dry land on earth would be volcanic islands, where rainwater would form ponds where lipids could form the first stages towards cell membranes. These predecessors of true cells are assumed to have behaved more like a superorganism rather than individuals structures, where the porous membranes would house molecules which would leak out and enter other protocells. Only when true cells had evolved would they gradually adapt to saltier environments and enter the ocean.[271]

9.8 See also

- Anthropic principle
- Artificial cell
- Astrochemistry
- Biological immortality
- Common descent
- Emergence
- Entropy and life
- GADV protein world
- Mediocrity principle
- Mycoplasma laboratorium
- Nexus for Exoplanet System Science
- Planetary habitability
- Rare Earth hypothesis
- Shadow biosphere
- Stromatolite

9.9 Notes

[1] The reactions are:

$$FeS + H_2S \rightarrow FeS_2 + 2H^+ + 2e^-$$
$$FeS + H_2S + CO_2 \rightarrow FeS_2 + HCOOH$$

[2] The reactions are:
Reaction 1: *Fayalite + water → magnetite + aqueous silica + hydrogen*

$$3Fe_2SiO_4 + 2H_2O \rightarrow 2Fe_3O_4 + 3SiO_2 + 2H_2$$

Reaction 2: *Forsterite + aqueous silica → serpentine*

$$3Mg_2SiO_4 + SiO_2 + 4H_2O \rightarrow 2Mg_3Si_2O_5(OH)_4$$

Reaction 3: *Forsterite + water → serpentine + brucite*

$$2Mg_2SiO_4 + 3H_2O \rightarrow Mg_3Si_2O_5(OH)_4 + Mg(OH)_2$$

Reaction 3 describes the hydration of olivine with water only to yield serpentine and $Mg(OH)_2$ (brucite). Serpentine is stable at high pH in the presence of brucite like calcium silicate hydrate, (C-S-H) phases formed along with portlandite ($Ca(OH)_2$) in hardened Portland cement paste after the hydration of belite (Ca_2SiO_4), the artificial calcium equivalent of forsterite. Analogy of reaction 3 with belite hydration in ordinary Portland cement: *Belite + water → C-S-H phase + portlandite*

$$2\ Ca_2SiO_4 + 4\ H_2O \rightarrow 3\ CaO \cdot 2\ SiO_2 \cdot 3\ H_2O + Ca(OH)_2$$

9.10   References

[1] Pronunciation: "/ˌeɪbaɪə(ʊ)ˈdʒɛnɪsɪs/". Pearsall, Judy; Hanks, Patrick, eds. (1998). "abiogenesis". *The New Oxford Dictionary of English* (1st ed.). Oxford, UK: Oxford University Press. p. 3. ISBN 0-19-861263-X.

[2] OED On-line (2003)

[3] Bernal 1960, p. 30

[4] Oparin 1953, p. vi

[5] Warmflash, David; Warmflash, Benjamin (November 2005). "Did Life Come from Another World?". *Scientific American* (Stuttgart: Georg von Holtzbrinck Publishing Group) **293** (5): 64–71. doi:10.1038/scientificamerican1105-64. ISSN 0036-8733.

[6] Yarus 2010, p. 47

[7] Peretó, Juli (2005). "Controversies on the origin of life" (PDF). *International Microbiology* (Barcelona: Spanish Society for Microbiology) **8** (1): 23–31. ISSN 1139-6709. PMID 15906258. Retrieved 2015-06-01.

[8] Elizabeth A. Bell. "Potentially biogenic carbon preserved in a 4.1 billion-year-old zircon".

[9] Voet & Voet 2004, p. 29

[10] Dyson 1999

[11]
  • Copley, Shelley D.; Smith, Eric; Morowitz, Harold J. (December 2007). "The origin of the RNA world: Co-evolution of genes and metabolism" (PDF). *Bioorganic Chemistry* (Amsterdam, the Netherlands: Elsevier) **35** (6): 430–443. doi:10.1016/j.bioorg.2007.08.001. ISSN 0045-2068. PMID 17897696. Retrieved 2015-06-08. The proposal that life on Earth arose from an RNA world is widely accepted.

  • Orgel, Leslie E. (April 2003). "Some consequences of the RNA world hypothesis". *Origins of Life and Evolution of the Biosphere* (Kluwer Academic Publishers) **33** (2): 211–218. doi:10.1023/A:1024616317965. ISSN 0169-6149. PMID 12967268. It now seems very likely that our familiar DNA/RNA/protein world was preceded by an RNA world...

  • Robertson & Joyce 2012: "There is now strong evidence indicating that an RNA World did indeed exist before DNA- and protein-based life."

  • Neveu, Kim & Benner 2013: "[The RNA world's existence] has broad support within the community today."

[12] Robertson, Michael P.; Joyce, Gerald F. (May 2012). "The origins of the RNA world". *Cold Spring Harbor Perspectives in Biology* (Cold Spring Harbor, NY: Cold Spring Harbor Laboratory Press) **4** (5): a003608. doi:10.1101/cshperspect.a003608. ISSN 1943-0264. PMC 3331698. PMID 20739415.

[13] Cech, Thomas R. (July 2012). "The RNA Worlds in Context". *Cold Spring Harbor Perspectives in Biology* (Cold Spring Harbor, NY: Cold Spring Harbor Laboratory Press) **4** (7): a006742. doi:10.1101/cshperspect.a006742. ISSN 1943-0264. PMC 3385955. PMID 21441585.

[14] Keller, Markus A.; Turchyn, Alexandra V.; Ralser, Markus (25 March 2014). "Non-enzymatic glycolysis and pentose phosphate pathway-like reactions in a plausible Archean ocean". *Molecular Systems Biology* (Heidelberg, Germany: EMBO Press on behalf of the European Molecular Biology Organization) **10** (725). doi:10.1002/msb.20145228. ISSN 1744-4292. PMC 4023395. PMID 24771084.

[15] Ehrenfreund, Pascale; Cami, Jan (December 2010). "Cosmic carbon chemistry: from the interstellar medium to the early Earth.". *Cold Spring Harbor Perspectives in Biology* (Cold Spring Harbor, NY: Cold Spring Harbor Laboratory Press) **2** (12): a002097. doi:10.1101/cshperspect.a002097. ISSN 1943-0264. PMC 2982172. PMID 20554702.

[16] Perkins, Sid (8 April 2015). "Organic molecules found circling nearby star". *Science* (News) (Washington, D.C.: American Association for the Advancement of Science). ISSN 1095-9203. Retrieved 2015-06-02.

[17] King, Anthony (14 April 2015). "Chemicals formed on meteorites may have started life on Earth". *Chemistry World* (News) (London: Royal Society of Chemistry). ISSN 1473-7604. Retrieved 2015-04-17.

[18] Saladino, Raffaele; Carota, Eleonora; Botta, Giorgia; et al. (13 April 2015). "Meteorite-catalyzed syntheses of nucleosides and of other prebiotic compounds from formamide under proton irradiation". *Proc. Natl. Acad. Sci. U.S.A.* (Washington, D.C.: National Academy of Sciences) **112** (21): E2746–E2755. doi:10.1073/pnas.1422225112. ISSN 1091-6490. PMID 25870268.

[19] Rampelotto, Pabulo Henrique (26 April 2010). *Panspermia: A Promising Field Of Research* (PDF). Astrobiology Science Conference 2010. Houston, TX: Lunar and Planetary Institute. p. 5224. Bibcode:2010LPICo1538.5224R. Retrieved 2014-12-03. Conference held at League City, TX

[20] Loeb, Abraham (October 2014). "The habitable epoch of the early Universe". *International Journal of Astrobiology* (Cambridge, UK: Cambridge University Press) **13** (4): 337–339. arXiv:1312.0613. Bibcode:2014IJAsB..13..337L. doi:10.1017/S1473550414000196.ISSN 1473-5504.

   • Loeb, Abraham (3 June 2014). "The Habitable Epoch of the Early Universe". arXiv:1312.0613v3 [astro-ph.CO].

[21] Dreifus, Claudia (2 December 2014). "Much-Discussed Views That Go Way Back". *The New York Times* (New York: The New York Times Company). p. D2. ISSN 0362-4331. Retrieved 2014-12-03.

[22] Graham, Robert W. (February 1990). "Extraterrestrial Life in the Universe" (PDF) (NASA Technical Memorandum 102363). Lewis Research Center, Cleveland, Ohio: NASA. Retrieved 2015-06-02.

[23] Altermann 2009, p. xvii

[24] "Age of the Earth". United States Geological Survey. 9 July 2007. Retrieved 2006-01-10.

[25] Dalrymple 2001, pp. 205–221

[26] Manhesa, Gérard; Allègre, Claude J.; Dupréa, Bernard; Hamelin, Bruno (May 1980). "Lead isotope study of basic-ultrabasic layered complexes: Speculations about the age of the earth and primitive mantle characteristics". *Earth and Planetary Science Letters* (Amsterdam, the Netherlands: Elsevier) **47** (3): 370–382. Bibcode:1980E&PSL...47..370M. doi:10.1016/0012-821X(80)90024-2. ISSN 0012-821X.

[27] Schopf, J. William; Kudryavtsev, Anatoliy B.; Czaja, Andrew D.; Tripathi, Abhishek B. (5 October 2007). "Evidence of Archean life: Stromatolites and microfossils". *Precambrian Research* (Amsterdam, the Netherlands: Elsevier) **158** (3–4): 141–155. doi:10.1016/j.precamres.2007.04.009. ISSN 0301-9268.

[28] Schopf, J. William (29 June 2006). "Fossil evidence of Archaean life". *Philosophical Transactions of the Royal Society B* (London: Royal Society) **361** (1470): 869–885. doi:10.1098/rstb.2006.1834. ISSN 0962-8436. PMC 1578735. PMID 16754604.

[29] Raven & Johnson 2002, p. 68

[30] Borenstein, Seth (13 November 2013). "Oldest fossil found: Meet your microbial mom". *Excite* (Yonkers, NY: Mindspark Interactive Network). Associated Press. Retrieved 2015-06-02.

[31] Pearlman, Jonathan (13 November 2013). "Oldest signs of life on Earth found". *The Daily Telegraph* (London: Telegraph Media Group). Retrieved 2014-12-15.

[32] Noffke, Nora; Christian, Daniel; Wacey, David; Hazen, Robert M. (16 November 2013). "Microbially Induced Sedimentary Structures Recording an Ancient Ecosystem in the *ca.* 3.48 Billion-Year-Old Dresser Formation, Pilbara, Western Australia". *Astrobiology* (New Rochelle, NY: Mary Ann Liebert, Inc.) **13** (12): 1103–1124. Bibcode:2013AsBio..13.1103N. doi:10.1089/ast.2013.1030. ISSN 1531-1074. PMC 3870916. PMID 24205812.

[33] Ohtomo, Yoko; Kakegawa, Takeshi; Ishida, Akizumi; et al. (January 2014). "Evidence for biogenic graphite in early Archaean Isua metasedimentary rocks". *Nature Geoscience* (London: Nature Publishing Group) **7** (1): 25–28. Bibcode:2014NatGe...7...25O.doi:10.1038/ngeo2025. ISSN 1752-0894.

[34] Borenstein, Seth (19 October 2015). "Hints of life on what was thought to be desolate early Earth". *Excite* (Yonkers, NY: Mindspark Interactive Network). Associated Press. Retrieved 2015-10-20.

[35] Bell, Elizabeth A.; Boehnike, Patrick; Harrison, T. Mark; et al. (19 October 2015). "Potentially biogenic carbon preserved in a 4.1 billion-year-old zircon" (PDF). *Proc. Natl. Acad. Sci. U.S.A.* (Washington, D.C.: National Academy of Sciences). doi:10.1073/pnas.1517557112. ISSN 1091-6490. Retrieved 2015-10-20. Early edition, published online before print.

[36] Moskowitz, Clara (29 March 2012). "Life's Building Blocks May Have Formed in Dust Around Young Sun". *Space.com* (Salt Lake City, UT: Purch). Retrieved 2012-03-30.

[37] Fesenkov 1959, p. 9

[38] Kasting, James F. (12 February 1993). "Earth's Early Atmosphere" (PDF). *Science* (Washington, D.C.: American Association for the Advancement of Science) **259** (5097): 922. doi:10.1126/science.11536547. ISSN 0036-8075. PMID 11536547. Retrieved 2015-07-28.

[39] Russell 2010

[40] Follmann, Hartmut; Brownson, Carol (November 2009). "Darwin's warm little pond revisited: from molecules to the origin of life". *Naturwissenschaften* (Berlin: Springer-Verlag) **96** (11): 1265–1292. Bibcode:doi=10.1007/s00114-009-0602-1 2009NW.....96.1265F doi=10.1007/s00114-009-0602-1. ISSN 0028-1042. PMID 19760276.

[41] Morse, John W.; MacKenzie, Fred T. (1998). "Hadean Ocean Carbonate Geochemistry". *Aquatic Geochemistry* (Kluwer Academic Publishers) **4** (3–4): 301–319. doi:10.1023/A:1009632230875. ISSN 1380-6165.

[42] Wilde, Simon A.; Valley, John W.; Peck, William H.; Graham, Colin M. (11 January 2001). "Evidence from detrital zircons for the existence of continental crust and oceans on the Earth 4.4 Gyr ago" (PDF). *Nature* (London: Nature Publishing Group) **409** (6817): 175–178. doi:10.1038/35051550. ISSN 0028-0836. PMID 11196637. Retrieved 2015-06-03.

[43] Rosing, Minik T.; Bird, Dennis K.; Sleep, Norman H.; et al. (22 March 2006). "The rise of continents—An essay on the geologic consequences of photosynthesis" (PDF). *Palaeogeography, Palaeoclimatology, Palaeoecology* (Amsterdam, the Netherlands: Elsevier) **232** (2–4): 99–113. doi:10.1016/j.palaeo.2006.01.007. ISSN 0031-0182. Retrieved 2015-06-08.

[44] Sleep, Norman H.; Zahnle, Kevin J.; Kasting, James F.; et al. (9 November 1989). "Annihilation of ecosystems by large asteroid impacts on early Earth". *Nature* (London: Nature Publishing Group) **342** (6246): 139–142. Bibcode:1989Natur.342..139S. doi:10.1038/342139a0. ISSN 0028-0836. PMID 11536616.

[45] Davies 1999

[46] O'Donoghue, James (21 August 2011). "Oldest reliable fossils show early life was a beach". *New Scientist* (London: Reed Business Information). ISSN 0262-4079. Retrieved 2014-10-13.

   • Wacey, David; Kilburn, Matt R.; Saunders, Martin; et al. (October 2011). "Microfossils of sulphur-metabolizing cells in 3.4-billion-year-old rocks of Western Australia". *Nature Geoscience* (London: Nature Publishing Group) **4** (10): 698–702. Bibcode:2011NatGe...4..698W. doi:10.1038/ngeo1238. ISSN 1752-0894.

[47] Wolpert, Stuart (19 October 2015). "Life on Earth likely started at least 4.1 billion years ago — much earlier than scientists had thought". UCLA. Retrieved 20 October 2015.

[48] Gomes, Rodney; Levison, Hal F.; Tsiganis, Kleomenis; Morbidelli, Alessandro (26 May 2005). "Origin of the cataclysmic Late Heavy Bombardment period of the terrestrial planets". *Nature* (London: Nature Publishing Group) **435** (7041): 466–469. Bibcode:2005Natur.435..466G. doi:10.1038/nature03676. ISSN 0028-0836. PMID 15917802.

[49] Davies 2007, pp. 61–73

[50] Maher, Kevin A.; Stevenson, David J. (18 February 1988). "Impact frustration of the origin of life". *Nature* (London: Nature Publishing Group) **331** (6157): 612–614. Bibcode:1988Natur.331..612M. doi:10.1038/331612a0. ISSN 0028-0836. PMID 11536595.

[51] Sheldon 2005

[52] Vartanian 1973, pp. 307–312

[53] Lennox 2001, pp. 229–258

[54] Balme, D. M. (1962). "Development of Biology in Aristotle and Theophrastus: Theory of Spontaneous Generation". *Phronesis* (Leiden, the Netherlands: Brill Publishers) **7** (1–2): 91–104. doi:10.1163/156852862X00052. ISSN 0031-8868.

[55] Ross 1652

[56] Dobell 1960

[57] Bondeson 1999

[58] Bernal 1967

[59] "Biogenesis". *Hmolpedia*. Ancaster, Ontario, Canada: WikiFoundry, Inc. Retrieved 2014-05-19.

[60] Huxley 1968

[61] Bastian 1871

[62] Bastian 1871, p. xi–xii

[63] Oparin 1953, p. 196

[64] Tyndall 1905, IV, XII (1876), XIII (1878)

[65] Bernal 1967, p. 139

[66] Priscu, John C. "Origin and Evolution of Life on a Frozen Earth". Arlington County, VA: National Science Foundation. Retrieved 2014-03-01.

[67] Darwin 1887, p. 18: "It is often said that all the conditions for the first production of a living organism are now present, which could ever have been present. But if (and oh! what a big if!) we could conceive in some warm little pond, with all sorts of ammonia and phosphoric salts, light, heat, electricity, &c., present, that a proteine compound was chemically formed ready to undergo still more complex changes, at the present day such matter would be instantly devoured or absorbed, which would not have been the case before living creatures were formed." — Charles Darwin, 1 February 1871

[68] Bernal 1967, *The Origin of Life* (A. I. Oparin, 1924), pp. 199–234

[69] Oparin 1953

[70] Shapiro 1987, p. 110

[71] Bryson 2004, pp. 300–302

[72] Miller, Stanley L. (15 May 1953). "A Production of Amino Acids Under Possible Primitive Earth Conditions". *Science* (Washington, D.C.: American Association for the Advancement of Science) 117 (3046): 528–529. Bibcode:1953Sci...117..528M. doi:10.1126/science.117.3046.528. ISSN 0036-8075. PMID 13056598.

[73] Parker, Eric T.; Cleaves, Henderson J.; Dworkin, Jason P.; et al. (5 April 2011). "Primordial synthesis of amines and amino acids in a 1958 Miller H_2S-rich spark discharge experiment" (PDF). *Proc. Natl. Acad. Sci. U.S.A.* (Washington, D.C.: National Academy of Sciences) 108 (14): 5526–5531. Bibcode:2011PNAS..108.5526P. doi:10.1073/pnas.1019191108. ISSN 0027-8424. PMC 3078417. PMID 21422282. Retrieved 2015-06-08.

[74] Bernal 1967, p. 143

[75] Walsh, J. Bruce (1995). "Part 4: Experimental studies of the origins of life". *Origins of life* (Lecture notes). Tucson, AZ: University Of Arizona. Archived from the original on 2008-01-13. Retrieved 2015-06-08.

[76] Woodward 1969, p. 287

[77] Bahadur, Krishna (1973). "Photochemical Formation of Self–sustaining Coacervates" (PDF). *Proceedings of the Indian National Science Academy* (New Delhi: Indian National Science Academy) 39B (4): 455–467. ISSN 0370-0046.

- Bahadur, Krishna (1975). "Photochemical Formation of Self-Sustaining Coacervates". *Zentralblatt für Bakteriologie, Parasitenkunde, Infektionskrankheiten und Hygiene* (Jena, Germany: Gustav Fischer Verlag) 130 (3): 211–218. doi:10.1016/S0044-4057(75)80076-1. OCLC 641018092. PMID 1242552.

[78] Kasting 1993, p. 922

[79] Kasting 1993, p. 920

[80] Bernal 1951

[81] Bernal, John Desmond (September 1949). "The Physical Basis of Life". *Proceedings of the Physical Society. Section A* (Bristol, UK: Physical Society) **62** (9): 537–558. Bibcode:1949PPSA...62..537B. doi:10.1088/0370-1298/62/9/301. ISSN 0370-1298.

[82] Kauffman 1995

[83] Oehlenschläger, Frank; Eigen, Manfred (December 1997). "30 Years Later – a New Approach to Sol Spiegelman's and Leslie Orgel's in vitro EVOLUTIONARY STUDIES Dedicated to Leslie Orgel on the occasion of his 70th birthday". *Origins of Life and Evolution of Biospheres* (Kluwer Academic Publishers) **27** (5-6): 437–457. doi:10.1023/A:1006501326129. ISSN 0169-6149. PMID 9394469.

[84] McCollom, Thomas; Mayhew, Lisa; Scott, Jim (7 October 2014). "NASA awards CU-Boulder-led team $7 million to study origins, evolution of life in universe" (Press release). Boulder, CO: University of Colorado Boulder. Retrieved 2015-06-08.

[85] Gibson, Daniel G.; Glass, John I.; Lartigue, Carole; et al. (2 July 2010). "Creation of a Bacterial Cell Controlled by a Chemically Synthesized Genome". *Science* (Washington, D.C.: American Association for the Advancement of Science) **329** (5987): 52–56. Bibcode:2010Sci...329...52G. doi:10.1126/science.1190719. ISSN 0036-8075. PMID 20488990.

[86] Swaby, Rachel (20 May 2010). "Scientists Create First Self-Replicating Synthetic Life". *Wired* (New York: Condé Nast). Retrieved 2015-06-08.

[87] Gawlowicz, Susan (6 November 2011). "Carbon-based organic 'carriers' in interstellar dust clouds? Newly discovered diffuse interstellar bands". *Science Daily* (Rockville, MD: ScienceDaily, LLC). Retrieved 2015-06-08. Post is reprinted from materials provided by the Rochester Institute of Technology.

   • Geballe, Thomas R.; Najarro, Francisco; Figer, Donald F.; et al. (10 November 2011). "Infrared diffuse interstellar bands in the Galactic Centre region". *Nature* (London: Nature Publishing Group) **479** (7372): 200–202. arXiv:1111.0613. Bibcode:2011Natur.479..200G. doi:10.1038/nature10527. ISSN 0028-0836. PMID 22048316.

[88] Klyce 2001

[89] Chyba, Christopher; Sagan, Carl (9 January 1992). "Endogenous production, exogenous delivery and impact-shock synthesis of organic molecules: an inventory for the origins of life". *Nature* (London: Nature Publishing Group) **355** (6356): 125–132. Bibcode:1992Natur.355..125C. doi:10.1038/355125a0. ISSN 0028-0836. PMID 11538392.

[90] Furukawa, Yoshihiro; Sekine, Toshimori; Oba, Masahiro; et al. (January 2009). "Biomolecule formation by oceanic impacts on early Earth". *Nature Geoscience* (London: Nature Publishing Group) **2** (1): 62–66. Bibcode:2009NatGe...2...62F. doi:10.1038/NGEO383. ISSN 1752-0894.

[91] Davies 1999, p. 155

[92] Bock & Goode 1996

[93] Palasek, Stan (23 May 2013). "Primordial RNA Replication and Applications in PCR Technology". arXiv:1305.5581v1 [q-bio.BM].

[94] Koonin, Eugene V.; Senkevich, Tatiana G.; Dolja, Valerian V. (19 September 2006). "The ancient Virus World and evolution of cells". *Biology Direct* (London: BioMed Central) **1**: 29. doi:10.1186/1745-6150-1-29. ISSN 1745-6150. PMC 1594570. PMID 16984643.

[95] Vlassov, Alexander V.; Kazakov, Sergei A.; Johnston, Brian H.; et al. (August 2005). "The RNA World on Ice: A New Scenario for the Emergence of RNA Information". *Journal of Molecular Evolution* (Berlin: Springer-Verlag) **61** (2): 264–273. doi:10.1007/s00239-004-0362-7. ISSN 0022-2844. PMID 16044244.

[96] Nussinov, Mark D.; Otroshchenko, Vladimir A.; Santoli, Salvatore (1997). "The emergence of the non-cellular phase of life on the fine-grained clayish particles of the early Earth's regolith". *BioSystems* (Amsterdam, the Netherlands: Elsevier) **42** (2–3): 111–118. doi:10.1016/S0303-2647(96)01699-1. ISSN 0303-2647. PMID 9184757.

[97] Saladino, Raffaele; Crestini, Claudia; Pino, Samanta; et al. (March 2012). "Formamide and the origin of life.". *Physics of Life Reviews* (Amsterdam, the Netherlands: Elsevier) **9** (1): 84–104. Bibcode:2012PhLRv...9...84S. doi:10.1016/j.plrev.2011.12.002. ISSN 1571-0645. PMID 22196896.

[98] Saladino, Raffaele; Botta, Giorgia; Pino, Samanta; et al. (July 2012). "From the one-carbon amide formamide to RNA all the steps are prebiotically possible". *Biochimie* (Amsterdam, the Netherlands: Elsevier) **94** (7): 1451–1456. doi:10.1016/j.biochi.2012.02.018. ISSN 0300-9084. PMID 22738728.

[99] Oró, Joan (16 September 1961). "Mechanism of Synthesis of Adenine from Hydrogen Cyanide under Possible Primitive Earth Conditions". *Nature* (London: Nature Publishing Group) **191** (4794): 1193–1194. Bibcode:1961Natur.191.1193O. doi:10.1038/1911193a0. ISSN 0028-0836. PMID 13731264.

[100] Basile, Brenda; Lazcano, Antonio; Oró, Joan (1984). "Prebiotic syntheses of purines and pyrimidines". *Advances in Space Research* (Amsterdam, the Netherlands: Elsevier) **4** (12): 125–131. Bibcode:1984AdSpR...4..125B. doi:10.1016/0273-1177(84)90554-4. ISSN 0273-1177. PMID 11537766.

[101] Orgel, Leslie E. (August 2004). "Prebiotic Adenine Revisited: Eutectics and Photochemistry". *Origins of Life and Evolution of Biospheres* (Kluwer Academic Publishers) **34** (4): 361–369. Bibcode:2004OLEB...34..361O. doi:10.1023/B:ORIG.0000029882.52156.c2. ISSN 0169-6149. PMID 15279171.

[102] Robertson, Michael P.; Miller, Stanley L. (29 June 1995). "An efficient prebiotic synthesis of cytosine and uracil". *Nature* (London: Nature Publishing Group) **375** (6534): 772–774. Bibcode:1995Natur.375..772R. doi:10.1038/375772a0. ISSN 0028-0836. PMID 7596408.

[103] Fox, Douglas (February 2008). "Did Life Evolve in Ice?". *Discover* (Waukesha, WI: Kalmbach Publishing). ISSN 0274-7529. Retrieved 2008-07-03.

[104] Levy, Matthew; Miller, Stanley L.; Brinton, Karen; Bada, Jeffrey L. (June 2000). "Prebiotic Synthesis of Adenine and Amino Acids Under Europa-like Conditions". *Icarus* (Amsterdam, the Netherlands: Elsevier) **145** (2): 609–613. Bibcode:2000Icar..145..609L. doi:10.1006/icar.2000.6365. ISSN 0019-1035. PMID 11543508.

[105] Menor-Salván, César; Ruiz-Bermejo, Marta; Guzmán, Marcelo I.; Osuna-Esteban, Susana; Veintemillas-Verdaguer, Sabino (20 April 2009). "Synthesis of Pyrimidines and Triazines in Ice: Implications for the Prebiotic Chemistry of Nucleobases". *Chemistry: A European Journal* (Weinheim, Germany: Wiley-VCH on behalf of ChemPubSoc Europe) **15** (17): 4411–4418. doi:10.1002/chem.200802656. ISSN 0947-6539. PMID 19288488.

[106] Roy, Debjani; Najafian, Katayoun; von Ragué Schleyer, Paul (30 October 2007). "Chemical evolution: The mechanism of the formation of adenine under prebiotic conditions". *Proc. Natl. Acad. Sci. U.S.A.* (Washington, D.C.: National Academy of Sciences) **104** (44): 17272–17277. Bibcode:2007PNAS..10417272R. doi:10.1073/pnas.0708434104. ISSN 0027-8424. PMC 2077245. PMID 17951429.

[107] Cleaves, H. James; Chalmers, John H.; Lazcano, Antonio; et al. (April 2008). "A Reassessment of Prebiotic Organic Synthesis in Neutral Planetary Atmospheres". *Origins of Life and Evolution of Biospheres* (Dordrecht, the Netherlands: Springer) **38** (2): 105–115. Bibcode:2008OLEB...38..105C. doi:10.1007/s11084-007-9120-3. ISSN 0169-6149. PMID 18204914.

[108] Chyba, Christopher F. (13 May 2005). "Rethinking Earth's Early Atmosphere". *Science* (Washington, D.C.: American Association for the Advancement of Science) **308** (5724): 962–963. doi:10.1126/science.1113157. ISSN 0036-8075. PMID 15890865.

[109] Barton et al. 2007, pp. 93–95

[110] Bada & Lazcano 2009, pp. 56–57

[111] Bada, Jeffrey L.; Lazcano, Antonio (2 May 2003). "Prebiotic Soup--Revisiting the Miller Experiment" (PDF). *Science* (Washington, D.C.: American Association for the Advancement of Science) **300** (5620): 745–746. doi:10.1126/science.1085145. ISSN 0036-8075. PMID 12730584. Retrieved 2015-06-13.

[112] Oró, Joan; Kimball, Aubrey P. (February 1962). "Synthesis of purines under possible primitive earth conditions: II. Purine intermediates from hydrogen cyanide". *Archives of Biochemistry and Biophysics* (Amsterdam, the Netherlands: Elsevier) **96** (2): 293–313. doi:10.1016/0003-9861(62)90412-5. ISSN 0003-9861. PMID 14482339.

[113] Ahuja, Mukesh, ed. (2006). "Origin of Life". *Life Science* **1**. Delhi: Isha Books. p. 11. ISBN 81-8205-386-2. OCLC 297208106.

[114] Service, Robert F. (16 March 2015). "Researchers may have solved origin-of-life conundrum". *Science* (News) (Washington, D.C.: American Association for the Advancement of Science). ISSN 1095-9203. Retrieved 2015-07-26.

[115] Patel, Bhavesh H.; Percivalle, Claudia; Ritson, Dougal J.; Duffy, Colm D.; Sutherland, John D. (April 2015). "Common origins of RNA, protein and lipid precursors in a cyanosulfidic protometabolism". *Nature Chemistry* (London: Nature Publishing Group) **7** (4): 301–307. Bibcode:2015NatCh...7..301P. doi:10.1038/nchem.2202. ISSN 1755-4330. PMID 25803468. Retrieved 2015-07-22.

[136] Chen, Irene A. (8 December 2006). "The Emergence of Cells During the Origin of Life". *Science* (Washington, D.C.: American Association for the Advancement of Science) **314** (5805): 1558–1559. doi:10.1126/science.1137541. ISSN 0036-8075. PMID 17158315. Retrieved 2015-06-15.

[137] Zimmer, Carl (26 June 2004). "What Came Before DNA?". *Discover* (Waukesha, WI: Kalmbach Publishing). ISSN 0274-7529.

[138] Shapiro, Robert (June 2007). "A Simpler Origin for Life". *Scientific American* (Stuttgart: Georg von Holtzbrinck Publishing Group) **296** (6): 46–53. doi:10.1038/scientificamerican0607-46. ISSN 0036-8733. PMID 17663224. Retrieved 2015-06-15.

[139] Chang 2007

[140] Switek, Brian (13 February 2012). "Debate bubbles over the origin of life". *Nature* (London: Nature Publishing Group). doi:10.1038/nature.2012.10024. ISSN 0028-0836.

[141] Grote, Mathias (September 2011). "*Jeewanu*, or the 'particles of life'" (PDF). *Journal of Biosciences* (Bangalore, India: Indian Academy of Sciences; Springer) **36** (4): 563–570. doi:10.1007/s12038-011-9087-0. ISSN 0250-5991. PMID 21857103. Retrieved 2015-06-15.

[142] Gupta, V. K.; Rai, R. K. (August 2013). "Histochemical localisation of RNA-like material in photochemically formed self-sustaining, abiogenic supramolecular assemblies 'Jeewanu'". *International Research Journal of Science & Engineering* (Amravati, India) **1** (1): 1–4. ISSN 2322-0015. Retrieved 2015-06-15.

[143] Welter, Kira (10 August 2015). "Peptide glue may have held first protocell components together". *Chemistry World* (News) (London: Royal Society of Chemistry). ISSN 1473-7604. Retrieved 2015-08-29.

- Kamat, Neha P.; Tobé, Sylvia; Hill, Ian T.; Szostak, Jack W. (29 July 2015). "Electrostatic Localization of RNA to Protocell Membranes by Cationic Hydrophobic Peptides". *Angewandte Chemie International Edition* (Weinheim, Germany: Wiley-VCH on behalf of the German Chemical Society). doi:10.1002/anie.201505742. ISSN 1433-7851. "Early View (Online Version of Record published before inclusion in an issue)"

[144] Wimberly, Brian T.; Brodersen, Ditlev E.; Clemons, William M., Jr.; et al. (21 September 2000). "Structure of the 30S ribosomal subunit". *Nature* (London: Nature Publishing Group) **407** (6802): 327–339. doi:10.1038/35030006. ISSN 0028-0836. PMID 11014182.

[145] Zimmer, Carl (25 September 2014). "A Tiny Emissary From the Ancient Past". *The New York Times* (New York: The New York Times Company). ISSN 0362-4331. Retrieved 2014-09-26.

[146] Wade, Nicholas (4 May 2015). "Making Sense of the Chemistry That Led to Life on Earth". *The New York Times* (New York: The New York Times Company). ISSN 0362-4331. Retrieved 2015-05-10.

[147] Yarus, Michael (April 2011). "Getting Past the RNA World: The Initial Darwinian Ancestor". *Cold Spring Harbor Perspectives in Biology* (Cold Spring Harbor, NY: Cold Spring Harbor Laboratory Press) **3** (4): a003590. doi:10.1101/cshperspect.a003590. ISSN 1943-0264. PMC 3062219. PMID 20719875.

[148] Neveu, Marc; Kim, Hyo-Joong; Benner, Steven A. (22 April 2013). "The 'Strong' RNA World Hypothesis: Fifty Years Old". *Astrobiology* (New Rochelle, NY: Mary Ann Liebert, Inc.) **13** (4): 391–403. Bibcode:2013AsBio..13..391N. doi:10.1089/ast.2012.0868. ISSN 1531-1074. PMID 23551238.

[149] Gilbert, Walter (20 February 1986). "Origin of life: The RNA world". *Nature* (London: Nature Publishing Group) **319** (6055): 618. Bibcode:1986Natur.319..618G. doi:10.1038/319618a0. ISSN 0028-0836.

[150] Noller, Harry F. (April 2012). "Evolution of protein synthesis from an RNA world.". *Cold Spring Harbor Perspectives in Biology* (Cold Spring Harbor, NY: Cold Spring Harbor Laboratory Press) **4** (4): a003681. doi:10.1101/cshperspect.a003681. ISSN 1943-0264. PMC 3312679. PMID 20610545.

[151] Koonin, Eugene V. (31 May 2007). "The cosmological model of eternal inflation and the transition from chance to biological evolution in the history of life". *Biology Direct* (London: BioMed Central) **2**: 15. doi:10.1186/1745-6150-2-15. ISSN 1745-6150. PMC 1892545. PMID 17540027.

[152] Yates, Diana (25 September 2015). "Study adds to evidence that viruses are alive" (Press release). Champaign, IL: University of Illinois at Urbana–Champaign. Retrieved 2015-10-20.

[153] Caetano-Anollés, Gustavo; Nasir, Arshan (6 September 2012). "Benefits of using molecular structure and abundance in phylogenomic analysis". *Frontiers in Genetics* (Lausanne, Switzerland: Frontiers Media) **3** (172). doi:10.3389/fgene.2012.00172. ISSN 1664-8021. PMC 3434437.

[154] Arshan, Nasir; Caetano-Anollés, Gustavo (25 September 2015). "A phylogenomic data-driven exploration of viral origins and evolution". *Science Advances* (Washington, D.C.: American Association for the Advancement of Science) **1** (8): e1500527. doi:10.1126/sciadv.1500527. ISSN 2375-2548.

[155] Nasir, Arshan; Naeem, Aisha; Jawad Khan, Muhammad; et al. (December 2011). "Annotation of Protein Domains Reveals Remarkable Conservation in the Functional Make up of Proteomes Across Superkingdoms". *Genes* (Basel, Switzerland: MDPI) **2** (4): 869–911. doi:10.3390/genes2040869. ISSN 2073-4425. PMC 3927607. PMID 24710297.

[156] Zimmer, Carl (12 September 2013). "A Far-Flung Possibility for the Origin of Life". *The New York Times* (New York: The New York Times Company). ISSN 0362-4331. Retrieved 2015-06-15.

[157] Webb, Richard (29 August 2013). "Primordial broth of life was a dry Martian cup-a-soup". *New Scientist* (London: Reed Business Information). ISSN 0262-4079. Retrieved 2015-06-16.

[158] Wentao Ma; Chunwu Yu; Wentao Zhang; et al. (November 2007). "Nucleotide synthetase ribozymes may have emerged first in the RNA world". *RNA* (Cold Spring Harbor, NY: Cold Spring Harbor Laboratory Press on behalf of the RNA Society) **13** (11): 2012–2019. doi:10.1261/rna.658507. ISSN 1355-8382. PMC 2040096. PMID 17878321.

[159] Orgel, Leslie E. (October 1994). "The origin of life on Earth". *Scientific American* (Stuttgart: Georg von Holtzbrinck Publishing Group) **271** (4): 76–83. doi:10.1038/scientificamerican1094-76. ISSN 0036-8733. PMID 7524147.

[160] Johnston, Wendy K.; Unrau, Peter J.; Lawrence, Michael S.; et al. (18 May 2001). "RNA-Catalyzed RNA Polymerization: Accurate and General RNA-Templated Primer Extension". *Science* (Washington, D.C.: American Association for the Advancement of Science) **292** (5520): 1319–1325. Bibcode:2001Sci...292.1319J. doi:10.1126/science.1060786. ISSN 0036-8075. PMID 11358999.

[161] Szostak, Jack W. (5 February 2015). "The Origins of Function in Biological Nucleic Acids, Proteins, and Membranes". Chevy Chase (CDP), MD: Howard Hughes Medical Institute. Retrieved 2015-06-16.

[162] Lincoln, Tracey A.; Joyce, Gerald F. (27 February 2009). "Self-Sustained Replication of an RNA Enzyme". *Science* (Washington, D.C.: American Association for the Advancement of Science) **323** (5918): 1229–1232. Bibcode:2009Sci...323.1229L. doi:10.1126/science.1167856. ISSN 0036-8075. PMC 2652413. PMID 19131595.

[163] Joyce, Gerald F. (2009). "Evolution in an RNA world" (PDF). *Cold Spring Harbor Perspectives in Biology* (Cold Spring Harbor, NY: Cold Spring Harbor Laboratory Press) **74** (Evolution: The Molecular Landscape): 17–23. doi:10.1101/sqb.2009.74.004. ISSN 1943-0264. PMC 2891321. PMID 19667013. Retrieved 2015-06-16.

[164] Bernstein, Harris; Byerly, Henry C.; Hopf, Frederick A.; et al. (June 1983). "The Darwinian Dynamic". *The Quarterly Review of Biology* (Chicago, IL: University of Chicago Press) **58** (2): 185–207. doi:10.1086/413216. ISSN 0033-5770. JSTOR 2828805.

[165] Michod 1999

[166] Orgel, Leslie E. (17 November 2000). "A Simpler Nucleic Acid". *Science* (Washington, D.C.: American Association for the Advancement of Science) **290** (5495): 1306–1307. doi:10.1126/science.290.5495.1306. ISSN 0036-8075. PMID 11185405.

[167] Nelson, Kevin E.; Levy, Matthew; Miller, Stanley L. (11 April 2000). "Peptide nucleic acids rather than RNA may have been the first genetic molecule". *Proc. Natl. Acad. Sci. U.S.A.* (Washington, D.C.: National Academy of Sciences) **97** (8): 3868–3871. Bibcode:2000PNAS...97.3868N. doi:10.1073/pnas.97.8.3868. ISSN 0027-8424. PMC 18108. PMID 10760258.

[168] Larralde, Rosa; Robertson, Michael P.; Miller, Stanley L. (29 August 1995). "Rates of Decomposition of Ribose and Other Sugars: Implications for Chemical Evolution" (PDF). *Proc. Natl. Acad. Sci. U.S.A.* (Washington, D.C.: National Academy of Sciences) **92** (18): 8158–8160. Bibcode:1995PNAS...92.8158L. doi:10.1073/pnas.92.18.8158. ISSN 0027-8424. PMC 41115. PMID 7667262.

[169] Lindahl, Tomas (22 April 1993). "Instability and decay of the primary structure of DNA". *Nature* (London: Nature Publishing Group) **362** (6422): 709–715. Bibcode:1993Natur.362..709L. doi:10.1038/362709a0. ISSN 0028-0836. PMID 8469282.

[170] Anastasi, Carole; Crowe, Michael A.; Powner, Matthew W.; Sutherland, John D. (18 September 2006). "Direct Assembly of Nucleoside Precursors from Two- and Three-Carbon Units". *Angewandte Chemie International Edition* (Weinheim, Germany: Wiley-VCH on behalf of the German Chemical Society) **45** (37): 6176–6179. doi:10.1002/anie.200601267. ISSN 1433-7851. PMID 16917794.

[171] Powner, Matthew W.; Sutherland, John D. (13 October 2008). "Potentially Prebiotic Synthesis of Pyrimidine β-D-Ribonucleotides by Photoanomerization/Hydrolysis of α-D-Cytidine-2'-Phosphate". *ChemBioChem* (Weinheim, Germany: Wiley-VCH) **9** (15): 2386–2387. doi:10.1002/cbic.200800391. ISSN 1439-4227. PMID 18798212.

[172] Powner, Matthew W.; Gerland, Béatrice; Sutherland, John D. (14 May 2009). "Synthesis of activated pyrimidine ribonucleotides in prebiotically plausible conditions". *Nature* (London: Nature Publishing Group) **459** (7244): 239–242. Bibcode:2009Natur.459..239P.doi:10.1038/nature08013. ISSN 0028-0836. PMID19444213.

[173] Senthilingam, Meera (25 April 2014). "Metabolism May Have Started in Early Oceans Before the Origin of Life" (Press release). Wellcome Trust. EurekAlert!. Retrieved 2015-06-16.

[174] Yue-Ching Ho, Eugene (July–September 1990). "Evolutionary Epistemology and Sir Karl Popper's Latest Intellectual Interest: A First-Hand Report". *Intellectus* (Hong Kong: Hong Kong Institute of Economic Science) **15**: 1–3. OCLC 26878740. Retrieved 2012-08-13.

[175] Wade, Nicholas (22 April 1997). "Amateur Shakes Up Ideas on Recipe for Life". *The New York Times* (New York: The New York Times Company). ISSN 0362-4331. Retrieved 2015-06-16.

[176] Popper, Karl R. (29 March 1990). "Pyrite and the origin of life". *Nature* (London: Nature Publishing Group) **344** (6265): 387. Bibcode:1990Natur.344..387P. doi:10.1038/344387a0. ISSN 0028-0836.

[177] Huber, Claudia; Wächtershäuser, Günter (31 July 1998). "Peptides by Activation of Amino Acids with CO on (Ni,Fe)S Surfaces: Implications for the Origin of Life". *Science* (Washington, D.C.: American Association for the Advancement of Science) **281** (5377): 670–672. Bibcode:1998Sci...281..670H. doi:10.1126/science.281.5377.670. ISSN 0036-8075. PMID 9685253.

[178] Lane 2009

[179] Musser, George (23 September 2011). "How Life Arose on Earth, and How a Singularity Might Bring It Down". *Observations* (Blog). *Scientific American*. ISSN 0036-8733. Retrieved 2015-06-17.

[180] Carroll, Sean (10 March 2010). "Free Energy and the Meaning of Life". *Cosmic Variance* (Blog). *Discover*. ISSN 0274-7529. Retrieved 2015-06-17.

[181] Wolchover, Natalie (22 January 2014). "A New Physics Theory of Life". *Quanta Magazine* (New York: Simons Foundation). Retrieved 2015-06-17.

[182] England, Jeremy L. (28 September 2013). "Statistical physics of self-replication" (PDF). *Journal of Chemical Physics* (College Park, MD: American Institute of Physics) **139**: 121923. arXiv:1209.1179. Bibcode:2013JChPh.139l1923E. doi:10.1063/1.4818538.ISSN 0021-9606. Retrieved2015-06-18.

[183] Orgel, Leslie E. (7 November 2000). "Self-organizing biochemical cycles". *Proc. Natl. Acad. Sci. U.S.A.* (Washington, D.C.: National Academy of Sciences) **97** (23): 12503–12507. Bibcode:2000PNAS...9712503O. doi:10.1073/pnas.220406697. ISSN 0027-8424. PMC 18793. PMID 11058157.

[184] Mulkidjanian, Armen Y. (24 August 2009). "On the origin of life in the zinc world: 1. Photosynthesizing, porous edifices built of hydrothermally precipitated zinc sulfide as cradles of life on Earth". *Biology Direct* (London: BioMed Central) **4**: 26. doi:10.1186/1745-6150-4-26. ISSN 1745-6150.

[185] Wächtershäuser, Günter (December   ). "Before Enzymes and Templates: Theory of Surface Metabolism". *Microbiological Reviews* (Washington, D.C.: American Society for Microbiology) **52** (4   . ISSN   -   . PMC 373159. PMID 3070320.

[186] Mulkidjanian, Armen Y.; Galperin, Michael Y. (24 August 2009). "On the origin of life in the zinc world. 2. Validation of the hypothesis on the photosynthesizing zinc sulfide edifices as cradles of life on Earth". *Biology Direct* (London: BioMed Central) **4**: 27. doi:10.1186/1745-6150-4-27. ISSN 1745-6150.

[187] Macallum, A. B. (1 April 1926). "The Paleochemistry of the body fluids and tissues". *Physiological Reviews* (Bethesda, MD: American Physiological Society) **6** (2): 316–357. ISSN 0031-9333. Retrieved 2015-06-18.

[188] Mulkidjanian, Armen Y.; Bychkov, Andrew Yu.; Dibrova, Daria V.; et al. (3 April 2012). "Origin of first cells at terrestrial, anoxic geothermal fields". *Proc. Natl. Acad. Sci. U.S.A.* (Washington, D.C.: National Academy of Sciences) **109** (14): E821–E830. Bibcode:2012PNAS..109E.821M. doi:10.1073/pnas.1117774109. ISSN 1091-6490. PMC 3325685. PMID 22331915.

[189] For a deeper integrative version of this hypothesis, see in particular Lankenau 2011, pp. 225–286, interconnecting the "Two RNA worlds" concept and other detailed aspects; and Davidovich, Chen; Belousoff, Matthew; Bashan, Anat; Yonath, Ada (September 2009). "The evolving ribosome: from non-coded peptide bond formation to sophisticated translation machinery". *Research in Microbiology* (Amsterdam, the Netherlands: Elsevier) **160** (7): 487–492. doi:10.1016/j.resmic.2009.07.004. ISSN 1769-7123. PMID 19619641.

[190] Schirber, Michael (24 June 2014). "Hydrothermal Vents Could Explain Chemical Precursors to Life". *NASA Astrobiology: Life in the Universe*. NASA. Retrieved 2015-06-19.

[191] Martin, William; Russell, Michael J. (29 January 2003). "On the origins of cells: a hypothesis for the evolutionary transitions from abiotic geochemistry to chemoautotrophic prokaryotes, and from prokaryotes to nucleated cells". *Philosophical Transactions of the Royal Society B* (London: Royal Society) **358** (1429): 59–83; discussion 83–85. doi:10.1098/rstb.2002.1183. ISSN 0962-8436. PMC 1693102. PMID 12594918.

[192] Ignatov, Ignat; Mosin, Oleg V. (2013). "Possible Processes for Origin of Life and Living Matter with modeling of Physiological Processes of Bacterium *Bacillus Subtilis* in Heavy Water as Model System". *Journal of Natural Sciences Research* (New York: International Institute for Science, Technology and Education) **3** (9): 65–76. ISSN 2225-0921.

[193] Calvin 1969

[194] Schirber, Michael (1 March 2010). "First Fossil-Makers in Hot Water". *Astrobiology Magazine* (New York: NASA). Retrieved 2015-06-19.

[195] Kurihara, Kensuke; Tamura, Mieko; Shohda, Koh-ichiroh; et al. (October 2011). "Self-Reproduction of supramolecular giant vesicles combined with the amplification of encapsulated DNA". *Nature Chemistry* (London: Nature Publishing Group) **3** (10): 775–781. Bibcode:2011NatCh...3..775K. doi:10.1038/nchem.1127. ISSN 1755-4330. PMID 21941249.

[196] Usher, Oli (27 April 2015). "Chemistry of seabed's hot vents could explain emergence of life" (Press release). University College London. Retrieved 2015-06-19.

[197] Roldan, Alberto; Hollingsworth, Nathan; Roffey, Anna; Islam, Husn-Ubayda; et al. (May 2015). "Bio-inspired CO2 conversion by iron sulfide catalysts under sustainable conditions" (PDF). *Chemical Communications* (London: Royal Society of Chemistry) **51** (35): 7501–7504. doi:10.1039/C5CC02078E. ISSN 1359-7345. PMID 25835242. Retrieved 2015-06-19.

[198] Muller, Anthonie W. J. (7 August 1985). "Thermosynthesis by biomembranes: Energy gain from cyclic temperature changes". *Journal of Theoretical Biology* (Amsterdam, the Netherlands: Elsevier) **115** (3): 429–453. doi:10.1016/S0022-5193(85)80202-2. ISSN 0022-5193. PMID 3162066.

[199] Muller, Anthonie W. J. (1995). "Were the first organisms heat engines? A new model for biogenesis and the early evolution of biological energy conversion". *Progress in Biophysics and Molecular Biology* (Oxford, UK; New York: Pergamon Press) **63** (2): 193–231. doi:10.1016/0079-6107(95)00004-7. ISSN 0079-6107. PMID 7542789.

[200] Muller, Anthonie W. J.; Schulze-Makuch, Dirk (1 April 2006). "Sorption heat engines: Simple inanimate negative entropy generators". *Physica A: Statistical Mechanics and its Applications* (Utrecht, the Netherlands: Elsevier) **362** (2): 369–381. arXiv:physics/0507173. Bibcode:2006PhyA..362..369M. doi:10.1016/j.physa.2005.12.003. ISSN 0378-4371.

[201] Orgel 1987, pp. 9–16

[202] Perry, Caroline (7 February 2011). "Clay-armored bubbles may have formed first protocells" (Press release). Cambridge, MA: Harvard University. EurekAlert!. Retrieved 2015-06-20.

[203] Dawkins 1996, pp. 148–161

[204] Wenhua Huang; Ferris, James P. (12 July 2006). "One-Step, Regioselective Synthesis of up to 50-mers of RNA Oligomers by Montmorillonite Catalysis". *Journal of the American Chemical Society* (Washington, D.C.: American Chemical Society) **128** (27): 8914–8919. doi:10.1021/ja061782k. ISSN 0002-7863. PMID 16819887.

[205] Moore, Caroline (16 July 2007). "Crystals as genes?". *Highlights in Chemical Science* (London: Royal Society of Chemistry). ISSN 2041-5818. Retrieved 2015-06-21.

- Bullard, Theresa; Freudenthal, John; Avagyan, Serine; et al. (2007). "Test of Cairns-Smith's 'crystals-as-genes' hypothesis". *Faraday Discussions* **136**: 231–245. Bibcode:2007FaDi..136..231B. doi:10.1039/b616612c. ISSN 1359-6640.

[206] Clark, Stuart (25 September 2002). "Tough Earth bug may be from Mars". *New Scientist* (London: Reed Business Information). ISSN 0262-4079. Retrieved 2015-06-21.

[207] "Exobiology and Radiation Assembly (ERA)". *NSSDC Master Catalog*. NASA. NSSDC ID: 1992-049B-03. Retrieved 2015-06-21. Experiment carried on board the European Retrievable Carrier (EURECA).

[208] Horneck, Gerda; Klaus, David M.; Mancinelli, Rocco L. (March 2010). "Space Microbiology". *Microbiology and Molecular Biology Reviews* (Washington, D.C.: American Society for Microbiology) **74** (1): 121–156. doi:10.1128/MMBR.00016-09. ISSN 1092-2172. PMC 2832349. PMID 20197502.

[209] Clancy, Brack & Horneck 2005

[210] Rabbow, Elke; Horneck, Gerda; Rettberg, Petra; et al. (December 2009). "EXPOSE, an Astrobiological Exposure Facility on the International Space Station – from Proposal to Flight". *Origins of Life and Evolution of Biospheres* (Dordrecht, the Netherlands: Springer) **39** (6): 581–598. Bibcode:2009OLEB...39..581R. doi:10.1007/s11084-009-9173-6. ISSN 0169-6149. PMID 19629743.

[211] Onofri, Silvano; de la Torre, Rosa; de Vera, Jean-Pierre; et al. (May 2012). "Survival of Rock-Colonizing Organisms After 1.5 Years in Outer Space". *Astrobiology* (New Rochelle, NY: Mary Ann Liebert, Inc.) **12** (5): 508–516. Bibcode:2012AsBio..12..508O. doi:10.1089/ast.2011.0736. ISSN 1531-1074. PMID 22680696.

[212] Amos, Jonathan (23 August 2010). "Beer microbes live 553 days outside ISS". *BBC News* (London: BBC). Retrieved 2015-06-22.

[213] "biological abundance of elements". *Encyclopedia of Science*. Dundee, Scotland: David Darling Enterprises. Retrieved 2008-10-09.

[214] Hoover, Rachel (21 February 2014). "Need to Track Organic Nano-Particles Across the Universe? NASA's Got an App for That". *Ames Research Center*. Mountain View, CA: NASA. Retrieved 2015-06-22.

[215] Chang, Kenneth (18 August 2009). "From a Distant Comet, a Clue to Life". *The New York Times* (New York: The New York Times Company). p. A18. ISSN 0362-4331. Retrieved 2015-06-22.

[216] Goncharuk, Vladislav V.; Zui, O. V. (February 2015). "Water and carbon dioxide as the main precursors of organic matter on Earth and in space". *Journal of Water Chemistry and Technology* (Dordrecht, the Netherlands: Springer on behalf of Allerton Press) **37** (1): 2–3. doi:10.3103/S1063455X15010026. ISSN 1063-455X.

[217] Abou Mrad, Ninette; Vinogradoff, Vassilissa; Duvernay, Fabrice; et al. (2015). "Laboratory experimental simulations: Chemical evolution of the organic matter from interstellar and cometary ice analogs" (PDF). *Bulletin de la Société Royale des Sciences de Liège* (Liège, Belgium: Société royale des sciences de Liège) **84**: 21–32. Bibcode:2015BSRSL..84...21A. ISSN 0037-9565. Retrieved 2015-04-06.

[218] Gallori, Enzo (June 2011). "Astrochemistry and the origin of genetic material". *Rendiconti Lincei* (Milan, Italy: Springer) **22** (2): 113–118. doi:10.1007/s12210-011-0118-4. ISSN 2037-4631. "Paper presented at the Symposium 'Astrochemistry: molecules in space and time' (Rome, 4–5 November 2010), sponsored by Fondazione 'Guido Donegani', Accademia Nazionale dei Lincei."

[219] Martins, Zita (February 2011). "Organic Chemistry of Carbonaceous Meteorites". *Elements* (Chantilly, VA: Mineralogical Society of America et al.) **7** (1): 35–40. doi:10.2113/gselements.7.1.35. ISSN 1811-5209.

[220] Martins, Zita; Botta, Oliver; Fogel, Marilyn L.; et al. (15 June 2008). "Extraterrestrial nucleobases in the Murchison meteorite". *Earth and Planetary Science Letters* (Amsterdam, the Netherlands: Elsevier) **270** (1–2): 130–136. arXiv:0806.2286. Bibcode:2008E&PSL.270..130M. doi:10.1016/j.epsl.2008.03.026. ISSN 0012-821X.

[221] "We may all be space aliens: study". *ABC News* (Sydney: Australian Broadcasting Corporation). AFP. 14 June 2008. Retrieved 2015-06-22.

[222] Callahan, Michael P.; Smith, Karen E.; Cleaves, H. James, II; et al. (23 August 2011). "Carbonaceous meteorites contain a wide range of extraterrestrial nucleobases". *Proc. Natl. Acad. Sci. U.S.A.* (Washington, D.C.: National Academy of Sciences) **108** (34): 13995–13998. Bibcode:2011PNAS..10813995C. doi:10.1073/pnas.1106493108. ISSN 0027-8424. PMC 3161613. PMID 21836052.

[223] Steigerwald, John (8 August 2011). "NASA Researchers: DNA Building Blocks Can Be Made in Space". *Goddard Space Flight Center*. Greenbelt, MD: NASA. Retrieved 2015-06-23.

[224] Chow, Denise (26 October 2011). "Discovery: Cosmic Dust Contains Organic Matter from Stars". *Space.com* (Ogden, UT: Purch). Retrieved 2015-06-23.

[225] "Astronomers Discover Complex Organic Matter Exists Throughout the Universe". Rockville, MD: ScienceDaily, LLC. 26 October 2011. Retrieved 2015-06-23. Post is reprinted from materials provided by The University of Hong Kong.

[226] Sun Kwok; Yong Zhang (3 November 2011). "Mixed aromatic–aliphatic organic nanoparticles as carriers of unidentified infrared emission features". *Nature* (London: Nature Publishing Group) **479** (7371): 80–83. Bibcode:2011Natur.479...80K. doi:10.1038/nature10542. ISSN 0028-0836. PMID 22031328.

[227] Clemence, Lara; Cohen, Jarrett (7 February 2005). "Space Sugar's a Sweet Find". *Goddard Space Flight Center*. Greenbelt, MD: NASA. Retrieved 2015-06-23.

[228] Than, Ker (30 August 2012). "Sugar Found In Space: A Sign of Life?". *National Geographic News* (Washington, D.C.: National Geographic Society). Retrieved 2015-06-23.

[229] "Sweet! Astronomers spot sugar molecule near star". *Excite* (Yonkers, NY: Mindspark Interactive Network). Associated Press. 29 August 2012. Retrieved 2015-06-23.

[230] "Building blocks of life found around young star". *News & Events*. Leiden, the Netherlands: Leiden University. 30 September 2012. Retrieved 2013-12-11.

[231] Jørgensen, Jes K.; Favre, Cécile; Bisschop, Suzanne E.; et al. (20 September 2012). "Detection of the simplest sugar, glycolaldehyde, in a solar-type protostar with ALMA" (PDF). *The Astrophysical Journal Letters* (Bristol, England: IOP Publishing for the American Astronomical Society) **757** (1). arXiv:1208.5498. Bibcode:2012ApJ...757L...4J. doi:10.1088/2041-8205/757/1/L4. ISSN 2041-8213. L4. Retrieved 2015-06-23.

[232] "'Life chemical' detected in comet". *BBC News* (London: BBC). 18 August 2009. Retrieved 2015-06-23.

[233] Thompson, William Reid; Murray, B. G.; Khare, Bishun Narain; Sagan, Carl (30 December 1987). "Coloration and darkening of methane clathrate and other ices by charged particle irradiation: Applications to the outer solar system". *Journal of Geophysical Research* (Washington, D.C.: American Geophysical Union) **92** (A13): 14933–14947. Bibcode:1987JGR....9214933T. doi:10.1029/JA092iA13p14933. ISSN 0148-0227. PMID 11542127.

[234] Stark, Anne M. (5 June 2013). "Life on Earth shockingly comes from out of this world". Livermore, CA: Lawrence Livermore National Laboratory. Retrieved 2015-06-23.

[235] Goldman, Nir; Tamblyn, Isaac (20 June 2013). "Prebiotic Chemistry within a Simple Impacting Icy Mixture". *Journal of Physical Chemistry A* (Washington, D.C.: American Chemical Society) **117** (24): 5124–5131. doi:10.1021/jp402976n. ISSN 1089-5639. PMID 23639050.

[236] Carey, Bjorn (18 October 2005). "Life's Building Blocks 'Abundant in Space'". *Space.com* (Watsonville, CA: Imaginova). Retrieved 2015-06-23.

[237] Hudgins, Douglas M.; Bauschlicher, Charles W., Jr.; Allamandola, Louis J. (10 October 2005). "Variations in the Peak Position of the 6.2 μm Interstellar Emission Feature: A Tracer of N in the Interstellar Polycyclic Aromatic Hydrocarbon Population" (PDF). *The Astrophysical Journal* (Bristol, England: IOP Publishing for the American Astronomical Society) **632** (1): 316–332. Bibcode:2005ApJ...632..316H. doi:10.1086/432495. ISSN 0004-637X.

[238] Des Marais, David J.; Allamandola, Louis J.; Sandford, Scott; et al. (2009). "Cosmic Distribution of Chemical Complexity". *Ames Research Center*. Mountain View, CA: NASA. Retrieved 2015-06-24. See the Ames Research Center 2009 annual team report to the NASA Astrobiology Institute here.

[239] Garcia-Hernández, Domingo. A.; Manchado, Arturo; Garcia-Lario, Pedro; et al. (20 November 2010). "Formation of Fullerenes in H-Containing Planetary Nebulae". *The Astrophysical Journal Letters* (Bristol, England: IOP Publishing for the American Astronomical Society) **724** (1): L39–L43. arXiv:1009.4357. Bibcode:2010ApJ...724L..39G. doi:10.1088/2041-8205/724/1/L39. ISSN 2041-8213.

[240] Atkinson, Nancy (27 October 2010). "Buckyballs Could Be Plentiful in the Universe". *Universe Today* (Courtenay, British Columbia: Fraser Cain). Retrieved 2015-06-24.

[241] Marlaire, Ruth, ed. (3 March 2015). "NASA Ames Reproduces the Building Blocks of Life in Laboratory". *Ames Research Center*. Moffett Field, CA: NASA. Retrieved 2015-03-05.

[242] Lancet, Doron (30 December 2014). "Systems Prebiology-Studies of the origin of Life". *The Lancet Lab*. Rehovot, Israel: Department of Molecular Genetics; Weizmann Institute of Science. Retrieved 2015-06-26.

[243] Segré, Daniel; Ben-Eli, Dafna; Deamer, David W.; Lancet, Doron (February 2001). "The Lipid World" (PDF). *Origins of Life and Evolution of the Biosphere* (Kluwer Academic Publishers) **31** (1–2): 119–145. doi:10.1023/A:1006746807104. ISSN 0169-6149. PMID 11296516. Retrieved 2008-09-11.

[244] Eigen, Manfred; Schuster, Peter (November 1977). "The Hypercycle. A Principle of Natural Self-Organization. Part A: Emergence of the Hypercycle" (PDF). *Naturwissenschaften* (Berlin: Springer-Verlag) **64** (11): 541–565. Bibcode:1977NW.....64..541E. doi:10.1007/bf00450633. ISSN 0028-1042. PMID593400. Retrieved2015-06-13.

  - Eigen, Manfred; Schuster, Peter (1978). "The Hypercycle. A Principle of Natural Self-Organization. Part B: The Abstract Hypercycle" (PDF). *Naturwissenschaften* (Berlin: Springer-Verlag) **65**: 7–41. Bibcode:1978NW.....65....7E. doi:10.1007/bf00420631. ISSN 0028-1042. Retrieved 2015-06-13.
  - Eigen, Manfred; Schuster, Peter (July 1978). "The Hypercycle. A Principle of Natural Self-Organization. Part C: The Realistic Hypercycle" (PDF). *Naturwissenschaften* (Berlin: Springer-Verlag) **65** (7): 341–369. Bibcode:1978NW.....65..341E. doi:10.1007/bf00439699. ISSN 0028-1042. Retrieved 2015-06-13.

[245] Markovitch, Omer; Lancet, Doron (Summer 2012). "Excess Mutual Catalysis Is Required for Effective Evolvability" (PDF). *Artificial Life* (Cambridge, MA: MIT Press) **18** (3): 243–266. doi:10.1162/artl_a_00064. ISSN 1064-5462. PMID 22662913. Retrieved 2015-06-26.

[246] Tessera, Marc (2011). "Origin of Evolution *versus* Origin of Life: A Shift of Paradigm". *International Journal of Molecular Sciences* (Basel, Switzerland: MDPI) **12** (6): 3445–3458. doi:10.3390/ijms12063445. ISSN 1422-0067. PMC 3131571. PMID 21747687. Special Issue: "Origin of Life 2011"

[247] Brown, Michael R. W.; Kornberg, Arthur (16 November 2004). "Inorganic polyphosphate in the origin and survival of species". *Proc. Natl. Acad. Sci. U.S.A.* (Washington, D.C.: National Academy of Sciences) **101** (46): 16085–16087. Bibcode:2004PNAS..10116085B. doi:10.1073/pnas.0406909101. ISSN 0027-8424. PMC 528972. PMID 15520374.

[248] Clark, David P. (3 August 1999). "The Origin of Life". *Microbiology 425: Biochemistry and Physiology of Microorganism* (Lecture). Carbondale, IL: College of Science; Southern Illinois University Carbondale. Archived from the original on 2000-10-02. Retrieved 2015-06-26.

[249] Pasek, Matthew A. (22 January 2008). "Rethinking early Earth phosphorus geochemistry". *Proc. Natl. Acad. Sci. U.S.A.* (Washington, D.C.: National Academy of Sciences) **105** (3): 853–858. Bibcode:2008PNAS..105..853P. doi:10.1073/pnas.0708205105. ISSN 0027-8424. PMC2242691. PMID 18195373.

[250] Witt, Adolf N.; Vijh, Uma P.; Gordon, Karl D. (2003). "Discovery of Blue Fluorescence by Polycyclic Aromatic Hydrocarbon Molecules in the Red Rectangle". *Bulletin of the American Astronomical Society* (Washington, D.C.: American Astronomical Society) **35**: 1381. Bibcode:2003AAS...20311017W. Archived from the original on 2003-12-19. Retrieved 2015-06-26. American Astronomical Society Meeting 203, #110.17, January 2004.

[251] "NASA Cooks Up Icy Organics to Mimic Life's Origins". *Space.com*. Ogden, UT: Purch. 20 September 2012. Retrieved 2015-06-26.

[252] Gudipati, Murthy S.; Rui Yang (1 September 2012). "In-situ Probing of Radiation-induced Processing of Organics in Astrophysical Ice Analogs—Novel Laser Desorption Laser Ionization Time-of-flight Mass Spectroscopic Studies". *The Astrophysical Journal Letters* (Bristol, England: IOP Publishing for the American Astronomical Society) **756** (1). Bibcode:2012ApJ...756L..24G. doi:10.1088/2041-8205/756/1/L24. ISSN 2041-8213. L24.

[253] "NASA Ames PAH IR Spectroscopic Database". NASA. Retrieved 2015-06-17.

[254] Dartnell, Lewis (12 January 2008). "Did life begin on a radioactive beach?". *New Scientist* (London: Reed Business Information) (2638): 8. ISSN 0262-4079. Retrieved 2015-06-26.

[255] Adam, Zachary (2007). "Actinides and Life's Origins". *Astrobiology* (New Rochelle, NY: Mary Ann Liebert, Inc.) **7** (6): 852–872. Bibcode:2007AsBio...7..852A. doi:10.1089/ast.2006.0066. ISSN 1531-1074. PMID 18163867.

[256] Parnell, John (December 2004). "Mineral Radioactivity in Sands as a Mechanism for Fixation of Organic Carbon on the Early Earth". *Origins of Life and Evolution of Biospheres* (Kluwer Academic Publishers) **34** (6): 533–547. Bibcode:2004OLEB...34..533P. doi:10.1023/B:ORIG.0000043132.23966.a1. ISSN 0169-6149. PMID15570707.

[257] Michaelian, Karo (30 June 2009). "Thermodynamic Function of Life". arXiv:0907.0040 [physics.gen-ph].

[258] Michaelian, Karo (25 January 2011). "Biological catalysis of the hydrological cycle: life's thermodynamic function". *Hydrology and Earth System Sciences Discussions* (Göttingen, Germany: Copernicus Publications on behalf of the European Geosciences Union) **8**: 1093–1123. Bibcode:2011HESSD...8.1093M. doi:10.5194/hessd-8-1093-2011. ISSN 1812-2116.

[259] Michaelian, Karo (11 March 2011). "Thermodynamic Dissipation Theory for the Origin of Life" (PDF). *Earth System Dynamics* (Göttingen, Germany: Copernicus Publications on behalf of the European Geosciences Union) **2**: 37–51. arXiv:0907.0042. Bibcode:2011ESD.....2...37M. doi:10.5194/esd-2-37-2011. ISSN 2190-4987. Retrieved 2015-06-28.

[260] Cnossen, Ingrid; Sanz-Forcada, Jorge; Favata, Fabio; et al. (February 2007). "Habitat of early life: Solar X-ray and UV radiation at Earth's surface 4–3.5 billion years ago". *Journal of Geophysical Research* (Washington, D.C.: American Geophysical Union) **112** (E2): E02008. arXiv:astro-ph/0702529. Bibcode:2007JGRE..112.2008C. doi:10.1029/2006JE002784. ISSN 0148-0227.

[261] Sagan, Carl (April 1973). "Ultraviolet Selection Pressure on the Earliest Organisms". *Journal of Theoretical Biology* (Amsterdam, the Netherlands: Elsevier) **39** (1): 195–200. doi:10.1016/0022-5193(73)90216-6. ISSN 0022-5193. PMID 4741712.

[262] Michaelian, Karo; Simeonov, Aleksandar (16 May 2014). "Fundamental Molecules of Life are Pigments which Arose and Evolved to Dissipate the Solar Spectrum". arXiv:1405.4059 [physics.bio-ph].

[263] Michaelian, Karo (2013). "A non-linear irreversible thermodynamic perspective on organic pigment proliferation and biological evolution" (PDF). *Journal of Physics: Conference Series* (Bristol, England: IOP Publishing) **475** (conference 1): 012010. arXiv:1307.5924. Bibcode:2013JPhCS.475a2010M. doi:10.1088/1742-6596/475/1/012010. ISSN 1742-6596. "4th National Meeting in Chaos, Complex System and Time Series 29 November to 2 December 2011, Xalapa, Veracruz, Mexico"

[264] Knauth 1992, pp. 123–152

[265] Knauth, L. Paul; Lowe, Donald R. (May 2003). "High Archean climatic temperature inferred from oxygen isotope geochemistry of cherts in the 3.5 Ga Swaziland group, South Africa". *Geological Society of America Bulletin* (Boulder, CO: Geological Society of America) **115**: 566–580. Bibcode:2003GSAB..115..566K. doi:10.1130/0016-7606(2003)115<0566:hactif>2.0.co;2. ISSN 0016-7606.

[266] Lowe, Donald R.; Tice, Michael M. (June 2004). "Geologic evidence for Archean atmospheric and climatic evolution: Fluctuating levels of CO_2, CH_4, and O_2 with an overriding tectonic control". *Geology* (Boulder, CO: Geological Society of America) **32** (6): 493–496. Bibcode:2004Geo....32..493L. doi:10.1130/G20342.1. ISSN 0091-7613.

[267] Michaelian, Karo; Santillán Padilla, Norberto (24 November 2014). "DNA Denaturing through UV-C Photon Dissipation: A Possible Route to Archean Non-enzymatic Replication" (PDF). *bioRxiv* (Cold Spring Harbor, NY: Cold Spring Harbor Laboratory). doi:10.1101/009126. Retrieved 2015-06-29.

[268] Michaelian, Karo (2010). "Homochirality Through Photon-Induced Melting of RNA/DNA: Thermodynamic Dissipation Theory Of The Origin Of Life". *WebmedCentral* (Durham, UK: Webmed Limited, UK) **1** (10): WMC00924. doi:10.9754/journal.

[269] Davies, Paul (December 2007). "Are Aliens Among Us?" (PDF). *Scientific American* (Stuttgart: Georg von Holtzbrinck Publishing Group) **297** (6): 62–69. doi:10.1038/scientificamerican1207-62. ISSN 0036-8733. Retrieved 2015-07-16. ...if life does emerge readily under terrestrial conditions, then perhaps it formed many times on our home planet. To pursue this possibility, deserts, lakes and other extreme or isolated environments have been searched for evidence of "alien" life-forms—organisms that would differ fundamentally from known organisms because they arose independently.

[270] Hartman, Hyman (October 1998). "Photosynthesis and the Origin of Life". *Origins of Life and Evolution of Biospheres* (Kluwer Academic Publishers) **28** (4–6): 515–521. Bibcode:1998OLEB...28..515H. doi:10.1023/A:1006548904157. ISSN 0169-6149. PMID 11536891.

[271] Damer, Bruce; Deamer, David (13 March 2015). "Coupled Phases and Combinatorial Selection in Fluctuating Hydrothermal Pools: A Scenario to Guide Experimental Approaches to the Origin of Cellular Life". *Life* (Basel, Switzerland: MDPI) **5** (1): 872–887. doi:10.3390/life5010872. ISSN 2075-1729. PMC 4390883. PMID 25780958.

9.11 Bibliography

- Altermann, Wladyslaw (2009). "From Fossils to Astrobiology – A Roadmap to Fata Morgana?" (PDF). In Seckbach, Joseph; Walsh, Maud. *From Fossils to Astrobiology: Records of Life on Earth and the Search for Extraterrestrial Biosignatures*. Cellular Origin, Life in Extreme Habitats and Astrobiology **12**. Dordrecht, the Netherlands; London: Springer Science+Business Media. ISBN 978-1-4020-8836-0. LCCN 2008933212. Retrieved 2015-06-05.

- Bada, Jeffrey L.; Lazcano, Antonio (2009). "The Origin of Life". In Ruse, Michael; Travis, Joseph. *Evolution: The First Four Billion Years*. Foreword by Edward O. Wilson. Cambridge, MA: Belknap Press of Harvard University Press. ISBN 978-0-674-03175-3. LCCN 2008030270. OCLC 225874308.

- Barton, Nicholas H.; Briggs, Derek E. G.; Eisen, Jonathan A.; et al. (2007). *Evolution*. Cold Spring Harbor, NY: Cold Spring Harbor Laboratory Press. ISBN 978-0-87969-684-9. LCCN 2007010767. OCLC 86090399.

- Bastian, H. Charlton (1871). *The Modes of Origin of Lowest Organisms*. London; New York: Macmillan and Company. LCCN 11004276. OCLC 42959303. Retrieved 2015-06-06.

- Bernal, J. D. (1951). *The Physical Basis of Life*. London: Routledge & Kegan Paul. LCCN 51005794.

- Bernal, J. D. (1960). "The Problem of Stages in Biopoesis". In Florkin, M. *Aspects of the Origin of Life*. International Series of Monographs on Pure and Applied Biology. Oxford, UK; New York: Pergamon Press. ISBN 978-1-4831-3587-8. LCCN 60013823.

- Bernal, J. D. (1967) [Reprinted work by A. I. Oparin originally published 1924; Moscow: The Moscow Worker]. *The Origin of Life*. The Weidenfeld and Nicolson Natural History. Translation of Oparin by Ann Synge. London: Weidenfeld & Nicolson. LCCN 67098482.

- Bock, Gregory R.; Goode, Jamie A., eds. (1996). *Evolution of Hydrothermal Ecosystems on Earth (and Mars?)*. Ciba Foundation Symposium **202**. Chichester, UK; New York: John Wiley & Sons. ISBN 0-471-96509-X. LCCN 96031351.

- Bondeson, Jan (1999). *The Feejee Mermaid and Other Essays in Natural and Unnatural History*. Ithaca, NY: Cornell University Press. ISBN 0-8014-3609-5. LCCN 98038295.

- Bryson, Bill (2004). *A Short History of Nearly Everything*. London: Black Swan. ISBN 978-0-552-99704-1. OCLC 55589795.

- Calvin, Melvin (1969). *Chemical Evolution: Molecular Evolution Towards the Origin of Living Systems on the Earth and Elsewhere*. Oxford, UK: Clarendon Press. ISBN 0-19-855342-0. LCCN 70415289. OCLC 25220.

- Chaichian, Masud; Rojas, Hugo Perez; Tureanu, Anca (2014). "Physics and Life". *Basic Concepts in Physics: From the Cosmos to Quarks*. Undergraduate Lecture Notes in Physics. Berlin; Heidelberg: Springer Berlin Heidelberg. doi:10.1007/978-3-642-19598-3_12. ISBN 978-3-642-19597-6. ISSN 2192-4791. LCCN 2013950482. OCLC 900189038.

- Chang, Thomas Ming Swi (2007). *Artificial Cells: Biotechnology, Nanomedicine, Regenerative Medicine, Blood Substitutes, Bioencapsulation, and Cell/Stem Cell Therapy*. Regenerative Medicine, Artificial Cells and Nanomedicine **1**. Hackensack, NJ: World Scientific. ISBN 978-981-270-576-1. LCCN 2007013738. OCLC 173522612.

- Clancy, Paul; Brack, André; Horneck, Gerda (2005). *Looking for Life, Searching the Solar System*. Cambridge, UK: Cambridge University Press. ISBN 978-0-521-82450-7. LCCN 2006271630. OCLC 57574490.

- Dalrymple, G. Brent (2001). "The age of the Earth in the twentieth century: a problem (mostly) solved". In Lewis, C. L. E.; Knell, S. J. *The Age of the Earth: from 4004 BC to AD 2002*. Geological Society Special Publication **190**. London: Geological Society of London. Bibcode:2001GSLSP.190..205D. doi:10.1144/gsl.sp.2001.190.01.14. ISBN 1-86239-093-2. ISSN 0305-8719. LCCN 2003464816. OCLC 48570033.

- Darwin, Charles (1887). Darwin, Francis, ed. *The Life and Letters of Charles Darwin, Including an Autobiographical Chapter* **3** (3rd ed.). London: John Murray. OCLC 834491774.

- Davies, Geoffrey F. (2007). "Chapter 2.3 Dynamics of the Hadean and Archaean Mantle". In van Kranendonk, Martin J.; Smithies, R. Hugh; Bennett, Vickie C. *Earth's Oldest Rocks*. Developments in Precambrian Geology **15**. Amsterdam, the Netherlands; Boston: Elsevier. doi:10.1016/S0166-2635(07)15023-4. ISBN 978-0-444-52810-0. LCCN 2009525003.

- Davies, Paul (1999). *The Fifth Miracle: The Search for the Origin of Life*. London: Penguin Books. ISBN 0-14-028226-2.

- Dawkins, Richard (1996). *The Blind Watchmaker* (Reissue with a new introduction ed.). New York: W. W. Norton & Company. ISBN 0-393-31570-3. LCCN 96229669. OCLC 35648431.

- Dawkins, Richard (2004). *The Ancestor's Tale: A Pilgrimage to the Dawn of Evolution*. Boston, MA: Houghton Mifflin. ISBN 0-618-00583-8. LCCN 2004059864. OCLC 56617123.

- Dobell, Clifford (1960) [Originally published 1932; New York: Harcourt, Brace & Company]. *Antony van Leeuwenhoek and His 'Little Animals'*. New York: Dover Publications. LCCN 60002548.

- Dyson, Freeman (1999). *Origins of Life* (Revised ed.). Cambridge, UK; New York: Cambridge University Press. ISBN 0-521-62668-4. LCCN 99021079.

- Eigen, M.; Schuster, P. (1979). *The Hypercycle: A Principle of Natural Self-Organization*. Berlin; New York: Springer-Verlag. ISBN 0-387-09293-5. LCCN 79001315. OCLC 4665354.

- Fesenkov, V. G. (1959). "Some Considerations about the Primaeval State of the Earth". In Oparin, A. I.; et al. *The Origin of Life on the Earth*. I.U.B. Symposium Series **1**. Edited for the International Union of Biochemistry by Frank Clark and R. L. M. Synge (English-French-German ed.). London; New York: Pergamon Press. ISBN 978-1-4832-2240-0. LCCN 59012060. Retrieved 2015-06-03. International Symposium on the Origin of Life on the Earth (held at Moscow, 19–24 August 1957)

- Hazen, Robert M. (2005). *Genesis: The Scientific Quest for Life's Origin*. Washington, D.C.: Joseph Henry Press. ISBN 0-309-09432-1. LCCN 2005012839. OCLC 60321860.

- Huxley, Thomas Henry (1968) [Originally published 1897]. "VIII Biogenesis and Abiogenesis [1870]". *Discourses, Biological and Geological*. Collected Essays **VIII** (Reprint ed.). New York: Greenwood Press. LCCN 70029958. Retrieved 2014-05-19.

- Kauffman, Stuart (1993). *The Origins of Order: Self-Organization and Selection in Evolution*. New York: Oxford University Press. ISBN 978-0-19-507951-7. LCCN 91011148. OCLC 23253930.

- Kauffman, Stuart (1995). *At Home in the Universe: The Search for Laws of Self-Organization and Complexity*. New York: Oxford University Press. ISBN 0-19-509599-5. LCCN 94025268.

- Klyce, Brig (22 January 2001). Kingsley, Stuart A.; Bhathal, Ragbir, eds. *Panspermia Asks New Questions*. The Search for Extraterrestrial Intelligence (SETI) in the Optical Spectrum III. Bellingham, WA: SPIE. doi:10.1117/12.435366. ISBN 0-8194-3951-7. LCCN 2001279159. Retrieved 2015-06-09. Proceedings of the SPIE held at San Jose, CA. 22–24 January 2001

- Knauth, L. Paul (1992). "Origin and diagenesis of cherts: An isotopic perspective". In Clauer, Norbert; Chaudhuri, Sambhu. *Isotopic Signatures and Sedimentary Records*. Lecture Notes in Earth Sciences **43**. Berlin; New York: Springer-Verlag. doi:10.1007/BFb0009863. ISBN 3-540-55828-4. ISSN 0930-0317. LCCN 92025372. OCLC 26262469.

- Lane, Nick (2009). *Life Ascending: The 10 Great Inventions of Evolution* (1st American ed.). New York: W. W. Norton & Company. ISBN 978-0-393-06596-1. LCCN 2009005046. OCLC 286488326.

- Lankenau, Dirk-Henner (2011). "Two RNA Worlds: Toward the Origin of Replication, Genes, Recombination and Repair". In Egel, Richard; Lankenau, Dirk-Henner; Mulkidjanian., Armen Y. *Origins of Life: The Primal Self-Organization*. Heidelberg: Springer. doi:10.1007/978-3-642-21625-1. ISBN 978-3-642-21624-4. LCCN 2011935879. OCLC 733245537.

- Lennox, James G. (2001). *Aristotle's Philosophy of Biology: Studies in the Origins of Life Science*. Cambridge Studies in Philosophy and Biology. Cambridge, UK; New York: Cambridge University Press. ISBN 0-521-65976-0. LCCN 00026070.

- McKinney, Michael L. (1997). "How do rare species avoid extinction? A paleontological view". In Kunin, William E.; Gaston, Kevin J. *The Biology of Rarity: Causes and consequences of rare—common differences* (1st ed.). London; New York: Chapman & Hall. ISBN 0-412-63380-9. LCCN 96071014. OCLC 36442106.

- Michod, Richard E. (1999). "Darwinian Dynamics: Evolutionary Transitions in Fitness and Individuality". Princeton, NJ: Princeton University Press. ISBN 0-691-02699-8. LCCN 98004166. OCLC 38948118.

- Miller, G. Tyler; Spoolman, Scott E. (2012). *Environmental Science* (14th ed.). Belmont, CA: Brooks/Cole. ISBN 978-1-111-98893-7. LCCN 2011934330. OCLC 741539226.

- Oparin, A. I. (1953) [Originally published 1938; New York: The Macmillan Company]. *The Origin of Life*. Translation and new introduction by Sergius Morgulis (2nd ed.). Mineola, NY: Dover Publications. ISBN 0-486-49522-1. LCCN 53010161.

- Orgel, Leslie E. (1987). "Evolution of the Genetic Apparatus: A Review". *Evolution of Catalytic Function*. Cold Spring Harbor Symposia on Quantitative Biology **52**. Cold Spring Harbor, NY: Cold Spring Harbor Laboratory Press. doi:10.1101/SQB.1987.052.01.004. ISBN 0-87969-054-2. OCLC 19850881. "Proceedings of a symposium held at Cold Spring Harbor Laboratory in 1987"

- Raven, Peter H.; Johnson, George B. (2002). *Biology* (6th ed.). Boston, MA: McGraw-Hill. ISBN 0-07-112261-3. LCCN 2001030052. OCLC 45806501.

- Ross, Alexander (1652). *Arcana Microcosmi*. Book II. London. Retrieved 2015-07-07.

- Russell, Michael, ed. (2010). *Origins, Abiogenesis and the Search for Life*. Cambridge, MA: Cosmology Science Publishers. ISBN 978-0-9829552-1-5.

- Shapiro, Robert (1987). *Origins: A Skeptic's Guide to the Creation of Life on Earth*. Toronto; New York: Bantam Books. ISBN 0-553-34355-6.

- Sheldon, Robert B. (22 September 2005). Hoover, Richard B.; Levin, Gilbert V.; Rozanov, Alexei Y.; Gladstone, G. Randall, eds. *Historical Development of the Distinction between Bio- and Abiogenesis* (PDF). Astrobiology and Planetary Missions. Bellingham, WA: SPIE. doi:10.1117/12.663480. ISBN 978-0-8194-5911-4. LCCN 2005284378. Retrieved 2015-04-13. Proceedings of the SPIE held at San Diego, CA, 31 July–2 August 2005

- Stearns, Beverly Peterson; Stearns, Stephen C. (1999). *Watching, from the Edge of Extinction*. New Haven, CT: Yale University Press. ISBN 0-300-07606-1. LCCN 98034087. OCLC 47011675.

- Tyndall, John (1905) [Originally published 1871; London; New York: Longmans, Green & Co.; D. Appleton and Company]. *Fragments of Science* **2** (6th ed.). New York: P.F. Collier & Sons. OCLC 726998155. Retrieved 2015-06-06.

- Vartanian, Aram (1973). "Spontaneous Generation". In Wiener, Philip P. *Dictionary of the History of Ideas* **IV**. New York: Charles Scribner's Sons. ISBN 0-684-13293-1. LCCN 72007943. Retrieved 2015-06-05.

- Voet, Donald; Voet, Judith G. (2004). *Biochemistry* **1** (3rd ed.). New York: John Wiley & Sons. ISBN 0-471-19350-X. LCCN 2003269978.

- Woodward, Robert J., ed. (1969). *Our Amazing World of Nature: Its Marvels & Mysteries*. Pleasantville, NY: Reader's Digest Association. ISBN 0-340-13000-8. LCCN 69010418.

- Yarus, Michael (2010). *Life from an RNA World: The Ancestor Within*. Cambridge, MA: Harvard University Press. ISBN 978-0-674-05075-4. LCCN 2009044011.

9.12   Further reading

- Arrhenius, Gustaf O.; Sales, Brian C.; Mojzsis, Stephen J.; et al. (21 August 1997). "Entropy and Charge in Molecular Evolution—the Case of Phosphate" (PDF). *Journal of Theoretical Biology* (Amsterdam, the Netherlands: Elsevier) **187** (4): 503–522. doi:10.1006/jtbi.1996.0385. ISSN 0022-5193. PMID 9299295.

- Cavalier-Smith, Thomas (June 2006). "Cell evolution and Earth history: stasis and revolution". *Philosophical Transactions of the Royal Society B* (London: Royal Society) **361** (1470): 969–1006. doi:10.1098/rstb.2006.1842. ISSN 0962-8436. PMC 1578732. PMID 16754610.

- de Duve, Christian (1995). *Vital Dust: Life As A Cosmic Imperative* (1st ed.). New York: Basic Books. ISBN 0-465-09044-3. LCCN 94012964. OCLC 30624716.

- Fernando, Chrisantha T.; Rowe, Jonathan (7 July 2007). "Natural selection in chemical evolution". *Journal of Theoretical Biology* (Amsterdam, the Netherlands) **247** (1): 152–167. doi:10.1016/j.jtbi.2007.01.028. ISSN 0022-5193. PMID 17399743.

- Gribbin, John (1998). *The Case of the Missing Neutrinos: And other Curious Phenomena of the Universe* (1st Fromm International ed.). New York: Fromm International. ISBN 0-88064-199-1. LCCN 98027948. OCLC 39368356.

- Harris, Henry (2002). *Things Come to Life: Spontaneous Generation Revisited*. Oxford, UK; New York: Oxford University Press. ISBN 0-19-851538-3. LCCN 2001054856. OCLC 48100507.

- Horgan, John (February 1991). "In the Beginning...". *Scientific American* (Stuttgart: Georg von Holtzbrinck Publishing Group) **264** (2): 116–125. doi:10.1038/scientificamerican0291-116. ISSN 0036-8733.

- Ignatov, Ignat; Mosin, Oleg V. (2013). "Modeling of Possible Processes for Origin of Life and Living Matter in Hot Mineral and Seawater with Deuterium". *Journal of Environment and Earth Science* (New York: International Institute for Science, Technology and Education) **3** (14): 103–118. ISSN 2224-3216. Retrieved 2015-06-29.

- Jortner, Joshua (October 2006). "Conditions for the emergence of life on the early Earth: summary and reflections". *Philosophical Transactions of the Royal Society B* (London: Royal Society) **361** (1474): 1877–1891. doi:10.1098/rstb.2006.1909. ISSN 0962-8436. PMC 1664691. PMID 17008225.

- Klotz, Irene (24 February 2012). "Did Life Start in a Pond, Not Oceans?". *Discovery News* (Silver Spring, MD: Discovery Communications). Retrieved 2015-06-29.

- Knoll, Andrew H. (2003). *Life on a Young Planet: The First Three Billion Years of Evolution on Earth*. Princeton, NJ: Princeton University Press. ISBN 0-691-00978-3. LCCN 2002035484. OCLC 50604948.

- Luisi, Pier Luigi (2006). *The Emergence of Life: From Chemical Origins to Synthetic Biology*. Cambridge, UK: Cambridge University Press. ISBN 978-0-521-82117-9. LCCN 2006285720. OCLC 173609999.

- Maynard Smith, John; Szathmáry, Eörs (1999). *The Origins of Life: From the Birth of Life to the Origin of Language*. Oxford, UK; New York: Oxford University Press. ISBN 0-19-850493-4. LCCN 99230990. OCLC 40980149.

- Morowitz, Harold J. (1992). *Beginnings of Cellular Life: Metabolism Recapitulates Biogenesis*. New Haven, CT: Yale University Press. ISBN 0-300-05483-1. LCCN 92006849. OCLC 25316379.

- NASA Astrobiology Institute: Harrison, T. Mark; McKeegan, Kevin D.; Mojzsis, Stephen J. "Earth's Early Environment and Life: When did Earth become suitable for habitation?". Archived from the original on 2012-02-17. Retrieved 2015-06-30.

- NASA Specialized Center of Research and Training in Exobiology: Arrhenius, Gustaf O. (11 September 2002). "Arrhenius". Archived from the original on 2007-12-21. Retrieved 2015-06-30.

- "The physico-chemical basis of life". *What is Life*. Spring Valley, CA: Lukas K. Buehler. Retrieved 27 October 2005.

- Pitsch, Stefan; Krishnamurthy, Ramanarayanan; Arrhenius, Gustaf O. (6 September 2000). "Concentration of Simple Aldehydes by Sulfite-Containing Double-Layer Hydroxide Minerals: Implications for Biopoesis". *Helvetica Chimica Acta* (Hoboken, NJ: John Wiley & Sons) **83** (9): 2398–2411. doi:10.1002/1522-2675(20000906)83:9<2398::AID-HLCA2398>3.0.CO;2-5. ISSN 0018-019X. PMID 11543578.

- Pons, Marie-Laure; Quitté, Ghylaine; Fujii, Toshiyuki; et al. (25 October 2011). "Early Archean Serpentine Mud Volcanoes at Isua, Greenland, as a Niche for Early Life". *Proc. Natl. Acad. Sci. U.S.A.* (Washington, D.C.: National Academy of Sciences) **108** (43): 17639–17643. Bibcode:2011PNAS..10817639P. doi:10.1073/pnas.1108061108. ISSN 0027-8424. PMC 3203773. PMID 22006301.

- Pross, Addy (2012). *What is Life?: How Chemistry Becomes Biology* (1st ed.). Oxford, UK: Oxford University Press. ISBN 978-0-19-964101-7. LCCN 2012538842. OCLC 812020290.

- Roy, Debjani; Schleyer, Paul von Ragué (2010). "Chemical Origin of Life: How do Five HCN Molecules Combine to form Adenine under Prebiotic and Interstellar Conditions". In Matta, Chérif F. *Quantum Biochemistry*. Weinheim, Germany: Wiley-VCH. doi:10.1002/9783527629213.ch6. ISBN 978-3-527-62921-3. LCCN 2011499476. OCLC 905973537.

- Russell, Michael J.; Hall, A. J.; Cairns-Smith, Alexander Graham; et al. (10 November 1988). "Submarine hot springs and the origin of life". *Nature* (London: Nature Publishing Group) **336** (6195): 117. Bibcode:1988Natur.336..117R. doi:10.1038/336117a0. ISSN 0028-0836. PMID 11536607.

- Shock, Everett L. (25 October 1997). "High-temperature life without photosynthesis as a model for Mars" (PDF). *Journal of Geophysical Research* (Washington, D.C.: American Geophysical Union) **102** (E10): 23687–23694. Bibcode:1997JGR...10223687S. doi:10.1029/97je01087. ISSN 0148-0227.

9.13   External links

- "Exploring Life's Origins: A Virtual Exhibit". *Exploring Life's Origins: A Virtual Exhibit*. Arlington County, VA: National Science Foundation. Retrieved 2015-07-02.

- Fields, Helen (October 2010). "The Origins of Life". *Smithsonian* (Washington, D.C.: Smithsonian Institution). ISSN 0037-7333. Retrieved 2015-07-02.

- Fox, Douglas (28 March 2007). "Primordial Soup's On: Scientists Repeat Evolution's Most Famous Experiment". *Scientific American* (Stuttgart: Georg von Holtzbrinck Publishing Group). ISSN 0036-8733. Retrieved 2015-07-02.

- "The Geochemical Origins of Life by Michael J. Russell & Allan J. Hall". Glasgow, Scotland: University of Glasgow. 13 December 2008. Retrieved 2015-07-02.

- Kauffman, Stuart (8 August 1996). "Even peptides do it". *Nature* (London: Nature Publishing Group) **382** (6591): 496–497. Bibcode:1996Natur.382..496K. doi:10.1038/382496a0. ISSN 0028-0836. PMID 8700218. Archived from the original on 2006-10-15. Retrieved 2015-07-02.

- Malory, Marcia. "How life began on Earth". *Earth Facts*. Retrieved 2015-07-02.

- Nowak, Martin A.; Ohtsuki, Hisashi (30 September 2008). "Prevolutionary dynamics and the origin of evolution" (PDF). *Proc. Natl. Acad. Sci. U.S.A.* (Washington, D.C.: National Academy of Sciences) **105** (39): 14924–14927. Bibcode:2008PNAS..10514924N. doi:10.1073/pnas.0806714105. ISSN 0027-8424. PMC 2567469. PMID 18791073.

- "Possible Connections Between Interstellar Chemistry and the Origin of Life on the Earth". *Space Science and Astrobiology at Ames*. NASA. Archived from the original on 2009-07-31. Retrieved 2015-07-02.

- "Research Spotlight: Jack Szostak: Making Life from Scratch". *Origins of Life Initiative*. Cambridge, MA: Harvard University. Retrieved 2015-07-02.

- Schirber, Michael (9 June 2006). "How Life Began: New Research Suggests Simple Approach". *LiveScience* (Ogden, UT: Purch). Retrieved 2015-07-02.

- "Scientists Find Clues That Life Began in Deep Space". *NASA Astrobiology Institute*. Mountain View, CA: NASA. 30 January 2001. Archived from the original on 2013-04-29. Retrieved 2015-07-02.

- "Simple Artificial Cell Created From Scratch To Study Cell Complexity". *Science Daily* (Rockville, MD: ScienceDaily, LLC). 16 May 2008. Retrieved 2015-07-02. Post is reprinted from materials provided by Pennsylvania State University.

- Singer, Emily (19 July 2015). "Chemists Invent New Letters for Nature's Genetic Alphabet". *Wired*. New York: Condé Nast. Retrieved 2015-07-20.

- Swaminathan, Nikhil (10 June 2008). "Scientists Close to Reconstructing First Living Cell". *Scientific American* (News) (Stuttgart: Georg von Holtzbrinck Publishing Group). ISSN 0036-8733. Retrieved 2015-07-02.

- Vasas, Vera; Fernando, Chrisantha; Santos, Mauro; et al. (5 January 2012). "Evolution before genes" (PDF). *Biology Direct* (London: BioMed Central) **7**: 1. doi:10.1186/1745-6150-7-1. ISSN 1745-6150.

- Zlobin, Andrei E. (2013). "Tunguska similar impacts and origin of life". *Modern Scientific Researches and Innovations* (Moscow: International Centre of Science and Innovations Ltd.) (12). Retrieved 2015-07-02.

- Zlobin, Andrei E. (2014). "Symmetry infringement in mathematical metrics of hydrogen atom as illustration of ideas by V.I.Vernadsky concerning origin of life and biosphere" (PDF). *Acta Naturae* (Moscow: Park Media Ltd.) (Special Issue 1): 48. ISSN 2075-8251. Retrieved 2015-07-02.

9.13.1   Video resources

- Hazen, Robert M. (29 April 2014). *The Origins of Life* (Webcast). Baltimore, MD: Space Telescope Science Institute. Retrieved 2015-07-03. — A 2014 Spring Symposium webcast (video; 38 m)

- "The Origin of Life" on YouTube — A Royal Institution Discourse lecture given by John Maynard Smith in 1995 (video; 58 m)

- "Space Experts Discuss the Search for Life in the Universe at NASA" on YouTube — Panel discussion at NASA headquarters on 14 July 2014 (video; 87 m)

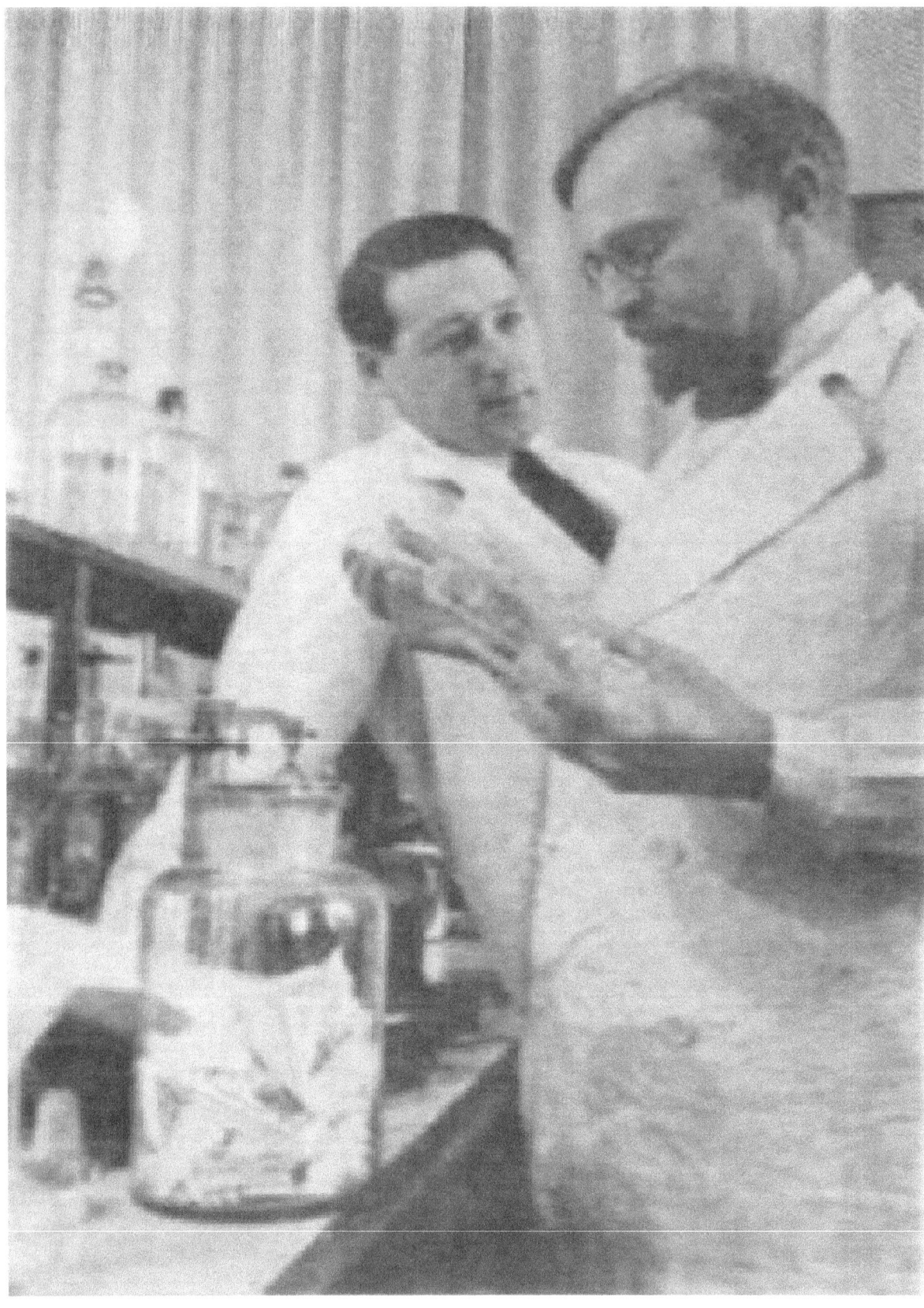

Alexander Oparin (right) at the laboratory

Phylogenetic Tree of Life

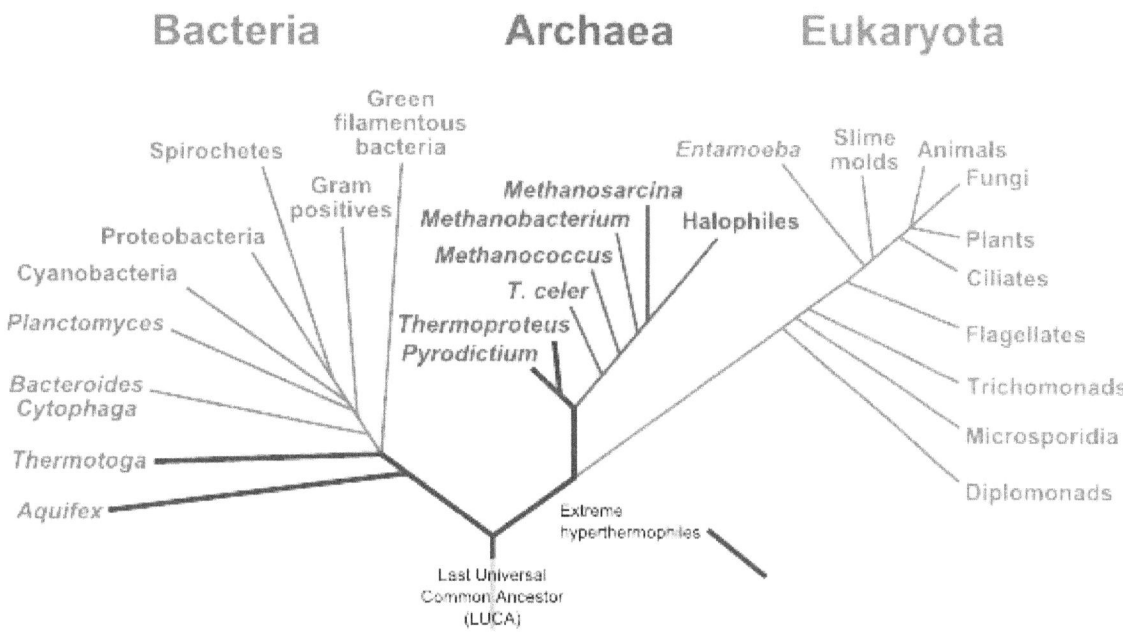

A cladogram demonstrating extreme hyperthermophiles at the base of the phylogenetic tree of life.

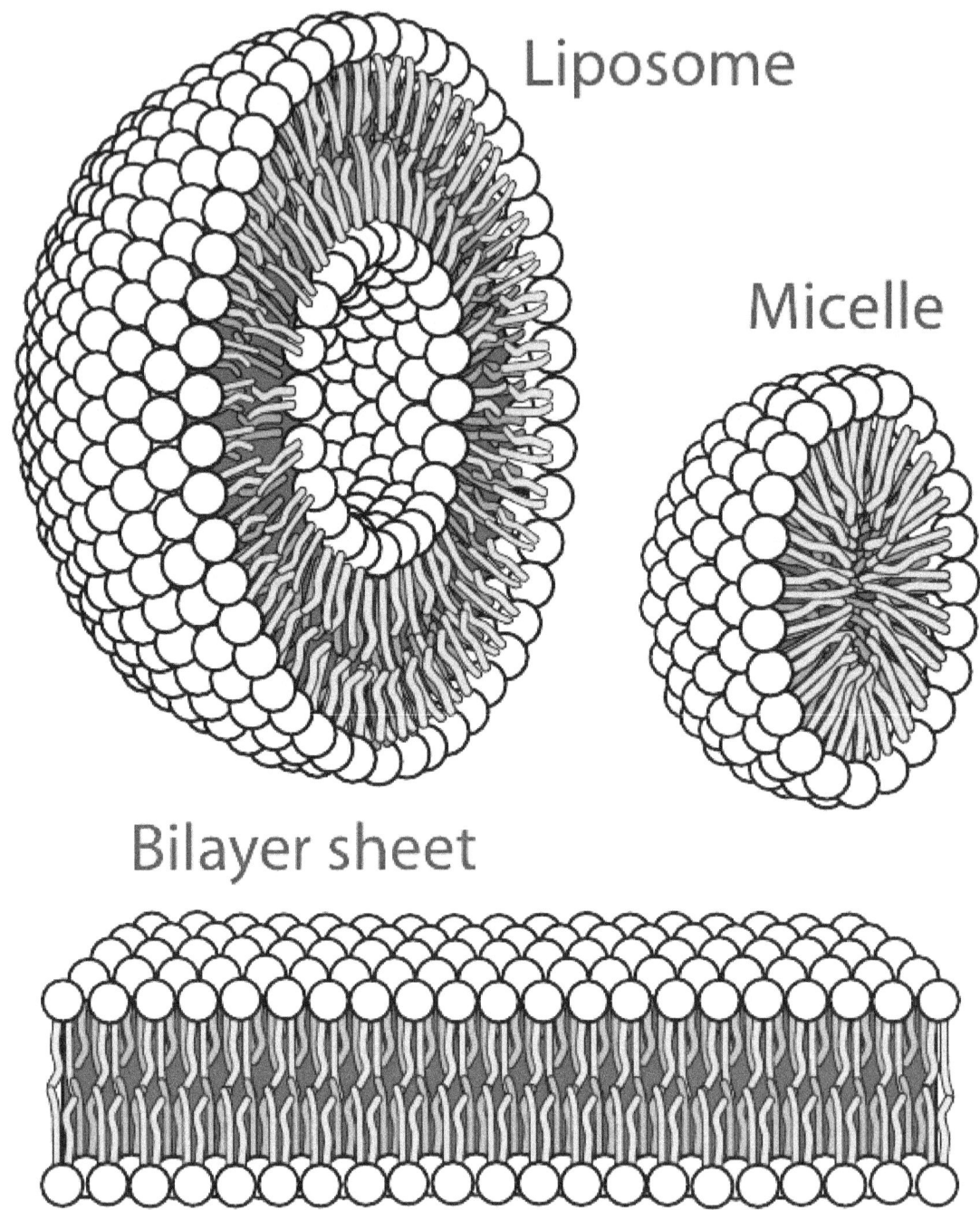

Liposome

Micelle

Bilayer sheet

The three main structures phospholipids form spontaneously in solution: the liposome (a closed bilayer), the micelle and the bilayer.

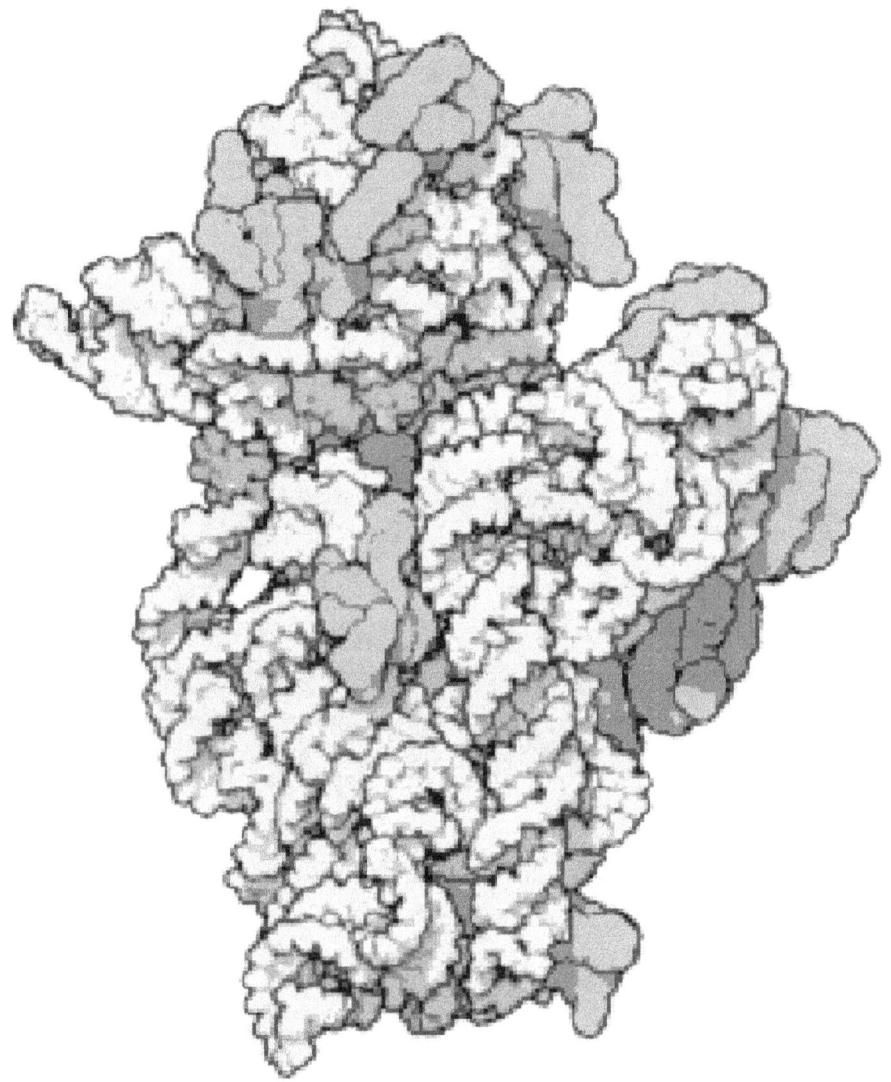

Molecular structure of the ribosome 30S subunit from Thermus thermophilus.[144] Proteins are shown in blue and the single RNA chain in orange.

Deep-sea hydrothermal vent or 'black smoker'

White smokers emitting liquid carbon dioxide (CO_2) at the Champagne vent, Marianas Trench Marine National Monument

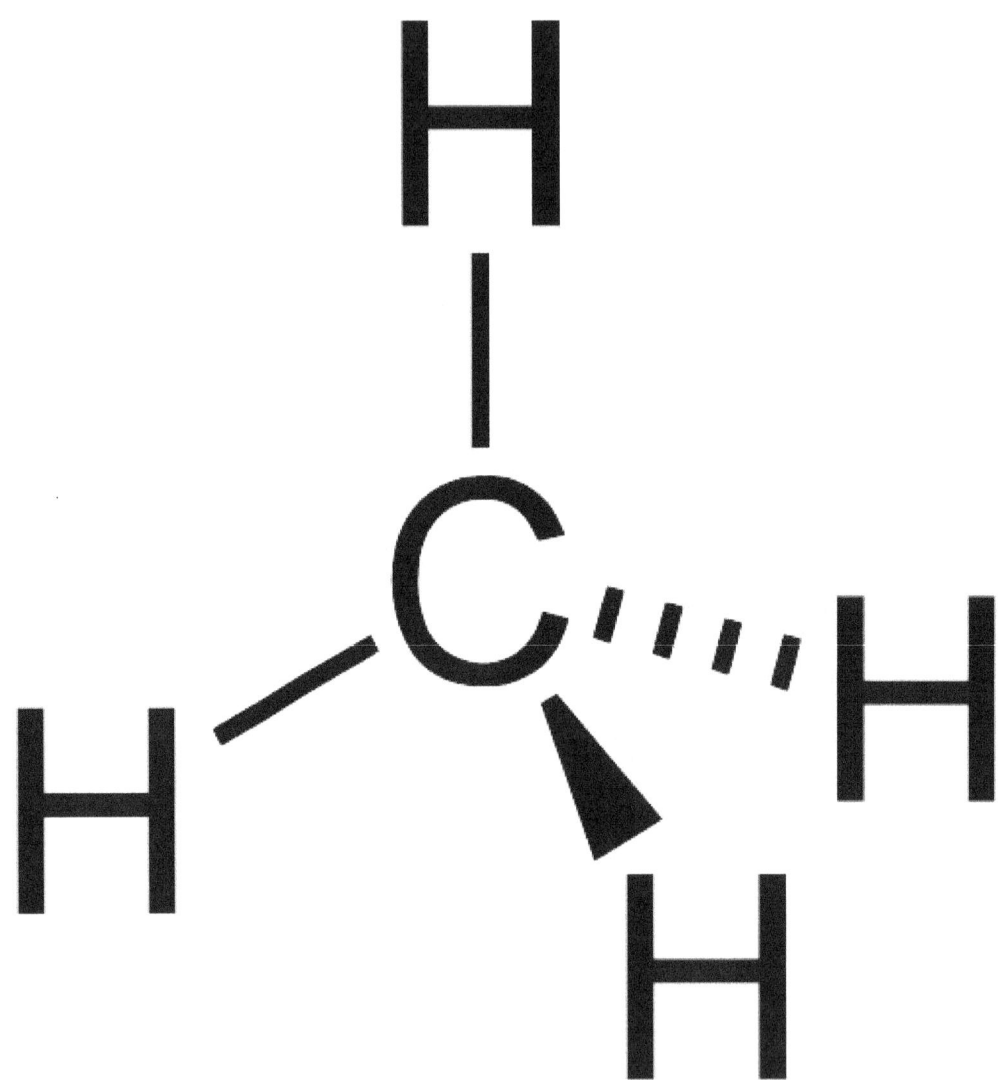

Methane is one of the simplest organic compounds

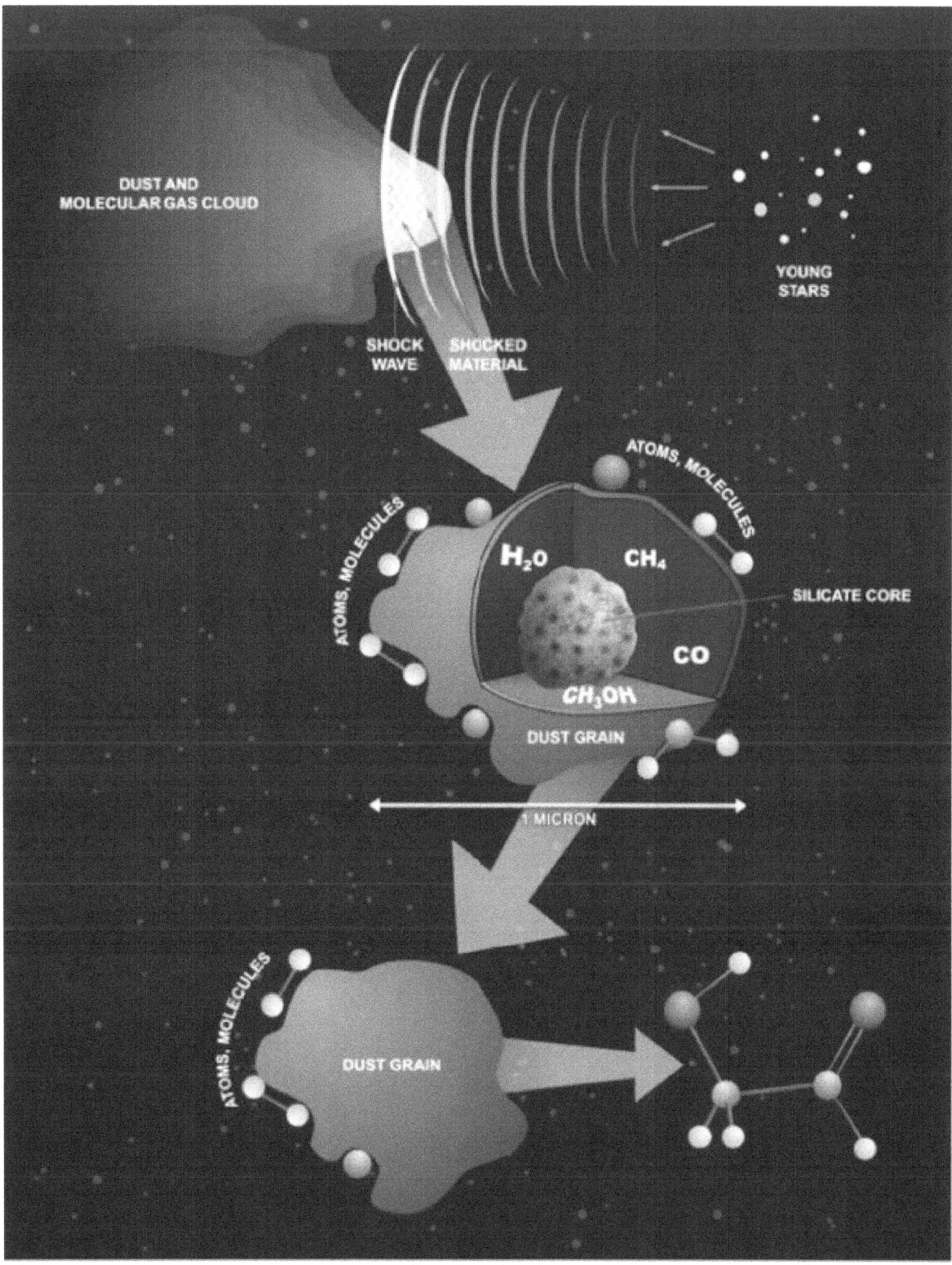

Formation of glycolaldehyde in stardust

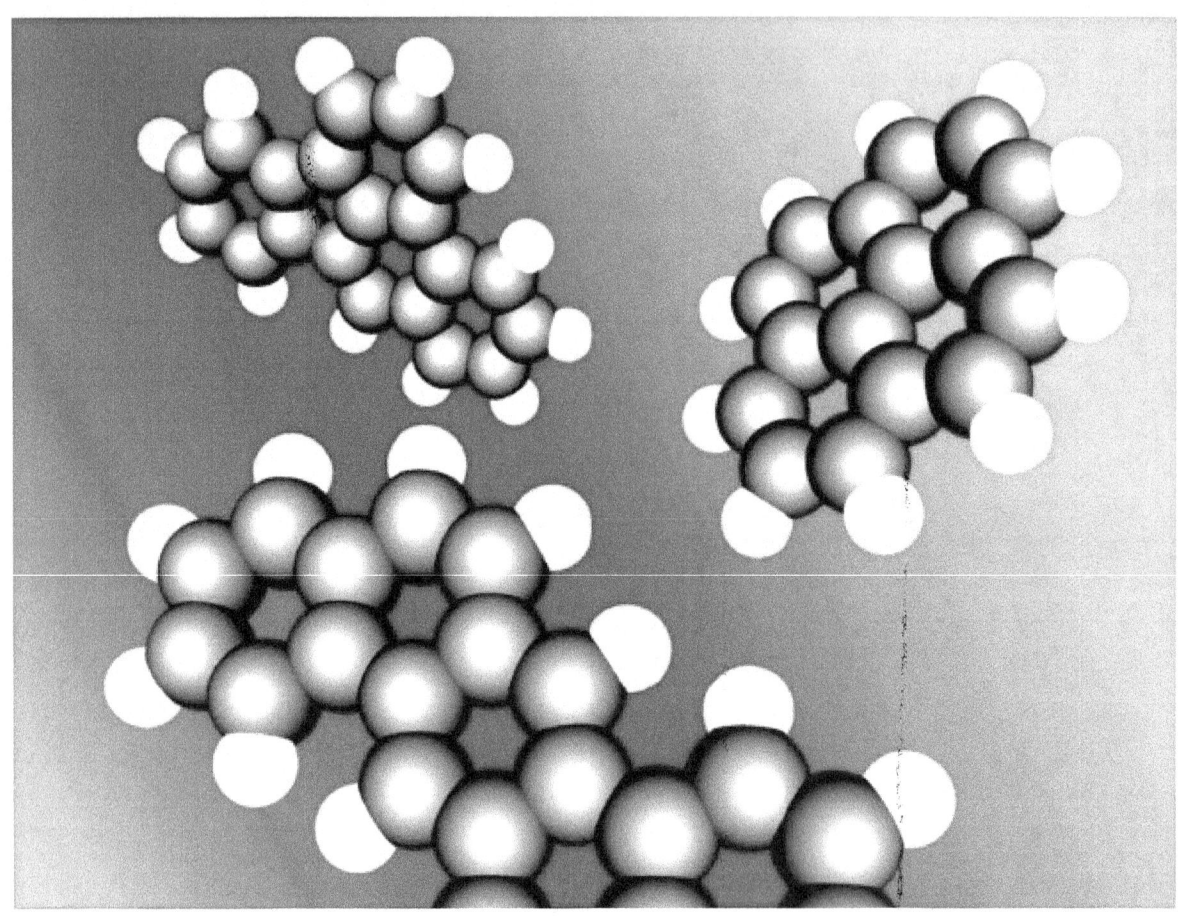

An illustration of typical polycyclic aromatic hydrocarbons. Clockwise from top left: benz(e)acephenanthrylene, pyrene and dibenz(ah)anthracene.

Chapter 10

Circumstellar habitable zone

"Goldilocks zone" redirects here. For the planet originally nicknamed "Goldilocks", see 70 Virginis b. For the more general Goldilocks principle, see Goldilocks principle.

"Habitable zone" redirects here. For the galactic zone, see Galactic habitable zone.

"Comfort zone (astronomy)" redirects here. For other uses, see Comfort zone (disambiguation).

In astronomy and astrobiology, the **circumstellar habitable zone (CHZ)**, or simply the **habitable zone**, is the region

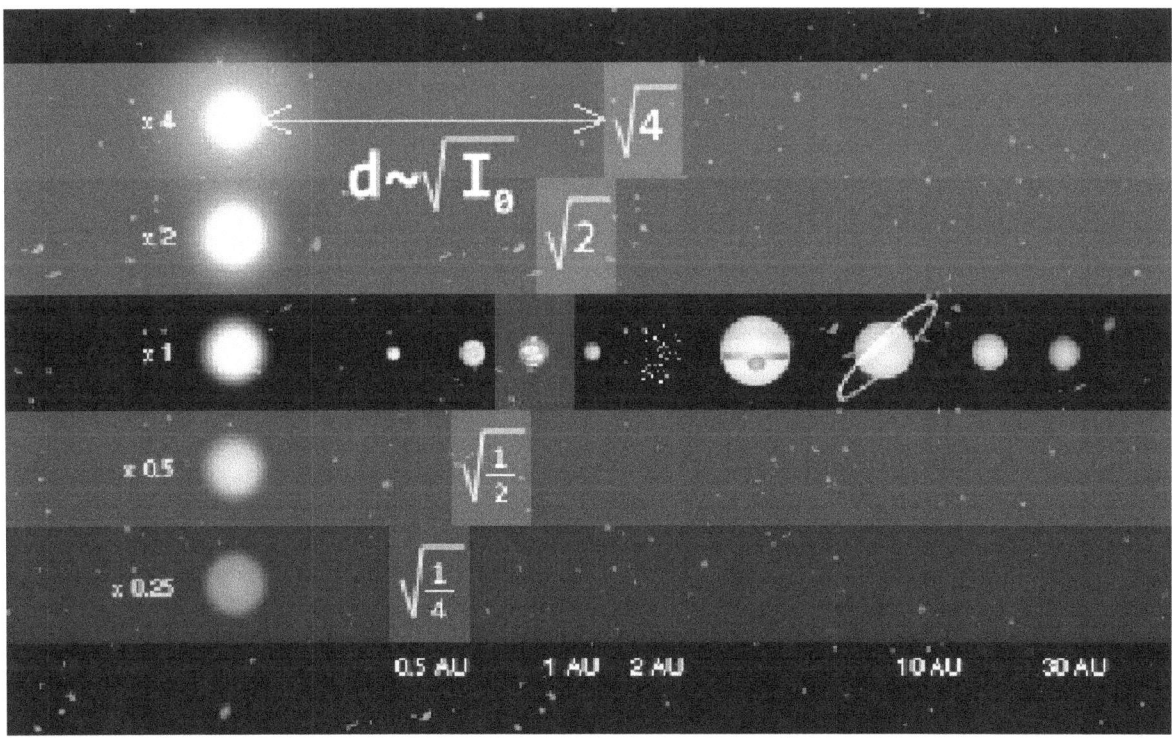

An example of a system based on stellar luminosity for predicting the location of the habitable zone around various types of stars. Planet sizes, star sizes, orbit lengths, and habitable zone sizes are not to scale.

around a star within which planetary-mass objects with sufficient atmospheric pressure can support liquid water at their surfaces.[1][2] The bounds of the CHZ are calculated using the known requirements of Earth's biosphere, its position in the Solar System and the amount of radiant energy it receives from the Sun. Due to the importance of liquid water to life as it exists on Earth, the nature of the CHZ and the objects within is believed to be instrumental in determining the scope and distribution of Earth-like extraterrestrial life and intelligence.

The habitable zone is also called the **Goldilocks zone**, a metaphor of the children's fairy tale of *Goldilocks and the Three Bears*, in which a little girl chooses from sets of three items, ignoring the ones that are too extreme (large or small, hot or cold, etc.), and settling on the one in the middle, which is "just right".

Since the concept was first presented in 1953,[3] stars have been confirmed to possess a CHZ planet, including some systems that consist of multiple CHZ planets.[4] Most such planets, being super-Earths or gas giants, are more massive than Earth, because such planets are easier to detect. On November 4, 2013, astronomers reported, based on *Kepler* data, that there could be as many as 40 billion Earth-sized planets orbiting in the habitable zones of Sun-like stars and red dwarfs in the Milky Way.[5][6] 11 billion of these may be orbiting Sun-like stars.[7] The nearest such planet may be 12 light-years away, according to the scientists.[5][6] The CHZ is also of particular interest to the emerging field of habitability of natural satellites, because planetary-mass moons in the CHZ might outnumber planets.[8]

In subsequent decades, the CHZ concept began to be challenged as a primary criterion for life. Since the discovery of evidence for extraterrestrial liquid water, substantial quantities of it are now believed to occur outside the circumstellar habitable zone. Sustained by other energy sources, such as tidal heating[9][10] or radioactive decay[11] or pressurized by other non-atmospheric means, the basic conditions for water-dependent life may be found even in interstellar space, on rogue planets, or their moons.[12] Liquid water can also exist at a wider range of temperatures and pressures as a solution, for example with sodium chlorides in seawater on Earth, chlorides and sulphates on Equatorial Mars,[13] or ammoniates,[14] due to its different colligative properties. In addition, other circumstellar zones, where non-water solvents favorable to hypothetical life based on alternative biochemistries could exist in liquid form at the surface, have been proposed.[15]

10.1 History

The concept of a Circumstellar Habitable Zone was first introduced in 1953 by Hubertus Strughold, who in his treatise *The Green and the Red Planet: A Physiological Study of the Possibility of Life on Mars* coined the term "ecosphere" and referred to various "zones" in which life could emerge.[3][16] In the same year, Harlow Shapley wrote "Liquid Water Belt", which described the same theory in further scientific detail. Both works stressed the importance of liquid water to life.[17] Su-Shu Huang, an American astrophysicist, first introduced the term "habitable zone" in 1959 to refer to the area around a star where liquid water could exist on a sufficiently large body, and was the first to introduce it in the context of planetary habitability and extraterrestrial life.[18][19] A major early contributor to habitable zone theory, Huang argued in 1960 that circumstellar habitable zones, and by extension extraterrestrial life, would be uncommon in multiple star systems, given the gravitational instabilities of those systems.[20]

The theory of habitable zones was further developed in 1964 by Stephen H. Dole in his book *Habitable Planets for Man*, in which he covered the circumstellar habitable zone itself as well as various other determinants of planetary habitability, eventually estimating the number of habitable planets in the Milky Way to be about 600 million.[21] At the same time, science-fiction author Isaac Asimov introduced the concept of a circumstellar habitable zone to the general public through his various explorations of space colonization.[22] The term "Goldilocks zone" emerged in the 1970s, referencing specifically a region around a star whose temperature is "just right" for water to be present in the liquid phase.[23] In 1993, astronomer James Kasting introduced the term "circumstellar habitable zone" to refer more precisely to the region then (and still) known as the habitable zone.[18]

An update to habitable-zone theory came in 2000, when astronomers Peter Ward and Donald Brownlee introduced the idea of the "galactic habitable zone", which they later developed with Guillermo Gonzalez.[24][25] The galactic habitable zone, defined as the region where life is most likely to emerge in a galaxy, encompasses those regions close enough to a galactic center that stars there are enriched with heavier elements, but not so close that star systems, planetary orbits, and the emergence of life would be frequently disrupted by the intense radiation and enormous gravitational forces commonly found at galactic centers.[24]

Subsequently, several planetary scientists have criticized the circumstellar habitable zone theory for its "carbon chauvinism", proposing that the concept be extended to other solvents, such as ammonia or methane, which could be the basis of life based on an alternative biochemistry.[15] In 2013, further developments in habitable zone theory were made with the proposal of a circum*planetary* habitable zone, also known as the "habitable edge", to encompass the region around a planet where the orbits of natural satellites would not be disrupted, and at the same time tidal heating from the planet would not cause liquid water to boil away.[26]

10.2   Determination of the circumstellar habitable zone

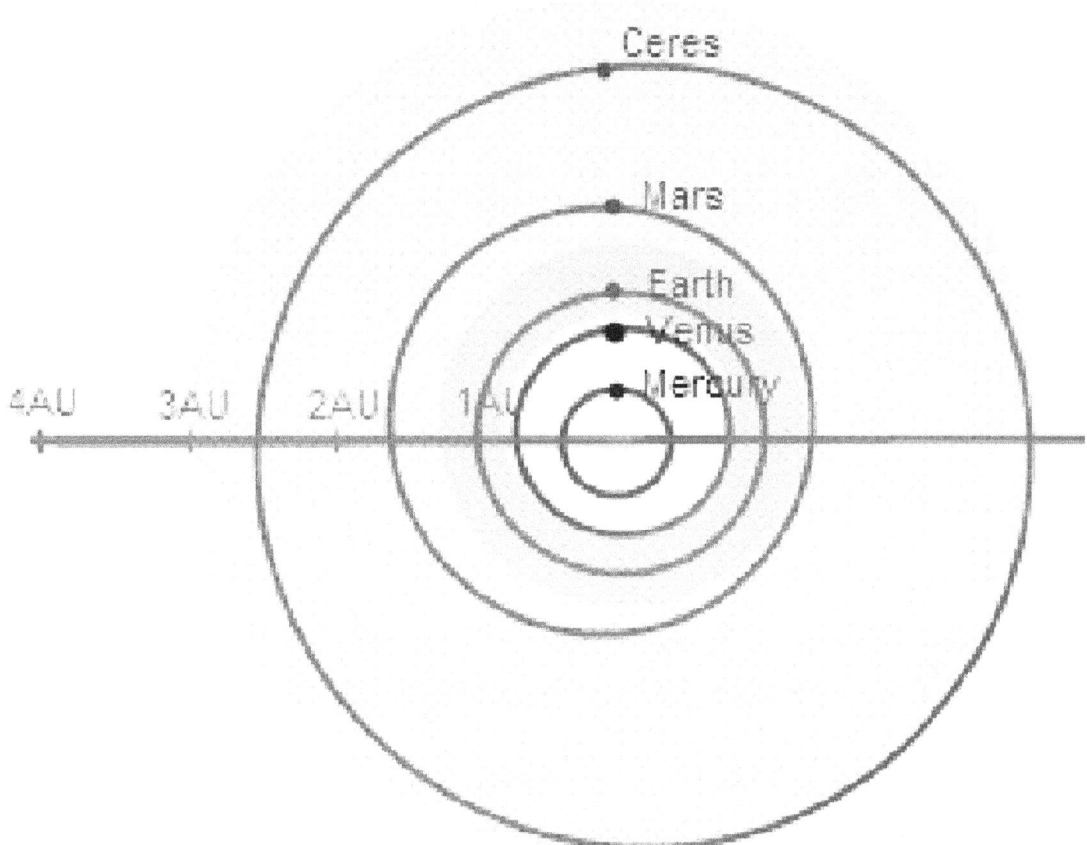

The range of published estimates for the extent of the Sun's CHZ. The conservative CHZ[26] is indicated by a dark-green band crossing the inner edge of the aphelion of Venus, whereas an extended CHZ,[27] extending to the orbit of the dwarf planet Ceres, is indicated by a light-green band.

Whether a body is in the circumstellar habitable zone of its host star is dependent on the radius of the planet's orbit (for natural satellites, the host planet's orbit), the mass of the body itself, and the radiative flux of the host star. Given the large spread in the masses of planets within a circumstellar habitable zone, coupled with the discovery of super-Earth planets which can sustain thicker atmospheres and stronger magnetic fields than Earth, circumstellar habitable zones are now split into two separate regions—a "conservative habitable zone" in which lower-mass planets like Earth or Venus can remain habitable, complemented by a larger "extended habitable zone" in which super-Earth planets, with stronger greenhouse effects, can have the right temperature for liquid water to exist at the surface.[28]

10.2.1   Solar System estimates

Estimates for the habitable zone within the Solar System range from 0.725 to 3.0 astronomical units, though arriving at these estimates has been challenging for a variety of reasons. Numerous planetary mass objects orbit within, or close to, this range and as such receive sufficient sunlight to raise temperatures above the freezing point of water. However their atmospheric conditions vary substantially. The aphelion of Venus, for example, touches the inner edge of the zone and while atmospheric pressure at the surface is sufficient for liquid water, a strong greenhouse effect raises surface temperatures to 462 °C (864 °F) at which water can only exist as vapour.[29] The entire orbits of the Moon,[30] Mars,[31] and numerous asteroids also lie within various estimates of the habitable zone. Only at Mars' lowest elevations (less than 30% of the planet's surface) is atmospheric pressure and temperature sufficient for water to, if present, exist in liquid form for short periods.[32] At Hellas Basin, for example, atmospheric pressures can reach 1,115 Pa and temperatures above zero (around the triple point for water) for 70 days in the Martian year.[32] Despite indirect evidence in the form of seasonal flows on warm Martian slopes,[33][34][35][36] no confirmation has been made of the presence of liquid water there. While other objects orbit partly within this zone, including comets, Ceres[37] is the only one of planetary mass. A combination of low mass and an inability to mitigate evaporation and atmosphere loss against the solar wind make it impossible for these bodies to sustain liquid water on their surface. Most estimates, therefore, are inferred from the effect that a repositioned orbit would have on the habitability of Earth or Venus.

According to extended habitable zone theory, planetary mass objects with atmospheres capable of inducing sufficient radiative forcing could possess liquid water farther out from the Sun. Such objects could include those whose atmospheres contain a high component of greenhouse gas and terrestrial planets much more massive than Earth (Super-Earth class planets), that have retained atmospheres with surface pressures of up to 100 kbar. There are no examples of such objects in the Solar System to study and not enough is known about the nature of atmospheres of these kinds of extrasolar objects and the net temperature effect of such atmospheres including induced albedo, anti-greenhouse or other possible heat sources cannot be determined by their position in the habitable zone.

10.2.2   Extrasolar extrapolation

See also: Habitability of red dwarf systems

Astronomers use stellar flux and the inverse-square law to extrapolate cirumstellar-habitable-zone models created for the Solar System to other stars. For example, although the Solar System has a circumstellar habitable zone centered at 1.34 AU from the Sun,[1] a star with 0.25 times the luminosity of the Sun would have a habitable zone centered at $\sqrt{0.25}$, or 0.5, the distance from the star, corresponding to a distance of 0.67 AU. Various complicating factors, though, including the individual characteristics of stars themselves, mean that extrasolar extrapolation of the CHZ concept is more complex.

Spectral types and star-system characteristics

Some scientists argue that the concept of a circumstellar habitable zone is actually limited to stars in certain types of systems or of certain spectral types. Binary systems, for example, have circumstellar habitable zones that differ from those of single-star planetary systems, in addition to the orbital-stability concerns inherent with a three-body configuration.[47] If the Solar System were such a binary system, the outer limits of the resulting circumstellar habitable zone could extend as far as 2.4 AU.[48][49]

With regard to spectral types, Zoltán Balog proposes that O-type stars cannot form planets due to the photoevaporation caused by their strong ultraviolet emissions.[50] Studying ultraviolet emissions, Andrea Buccino found that only 40 percent of stars studied (including the Sun) had overlapping liquid water and ultraviolet habitable zones.[51] Stars smaller than the Sun, on the other hand, have distinct impediments to habitability. Michael Hart, for example, proposed that only main-sequence stars of spectral class K0 or brighter could possess habitable zones, an idea which has evolved in modern times into the concept of a tidal locking radius for red dwarfs. Within this radius, which is coincidental with the red-dwarf habitable zone, it has been suggested that the volcanism caused by tidal heating could cause a "tidal Venus" planet with high temperatures and no ability to support life.[52]

Others maintain that circumstellar habitable zones are more common and that it is indeed possible for water to exist on planets orbiting cooler stars. Climate modelling from 2013 supports the idea that red dwarf stars can support planets with relatively constant temperatures over their surfaces in spite of tidal locking.[53] Astronomy professor Eric Agol argues that even white dwarfs may support a relatively brief habitable zone through planetary migration.[54] At the same time, others have written in similar support of semi-stable, temporary habitable zones around brown dwarfs.[52]

A video explaining the significance of the 2011 discovery of a planet in the circumbinary habitable zone of Kepler-47.

Natural defenses against space weather, such as the magnetosphere depicted in this artistic rendition, may be required for planets to sustain surface water for prolonged periods.

Stellar evolution

Circumstellar habitable zones change over time with stellar evolution. For example, hot O-type stars, which may remain on the main sequence for fewer than 10 million years,[55] would have rapidly changing habitable zones not conducive to the development of life. Red dwarf stars, on the other hand, which can live for hundreds of billions of years on the main sequence, would have planets with ample time for life to develop and evolve.[56][57] Even while stars are on the main sequence, though, their energy output steadily increases, pushing their habitable zones farther and farther out: our Sun, for example, was only 75 percent as bright in the Archaean as it is now,[58] and in the future continued increases in energy output will put Earth outside the Sun's habitable zone, even before it reaches the red giant phase.[59] In order to deal with this increase in luminosity, the concept of a *continuously habitable zone* has been introduced. As the name suggests, the continuously habitable zone is a region around a star in which planetary-mass bodies can sustain liquid water for a given period of time. Like the general circumstellar habitable zone, the continuously habitable zone of a star is divided into a conservative and extended region.[59]

In red dwarf systems, gigantic stellar flares which could double a star's brightness in minutes[60] and huge starspots which can cover 20 percent of the star's surface area,[61] have the potential to strip an otherwise habitable planet of its atmosphere and water.[62] As with more massive stars, though, stellar evolution changes their nature,[63] so by about 1.2 billion years of age, red dwarfs generally become sufficiently constant to allow for the development of life.[62][64]

Once a star has evolved sufficiently to become a red giant, its circumstellar habitable zone will change dramatically from its main-sequence size. For example, the Sun is expected to engulf the previously-habitable Earth as a red giant.[65] However, once a red giant star reaches the horizontal branch, it achieves a new equilibrium and can sustain a circumstellar habitable zone, which in the case of the Sun would range from 7 to 22 AU.[66] At such stage, Saturn's moon Titan would likely be habitable in Earth's sense.[67] Given that this new equilibrium lasts for about 1 Gyr, and because life on Earth emerged by 0.7 Gyr from the formation of the Solar System at latest, life could conceivably develop on planetary mass objects in the habitable zone of red giants.[66] However, around such a helium-burning star, important life processes like photosynthesis could only happen around planets where the atmosphere has been artificially seeded with carbon dioxide, as by the time a solar-mass star becomes a red giant, planetary-mass bodies would have already absorbed much of their free carbon dioxide.[68]

Desert planets

A planet's atmospheric conditions influence its ability to retain heat, so that the location of the habitable zone is also specific to each type of planet: desert planets (also known as dry planets), with very little water, will have less water vapor in the atmosphere than Earth and so have a reduced greenhouse effect, meaning that a desert planet could maintain oases of water closer to its star than Earth is to the Sun. The lack of water also means there is less ice to reflect heat into space, so the outer edge of desert-planet habitable zones is further out.[69][70]

Other considerations

See also: Planetary habitability and Natural satellite habitability

A planet cannot have a hydrosphere—a key ingredient for the formation of carbon-based life—unless there is a source for water within its stellar system. The origin of water on Earth is still not completely understood; possible sources include the result of impacts with icy bodies, outgassing, mineralization, leakage from hydrous minerals from the lithosphere, and photolysis.[71][72] For an extrasolar system, an icy body from beyond the frost line could migrate into the habitable zone of its star, creating an ocean planet with seas hundreds of kilometers deep[73] such as GJ 1214 b[74][75] or Kepler-22b may be.[76]

Maintenance of liquid surface water also requires a sufficiently thick atmosphere. Possible origins of terrestrial atmospheres are currently theorised to outgassing, impact degassing and ingassing.[77] Atmospheres are thought to be maintained through similar processes along with biogeochemical cycles and the mitigation of atmospheric escape.[78] In a 2013 study led by Italian astronomer Giovanni Vladilo, it was shown that the size of the circumstellar habitable zone increased with greater atmospheric pressure.[45] Below an atmospheric pressure of about 15 millibars, it was found that habitability could not be maintained[45] because even a small shift in pressure or temperature could render water unable to form a liquid.[79]

In the case of planets orbiting in the CHZs of red dwarf stars, the extremely close distances to the stars cause tidal locking, an important factor in habitability. For a tidally locked planet, the sidereal day is as long as the orbital period, causing one side to permanently face the host star and the other side to face away. In the past, such tidal locking was believed to cause

Earth's hydrosphere. Water covers 71% of Earth's surface, with the global ocean accounting for 97.3% of the water distribution on Earth.

extreme heat on the star-facing side and bitter cold on the opposite side, making many red dwarf planets uninhabitable; however, a 2013 paper written by geophysicist Jun Yang of the University of Chicago and collaborators, using three-dimensional climate models, showed that the side of a red dwarf planet facing the host star would have extensive cloud cover, increasing its Bond albedo and reducing significantly temperature differences between the two sides.[53]

Planetary-mass natural satellites have the potential to be habitable as well. However, these bodies need to fulfill additional parameters, in particular being located within the circumplanetary habitable zones of their host planets.[26] More specifically, planets need to be far enough from their host giant planets that they are not transformed by tidal heating into volcanic worlds like Io,[26] but must still remain within the Hill radius of the planet so that they are not pulled out of orbit of their host planet.[80] Red dwarfs that have masses less than 20 percent of that of the Sun cannot have habitable moons around giant planets, as the small size of the circumstellar habitable zone would put a habitable moon so close to a star that it would be stripped from its host planet. In such a system, a moon close enough to its host planet to maintain its orbit would have tidal heating so intense as to eliminate any prospects of habitability.[26]

A planetary object that orbits a star with high orbital eccentricity may spend only some of its year in the CHZ and experience a large variation in temperature and atmospheric pressure. This would result in dramatic seasonal phase shifts where liquid water may exist only intermittently. It is possible that subsurface habitats could be insulated from such changes and that extremophiles on or near the surface might survive through adaptions such as hibernation (cryptobiosis) and/or hyperthermostability. Tardigrades, for example, can survive in a dehydrated state temperatures between 0.150 K (−273 °C)[81] and 424 K (151 °C).[82] Life on a planetary object orbiting outside CHZ might hibernate on the cold side as the planet approaches the apastron where the planet is coolest and become active on approach to the periastron when the planet is sufficiently warm.[83]

10.3 Extrasolar discoveries

See also: List of potentially habitable exoplanets

Among exoplanets, a review in 2015 came to the conclusion that Kepler-62f, Kepler-186f and Kepler-442b were likely the best candidates for being potentially habitable.[84] These are at a distance of 1200, 490 and 1,120 light-years away, respectively. Of these, Kepler-186f is in similar size to Earth with its 1.2-Earth-radius measure, and it is located towards

Artists concept of a planet on an eccentric orbit that passes through the CHZ for only part of its year

the outer edge of the habitable zone around its red dwarf sun. Among nearest terrestrial exoplanet candidates, Tau Ceti e is merely 11.9 light-years away. It's in the inner edge of its solar system's habitable zone, giving it an estimated average surface temperature of 68 °C (154 °F).[85]

Studies that have attempted to estimate the number of terrestrial planets within the circumstellar habitable zone tend to reflect the availability of scientific data. A 2013 study by Ravi Kumar Kopparapu put ηe, the fraction of stars with planets in the CHZ, at 0.48,[1] meaning that there may be roughly 95–180 billion habitable planets in the Milky Way.[86] However, this is merely a statistical prediction; only a small fraction of these possible planets have yet been discovered.[87]

Previous studies have been more conservative. In 2011, Seth Borenstein concluded that there are roughly 500 million habitable planets in the Milky Way.[88] NASA's Jet Propulsion Laboratory 2011 study, based on observations from the *Kepler* mission, raised the number somewhat, concluding that about "1.4 to 2.7 percent" of all stars of spectral class F, G, and K are expected to have planets in their CHZs.[89][90]

10.3.1 Early findings

See also: Category:Gas giants in the habitable zone.

The first discoveries of extrasolar planets in the CHZ occurred just a few years after the first extrasolar planets were discovered. One of the first discoveries was 70 Virginis b, a gas giant initially nicknamed "Goldilocks" due to it being neither "too hot" nor "too cold." Later study revealed temperatures analogous to Venus ruling out any potential for liquid water.[91] 16 Cygni Bb, also discovered in 1996, has an extremely eccentric orbit that causes extreme seasonal effects on the planet's surface. In spite of this, simulations have suggested that it is possible for a terrestrial natural satellite to support water at its surface year-round.[92]

Gliese 876 b, discovered in 1998, and Gliese 876 c, discovered in 2001, are both gas giants discovered in the habitable zone around Gliese 876. Although they are not thought to themselves possess significant water at their surfaces, both may have habitable moons.[93] Upsilon Andromedae d, discovered in 1999, is a gas giant in its star's circumstellar habitable zone considered to be large enough to favor the formation of large, Earth-like moons.[94]

Announced on April 4, 2001, HD 28185 b is a gas giant found to orbit entirely within its star's circumstellar habitable zone[95] and has a low orbital eccentricity, comparable to that of Mars in the Solar System.[96] Tidal interactions suggest that HD 28185 b could harbor habitable Earth-mass satellites in orbit around it for many billions of years,[97] though it is unclear whether such satellites could form in the first place.[98]

HD 69830 d, a gas giant with 17 times the mass of Earth, was in 2006 found orbiting within the circumstellar habitable zone of HD 69830, 41 light years away from Earth.[99] The following year, 55 Cancri f was discovered within the CHZ of its host star 55 Cancri A.[100][101] Although conditions on this massive and dense planet are not conducive to the formation of water or life as we know it, a hypothetical moon of this planet with the proper mass and composition could be able to support liquid water at its surface.[102]

10.3.2   Habitable super-Earths

See also: Category:Super-Earths in the habitable zone.

 The 2007 discovery of Gliese 581 c, the first super-Earth in the circumstellar habitable zone, created significant interest in

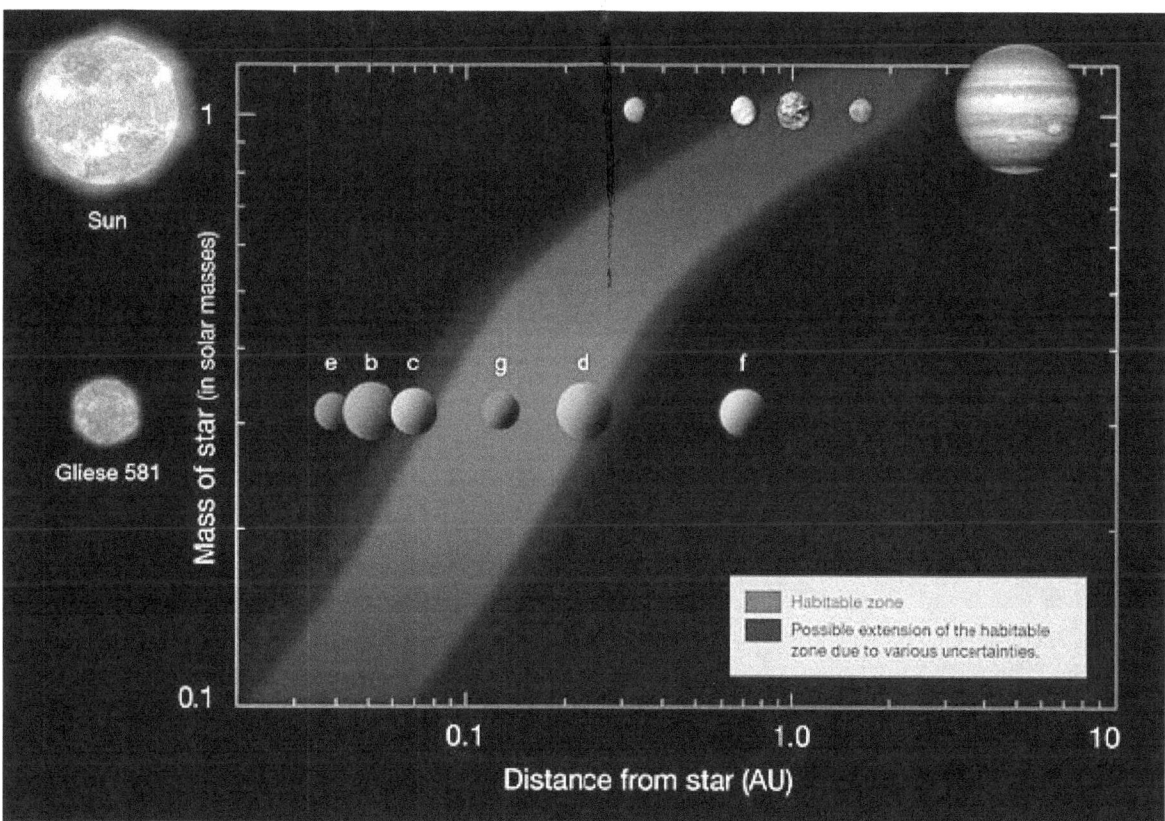

The habitable zone of Gliese 581 compared with our Solar System's habitable zone.

the system by the scientific community, although the planet was later found to have surface conditions that likely resemble Venus more than Earth.[103] Gliese 581 d, another planet in the same system and thought to be a better candidate for habitability, was also announced in 2007. Its existence was later disconfirmed in 2014. Gliese 581 g, yet another planet thought to have been discovered in the circumstellar habitable zone of the system, was considered to be more habitable than both Gliese 581 c and d. However, its existence was also disconfirmed in 2014.[104]

Discovered in August 2011, HD 85512 b was initially believed to be habitable,[105] but the new circumstellar-habitable-

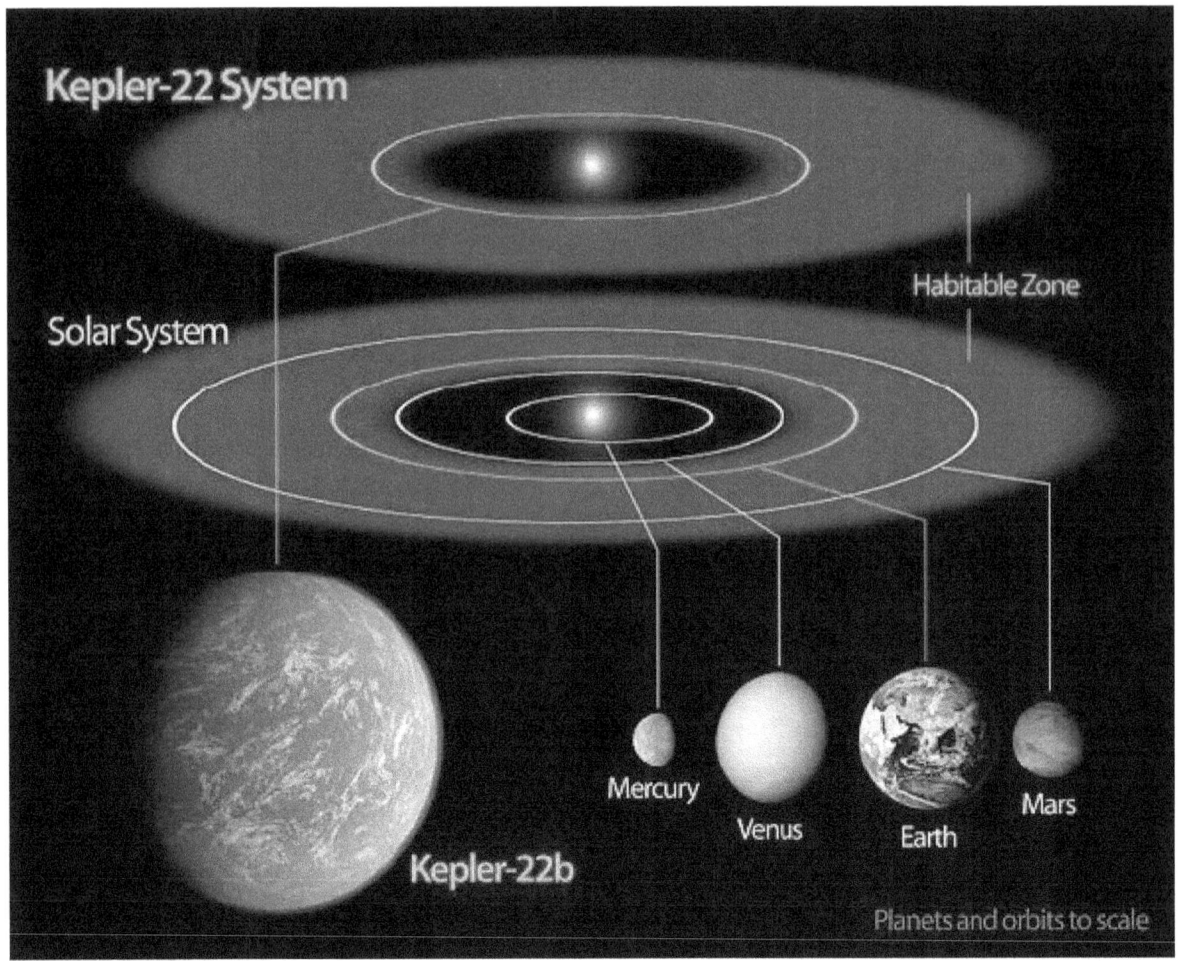

A diagram comparing size (artist's impression) and orbital position of planet Kepler-22b within Sun-like star Kepler 22's habitable zone and that of Earth in the Solar System

zone criteria devised by Kopparapu et al. in 2013 place the planet outside the circumstellar habitable zone.[87] With an increase in the intensity of exoplanet discovery, the Earth Similarity Index was devised in October 2011 as a way of comparing planetary properties, such as surface temperature and density, to those of Earth in order to better gauge the habitability of extrasolar bodies.[106]

Kepler-22 b, discovered in December 2011 by the *Kepler* space probe,[107] is the first transiting exoplanet discovered around a sunlike star. With a radius 2.4 times that of Earth, Kepler-22b has been predicted by some to be an ocean planet.[108] Gliese 667 Cc, discovered in 2011 but announced in 2012,[109] is a super-Earth orbiting in the circumstellar habitable zone of Gliese 667 C. Subsequently in June 2013, two other habitable super-Earths orbiting the same star, Gliese 667 Cf and Gliese 667 Ce, were discovered in the CHZ.[110]

Gliese 163 c, discovered in September 2012 in orbit around the red dwarf Gliese 163[111] is located 49 light years from Earth. The planet has 6.9 Earth masses and 1.8–2.4 Earth radii, and with its close orbit receives 40 percent more stellar radiation than Earth, leading to surface temperatures of about 60° C.[112][113][114] HD 40307 g, a candidate planet tentatively discovered in November 2012, is in the circumstellar habitable zone of HD 40307.[115] In December 2012, Tau Ceti e and Tau Ceti f were found in the circumstellar habitable zone of Tau Ceti, a sunlike star just 12 light years away.[116] Although more massive than Earth, they are among the least massive planets found to date orbiting in the zone;[117] however, Tau Ceti f, like HD 85512 b, did not fit the new circumstellar-habitable-zone criteria established by the 2013 Kopparapu study.[118]

10.3.3   Earth-sized planets and Solar analogs

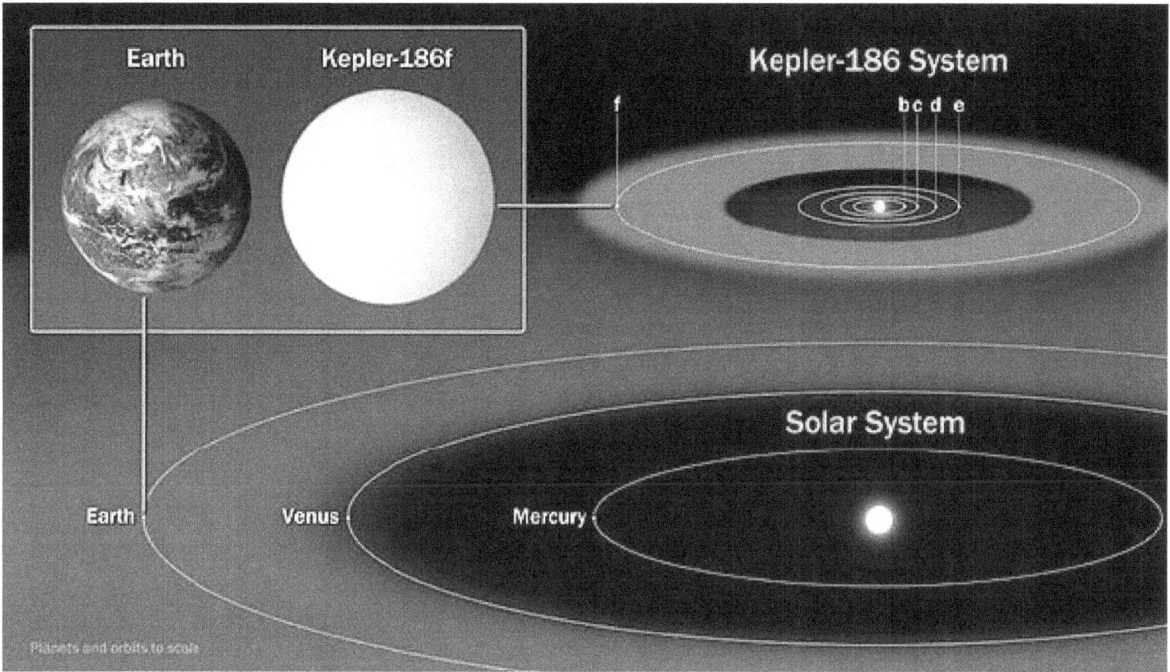

Comparison of the CHZ position of Earth-radius planet Kepler-186f and the Solar System (17 April 2014)

Recent discoveries have uncovered planets that are believed to be similar in many ways to the Earth (that is Earth analogs, or terrestrial planets relatively high Earth Similarity Indexes). While there is no universal definition of "Earth-sized", ranges are typically defined by mass. The lower range used in many definitions of the Super-Earth class is 1.9 Earth masses, likewise. Sub-Earths range up to the size of Venus (~0.815 Earth masses). An upper limit of 1.5 Earth radii is also considered, given that above 1.5 $R_\oplus$ the average planet density rapidly decreases with increasing radius, indicating that these planets have a large fraction of volatiles by volume overlying a rocky core.[119]

On 7 January 2013, astronomers from the *Kepler* team announced the discovery of Kepler-69c (formerly *KOI-172.02*), an Earth-like exoplanet candidate (1.7 times the radius of Earth) orbiting Kepler-69, a star similar to our Sun, in the CHZ and a "prime candidate to host alien life".[120][121][122][123] The discovery of two planets orbiting in the habitable zone of Kepler-62, by the Kepler team was announced on April 19, 2013. The planets, named Kepler-62e and Kepler-62f, are likely solid planets with sizes 1.6 and 1.4 times the radius of Earth, respectively.[122][124][125]

With a radius measured at 1.1 Earth, Kepler-186f, discovery announced in April 2014, is the closest yet size to Earth of an exoplanet confirmed by the transit method[126][127][128] though its mass remains unknown and its parent star is not a Solar analog.

On 6 January 2015, NASA announced the 1000th confirmed exoplanet discovered by the Kepler Space Telescope. Three of the newly confirmed exoplanets were found to orbit within habitable zones of their related stars: two of the three, Kepler-438b and Kepler-442b, are near-Earth-size and likely rocky; the third, Kepler-440b, is a super-Earth.[129] (Announced 16 January, EPIC 201367065 d is a planet of 1.5 Earth radii found to orbit within a habitable zone (as calculated by Selsis, Kasting et al.) of EPIC 201367065, receiving 1.4 times the intensity of visible light as Earth.[130]

Kepler-452b, publicly announced on 23 July 2015 is 50% bigger than Earth, likely rocky and takes approximately 385 Earth days orbit in the habitable zone of its G-class (solar analog) star Kepler-452.[131][132]

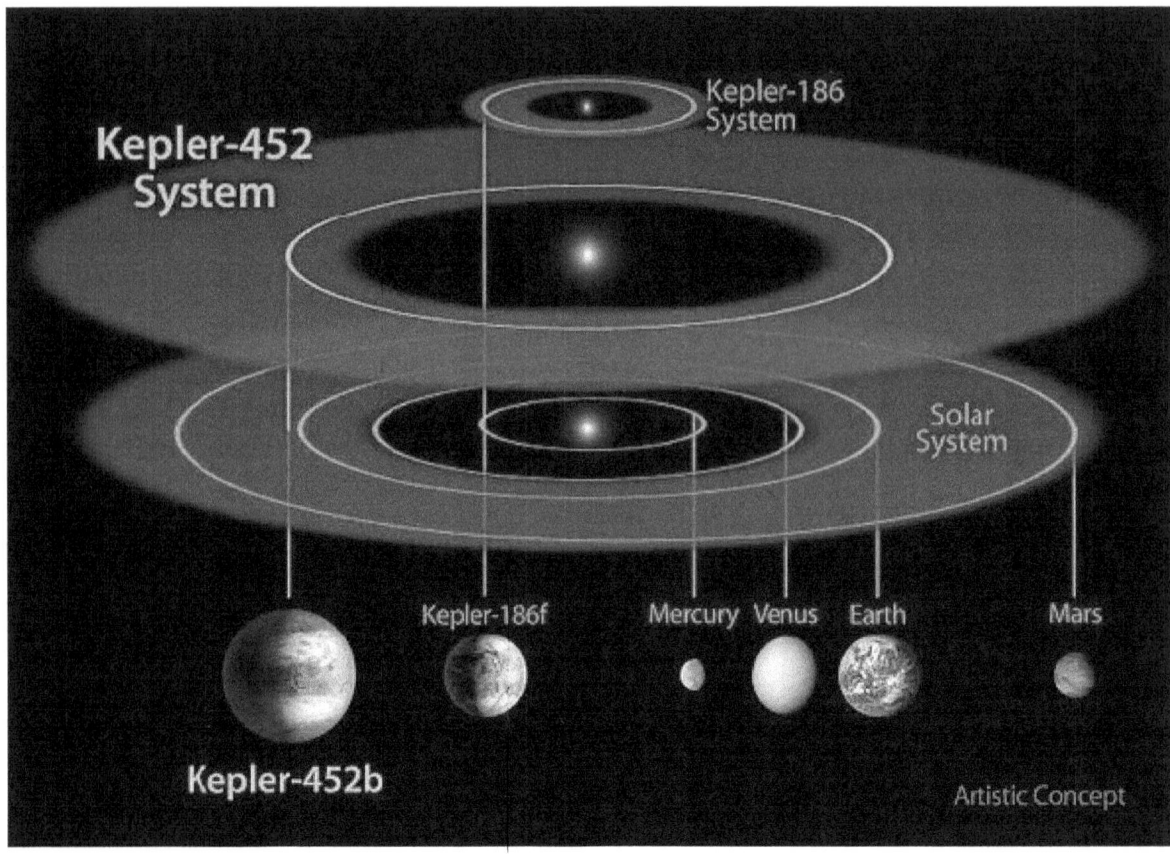

While larger than Kepler 186f, Kepler-452b's orbit and star are more similar to Earth's.

10.4   Habitability outside the CHZ

Liquid-water environments have been found to exist in the absence of atmospheric pressure, and at temperatures outside the CHZ temperature range. For example, Saturn's moon Titan and Jupiter's Europa, both outside the habitable zone, may hold large volumes of liquid water in subsurface oceans.[133]

Outside the CHZ, tidal heating and radioactive decay are two possible heat sources that could contribute to the existence of liquid water.[9][10] Abbot and Switzer (2011) put forward the possibility that subsurface water could exist on rogue planets as a result of radioactive decay-based heating and insulation by a thick surface layer of ice.[12]

With some theorising that life on Earth may have actually originated in stable, subsurface habitats,[134][135] it has been suggested that it may be common for wet subsurface extraterrestrial habitats such as these to 'teem with life'.[136] Indeed, on Earth itself living organisms may be found more than 6 kilometres below the surface.[137]

Another possibility is that outside the CHZ organisms may use alternative biochemistries that do not require water at all. Astrobiologists, including NASA's Christopher McKay, have suggested that methane may be a solvent conducive to the development of "cryolife", with the Sun's "methane habitable zone" being centered on 1,610,000,000 km (1.0×10^9 mi; 11 AU) from the star.[15] This distance is coincidental with the location of Titan, whose lakes and rain of methane make it an ideal location to find McKay's proposed cryolife.[15] In addition, testing of a number of organisms has found some are capable of surviving in extra-CHZ conditions.[138]

The discovery of hydrocarbon lakes on Saturn's moon Titan has begun to call into question the carbon chauvinism that underpins CHZ theory.

10.5 Significance for complex and intelligent life

The Rare Earth hypothesis argues that complex and intelligent life is uncommon and that the CHZ is one of many critical factors. According to Ward & Brownlee (2004) and others, not only is a CHZ orbit and surface water a primary requirement to sustain life but a requirement to support the secondary conditions required for multicellular life to emerge and evolve. The secondary habitability factors are both geological (the role of surface water in sustaining necessary plate tectonics)[24] and biochemical (the role of radiant energy in support photosynthesis for necessary atmospheric oxygenation).[139] But others, such as Ian Stewart and Jack Cohen in their 2002 book *Evolving the Alien* argue that complex intelligent life may arise outside the CHZ.[140] Intelligent life outside the CHZ may have evolved in subsurface environments, from alternative biochemistries[140] or even from nuclear reactions.[141].

On Earth, complex multicellular life has been found with the potential to survive the conditions that might exist outside the CHZ. An animal example of such a life form is the tardigrade, which can withstand both temperatures well above the boiling point of water and the vacuum of outer space.[142] In addition, the plant *Rhizocarpon geographicum* has been found to survive in an environment where the atmospheric pressure is far too low for surface liquid water and where the radiant energy is also much lower than that which most plants require to photosynthesize.[143][144] If the human race, however, is to colonize other planets, true Earth analogs in the CHZ are most likely to provide the closest natural habitats for human beings; this concept was the basis of Stephen H. Dole's 1964 study. With suitable temperature, gravity, atmospheric pressure and the presence of water, the necessity of spacesuits may be eliminated and complex Earth-life can be allowed to flourish.[21]

Planets in the CHZ remain of paramount interest to researchers looking for intelligent life elsewhere in the universe.[145] The 1961 Drake equation, still used as means of calculating the number of intelligent civilizations in our galaxy, contains a parameter ηe, which is generally considered to imply the fraction of stars that have planetary mass objects orbiting within the CHZ. A low value lends support to the Rare Earth hypothesis, which posits that intelligent life is a rarity in the Universe, whereas a high value provides evidence for the Copernican mediocrity principle, the view that habitability—and therefore life—is common throughout the Universe.[24] A 1971 NASA report by Drake and Bernard Oliver proposed the "waterhole", based on the spectral absorption lines of the hydrogen and hydroxyl components of water, as a good, obvious band for communication with extraterrestrial intelligence[146][147] that has since been widely adopted by astronomers involved in the search for extraterrestrial intelligence. According to Jill Tarter, Margaret Turnbull and many others, CHZ candidates are the priority targets to narrow waterhole searches[148][149] and the Allen Telescope Array now extends Project Phoenix to such candidates.[150]

Because the CHZ is considered the most likely habitat for intelligent life, METI efforts have also been focused on systems likely to have planets there. The 2001 Teen Age Message and the 2003 Cosmic Call 2, for example, were sent to the 47 Ursae Majoris system, known to contain three Jupiter-mass planets and possibly with a terrestrial planet in the CHZ.[151][152][153][154] The Teen Age Message, and the later Wow! reply, were also directed to the 55 Cancri system, which has a gas giant in its CHZ.[100] A Message to Earth in 2008, and Hello From Earth in 2009, were directed to the Gliese 581 system, containing three planets in the CHZ—Gliese 581 c, d, and the unconfirmed g.[155]

10.6 See also

- Hypothetical types of biochemistry

- Earth analog

- Earth Similarity Index

- Extraterrestrial liquid water

- Extraterrestrial life

- Galactic habitable zone

- Natural satellite habitability

- Planetary habitability

- Rare Earth hypothesis

- Venus zone

10.7   References

[1] Kopparapu, Ravi Kumar (2013).   "A revised estimate of the occurrence rate of terrestrial planets in the habitable zones around kepler m-dwarfs". *The Astrophysical Journal Letters* **767** (1): L8. arXiv:1303.2649. Bibcode:2013ApJ...767L...8K. doi:10.1088/2041-8205/767/1/L8.

[2] Cruz, Maria; Coontz, Robert (2013). "Exoplanets - Introduction to Special Issue".*Science***340**(6132):   565. doi:10.1126/scienc Retrieved 18 May 2013.

[3] Huggett, Richard J. (1995). *Geoecology: An Evolutionary Approach.* Routledge, Chapman & Hall. p. 10. ISBN 978-0-415-08689-9.

[4] Overbye, Dennis (January 6, 2015). "As Ranks of Goldilocks Planets Grow, Astronomers Consider What's Next". *New York Times.* Retrieved January 6, 2015.

[5] Overbye, Dennis (November 4, 2013). "Far-Off Planets Like the Earth Dot the Galaxy". *New York Times.* Retrieved November 5, 2013.

[6] Petigura, Eric A.; Howard, Andrew W.; Marcy, Geoffrey W. (October 31, 2013). "Prevalence of Earth-size planets orbiting Sun-like stars". *Proceedings of the National Academy of Sciences of the United States of America.* arXiv:1311.6806. Bibcode:2013PNAS..11019273P. doi:10.1073/pnas.1319909110. Retrieved November 5, 2013.

[7] Khan, Amina (November 4, 2013). "Milky Way may host billions of Earth-size planets". *Los Angeles Times.* Retrieved November 5, 2013.

[8] Schirber, Michael (26 Oct 2009). "Detecting Life-Friendly Moons". *Astrobiology Magazine.* NASA. Retrieved 9 May 2013.

[9] Cowen, Ron (2008-06-07). "A Shifty Moon". *Science News.*

[10] Bryner, Jeanna (24 June 2009). "Ocean Hidden Inside Saturn's Moon". *Space.com.* TechMediaNetwork. Retrieved 22 April 2013.

[11] Abbot, D. S.; Switzer, E. R. (2011). "The Steppenwolf: A Proposal for a Habitable Planet in Interstellar Space". *The Astrophysical Journal* **735** (2): L27. arXiv:1102.1108. Bibcode:2011ApJ...735L..27A. doi:10.1088/2041-8205/735/2/L27.

[12] "Rogue Planets Could Harbor Life in Interstellar Space, Say Astrobiologists". *MIT Technology Review.* MIT Technology Review. 9 February 2011. Retrieved 24 June 2013.

[13] Wall, Mike (28 September 2015). "Salty Water Flows on Mars Today, Boosting Odds for Life". *Space.com.* Retrieved 2015-09-28.

[14] Sun, Jiming; Clark, Bryan K.; Torquato, Salvatore; Car, Roberto (2015). "The phase diagram of high-pressure superionic ice". *Nature Communications* **6**: 8156. doi:10.1038/ncomms9156. ISSN 2041-1723.

[15] Villard, Ray (November 18, 2011). "Alien Life May Live in Various Habitable Zones : Discovery News". News.discovery.com. Discovery Communications LLC. Retrieved April 22, 2013.

[16] Strughold, Hubertus (1953). *The Green and Red Planet: A Physiological Study of the Possibility of Life on Mars.* University of New Mexico Press.

[17] Kasting, James (2010). *How to Find a Habitable Planet.* Princeton University Press. p. 127. ISBN 978-0-691-13805-3. Retrieved 4 May 2013.

[18] Kasting, James F.; Whitmire, Daniel P.; Reynolds, Ray T. (January 1993). "Habitable Zones around Main Sequence Stars". *Icarus* **101** (1): 108–118. Bibcode:1993Icar..101..108K. doi:10.1006/icar.1993.1010. PMID 11536936.

[19] Huang, Su-Shu (1966). *Extraterrestrial life: An Anthology and Bibliography.* National Research Council (U.S.). Study Group on Biology and the Exploration of Mars. Washington, D. C.: National Academy of Sciences. pp. 87–93.

[20] Huang, Su-Shu (April 1960). "Life-Supporting Regions in the Vicinity of Binary Systems". *Publications of the Astronomical Society of the Pacific* **72** (425): 106–114. Bibcode:1960PASP...72..106H. doi:10.1086/127489.

[21] Dole, Stephen H (1964). *Habitable Planets for Man*. Blaisdell Publishing Company. p. 103.

[22] Gilster, Paul (2004). *Centauri Dreams: Imagining and Planning Interstellar Exploration*. Springer. p. 40. ISBN 978-0-387-00436-5.

[23] "The Goldilocks Zone" (Press release). NASA. October 2, 2003. Retrieved April 22, 2013.

[24] Brownlee, Donald; Ward, Peter (2004). *Rare Earth: Why Complex Life Is Uncommon in the Universe*. New York: Copernicus. ISBN 0-387-95289-6.

[25] Gonzalez, Guillermo; Brownlee, Donald; Ward, Peter (July 2001). "The Galactic Habitable Zone I. Galactic Chemical Evolution". *Icarus* **152** (1): 185–200. arXiv:astro-ph/0103165. Bibcode:2001Icar..152..185G. doi:10.1006/icar.2001.6617.

[26] Hadhazy, Adam (April 3, 2013). "The 'Habitable Edge' of Exomoons". *Astrobiology Magazine*. NASA. Retrieved April 22, 2013.

[27] Fogg, M. J. (1992). "An Estimate of the Prevalence of Biocompatible and Habitable Planets". *Journal of the British Interplanetary Society* **45** (1): 3–12. Bibcode:1992JBIS...45....3F. PMID 11539465.

[28] Redd, Nola Taylor (25 August 2011). "Greenhouse Effect Could Extend Habitable Zone". *Astrobiology Magazine*. NASA. Retrieved 25 June 2013.

[29] "Venus". Case Western Reserve University. 13 September 2006. Retrieved 2011-12-21.

[30] Sharp, Tim. "Atmosphere of the Moon". *Space.com*. TechMediaNetwork. Retrieved April 23, 2013.

[31] Bolonkin, Alexander A. (2009). *Artificial Environments on Mars*. Berlin Heidelberg: Springer. pp. 599–625. ISBN 978-3-642-03629-3.

[32] Haberle, Robert M.; McKay, Christopher P.; Schaeffer, James; Cabrol, Nathalie A.; Grin, Edmon A.; Zent, Aaron P.; Quinn, Richard (2001). "On the possibility of liquid water on present-day Mars". *Journal of Geophysical Research* **106** (E10): 23317. Bibcode:2001JGR...10623317H. doi:10.1029/2000JE001360. ISSN 0148-0227.

[33] Mann, Adam (February 18, 2014). "Strange Dark Streaks on Mars Get More and More Mysterious". *Wired (magazine)*. Retrieved February 18, 2014.

[34] "NASA Finds Possible Signs of Flowing Water on Mars". voanews.com. Retrieved August 5, 2011.

[35] "Is Mars Weeping Salty Tears?". news.sciencemag.org. Retrieved August 5, 2011.

[36] Webster, Guy; Brown, Dwayne (December 10, 2013). "NASA Mars Spacecraft Reveals a More Dynamic Red Planet". *NASA*. Retrieved December 10, 2013.

[37] A'Hearn, Michael F.; Feldman, Paul D. (1992). "Water vaporization on Ceres". *Icarus* **98** (1): 54–60. Bibcode:1992Icar...98...54A. doi:10.1016/0019-1035(92)90206-M.

[38] Budyko, M. I. (1969). "The effect of solar radiation variations on the climate of the Earth". *Tellus* **21** (5): 611. doi:10.1111/j.2153-3490.1969.tb00466.x.

[39] Sellers, William D. (June 1969). "A Global Climatic Model Based on the Energy Balance of the Earth-Atmosphere System". *Journal of Applied Meteorology* **8**(3): 392–400. Bibcode:1969JApMe...8..392S.doi:10.1175/1520-0450(1969)008<0392:AGC

[40] North, Gerald R. (November 1975). "Theory of Energy-Balance Climate Models". *Journal of the Atmospheric Sciences* **32** (11): 2033–2043. Bibcode:1975JAtS...32.2033N. doi:10.1175/1520-0469(1975)032<2033:TOEBCM>2.0.CO;2.

[41] Rasool, I.; De Bergh, C. (Jun 1970). "The Runaway Greenhouse and the Accumulation of CO_2 in the Venus Atmosphere" (PDF). *Nature* **226** (5250): 1037–1039. Bibcode:1970Natur.226.1037R. doi:10.1038/2261037a0. ISSN 0028-0836. PMID 16057644.

[42] Hart, M. H. (1979). "Habitable zones about main sequence stars". *Icarus* **37**: 351–357. Bibcode:1979Icar...37..351H. doi:10.1016/0019-1035(79)90141-6.

[43] Spiegel, D. S.; Raymond, S. N.; Dressing, C. D.; Scharf, C. A.; Mitchell, J. L. (2010). "Generalized Milankovitch Cycles and Long-Term Climatic Habitability". *The Astrophysical Journal* **721** (2): 1308. arXiv:1002.4877. Bibcode:2010ApJ...721.1308S. doi:10.1088/0004-637X/721/2/1308.

[44] Abe, Y.; Abe-Ouchi, A.; Sleep, N. H.; Zahnle, K. J. (2011). "Habitable Zone Limits for Dry Planets". *Astrobiology* **11** (5): 443–460. Bibcode:2011AsBio..11..443A. doi:10.1089/ast.2010.0545. PMID 21707386.

[45] Vladilo, Giovanni; Murante, Giuseppe; Silva, Laura; Provenzale, Antonello; Ferri, Gaia; Ragazzini, Gregorio (March 2013). "The habitable zone of Earth-like planets with different levels of atmospheric pressure". *The Astrophysical Journal (accepted)* **767** (1): 65–?. arXiv:1302.4566. Bibcode:2013ApJ...767...65V. doi:10.1088/0004-637X/767/1/65.

[46] Zsom, Andras; Seager, Sara; De Wit, Julien (2013). "Towards the Minimum Inner Edge Distance of the Habitable Zone". arXiv:1304.3714 [astro-ph.EP].

[47] Cuntz, Manfred (2013). "S-Type and P-Type Habitability in Stellar Binary Systems: A Comprehensive Approach. I. Method and Applications". arXiv:1303.6645 [astro-ph.EP].

[48] Forget, F.; Pierrehumbert, RT (1997). "Warming Early Mars with Carbon Dioxide Clouds That Scatter Infrared Radiation". *Science* **278** (5341): 1273–6. Bibcode:1997Sci...278.1273F. doi:10.1126/science.278.5341.1273. PMID 9360920.

[49] Mischna, M; Kasting, JF; Pavlov, A; Freedman, R (2000). "Influence of Carbon Dioxide Clouds on Early Martian Climate". *Icarus* **145** (2): 546–54. Bibcode:2000Icar..145..546M. doi:10.1006/icar.2000.6380. PMID 11543507.

[50] Vu, Linda. "Planets Prefer Safe Neighborhoods" (Press release). Spitzer.caltech.edu. NASA/Caltech. Retrieved April 22, 2013.

[51] Buccino, Andrea P.; Lemarchand, Guillermo A.; Mauas, Pablo J.D. (2006). "Ultraviolet radiation constraints around the circumstellar habitable zones". *Icarus* **183** (2): 491–503. arXiv:astro-ph/0512291. Bibcode:2006Icar..183..491B. doi:10.1016/j.i

[52] Barnes, Rory; Heller, René (March 2013). "Habitable Planets Around White and Brown Dwarfs: The Perils of a Cooling Primary". *Astrobiology* **13** (3): 279–291. arXiv:1203.5104. Bibcode:2013AsBio..13..279B. doi:10.1089/ast.2012.0867. PMC 3612282. PMID 23537137.

[53] Yang, J.; Cowan, N. B.; Abbot, D. S. (2013). "Stabilizing Cloud Feedback Dramatically Expands the Habitable Zone of Tidally Locked Planets". *The Astrophysical Journal* **771** (2): L45. arXiv:1307.0515. Bibcode:2013ApJ...771L..45Y. doi:10.1088/2041-8205/771/2/L45.

[54] Agol, Eric (April 2011). "Transit Surveys for Earths in the Habitable Zones of White Dwarfs". *The Astrophysical Journal Letters* **731** (2): 1–5. arXiv:1103.2791. Bibcode:2011ApJ...731L..31A. doi:10.1088/2041-8205/731/2/L31.

[55] Carroll, Bradley; Ostlie, Dale (2007). *An Introduction to Modern Astrophysics* (2 ed.).

[56] Richmond, Michael (November 10, 2004). "Late stages of evolution for low-mass stars". Rochester Institute of Technology. Retrieved 2007-09-19.

[57] Guo, J.; Zhang, F.; Chen, X.; Han, Z. (2009). "Probability distribution of terrestrial planets in habitable zones around host stars". *Astrophysics and Space Science* **323** (4): 367. arXiv:1003.1368. Bibcode:2009Ap&SS.323..367G. doi:10.1007/s10509-009-0081-z.

[58] Kasting, J.F.; Ackerman, T.P. (1986). "Climatic Consequences of Very High Carbon Dioxide Levels in the Earth's Early Atmosphere". *Science* **234** (4782): 1383–1385. doi:10.1126/science.11539665. PMID 11539665.

[59] Franck, S.; von Bloh, W.; Bounama, C.; Steffen, M.; Schönberner, D.; Schellnhuber, H.-J. (2002). "Habitable Zones and the Number of Gaia's Sisters" (PDF). In Montesinos, Benjamin; Giménez, Alvaro; Guinan, Edward F. *ASP Conference Series*. The Evolving Sun and its Influence on Planetary Environments. Astronomical Society of the Pacific. pp. 261–272. Bibcode:2002ASPC..269..261F. ISBN 1-58381-109-5. Retrieved April 26, 2013.

[60] Croswell, Ken (January 27, 2001). "Red, willing and able" (Full reprint). *New Scientist*. Retrieved August 5, 2007.

[61] Alekseev, I. Y.; Kozlova, O. V. (2002). "Starspots and active regions on the emission red dwarf star LQ Hydrae". *Astronomy and Astrophysics* **396**: 203. Bibcode:2002A&A...396..203A. doi:10.1051/0004-6361:20021424.

[62] Alpert, Mark (November 7, 2005). "Red Star Rising". *Scientific American*. Retrieved January 19, 2013.

[63] Research Corporation (December 19, 2006). "Andrew West: 'Fewer flares, starspots for older dwarf stars'". *EarthSky*. Retrieved April 27, 2013.

[64] Cain, Fraser; Gay, Pamela (2007). "AstronomyCast episode 40: American Astronomical Society Meeting, May 2007". *Universe Today*. Retrieved 2007-06-17.

[65] Christensen, Bill (April 1, 2005). "Red Giants and Planets to Live On". *Space.com*. TechMediaNetwork. Retrieved April 27, 2013.

[66] Lopez, B.; Schneider, J.; Danchi, W. C. (2005). "Can Life Develop in the Expanded Habitable Zones around Red Giant Stars?". *The Astrophysical Journal* **627** (2): 974. arXiv:astro-ph/0503520. Bibcode:2005ApJ...627..974L. doi:10.1086/430416.

[67] Lorenz, Ralph D.; Lunine, Jonathan I.; McKay, Christopher P. (1997). "Titan under a red giant sun: A new kind of "habitable" moon". *Geophysical Research Letters* **24** (22): 2905–2908. Bibcode:1997GeoRL..24.2905L. doi:10.1029/97GL52843. ISSN 0094-8276. PMID 11542268.

[68] Voisey, Jon (February 23, 2011). "Plausibility Check – Habitable Planets around Red Giants". *Universe Today*. Retrieved April 27, 2013.

[69] Alien Life More Likely on 'Dune' Planets, 09/01/11, Charles Q. Choi, *Astrobiology Magazine*

[70] Habitable Zone Limits for Dry Planets, Yutaka Abe, Ayako Abe-Ouchi, Norman H. Sleep, and Kevin J. Zahnle. *Astrobiology*. June 2011, 11(5): 443–460. doi:10.1089/ast.2010.0545

[71] Drake, Michael J. (April 2005). "Origin of water in the terrestrial planets". *Meteoritics & Planetary Science* (John Wiley & Sons) **40** (4): 519–527. Bibcode:2005M&PS...40..519D. doi:10.1111/j.1945-5100.2005.tb00960.x.

[72] Drake, Michael J.; et al. (August 2005). "Origin of water in the terrestrial planets". *Asteroids, Comets, and Meteors (IAU S229)*. 229th Symposium of the International Astronomical Union **1** (4). Búzios, Rio de Janeiro, Brazil: Cambridge University Press. pp. 381–394. Bibcode:2006IAUS..229..381D. doi:10.1017/S1743921305006861. ISBN 978-0-521-85200-5.

[73] Kuchner, Marc (2003). "Volatile-rich Earth-Mass Planets in the Habitable Zone". *Astrophysical Journal* **596**: L105–L108. arXiv:astro-ph/0303186. Bibcode:2003ApJ...596L.105K. doi:10.1086/378397.

[74] Charbonneau, David; Zachory K. Berta; Jonathan Irwin; Christopher J. Burke; Philip Nutzman; Lars A. Buchhave; Christophe Lovis; Xavier Bonfils; et al. (2009). "A super-Earth transiting a nearby low-mass star". *Nature* **462** (17 December 2009): 891–894. arXiv:0912.3229. Bibcode:2009Natur.462..891C. doi:10.1038/nature08679. PMID 20016595. Retrieved 2009-12-15.

[75] Kuchner, Seager; Hier-Majumder, M.; Militzer, C. A. (2007). "Mass–radius relationships for solid exoplanets". *The Astrophysical Journal* **669** (2): 1279–1297. arXiv:0707.2895. Bibcode:2007ApJ...669.1279S. doi:10.1086/521346.

[76] Vastag, Brian (December 5, 2011). "Newest alien planet is just the right temperature for life". *The Washington Post*. Retrieved April 27, 2013.

[77] Robinson, Tyler D.; Catling, David C. (2012). "An Analytic Radiative-Convective Model for Planetary Atmospheres". *The Astrophysical Journal* **757** (1): 104. arXiv:1209.1833. Bibcode:2012ApJ...757..104R. doi:10.1088/0004-637X/757/1/104.

[78] Shizgal, B. D.; Arkos, G. G. (1996). "Nonthermal escape of the atmospheres of Venus, Earth, and Mars". *Reviews of Geophysics* **34** (4): 483–505. Bibcode:1996RvGeo..34..483S. doi:10.1029/96RG02213.

[79] Chaplin, Martin (April 8, 2013). "Water Phase Diagram". *Ices*. London South Bank University. Retrieved April 27, 2013.

[80] D.P. Hamilton & J.A. Burns (1992). "Orbital stability zones about asteroids. II - The destabilizing effects of eccentric orbits and of solar radiation". *Icarus* **96** (1): 43. Bibcode:1992Icar...96...43H. doi:10.1016/0019-1035(92)90005-R.

[81] Becquerel P. (1950). "La suspension de la vie au dessous de 1/20 K absolu par demagnetization adiabatique de l'alun de fer dans le vide les plus eléve". *C. R. Hebd. Séances Acad. Sci. Paris* (in French) **231**: 261–263.

[82] Horikawa, Daiki D. (2012). Alexander V. Altenbach, Joan M. Bernhard & Joseph Seckbach, ed. *Anoxia Evidence for Eukaryote Survival and Paleontological Strategies*. (21 ed.). Springer Netherlands. pp. 205–217. ISBN 978-94-007-1895-1. Retrieved 21 January 2012.

[83] Kane, Stephen R.; Gelino, Dawn M. (2012). "The Habitable Zone and Extreme Planetary Orbits". *Astrobiology* **12** (10): 940–945. arXiv:1205.2429. Bibcode:2012AsBio..12..940K. doi:10.1089/ast.2011.0798. PMID 23035897.

[84] Paul Gilster, Andrew LePage (2015-01-30). "A Review of the Best Habitable Planet Candidates". Centauri Dreams, Tau Zero Foundation. Retrieved 2015-07-24.

[85] Giovanni F. Bignami (2015). *The Mystery of the Seven Spheres: How Homo sapiens will Conquer Space*. Springer. ISBN 9783319170046., Page 110

[86] Wethington, Nicholos (September 16, 2008). "How Many Stars are in the Milky Way?". *UniverseToday*. Retrieved April 21, 2013.

[87] Torres, Abel Mendez (April 26, 2013). "Ten potentially habitable exoplanets now". *Habitable Exoplanets Catalog*. University of Puerto Rico. Retrieved April 29, 2013.

[88] Borenstein, Seth (19 February 2011). "Cosmic census finds crowd of planets in our galaxy". Associated Press. Retrieved 24 April 2011.

[89] Choi, Charles Q. (21 March 2011). "New Estimate for Alien Earths: 2 Billion in Our Galaxy Alone". Space.com. Retrieved 2011-04-24.

[90] Catanzarite, J.; Shao, M. (2011). "The Occurrence Rate of Earth Analog Planets Orbiting Sun-Like Stars". *The Astrophysical Journal* **738** (2): 151. arXiv:1103.1443. Bibcode:2011ApJ...738..151C. doi:10.1088/0004-637X/738/2/151.

[91] "70 Virginis b". *Extrasolar Planet Guide*. Extrasolar.net. Archived from the original on 2012-06-19. Retrieved 2009-04-02.

[92] Williams, D., Pollard, D. (2002). "Earth-like worlds on eccentric orbits: excursions beyond the habitable zone". *International Journal of Astrobiology* **1** (1): 61–69. Bibcode:2002IJAsB...1...61W. doi:10.1017/S1473550402001064.

[93] Sudarsky, David; et al. (2003). "Theoretical Spectra and Atmospheres of Extrasolar Giant Planets". *The Astrophysical Journal* **588** (2): 1121–1148. arXiv:astro-ph/0210216. Bibcode:2003ApJ...588.1121S. doi:10.1086/374331.

[94] Williams, D., Pollard, D. (2002). "Earth-like worlds on eccentric orbits: excursions beyond the habitable zone". *International Journal of Astrobiology*(Cambridge University Press)**1**(1): 61–69. Bibcode:2002IJAsB...1...61W.doi:10.1017/S14735504020

[95] Jones, B. W.; Sleep, P. N.; Underwood, D. R. (2006). "Habitability of Known Exoplanetary Systems Based on Measured Stellar Properties".*The Astrophysical Journal***649**(2): 1010. arXiv:astro-ph/0603200. Bibcode:2006ApJ...649.1010J.doi:10.1086/5

[96] Butler, R. P.; Wright, J. T.; Marcy, G. W.; Fischer, D. A.; Vogt, S. S.; Tinney, C. G.; Jones, H. R. A.; Carter, B. D.; Johnson, J. A.; McCarthy, C.; Penny, A. J. (2006). "Catalog of Nearby Exoplanets". *The Astrophysical Journal* **646**: 505. arXiv:astro-ph/0607493. Bibcode:2006ApJ...646..505B. doi:10.1086/504701.

[97] Barnes, J. W.; O'Brien, D. P. (2002). "Stability of Satellites around Close-in Extrasolar Giant Planets". *The Astrophysical Journal* **575**: 1087. arXiv:astro-ph/0205035. Bibcode:2002ApJ...575.1087B. doi:10.1086/341477.

[98] Canup, R. M.; Ward, W. R. (2006). "A common mass scaling for satellite systems of gaseous planets". *Nature* **441** (7095): 834–839. Bibcode:2006Natur.441..834C. doi:10.1038/nature04860. PMID 16778883.

[99] Lovis; et al. (2006). "An extrasolar planetary system with three Neptune-mass planets". *Nature* **441** (7091): 305–309. arXiv:astro-ph/0703024. Bibcode:2006Natur.441..305L. doi:10.1038/nature04828. PMID 16710412.

[100] "Astronomers Discover Record Fifth Planet Around Nearby Star 55 Cancri". Sciencedaily.com. November 6, 2007. Archived from the original on 26 September 2008. Retrieved 2008-09-14.

[101] Fischer, Debra A.; et al. (2008). "Five Planets Orbiting 55 Cancri".*The Astrophysical Journal***675**(1): 790–801. arXiv:0712.39 Bibcode:2008ApJ...675..790F. doi:10.1086/525512.

[102] Ian Sample, science correspondent (7 November 2007). "Could this be Earth's near twin? Introducing planet 55 Cancri f". London: *The Guardian*. Archived from the original on 2 October 2008. Retrieved 17 October 2008.

[103] Than, Ker (2007-02-24). "Planet Hunters Edge Closer to Their Holy Grail". space.com. Retrieved 2007-04-29.

[104] Robertson, Paul; Mahadevan, Suvrath; Endl, Michael; Roy, Arpita (3 July 2014). "Stellar activity masquerading as planets in the habitable zone of the M dwarf Gliese 581".*Science*. arXiv:1407.1049. Bibcode:2014Sci...345..440R.doi:10.1126/science.125

[105] "Researchers find potentially habitable planet" (in French). maxisciences.com. Retrieved 2011-08-31.

[106] Schulze-Makuch, D.; Méndez, A.; Fairén, A. G.; Von Paris, P.; Turse, C.; Boyer, G.; Davila, A. F.; António, M. R. D. S.; Catling, D.; Irwin, L. N. (2011). "A Two-Tiered Approach to Assessing the Habitability of Exoplanets". *Astrobiology* **11** (10): 1041–1052. Bibcode:2011AsBio..11.1041S. doi:10.1089/ast.2010.0592. PMID 22017274.

[107] "Kepler 22-b: Earth-like planet confirmed". BBC. December 5, 2011. Retrieved May 2, 2013.

[108] Scharf, Caleb A. (2011-12-08). "You Can't Always Tell an Exoplanet by Its Size". *Scientific American*. Retrieved 2012-09-20.: "If it [Kepler-22b] had a similar composition to Earth, then we're looking at a world in excess of about 40 Earth masses".

[109] Anglada-Escude, Guillem; Arriagada, Pamela; Vogt, Steven; Rivera, Eugenio J.; Butler, R. Paul; Crane, Jeffrey D.; Shectman, Stephen A.; Thompson, Ian B.; Minniti, Dante (2012). "A planetary system around the nearby M dwarf GJ 667C with at least one super-Earth in its habitable zone". arXiv:1202.0446 [astro-ph.EP].

[110] Anglada-Escudé, Guillem; Tuomi, Mikko; Gerlach, Enrico; Barnes, Rory; Heller, René; Jenkins, James S.; Wende, Sebastian; Vogt, Steven S.; Butler, R. Paul; Reiners, Ansgar; Jones, Hugh R. A. (2013-06-07). "A dynamically-packed planetary system around GJ 667C with three super-Earths in its habitable zone" (PDF). *Astronomy & Astrophysics*. arXiv:1306.6074. Bibcode:2013A&A...556A.126A. doi:10.1051/0004-6361/201321331. Retrieved 2013-06-25.

[111] Staff (September 20, 2012). "LHS 188 -- High proper-motion Star". Centre de données astronomiques de Strasbourg (Strasbourg astronomical Data Center). Retrieved September 20, 2012.

[112] Méndez, Abel (August 29, 2012). "A Hot Potential Habitable Exoplanet around Gliese 163". University of Puerto Rico at Arecibo (Planetary Habitability Laboratory). Retrieved September 20, 2012.

[113] Redd (September 20, 2012). "Newfound Alien Planet a Top Contender to Host Life". Space.com. Retrieved September 20, 2012.

[114] "A Hot Potential Habitable Exoplanet around Gliese 163". Spacedaily.com. Retrieved 2013-02-10.

[115] Tuomi, Mikko; Anglada-Escude, Guillem; Gerlach, Enrico; Jones, Hugh R. R.; Reiners, Ansgar; Rivera, Eugenio J.; Vogt, Steven S.; Butler, Paul (2012). "Habitable-zone super-Earth candidate in a six-planet system around the K2.5V star HD 40307". *Astronomy and Astrophysics (accepted)* **549**: A48. arXiv:1211.1617. Bibcode:2013A&A...549A..48T. doi:10.1051/0004-6361/201220268.

[116] Aron, Jacob (December 19, 2012). "Nearby Tau Ceti may host two planets suited to life". *New Scientist*. Reed Business Information. Retrieved April 1, 2013.

[117] Tuomi, M.; Jones, H. R. A.; Jenkins, J. S.; Tinney, C. G.; Butler, R. P.; Vogt, S. S.; Barnes, J. R.; Wittenmyer, R. A.; o'Toole, S.; Horner, J.; Bailey, J.; Carter, B. D.; Wright, D. J.; Salter, G. S.; Pinfield, D. (2013). "Signals embedded in the radial velocity noise". *Astronomy & Astrophysics* **551**: A79. arXiv:1212.4277. Bibcode:2013A&A...551A..79T. doi:10.1051/0004-6361/201220509.

[118] Torres, Abel Mendez (May 1, 2013). "The Habitable Exoplanets Catalog". *Habitable Exoplanets Catalog*. University of Puerto Rico. Retrieved May 1, 2013.

[119] Lauren M. Weiss, and Geoffrey W. Marcy. "The mass-radius relation for 65 exoplanets smaller than 4 Earth radii"

[120] Moskowitz, Clara (January 9, 2013). "Most Earth-Like Alien Planet Possibly Found". Space.com. Retrieved January 9, 2013.

[121] Barclay, Thomas; Burke, Christopher J.; Howell, Steve B.; Rowe, Jason F.; Huber, Daniel; Isaacson, Howard; Jenkins, Jon M.; Kolbl, Rea; Marcy, Geoffrey W. (2013). "A Super-Earth-Sized Planet Orbiting in or Near the Habitable Zone Around a Sun-Like Star". *The Astrophysical Journal* **768** (2): 101. arXiv:1304.4941. Bibcode:2013ApJ...768..101B. doi:10.1088/0004-637X/768/2/101.

[122] Johnson, Michele; Harrington, J.D. (18 April 2013). "NASA's Kepler Discovers Its Smallest 'Habitable Zone' Planets to Date". *NASA*. Retrieved 18 April 2013.

[123] Overbye, Dennis (18 April 2013). "Two Promising Places to Live, 1,200 Light-Years from Earth". *New York Times*. Retrieved 18 April 2013.

[124] Overbye, Dennis (18 April 2013). "2 Good Places to Live, 1,200 Light-Years Away". *New York Times*. Retrieved 18 April 2013.

[125] Borucki, William J.; et al. (18 April 2013). "Kepler-62: A Five-Planet System with Planets of 1.4 and 1.6 Earth Radii in the Habitable Zone". *Science Express* **340** (6132): 587. arXiv:1304.7387. Bibcode:2013Sci...340..587B. doi:10.1126/science.1234 702.Retrieved18April2013.

[126] Chang, Kenneth (17 April 2014). "Scientists Find an 'Earth Twin,' or Maybe a Cousin". *New York Times*. Retrieved 17 April 2014.

[127] Chang, Alicia (17 April 2014). "Astronomers spot most Earth-like planet yet". *AP News*. Retrieved 17 April 2014.

[128] Morelle, Rebecca (17 April 2014). "'Most Earth-like planet yet' spotted by Kepler". *BBC News*. Retrieved 17 April 2014.

[129] Clavin, Whitney; Chou, Felicia; Johnson, Michele (6 January 2015). "NASA's Kepler Marks 1,000th Exoplanet Discovery, Uncovers More Small Worlds in Habitable Zones". *NASA*. Retrieved 6 January 2015.

[130] Jensen, Mari N. (16 January 2015). "Three nearly Earth-size planets found orbiting nearby star: One in 'Goldilocks' zone". *Science Daily*. Retrieved 25 July 2015.

[131] Jenkins, Jon M.; Twicken, Joseph D.; Batalha, Natalie M.; Caldwell, Douglas A.; Cochran, William D.; Endl, Michael; Latham, David W.; Esquerdo, Gilbert A.; Seader, Shawn; Bieryla, Allyson; Petigura, Erik; Ciardi, David R.; Marcy, Geoffrey W.; Isaacson, Howard; Huber, Daniel; Rowe, Jason F.; Torres, Guillermo; Bryson, Stephen T.; Buchhave, Lars; Ramirez, Ivan; Wolfgang, Angie; Li, Jie; Campbell, Jennifer R.; Tenenbaum, Peter; Sanderfer, Dwight; Henze, Christopher E.; Catanzarite, Joseph H.; Gilliland, Ronald L.; Borucki, William J. (23 July 2015). "Discovery and Validation of Kepler-452b: A 1.6 R⊕ Super Earth Exoplanet in the Habitable Zone of a G2 Star". *The Astronomical Journal* **150** (2): 56. arXiv:1507.06723. Bibcode:2015AJ....150...56J. doi:10.1088/0004-6256/150/2/56. ISSN 1538-3881. Retrieved 24 July 2015.

[132] "NASA telescope discovers Earth-like planet in star's 'habitable zone'". *BNO News*, 23 July 2015. Retrieved 23 July 2015.

[133] Torres, Abel (2012-06-12). "Liquid Water in the Solar System". Retrieved 2013-12-15.

[134] Munro, Margaret (2013). "Miners deep underground in northern Ontario find the oldest water ever known", *National Post*, retrieved 2013-10-06

[135] Davies, Paul (2013), *The Origin of Life II: How did it begin?* (PDF), retrieved 2013-10-06

[136] Taylor, Geoffrey (1996), "Life Underground" (PDF), *Planetary Science Research*, retrieved 2013-10-06

[137] Doyle, Alister (4 March 2013), "Deep underground, worms and "zombie microbes" rule", *Reuters*, retrieved 2013-10-06

[138] Nicholson, W. L.; Moeller, R.; Horneck, G.; PROTECT Team (2012). "Transcriptomic Responses of GerminatingBacillus subtilisSpores Exposed to 1.5 Years of Space and Simulated Martian Conditions on the EXPOSE-E Experiment PROTECT". *Astrobiology* **12** (5): 469–86. Bibcode:2012AsBio..12..469N. doi:10.1089/ast.2011.0748. PMID 22680693.

[139] Decker, Heinz; Holde, Kensal E. (2011). "Oxygen and the Exploration of the Universe (article) (book:Oxygen and the Evolution of Life)". pp. 157–168. doi:10.1007/978-3-642-13179-0_9. ISBN 978-3-642-13178-3.

[140] Stewart, Ian; Cohen, Jack (2002). *Evolving the Alien*. Ebury Press. ISBN 978-0-09-187927-3.

[141] Goldsmith, Donald; Owen, Tobias (1992). *The Search for Life in the Universe* (2 ed.). Addison-Wesley. p. 247. ISBN 0-201-56949-3.

[142] Guidetti, R. & Jönsson, K.I. (2002). "Long-term anhydrobiotic survival in semi-terrestrial micrometazoans". *Journal of Zoology* **257** (2): 181–187. doi:10.1017/S0952836902000783X.

[143] Baldwin, Emily (26 April 2012). "Lichen survives harsh Mars environment". Skymania News. Retrieved 27 April 2012.

[144] de Vera, J.-P.; Kohler, Ulrich (26 April 2012). "The adaptation potential of extremophiles to Martian surface conditions and its implication for the habitability of Mars" (PDF). European Geosciences Union. Retrieved 27 April 2012.

[145] Palca, Joe (September 29, 2010). "'Goldilocks' Planet's Temperature Just Right For Life". *NPR* (NPR). Retrieved April 5, 2011.

[146] "Project Cyclops: A design study of a system for detecting extraterrestrial intelligent life" (PDF). NASA. 1971. Retrieved June 28, 2009.

[147] Joseph A. Angelo (2007). *Life in the Universe*. Infobase Publishing. p. 163. ISBN 978-1-4381-0892-6. Retrieved 26 June 2013.

[148] Turnbull, Margaret C.; Tarter, Jill C. (2003). "Target Selection for SETI. I. A Catalog of Nearby Habitable Stellar Systems". *The Astrophysical Journal Supplement Series* **145** (1): 181–198. arXiv:astro-ph/0210675. Bibcode:2003ApJS..145..181T. doi:10.1086/345779.

[149] Siemion, Andrew P. V.; Demorest, Paul; Korpela, Eric; Maddalena, Ron J.; Werthimer, Dan; Cobb, Jeff; Howard, Andrew W.; Langston, Glen; Lebofsky, Matt (2013). "A 1.1 to 1.9 GHz SETI Survey of the *Kepler* Field: I. A Search for Narrowband Emission from Select Targets". *The Astrophysical Journal* **767** (1): 94. arXiv:1302.0845. Bibcode:2013ApJ...767...94S. doi:10.1088/0004-637X/767/1/94.

[150] Wall, Mike (2011). "HabStars: Speeding Up In the Zone". Retrieved 2013-06-26

[151] Zaitsev, A. L. (June 2004). "Transmission and reasonable signal searches in the Universe". *Horizons of the Universe* Передача и поиски разумных сигналов во Вселенной. Plenary presentation at the National Astronomical Conference WAC-2004 "Horizons of the Universe", Moscow, Moscow State University, June 7, 2004 (in Russian). Moscow. Retrieved 2013-06-30.

[152] Grinspoon, David (12 December 2007). "Who Speaks for Earth?". Seedmagazine.com. Retrieved 2012-08-21.

[153] P. C. Gregory, D. A. Fischer (2010). "A Bayesian periodogram finds evidence for three planets in 47 Ursae Majoris". *Monthly Notices of the Royal Astronomical Society* **403** (2): 731–747. arXiv:1003.5549. Bibcode:2010MNRAS.403..731G. doi:10.1111/j.1365-2966.2009.16233.x.

[154] B. Jones; Underwood, David R.; et al. (2005). "Prospects for Habitable "Earths" in Known Exoplanetary Systems". *Astrophysical Journal* **622** (2): 1091–1101. arXiv:astro-ph/0503178. Bibcode:2005ApJ...622.1091J. doi:10.1086/428108.

[155] Moore, Matthew (October 9, 2008). "Messages from Earth sent to distant planet by Bebo". London: .telegraph.co.uk. Archived from the original on 11 October 2008. Retrieved 2008-10-09.

10.8 External links

- "Circumstellar Habitable Zone Simulator". Astronomy Education at the University of Nebraska-Lincoln.

- "The Habitable Exoplanets Catalog". PHL/University of Puerto Rico at Arecibo.

- "The Habitable Zone Gallery".

- "Stars and Habitable Planets". SolStation.

- Nikos Prantzos (2006). "On the Galactic Habitable Zone". *Space Science Reviews* **135**: 313–322. arXiv:astro-ph/0612316. Bibcode:2008SSRv..135..313P. doi:10.1007/s11214-007-9236-9.

- Interstellar Real Estate: Location, Location, Location – Defining the Habitable Zone

- "Exoplanets in relation to host star's current habitable zone". *www.planetarybiology.com*.

- "exoExplorer: a free Windows application for visualizing exoplanet environments in 3D". *www.planetarybiology.com*.

- Shiga, David (November 19, 2009). "Why the universe may be teeming with aliens". *NewScientist*.

- Simmons; et al. "The New Worlds Observer: a mission for high-resolution spectroscopy of extra-solar terrestrial planets" (PDF). *New Worlds*.

- Cockell, Charles S.; Herbst, Tom; Léger, Alain; Absil, O.; Beichman, Charles; Benz, Willy; Brack, Andre; Chazelas, Bruno; Chelli, Alain (2009). "Darwin – an experimental astronomy mission to search for extrasolar planets" (PDF). *Experimental Astronomy* **23**: 435–461. Bibcode:2009ExA....23..435C. doi:10.1007/s10686-008-9121-x.

- Atkinson, Nancy (March 19, 2009). "JWST Will Provide Capability to Search for Biomarkers on Earth-like Worlds". *Universe Today*.

Chapter 11

Planetary habitability

Planetary habitability is the measure of a planet's or a natural satellite's potential to develop and sustain life. Life may develop directly on a planet or satellite or be transferred to it from another body, a theoretical process known as panspermia. As the existence of life beyond Earth is unknown, planetary habitability is largely an extrapolation of conditions on Earth and the characteristics of the Sun and Solar System which appear favourable to life's flourishing—in particular those factors that have sustained complex, multicellular organisms and not just simpler, unicellular creatures. Research and theory in this regard is a component of planetary science and the emerging discipline of astrobiology.

An absolute requirement for life is an energy source, and the notion of planetary habitability implies that many other geophysical, geochemical, and astrophysical criteria must be met before an astronomical body can support life. In its astrobiology roadmap, NASA has defined the principal habitability criteria as "extended regions of liquid water,[1] conditions favourable for the assembly of complex organic molecules, and energy sources to sustain metabolism."[2]

In determining the habitability potential of a body, studies focus on its bulk composition, orbital properties, atmosphere, and potential chemical interactions. Stellar characteristics of importance include mass and luminosity, stable variability, and high metallicity. Rocky, terrestrial-type planets and moons with the potential for Earth-like chemistry are a primary focus of astrobiological research, although more speculative habitability theories occasionally examine alternative biochemistries and other types of astronomical bodies.

The idea that planets beyond Earth might host life is an ancient one, though historically it was framed by philosophy as much as physical science.[lower alpha 1] The late 20th century saw two breakthroughs in the field. The observation and robotic spacecraft exploration of other planets and moons within the Solar System has provided critical information on defining habitability criteria and allowed for substantial geophysical comparisons between the Earth and other bodies. The discovery of extrasolar planets, beginning in the early 1990s[3][4] and accelerating thereafter, has provided further information for the study of possible extraterrestrial life. These findings confirm that the Sun is not unique among stars in hosting planets and expands the habitability research horizon beyond the Solar System.

The chemistry of life may have begun shortly after the Big Bang, 13.8 billion years ago, during a habitable epoch when the Universe was only 10–17 million years old.[5][6] According to the panspermia hypothesis, microscopic life—distributed by meteoroids, asteroids and other small Solar System bodies—may exist throughout the universe.[7] Nonetheless, Earth is the only place in the universe known to harbor life.[8][9] Estimates of habitable zones around other stars,[10][11] along with the discovery of hundreds of extrasolar planets and new insights into the extreme habitats here on Earth, suggest that there may be many more habitable places in the universe than considered possible until very recently.[12] On 4 November 2013, astronomers reported, based on *Kepler* space mission data, that there could be as many as 40 billion Earth-sized planets orbiting in the habitable zones of Sun-like stars and red dwarfs within the Milky Way.[13][14] 11 billion of these estimated planets may be orbiting Sun-like stars.[15] The nearest such planet may be 12 light-years away, according to the scientists.[13][14]

Understanding planetary habitability is partly an extrapolation of the conditions on Earth, as this is the only planet known to support life.

11.1 Suitable star systems

An understanding of planetary habitability begins with stars. While bodies that are generally Earth-like may be plentiful, it is just as important that their larger system be agreeable to life. Under the auspices of SETI's Project Phoenix, scientists Margaret Turnbull and Jill Tarter developed the "HabCat" (or Catalogue of Habitable Stellar Systems) in 2002. The catalogue was formed by winnowing the nearly 120,000 stars of the larger Hipparcos Catalogue into a core group of 17,000 "HabStars", and the selection criteria that were used provide a good starting point for understanding which astrophysical factors are necessary to habitable planets.[16] According to research published in August 2015, very large galaxies may be more favorable to the creation and development of habitable planets than smaller galaxies, like the Milky Way galaxy.[17]

11.1.1 Spectral class

The spectral class of a star indicates its photospheric temperature, which (for main-sequence stars) correlates to overall mass. The appropriate spectral range for "HabStars" is considered to be "early F" or "G", to "mid-K". This corresponds to temperatures of a little more than 7,000 K down to a little more than 4,000 K (6,700 °C to 3,700 °C); the Sun, a G2 star at 5,777 K, is well within these bounds. "Middle-class" stars of this sort have a number of characteristics considered important to planetary habitability:

- They live at least a few billion years, allowing life a chance to evolve. More luminous main-sequence stars of the "O", "B", and "A" classes usually live less than a billion years and in exceptional cases less than 10 million.[18][lower-alpha 2]

- They emit enough high-frequency ultraviolet radiation to trigger important atmospheric dynamics such as ozone formation, but not so much that ionisation destroys incipient life.[19]

- Liquid water may exist on the surface of planets orbiting them at a distance that does not induce tidal locking. K Spectrum stars may be able to support life for long periods, far longer than the Sun.[20]

This spectral range probably accounts for between 5% and 10% of stars in the local Milky Way galaxy. Whether fainter late K and M class red dwarf stars are also suitable hosts for habitable planets is perhaps the most important open question in the entire field of planetary habitability given their prevalence (habitability of red dwarf systems). Gliese 581 c, a "super-Earth", has been found orbiting in the "habitable zone" of a red dwarf and may possess liquid water. However it is also possible that a greenhouse effect may render it too hot to support life, while its neighbor, Gliese 581 d, may be a more likely candidate for habitability.[21] In September 2010, the discovery was announced of another planet, Gliese 581 g, in an orbit between these two planets. However, reviews of the discovery have placed the existence of this planet in doubt, and it is listed as "unconfirmed". In September 2012, the discovery of two planets orbiting Gliese 163[22] was announced.[23][24] One of the planets, Gliese 163 c, about 6.9 times the mass of Earth and somewhat hotter, was considered to be within the habitable zone.[23][24]

A recent study suggests that cooler stars that emit more light in the infrared and near infrared may actually host warmer planets with less ice and incidence of snowball states. These wavelengths are absorbed by their planets' ice and greenhouse gases and remain warmer.[25][26]

11.1.2 A stable habitable zone

Main article: Habitable zone

The habitable zone (HZ, categorized by the Planetary Habitability Index) is a shell-shaped region of space surrounding a star in which a planet could maintain liquid water on its surface. After an energy source, liquid water is considered the most important ingredient for life, considering how integral it is to all life systems on Earth. This may reflect the known dependence of life on water; however, if life is discovered in the absence of water, the definition of an HZ may have to be greatly expanded.

A "stable" HZ implies two factors. First, the range of an HZ should not vary greatly over time. All stars increase in luminosity as they age, and a given HZ thus migrates outwards, but if this happens too quickly (for example, with a super-massive star) planets may only have a brief window inside the HZ and a correspondingly smaller chance of developing life. Calculating an HZ range and its long-term movement is never straightforward, as negative feedback loops such as the CNO cycle will tend to offset the increases in luminosity. Assumptions made about atmospheric conditions and geology thus have as great an impact on a putative HZ range as does stellar evolution: the proposed parameters of the Sun's HZ, for example, have fluctuated greatly.[27]

Second, no large-mass body such as a gas giant should be present in or relatively close to the HZ, thus disrupting the formation of Earth-like bodies. The matter in the asteroid belt, for example, appears to have been unable to accrete into a planet due to orbital resonances with Jupiter; if the giant had appeared in the region that is now between the orbits of Venus and Mars, Earth would almost certainly not have developed in its present form. However a gas giant inside the HZ might have habitable moons under the right conditions.[28]

In the Solar System, the inner planets are terrestrial, and the outer ones are gas giants, but discoveries of extrasolar planets suggest that this arrangement may not be at all common: numerous Jupiter-sized bodies have been found in close orbit about their primary, disrupting potential HZs. However, present data for extrasolar planets is likely to be skewed towards that type (large planets in close orbits) because they are far easier to identify; thus it remains to be seen which type of planetary system is the norm, or indeed if there is one.

11.1.3   Low stellar variation

Main article: Variable star

Changes in luminosity are common to all stars, but the severity of such fluctuations covers a broad range. Most stars are relatively stable, but a significant minority of variable stars often undergo sudden and intense increases in luminosity and consequently in the amount of energy radiated toward bodies in orbit. These stars are considered poor candidates for hosting life-bearing planets, as their unpredictability and energy output changes would negatively impact organisms: living things adapted to a specific temperature range could not survive too great a temperature variation. Further, upswings in luminosity are generally accompanied by massive doses of gamma ray and X-ray radiation which might prove lethal. Atmospheres do mitigate such effects, but their atmosphere might not be retained by planets orbiting variables, because the high-frequency energy buffeting these planets would continually strip them of their protective covering.

The Sun, in this respect as in many others, is relatively benign: the variation between its maximum and minimum energy output is roughly 0.1% over its 11-year solar cycle. There is strong (though not undisputed) evidence that even minor changes in the Sun's luminosity have had significant effects on the Earth's climate well within the historical era: the Little Ice Age of the mid-second millennium, for instance, may have been caused by a relatively long-term decline in the Sun's luminosity.[29] Thus, a star does not have to be a true variable for differences in luminosity to affect habitability. Of known solar analogs, one that closely resembles the Sun is considered to be 18 Scorpii; unfortunately for the prospects of life existing in its proximity, the only significant difference between the two bodies is the amplitude of the solar cycle, which appears to be much greater for 18 Scorpii.[30]

11.1.4   High metallicity

See also: Metallicity

While the bulk of material in any star is hydrogen and helium, there is a great variation in the amount of heavier elements (metals) that stars contain. A high proportion of metals in a star correlates to the amount of heavy material initially available in the protoplanetary disk. A smaller amount of metal makes the formation of planets much less likely, under the solar nebula theory of planetary system formation. Any planets that did form around a metal-poor star would probably be low in mass, and thus unfavorable for life. Spectroscopic studies of systems where exoplanets have been found to date confirm the relationship between high metal content and planet formation: "Stars with planets, or at least with planets similar to the ones we are finding today, are clearly more metal rich than stars without planetary companions."[31] This relationship between high metallicity and planet formation also means that habitable systems are more likely to be found around younger stars, since stars that formed early in the universe's history have low metal content.

11.2   Planetary characteristics

The chief assumption about habitable planets is that they are terrestrial. Such planets, roughly within one order of magnitude of Earth mass, are primarily composed of silicate rocks, and have not accreted the gaseous outer layers of hydrogen and helium found on gas giants. That life could evolve in the cloud tops of giant planets has not been decisively ruled out,[lower alpha 3] though it is considered unlikely, as they have no surface and their gravity is enormous.[35] The natural satellites of giant planets, meanwhile, remain valid candidates for hosting life.[32]

In February 2011 the Kepler Space Observatory Mission team released a list of 1235 extrasolar planet candidates, in-

The moons of some gas giants could potentially be habitable.[32]

cluding 54 that may be in the habitable zone.[36][37] Six of the candidates in this zone are smaller than twice the size of Earth.[36] A more recent study found that one of these candidates (KOI 326.01) is much larger and hotter than first reported.[38] Based on the findings, the Kepler team estimated there to be "at least 50 billion planets in the Milky Way" of which "at least 500 million" are in the habitable zone.[39]

In analyzing which environments are likely to support life, a distinction is usually made between simple, unicellular organisms such as bacteria and archaea and complex metazoans (animals). Unicellularity necessarily precedes multicellularity in any hypothetical tree of life, and where single-celled organisms do emerge there is no assurance that greater complexity will then develop.[lower alpha 4] The planetary characteristics listed below are considered crucial for life generally, but in every case multicellular organisms are more picky than unicellular life.

11.2.1   Mass

Low-mass planets are poor candidates for life for two reasons. First, their lesser gravity makes atmosphere retention difficult. Constituent molecules are more likely to reach escape velocity and be lost to space when buffeted by solar wind or stirred by collision. Planets without a thick atmosphere lack the matter necessary for primal biochemistry, have little insulation and poor heat transfer across their surfaces (for example, Mars, with its thin atmosphere, is colder than the Earth would be if it were at a similar distance from the Sun), and provide less protection against meteoroids and high-frequency radiation. Further, where an atmosphere is less dense than 0.006 Earth atmospheres, water cannot exist in liquid form as the required atmospheric pressure, 4.56 mm Hg (608 Pa) (0.18 inch Hg), does not occur. The temperature range at which water is liquid is smaller at low pressures generally.

Secondly, smaller planets have smaller diameters and thus higher surface-to-volume ratios than their larger cousins. Such

Mars, with its rarefied atmosphere, is colder than the Earth would be if it were at a similar distance from the Sun.

bodies tend to lose the energy left over from their formation quickly and end up geologically dead, lacking the volcanoes, earthquakes and tectonic activity which supply the surface with life-sustaining material and the atmosphere with temperature moderators like carbon dioxide. Plate tectonics appear particularly crucial, at least on Earth: not only does the process recycle important chemicals and minerals, it also fosters bio-diversity through continent creation and increased environmental complexity and helps create the convective cells necessary to generate Earth's magnetic field.[40]

"Low mass" is partly a relative label: the Earth is low mass when compared to the Solar System's gas giants, but it is the largest, by diameter and mass, and the densest of all terrestrial bodies.[lower-alpha 5] It is large enough to retain an atmosphere through gravity alone and large enough that its molten core remains a heat engine, driving the diverse geology of the surface (the decay of radioactive elements within a planet's core is the other significant component of planetary heating). Mars, by contrast, is nearly (or perhaps totally) geologically dead and has lost much of its atmosphere.[41] Thus it would be fair to infer that the lower mass limit for habitability lies somewhere between that of Mars and that of Earth or Venus: 0.3 Earth masses has been offered as a rough dividing line for habitable planets.[42] However, a 2008 study by the Harvard-Smithsonian Center for Astrophysics suggests that the dividing line may be higher. Earth may in fact lie on the lower boundary of habitability: if it were any smaller, plate tectonics would be impossible. Venus, which has 85% of

Earth's mass, shows no signs of tectonic activity. Conversely, "super-Earths", terrestrial planets with higher masses than Earth, would have higher levels of plate tectonics and thus be firmly placed in the habitable range.[43]

Exceptional circumstances do offer exceptional cases: Jupiter's moon Io (which is smaller than any of the terrestrial planets) is volcanically dynamic because of the gravitational stresses induced by its orbit, and its neighbor Europa may have a liquid ocean or icy slush underneath a frozen shell also due to power generated from orbiting a gas giant.

Saturn's Titan, meanwhile, has an outside chance of harbouring life, as it has retained a thick atmosphere and has liquid methane seas on its surface. Organic-chemical reactions that only require minimum energy are possible in these seas, but whether any living system can be based on such minimal reactions is unclear, and would seem unlikely. These satellites are exceptions, but they prove that mass, as a criterion for habitability, cannot necessarily be considered definitive at this stage of our understanding.

A larger planet is likely to have a more massive atmosphere. A combination of higher escape velocity to retain lighter atoms, and extensive outgassing from enhanced plate tectonics may greatly increase the atmospheric pressure and temperature at the surface compared to Earth. The enhanced greenhouse effect of such a heavy atmosphere would tend to suggest that the habitable zone should be further out from the central star for such massive planets.

Finally, a larger planet is likely to have a large iron core. This allows for a magnetic field to protect the planet from stellar wind and cosmic radiation, which otherwise would tend to strip away planetary atmosphere and to bombard living things with ionized particles. Mass is not the only criterion for producing a magnetic field—as the planet must also rotate fast enough to produce a dynamo effect within its core[44]—but it is a significant component of the process.

11.2.2   Orbit and rotation

As with other criteria, stability is the critical consideration in evaluating the effect of orbital and rotational characteristics on planetary habitability. Orbital eccentricity is the difference between a planet's farthest and closest approach to its parent star divided by the sum of said distances. It is a ratio describing the shape of the elliptical orbit. The greater the eccentricity the greater the temperature fluctuation on a planet's surface. Although they are adaptive, living organisms can stand only so much variation, particularly if the fluctuations overlap both the freezing point and boiling point of the planet's main biotic solvent (e.g., water on Earth). If, for example, Earth's oceans were alternately boiling and freezing solid, it is difficult to imagine life as we know it having evolved. The more complex the organism, the greater the temperature sensitivity.[45] The Earth's orbit is almost wholly circular, with an eccentricity of less than 0.02; other planets in the Solar System (with the exception of Mercury) have eccentricities that are similarly benign.

Data collected on the orbital eccentricities of extrasolar planets has surprised most researchers: 90% have an orbital eccentricity greater than that found within the Solar System, and the average is fully 0.25.[46] This means that the vast majority of planets have highly eccentric orbits and of these, even if their average distance from their star is deemed to be within the HZ, they nonetheless would be spending only a small portion of their time within the zone.

A planet's movement around its rotational axis must also meet certain criteria if life is to have the opportunity to evolve. A first assumption is that the planet should have moderate seasons. If there is little or no axial tilt (or obliquity) relative to the perpendicular of the ecliptic, seasons will not occur and a main stimulant to biospheric dynamism will disappear. The planet would also be colder than it would be with a significant tilt: when the greatest intensity of radiation is always within a few degrees of the equator, warm weather cannot move poleward and a planet's climate becomes dominated by colder polar weather systems.

If a planet is radically tilted, meanwhile, seasons will be extreme and make it more difficult for a biosphere to achieve homeostasis. The axial tilt of the Earth is higher now (in the Quaternary) than it has been in the past, coinciding with reduced polar ice, warmer temperatures and *less* seasonal variation. Scientists do not know whether this trend will continue indefinitely with further increases in axial tilt (see Snowball Earth).

The exact effects of these changes can only be computer modelled at present, and studies have shown that even extreme tilts of up to 85 degrees do not absolutely preclude life "provided it does not occupy continental surfaces plagued seasonally by the highest temperature."[47] Not only the mean axial tilt, but also its variation over time must be considered. The Earth's tilt varies between 21.5 and 24.5 degrees over 41,000 years. A more drastic variation, or a much shorter periodicity, would induce climatic effects such as variations in seasonal severity.

Other orbital considerations include:

- The planet should rotate relatively quickly so that the day-night cycle is not overlong. If a day takes years, the temperature differential between the day and night side will be pronounced, and problems similar to those noted with extreme orbital eccentricity will come to the fore.

- The planet also should rotate quickly enough so that a magnetic dynamo may be started in its iron core to produce a magnetic field.

- Change in the direction of the axis rotation (precession) should not be pronounced. In itself, precession need not affect habitability as it changes the direction of the tilt, not its degree. However, precession tends to accentuate variations caused by other orbital deviations; see Milankovitch cycles. Precession on Earth occurs over a 26,000-year cycle.

The Earth's Moon appears to play a crucial role in moderating the Earth's climate by stabilising the axial tilt. It has been suggested that a chaotic tilt may be a "deal-breaker" in terms of habitability—i.e. a satellite the size of the Moon is not only helpful but required to produce stability.[48] This position remains controversial.[lower-alpha 6]

11.2.3 Geochemistry

Main article: Geochemistry

It is generally assumed that any extraterrestrial life that might exist will be based on the same fundamental biochemistry as found on Earth, as the four elements most vital for life, carbon, hydrogen, oxygen, and nitrogen, are also the most common chemically reactive elements in the universe. Indeed, simple biogenic compounds, such as very simple amino acids such as glycine, have been found in meteorites and in the interstellar medium.[49] These four elements together comprise over 96% of Earth's collective biomass. Carbon has an unparalleled ability to bond with itself and to form a massive array of intricate and varied structures, making it an ideal material for the complex mechanisms that form living cells. Hydrogen and oxygen, in the form of water, compose the solvent in which biological processes take place and in which the first reactions occurred that led to life's emergence. The energy released in the formation of powerful covalent bonds between carbon and oxygen, available by oxidizing organic compounds, is the fuel of all complex life-forms. These four elements together make up amino acids, which in turn are the building blocks of proteins, the substance of living tissue. In addition, neither sulfur, required for the building of proteins, nor phosphorus, needed for the formation of DNA, RNA, and the adenosine phosphates essential to metabolism, is rare.

Relative abundance in space does not always mirror differentiated abundance within planets; of the four life elements, for instance, only oxygen is present in any abundance in the Earth's crust.[50] This can be partly explained by the fact that many of these elements, such as hydrogen and nitrogen, along with their simplest and most common compounds, such as carbon dioxide, carbon monoxide, methane, ammonia, and water, are gaseous at warm temperatures. In the hot region close to the Sun, these volatile compounds could not have played a significant role in the planets' geological formation. Instead, they were trapped as gases underneath the newly formed crusts, which were largely made of rocky, involatile compounds such as silica (a compound of silicon and oxygen, accounting for oxygen's relative abundance). Outgassing of volatile compounds through the first volcanoes would have contributed to the formation of the planets' atmospheres. The Miller–Urey experiment showed that, with the application of energy, simple inorganic compounds exposed to a primordial atmosphere can react to synthesize amino acids.[51]

Even so, volcanic outgassing could not have accounted for the amount of water in Earth's oceans.[52] The vast majority of the water —and arguably carbon— necessary for life must have come from the outer Solar System, away from the Sun's heat, where it could remain solid. Comets impacting with the Earth in the Solar System's early years would have deposited vast amounts of water, along with the other volatile compounds life requires onto the early Earth, providing a kick-start to the origin of life.

Thus, while there is reason to suspect that the four "life elements" ought to be readily available elsewhere, a habitable system probably also requires a supply of long-term orbiting bodies to seed inner planets. Without comets there is a possibility that life as we know it would not exist on Earth.

11.2.4   Microenvironments and extremophiles

One important qualification to habitability criteria is that only a tiny portion of a planet is required to support life. Astrobiologists often concern themselves with "micro-environments", noting that "we lack a fundamental understanding of how evolutionary forces, such as mutation, selection, and genetic drift, operate in micro-organisms that act on and respond to changing micro-environments."[53] Extremophiles are Earth organisms that live in niche environments under severe conditions generally considered inimical to life. Usually (although not always) unicellular, extremophiles include acutely alkaliphilic and acidophilic organisms and others that can survive water temperatures above 100 °C in hydrothermal vents.

The discovery of life in extreme conditions has complicated definitions of habitability, but also generated much excitement amongst researchers in greatly broadening the known range of conditions under which life can persist. For example, a planet that might otherwise be unable to support an atmosphere given the solar conditions in its vicinity, might be able to do so within a deep shadowed rift or volcanic cave.[54] Similarly, craterous terrain might offer a refuge for primitive life. The Lawn Hill crater has been studied as an astrobiological analog, with researchers suggesting rapid sediment infill created a protected microenvironment for microbial organisms; similar conditions may have occurred over the geological history of Mars.[55]

Earth environments that *cannot* support life are still instructive to astrobiologists in defining the limits of what organisms can endure. The heart of the Atacama desert, generally considered the driest place on Earth, appears unable to support life, but it has been subject to study by NASA and ESA for that reason: it provides a Mars analog and the moisture gradients along its edges are ideal for studying the boundary between sterility and habitability.[56] The Atacama was the subject of study in 2003 that partly replicated experiments from the Viking landings on Mars in the 1970s; no DNA could be recovered from two soil samples, and incubation experiments were also negative for biosignatures.[57]

11.2.5   Ecological factors

The two current ecological approaches for predicting the potential habitability use 19 or 20 environmental factors, with emphasis on water availability, temperature, presence of nutrients, an energy source, and protection from solar ultraviolet and galactic cosmic radiation.[58][59]

11.3   Uninhabited habitats

An important distinction in habitability is between habitats that contain active life (inhabited habitats) and habitats that are habitable for life, but uninhabited.[60] Uninhabited (or vacant) habitats could arise on a planet where there was no origin of life (and no transfer of life to the planet from another, inhabited, planet), but where habitable environments exist. They might also occur on a planet that is inhabited, but the lack of connectivity between habitats might mean that many habitats remain uninhabited. Uninhabited habitats underline the importance of decoupling habitability and the presence of life. Charles Cockell and co-workers discuss Mars as one plausible world that might harbor uninhabited habitats. Other stellar systems might host planets that are habitable, but devoid of life.

11.4   Alternative star systems

In determining the feasibility of extraterrestrial life, astronomers had long focused their attention on stars like the Sun. However, since planetary systems that resemble the Solar System are proving to be rare, they have begun to explore the possibility that life might form in systems very unlike our own.

11.4.1   Binary systems

Main article: Habitability of binary star systems

Typical estimates often suggest that 50% or more of all stellar systems are binary systems. This may be partly sample bias, as massive and bright stars tend to be in binaries and these are most easily observed and catalogued; a more precise analysis has suggested that the more common fainter stars are usually singular, and that up to two thirds of all stellar systems are therefore solitary.[61]

The separation between stars in a binary may range from less than one astronomical unit (AU, the average Earth–Sun distance) to several hundred. In latter instances, the gravitational effects will be negligible on a planet orbiting an otherwise suitable star and habitability potential will not be disrupted unless the orbit is highly eccentric (see Nemesis, for example). However, where the separation is significantly less, a stable orbit may be impossible. If a planet's distance to its primary exceeds about one fifth of the closest approach of the other star, orbital stability is not guaranteed.[62] Whether planets might form in binaries at all had long been unclear, given that gravitational forces might interfere with planet formation. Theoretical work by Alan Boss at the Carnegie Institution has shown that gas giants can form around stars in binary systems much as they do around solitary stars.[63]

One study of Alpha Centauri, the nearest star system to the Sun, suggested that binaries need not be discounted in the search for habitable planets. Centauri A and B have an 11 AU distance at closest approach (23 AU mean), and both should have stable habitable zones. A study of long-term orbital stability for simulated planets within the system shows that planets within approximately three AU of either star may remain rather stable (i.e. the semi-major axis deviating by less than 5% during 32 000 binary periods). The HZ for Centauri A is conservatively estimated at 1.2 to 1.3 AU and Centauri B at 0.73 to 0.74—well within the stable region in both cases.[64]

11.4.2   Red dwarf systems

Main article: Habitability of red dwarf systems
 Determining the habitability of red dwarf stars could help determine how common life in the universe might be, as red dwarfs make up between 70 to 90% of all the stars in the galaxy.

Size

Astronomers for many years ruled out red dwarfs as potential abodes for life. Their small size (from 0.08 to 0.45 solar masses) means that their nuclear reactions proceed exceptionally slowly, and they emit very little light (from 3% of that produced by the Sun to as little as 0.01%). Any planet in orbit around a red dwarf would have to huddle very close to its parent star to attain Earth-like surface temperatures; from 0.3 AU (just inside the orbit of Mercury) for a star like Lacaille 8760, to as little as 0.032 AU for a star like Proxima Centauri[65] (such a world would have a year lasting just 6.3 days). At those distances, the star's gravity would cause tidal locking. One side of the planet would eternally face the star, while the other would always face away from it. The only ways in which potential life could avoid either an inferno or a deep freeze would be if the planet had an atmosphere thick enough to transfer the star's heat from the day side to the night side, or if there was a gas giant in the habitable zone, with a habitable moon, which would be locked to the planet instead of the star, allowing a more even distribution of radiation over the planet. It was long assumed that such a thick atmosphere would prevent sunlight from reaching the surface in the first place, preventing photosynthesis.

This pessimism has been tempered by research. Studies by Robert Haberle and Manoj Joshi of NASA's Ames Research Center in California have shown that a planet's atmosphere (assuming it included greenhouse gases CO_2 and H_2O) need only be 100 mbs, or 10% of Earth's atmosphere, for the star's heat to be effectively carried to the night side.[66] This is well within the levels required for photosynthesis, though water would still remain frozen on the dark side in some of their models. Martin Heath of Greenwich Community College, has shown that seawater, too, could be effectively circulated without freezing solid if the ocean basins were deep enough to allow free flow beneath the night side's ice cap. Further research—including a consideration of the amount of photosynthetically active radiation—suggested that tidally locked planets in red dwarf systems might at least be habitable for higher plants.[67]

Other factors limiting habitability

Size is not the only factor in making red dwarfs potentially unsuitable for life, however. On a red dwarf planet, photosynthesis on the night side would be impossible, since it would never see the sun. On the day side, because the sun does not rise or set, areas in the shadows of mountains would remain so forever. Photosynthesis as we understand it would be complicated by the fact that a red dwarf produces most of its radiation in the infrared, and on the Earth the process depends on visible light. There are potential positives to this scenario. Numerous terrestrial ecosystems rely on chemosynthesis rather than photosynthesis, for instance, which would be possible in a red dwarf system. A static primary star position removes the need for plants to steer leaves toward the sun, deal with changing shade/sun patterns, or change from photosynthesis to stored energy during night. Because of the lack of a day-night cycle, including the weak light of morning and evening, far more energy would be available at a given radiation level.

Red dwarfs are far more variable and violent than their more stable, larger cousins. Often they are covered in starspots that can dim their emitted light by up to 40% for months at a time, while at other times they emit gigantic flares that can double their brightness in a matter of minutes.[68] Such variation would be very damaging for life, as it would not only destroy any complex organic molecules that could possibly form biological precursors, but also because it would blow off sizeable portions of the planet's atmosphere.

For a planet around a red dwarf star to support life, it would require a rapidly rotating magnetic field to protect it from the flares. However, a tidally locked planet rotates only very slowly, and so cannot produce a geodynamo at its core. However, the violent flaring period of a red dwarf's life cycle is estimated to only last roughly the first 1.2 billion years of its existence. If a planet forms far away from a red dwarf so as to avoid tidal locking, and then migrates into the star's habitable zone after this turbulent initial period, it is possible that life may have a chance to develop.[69]

Longevity and ubiquity

There is, however, one major advantage that red dwarfs have over other stars as abodes for life: they live a long time. It took 4.5 billion years before humanity appeared on Earth, and life as we know it will see suitable conditions for 1[70] to 2.3[71] billion years more. Red dwarfs, by contrast, could live for trillions of years because their nuclear reactions are far slower than those of larger stars, meaning that life would have longer to evolve and survive.

While the odds of finding a planet in the habitable zone around any specific red dwarf are slim, the total amount of habitable zone around all red dwarfs combined is equal to the total amount around Sun-like stars given their ubiquity.[72] Furthermore, this total amount of habitable zone will last longer, because red dwarf stars live for hundreds of billions of years or even longer on the main sequence.[73]

11.4.3   Massive stars

Recent research suggests that very large stars, greater than ~100 solar masses, could have planetary systems consisting of hundreds of Mercury-sized planets within the habitable zone. Such systems could also contain brown dwarfs and low-mass stars (~0.1–0.3 solar masses).[74] However the very short lifespans of stars of more than a few solar masses would scarcely allow time for a planet to cool, let alone the time needed for a stable biosphere to develop. Massive stars are thus eliminated as possible abodes for life.[75]

However, a massive-star system could be a progenitor of life in another way – the supernova explosion of the massive star in the central part of the system. This supernova will disperse heavier elements throughout its vicinity, created during the phase when the massive star has moved off of the main sequence, and the systems of the potential low-mass stars (which are still on the main sequence) within the former massive-star system may be enriched with the relatively large supply of the heavy elements so close to a supernova explosion. However, this states nothing about what types of planets would form as a result of the supernova material, or what their habitability potential would be.

11.5   The galactic neighborhood

Along with the characteristics of planets and their star systems, the wider galactic environment may also impact habitability. Scientists considered the possibility that particular areas of galaxies (galactic habitable zones) are better suited to life than others; the Solar System in which we live, in the Orion Spur, on the Milky Way galaxy's edge is considered to be in a life-favorable spot.[76]

- It is not in a globular cluster where immense star densities are inimical to life, given excessive radiation and gravitational disturbance. Globular clusters are also primarily composed of older, probably metal-poor, stars. Furthermore, in globular clusters, the great ages of the stars would mean a large amount of stellar evolution by the host or other nearby stars, which due to their proximity may cause extreme harm to life on any planets, provided that they can form.

- It is not near an active gamma ray source.

- It is not near the galactic center where once again star densities increase the likelihood of ionizing radiation (e.g., from magnetars and supernovae). A supermassive black hole is also believed to lie at the middle of the galaxy which might prove a danger to any nearby bodies.

- The circular orbit of the Sun around the galactic center keeps it out of the way of the galaxy's spiral arms where intense radiation and gravitation may again lead to disruption.[77]

Thus, relative isolation is ultimately what a life-bearing system needs. If the Sun were crowded amongst other systems, the chance of being fatally close to dangerous radiation sources would increase significantly. Further, close neighbors might disrupt the stability of various orbiting bodies such as Oort cloud and Kuiper belt objects, which can bring catastrophe if knocked into the inner Solar System.

While stellar crowding proves disadvantageous to habitability, so too does extreme isolation. A star as metal-rich as the Sun would probably not have formed in the very outermost regions of the Milky Way given a decline in the relative abundance of metals and a general lack of star formation. Thus, a "suburban" location, such as the Solar System enjoys, is preferable to a Galaxy's center or farthest reaches.[78]

11.6   Other considerations

11.6.1   Alternative biochemistries

Main article: Hypothetical types of biochemistry

While most investigations of extraterrestrial life start with the assumption that advanced life-forms must have similar requirements for life as on Earth, the hypothesis of other types of biochemistry suggests the possibility of lifeforms evolving around a different metabolic mechanism. In *Evolving the Alien*, biologist Jack Cohen and mathematician Ian Stewart argue astrobiology, based on the Rare Earth hypothesis, is restrictive and unimaginative. They suggest that Earth-like planets may be very rare, but non-carbon-based complex life could possibly emerge in other environments. The most frequently mentioned alternative to carbon is silicon-based life, while ammonia and hydrocarbons are sometimes suggested as alternative solvents to water. The astrobiologist Dirk Schulze-Makuch and other scientists have proposed a Planet Habitability Index whose criteria include "potential for holding a liquid solvent" that is not necessarily restricted to water.[79][80]

More speculative ideas have focused on bodies altogether different from Earth-like planets. Astronomer Frank Drake, a well-known proponent of the search for extraterrestrial life, imagined life on a neutron star: submicroscopic "nuclear molecules" combining to form creatures with a life cycle millions of times quicker than Earth life.[81] Called "imaginative and tongue-in-cheek", the idea gave rise to science fiction depictions.[82] Carl Sagan, another optimist with regards to extraterrestrial life, considered the possibility of organisms that are always airborne within the high atmosphere of Jupiter in a 1976 paper.[33][34] Cohen and Stewart also envisioned life in both a solar environment and in the atmosphere of a gas giant.

11.6.2   "Good Jupiters"

"Good Jupiters" are gas giants, like the Solar System's Jupiter, that orbit their stars in circular orbits far enough away from the habitable zone not to disturb it but close enough to "protect" terrestrial planets in closer orbit in two critical ways.

First, they help to stabilize the orbits, and thereby the climates, of the inner planets. Second, they keep the inner Solar System relatively free of comets and asteroids that could cause devastating impacts.[83] Jupiter orbits the Sun at about five times the distance between the Earth and the Sun. This is the rough distance we should expect to find good Jupiters elsewhere. Jupiter's "caretaker" role was dramatically illustrated in 1994 when Comet Shoemaker–Levy 9 impacted the giant; had Jovian gravity not captured the comet, it may well have entered the inner Solar System.

However, the story is not quite so clear cut. Research has shown that Jupiter's role in determining the rate at which objects hit the Earth is, at the very least, significantly more complicated than once thought.[84][85][86][87] Whilst for the long-period comets (which contribute only a small fraction of the impact risk to the Earth) it is true that Jupiter acts as a shield, it actually seems to increase the rate at which asteroids and short-period comets are flung towards our planet. Were Jupiter absent, it seems likely that the Earth would actually experience significantly fewer impacts from potentially hazardous objects. By extension, it is becoming clear that the presence of Jupiter-like planets is no longer required as a pre-requisite for planetary habitability – indeed, our first searches for life beyond the Solar System might be better directed to systems where no such planet has formed, since in those systems, less material will be directed to impact on the potentially inhabited planets.

The role of Jupiter in the early history of the Solar System is somewhat better established, and the source of significantly less debate. Early in the Solar System's history, Jupiter is accepted as having played an important role in the hydration of our planet: it increased the eccentricity of asteroid belt orbits and enabled many to cross Earth's orbit and supply the planet with important volatiles. Before Earth reached half its present mass, icy bodies from the Jupiter–Saturn region and small bodies from the primordial asteroid belt supplied water to the Earth due to the gravitational scattering of Jupiter and, to a lesser extent, Saturn.[88] Thus, while the gas giants are now helpful protectors, they were once suppliers of critical habitability material.

In contrast, Jupiter-sized bodies that orbit too close to the habitable zone but not in it (as in 47 Ursae Majoris), or have a highly elliptical orbit that crosses the habitable zone (like 16 Cygni B) make it very difficult for an independent Earth-like planet to exist in the system. See the discussion of a stable habitable zone above. However, during the process of migrating into a habitable zone, a Jupiter-size planet may capture a terrestrial planet as a moon. Even if such a planet is initially loosely bound and following a strongly inclined orbit, gravitational interactions with the star can stabilize the new moon into a close, circular orbit that is coplanar with the planet's orbit around the star.[89]

11.6.3   Life's impact on habitability

A supplement to the factors that support life's emergence is the notion that life itself, once formed, becomes a habitability factor in its own right. An important Earth example was the production of oxygen by ancient cyanobacteria, and eventually photosynthesizing plants, leading to a radical change in the composition of Earth's atmosphere. This oxygen would prove fundamental to the respiration of later animal species. The Gaia hypothesis, a class of scientific models of the geo-biosphere pioneered by Sir James Lovelock in 1975, argues that life as a whole fosters and maintains suitable conditions for itself by helping to create a planetary environment suitable for its continuity. Similarly, David Grinspoon has suggested a "living worlds hypothesis" in which our understanding of what constitutes habitability cannot be separated from life already extant on a planet. Planets that are geologically and meteorologically alive are much more likely to be biologically alive as well and "a planet and its life will co-evolve."[90]

11.7   See also

- *Alien Planet*
- Class M planet
- Circumstellar habitable zone
- Darwin mission
- Definition of planet
- Extraterrestrial liquid water

- *Habitable Planets for Man*

- List of potentially habitable exoplanets

- Natural satellite habitability

- Neocatastrophism

- Mars Science Laboratory

- Rare Earth hypothesis

- Space colonization

- Terraforming

- Terrestrial Planet Finder

- Superhabitable Exoplanet

11.8 Notes

[1] This article is an analysis of planetary habitability from the perspective of contemporary physical science. A historical viewpoint on the possibility of habitable planets can be found at Beliefs in extraterrestrial life and Cosmic pluralism. For a discussion of the probability of alien life see the Drake equation and Fermi paradox. Habitable planets are also a staple of fiction; see Planets in science fiction.

[2] Life appears to have emerged on Earth approximately 500 million years after the planet's formation. "A" class stars (which shine for between 600 million and 1.2 billion years) and a small fraction of "B" class stars (which shine 10+ million to 600 million) fall within this window. At least theoretically life could emerge in such systems but it would almost certainly not reach a sophisticated level given these time-frames and the fact that increases in luminosity would occur quite rapidly. Life around "O" class stars is exceptionally unlikely, as they shine for less than ten million years.

[3] In *Evolving the Alien*, Jack Cohen and Ian Stewart evaluate plausible scenarios in which life might form in the cloud-tops of Jovian planets. Similarly, Carl Sagan suggested that the clouds of Jupiter might host life.[13][14]

[4] There is an emerging consensus that single-celled micro-organisms may in fact be common in the universe, especially since Earth's extremophiles flourish in environments that were once considered hostile to life. The potential occurrence of complex multi-celled life remains much more controversial. In their work *Rare Earth: Why Complex Life Is Uncommon in the Universe*, Peter Ward and Donald Brownlee argue that microbial life is probably widespread while complex life is very rare and perhaps even unique to Earth. Current knowledge of Earth's history partly buttresses this theory: multi-celled organisms are believed to have emerged at the time of the Cambrian explosion close to 600 million years ago, but more than 3 billion years after life first appeared. That Earth life remained unicellular for so long underscores that the decisive step toward complex organisms need not necessarily occur.

[5] There is a "mass-gap" in the Solar System between Earth and the two smallest gas giants, Uranus and Neptune, which are 13 and 17 Earth masses. This is probably just chance, as there is no geophysical barrier to the formation of intermediate bodies (see for instance OGLE-2005-BLG-390Lb and Super-Earth) and we should expect to find planets throughout the galaxy between two and twelve Earth masses. If the star system is otherwise favourable, such planets would be good candidates for life as they would be large enough to remain internally dynamic and to retain an atmosphere for billions of years but not so large as to accrete a gaseous shell which limits the possibility of life formation.

[6] According to prevailing theory, the formation of the Moon commenced when a Mars-sized body struck the Earth in a glancing collision late in its formation, and the ejected material coalesced and fell into orbit (see giant impact hypothesis). In *Rare Earth* Ward and Brownlee emphasize that such impacts ought to be rare, reducing the probability of other Earth-Moon type systems and hence the probability of other habitable planets. Other moon formation processes are possible, however, and the proposition that a planet may be habitable in the absence of a moon has not been disproven.

11.9   References

[1] Dyches, Preston; Chou, Felcia (7 April 2015). "The Solar System and Beyond is Awash in Water". *NASA*. Retrieved 8 April 2015.

[2] "Goal 1: Understand the nature and distribution of habitable environments in the Universe". *Astrobiology: Roadmap*. NASA. Retrieved 11 August 2007.

[3] Wolszczan, A.; Frail, D. A. (9 January 1992). "A planetary system around the millisecond pulsar PSR1257 + 12". *Nature* **355** (6356): 145–147. Bibcode:1992Natur.355..145W. doi:10.1038/355145a0.

[4] Wolszczan, A (1994). "Confirmation of Earth Mass Planets Orbiting the Millisecond Pulsar PSR:B1257+12". *Science* **264** (5158): 538–42. Bibcode:1994Sci...264..538W. doi:10.1126/science.264.5158.538. JSTOR 2883699. PMID 17732735.

[5] Loeb, Abraham (October 2014). "The Habitable Epoch of the Early Universe". *International Journal of Astrobiology* **13** (04): 337–339. arXiv:1312.0613. doi:10.1017/S1473550414000196.

[6] Dreifus, Claudia (2 December 2014). "Much-Discussed Views That Go Way Back – Avi Loeb Ponders the Early Universe, Nature and Life". *New York Times*. Retrieved 3 December 2014.

[7] Rampelotto, P.H. (2010). "Panspermia: A Promising Field Of Research" (PDF). *Astrobiology Science Conference*. Retrieved 3 December 2014. External link in |work= (help)

[8] Graham, Robert W. (February 1990). "NASA Technical Memorandum 102363 – Extraterrestrial Life in the Universe" (PDF). *NASA* (Lewis Research Center, Ohio). Retrieved 7 July 2014.

[9] Altermann, Wladyslaw (2008). "From Fossils to Astrobiology – A Roadmap to Fata Morgana?". In Seckbach, Joseph; Walsh, Maud. *From Fossils to Astrobiology: Records of Life on Earth and the Search for Extraterrestrial Biosignatures* **12**. p. xvii. ISBN 1-4020-8836-1.

[10] Horneck, Gerda; Petra Rettberg (2007). *Complete Course in Astrobiology*. Wiley-VCH. ISBN 3-527-40660-3.

[11] Davies, Paul (18 November 2013). "Are We Alone in the Universe?". *New York Times*. Retrieved 20 November 2013.

[12] Overbye, Dennis (6 January 2015). "As Ranks of Goldilocks Planets Grow, Astronomers Consider What's Next". *New York Times*. Retrieved 6 January 2015.

[13] Overbye, Dennis (4 November 2013). "Far-Off Planets Like the Earth Dot the Galaxy". *New York Times*. Retrieved 5 November 2013.

[14] Petigura, Eric A.; Howard, Andrew W.; Marcy, Geoffrey W. (31 October 2013). "Prevalence of Earth-size planets orbiting Sun-like stars". *Proceedings of the National Academy of Sciences of the United States of America*. arXiv:1311.6806. Bibcode:2013PNAS..11019273P. doi:10.1073/pnas.1319909110. Retrieved 5 November 2013.

[15] Khan, Amina (4 November 2013). "Milky Way may host billions of Earth-size planets". *Los Angeles Times*. Retrieved 5 November 2013.

[16] Turnbull, Margaret C.; Tarter, Jill C. (March 2003). "Target selection for SETI: A catalog of nearby habitable stellar systems" (PDF). *The Astrophysical Journal Supplement Series* **145**: 181–198. arXiv:astro-ph/0210675. Bibcode:2003ApJS..145..181T. doi:10.1086/345779. Habitability criteria defined—the foundational source for this article.

[17] Choi, Charles Q. (21 August 2015). "Giant Galaxies May Be Better Cradles for Habitable Planets". *Space.com*. Retrieved 24 August 2015.

[18] "Star tables". California State University, Los Angeles. Archived from the original on 14 June 2008. Retrieved 12 August 2010.

[19] Kasting, James F.; Whittet, DC; Sheldon, WR (August 1997). "Ultraviolet radiation from F and K stars and implications for planetary habitability". *Origins of Life and Evolution of Biospheres* **27** (4): 413–420. doi:10.1023/A:1006596806012. PMID 11536831.

[20] Guinan, Edward; Cuntz, Manfred (10 August 2009). "The violent youth of solar proxies steer course of genesis of life". International Astronomical Union. Retrieved 27 August 2009.

[21] "Gliese 581: one planet might indeed be habitable" (Press release). Astronomy & Astrophysics. 13 December 2007. Retrieved 7 April 2008.

[22] Staff (20 September 2012). "LHS 188 – High proper-motion Star". Centre de données astronomiques de Strasbourg (Strasbourg astronomical Data Center). Retrieved 20 September 2012.

[23] Méndez, Abel (29 August 2012). "A Hot Potential Habitable Exoplanet around Gliese 163". University of Puerto Rico at Arecibo (Planetary Habitability Laboratory). Retrieved 20 September 2012.

[24] Redd, Nola Taylor (20 September 2012). "Newfound Alien Planet a Top Contender to Host Life". Space.com. Retrieved 20 September 2012.

[25] "Planets May Keep Warmer In A Cool Star System". Redorbit. 19 July 2013.

[26] Shields, A. L.; Meadows, V. S.; Bitz, C. M.; Pierrehumbert, R. T.; Joshi, M. M.; Robinson, T. D. (2013). "The Effect of Host Star Spectral Energy Distribution and Ice-Albedo Feedback on the Climate of Extrasolar Planets". *Astrobiology* **13** (8): 715. Bibcode:2013AsBio..13..715S. doi:10.1089/ast.2012.0961. PMID 23855332.

[27] Kasting, James F.; Whitmore, Daniel P.; Reynolds, Ray T. (1993). "Habitable Zones Around Main Sequence Stars" (PDF). *Icarus* **101** (1): 108–128. Bibcode:1993Icar..101..108K. doi:10.1006/icar.1993.1010. PMID 11536936. Retrieved 6 August 2007.

[28] Williams, Darren M.; Kasting, James F.; Wade, Richard A. (January 1997). "Habitable moons around extrasolar giant planets". *Nature* **385** (6613): 234–236. Bibcode:1996DPS....28.1221W. doi:10.1038/385234a0. PMID 9000072.

[29] "The Little Ice Age". *Department of Atmospheric Science*. University of Washington. Retrieved 11 May 2007.

[30] "18 Scorpii". *www.solstation.com*. Sol Company. Retrieved 11 May 2007.

[31] Santos, Nuno C.; Israelian, Garik; Mayor, Michael (2003). "Confirming the Metal-Rich Nature of Stars with Giant Planets" (PDF). *Proceedings of 12th Cambridge Workshop on Cool Stars, Stellar Systems, and The Sun*. University of Colorado. Retrieved 11 August 2007.

[32] "An interview with Dr. Darren Williams". *Astrobiology: The Living Universe*. 2000. Retrieved 5 August 2007.

[33] Sagan, C.; Salpeter, E. E. (1976). "Particles, environments, and possible ecologies in the Jovian atmosphere". *The Astrophysical Journal Supplement Series* **32**: 737. Bibcode:1976ApJS...32..737S. doi:10.1086/190414.

[34] Darling, David. "Jupiter, life on". The Encyclopedia of Astrobiology, Astronomy, and Spaceflight. Retrieved 6 August 2007.

[35] "Could there be life in the outer solar system?". *Millennium Mathematics Project, Videoconferences for Schools*. University of Cambridge. 2002. Retrieved 5 August 2007.

[36] Borucki, William J.; Koch, David G.; Basri, Gibor; Batalha, Natalie; Brown, Timothy M.; Bryson, Stephen T.; Caldwell, Douglas; Christensen-Dalsgaard, Jørgen; Cochran, William D.; Devore, Edna; Dunham, Edward W.; Gautier, Thomas N.; Geary, John C.; Gilliland, Ronald; Gould, Alan; Howell, Steve B.; Jenkins, Jon M.; Latham, David W.; Lissauer, Jack J.; Marcy, Geoffrey W.; Rowe, Jason; Sasselov, Dimitar; Boss, Alan; Charbonneau, David; Ciardi, David; Doyle, Laurance; Dupree, Andrea K.; Ford, Eric B.; Fortney, Jonathan; et al. (2011). "Characteristics of planetary candidates observed by Kepler, II: Analysis of the first four months of data". *The Astrophysical Journal* **736** (1): 19. arXiv:1102.0541. Bibcode:2011ApJ...736...19B. doi:10.1088/0004-637X/736/1/19.

[37] "NASA Finds Earth-size Planet Candidates in Habitable Zone, Six Planet System". NASA. 2 February 2011. Retrieved 2 February 2011.

[38] Grant, Andrew (8 March 2011). "Exclusive: "Most Earth-Like" Exoplanet Gets Major Demotion—It Isn't Habitable". Discover Magazine. Retrieved 9 March 2011.

[39] Borenstein, Seth (19 February 2011). "Cosmic census finds crowd of planets in our galaxy". Associated Press. Retrieved 19 February 2011.

[40] Ward, pp. 191–220

[41] "The Heat History of the Earth". *Geolab*. James Madison University. Retrieved 11 May 2007.

[42] Raymond, Sean N.; Quinn, Thomas; Lunine, Jonathan I. (January 2007). "High-resolution simulations of the final assembly of Earth-like planets 2: water delivery and planetary habitability". *Astrobiology* **7** (1): 66–84. arXiv:astro-ph/0510285. Bibcode:2007AsBio...7...66R. doi:10.1089/ast.2006.06-0126. PMID 17407404.

[43] "Earth: A Borderline Planet for Life?". *Harvard-Smithsonian Center for Astrophysics*. 2008. Retrieved 4 June 2008.

[44] Nave, C. R. "Magnetic Field of the Earth". *HyperPhysics*. Georgia State University. Retrieved 11 May 2007.

[45] Ward, pp. 122–123.

[46] Bortman, Henry (22 June 2005). "Elusive Earths". Astrobiology Magazine. Retrieved 11 May 2007.

[47] "Planetary Tilt Not A Spoiler For Habitation" (Press release). Penn State University. 25 August 2003. Retrieved 11 May 2007.

[48] Lasker, J.; Joutel, F.; Robutel, P. (July 1993). "Stabilization of the earth's obliquity by the moon". *Nature* **361** (6413): 615–617. Bibcode:1993Natur.361..615L. doi:10.1038/361615a0.

[49] "Organic Molecule, Amino Acid-Like, Found In Constellation Sagittarius". ScienceDaily. 2008. Retrieved 20 December 2008.

[50] Darling, David. "Elements, biological abundance". The Encyclopedia of Astrobiology, Astronomy, and Spaceflight. Retrieved 11 May 2007.

[51] "How did chemistry and oceans produce this?". *The Electronic Universe Project*. University of Oregon. Retrieved 11 May 2007.

[52] "How did the Earth Get to Look Like This?". *The Electronic Universe Project*. University of Oregon. Retrieved 11 May 2007.

[53] "Understand the evolutionary mechanisms and environmental limits of life". *Astrobiology: Roadmap*. NASA. September 2003. Retrieved 6 August 2007.

[54] Hart, Stephen (17 June 2003). "Cave Dwellers: ET Might Lurk in Dark Places". Space.com. Archived from the original on 20 June 2003. Retrieved 6 August 2007.

[55] Lindsay, J; Brasier, M (2006). "Impact Craters as biospheric microenvironments, Lawn Hill Structure, Northern Australia". *Astrobiology* **6** (2): 348–363. Bibcode:2006AsBio...6..348L. doi:10.1089/ast.2006.6.348. PMID 16689651.

[56] McKay, Christopher (June 2002). "Too Dry for Life: The Atacama Desert and Mars" (PDF). *Ames Research Center*. NASA. Retrieved 26 August 2009.

[57] Navarro-González, Rafael; McKay, Christopher P. (7 November 2003). "Mars-Like Soils in the Atacama Desert, Chile, and the Dry Limit of Microbial Life". *Science* **302** (5647): 1018–1021. Bibcode:2003Sci...302.1018N. doi:10.1126/science.1089143. PMID 14605363.

[58] Schuerger, Andrew C.; Golden, D. C.; Ming, Doug W. (20 July 2012). "Biotoxicity of Mars soils:1. Dry deposition of analog soils on microbial colonies and survival under Martian conditions" (PDF). *Elsevier -Planetary and Space Science*. Retrieved 6 June 2013.

[59] Beaty, David W.; et al. (14 July 2006). "MEPAG SR-SAG (2006) Unpublished white paper", in the Mars Exploration Program Analysis Group (MEPAG). *Findings of the Mars Special Regions Science Analysis Group* (PDF). Jet Propulsion Laboratory – NASA. p. 17, retrieved 6 June 2013

[60] Cockell, Charles S.; Balme, Matt; Bridges, John C.; Davila, Alfsonso; Schwenzer, Susanne P. (January 2012). "Uninhabited habitats on Mars" (PDF). *Icarus* **217**: 184–193. Bibcode:2012Icar..217..184C. doi:10.1016/j.icarus.2011.10.025.

[61] "Most Milky Way Stars Are Single" (Press release). Harvard-Smithsonian Center for Astrophysics. 30 January 2006. Archived from the original on 13 August 2007. Retrieved 5 June 2007.

[62] "Stars and Habitable Planets". *www.solstation.com*. Sol Company. Retrieved 5 June 2007.

[63] Boss, Alan (January 2006). "Planetary Systems can from around Binary Stars" (Press release). Carnegie Institution. Retrieved 5 June 2007.

[64] Wiegert, Paul A.; Holman, Matt J. (April 1997). "The stability of planets in the Alpha Centauri system". *The Astronomical Journal* **113** (4): 1445–1450. arXiv:astro-ph/9609106. Bibcode:1997AJ....113.1445W. doi:10.1086/118360.

[65] "Habitable zones of stars". *NASA Specialized Center of Research and Training in Exobiology*. University of Southern California, San Diego. Archived from the original on 21 November 2000. Retrieved 11 May 2007.

[66] Joshi, M. M.; Haberle, R. M.; Reynolds, R. T. (October 1997). "Simulations of the Atmospheres of Synchronously Rotating Terrestrial Planets Orbiting M Dwarfs: Conditions for Atmospheric Collapse and the Implications for Habitability" (PDF). *Icarus* **129** (2): 450–465. Bibcode:1997Icar..129..450J. doi:10.1006/icar.1997.5793.

[67] Heath, Martin J.; Doyle, Laurance R.; Joshi, Manoj M.; Haberle, Robert M. (1999). "Habitability of Planets Around Red Dwarf Stars" (PDF). *Origins of Life and Evolution of the Biosphere* **29** (4): 405–424. doi:10.1023/A:1006596718708. PMID 10472629. Retrieved 11 August 2007.

[68] Croswell, Ken (27 January 2001). "Red, willing and able" (Full reprint). New Scientist. Retrieved 5 August 2007.

[69] Cain, Fraser; Gay, Pamela (2007). "AstronomyCast episode 40: American Astronomical Society Meeting, May 2007". *Universe Today*. Retrieved 17 June 2007.

[70] Hines, Sandra (13 January 2003). "'The end of the world' has already begun, UW scientists say" (Press release). University of Washington. Retrieved 5 June 2007.

[71] Li, King-Fai; Pahlevan, Kaveh; Kirschvink, Joseph L.; Yung, Yuk L. (2009). "Atmospheric pressure as a natural climate regulator for a terrestrial planet with a biosphere" (PDF). *Proceedings of the National Academy of Sciences* **106** (24): 9576–9579. Bibcode:2009PNAS..106.9576L. doi:10.1073/pnas.0809436106. PMC 2701016. PMID 19487662. Retrieved 19 July 2009.

[72] "M Dwarfs: The Search for Life is On, Interview with Todd Henry". Astrobiology Magazine. 29 August 2005. Retrieved 5 August 2007.

[73] Cain, Fraser (4 February 2009). "Red Dwarf Stars". Universe Today.

[74] Kashi, Amit; Soker, Noam (2011). "The outcome of the protoplanetary disk of very massive stars, January 2011". *New Astronomy* **16**: 27–32. arXiv:1002.4693. Bibcode:2011NewA...16...27K. doi:10.1016/j.newast.2010.06.003.

[75] Stellar mass#Age

[76] Mullen, Leslie (18 May 2001). "Galactic Habitable Zones". Astrobiology Magazine. Retrieved 5 August 2007.

[77] Ward, pp. 26–29.

[78] Dorminey, Bruce (July 2005). "Dark Threat". *Astronomy* **33**: 40–45. Bibcode:2005Ast.....33g..40D.

[79] Alan Boyle (2011-11-22). "Which alien worlds are most livable?". msnbc.com. Retrieved 2015-03-20.

[80] Dirk Schulze-Makuch; et al. (Dec 2011). "A Two-Tiered Approach to Assessing the Habitability of Exoplanets". *Astrobiology* **11** (10): 1041–1052. doi:10.1089/ast.2010.0592.

[81] Drake, Frank (1973). "Life on a Neutron Star". *Astronomy* **1** (5): 5.

[82] Darling, David. "Neutron star, life on". The Encyclopedia of Astrobiology, Astronomy, and Spaceflight. Retrieved 5 September 2009.

[83] Bortman, Henry (29 September 2004). "Coming Soon: "Good" Jupiters". Astrobiology Magazine. Retrieved 5 August 2007.

[84] Horner, Jonathan; Jones, Barrie (December 2010). "Jupiter – Friend or Foe? An answer". *Astronomy and Geophysics* **51** (6): 16–22. Bibcode:2010A&G....51f..16H. doi:10.1111/j.1468-4004.2010.51616.x.

[85] Horner, Jonathan; Jones, B. W. (October 2008). "Jupiter – Friend or Foe? I: The Asteroids". *International Journal of Astrobiology* **7** (3–4): 251–261. arXiv:0806.2795. Bibcode:2008IJAsB...7..251H. doi:10.1017/S1473550408004187.

[86] Horner, Jonathan; Jones, B. W. (April 2009). "Jupiter – friend or foe? II: the Centaurs". *International Journal of Astrobiology* **8** (2): 75–80. arXiv:0903.3305. Bibcode:2009IJAsB...8...75H. doi:10.1017/S1473550408004357.

[87] Horner, Jonathan; Jones, B. W.; Chambers, J. (January 2010). "Jupiter – friend or foe? III: the Oort cloud comets". *International Journal of Astrobiology* **9** (1): 1–10. arXiv:0911.4381. Bibcode:2010IJAsB...9....1H. doi:10.1017/S1473550409990346.

[88] Lunine, Jonathan I. (30 January 2001). "The occurrence of Jovian planets and the habitability of planetary systems". *Proceedings of the National Academy of Sciences* **98** (3): 809–814. Bibcode:2001PNAS...98..809L. doi:10.1073/pnas.98.3.809. PMC 14664. PMID 11158551.

[89] Porter, Simon B.; Grundy, William M. (July 2011). "Post-capture Evolution of Potentially Habitable Exomoons". *The Astrophysical Journal Letters* **736** (1): L14. arXiv:1106.2800. Bibcode:2011ApJ...736L..14P. doi:10.1088/2041-8205/736/1/L14

[90] "The Living Worlds Hypothesis". Astrobiology Magazine. 22 September 2005. Retrieved 6 August 2007.

11.10   Bibliography

- Ward, Peter; Brownlee, Donald (2000). *Rare Earth: Why Complex Life is Uncommon in the Universe*. Springer. ISBN 0-387-98701-0.

11.11   Further reading

- Cohen, Jack and Ian Stewart. *Evolving the Alien: The Science of Extraterrestrial Life*, Ebury Press, 2002. ISBN 0-09-187927-2

- Dole, Stephen H. (1965). *Habitable Planets for Man* (1st ed.). Rand Corporation. ISBN 0-444-00092-5.

- Fogg, Martyn J., ed. "Terraforming" (entire special issue) *Journal of the British Interplanetary Society*, April 1991

- Fogg, Martyn J. *Terraforming: Engineering Planetary Environments*, SAE International, 1995. ISBN 1-56091-609-5

- Gonzalez, Guillermo and Richards, Jay W. *The Privileged Planet*, Regnery, 2004. ISBN 0-89526-065-4

- Grinspoon, David. *Lonely Planets: The Natural Philosophy of Alien Life*, HarperCollins, 2004.

- Lovelock, James. *Gaia: A New Look at Life on Earth*. ISBN 0-19-286218-9

- Schmidt, Stanley and Robert Zubrin, eds. *Islands in the Sky*, Wiley, 1996. ISBN 0-471-13561-5

- Webb, Stephen *If The Universe Is Teeming With Aliens ... Where Is Everybody? Fifty Solutions to the Fermi Paradox and the Problem of Extraterrestrial Life* New York: January 2002 Springer-Verlag ISBN 978-0-387-95501-8

11.12   External links

- Planetary Sciences and Habitability Group, Spanish Research Council

- The Habitable Zone Gallery

- Planetary Habitability Laboratory (PHL/UPR Arecibo)

- The Habitable Exoplanets Catalog (PHL/UPR Arecibo)

- David Darling encyclopedia

- General interest astrobiology

- Sol Station

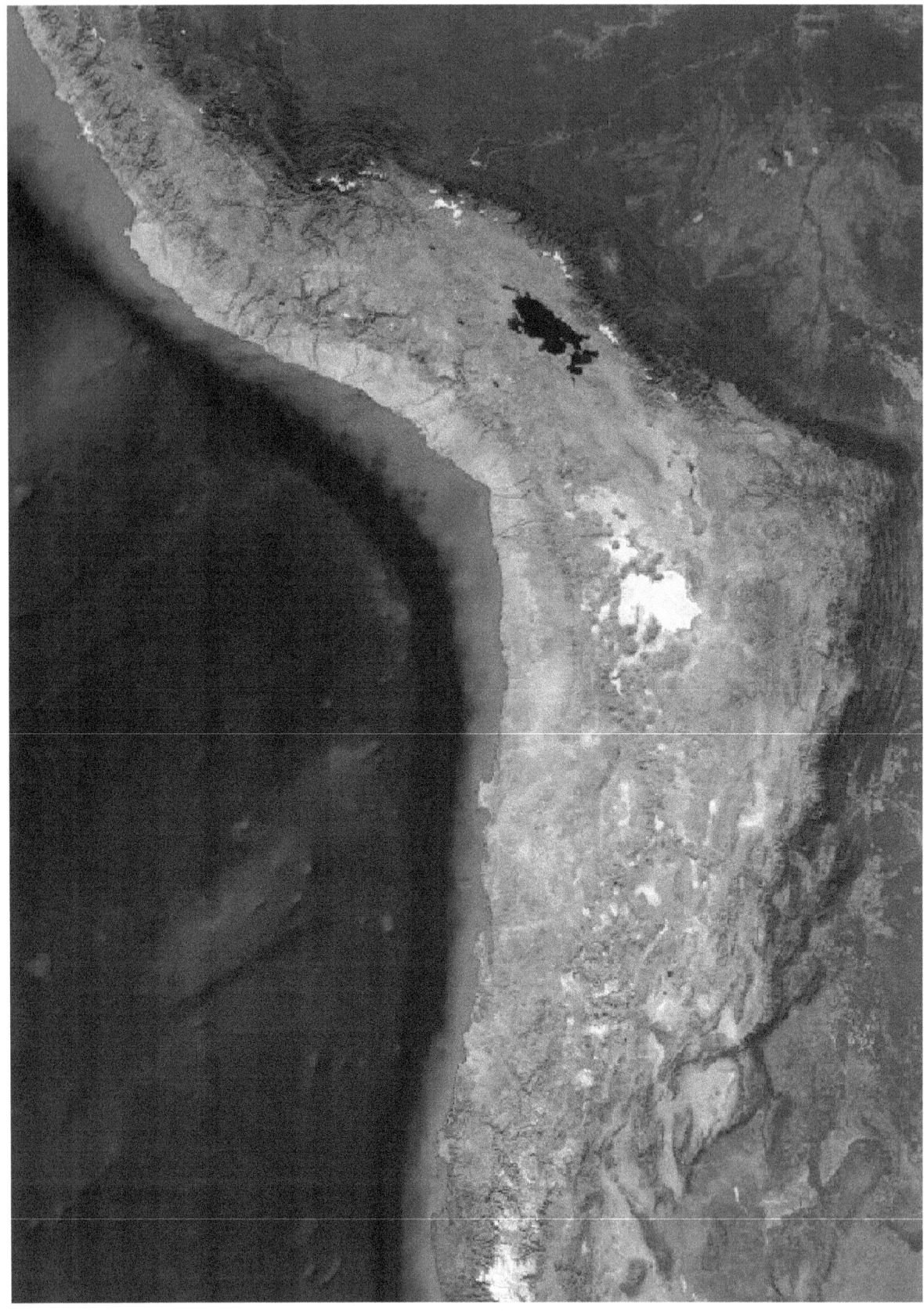

The Atacama Desert provides an analog to Mars and an ideal environment to study the boundary between sterility and habitability.

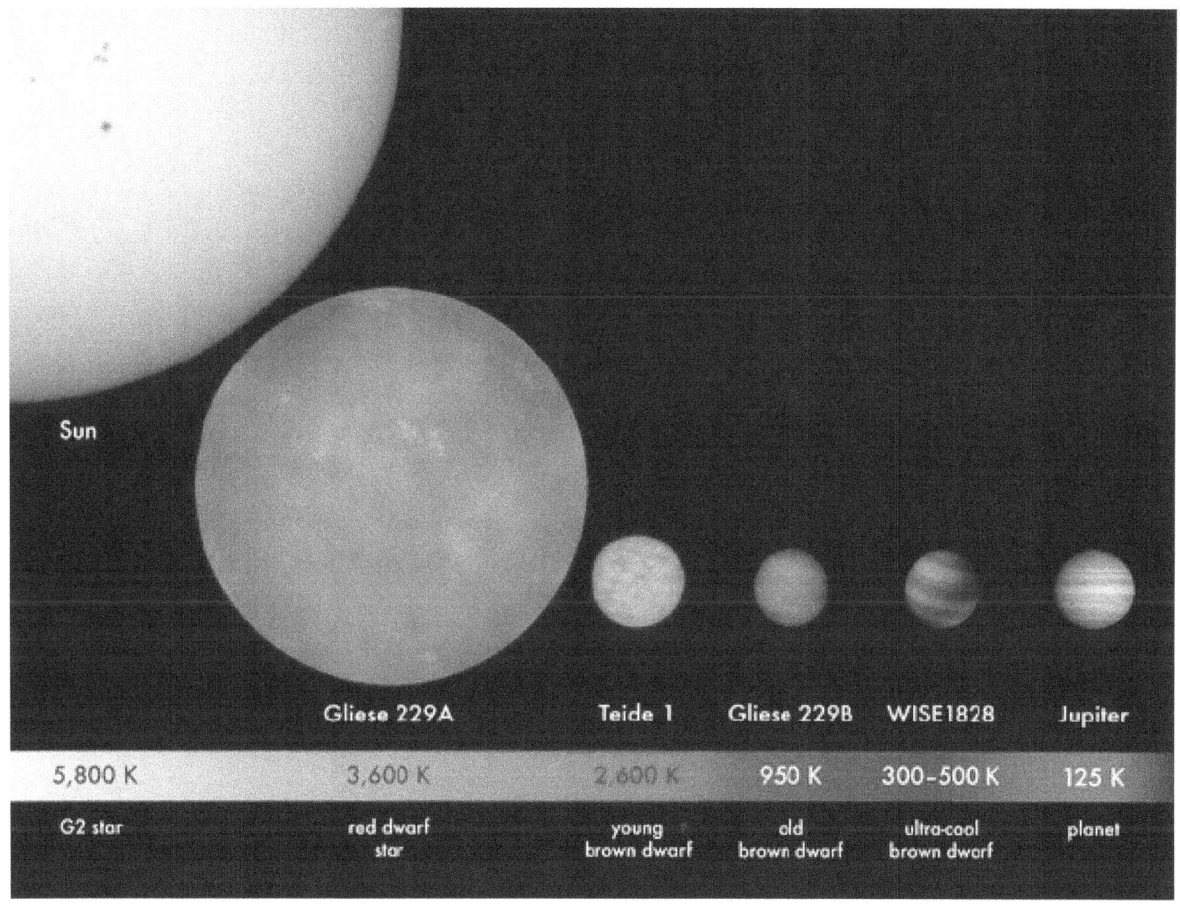

Relative star sizes and photospheric temperatures. Any planet around a red dwarf such as the one shown here (Gliese 229A) would have to huddle close to achieve Earth-like temperatures, probably inducing tidal locking. See Aurelia. Credit: MPIA/V. Joergens.

An artist's impression of GJ 667 Cc, a potentially habitable planet orbiting a red dwarf constituent in a trinary star system.

Chapter 12

Habitability of natural satellites

Artist's impression of a hypothetical habitable moon of Upsilon Andromedae d

Natural satellite habitability is the measure of a natural satellite's potential to sustain life.[1] It is an emerging study which is considered important to astrobiology for several reasons, foremost being that natural satellites are predicted to greatly outnumber planets and that it is hypothesized that habitability factors are likely to be similar to those of planets.[2][3] There are, however, key environmental differences which have a bearing on moons as potential sites for extraterrestrial life.

The strongest candidates for natural satellite habitability are currently icy satellites[4] such as those of Jupiter and Saturn—Europa[5] and Enceladus[6] respectively, although if life exists in either place, it would probably be confined to subsurface habitats. Historically, life on Earth was thought to be strictly a surface phenomenon, but recent studies have shown that up to half of Earth life could live below the surface.[7] Europa and Enceladus exist outside the circumstellar habitable zone which has historically defined the limits of life within the Solar System as the zone in which water can exist as liquid at the surface. In the Solar System's habitable zone there are only three natural satellites—the Moon, and Mars's moons Phobos and Deimos (although some estimates show Mars and its moons to be slightly outside the habitable zone)[8]—none of which sustain an atmosphere or water in liquid form. Tidal forces[9][10] are likely to play as significant a role as stellar radiation in the potential habitability of natural satellites.

Extrasolar moons are not yet confirmed to exist. Detecting them is extremely difficult, because current methods are limited to transit timing.[11] It is possible that some of their attributes could be determined by similar methods as those of transiting planets.[12] Despite this some scientists estimate that there are as many habitable exomoons as habitable exoplanets.[2]

12.1 Presumed conditions

The conditions of habitability for natural satellites are similar to those of planetary habitability. However, there are several factors which differentiate natural satellite habitability and additionally extend their habitability outside the planetary habitable zone.[13]

12.1.1 Liquid water

Main article: Extraterrestrial liquid water

Liquid water is suggested by many astrobiologists as a prerequisite for extraterrestrial life. There is growing evidence of subsurface liquid water on several moons in the Solar System orbiting the gas giants Jupiter, Saturn, Uranus, and Neptune. However, none of these subsurface bodies of water has received final confirmation to date.

12.1.2 Orbital stability

For a stable orbit the ratio between the moon's orbital period P_s around its primary and that of the primary around its star P_p must be < 1/9, e.g. if a planet takes 90 days to orbit its star, the maximum stable orbit for a moon of that planet is less than 10 days.[14][15] Simulations suggest that a moon with an orbital period less than about 45 to 60 days will remain safely bound to a massive giant planet or brown dwarf that orbits 1 AU from a Sun-like star.[16]

12.1.3 Atmosphere

An atmosphere is considered by astrobiologists to be important in developing primal biochemistry, sustaining life and for surface water to exist. Most natural satellites in the Solar System lack significant atmospheres, the sole exception being Saturn's moon, Titan.

Sputtering, a process whereby atoms are ejected from a solid target material due to bombardment of the target by energetic particles, presents a significant problem for natural satellites. All the gas giants in the Solar System, and likely those orbiting other stars, have magnetospheres with radiation belts potent enough to completely erode an atmosphere of an Earth-like moon in just a few hundred million years. Strong stellar winds can also strip gas atoms from the top of an atmosphere causing them to be lost to space.

To support an Earth-like atmosphere for around 4.6 billion years (Earth's current age), a moon with a Mars-like density is estimated to need at least 7% of Earth's mass.[17] One way to decrease loss from sputtering is for the moon to have strong magnetic field which can deflect stellar wind and radiation belts. NASA's Galileo's measurements hints large moons can have magnetic fields; it found Ganymede has its own magnetosphere, even though its mass is only 2.5% of Earth's.[16] An exception is if the moon's atmosphere is constantly replenished by gases from subsurface sources (as believed by some scientists to be the case with Titan).

12.1.4   Tidal effects

While the effects of tidal acceleration are relatively modest on planets, it can be a significant source of energy for natural satellites and an alternative energy source for sustaining life.

Moons orbiting gas giants or brown dwarfs are likely to be tidally locked to their primary: that is, their days are as long as their orbits. While tidal locking may adversely affect planets within habitable zones by interfering with the distribution of stellar radiation, it may work in favour of satellite habitability by allowing tidal heating. Monoj Joshi and Robert Haberle (NASA/Ames Research Center) and their colleagues modelled the temperature on tide-locked exoplanets in the habitability zone of red dwarfs. They found that an atmosphere with a carbon-dioxide pressure of only 1 to 1.5 atmospheres not only allows habitable temperatures but allows liquid water on the dark side. The temperature range of a moon that is tidally locked to a gas giant could be less extreme than with a planet locked to a sun. Even though no studies have been done on the subject, modest amounts of CO_2 would make the temperature habitable.[16]

Furthermore, tidal effects could also allow a moon to sustain plate tectonics, which would cause volcanic activity to regulate the moon's temperature[18][19] and create a geodynamo effect which would give the satellite a strong magnetic field.[20]

Axial tilt and climate

Provided gravitational interaction of a moon with other satellites can be neglected, moons tend to be tidally locked with their planets. In addition to the rotational locking mentioned above, there will also be a process termed 'tilt erosion', which has originally been coined for the tidal erosion of planetary obliquity against a planet's orbit around its host star.[21] The final spin state of a moon then consists of a rotational period equal to its orbital period around the planet and a rotational axis that is perpendicular to the orbital plane.

If the moon's mass is not too low compared to the planet, it may in turn stabilize the planet's axial tilt, i.e. its obliquity against the orbit around the star. On Earth, the Moon has played an important role in stabilizing the axial tilt of the Earth, thereby reducing the impact of gravitational perturbations from the other planets and ensuring only moderate climate variations throughout the planet.[22] On Mars, however, a planet without significant tidal effects from its relatively low-mass moons Phobos and Deimos, axial tilt can undergo extreme changes from 13° to 40° on timescales of 5 to 10 million years.[23][24]

Being tidally locked to a giant planet or sub-brown dwarf would allow for more moderate climates on a moon than there would be if the moon were a similar-sized planet orbiting in locked rotation in the habitable zone of the star.[25] This is especially true of red dwarf systems, where comparatively high gravitational forces and low luminosities leave the habitable zone in an area where tidal locking would occur. If tidally locked, one rotation about the axis may take a long time relative to a planet (for example, ignoring the slight axial tilt of earth's moon and topographical shadowing, any given point on it has two weeks – in Earth time – of sunshine and two weeks of night in its lunar day) but these long periods of light and darkness are not as challenging for habitability as the eternal days and eternal nights on a planet tidally locked to its star.

12.2   Possible origins

Complex conditions thought to be required for abiogenesis are not known to exist anywhere within the solar system. However several candidates beyond Sol's habitable zone have been identified that have some of the ingredients thought necessary for life to exist. The alternate theory of panspermia suggests life may have been introduced to such environments.

There is also the theoretical possibility of extraterrestrial biochemistries exotic beyond current human speculation.

Deliberate or accidental future forward-contamination by organisms originating from Earth is a distinct possibility in these potentially habitable environments. Such cases would make it difficult to determine where the origin of life was.

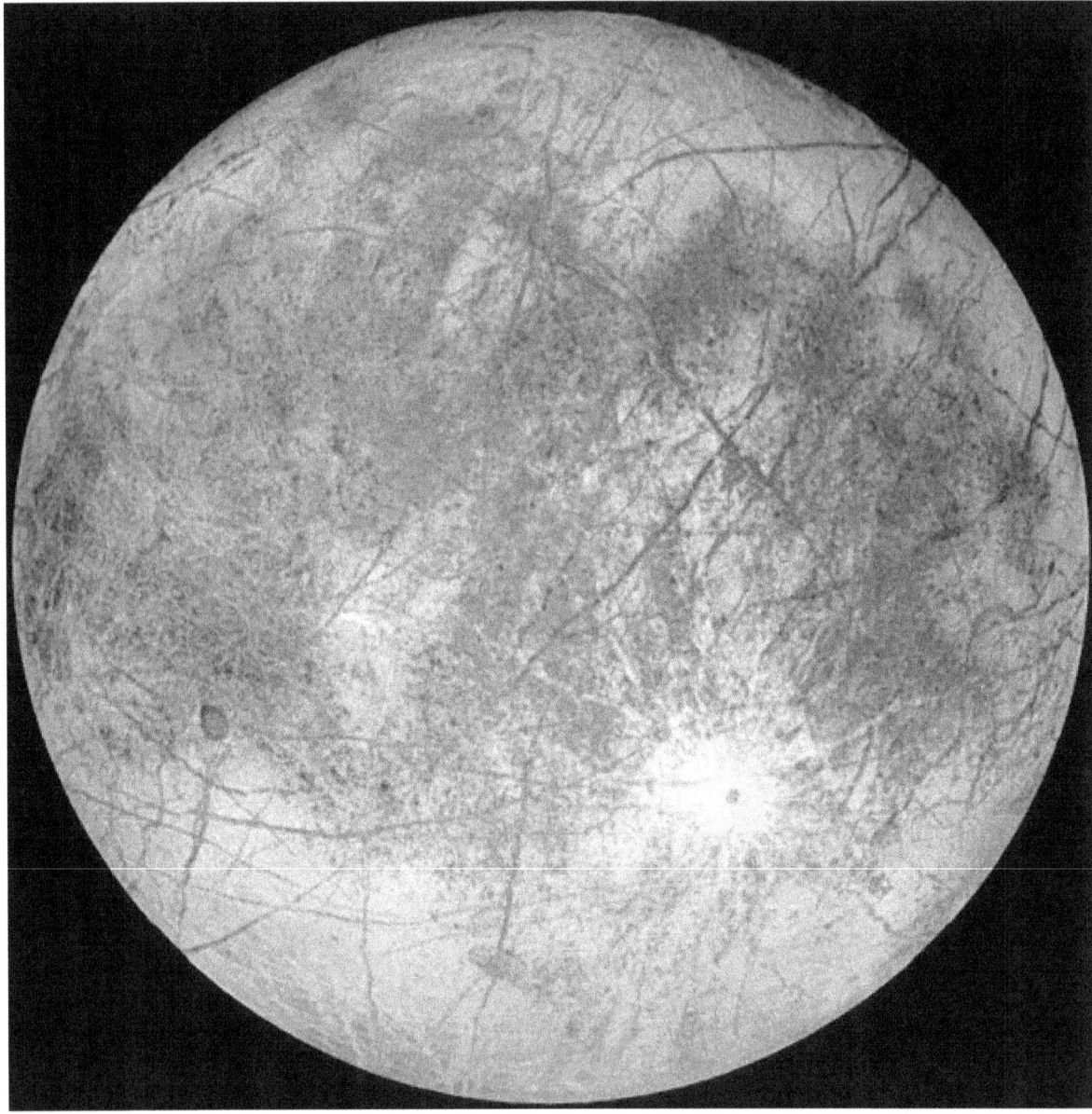

Europa, a moon of Jupiter, with a possibility of having life

12.3 In the Solar System

The following is a list of natural satellites and environments in the Solar System with a possibility of harboring extraterrestrial life.

12.4 Extrasolar

Further information: Extrasolar moon
See also: Category:Giant planets in the habitable zone.

No extrasolar natural satellites have yet been detected. Large planets in the Solar System like Jupiter and Saturn are known to have large moons with some of the conditions for life. Therefore some scientists speculate that large extrasolar

Artist's impression of a hypothetical moon around a Saturn-like exoplanet that could be habitable.

planets (and double planets) may have similarly large moons that are potentially habitable. A moon with sufficient mass may support an atmosphere like Titan and may also sustain liquid water on the surface.

Massive exoplanets known to be located within a habitable zone (such as Gliese 876 b, 55 Cancri f, Upsilon Andromedae d, 47 Ursae Majoris b, HD 28185 b and HD 37124 c) are of particular interest as they may potentially possess natural satellites with liquid water on the surface.

Habitability of extrasolar moons will depend on stellar and planetary illumination on moons as well as the effect of eclipses on their orbit-averaged surface illumination.[35] Beyond that, tidal heating might play a role for a moon's habitability. In Section 4 in their paper, Heller & Barnes[35] introduced a concept to define the habitable orbits of moons. Referring to the concept of the circumstellar habitable zone for planets, they define an inner border for a moon to be habitable around a certain planet and call it the circumplanetary "habitable edge". Moons closer to their planet than the habitable edge are uninhabitable. When effects of eclipses as well as constraints from a satellite's orbital stability are included into this concept, one finds that — depending on a moon's orbital eccentricity — there is a minimum mass of roughly 0.2 solar masses for stars to host habitable moons within the stellar HZ.[36] The magnetic environment of exomoons, which is critically triggered by the intrinsic magnetic field of the host planet, has been identified as another effect on exomoon habitability.[37] Most notably, it was found that moons at distances between about 5 and 20 planetary radii from a giant planet can be habitable from an illumination and tidal heating point of view, but still the planetary magnetosphere would critically influence their habitability.

12.5   See also

- Earth analog

- List of potentially habitable moons

12.6 References

[1] Dyches, Preston; Chou, Felcia (7 April 2015). "The Solar System and Beyond is Awash in Water". *NASA*. Retrieved 8 April 2015.

[2] Shriber, Michael (26 Oct 2009). "Detecting Life-Friendly Moons". *Astrobiology Magazine*. Retrieved 9 May 2013.

[3] Woo, Marcus (27 January 2015). "Why We're Looking for Alien Life on Moons, Not Just Planets". *Wired*. Retrieved 27 January 2015.

[4] Castillo, Julie; Vance, Steve (2008). "Session 13. The Deep Cold Biosphere? Interior Processes of Icy Satellites and Dwarf Planets". *Astrobiology* **8** (2): 344–346. Bibcode:2008AsBio...8..344C. doi:10.1089/ast.2008.1237. ISSN 1531-1074.

[5] Greenberg, Richard (2011). "Exploration and Protection of Europa's Biosphere: Implications of Permeable Ice". *Astrobiology* **11** (2): 183–191. Bibcode:2011AsBio..11..183G. doi:10.1089/ast.2011.0608. ISSN 1531-1074.

[6] Parkinson, Christopher D.; Liang, Mao-Chang; Yung, Yuk L.; Kirschivnk, Joseph L. (2008). "Habitability of Enceladus: Planetary Conditions for Life". *Origins of Life and Evolution of Biospheres* **38** (4): 355–369. Bibcode:2008OLEB...38..355P. doi:10.1007/s11084-008-9135-4. ISSN 0169-6149.

[7] BOYD, ROBERT S.; . Buried alive: Half of Earth\'s life may lie below land, sea. McClatchy DC. 2014-04-24. URL:http://www.mcclatchydc.com/2010/03/08/90020/buried-alive-half-of-earths-life.html. Accessed: 2014-04-24.(Archived by WebCite® at http://www.webcitation.org/6P5GlqXem)

[8] University of Arizona; NASA; NASA Jet Propulsion Laboratory. Phoenix Mars Mission – Habitability and Biology. University of Arizona. 2014-04-24. URL:http://phoenix.lpl.arizona.edu/mars141.php#1. Accessed: 2014-04-24.(Archived by WebCite® at http://www.webcitation.org/6P58Tsw9q)

[9] Cowen, Ron (2008-06-07). "A Shifty Moon". *Science News*.

[10] Bryner, Jeanna (24 June 2009). "Ocean Hidden Inside Saturn's Moon". *Space.com*. TechMediaNetwork. Retrieved 22 April 2013.

[11] Kipping, David M.; Fossey, Stephen J.; Campanella, Giammarco (2009). "On the detectability of habitable exomoons with Kepler-class photometry". *Monthly Notices of the Royal Astronomical Society* **400** (1): 398–405. arXiv:0907.3909. Bibcode:2009MNRAS.400..398K.doi:10.1111/j.1365-2966.2009.15472.x. ISSN 0035-8711.

[12] Kaltenegger, L. (2010). "CHARACTERIZING HABITABLE EXOMOONS". *The Astrophysical Journal* **712** (2): L125–L130. arXiv:0912.3484. Bibcode:2010ApJ...712L.125K. doi:10.1088/2041-8205/712/2/L125. ISSN 2041-8205.

[13] Scharf, Caleb Exomoons Ever Closer Scientific American. October 4, 2011

[14] Kipping, David (2009). "Transit timing effects due to an exomoon". *Monthly Notes of the Royal Astronomical Society* **392**: 181–189. arXiv:0810.2243. Bibcode:2009MNRAS.392..181K. doi:10.1111/j.1365-2966.2008.13999.x. Retrieved 22 February 2012.

[15] Heller, R. (2012). "Exomoon habitability constrained by energy flux and orbital stability". *Astronomy & Astrophysics* **545**: L8. arXiv:1209.0050. Bibcode:2012A&A...545L...8H. doi:10.1051/0004-6361/201220003. ISSN 0004-6361.

[16] Andrew J. LePage. "Habitable Moons:What does it take for a moon — or any world — to support life?". SkyandTelescope.com. Retrieved 2011-07-11.

[17] "In Search Of Habitable Moons". Pennsylvania State University. Retrieved 2011-07-11.

[18] Glatzmaier, Gary A. "How Volcanoes Work – Volcano Climate Effects". *How Volcanoes Work*. Retrieved 29 February 2012.

[19] "Solar System Exploration:Planets:Jupiter:Moons:Io". *Solar System Exploration*. NASA. Retrieved 29 February 2012.

[20] Nave, R. "Magnetic Field of the Earth". Retrieved 29 February 2012.

[21] Heller, René; Barnes, Rory; Leconte, Jérémy (April 2011). "Tidal obliquity evolution of potentially habitable planets". *Astronomy and Astrophysics* **528**: A27. arXiv:1101.2156. Bibcode:2011A&A...528A..27H. doi:10.1051/0004-6361/201015809.

[22] Henney, Paul. "How Earth and the Moon interact". *Astronomy Today*. Retrieved 25 December 2011.

[23] "Mars 101 – Overview". *Mars 101*. NASA. Retrieved 25 December 2011.

[24] Armstrong, John C.; Leovy, Conway B.; Quinn, Thomas (October 2004). "A 1 Gyr climate model for Mars: new orbital statistics and the importance of seasonally resolved polar processes". *Icarus* **171**: 255–271. Bibcode:2004Icar..171..255A. doi:10.1016/j.icarus.2004.05.007. Retrieved 22 February 2012.

[25] Choi, Charles Q. (27 December 2009). "Moons Like Avatar's Pandora Could Be Found". *Space.com*. Retrieved 16 January 2012.

[26] Greenberg, R.; Hoppa, G. V.; Tufts, B. R.; Geissler, P.; Riley, J.; Kadel, S. (October 1999). "Chaos on Europa". *Icarus* **141**: 263–286. Bibcode:1999Icar..141..263G. doi:10.1006/icar.1999.6187.

[27] Schmidt, B. E.; Blankenship, D. D.; Patterson, G. W. (November 2011). "Active formation of 'chaos terrain' over shallow subsurface water on Europa". *Nature* **479**: 502–505. Bibcode:2011Natur.479..502S. doi:10.1038/nature10608. PMID 22089135.

[28] "Moon of Jupiter could support life:Europa has a liquid ocean that lies beneath several miles of ice". msnbc.com. Retrieved 2011-07-10.

[29] "Liquid water on Saturn moon could support life:Cassini spacecraft sees signs of geysers on icy Enceladus". msnbc.com. Retrieved 2011-07-10.

[30] "Life On Titan? New Clues to What's Consuming Hydrogen, Acetylene On Saturn's Moon". Science Daily. 2010-06-07. Retrieved 2011-07-10.

[31] Phillips, T. (1998-10-23). "Callisto makes a big splash". Science@NASA.

[32] Charles Q. Choi (2010-06-07). "Chance For Life On Io". Science Daily. Retrieved 2011-07-10.

[33] Louis Neal Irwin; Dirk Schulze-Makuch (June 2001). "Assessing the Plausibility of Life on Other Worlds". *Astrobiology* **1** (2): 143–60. Bibcode:2001AsBio...1..143I. doi:10.1089/153110701753198918. PMID 12467118.

[34] "Water on Pluto moon". *The Sydney Morning Herald*. 2007-07-19.

[35] Heller, René; Rory Barnes (2012). "Exomoon habitability constrained by illumination and tidal heating". *Astrobiology*. arXiv:1209.5323. doi:10.1089/ast.2012.0859.

[36] Heller, René (September 2012). "Exomoon habitability constrained by energy flux and orbital stability". *Astronomy and Astrophysics* **545**: L8. arXiv:1209.0050. Bibcode:2012A&A...545L...8H. doi:10.1051/0004-6361/201220003.

[37] Heller, René (September 2013). "Magnetic shielding of exomoons beyond the circumplanetary habitable edge". *The Astrophysical Journal Letters*. arXiv:1309.0811. Bibcode:2013ApJ...776L..33H. doi:10.1088/2041-8205/776/2/L33.

Chapter 13

Extremophile

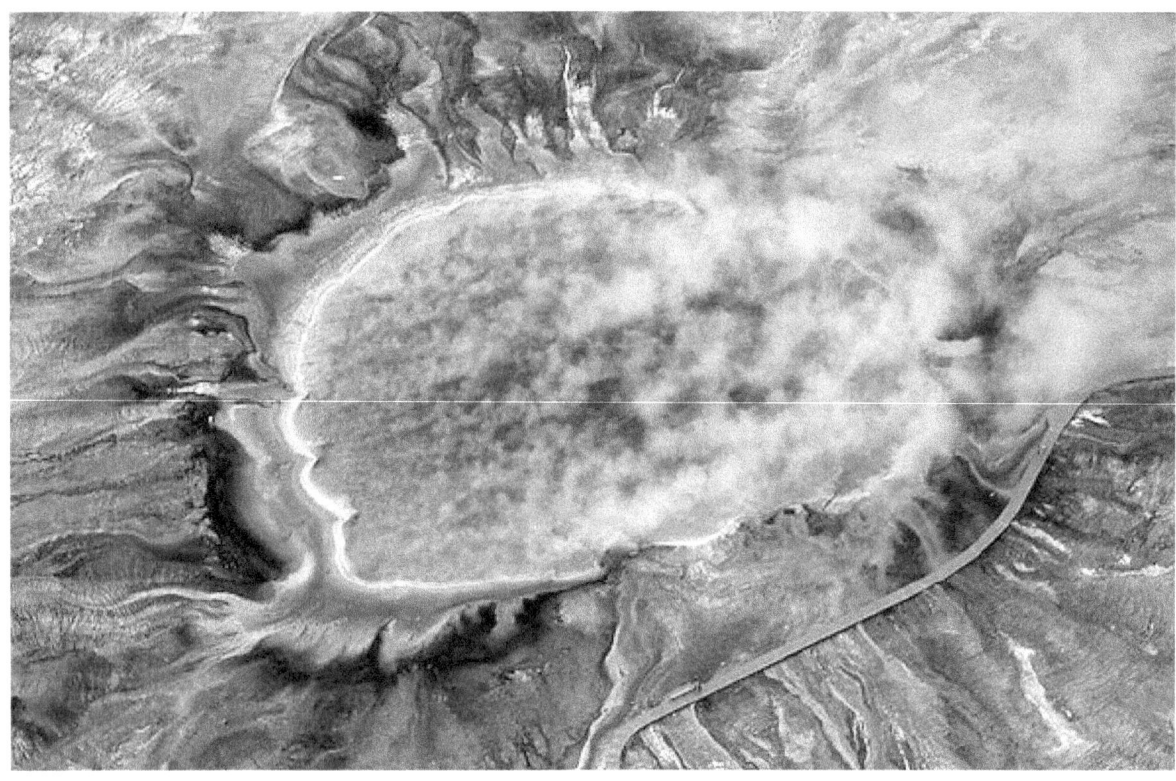

Thermophiles, a type of extremophile, produce some of the bright colors of Grand Prismatic Spring, Yellowstone National Park

An **extremophile** (from Latin *extremus* meaning "extreme" and Greek *philiā* (φιλία) meaning "love") is an organism that thrives in physically or geochemically extreme conditions that are detrimental to most life on Earth.[1][2] In contrast, organisms that live in more moderate environments may be termed mesophiles or neutrophiles.

13.1 Characteristics

In the 1980s and 1990s, biologists found that microbial life has an amazing flexibility for surviving in extreme environments — niches that are extraordinarily hot, or acidic, for example — that would be completely inhospitable to complex organisms. Some scientists even concluded that life may have begun on Earth in hydrothermal vents far under the ocean's surface.[3] According to astrophysicist Dr. Steinn Sigurdsson, "There are viable bacterial spores that have been found that are 40 million years old on Earth — and we know they're very hardened to radiation."[4] On 6 February 2013, scientists reported that bacteria were found living in the cold and dark in a lake buried a half-mile deep under the ice in Antarctica.[5] On 17 March 2013, researchers reported data that suggested microbial life forms thrive in the Mariana Trench, the deepest spot on the Earth.[6][7] Other researchers reported related studies that microbes thrive inside rocks up to 1900 feet below the sea floor under 8500 feet of ocean off the coast of the northwestern United States.[6][8] According to one of the researchers,"You can find microbes everywhere — they're extremely adaptable to conditions, and survive wherever they are."[6]

13.2   Morphology

Most known extremophiles are microbes. The domain Archaea contains renowned examples, but extremophiles are present in numerous and diverse genetic lineages of bacteria and archaeans. Furthermore, it is erroneous to use the term extremophile to encompass all archaeans, as some are mesophilic. Neither are all extremophiles unicellular; protostome animals found in similar environments include the Pompeii worm, the psychrophilic Grylloblattidae (insects) and Antarctic krill (a crustacean). Many would also classify tardigrades (water bears) as extremophiles but while tardigrades can survive in extreme environments, they are not considered extremophiles because they are not adapted to live in these conditions. Their chances of dying increase the longer they are exposed to the extreme environment.

13.3   Classifications

There are many classes of extremophiles that range all around the globe, each corresponding to the way its environmental niche differs from mesophilic conditions. These classifications are not exclusive. Many extremophiles fall under multiple categories and are classified as polyextremophiles. For example, organisms living inside hot rocks deep under Earth's surface are thermophilic and barophilic such as *Thermococcus barophilus*.[9] A polyextremophile living at the summit of a mountain in the Atacama Desert might be a radioresistant xerophile, a psychrophile, and an oligotroph. Polyextremophiles are well known for their ability to tolerate both high and low pH levels.

13.3.1   Terms

Acidophile An organism with optimal growth at pH levels of 3 or below

Alkaliphile An organism with optimal growth at pH levels of 9 or above

Anaerobe An organism that does not require oxygen for growth such as *Spinoloricus Cinzia*. Two sub-types exist: facultative anaerobe and obligate anaerobe. A facultative anaerobe can tolerate anaerobic and aerobic conditions; however, an obligate anaerobe would die in presence of even trace levels of oxygen.

Cryptoendolith An organism that lives in microscopic spaces within rocks, such as pores between aggregate grains; these may also be called Endolith, a term that also includes organisms populating fissures, aquifers, and faults filled with groundwater in the deep subsurface.

Halophile An organism requiring at least 0.2M concentrations of salt (NaCl) for growth[10]

Hyperthermophile An organism that can thrive at temperatures above 80 °C, such as those found in hydrothermal systems

Hypolith An organism that lives underneath rocks in cold deserts

Lithoautotroph An organism (usually bacteria) whose sole source of carbon is carbon dioxide and exergonic inorganic oxidation (chemolithotrophs) such as *Nitrosomonas europaea*; these organisms are capable of deriving energy from reduced mineral compounds like pyrites, and are active in geochemical cycling and the weathering of parent bedrock to form soil

Metallotolerant capable of tolerating high levels of dissolved heavy metals in solution, such as copper, cadmium, arsenic, and zinc; examples include *Ferroplasma sp.*, *Cupriavidus metallidurans* and GFAJ-1.[11][12][13]

Oligotroph An organism capable of growth in nutritionally limited environments

Osmophile An organism capable of growth in environments with a high sugar concentration

Piezophile (Also referred to as barophile). An organism that lives optimally at high pressures such as those deep in the ocean or underground;[14] common in the deep terrestrial subsurface, as well as in oceanic trenches

Polyextremophile A **polyextremophile** (faux Ancient Latin/Greek for 'affection for many extremes') is an organism that qualifies as an extremophile under more than one category.

Psychrophile/Cryophile An organism capable of survival, growth or reproduction at temperatures of −15 °C or lower for extended periods; common in cold soils, permafrost, polar ice, cold ocean water, and in or under alpine snowpack

Radioresistant Organisms resistant to high levels of ionizing radiation, most commonly ultraviolet radiation, but also including organisms capable of resisting nuclear radiation

Thermophile An organism that can thrive at temperatures between 45–122 °C

Thermoacidophile Combination of thermophile and acidophile that prefer temperatures of 70–80 °C and pH between 2 and 3

Xerophile An organism that can grow in extremely dry, desiccating conditions; this type is exemplified by the soil microbes of the Atacama Desert

13.4 In astrobiology

Astrobiology is the field concerned with forming theories, such as panspermia, about the distribution, nature, and future of life in the universe. In it, microbial ecologists, astronomers, planetary scientists, geochemists, philosophers, and explorers cooperate constructively to guide the search for life on other planets. Astrobiologists are particularly interested in studying extremophiles, as many organisms of this type are capable of surviving in environments similar to those known to exist on other planets. For example, Mars may have regions in its deep subsurface permafrost that could harbor endolith communities. The subsurface water ocean of Jupiter's moon Europa may harbor life, especially at hypothesized hydrothermal vents at the ocean floor.

Recent research carried out on extremophiles in Japan involved a variety of bacteria including *Escherichia coli* and *Paracoccus denitrificans* being subject to conditions of extreme gravity. The bacteria were cultivated while being rotated in an ultracentrifuge at high speeds corresponding to 403,627 g (i.e. 403,627 times the gravity experienced on Earth). *Paracoccus denitrificans* was one of the bacteria which displayed not only survival but also robust cellular growth under these conditions of hyperacceleration which are usually found only in cosmic environments, such as on very massive stars or in the shock waves of supernovas. Analysis showed that the small size of prokaryotic cells is essential for successful growth under hypergravity. The research has implications on the feasibility of panspermia.[15][16]

On 26 April 2012, scientists reported that lichen survived and showed remarkable results on the adaptation capacity of photosynthetic activity within the simulation time of 34 days under Martian conditions in the Mars Simulation Laboratory (MSL) maintained by the German Aerospace Center (DLR).[17][18]

On 29 April 2013, scientists at Rensselaer Polytechnic Institute, funded by NASA, reported that, during spaceflight on the International Space Station, microbes seem to adapt to the space environment in ways "not observed on Earth" and in ways that "can lead to increases in growth and virulence".[19]

On 19 May 2014, scientists announced that numerous microbes, like *Tersicoccus phoenicis*, may be resistant to methods usually used in spacecraft assembly clean rooms. It's not currently known if such resistant microbes could have withstood space travel and are present on the *Curiosity* rover now on the planet Mars.[20]

On 20 August 2014, scientists confirmed the existence of microorganisms living half a mile below the ice of Antarctica.[21]

13.5   Examples

New sub-types of -philes are identified frequently and the sub-category list for extremophiles is always growing. For example, microbial life lives in the liquid asphalt lake, Pitch Lake. Research indicates that extremophiles inhabit the asphalt lake in populations ranging between 10^6 to 10^7 cells/gram.[23][24] Likewise, until recently boron tolerance was unknown but a strong borophile was discovered in bacteria. With the recent isolation of *Bacillus boroniphilus*, borophiles came into discussion.[25] Studying these borophiles may help illuminate the mechanisms of both boron toxicity and boron deficiency.

13.6   Industrial uses

The thermoalkaliphilic catalase, which initiates the breakdown of hydrogen peroxide into oxygen and water, was isolated from an organism, *Thermus brockianus*, found in Yellowstone National Park by Idaho National Laboratory researchers. The catalase operates over a temperature range from 30 °C to over 94 °C and a pH range from 6-10. This catalase is extremely stable compared to other catalases at high temperatures and pH. In a comparative study, the *T. brockianus* catalase exhibited a half life of 15 days at 80 °C and pH 10 while a catalase derived from *Aspergillus niger* had a half life of 15 seconds under the same conditions. The catalase will have applications for removal of hydrogen peroxide in industrial processes such as pulp and paper bleaching, textile bleaching, food pasteurization, and surface decontamination of food packaging.[26]

DNA modifying enzymes such as *Taq* DNA polymerase and some *Bacillus* enzymes used in clinical diagnostics and starch liquefaction are produced commercially by several biotechnology companies.[27]

13.7   DNA transfer

Over 65 prokaryotic species are known to be naturally competent for genetic transformation, the ability to transfer DNA from one cell to another cell followed by integration of the donor DNA into the recipient cell's chromosome.[28] Several extremophiles are able to carry out species-specific DNA transfer, as described below. However, it is not yet clear how common such a capability is among extremophiles.

The bacterium *Deinococcus radiodurans* is one of the most radioresistant organisms known. This bacterium can also survive cold, dehydration, vacuum and acid and is thus known as a polyextremophile. *D. radiodurans* is competent to perform genetic transformation.[29] Recipient cells are able to repair DNA damage in donor transforming DNA that had been UV irradiated as efficiently as they repair cellular DNA when the cells themselves are irradiated. The extreme thermophilic bacterium *Thermus thermophilus* and other related *Thermus* species are also capable of genetic transformation.[30]

Halobacterium volcanii, an extreme halophilic (saline tolerant) archaeon, is capable of natural genetic transformation. Cytoplasmic bridges are formed between cells that appear to be used for DNA transfer from one cell to another in either direction.[31]

Sulfolobus solfataricus and *Sulfolobus acidocaldarius* are hyperthermophilic archaea. Exposure of these organisms to the DNA damaging agents UV irradiation, bleomycin or mitomycin C induces species-specific cellular aggregation.[32][33] UV-induced cellular aggregation of *S. acidocaldarius* mediates chromosomal marker exchange with high frequency.[33] Recombination rates exceed those of uninduced cultures by up to three orders of magnitude. Frols et al.[32] and Ajon et al.[33] hypothesized that cellular aggregation enhances species-specific DNA transfer between *Sulfolobus* cells in order to repair damaged DNA by means of homologous recombination. Van Wolferen et al.[34] noted that this DNA exchange process may be crucial under DNA damaging conditions such as high temperatures. It has also been suggested that DNA transfer in *Sulfolobus* may be an early form of sexual interaction similar to the more well-studied bacterial transformation systems that involve species-specific DNA transfer leading to homologous recombinational repair of DNA damage[35] (and see Transformation (genetics)).

Extracellular membrane vesicles (MVs) might be involved in DNA transfer between different hyperthermophilic archaeal species.[36] It has been shown that both plasmids[37] and viral genomes[36] can be transferred via MVs. Notably, a horizontal plasmid transfer has been documented between hyperthermophilic *Thermococcus* and *Methanocaldococcus* species, respectively belonging to the orders *Thermococcales* and *Methanococcales*.[138]

13.8   See also

- *Deinococcus radiodurans*
- Extremotroph
- List of microorganisms tested in outer space
- Tardigrade
- Thermophile

13.9   References

[1] Rampelotto, P. H. (2010). "Resistance of microorganisms to extreme environmental conditions and its contribution to Astrobiology". *Sustainability* **2** (6): 1602–1623. Bibcode:2010Sust....2.1602R. doi:10.3390/su2061602.

[2] Rothschild, L.J.; Mancinelli, R.L. (2001). "Life in extreme environments".*Nature***409**(6823): 1092–1101. Bibcode:2001Natur doi:10.1038/35059215. PMID 11234023.

[3] "Mars Exploration - Press kit" (PDF). NASA. June 2003. Retrieved 14 July 2009.

[4] BBC Staff (23 August 2011). "Impacts 'more likely' to have spread life from Earth". BBC. Retrieved 24 August 2011.

[5] Gorman, James (6 February 2013). "Bacteria Found Deep Under Antarctic Ice, Scientists Say". *New York Times*. Retrieved 6 February 2013.

[6] Choi, Charles Q. (17 March 2013). "Microbes Thrive in Deepest Spot on Earth". LiveScience. Retrieved 17 March 2013.

[7] Glud, Ronnie; Wenzhöfer, Frank; Middleboe, Mathias; Oguri, Kazumasa; Turnewitsch, Robert; Canfield, Donald E.; Kitazato, Hiroshi (17 March 2013). "High rates of microbial carbon turnover in sediments in the deepest oceanic trench on Earth". *Nature Geoscience* **6** (4): 284–288. Bibcode:2013NatGe...6..284G. doi:10.1038/ngeo1773. Retrieved 17 March 2013.

[8] Oskin, Becky (14 March 2013). "Intraterrestrials: Life Thrives in Ocean Floor". LiveScience. Retrieved 17 March 2013.

[9] Thermococcus barophilus sp. nov., a new barophilic and hyperthermophilic archaeon isolated under high hydrostatic pressure from a deep-sea hydrothermal vent. *IJSEM*, p. 351-359, 49, 1999.

[10] Cavicchioli, R. & Thomas, T. 2000. Extremophiles. In: J. Lederberg. (ed.) Encyclopedia of Microbiology, Second Edition, Vol. 2, pp. 317–337. Academic Press, San Diego.

[11] "Studies refute arsenic bug claim". *BBC News*. 9 July 2012. Retrieved 10 July 2012.

[12] Erb, Tobias J.; Kiefer, Patrick; Hattendorf, Bodo; Günther, Detlef; Vorholt, Julia A. (8 July 2012). "GFAJ-1 Is an Arsenate-Resistant, Phosphate-Dependent Organism". *Science* **337** (6093): 467–70. Bibcode:2012Sci...337..467E. doi:10.1126/science. 1218455.PMID22773139. Retrieved10 July2012.

[13] Reaves, Marshall Louis; Sinha, Sunita; Rabinowitz, Joshua D.; Kruglyak, Leonid; Redfield, Rosemary J. (8 July 2012). "Absence of Detectable Arsenate in DNA from Arsenate-Grown GFAJ-1 Cells". *Science* **337** (6093): 470–3. arXiv:1201.6643. Bibcode: 2012Sci...337..470R.doi:10.1126/science.1219861. PMC3845625. PMID22773140. Retrieved10July2012.

[14] Dworkin, Martin; Falkow, Stanley (13 July 2006). *The Prokaryotes: Vol. 1: Symbiotic Associations, Biotechnology, Applied Microbiology*. Springer. p. 94. ISBN 978-0-387-25476-0.

[15] Than, Ker (25 April 2011). "Bacteria Grow Under 400,000 Times Earth's Gravity". *National Geographic- Daily News*. National Geographic Society. Retrieved 28 April 2011.

[16] Deguchi, Shigeru; Hirokazu Shimoshige, Mikiko Tsudome, Sada-atsu Mukai, Robert W. Corkery, Susumu Ito, and Koki Horikoshi; Tsudome, M.; Mukai, S.-a.; Corkery, R. W.; Ito, S.; Horikoshi, K. (2011). "Microbial growth at hyperaccelerations up to 403,627 xg". *Proceedings of the National Academy of Sciences* **108** (19): 7997–8002. Bibcode:2011PNAS..108.7997D. doi:10.1073/pnas.1018027108. Retrieved 28 April 2011.

[17] Baldwin, Emily (26 April 2012). "Lichen survives harsh Mars environment". Skymania News. Retrieved 27 April 2012.

[18] de Vera, J.-P.; Kohler, Ulrich (26 April 2012). "The adaptation potential of extremophiles to Martian surface conditions and its implication for the habitability of Mars" (PDF). European Geosciences Union. Retrieved 27 April 2012.

[19] Kim W; et al. (29 April 2013). "Spaceflight Promotes Biofilm Formation by Pseudomonas aeruginosa". *Plos One* **8** (4): e6237. Bibcode:2013PLoSO...862437K. doi:10.1371/journal.pone.0062437. Retrieved 5 July 2013.

[20] Madhusoodanan, Jyoti (19 May 2014). "Microbial stowaways to Mars identified".*Nature (journal)*. doi:10.1038/nature.2014.15 Retrieved 23 May 2014.

[21] Fox, Douglas (20 August 2014). "Lakes under the ice: Antarctica's secret garden". *Nature (journal)* **512** (7514): 244–246. Bibcode:2014Natur.512..244F. doi:10.1038/512244a. Retrieved 21 August 2014.

[22] Mack, Eric (20 August 2014). "Life Confirmed Under Antarctic Ice; Is Space Next?". *Forbes*. Retrieved 21 August 2014.

[23] Microbial Life Found in Hydrocarbon Lake. *the physics arXiv blog* 15 April 2010.

[24] Schulze-Makuch, Haque, Antonio, Ali, Hosein, Song, Yang, Zaikova, Beckles, Guinan, Lehto, Hallam. Microbial Life in a Liquid Asphalt Desert.

[25] Ahmed, Iftikhar; Yokota, Akira; Fujiwara, Toru (2006). "A novel highly boron tolerant bacterium, Bacillus boroniphilus sp. nov., isolated from soil, that requires boron for its growth". *Extremophiles* **11** (2): 217–224. doi:10.1007/s00792-006-0027-0. PMID 17072687.

[26] "Bioenergy and Industrial Microbiology". *Idaho National Laboratory*. U.S. Department of Energy. Retrieved 3 February 2014.

[27] Anitori, RP (editor) (2012). *Extremophiles: Microbiology and Biotechnology*. Caister Academic Press. ISBN 978-1-904455-98-1.

[28] Johnsborg, O; Eldholm, V; Håvarstein, LS. (2007). "Natural genetic transformation: prevalence, mechanisms and function". *Res Microbiol* **158** (10): 767–78. doi:10.1016/j.resmic.2007.09.004. PMID 17997281.

[29] Moseley, BE; Setlow, JK. (1968). "Transformation in Micrococcus radiodurans and the ultraviolet sensitivity of its transforming DNA". *Proc Natl Acad Sci U S A* **61** (1): 176–83. Bibcode:1968PNAS...61..176M. doi:10.1073/pnas.61.1.176. PMID 5303325.

[30] Koyama, Y; Hoshino, T; Tomizuka, N; Furukawa, K. (1986). "Genetic transformation of the extreme thermophile Thermus thermophilus and of other Thermus spp". *J Bacteriol* **166** (1): 338–40. PMID 3957870.

[31] Rosenshine, I; Tchelet, R; Mevarech, M. (1989). "The mechanism of DNA transfer in the mating system of an archaebacterium". *Science* **245** (4924): 1387–9. Bibcode:1989Sci...245.1387R. doi:10.1126/science.2818746. PMID 2818746.

[32] Fröls, S; Ajon, M; Wagner, M; Teichmann, D; Zolghadr, B; Folea, M; Boekema, EJ; Driessen, AJ; Schleper, C; et al. (2008). "UV-inducible cellular aggregation of the hyperthermophilic archaeon Sulfolobus solfataricus is mediated by pili formation". *Mol Microbiol* **70** (4): 938–52. doi:10.1111/j.1365-2958.2008.06459.x. PMID 18990182.

[33] Ajon, M; Fröls, S; van Wolferen, M; Stoecker, K; Teichmann, D; Driessen, AJ; Grogan, DW; Albers, SV; Schleper, C.; et al. (2011). "UV-inducible DNA exchange in hyperthermophilic archaea mediated by type IV pili". *Mol Microbiol* **82** (4): 807–17. doi:10.1111/j.1365-2958.2011.07861.x. PMID 21999488.

[34] Van Wolferen, M; Ajon, M; Driessen, AJ; Albers, SV. (2013). "How hyperthermophiles adapt to change their lives: DNA exchange in extreme conditions". *Extremophiles* **17** (4): 545–63. doi:10.1007/s00792-013-0552-6. PMID 23712907.

[35] Bernstein H and Bernstein C (2013). Evolutionary Origin and Adaptive Function of Meiosis. Meiosis, Dr. Carol Bernstein (Ed.), ISBN978-953-51-1197-9.InTech.http://www.intechopen.com/books/meiosis/evolutionary-origin-and-adaptive-function-

[36] Gaudin M, Krupovic M, Marguet E, Gauliard E, Cvirkaite-Krupovic V, Le Cam E, Oberto J, Forterre P; Krupovic; Marguet; Gauliard; Cvirkaite-Krupovic; Le Cam; Oberto; Forterre (2014). "Extracellular membrane vesicles harbouring viral genomes". *Environ Microbiol* **16** (4): 1167–75. doi:10.1111/1462-2920.12235. PMID 24034793.

[37] Gaudin M, Gauliard E, Schouten S, Houel-Renault L, Lenormand P, Marguet E, Forterre P.; Gauliard; Schouten; Houel-Renault; Lenormand; Marguet; Forterre (2013). "Hyperthermophilic archaea produce membrane vesicles that can transfer DNA". *Environ Microbiol Rep* **5** (1): 109–16. doi:10.1111/j.1758-2229.2012.00348.x. PMID 23757139.

[38] Krupovic M, Gonnet M, Hania W B, Forterre P, Erauso G; Gonnet; Hania; Forterre; Erauso (2013). "Insights into dynamics of mobile genetic elements in hyperthermophilic environments from five new Thermococcus plasmids". *PLoS ONE* **8** (1): e49044. Bibcode:2013PLoSO...849044K. doi:10.1371/journal.pone.0049044. PMC 3543421. PMID 23326305.

13.10   Further reading

- Wilson, Z. E. and Brimble, M. A. (January 2009). "Molecules derived from the extremes of life". *Nat. Prod. Rep.* **26** (1): 44–71. doi:10.1039/b800164m. PMID 19374122.

- Rossi M; et al. (July 2003). "Extremophiles 2002". *J Bacteriol.* **185** (13): 3683–9. doi:10.1128/JB.185.13.3683-3689.2003. PMC 161588. PMID 12813059.

- C.Michael Hogan (2010). "Extremophile". *Encyclopedia of Earth, National Council of Science & the Environment,* eds. E.Monosson & C.Cleveland.

- Joseph Seckbach, et al.: *Polyextremophiles: life under multiple forms of stress.* Springer, Dordrecht 2013. ISBN 978-94-007-6488-0.

13.11   External links

- Extreme Environments - Science Education Resource Center

- Extremophile Research

- Eukaryotes in extreme environments

- The Research Center of Extremophiles

- DaveDarling's Encyclopedia of Astrobiology, Astronomy, and Spaceflight

- The International Society for Extremophiles

- Idaho National Laboratory

- Polyextremophile on David Darling's *Encyclopedia of Astrobiology, Astronomy, and Spaceflight*

Chapter 14

Biosignature

For other uses, see Biomarker (disambiguation).

A **biosignature** is any substance – such as an element, isotope, molecule, or phenomenon – that provides scientific evidence of past or present life.[1][2][3] Measurable attributes of life include its complex physical and chemical structures and also its utilization of free energy and the production of biomass and wastes. Due to its unique characteristics, a biosignature can be interpreted as having been produced by living organisms; however, it is important that they not be considered definitive because there is no way of knowing in advance which ones are universal to life and which ones are unique to the peculiar circumstances of life on Earth.[4] Nonetheless, life forms are known to shed unique chemicals, including DNA, into the environment as evidence of their presence in a particular location.[5]

14.1 In geomicrobiology

The ancient record on Earth provides an opportunity to see what geochemical signatures are produced by microbial life and how these signatures are preserved over geologic time. Some related disciplines such as geochemistry, geobiology, and geomicrobiology often use biosignatures to determine if living organisms are or were present in a sample. These possible biosignatures include: (a) microfossils and stromatolites; (b) molecular structures (biomarkers) and isotopic compositions of carbon, nitrogen and hydrogen in organic matter; (c) multiple sulfur and oxygen isotope ratios of minerals; and (d) abundance relationships and isotopic compositions of redox sensitive metals (e.g., Fe, Mo, Cr, and rare earth elements).[6][7]

For example, the particular fatty acids measured in a sample can indicate which types of bacteria and archaea live in that environment. Another example are the long-chain fatty alcohols with more than 23 atoms that are produced by planktonic bacteria.[8] When used in this sense, geochemists often prefer the term biomarker. An other example is the presence of straight-chain lipids in the form of alkanes, alcohols an fatty acids with 20-36 carbon atoms in soils or sediments. Peat deposits are an indication of originating from the epicuticular wax of higher plants.

Life processes may produce a range of biosignatures such as nucleic acids, lipids, proteins, amino acids, kerogen-like material and various morphological features that are detectable in rocks and sediments.[9] Microbes often interact with geochemical processes, leaving features in the rock record indicative of biosignatures. For example, bacterial micrometer-sized pores in carbonate rocks resemble inclusions under transmitted light, but have distinct size, shapes and patterns (swirling or dendritic) and are distributed differently from common fluid inclusions.[10] A potential biosignature is a phenomenon that *may* have been produced by life, but for which alternate abiotic origins may also be possible.

14.2 In astrobiology

Astrobiological exploration is founded upon the premise that biosignatures encountered in space will be recognizable as

Electron micrograph of microfossils from a sediment core obtained by the Deep Sea Drilling Program

extraterrestrial life. The usefulness of a biosignature is determined, not only by the probability of life creating it, but also by the improbability of nonbiological (abiotic) processes producing it.[13] An example of such a biosignature might be complex organic molecules and/or structures whose formation is virtually unachievable in the absence of life. For example, some categories of biosignatures can include the following: cellular and extracellular morphologies, biogenic substance in rocks, bio-organic molecular structures, chirality, biogenic minerals, biogenic stable isotope patterns in minerals and organic compounds, atmospheric gases, and remotely detectable features on planetary surfaces, such as photosynthetic pigments, etc.[13]

Biosignatures need not be chemical, however, and can also be suggested by a distinctive magnetic biosignature.[14] Another possible biosignature might be morphology since the shape and size of certain objects may potentially indicate the presence of past or present life. For example, microscopic magnetite crystals in the Martian meteorite ALH84001 were the longest-debated of several potential biosignatures in that specimen because it was believed until recently that only bacteria could create crystals of their specific shape. However, anomalous features discovered that are "possible biosignatures" for life forms would be investigated as well. Such features constitute a working hypothesis, not a confirmation of detection of life. Concluding that evidence of an extraterrestrial life form (past or present) has been discovered, requires proving that a possible biosignature was produced by the activities or remains of life.[1] For example, the possible biomineral studied in the Martian ALH84001 meteorite includes putative microbial fossils, tiny rock-like structures whose shape was a potential biosignature because it resembled known bacteria. Most scientists ultimately concluded that these were far too small to be fossilized cells. A consensus that has emerged from these discussions, and is now seen as a critical requirement, is the demand for further lines of evidence in addition to any morphological data that supports such extraordinary claims.[1]

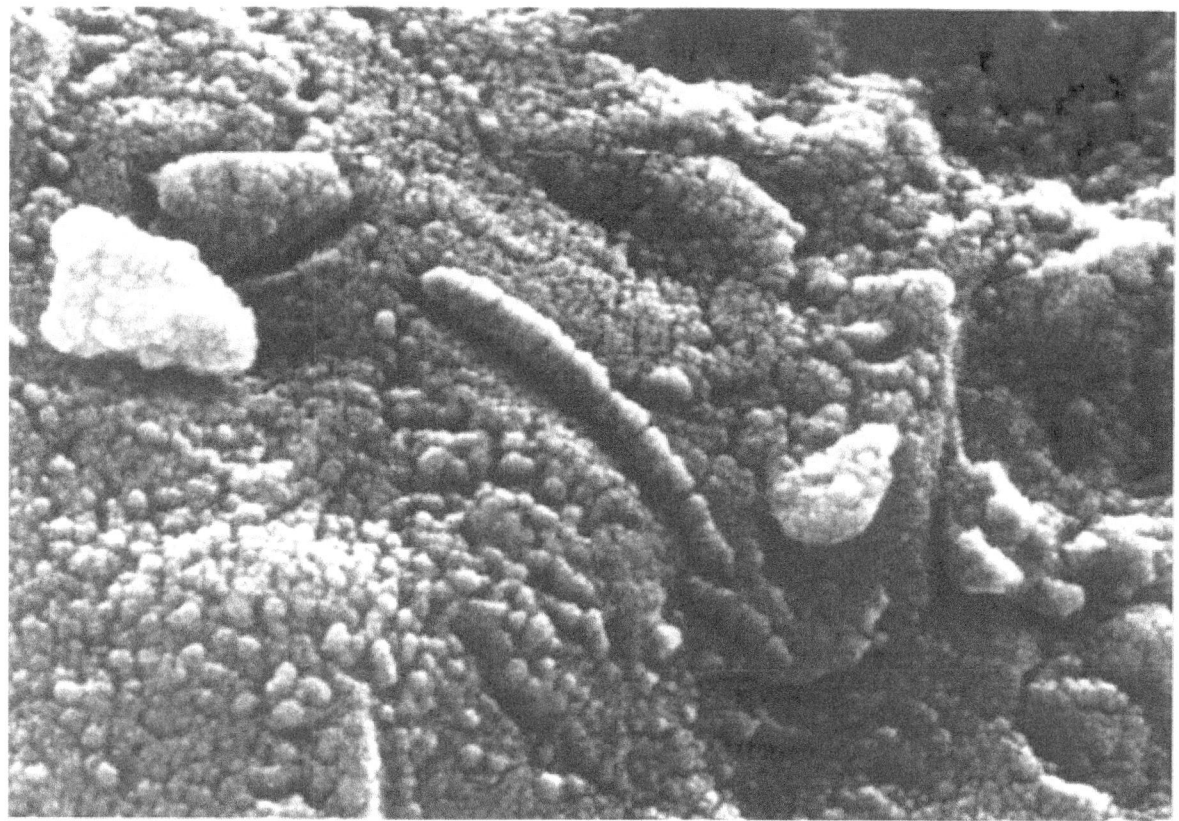

Some researchers suggested that these microscopic structures on the Martian ALH84001 meteorite could be fossilized bacteria.[11][12]

Scientific observations include the possible identification of biosignatures through indirect observation. For example, electromagnetic information through infrared radiation telescopes, radio-telescopes, space telescopes, etc.[15][16] From this discipline, the hypothetical electromagnetic radio signatures that SETI scans for would be a biosignature, since a message from intelligent aliens would certainly demonstrate the existence of extraterrestrial life.

On Mars, surface oxidants and UV radiation will have altered or destroyed organic molecules at or near the surface.[3] One issue that may add ambiguity in such a search is the fact that, throughout Martian history, abiogenic organic-rich chondritic meteorites have undoubtedly rained upon the Martian surface. At the same time, strong oxidants in Martian soil along with exposure to ionizing radiation might alter or destroy molecular signatures from meteorites or organisms.[3] An alternative approach would be to seek concentrations of buried crystalline minerals, such as clays and evaporites, which may protect organic matter from the destructive effects of ionizing radiation and strong oxidants.[3] The search for Martian biosignatures has become more promising due to the discovery that surface and near-surface aqueous environments existed on Mars at the same time when biological organic matter was being preserved in ancient aqueous sediments on Earth.[3]

Atmosphere

Over billions of years, the processes of life on a planet would result in a mixture of chemicals unlike anything that could form in an ordinary chemical equilibrium.[17] For example, large amounts of oxygen and small amounts of methane are generated by life on Earth. The presence of methane in the atmosphere of Mars indicates that there must be an active source on the planet, as it is an unstable gas. Furthermore, current photochemical models cannot explain the presence of methane in the atmosphere of Mars and its reported rapid variations in space and time. Neither its fast appearance nor disappearance can be explained yet.[18] To rule out a biogenic origin for the methane, a future probe or lander hosting a mass spectrometer will be needed, as the isotopic proportions of carbon-12 to carbon-14 in methane could distinguish between a biogenic and non-biogenic origin.[19] In June, 2012, scientists reported that measuring the ratio of hydrogen and methane levels on Mars may help determine the likelihood of life on Mars.[20][21] According to the scientists, "...low H_2/CH_4 ratios (less than approximately 40) indicate that life is likely present and active."[20] Other scientists have recently

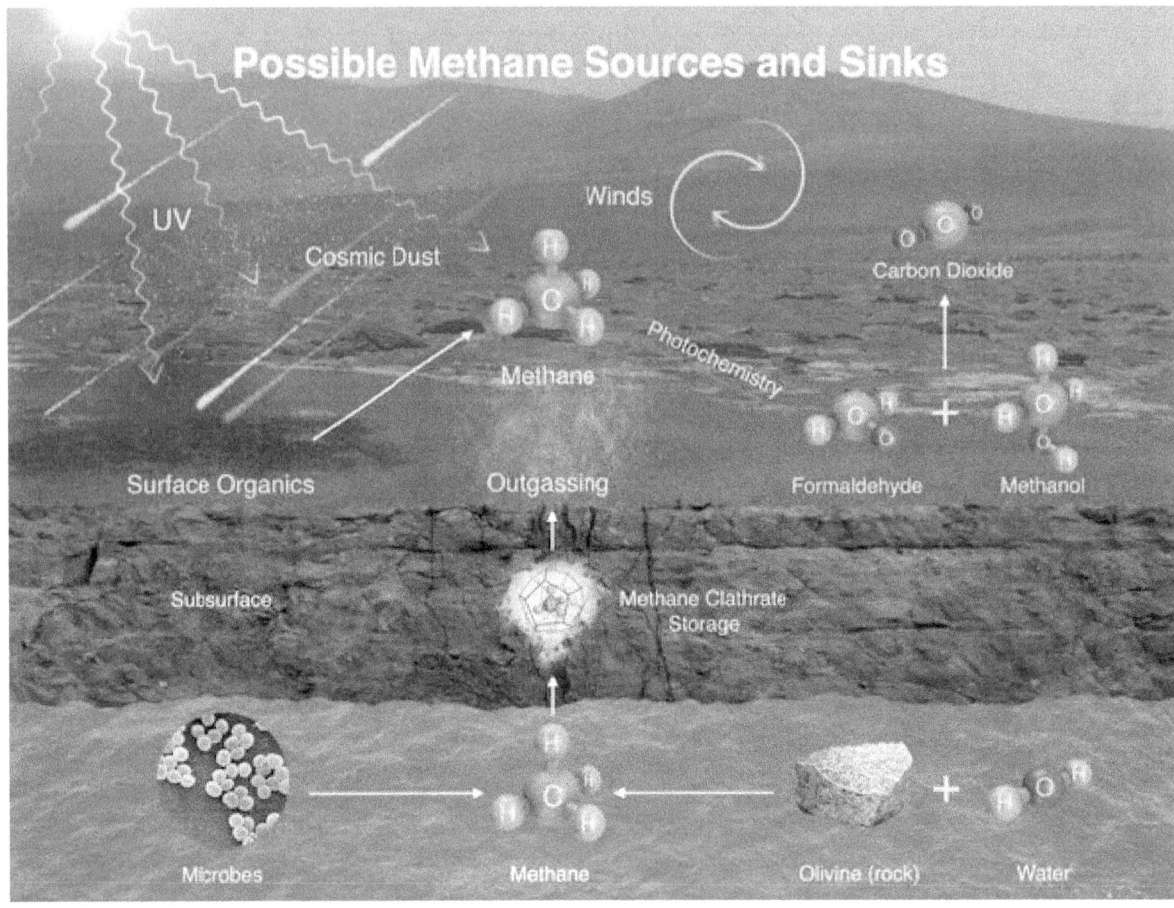

Methane (CH_4) on Mars - potential sources and sinks.

reported methods of detecting hydrogen and methane in extraterrestrial atmospheres.[22][23] The planned ExoMars Trace Gas Orbiter to be launched in 2016 to Mars, will study atmospheric trace gases and will attempt to characterize potential biochemical and geochemical processes at work.[24]

14.2.1 The Viking missions to Mars

Main article: Viking biological experiments
The Viking missions to Mars in the 1970s conducted the first experiments which were explicitly designed to look for biosignatures on another planet. Each of the two Viking landers carried three life-detection experiments which looked for signs of metabolism; however, the results were declared 'inconclusive'.[9][25][26][27][28]

14.2.2 Mars Science Laboratory

Main article: Timeline of Mars Science Laboratory

The *Curiosity* rover from the Mars Science Laboratory mission, is currently assessing the potential past and present habitability of the Martian environment and is attempting to detect biosignatures on the surface of Mars.[3] Considering the MSL instrument payload package, the following classes of biosignatures are within the MSL detection window: organism morphologies (cells, body fossils, casts), biofabrics (including microbial mats), diagnostic organic molecules, isotopic signatures, evidence of biomineralization and bioalteration, spatial patterns in chemistry, and biogenic gases. Of these, biogenic organic molecules and biogenic atmospheric gases are considered the most definitive and most readily

Carl Sagan with a model of the Viking lander

detectable by MSL.[3] The *Curiosity* rover targets outcrops to maximize the probability of detecting 'fossilized' organic matter preserved in sedimentary deposits.

On January 24, 2014, NASA reported that current studies by the *Curiosity* and *Opportunity* rovers on the planet Mars will now be searching for evidence of ancient life, including a biosphere based on autotrophic, chemotrophic and/or chemolithoautotrophic microorganisms, as well as ancient water, including fluvio-lacustrine environments (plains related to ancient rivers or lakes) that may have been habitable.[29][30][31][32] The search for evidence of habitability, taphonomy (related to fossils), and organic carbon on the planet Mars is now a primary NASA objective.[29][30]

14.2.3 ExoMars

The 2016 Trace Gas Orbiter (TGO) will be a Mars telecommunications orbiter and atmospheric gas analyzer mission. It will deliver the ExoMars EDM lander and then proceed to map the sources of methane on Mars and other gases, and in doing so, help select the landing site for the ExoMars rover to be launched on 2018.[33] The primary objective of the 2018 ExoMars rover mission is the search for biosignatures on the surface and subsurface by using a drill able to collect samples down to a depth of 2 metres (6.6 ft).[28][34]

14.3   See also

- Bioindicator

- Biomarker

- Planetary habitability

- Taphonomy

- Technosignature

14.4 References

[1] Steele; Beaty; et al. (September 26, 2006). "Final report of the MEPAG Astrobiology Field Laboratory Science Steering Group (AFL-SSG)" (.doc). *The Astrobiology Field Laboratory*. U.S.A.: the Mars Exploration Program Analysis Group (MEPAG) - NASA. p. 72.

[2] "Biosignature - definition". *Science Dictionary*. 2011. Retrieved 2011-01-12.

[3] Summons, Roger E.; Jan P. Amend; David Bish; Roger Buick; George D. Cody; David J. Des Marais; Dromart, G; Eigenbrode, J. L.; Knoll, A. H.; Sumner, D. Y. (23 February 2011). "Preservation of Martian Organic and Environmental Records: Final Report of the Mars Biosignature Working Group" (PDF). *The Astrobiology Journal* **11** (2): 157–81. Bibcode:2011AsBio..11..157S. doi:10.1089/ast.2010.0506. PMID 21417945. Retrieved 2013-06-22.

[4] Carol Cleland; Gamelyn Dykstra; Ben Pageler (2003). "Philosophical Issues in Astrobiology". NASA Astrobiology Institute. Retrieved 2011-04-15.

[5] Zimmer, Carl (January 22, 2015). "Even Elusive Animals Leave DNA, and Clues, Behind". *New York Times*. Retrieved January 23, 2015.

[6] "SIGNATURES OF LIFE FROM EARTH AND BEYOND". *Penn State Astrobiology Research Center (PSARC)*. Penn State. 2009. Retrieved 2011-01-14.

[7] Tenenbaum, David (July 30, 2008). "Reading Archaean Biosignatures". NASA. Retrieved 2014-11-23.

[8] Fatty alcohols

[9] Beegle, Luther W.; Abilleira, Fernando; Jordan, James F.; Wilson, Gregory R.; et al. (August 2007). "A Concept for NASA's Mars 2016 Astrobiology Field Laboratory". *Astrobiology* **7** (4): 545–577. Bibcode:2007AsBio...7..545B. doi:10.1089/ast.2007.0153. PMID17723090. Retrieved2009-07-20. Missing|last2=in Authors list(help)

[10] Bosak, Tanja Bosak; Virginia Souza-Egipsy; Frank A. Corsetti and Dianne K. Newman; Corsetti, Frank A.; Newman, Dianne K. (May 18, 2004). "Micrometer-scale porosity as a biosignature in carbonate crusts". *Geology* **32** (9): 781–784. Bibcode:2004Geo.....32..781B. doi:10.1130/G20681.1. Retrieved 2011-01-14.

[11] Crenson, Matt (2006-08-06). "After 10 years, few believe life on Mars". Associated Press (on usatoday.com). Retrieved 2009-12-06.

[12] McKay, David S.; Thomas-Keprta, Kathie L.; Vali, Hojatollah; Romanek, Christopher S.; Clemett, Simon J.; Chillier, Xavier D. F.; Maechling, Claude R.; Zare, Richard N.; et al. (1996). "Search for Past Life on Mars: Possible Relic Biogenic Activity in Martian Meteorite ALH84001". *Science* **273** (5277): 924–930. Bibcode:1996Sci...273..924M. doi:10.1126/science.273.5277.924.PMID8688069. Missing|last2=in Authors list(help)

[13] Rothschild, Lynn (September 2003). "Understand the evolutionary mechanisms and environmental limits of life". NASA. Retrieved 2009-07-13.

[14] Wall, Mike (13 December 2011). "Mars Life Hunt Could Look for Magnetic Clues". Space.com. Retrieved 2011-12-15.

[15] Gardner, James N. (February 28, 2006). "The Physical Constants as Biosignature: An anthropic retrodiction of the Selfish Biocosm Hypothesis". Kurzweil. Retrieved 2011-01-14.

[16] "Astrobiology". Biology Cabinet. September 26, 2006. Retrieved 2011-01-17.

[17] "Artificial Life Shares Biosignature With Terrestrial Cousins". *The Physics arXiv Blog*. MIT. 10 January 2011. Retrieved 2011-01-14.

[18] Mars Trace Gas Mission (September 10, 2009)

[19] Remote Sensing Tutorial, Section 19-13a - Missions to Mars during the Third Millennium, Nicholas M. Short, Sr., et al., NASA

[20] Oze, Christopher; Jones, Camille; Goldsmith, Jonas I.; Rosenbauer, Robert J. (June 7, 2012). "Differentiating biotic from abiotic methane genesis in hydrothermally active planetary surfaces". *PNAS* **109** (25): 9750–9754. Bibcode:2012PNAS..109.9750O. doi:10.1073/pnas.1205223109. PMC 3382529. PMID 22679287. Retrieved June 27, 2012.

[21] Staff (June 25, 2012). "Mars Life Could Leave Traces in Red Planet's Air: Study". Space.com. Retrieved June 27, 2012.

[22] Brogi, Matteo; Snellen, Ignas A. G.; de Krok, Remco J.; Albrecht, Simon; Birkby, Jayne; de Mooij, Ernest J. W. (June 28, 2012). "The signature of orbital motion from the dayside of the planet t Boötis b". *Nature* **486** (7404): 502–504. arXiv:1206.6109. Bibcode:2012Natur.486..502B. doi:10.1038/nature11161. Retrieved June 28, 2012.

[23] Mann, Adam (June 27, 2012). "New View of Exoplanets Will Aid Search for E.T.". Wired (magazine). Retrieved June 28, 2012.

[24] Mark Allen; = Olivier Witasse (June 16, 2011). "2016 ESA/NASA ExoMars Trace Gas Orbiter" (PDF). *MEPAG June 2011*. Jet Propulsion Laboratory (PDF)

[25] Levin, G and P. Straaf. 1976. Viking Labeled Release Biology Experiment: Interim Results. Science: vol: 194. pp: 1322-1329.

[26] Chambers, Paul (1999). *Life on Mars; The Complete Story*. London: Blandford. ISBN 0-7137-2747-0.

[27] Klein, Harold P.; Levin, Gilbert V.; Levin, Gilbert V.; Oyama, Vance I.; Lederberg, Joshua; Rich, Alexander; Hubbard, Jerry S.; Hobby, George L.; Straat, Patricia A.; Berdahl, Bonnie J.; Carle, Glenn C.; Brown, Frederick S.; Johnson, Richard D. (1976-10-01). "The Viking Biological Investigation: Preliminary Results". *Science* **194** (4260): 99–105. Bibcode:1976Sci...194...99K. doi:10.1126/science.194.4260.99. PMID 17793090. Retrieved 2008-08-15.

[28] ExoMars rover

[29] Grotzinger, John P. (January 24, 2014). "Introduction to Special Issue - Habitability, Taphonomy, and the Search for Organic Carbon on Mars". *Science* **343** (6169): 386–387. Bibcode:2014Sci...343..386G. doi:10.1126/science.1249944. Retrieved January 24, 2014.

[30] Various (January 24, 2014). "Special Issue - Table of Contents - Exploring Martian Habitability". *Science* **343** (6169): 345–452. Retrieved January 24, 2014.

[31] Various (January 24, 2014). "Special Collection - Curiosity - Exploring Martian Habitability". *Science*. Retrieved January 24, 2014.

[32] Grotzinger, J. P.; et al. (January 24, 2014). "A Habitable Fluvio-Lacustrine Environment at Yellowknife Bay, Gale Crater, Mars". *Science* **343** (6169): 1242777. Bibcode:2014Sci...343A.386G. doi:10.1126/science.1242777. Retrieved January 24, 2014.

[33] Pavlishchev, Boris (Jul 15, 2012). "ExoMars program gathers strength". *The Voice of Russia*. Retrieved 2012-07-15.

[34] "Mars Science Laboratory: Mission". NASA/JPL. Retrieved 2010-03-12.

Chapter 15

List of multiplanetary systems

See also: List of exoplanets

From the total of 1288 stars known to have exoplanets (as of December 11, 2015), there are a total of **502** known

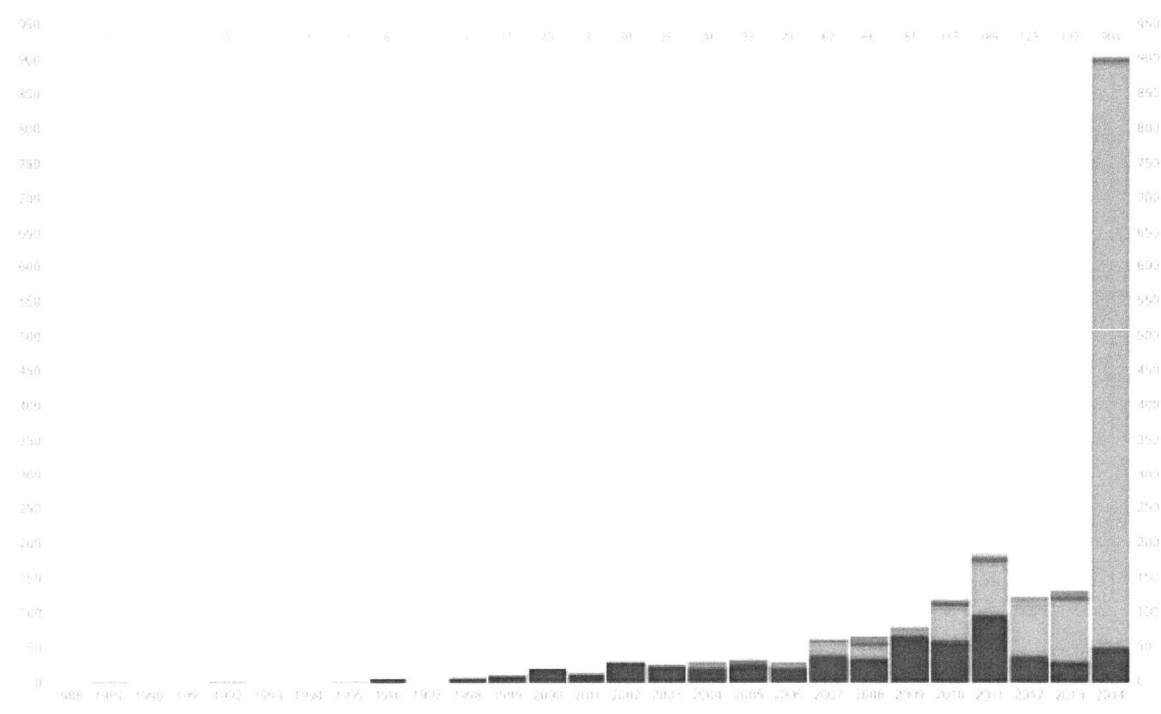

Number of extrasolar planet discoveries per year through September 2014. Colors indicate method of detection.

multiplanetary systems,[1] or stars with at least two confirmed planets, beyond our Solar System. About 280 of these have only two confirmed exoplanets, but some have a significantly larger number. The star with the most confirmed planets is our Sun with 8 confirmed planets, while the stars with the most confirmed exoplanets are Kepler-90 and HD 10180 with 7 confirmed planets each; in 2012, two more exoplanet candidates have been suggested for HD 10180, which would bring the total to 9 exoplanets in that system.

The 502 multiplanetary systems are listed below according to the star's distance from Earth. Gliese 876, with 4 confirmed exoplanets, is the closest multiplanetary system at 15 light years from our Solar System. A total of 12 systems are known that are closer than 50 light years away, but most are much farther away. The farthest confirmed multiplanetary system is OGLE-2012-BLG-0026L, at 13,300 ly away.

The table below contains information about the coordinates, spectral and physical properties, and number of confirmed planets. The two most important stellar properties are mass and metallicity because it determines how these planetary systems form. The higher mass and metallicity tend to have more planets and more massive planets.

15.1   Multiplanetary systems

- Need to be added: Gliese 433, Gliese 785, HD 136352, Gliese 439, Gliese 253, HD 7449, HIP 49067, UZ Fornacis

15.2   Stars orbited by both planets and brown dwarfs

Stars orbited by objects on both sides of the 13 Jupiter mass dividing line.

- 54 Piscium (HD 3651)
- HD 168443

15.3   Planetary system statistics

15.4   See also

- List of exoplanets
- Methods of detecting extrasolar planets
- List of extrasolar planet firsts
- List of extrasolar planet extremes
- List of brown dwarfs
- Lists of stars
- List of nearest stars
- List of stars with proplyds

For links to specific lists of extrasolar planets see:

- List of exoplanetary host stars
- List of extrasolar planets detected by radial velocity
- List of transiting extrasolar planets
- List of extrasolar planets detected by microlensing
- List of extrasolar planets that were directly imaged
- List of extrasolar planets detected by timing
- List of nearest terrestrial exoplanet candidates

Online archives:

- NASA Exoplanet Archive

15.5 References

[1] Schneider, Jean (10 September 2011). "Interactive Extra-solar Planets Catalog". *The Extrasolar Planets Encyclopaedia*. Retrieved 2012-01-30.

[2] "[1310.5771] Characterization of the KOI-94 System with Transit Timing Variation Analysis: Implication for the Planet-Planet Eclipse". *arxiv.org*. line feed character in |title= at position 80 (help)

[3] http://www.openexoplanetcatalogue.com/system.html?id=Kepler-65[]

[4] "[1501.06227] An ancient extrasolar system with five sub-Earth-size planets". *arxiv.org*.

Chapter 16

Terrestrial planet

The terrestrial planets Mercury, Venus, Earth, and Mars, sizes to scale

A **terrestrial planet**, **telluric planet** or **rocky planet** is a planet that is composed primarily of silicate rocks or metals. Within the Solar System, the terrestrial planets are the inner planets closest to the Sun, i.e. Mercury, Venus, Earth, and Mars. The terms "terrestrial planet" and "telluric planet" are derived from Latin words for Earth (*Terra* and *Tellus*), as these planets are, in terms of composition, "Earth-like".

Terrestrial planets have a solid planetary surface, making them substantially different from the larger giant planets, which are composed mostly of some combination of hydrogen, helium, and water existing in various physical states.

16.1 Structure

All terrestrial planets have approximately the same type of structure: a central metallic core, mostly iron, with a surrounding silicate mantle. The Moon is similar, but has a much smaller iron core. Io and Europa are also satellites that have internal structures similar to that of terrestrial planets. Terrestrial planets can have canyons, craters, mountains, volcanoes, and other surface structures, depending on the presence of water and tectonic activity. Terrestrial planets have secondary atmospheres, generated through volcanism or comet impacts, in contrast to the giant planets, whose atmospheres are primary, captured directly from the original solar nebula.[1]

16.2   Solar terrestrial planets

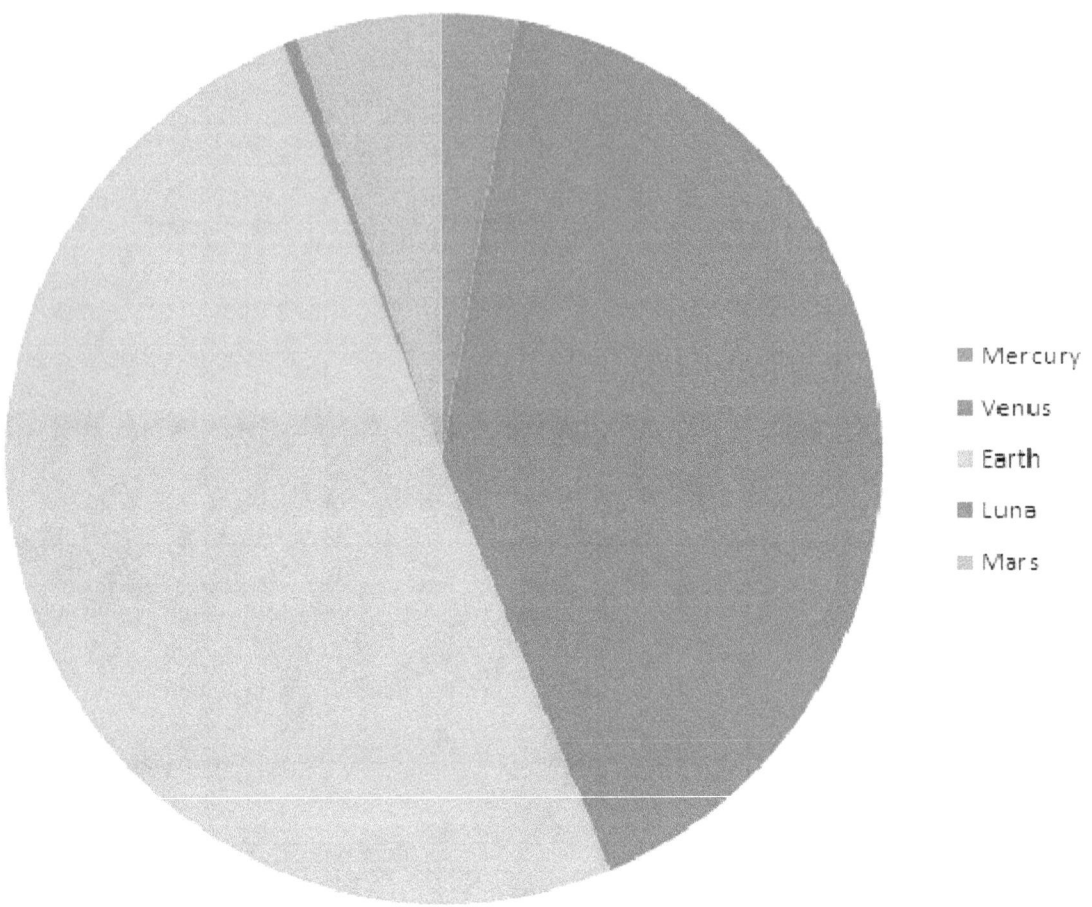

Relative masses of the terrestrial planets of the Solar System, including the Moon (designated here as "Luna")

The Solar System has four terrestrial planets: Mercury, Venus, Earth, and Mars. Only one terrestrial planet, Earth, is known to have an active hydrosphere.

During the formation of the Solar System, there were probably many more "terrestrial" planetesimals, but most merged with or were ejected by the four terrestrial planets.

Dwarf planets, such as Ceres and Pluto, and large small Solar System bodies are similar to terrestrial planets in the fact that they do have a solid surface, but are, on average, composed of more icy materials (Ceres and Pluto have densities 2.17 and 1.87 g cm^{-3}, respectively, and Haumea's density is similar to Pallas's 2.8 g cm^{-3}).

16.2.1   Density trends

The uncompressed density of a terrestrial planet is the average density its materials would have at zero pressure. A greater uncompressed density indicates greater metal content. Uncompressed density differs from the true average density because compression within planet cores increases their density; the average density depends on planet size as well as composition.

The density of terrestrial planets trends towards lower values as the distance from the Sun increases. The rocky minor planet Vesta orbiting outside of Mars is less dense than Mars still, at 3.4 g cm^{-3}.

It is unknown whether extrasolar terrestrial planets in general will also follow this trend.

16.3   Extrasolar terrestrial planets

See also: Super-Earth and List of nearest terrestrial exoplanet candidates

Most of the planets found outside the Solar System are giant planets, because they are more easily detectable.[2][3][4] But since 2005, hundreds of potentially terrestrial extrasolar planets have been found, with several being confirmed as terrestrial. Most of these are super-Earths, i.e. planets with masses between Earth's and Neptune's; super-Earths may be gas planets or terrestrial, depending on their mass and other parameters.

During the early 1990s, the first extrasolar planets were discovered orbiting the pulsar PSR B1257+12, with masses of 0.02, 4.3, and 3.9 times that of Earth's, by pulsar timing.

When 51 Pegasi b, the first planet found around a star still undergoing fusion, was discovered, many astronomers assumed it to be a gigantic terrestrial, because it was assumed no gas giant could exist as close to its star (0.052 AU) as 51 Pegasi b did. It was later found to be a gas giant.

In 2005, the first planets around main-sequence stars that may be terrestrial were found: Gliese 876 d, has a mass 7 to 9 times that of Earth and an orbital period of just two Earth days. It orbits the red dwarf Gliese 876, 15 light years from Earth. OGLE-2005-BLG-390Lb, about 5.5 times the mass of Earth, orbits a star about 21,000 light years away in the constellation Scorpius. From 2007 to 2010, three (possibly four) potential terrestrial planets were found orbiting the red dwarf Gliese 581. The smallest, Gliese 581 e, is only about 1.9 Earth mass,[5] but orbits very close to the star. An ideal terrestrial planet would be 2 Earth masses with a 25-day orbital period around a red dwarf.[6] Two others, Gliese 581 c and Gliese 581 d, as well as a disputed planet, Gliese 581 g, are more-massive super-Earths orbiting in or close to the habitable zone of the star, so they could potentially be habitable, with Earth-like temperatures.

Another possibly terrestrial planet, HD 85512 b, was discovered in 2011; it has at least 3.6 times the mass of Earth.[7] But the radius and composition of all these planets are unknown.

The first confirmed terrestrial exoplanet, Kepler-10b, was found in 2011 by the Kepler Mission, specifically designed to discover Earth-like planets around other stars using the transit method.[8]

In the same year, the Kepler Space Observatory Mission team released a list of 1235 extrasolar planet candidates, including six that are "Earth-size" or "super-Earth-size" (i.e. they have a radius less than 2 Earth radii)[9] and in the habitable zone.[10] Since then, Kepler has discovered hundreds of planets ranging from Moon-sized to super-Earths, with many more candidates in this size range (see image).

16.3.1   List of terrestrial exoplanets

The following exoplanets have a density of at least 5 g/cm³ and a mass below Neptune's and are thus very likely terrestrial: Kepler-10b, Kepler-20b, Kepler-36b, Kepler-48b, Kepler-78b, Kepler-89b, Kepler-97b, Kepler-99b, Kepler-131b.

The Neptune-mass planet Kepler-10c also has a density >5 g/cm³ and is thus very likely terrestrial.

16.3.2   Frequency

In 2013, astronomers reported, based on *Kepler* space mission data, that there could be as many as 40 billion Earth- and super-Earth-sized planets orbiting in the habitable zones of Sun-like stars and red dwarfs within the Milky Way.[11][32][13] 11 billion of these estimated planets may be orbiting Sun-like stars.[14] The nearest such planet may be 12 light-years away, according to the scientists.[11][12] However, this does not give estimates for the number of extrasolar terrestrial planets, because there are planets as small as Earth that have been shown to be gas planets (see KOI-314c).[15]

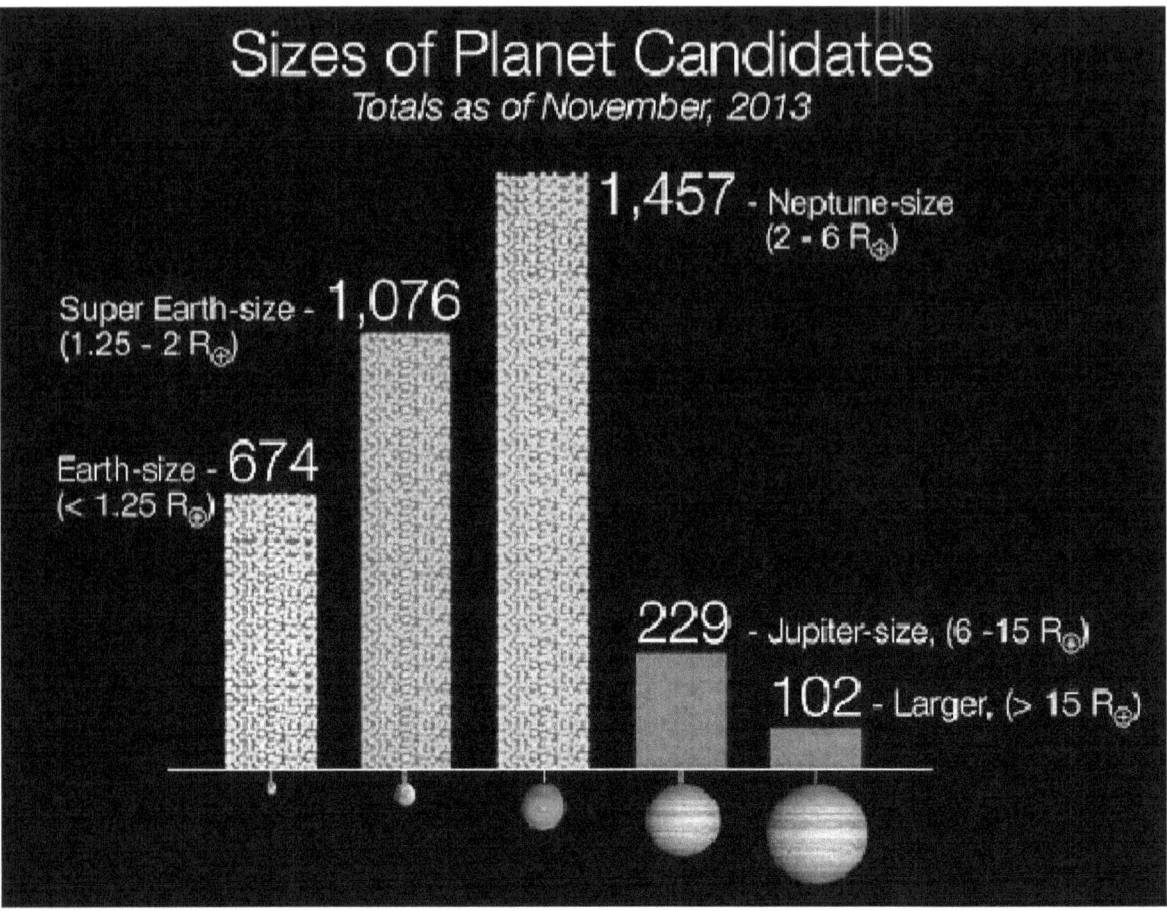

Sizes of Kepler planet candidates based on 2,740 candidates orbiting 2,036 stars as of November 4, 2013 (NASA).

16.4   Types

Several possible classifications for terrestrial planets have been proposed:[16]

Silicate planet   The standard type of terrestrial planet seen in the Solar System, made primarily of silicon-based rocky mantle with a metallic (iron) core.

Iron planet   A theoretical type of terrestrial planet that consists almost entirely of iron and therefore has a greater density and a smaller radius than other terrestrial planets of comparable mass. Mercury in the Solar System has a metallic core equal to 60–70% of its planetary mass. Iron planets are believed to form in the high-temperature regions close to a star, like Mercury, and if the protoplanetary disk is rich in iron.

Coreless planet   A theoretical type of terrestrial planet that consists of silicate rock but has no metallic core, i.e. the opposite of an iron planet. Although the Solar System contains no coreless planets, chondrite asteroids and meteorites are common in the Solar System. Coreless planets are believed to form farther from the star where volatile oxidizing material is more common.

Carbon planet (also called "diamond planet")   A theoretical class of planets, composed of a metal core surrounded by primarily carbon-based minerals. They may be considered a type of terrestrial planet if the metal content dominates. The Solar System contains no carbon planets, but does have carbonaceous asteroids.

Artist's impression of a carbon planet

16.5   See also

- Earth analog

- Chthonian planet

- Planetary habitability

- List of potentially habitable exoplanets

- Venus zone

16.6   References

[1] Dr. James Schombert (2004). "Primary Atmospheres (Astronomy 121: Lecture 14 Terrestrial Planet Atmospheres)". Department of Physics University of Oregon. Retrieved 22 December 2009.

[2] Carole Haswell, Transiting Exoplanets

[3] Michael Perryman, The Exoplanet Handbook

[4] Sara Seager, Exoplanets

[5] "Lightest exoplanet yet discovered". ESO (ESO 15/09 – Science Release). 21 April 2009. Retrieved 15 July 2009.

[6] M. Mayor; X. Bonfils; T. Forveille; X. Delfosse; S. Udry; J.-L. Bertaux; H. Beust; F. Bouchy; C. Lovis; F. Pepe; C. Perrier; D. Queloz; N. C. Santos (2009). "The HARPS search for southern extra-solar planets,XVIII. An Earth-mass planet in the GJ 581 planetary system". arXiv:0906.2780 [astro-ph].

[7] Kaufman, Rachel (30 August 2011). "New Planet May Be Among Most Earthlike – Weather Permitting, Alien world could host liquid water if it has 50 percent cloud cover, study says". National Geographic News. Retrieved 5 September 2011.

[8] http://www.bbc.com/news/science-environment-17454005

[9] Namely: KOI 326.01 [Rp=0.85], KOI 701.03 [Rp=1.73], KOI 268.01 [Rp=1.75], KOI 1026.01 [Rp=1.77], KOI 854.01 [Rp=1.91], KOI 70.03 [Rp=1.96] – Table 6). A more recent study found that one of these candidates (KOI 326.01) is in fact much larger and hotter than first reported. Grant, Andrew (8 March 2011). "Exclusive: "Most Earth-Like" Exoplanet Gets Major Demotion—It Isn't Habitable". *80beats*. Discover Magazine. Retrieved 9 March 2011. External link in |work= (help)

[10] Borucki, William J.; Koch, David G; Basri, Gibor; Batalha, Natalie; Brown, Timothy M.; et al. (1 February 2011). "Characteristics of planetary candidates observed by Kepler, II: Analysis of the first four months of data" (PDF). arXiv. Retrieved 16 February 2011.

[11] Overbye, Dennis (4 November 2013). "Far-Off Planets Like the Earth Dot the Galaxy". *New York Times*. Retrieved 5 November 2013.

[12] Petigura, Eric A.; Howard, Andrew W.; Marcy, Geoffrey W. (31 October 2013). "Prevalence of Earth-size planets orbiting Sun-like stars". *Proceedings of the National Academy of Sciences of the United States of America*. arXiv:1311.6806. Bibcode:2013PNAS..11019273P. doi:10.1073/pnas.1319909110. Retrieved 5 November 2013.

[13] Staff (January 7, 2013). "17 Billion Earth-Size Alien Planets Inhabit Milky Way". Space.com. Retrieved January 8, 2013.

[14] Khan, Amina (4 November 2013). "Milky Way may host billions of Earth-size planets". *Los Angeles Times*. Retrieved 5 November 2013.

[15] http://www.cfa.harvard.edu/news/2014-01

[16] Naeye, Bob (24 September 2007). "Scientists Model a Cornucopia of Earth-sized Planets". NASA, Goddard Space Flight Center. Retrieved 23 October 2013.

16.7 External links

- SPACE.com: Q&A: The IAU's Proposed Planet Definition 16 August 2006

- BBC News: Q&A New planets proposal Wednesday, 16 August 2006

Chapter 17

Planetary system

"Solar systems" redirects here. For the planetary system of the Sun, see Solar System. For the solar power company, see Solar Systems (company).

A **planetary system** is a set of gravitationally bound non-stellar objects in orbit around a star or star system. Generally

An artist's concept of a planetary system

speaking, systems with one or more planets constitute a planetary system, although such systems may also consist of bodies such as dwarf planets, asteroids, natural satellites, meteoroids, comets, planetesimals[1][2] and circumstellar disks. The

278

Sun together with its planetary system, which includes Earth, is known as the Solar System.[3][4] The term **exoplanetary system** is sometimes used in reference to other planetary systems.

A total of 2030 exoplanets (in 1288 planetary systems, including 502 multiple planetary systems) have been identified as of 11 December 2015.[5]

Of particular interest to astrobiology is the habitable zone of planetary systems where planets could have surface liquid water.

17.1 History

17.1.1 Heliocentrism

Historically, heliocentrism (the doctrine that the Sun is the centre of the universe) was opposed to geocentrism (placing the Earth at the center of the universe).

The notion of a heliocentric Solar System, with the Sun at the center, is possibly first suggested in the Vedic literature of ancient India, which often refer to the Sun as the "centre of spheres". Some interpret Aryabhatta's writings in Āryabhaṭīya as implicitly heliocentric.

The idea was first proposed in western philosophy and Greek astronomy as early as the 3rd century BC by Aristarchus of Samos,[6] but had received no support from most other ancient astronomers.

17.1.2 Discovery of the Solar System

Main article: Discovery and exploration of the Solar System

De revolutionibus orbium coelestium by Nicolaus Copernicus, published in 1543, was the first mathematically predictive heliocentric model of a planetary system. 17th-century successors Galileo Galilei, Johannes Kepler, and Isaac Newton developed an understanding of physics which led to the gradual acceptance of the idea that the Earth moves round the Sun and that the planets are governed by the same physical laws that governed the Earth.

17.1.3 Speculation on extrasolar planetary systems

In the 16th century the Italian philosopher Giordano Bruno, an early supporter of the Copernican theory that the Earth and other planets orbit the Sun, put forward the view that the fixed stars are similar to the Sun and are likewise accompanied by planets. He was burned at the stake for his ideas by the Roman Inquisition.[7]

In the 18th century the same possibility was mentioned by Isaac Newton in the "General Scholium" that concludes his *Principia*. Making a comparison to the Sun's planets, he wrote "And if the fixed stars are the centers of similar systems, they will all be constructed according to a similar design and subject to the dominion of *One*."[8]

His theories gained traction through the 19th and 20th centuries despite a lack of supporting evidence. Long before their confirmation by astronomers, conjecture on the nature of planetary systems had been a focus of the search for extraterrestrial intelligence and has been a prevalent theme in fiction, particularly science fiction.

17.1.4 Detection of exoplanets

The first confirmed detection of an exoplanet was in 1992, with the discovery of several terrestrial-mass planets orbiting the pulsar PSR B1257+12. The first confirmed detection of exoplanets of a main-sequence star was made in 1995, when a giant planet, 51 Pegasi b, was found in a four-day orbit around the nearby G-type star 51 Pegasi. The frequency of detections has increased since then, particularly through advancements in methods of detecting extrasolar planets and dedicated planet finding programs such as the Kepler mission.

17.2 Origin and evolution

See also: Nebular hypothesis, Planetary migration and Formation and evolution of the Solar System
Planetary systems come from protoplanetary disks that form around stars as part of the process of star formation.

During formation of a system much material is gravitationally scattered into far-flung orbits and some planets are ejected completely from the system becoming rogue planets.

17.2.1 Evolved systems

High-mass stars

Planets orbiting pulsars have been discovered, and pulsars are the remnants of the supernova explosions of high-mass stars. A planetary system that existed before the supernova would likely be mostly destroyed — planets would either evaporate, be pushed off of their orbits by the masses of gas from the exploding star, or the sudden loss of most of the mass of the central star would see them escape the gravitational hold of the star, or in some cases the supernova would kick the pulsar itself out of the system at high velocity so any planets that had survived the explosion would be left behind as free-floating objects. Planets found around pulsars may have formed as a result of pre-existing stellar companions that were almost entirely evaporated by the supernova blast, leaving behind planet-sized bodies. Alternatively, planets may form in an accretion disk of fallback matter surrounding a pulsar.[9] Fallback disks of matter that failed to escape orbit during a supernova may also form planets around black holes.[10]

Lower-mass stars

As stars evolve and turn into red giants, asymptotic giant branch stars and planetary nebulae they engulf the inner planets, evaporating or partially evaporating them depending on how massive they are. As the star loses mass, planets that are not engulfed move further out from the star.

If an evolved star is in a binary or multiple system then the mass it loses can transfer to another star, creating new protoplanetary disks and second- and third-generation planets which may differ in composition from the original planets which may also be affected by the mass transfer.

- Planets in evolved binary systems, Hagai B. Perets, 13 Jan 2011

- Can Planets survive Stellar Evolution?, Eva Villaver, Mario Livio, Feb 2007

- The Orbital Evolution of Gas Giant Planets around Giant Stars, Eva Villaver, Mario Livio, 13 Oct 2009

- On the survival of brown dwarfs and planets engulfed by their giant host star, Jean-Claude Passy, Mordecai-Mark Mac Low, Orsola De Marco, 2 Oct 2012

- Foretellings of Ragnarök: World-engulfing Asymptotic Giants and the Inheritance of White Dwarfs, Alexander James Mustill, Eva Villaver, 5 Dec 2012

17.3 System architecture

The Solar System consists of an inner region of small rocky planets and outer region of large gas giants however other planetary systems can have quite different architectures. Many systems with a hot Jupiter gas giant very close to the star have been found. Theories, such as planetary migration or scattering, have been proposed for the formation of large planets close to their parent stars.[11] At present, few systems have been found to be analogous to the Solar System with terrestrial planets close to the parent star. More commonly, systems consisting of multiple Super-Earths have been detected.[12]

17.3.1 Orbital properties

Unlike the Solar System, which has orbits that are nearly circular, many of the known planetary systems display much higher orbital eccentricity.[13] An example of such a system is 16 Cygni.

Mutual inclination

The mutual inclination between two planets is the angle between their orbital planes. Many compact systems with multiple close-in planets interior to the equivalent orbit of Venus are expected to have very low mutual inclinations, so the system (at least the close-in part) would be even flatter than the solar system. Captured planets could be captured into any arbitrary angle to the rest of the system. The only system where mutual inclinations have actually been measured is the Upsilon Andromedae system: the planets, c and d, have a mutual inclination of about 30 degrees.[14][15]

Orbital dynamics

Planetary systems can be categorized according to their orbital dynamics as resonant, non-resonant-interacting, hierarchical, or some combination of these. In resonant systems the orbital periods of the planets are in integer ratios. The Kepler-223 system contains four planets in an 8:6:4:3 orbital resonance.[16] Giant planets are found in mean-motion resonances more often than smaller planets.[17] In interacting systems the planets orbits are close enough together that they perturb the orbital parameters. The Solar System could be described as weakly interacting. In strongly interacting systems Kepler's laws do not hold.[18] In hierarchical systems the planets are arranged so that the system can be gravitationally considered as a nested system of two-bodies, e.g. in a star with a close-in hot jupiter with another gas giant much further out, the star and hot jupiter form a pair that appears as a single object to another planet that is far enough out.

Other, as yet unobserved, orbital possibilities include: double planets; various co-orbital planets such as quasi-satellites, trojans and exchange orbits; and interlocking orbits maintained by precessing orbital planes.[19]

- Extrasolar Binary Planets I: Formation by tidal capture during planet-planet scattering, H. Ochiai, M. Nagasawa, S. Ida, 26 Jun 2014

- Disruption of co-orbital (1:1) planetary resonances during gas-driven orbital migration, Arnaud Pierens, Sean Raymond, 19 May 2014

17.3.2 Captured planets

Free-floating planets in open clusters have similar velocities to the stars and so can be recaptured. They are typically captured into wide orbits between 100 and 10^5 AU. The capture efficiency decreases with increasing cluster size, and for a given cluster size it increases with the host/primary mass. It is almost independent of the planetary mass. Single and multiple planets could be captured into arbitrary unaligned orbits, non-coplanar with each other or with the stellar host spin, or pre-existing planetary system. Some planet–host metallicity correlation may still exist due to the common origin of the stars from the same cluster. Planets would be unlikely to be captured around neutron stars because these are likely to be ejected from the cluster by a pulsar kick when they form. Planets could even be captured around other planets to form free-floating planet binaries. After the cluster has dispersed some of the captured planets with orbits larger than 10^6 AU would be slowly disrupted by the galactic tide and likely become free-floating again through encounters with other field stars or giant molecular clouds.[20]

17.3.3 Number of planets, relative parameters and spacings

- On The Relative Sizes of Planets Within Kepler Multiple Candidate Systems, David R. Ciardi et al. 9 Dec 2012

- The Kepler Dichotomy among the M Dwarfs: Half of Systems Contain Five or More Coplanar Planets, Sarah Ballard, John Asher Johnson, 15 Oct 2014

- Exoplanet Predictions Based on the Generalised Titius-Bode Relation, Timothy Bovaird, Charles H. Lineweaver, 1 Aug 2013

- The Solar System and the Exoplanet Orbital Eccentricity - Multiplicity Relation, Mary Anne Limbach, Edwin L. Turner, 9 Apr 2014

- The period ratio distribution of Kepler's candidate multiplanet systems, Jason H. Steffen, Jason A. Hwang, 11 Sep 2014

- Are Planetary Systems Filled to Capacity? A Study Based on Kepler Results, Julia Fang, Jean-Luc Margot, 28 Feb 2013

17.4   Planet-hosting stars

17.4.1   Proportion of stars with planets

There is at least one planet on average per star.[21] Around 1 in 5 Sun-like stars[lower alpha 1] have an "Earth-sized"[lower alpha 2] planet in the habitable zone.

Most stars have planets but exactly what proportion of stars have planets is uncertain because not all planets can yet be detected. The radial-velocity method and the transit method (which between them are responsible for the vast majority of detections) are most sensitive to large planets in small orbits. Thus many known exoplanets are "hot Jupiters": planets of Jovian mass or larger in very small orbits with periods of only a few days. A 2005 survey of radial-velocity-detected planets found that about 1.2% of Sun-like stars have a hot jupiter, where "Sun-like star" refers to any main-sequence star of spectral classes late-F, G, or early-K without a close stellar companion.[22] This 1.2% is more than double the frequency of hot jupiters detected by the Kepler spacecraft, which may be because the Kepler field of view covers a different region of the Milky Way where the metallicity of stars is different.[23] It is further estimated that 3% to 4.5% of Sun-like stars possess a giant planet with an orbital period of 100 days or less, where "giant planet" means a planet of at least 30 Earth masses.[24]

It is known that small planets (of roughly Earth-like mass or somewhat larger) are more common than giant planets.[25] It also appears that there are more planets in large orbits than in small orbits. Based on this, it is estimated that perhaps 20% of Sun-like stars have at least one giant planet whereas at least 40% may have planets of lower mass.[24][26][27] A 2012 study of gravitational microlensing data collected between 2002 and 2007 concludes the proportion of stars with planets is much higher and estimates an average of 1.6 planets orbiting between 0.5–10 AU per star in the Milky Way, the authors of this study conclude that "stars are orbited by planets as a rule, rather than the exception".[21] In November 2013 it was announced that 22±8% of Sun-like[lower alpha 1] stars have an Earth-sized[lower alpha 2] planet in the habitable[lower alpha 3] zone.[28][29]

Whatever the proportion of stars with planets, the total number of exoplanets must be very large. Because the Milky Way has at least 200 billion stars, it must also contain tens or hundreds of billions of planets.

17.4.2   Type of star, spectral classification

See also planets orbiting Herbig Ae/Be stars, T Tauri stars, Subgiants, Yellow giants, Red giants, B-type stars, A-type stars, Subdwarf B stars, F-type stars, G-type stars, K-type stars, Red dwarfs, Brown dwarfs, White dwarfs, or Pulsars.
  Most known exoplanets orbit stars roughly similar to the Sun, that is, main-sequence stars of spectral categories F, G, or K. One reason is that planet-search programs have tended to concentrate on such stars. In addition, statistical analyses indicate that lower-mass stars (red dwarfs, of spectral category M) are less likely to have planets massive enough to be detected by the radial-velocity method.[24][30] Nevertheless, several tens of planets around red dwarfs have been discovered by the Kepler spacecraft by the transit method, which can detect smaller planets.

Stars of spectral category A typically rotate very quickly, which makes it very difficult to measure the small Doppler shifts induced by orbiting planets because the spectral lines are very broad. However, this type of massive star eventually evolves into a cooler red giant that rotates more slowly and thus can be measured using the radial-velocity method. A few tens of planets have been found around red giants.

Observations using the Spitzer Space Telescope indicate that extremely massive stars of spectral category O, which are much hotter than the Sun, produce a photo-evaporation effect that inhibits planetary formation.[31] When the O-type star goes supernova any planets that had formed would become free-floating due to the loss of stellar mass unless the natal kick of the resulting remnant pushes it in the same direction as an escaping planet.[32] Fallback disks of matter that failed to escape orbit during a supernova may form planets around neutron stars and black holes.[10]

Doppler surveys around a wide variety of stars indicate about 1 in 6 stars having twice the mass of the Sun are orbited by one or more Jupiter-sized planets, vs. 1 in 16 for Sun-like stars and only 1 in 50 for red dwarfs. On the other hand, microlensing surveys indicate that long-period Neptune-mass planets are found around 1 in 3 red dwarfs. [33] Kepler Space Telescope observations of planets with up to one year periods show that occurrence rates of Earth- to Neptune-sized planets (1 to 4 Earth radii) around M, K, G, and F stars are successively higher towards cooler, less massive stars.[34]

At the low-mass end of star-formation are sub-stellar objects that don't fuse hydrogen: the brown dwarfs and sub-brown dwarfs, of spectral classification L,T and Y. Planets and protoplanetary disks have been discovered around brown dwarfs, and disks have been found around sub-brown dwarfs (e.g. OTS 44).

Rogue planets ejected from their system could retain a system of satellites.[35]

17.4.3 Metallicity

Stars with a higher metallicity than the Sun are more likely to have planets, especially giant planets, than stars with lower metallicity.

Ordinary stars are composed mainly of the light elements hydrogen and helium. They also contain a small proportion of heavier elements, and this fraction is referred to as a star's metallicity (even if the elements are not metals in the traditional sense),[22] denoted [m/H] and expressed on a logarithmic scale where zero is the Sun's metallicity.

A 2012 study of the Kepler spacecraft data found that smaller planets, with radii smaller than Neptune's were found around stars with metallicities in the range $-0.6 < [m/H] < +0.5$ (about four times less than that of the Sun to three times more),[lower alpha 4] whereas larger planets were found mostly around stars with metallicities at the higher end of this range (at solar metallicity and above). In this study small planets occurred about three times as frequently as large planets around stars of metallicity greater than that of the Sun, but they occurred around six times as frequently for stars of metallicity less than that of the Sun. The lack of gas giants around low-metallicity stars could be because the metallicity of protoplanetary disks affects how quickly planetary cores can form and whether they accrete a gaseous envelope before the gas dissipates. However, Kepler can only observe planets very close to their star and the detected gas giants probably migrated from further out, so a decreased efficiency of migration in low-metallicity disks could also partly explain these findings.[36]

A 2014 study found that not only giant planets, but planets of all sizes have an increased occurrence rate around metal-rich stars compared to metal-poor, although the larger the planet, the greater this increase as the metallicity increases. The study divided planets into three groups based on radius: gas giants, gas dwarfs, and terrestrial planets with the dividing lines at 1.7 and 3.9 Earth radii. For these three groups the planet occurrence rates are 9.30, 2.03, and 1.72 times higher for metal-rich stars than for metal-poor stars, respectively. There is a bias against detecting smaller planets because metal-rich stars tend to be larger, making it more difficult to detect smaller planets, which means that these increases in occurrence rates are lower limits.[37]

It has also been shown that stars with planets are more likely to be deficient in lithium.[38]

17.4.4 Multiple stars

Stellar multiplicity increases with stellar mass: the likelihood of stars being in multiple systems is about 25% for red dwarfs, about 45% for Sun-like stars, and rises to about 80% for the most massive stars. Of the multiple stars about 75% are binaries and the rest are higher-order multiplicities.[39]

Some planets have been discovered orbiting one member of a binary star system (e.g. 55 Cancri, possibly Alpha Centauri Bb),[40] and several circumbinary planets have been discovered which orbit around both members of binary star (e.g. PSR B1620-26 b, Kepler-16b). A few planets in triple star systems are known (e.g. 16 Cygni Bb)[41] and one in the quadruple system Kepler 64.

The Kepler results indicate circumbinary planetary systems are relatively common (as of October 2013 the spacecraft had found seven circumbinary planets out of roughly 1000 eclipsing binaries searched). One puzzling finding is that although half of the binaries have an orbital period of 2.7 days or less, none of the binaries with circumbinary planets have a period less than 7.4 days. Another surprising Kepler finding is circumbinary planets tend to orbit their stars close to the critical instability radius (theoretical calculations indicate the minimum stable separation is roughly two to three times the size of the stars' separation).[42]

In 2014, from statisitcal studies of searches for companion stars, it was inferred that around half of exoplanet host stars have a companion star, usually within 100AU.[43][44] This means that many exoplanet host stars that were thought to be single are binaries, so in many cases it is not known which of the stars a planet actually orbits, and the published parameters of transiting planets could be significantly incorrect because the *planet radius* and *distance from star* are derived from the stellar parameters. Follow-up studies with imaging (such as speckle imaging) are needed to find or rule out companions (and radial velocity techniques would be required to detect binaries really close together) and this has not yet been done for most exoplanet host stars. Examples of known binary stars where it is not known which of the stars a planet orbits are Kepler-132 and Kepler-296.[45]

17.4.5   Open clusters

Most stars form in open clusters, but very few planets have been found in open clusters and this led to the hypothesis that the open-cluster environment hinders planet formation. However, a 2011 study concluded that there have been an insufficient number of surveys of clusters to make such a hypothesis.[46] The lack of surveys was because there are relatively few suitable open clusters in the Milky Way. Recent discoveries of both giant planets[47] and low-mass planets[48] in open clusters are consistent with there being similar planet occurrence rates in open clusters as around field stars.

The open cluster NGC 6811 contains two known planetary systems Kepler-66 and Kepler-67.

17.4.6   Age

- The Ages of Stars, David R. Soderblom, 31 Mar 2010

- Towards asteroseismically calibrated age-rotation-activity relations for Kepler solar-like stars, R.A. Garcia et al. 27 Mar 2014

- Accurate parameters of the oldest known rocky-exoplanet hosting system: Kepler-10 revisited, Alexandra Fogtmann-Schulz et al. 5 Dec 2013

17.4.7   Asteroseismology

- The importance of asteroseismology in exoplanetary science, F Borsa, E Poretti - sait.oat.ts.astro.it

- What asteroseismology can do for exoplanets: Kepler-410A b is a Small Neptune around a bright star, in an eccentric orbit consistent with low obliquity, Vincent Van Eylen et al. 17 Dec 2013

- Pulsations and planets: the asteroseismology-extrasolar-planet connection, Sonja Schuh, 19 May 2010

17.4.8   Stellar activity

- How stellar activity affects the size estimates of extrasolar planets, S. Czesla, K. F. Huber, U. Wolter, S. Schröter, J. H. M. M. Schmitt, 19 Jun 2009

- Hot Jupiters and stellar magnetic activity, A. F. Lanza, 20 May 2008

- Extrasolar Giant Planets and X-ray Activity, Vinay L. Kashyap, Jeremy J. Drake, Steven H. Saar, 21 Jul 2008

- Mass loss of "Hot Jupiters"—Implications for CoRoT discoveries. Part I: The importance of magnetospheric protection of a planet against ion loss caused by coronal mass ejections, Khodachenko et al. April 2007

17.5 Zones

17.5.1 Habitable zone

Main article: Habitable zone

The habitable zone around a star is the region where the temperature is just right to allow liquid water to exist on a planet; that is, not too close to the star for the water to evaporate and not too far away from the star for the water to freeze. The heat produced by stars varies depending on the size and age of the star so that the habitable zone can be at different distances. Also, the atmospheric conditions on the planet influence the planet's ability to retain heat so that the location of the habitable zone is also specific to each type of planet.

Habitable zones have usually been defined in terms of surface temperature, however over half of Earth's biomass is from subsurface microbes,[49] and the temperature increases as you go deeper underground, so the subsurface can be conducive for life when the surface is frozen and if this is considered, the habitable zone extends much further from the star,[50] even rogue planets could have liquid water at sufficient depths underground.[51]

17.5.2 Venus zone

The **Venus zone** is the region around a star where a terrestrial planet would have runaway greenhouse conditions like Venus, but not so near the star that the atmosphere completely evaporates. As with the habitable zone, the location of the Venus zone depends on several factors including the type of star and properties of the planets such as mass, rotation rate and atmospheric clouds. Studies of the Kepler spacecraft data indicate that 32% of red dwarfs have potentially Venus-like planets based on planet size and distance from star, rising to 45% for K-type and G-type stars. Several candidates have been identified but spectroscopic follow-up studies of their atmospheres will be required to see if they really are like Venus.[52][53]

17.6 Galactic distribution of planets

See also: Galactic habitable zone, Extragalactic planet and Planets in globular clusters

The Milky Way is 100,000 light-years across, but 90% of planets with known distances lie within about 2000 light years of Earth, as of July 2014. One method that can detect planets much further away is microlensing. The WFIRST spacecraft could use microlensing to measure the relative frequency of planets in the galactic bulge vs. galactic disk.[54] So far, the indications are that planets are more common in the disk than the bulge.[55] Estimates of the distance of microlensing events is difficult: the first planet considered with high probability of being in the bulge is MOA-2011-BLG-293Lb at a distance of 7.7 kiloparsecs (about 25,000 light years).[56]

Population I, or **metal-rich stars**, are those young stars whose metallicity is highest. The high metallicity of population I stars makes them more likely to possess planetary systems than older populations, because planets form by the accretion of metals. The Sun is an example of a metal-rich star. These are common in the spiral arms of the Milky Way. Generally, the youngest stars, the extreme population I, are found farther in and intermediate population I stars are farther out, etc. The Sun is considered an intermediate population I star. Population I stars have regular elliptical orbits around the Galactic Center, with a low relative velocity.[57]

Population II, or **metal-poor stars**, are those with relatively low metallicity which can have hundreds (e.g. BD +17° 3248) or thousands (e.g. Sneden's Star) times less metallicity than the Sun. These objects formed during an earlier time of the universe. Intermediate population II stars are common in the bulge near the center of the Milky Way, whereas Population II stars found in the galactic halo are older and thus more metal-poor. Globular clusters also contain high numbers of population II stars.[58] In 2014 the first planets around a halo star were announced around Kapteyn's star, the nearest halo star to Earth, around 13 light years away. However later research suggests that Kapteyn b is just an artefact of stellar activity and that Kapteyn c needs more study to be confirmed.[59] The metallicity of Kapteyn's star is estimated to be about $8^{[lower\ alpha\ 5]}$ times less than the Sun.[60]

Different types of galaxies have different histories of star formation and hence planet formation. Planet formation is affected by the ages, metallicities, and orbits of stellar populations within a galaxy. Distribution of stellar populations within a galaxy varies between the different types of galaxies.[61] Stars in elliptical galaxies are much older than stars in spiral galaxies. Most elliptical galaxies contain mainly low-mass stars, with minimal star-formation activity.[62] The distribution of the different types of galaxies in the universe depends on their location within galaxy clusters, with elliptical galaxies found mostly close to their centers.[63]

17.7 See also

- Protoplanetary disc

- List of exoplanets

- List of planetary systems

- List of exoplanetary host stars

- Exomoon

- Exocomet

- Interplanetary dust cloud

- Zodiacal dust

- Exozodiacal dust

- Debris disk

- Asteroid belt

- Kuiper belt

- Oort cloud

17.8 References

[1] For the purpose of this 1 in 5 statistic, "Sun-like" means G-type star. Data for Sun-like stars wasn't available so this statistic is an extrapolation from data about K-type stars

[2] For the purpose of this 1 in 5 statistic, Earth-sized means 1–2 Earth radii

[3] For the purpose of this 1 in 5 statistic, "habitable zone" means the region with 0.25 to 4 times Earth's stellar flux (corresponding to 0.5–2 AU for the Sun).

[4] Converting log scale [m/H] to multiple of solar metallicity: $[(10^{-0.6} \approx 1/4), (10^{0.5} \approx 3)]$

[5] Metallicity of Kapteyn's star estimated at [Fe/H]= −0.89. $10^{-0.89} \approx 1/8$

[1] p. 394, *The Universal Book of Astronomy, from the Andromeda Galaxy to the Zone of Avoidance*, David J. Dsrling, Hoboken, New Jersey: Wiley, 2004. ISBN 0-471-26569-1.

[2] p. 314, *Collins Dictionary of Astronomy*, Valerie Illingworth, London: Collins, 2000. ISBN 0-00-710297-6.

[3] p. 382, *Collins Dictionary of Astronomy*.

[4] p. 420, *A Dictionary of Astronomy*, Ian Ridpath, Oxford, New York: Oxford University Press, 2003. ISBN 0-19-860513-7.

[5] Schneider, Jean (10 September 2011). "Interactive Extra-solar Planets Catalog". *The Extrasolar Planets Encyclopedia*. Retrieved 2012-07-13.

[6] Dreyer (1953), pp.135–48; Linton (2004), pp.38–9). The work of Aristarchus's in which he proposed his heliocentric system has not survived. We only know of it now from a brief passage in Archimedes's *The Sand Reckoner*.

[7] "Cosmos" in *The New Encyclopædia Britannica* (15th edition, Chicago, 1991) **16**:787:2a. "For his advocacy of an infinity of suns and earths, he was burned at the stake in 1600."

[8] Newton, Isaac; Cohen, I. Bernard; Whitman, Anne (1999) [First published 1713]. *The Principia: A New Translation and Guide*. University of California Press. p. 940. ISBN 0-520-20217-1.

[9] Podsiadlowski, Philipp (1993). "Planet formation scenarios". *In: Planets around pulsars; Proceedings of the Conference* **36**: 149. Bibcode:1993ASPC...36..149P.

[10] The fate of fallback matter around newly born compact objects, Rosalba Perna, Paul Duffell, Matteo Cantiello, Andrew MacFadyen, (Submitted on 17 Dec 2013)

[11] Stuart J. Weidenschilling & Francesco Marzari (1996). "Gravitational scattering as a possible origin for giant planets at small stellar distances". *Nature* **384** (6610): 619–621. Bibcode:1996Natur.384..619W. doi:10.1038/384619a0. PMID 8967949.

[12] Types and Attributes at Astro Washington.com.

[13] Dvorak R,Pilat-Lohinger E,Bois E,Schwarz R,Funk B,Beichman C,Danchi W,Eiroa C,Fridlund M,Henning T,Herbst T, Kaltenegger L,Lammer H,Léger A,Liseau R,Lunine J,Paresce F,Penny A,Quirrenbach A,Röttgering H,Selsis F,Schneider J,Stam D, Tinetti G,White G. "Dynamical habitability of planetary systems" Institute for Astronomy, University of Vienna, Vienna , Austria.January 2010

[14] The 3-dimensional architecture of the Upsilon Andromedae planetary system, Russell Deitrick, Rory Barnes, Barbara McArthur, Thomas R. Quinn, Rodrigo Luger, Adrienne Antonsen, G. Fritz Benedict, (Submitted on 4 Nov 2014)

[15] "NASA – Out of Whack Planetary System Offers Clues to a Disturbed Past". Nasa.gov. 2010-05-25. Retrieved 2012-08-17.

[16] Emspak, Jesse. "Kepler Finds Bizarre Systems". *International Business Times*. International Business Times Inc. Retrieved 2 March 2011.

[17] The Occurrence and Architecture of Exoplanetary Systems, Joshua N. Winn (MIT), Daniel C. Fabrycky (U. Chicago), (Submitted on 15 Oct 2014)

[18] Fabrycky, Daniel C. (2010). "Non-Keplerian Dynamics", arXiv:1006.3834 [astro-ph.EP].

[19] Equilibria in the secular, non-coplanar two-planet problem, Cezary Migaszewski, Krzysztof Gozdziewski, 2 Feb 2009

[20] On the origin of planets at very wide orbits from the recapture of free-floating planets, Hagai B. Perets, M. B. N. Kouwenhoven, 2012

[21] Cassan, A.; Kubas, D.; Beaulieu, J. P.; Dominik, M; et al. (2012). "One or more bound planets per Milky Way star from microlensing observations". *Nature* **481** (7380): 167–169. arXiv:1202.0903. Bibcode:2012Natur.481..167C. doi:10.1038/nature 10684.PMID 22237108.

[22] Marcy, G.; et al. (2005). "Observed Properties of Exoplanets: Masses, Orbits and Metallicities". *Progress of Theoretical Physics Supplement* **158**: 24–42. arXiv:astro-ph/0505003. Bibcode:2005PThPS.158...24M. doi:10.1143/PTPS.158.24.

[23] The Frequency of Hot Jupiters Orbiting Nearby Solar-Type Stars, J. T. Wright, G. W. Marcy, A. W. Howard, John Asher Johnson, T. Morton, D. A. Fischer, (Submitted on 10 May 2012)

[24] Andrew Cumming; R. Paul Butler; Geoffrey W. Marcy; et al. (2008). "The Keck Planet Search: Detectability and the Minimum Mass and Orbital Period Distribution of Extrasolar Planets". *Publications of the Astronomical Society of the Pacific* **120** (867): 531–554. arXiv:0803.3357. Bibcode:2008PASP..120..531C. doi:10.1086/588487.

[25] Planet Occurrence within 0.25 AU of Solar-type Stars from Kepler, Andrew W. Howard et al. (Submitted on 13 Mar 2011)

[26] Amos, Jonathan (19 October 2009). "Scientists announce planet bounty". *BBC News*. Retrieved 2010-03-31.

[27] David P. Bennett; Jay Anderson; Ian A. Bond; Andrzej Udalski; et al. (2006). "Identification of the OGLE-2003-BLG-235/MOA-2003-BLG-53 Planetary Host Star". *Astrophysical Journal Letters* **647** (2): L171–L174. arXiv:astro-ph/0606038. Bibcode:2006ApJ...647L.171B. doi:10.1086/507585.

[28] Sanders, R. (4 November 2013). "Astronomers answer key question: How common are habitable planets?".*newscenter.berkeley*

[29] Petigura, E. A.; Howard, A. W.; Marcy, G. W. (2013). "Prevalence of Earth-size planets orbiting Sun-like stars". *Proceedings of the National Academy of Sciences***110**(48): 19273. arXiv:1311.6806. Bibcode:2013PNAS..11019273P.doi:10.1073/pnas.13

[30] Bonfils, X.; et al. (2005). "The HARPS search for southern extra-solar planets: VI. A Neptune-mass planet around the nearby M dwarf Gl 581". *Astronomy & Astrophysics* **443** (3): L15–L18. arXiv:astro-ph/0509211. Bibcode:2005A&A...443L..15B. doi:10.1051/0004-6361:200500193.

[31] L. Vu (3 October 2006). "Planets Prefer Safe Neighborhoods". Spitzer Science Center. Archived from the original on 13 July 2007. Retrieved 2007-09-01.

[32] Limits on Planets Orbiting Massive Stars from Radio Pulsar Timing, Thorsett, S.E. Dewey, R.J. 16-Sep-1993

[33] J. A. Johnson (2011). "The Stars that Host Planets". *Sky & Telescope* (April): 22–27.

[34] A stellar-mass-dependent drop in planet occurrence rates, Gijs D. Mulders, Ilaria Pascucci, Daniel Apai. (Submitted on 28 Jun 2014)

[35] The Survival Rate of Ejected Terrestrial Planets with Moons by J. H. Debes, S. Sigurdsson

[36] Buchhave, L. A.; et al. (2012). "An abundance of small exoplanets around stars with a wide range of metallicities". *Nature*. Bibcode:2012Natur.486..375B. doi:10.1038/nature11121.

[37] Revealing A Universal Planet-Metallicity Correlation For Planets of Different Sizes Around Solar-Type Stars, Ji Wang, Debra A. Fischer. (Submitted on 29 Oct 2013 (v1), last revised 16 Oct 2014 (this version, v3))

[38] Israelian, G.; et al. (2009). "Enhanced lithium depletion in Sun-like stars with orbiting planets". *Nature* **462** (7270): 189–191. arXiv:0911.4198. Bibcode:2009Natur.462..189I. doi:10.1038/nature08483. PMID 19907489.

[39] Stellar Multiplicity, Gaspard Duchêne (1,2), Adam Kraus (3) ((1) UC Berkeley, (2) Institut de Planetologie et d'Astrophysique de Grenoble, (3) Harvard-Smithsonian CfA). (Submitted on 12 Mar 2013)

[40] BINARY CATALOGUE OF EXOPLANETS, Maintained by Richard Schwarz], retrieved 28 Sept 2013

[41] http://www.univie.ac.at/adg/schwarz/multi.html

[42] Welsh, William F.; Doyle, Laurance R. (2013). "Worlds with Two Suns".*Scientific American***309**(5): 40. doi:10.1038/scientifi 40.

[43] One Planet, Two Stars: A System More Common Than Previously Thought, www.universetoday.com, by Shannon Hall on September 4, 2014

[44] Most Sub-Arcsecond Companions of Kepler Exoplanet Candidate Host Stars are Gravitationally Bound, Elliott P. Horch, Steve B. Howell, Mark E. Everett, David R. Ciardi. 3 Sep 2014

[45] Validation of Kepler's Multiple Planet Candidates. II: Refined Statistical Framework and Descriptions of Systems of Special Interest. Jack J. Lissauer, Geoffrey W. Marcy, Stephen T. Bryson, Jason F. Rowe, Daniel Jontof-Hutter, Eric Agol, William J. Borucki, Joshua A. Carter, Eric B. Ford, Ronald L. Gilliland, Rea Kolbl, Kimberly M. Star, Jason H. Steffen, Guillermo Torres. (Submitted on 25 Feb 2014)

[46] Ensemble analysis of open cluster transit surveys: upper limits on the frequency of short-period planets consistent with the field. Jennifer L. van Saders, B. Scott Gaudi. (Submitted on 15 Sep 2010)

[47] Three planetary companions around M67 stars, A. Brucalassi (1,2), L. Pasquini (3), R. Saglia (1,2), M. T. Ruiz (4), P. Bonifacio (5), L. R. Bedin (6), K. Biazzo (7), C. Melo (8), C. Lovis (9), S. Randich (10) ((1) MPI Munich, (2) UOM-LMU Munchen, (3) ESO Garching, (4) Astron. Dpt. Univ. de Chile, (5) GEPI Paris, (6) INAF-OAPD, (7) INAF-OACT, (8) ESO Santiago, (9) Obs. de Geneve, (10) INAF-OAFI) (Submitted on 20 Jan 2014)

[48] The same frequency of planets inside and outside open clusters of stars, Søren Meibom, Guillermo Torres, Francois Fressin, David W. Latham, Jason F. Rowe, David R. Ciardi, Steven T. Bryson, Leslie A. Rogers, Christopher E. Henze, Kenneth Janes, Sydney A. Barnes, Geoffrey W. Marcy, Howard Isaacson, Debra A. Fischer, Steve B. Howell, Elliott P. Horch, Jon M. Jenkins, Simon C. Schuler & Justin Crepp Nature 499, 55–58 (04 July 2013) doi:10.1038/nature12279 Received 06 November 2012 Accepted 02 May 2013 Published online 26 June 2013

[49] Amend, J. P., & Teske, A. (2005). Expanding frontiers in deep subsurface microbiology. Palaeogeography, Palaeoclimatology, Palaeoecology, 219(1–2), 131–155, Elsevier.

[50] Further away planets 'can support life' say researchers, BBC, 7 January 2014 Last updated at 12:40

[51] The Steppenwolf: A proposal for a habitable planet in interstellar space, Dorian S. Abbot, Eric R. Switzer, (Submitted on 5 Feb 2011 (v1), last revised 2 Jun 2011 (this version, v2))

[52] Habitable Zone Gallery - Venus

[53] On the Frequency of Potential Venus Analogs from Kepler Data, Stephen R. Kane, Ravi Kumar Kopparapu, Shawn D. Domagal-Goldman, (Submitted on 9 Sep 2014)

[54] SAG 11: Preparing for the WFIRST Microlensing Survey, Jennifer Yee

[55] Toward a New Era in Planetary Microlensing, Andy Gould, September 21, 2010

[56] MOA-2011-BLG-293Lb: First Microlensing Planet possibly in the Habitable Zone, V. Batista, J.-P. Beaulieu, A. Gould, D.P. Bennett, J.C Yee, A. Fukui, B.S. Gaudi, T. Sumi, A. Udalski, (Submitted on 14 Oct 2013 (v1), last revised 30 Oct 2013 (this version, v3))

[57] Charles H. Lineweaver (2000). "An Estimate of the Age Distribution of Terrestrial Planets in the Universe: Quantifying Metallicity as a Selection Effect".*Icarus***151**(2): 307–313. arXiv:astro-ph/0012399. Bibcode:2001Icar..151..307L.doi:10.1006/ica

[58] T. S. van Albada; Norman Baker (1973). "On the Two Oosterhoff Groups of Globular Clusters". *Astrophysical Journal* **185**: 477–498. Bibcode:1973ApJ...185..477V. doi:10.1086/152434.

[59] Stellar activity mimics a habitable-zone planet around Kapteyn's star, Paul Robertson (1 and 2), Arpita Roy (1 and 2 and 3), Suvrath Mahadevan (1 and 2 and 3) ((1) Dept. of Astronomy and Astrophysics, Penn State University, (2) Center for Exoplanets & Habitable Worlds, Penn State University, (3) The Penn State Astrobiology Research Center), (Submitted on 11 May 2015 (v1), last revised 1 Jun 2015 (this version, v2))

[60] Two planets around Kapteyn's star : a cold and a temperate super-Earth orbiting the nearest halo red-dwarf, Guillem Anglada-Escudé, Pamela Arriagada, Mikko Tuomi, Mathias Zechmeister, James S. Jenkins, Aviv Ofir, Stefan Dreizler, Enrico Gerlach, Chris J. Marvin, Ansgar Reiners, Sandra V. Jeffers, R. Paul Butler, Steven S. Vogt, Pedro J. Amado, Cristina Rodríguez-López, Zaira M. Berdiñas, Julian Morin, Jeff D. Crane, Stephen A. Shectman, Ian B. Thompson, Mateo Diaz, Eugenio Rivera, Luis F. Sarmiento, Hugh R.A. Jones, (Submitted on 3 Jun 2014)

[61] Habitable Zones in the Universe, G. Gonzalez, (Submitted on 14 Mar 2005 (v1), last revised 21 Mar 2005 (this version, v2))

[62] John, D. (2006), *Astronomy*, ISBN 1-4054-6314-7, p. 224-225

[63] Dressler, A. (March 1980). "Galaxy morphology in rich clusters - Implications for the formation and evolution of galaxies.". *The Astrophysical Journal* **236**: 351–365. Bibcode:1980ApJ...236..351D. doi:10.1086/157753.

17.9 Further reading

- On the Relationship Between Debris Disks and Planets, Ágnes Kóspál, David R. Ardila, Attila Moór, Péter Ábrahám, 30 Jun 2009

- Signatures of exosolar planets in dust debris disks, Leonid M. Ozernoy, Nick N. Gorkavyi, John C. Mather, Tanya Taidakova, 4 Jul 2000

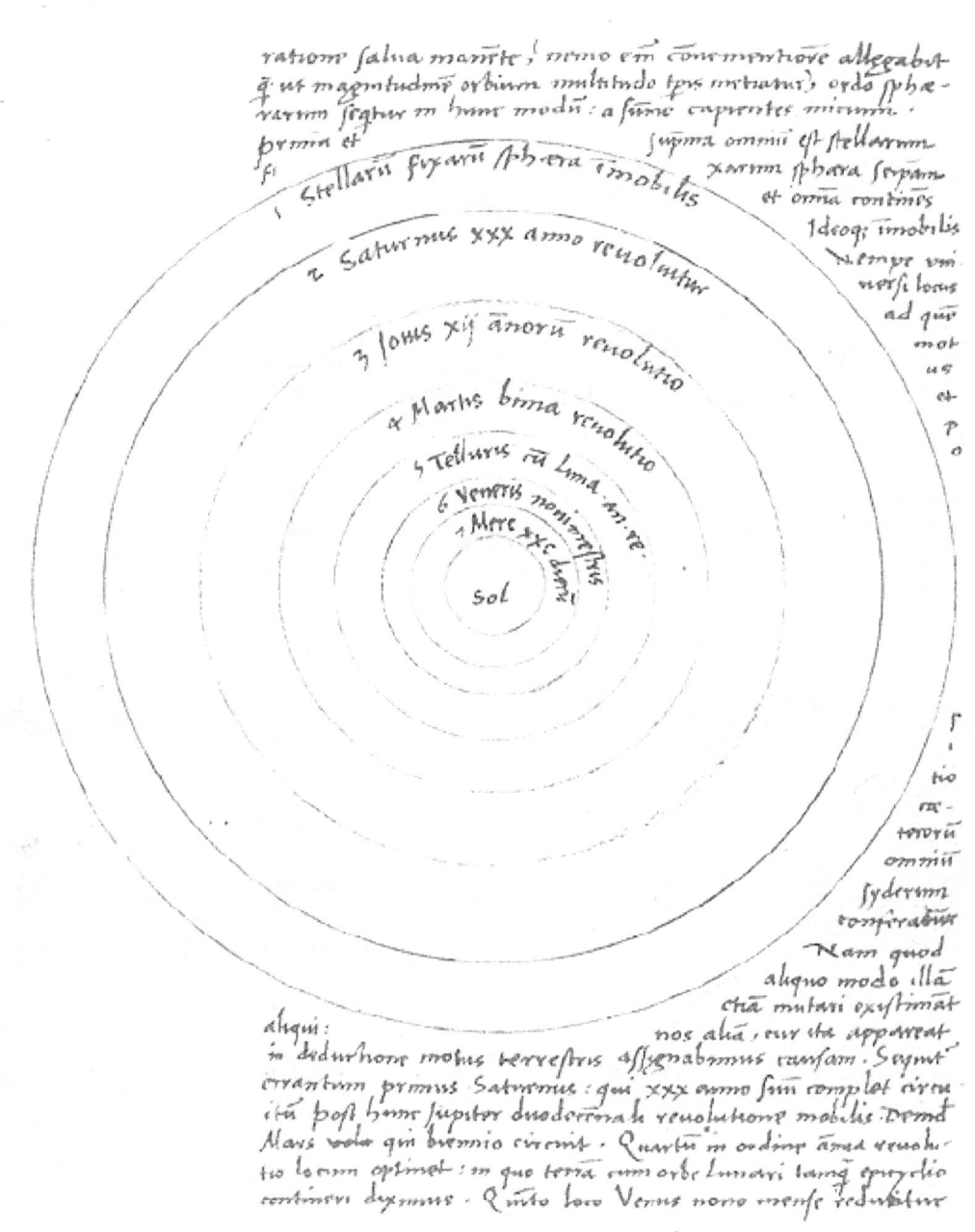

Heliocentric model of the Solar System in Copernicus' manuscript

An artist's concept of a protoplanetary disk

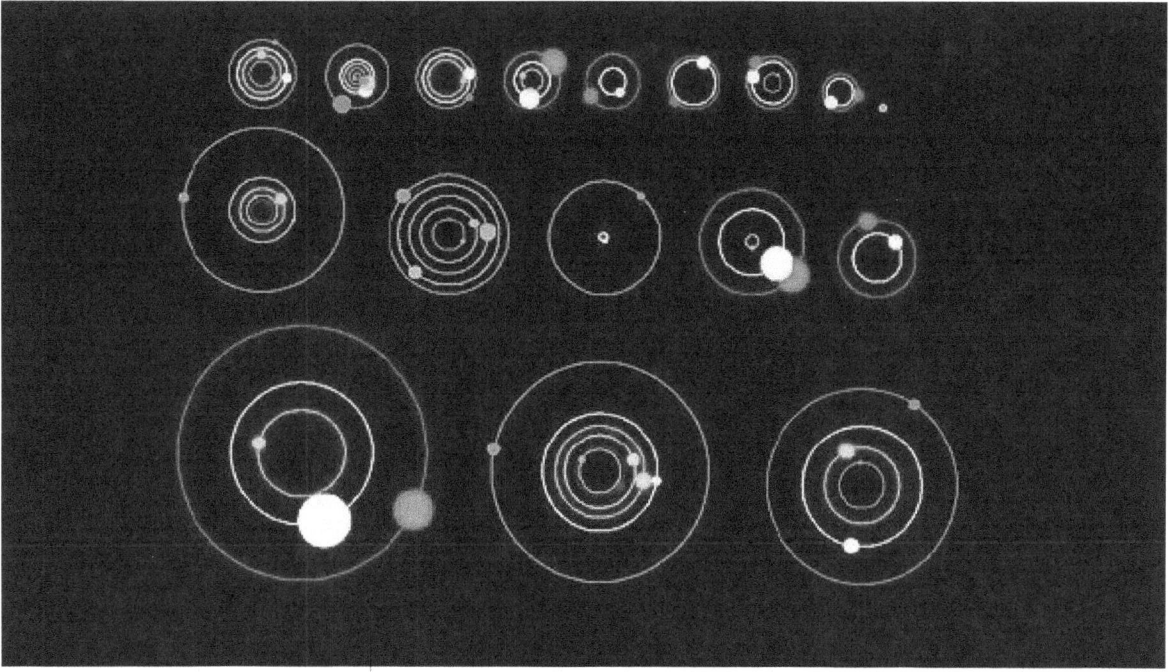

The spacings between orbits vary widely amongst the different systems discovered by the Kepler spacecraft.

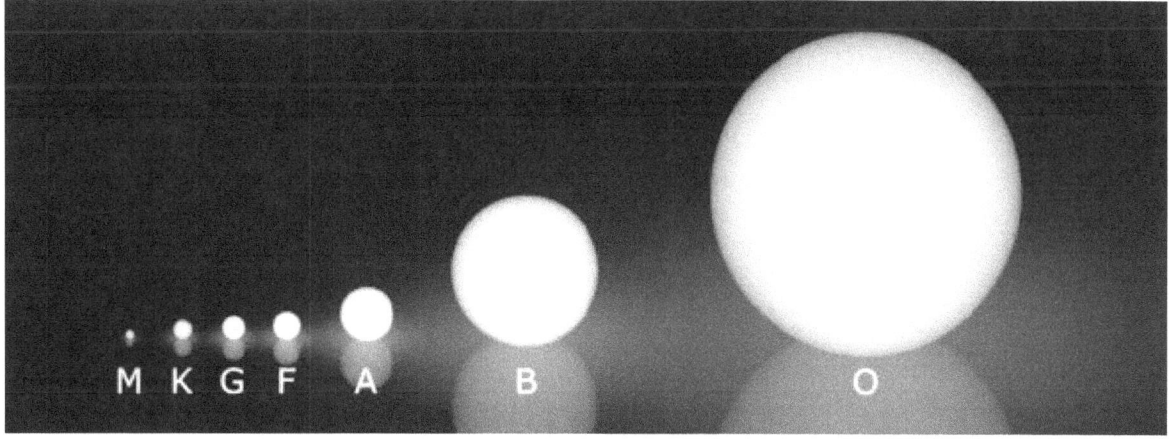

The Morgan-Keenan spectral classification

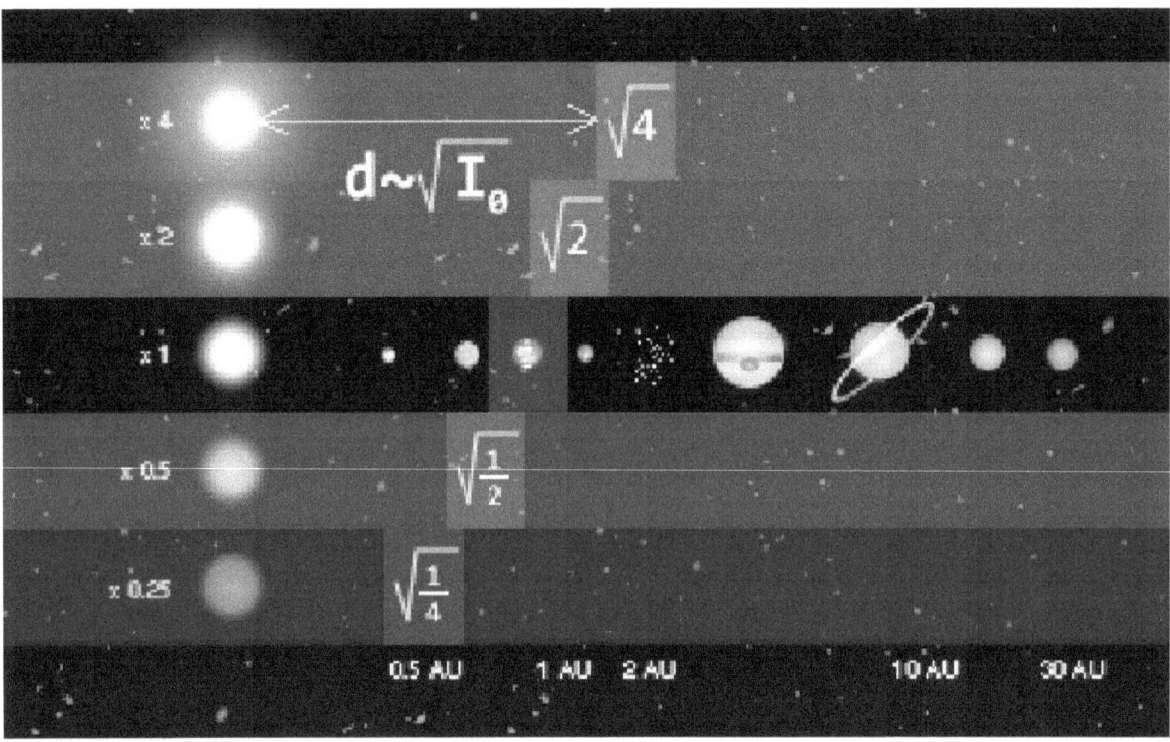

Location of habitable zone around different types of stars

90% of planets with known distances lie within about 2000 light years of Earth, as of July 2014.

Chapter 18

Extragalactic planet

Artist's impression of HIP 13044 b, the first candidate planet of extragalactic origin. This planet was detected using the MPG/ESO 2.2-metre telescope at ESO's La Silla Observatory.[1]

An **extragalactic planet**, also known as an **extragalactic exoplanet**, is a planet that is outside the Milky Way.[2] Of all the exoplanets discovered to date, only one is of extragalactic origin. Due to the huge distance of those worlds, they are very hard to detect.

18.1   HIP 13044 b

A planet with a mass of at least 1.25 times that of Jupiter has been discovered by the European Southern Observatory (ESO) orbiting a star of extragalactic origin, even though the star now finds itself within our own galaxy. The Jupiter-like planet is particularly unusual, as it is orbiting a star nearing the end of its life and could be about to be engulfed by it, giving tantalizing clues about the fate of our own planetary system in the distant future. The star is called HIP 13044 and it lies about 2000 light-years from Earth in the southern constellation of Fornax.[1]

18.2   Twin Quasar-related planet

A microlensing event in the Twin Quasar gravitational lensing system was observed in 1996, by R. E. Schild, in the "A" lobe of the lensed quasar. It is predicted that a 3-Earth-mass planet in the lensing galaxy, YGKOW G1, caused the event. This was the first extragalactic planet candidate announced. This, however, is not a repeatable observation, as it was a one-time chance alignment. This predicted planet lies 4 billion light years away.[3][4]

18.3   Andromeda Galaxy planets

A team of scientists has used gravitational microlensing to come up with a tentative detection of an extragalactic exoplanet in Andromeda, our nearest large galactic neighbor. The lensing pattern fits a star with a smaller companion weighing just 6 or 7 times the mass of Jupiter. This suspected planet is the first announced in the Andromeda Galaxy.[5][6]

18.4   See also

- Exoplanet
- Planet
- Solar System planets

18.5   References

[1] "Planet from another galaxy discovered". *ESO Press Release*. 18 November 2010. Retrieved 17 November 2011.

[2] Extrasolar Visions, "Extragalactic Worlds" (accessed 1 September 2009)

[3] New Scientist (issue 2037), Do alien worlds throng faraway galaxy? *Govert Schilling* 06 July 1996

[4] Extrasolar Visions, "The Q0957+561 Planet" (accessed 1 September 2009)

[5] Thaindian News, First extragalactic exoplanet may have been found by gravitational microlensing, 11 June 2009

[6] New Scientist, First extragalactic exoplanet may have been found, 10 June 2009

Chapter 19

Drake equation

This article is about Frank Drake's equation. For other uses, see Drake equation (disambiguation).

The **Drake equation** is a probabilistic argument used to arrive at an estimate of the number of active, communicative extraterrestrial civilizations in the Milky Way galaxy.[1][2] The number of such civilizations, N, is assumed to be equal to the mathematical product of (i) the average rate of star formation, $R*$, in our galaxy, (ii) the fraction of formed stars, fp, that have planets, (iii) the average number of planets per star that has planets, ne, that can potentially support life, (iv) the fraction of those planets, fl, that actually develop life, (v) the fraction of planets bearing life on which intelligent, civilized life, fi, has developed, (vi) the fraction of these civilizations that have developed communications, fc, i.e., technologies that release detectable signs into space, and (vii) the length of time, L, over which such civilizations release detectable signals, for a combined expression of:

$$N = R_* \cdot f_\mathrm{p} \cdot n_e \cdot f_l \cdot f_i \cdot f_c \cdot L$$

The equation was written in 1961 by Frank Drake, not for purposes of quantifying the number of civilizations,[3] but as a way to stimulate scientific dialogue at a meeting on the search for extraterrestrial intelligence (SETI). The equation summarizes the main concepts which scientists must contemplate when considering the question of other radio-communicative life.[3] Criticism of the Drake equation follows from the fact that several of its terms are conjectural, the net result being that the error associated with any derived value is very large such that the equation cannot be used to draw firm conclusions.

19.1 History

In September 1959, physicists Giuseppe Cocconi and Philip Morrison published an article in the journal *Nature* with the provocative title "Searching for Interstellar Communications."[4][5] Cocconi and Morrison argued that radio telescopes had become sensitive enough to pick up transmissions that might be broadcast into space by civilizations orbiting other stars. Such messages, they suggested, might be transmitted at a wavelength of 21 centimeters (1,420.4 megahertz). This is the wavelength of radio emission by neutral hydrogen, the most common element in the universe, and they reasoned that other intelligences might see this as a logical landmark in the radio spectrum.

Two months later, Harvard University astronomy professor Harlow Shapley speculated on the number of inhabited planets in the universe, saying "The universe has 10 million, million, million suns (10 followed by 18 zeros) similar to our own. One in a million has planets around it. Only one in a million million has the right combination of chemicals, temperature, water, days and nights to support planetary life as we know it. This calculation arrives at the estimated figure of 100 million worlds where life has been forged by evolution."[6]

Seven months after Cocconi and Morrison published their article, Drake made the first systematic search for signals from extraterrestrial intelligent beings. Using the 25 meter dish of the National Radio Astronomy Observatory in Green Bank, West Virginia, Drake monitored two nearby Sun-like stars: Epsilon Eridani and Tau Ceti. In this project, which he called

Dr. Frank Drake

Project Ozma. he slowly scanned frequencies close to the 21 cm wavelength for six hours a day from April to July 1960.[5] The project was well designed. inexpensive. and simple by today's standards. It was also unsuccessful.

Soon thereafter. Drake hosted a "search for extraterrestrial intelligence" meeting on detecting their radio signals. The meeting was held at the Green Bank facility in 1961. The equation that bears Drake's name arose out of his preparations for the meeting.[7]

> As I planned the meeting. I realized a few day[s] ahead of time we needed an agenda. And so I wrote down all the things you needed to know to predict how hard it's going to be to detect extraterrestrial life. And looking at them it became pretty evident that if you multiplied all these together. you got a number, N,

which is the number of detectable civilizations in our galaxy. This was aimed at the radio search, and not to search for primordial or primitive life forms. —Frank Drake.

The ten attendees were conference organizer J. Peter Pearman, Frank Drake, Philip Morrison, businessman and radio amateur Dana Atchley, chemist Melvin Calvin, astronomer Su-Shu Huang, neuroscientist John C. Lilly, inventor Barney Oliver, astronomer Carl Sagan and radio-astronomer Otto Struve.[8] These participants dubbed themselves "The Order of the Dolphin" (because of Lilly's work on dolphin communication), and commemorated their first meeting with a plaque at the observatory hall.[9][10]

19.2 Equation

The Drake equation is:

$$N = R_* \cdot f_p \cdot n_e \cdot f_l \cdot f_i \cdot f_c \cdot L$$

where:

N = the number of civilizations in our galaxy with which communication might be possible (i.e. which are on our current past light cone);

and

R^* = the average rate of star formation in our galaxy

f_p = the fraction of those stars that have planets

n_e = the average number of planets that can potentially support life per star that has planets

f_l = the fraction of planets that could support life that actually develop life at some point

f_i = the fraction of planets with life that actually go on to develop intelligent life (civilizations)

f_c = the fraction of civilizations that develop a technology that releases detectable signs of their existence into space

L = the length of time for which such civilizations release detectable signals into space[11][12]

19.3 Usefulness

The Drake equation amounts to a summary of the factors affecting the likelihood that we might detect radio-communication from intelligent extraterrestrial life.[11][11][13] The last four parameters, f_l, f_i, f_c, and L, are not known and are very hard to estimate, with values ranging over many orders of magnitude (see criticism). Therefore, the usefulness of the Drake equation is not in the solving, but rather in the contemplation of all the various concepts which scientists must incorporate when considering the question of life elsewhere,[11][13] and gives the question of life elsewhere a basis for scientific analysis. The Drake equation is a statement that stimulates intellectual curiosity about the universe around us, for helping us to understand that life as we know it is the end product of a natural, cosmic evolution, and for helping us realize how much we are a part of that universe.[12] What the equation and the search for life has done is focus science on some of the other questions about life in the universe, specifically abiogenesis, the development of multi-cellular life and the development of intelligence itself.[14]

Within the limits of our existing technology, any practical search for distant intelligent life must necessarily be a search for some manifestation of a distant technology. After about 50 years, the Drake equation is still of seminal importance because it is a 'road map' of what we need to learn in order to solve this fundamental existential question.[1] It also formed the backbone of astrobiology as a science; although speculation is entertained to give context, astrobiology concerns

The Allen Telescope Array for SETI

itself primarily with hypotheses that fit firmly into existing scientific theories. Some 50 years of SETI have failed to find anything, even though radio telescopes, receiver techniques, and computational abilities have improved enormously since the early 1960s, but it has been discovered, at least, that our galaxy is not teeming with very powerful alien transmitters continuously broadcasting near the 21 cm hydrogen frequency. No one could say this in 1961.[15]

19.4 Modifications

As many observers have pointed out, the Drake equation is a very simple model that does not include potentially relevant parameters,[16] and many changes and modifications to the equation have been proposed. One line of modification, for example, attempts to account for the uncertainty inherent in many of the terms.[17]

Others note that the Drake equation ignores many concepts that might be relevant to the odds of contacting other civilizations. For example, David Brin states: "The Drake equation merely speaks of the number of sites at which ETIs spontaneously arise. The equation says nothing directly about the contact cross-section between an ETIS and contemporary human society".[18] Because it is the contact cross-section that is of interest to the SETI community, many additional factors and modifications of the Drake equation have been proposed.

Colonization

It has been proposed to generalize the Drake equation to include additional effects of alien civilizations colonizing other star systems. Each original site expands with an expansion velocity v, and establishes additional sites that survive for a lifetime L. The result is a more complex set of 3 equations.[18]

Reappearance factor

The Drake equation may furthermore be multiplied by *how many times* an intelligent civilization may occur on planets where it has happened once. Even if an intelligent civilization reaches the end of its lifetime after, for example, 10,000 years, life may still prevail on the planet for billions of years, permitting the next civilization to evolve. Thus, several civilizations may come and go during the lifespan of one and the same planet. Thus, if nr is the average number of times a new civilization *reappears* on the same planet where a previous civilization once has appeared and ended, then the total number of civilizations on such a planet would be $(1+nr)$, which is the actual *reappearance factor* added to the equation.

The factor depends on what generally is the cause of civilization extinction. If it is generally by temporary uninhabitability, for example a nuclear winter, then nr may be relatively high. On the other hand, if it is generally by permanent uninhabitability, such as stellar evolution, then nr may be almost zero. In the case of total life extinction, a similar factor may be applicable for ff, that is, *how many times* life may appear on a planet where it has appeared once.

METI factor

Alexander Zaitsev said that to be in a communicative phase and emit dedicated messages are not the same. For example, humans, although being in a communicative phase, are not a communicative civilization; we do not practise such activities as the purposeful and regular transmission of interstellar messages. For this reason, he suggested introducing the METI factor (Messaging to Extra-Terrestrial Intelligence) to the classical Drake equation.[19] He defined the factor as "the fraction of communicative civilizations with clear and non-paranoid planetary consciousness", or alternatively expressed, the fraction of communicative civilizations that actually engage in deliberate interstellar transmission.

The METI factor is somewhat misleading since active, purposeful transmission of messages by a civilization is not required for them to receive a broadcast sent by another that is seeking first contact. It is merely required they have capable and compatible receiver systems operational; however, this is a variable humans cannot accurately estimate.

Biogenic gases

Astronomer Sara Seager proposed a revised equation that focuses on the search for planets with biosignature gases. These gases are produced by living organisms that can accumulate in a planet atmosphere to levels that can be detected with remote space telescopes.[20]

The Seager equation looks like this:[20][lower alpha 1] $N = N_* * F_Q * F_{HZ} * F_O * F_L * F_S$ Where:

N = the number of planets with detectable signs of life

N^* = the number of stars observed

FQ = the fraction of stars that are quiet

FHZ = the fraction of stars with rocky planets in the habitable zone

FO = the fraction of those planets that can be observed

FL = the fraction that have life

FS = the fraction on which life produces a detectable signature gas

Seager stresses, "We're not throwing out the Drake Equation, which is really a different topic," explaining, "Since Drake came up with the equation, we have discovered thousands of exoplanets. We as a community have had our views revolutionized as to what could possibly be out there. And now we have a real question on our hands, one that's not related to intelligent life: Can we detect any signs of life in any way in the very near future?"[21]

19.5 Estimates

19.5.1 Original estimates

There is considerable disagreement on the values of these parameters, but the 'educated guesses' used by Drake and his colleagues in 1961 were:[22][23]

- R^* = 1/year (1 star formed per year, on the average over the life of the galaxy; this was regarded as conservative)

- f_p = 0.2-0.5 (one fifth to one half of all stars formed will have planets)

- n_e = 1-5 (stars with planets will have between 1 and 5 planets capable of developing life)

- f_l = 1 (100% of these planets will develop life)

- f_i = 1 (100% of which will develop intelligent life)

- f_c = 0.1-0.2 (10-20% of which will be able to communicate)

- L = 1000-100,000,000 years (which will last somewhere between 1000 and 100,000,000 years)

Inserting the above minimum numbers into the equation gives a minimum N of 20. Inserting the maximum numbers gives a maximum of 50,000,000. Drake states that given the uncertainties, the original meeting concluded that $N \approx L$, and there were probably between 1000 and 100,000,000 civilizations in the Milky Way galaxy.

19.5.2 Current estimates

This section discusses and attempts to list the best current estimates for the parameters of the Drake equation.

Rate of star creation in our galaxy, R^*

Latest calculations from NASA and the European Space Agency indicate that the current rate of star formation in our galaxy is about 7 per year.[24]

Fraction of those stars that have planets, *fp*

Recent analysis of Microlensing surveys has found that f_p may approach 1 -- that is, stars are orbited by planets as a rule, rather than the exception; and that there are one or more bound planets per Milky Way star.[25][26]

Average number of planets per star having planets that might support life, *ne*

Here it is understood that satellites might also serve as good candidates.

In November 2013, astronomers reported, based on *Kepler* space mission data, that there could be as many as 40 billion Earth-sized planets orbiting in the habitable zones of sun-like stars and red dwarf stars within the Milky Way Galaxy.[27][28] 11 billion of these estimated planets may be orbiting sun-like stars.[29] Since there are about 100 billion stars in the galaxy, this implies f_p*n_e is roughly 0.4. The nearest planet in the habitable zone may be as little as 12 light-years away, according to the scientists.[27][28]

Even if planets are in the habitable zone, however, the number of planets with the right proportion of elements is difficult to estimate.[30] Brad Gibson, Yeshe Fenner, and Charley Lineweaver determined that about 10% of star systems in the Milky Way galaxy are hospitable to life, by having heavy elements, being far from supernovae and being stable for a sufficient time.[31]

The discovery of numerous gas giants in close orbit with their stars has introduced doubt that life-supporting planets commonly survive the formation of their stellar systems. So-called hot Jupiters may migrate from distant orbits to near orbits, in the process disrupting the orbits of habitable planets.

In addition, most stars in our galaxy are red dwarfs, which flare violently, mostly in X-rays, a property not conducive to life as we know it. Simulations also suggest that these bursts erode planetary atmosphere.[32]

On the other hand, the variety of star systems that might have habitable zones is not just limited to solar-type stars and Earth-sized planets; it is now estimated that even tidally locked planets close to red dwarfs might have habitable zones.[33] The possibility of life on moons of gas giants (such as Jupiter's moon Europa, or Saturn's moon Titan) adds further uncertainty to this figure.

The authors of the rare Earth hypothesis propose a number of additional constraints on habitability for planets, including being in galactic zones with suitably low radiation, high star metallicity, and low enough density to avoid excessive asteroid bombardment. They also propose that it is necessary to have a planetary system with large gas giants which provide bombardment protection without a hot Jupiter; and a planet with plate tectonics, a large moon that creates tidal pools, and moderate axial tilt to generate seasonal variation.[34]

Fraction of the above that actually go on to develop life, *fl*

Geological evidence from the Earth suggests that f_l may be high; life on Earth appears to have begun around the same time as favorable conditions arose, suggesting that abiogenesis may be relatively common once conditions are right. However, this evidence only looks at the Earth (a single model planet), and contains anthropic bias, as the planet of study was not chosen randomly, but by the living organisms that already inhabit it (ourselves). From a classical hypothesis testing standpoint, there are zero degrees of freedom, permitting no valid estimates to be made. If life were to be found on Mars that developed independently from life on Earth it would imply a value for f_l close to one. While this would raise the degrees of freedom from zero to one, there would remain a great deal of uncertainty on any estimate due to the small sample size, and the chance they are not really independent.

Countering this argument is that there is no evidence for abiogenesis occurring more than once on the Earth — that is, all terrestrial life stems from a common origin. If abiogenesis were more common it would be speculated to have occurred more than once on the Earth. Scientists have searched for this by looking for bacteria that are unrelated to other life on Earth, but none have been found yet.[35] It is also possible that life arose more than once, but that other branches were out-competed, or died in mass extinctions, or were lost in other ways. Biochemists Francis Crick and Leslie Orgel laid special emphasis on this uncertainty: "At the moment we have no means at all of knowing" whether we are "likely to be alone in the galaxy (Universe)" or whether "the galaxy may be pullulating with life of many different forms."[36] As an alternative to abiogenesis on Earth, they proposed the hypothesis of directed panspermia, which states that Earth life began with "microorganisms sent here deliberately by a technological society on another planet, by means of a special long-range unmanned spaceship" (Crick and Orgel, *op.cit.*).

Fraction of the above that develops intelligent life, *fi*

This value remains particularly controversial. Those who favor a low value, such as the biologist Ernst Mayr, point out that of the billions of species that have existed on Earth, only one has become intelligent and from this, infer a tiny value for f_i.[37] Likewise, the Rare Earth hypothesis, notwithstanding their low value for *ne* above, also think a low value for *fi* dominates the analysis.[38] Those who favor higher values note the generally increasing complexity of life over time, concluding that the appearance of intelligence is almost inevitable,[39][40] implying an f_i approaching 1. Skeptics point out that the large spread of values in this factor and others make all estimates unreliable. (See Criticism).

In addition, while it appears that life developed soon after the formation of Earth, the Cambrian explosion, in which a large variety of multicellular life forms came into being, occurred a considerable amount of time after the formation of Earth, which suggests the possibility that special conditions were necessary. Some scenarios such as the Snowball Earth or research into the extinction events have raised the possibility that life on Earth is relatively fragile. Research on any past life on Mars is relevant since a discovery that life did form on Mars but ceased to exist might raise our estimate of f_i but would indicate that in half the known cases, intelligent life did not develop.

This model also has a large anthropic bias and there are still zero degrees of freedom. Note that the capacity and willingness to participate in extraterrestrial communication has come relatively recently, with the Earth having only an estimated 100,000 year history of intelligent human life, and less than a century of technological ability.

Estimates of f_i have been affected by discoveries that the Solar System's orbit is circular in the galaxy, at such a distance that it remains out of the spiral arms for tens of millions of years (evading radiation from novae). Also, Earth's large moon may aid the evolution of life by stabilizing the planet's axis of rotation.

Fraction of the above revealing their existence via signal release into space, *fc*

For deliberate communication, the one example we have (the Earth) does not do much explicit communication, though there are some efforts covering only a tiny fraction of the stars that might look for our presence. (See Arecibo message, for example). There is considerable speculation why an extraterrestrial civilization might exist but choose not to communicate. However, deliberate communication is not required, and calculations indicate that current or near-future Earth-level technology might well be detectable to civilizations not too much more advanced than our own.[41] By this standard, the Earth is a communicating civilization.

Another question is what percentage of civilizations in the galaxy are close enough for us to detect, assuming that they send out signals.

Lifetime of such a civilization wherein it communicates its signals into space, *L*

Michael Shermer estimated *L* as 420 years, based on the duration of sixty historical Earthly civilizations.[42] Using 28 civilizations more recent than the Roman Empire, he calculates a figure of 304 years for "modern" civilizations. It could also be argued from Michael Shermer's results that the fall of most of these civilizations was followed by later civilizations that carried on the technologies, so it is doubtful that they are separate civilizations in the context of the Drake equation. In the expanded version, including *reappearance number*, this lack of specificity in defining single civilizations does not matter for the end result, since such a civilization turnover could be described as an increase in the *reappearance number* rather than increase in *L*, stating that a civilization reappears in the form of the succeeding cultures. Furthermore, since none could communicate over interstellar space, the method of comparing with historical civilizations could be regarded as invalid.

David Grinspoon has argued that once a civilization has developed enough, it might overcome all threats to its survival. It will then last for an indefinite period of time, making the value for *L* potentially billions of years. If this is the case, then he proposes that the Milky Way galaxy may have been steadily accumulating advanced civilizations since it formed.[43] He proposes that the last factor *L* be replaced with $fIC*T$, where fIC is the fraction of communicating civilizations become "immortal" (in the sense that they simply do not die out), and *T* representing the length of time during which this process has been going on. This has the advantage that *T* would be a relatively easy to discover number, as it would simply be some fraction of the age of the universe.

It has also been hypothesized that once a civilization has learned of a more advanced one, its longevity could increase because it can learn from the experiences of the other.[44]

The astronomer Carl Sagan speculated that all of the terms, except for the lifetime of a civilization, are relatively high and the determining factor in whether there are large or small numbers of civilizations in the universe is the civilization lifetime, or in other words, the ability of technological civilizations to avoid self-destruction. In Sagan's case, the Drake equation was a strong motivating factor for his interest in environmental issues and his efforts to warn against the dangers of nuclear warfare.

Inserting these current estimates into the original equation, using a value of 0.1 wherever the text says someone has proposed an unspecified "low value," results in the range of N being from a low of 2 to a high of 280,000,000. As study of the concepts has gone on, the range has increased at both the minimum and maximum ends.

19.5.3   Range of results

As many skeptics have pointed out, the Drake equation can give a very wide range of values, depending on the assumptions. In particular,the result can be N<<1, meaning we are likely alone in the galaxy, or N>>1, implying there are many civilizations we might contact. One of the few points of wide agreement is that the presence of humanity implies a probability of intelligence arising of greater than zero.[46]Beyond this, however, the values one may attribute to each factor in this equation tell more about a person's beliefs than about scientific facts.[47]

As an example of a low estimate, combining NASA's star formation rates, the rare Earth hypothesis value of $f_p*n_e*f_l = 10^{-5}$,[48] Mayr's view on intelligence arising, Drake's view of communication, and Shermer's's estimate of lifetime:

$$R* = 7/\text{year},^{[24]} f_p*n_e*f_l = 10^{-5},^{[34]} f_i = 10^{-9},^{[37]} f_c = 0.2^{[\text{Drake, above}]}, \text{ and } L = 304 \text{ years}^{[42]}$$

gives:

$$N = 7 \times 10^{-5} \times 10^{-9} \times 0.2 \times 304 = 4 \times 10^{-12}$$

i.e., suggesting that we are probably alone in this galaxy, and possibly the observable universe.

On the other hand, with larger values for each of the parameters above, values of N can be derived that are greater than 1. The following higher values that have been proposed for each of the parameters:

$$R* = 7/\text{year},^{[24]} f_p = 1,^{[25]} n_e = 0.2,^{[49][50]} f_l = 0.13,^{[51]} f_i = 1,^{[39]} f_c = 0.2^{[\text{Drake, above}]}, \text{ and } L = 10^9 \text{ years}^{[43]}$$

Use of these parameters gives:

$$N = 7 \times 1 \times 0.2 \times 0.13 \times 1 \times 0.2 \times 10^9 = 36.4 \text{ million.}$$

Monte Carlo simulations of estimates of the Drake equation factors based on a stellar and planetary model of the Milky Way have resulted in the number of civilizations varying by a factor of 100.[52]

19.6   Criticism

Criticism of the Drake equation follows mostly from the observation that several terms in the equation are largely or entirely based on conjecture. Star formation rates are well-known, and the incidence of planets has a sound theoretical and observational basis, but the other terms in the equation become very speculative. The uncertainties revolve around our understanding of the evolution of life, intelligence, and civilization, not physics. No statistical estimates are possible for some of the parameters, where only one example is known. The net result is that the equation cannot be used to draw firm conclusions of any kind, and the resulting margin of error is huge, far beyond what some consider acceptable or meaningful.[53]

One reply to such criticisms[54] is that even though the Drake equation currently involves speculation about unmeasured parameters, it was intended as a way to stimulate dialogue on these topics. Then the focus becomes how to proceed experimentally. Indeed, Drake originally formulated the equation merely as an agenda for discussion at the Green Bank conference.[55]

19.6.1   Fermi paradox

Main article: Fermi paradox

The pessimists' most telling argument in the SETI debate stems not from theory or conjecture but from an actual observation: the presumed lack of extraterrestrial contact.[5] A civilization lasting for tens of millions of years would have plenty of time to travel anywhere in the galaxy, even at the slow speeds foreseeable with our own kind of technology. Furthermore, no confirmed signs of intelligence elsewhere have been spotted, either in our galaxy or the more than 80 billion other galaxies of the observable universe. According to this line of thinking, the tendency to fill up all available territory seems to be a universal trait of living things, so the Earth should have already been colonized, or at least visited, but no evidence of this exists. Hence Fermi's question "Where is everybody?".[56][57]

A large number of explanations have been proposed to explain this lack of contact; a recent book elaborated on 50 different explanations.[58] In terms of the Drake Equation, the explanations can be divided into three classes:

- Few intelligent civilizations ever arise. This is an argument that at least one of the first few terms, $R^* \cdot f_p \cdot n_e \cdot f_l \cdot f_i$, has a low value. The most common suspect is f_i, but explanations such as the rare Earth Hypothesis argue that n_e is the small term.

- Intelligent civilizations exist, but we see no evidence, meaning f_c is small. Typical arguments include that civilizations are too far apart, it is too expensive to spread throughout the galaxy, civilizations broadcast signals for only a brief period of time, it is dangerous to communicate, and many others.

- The lifetime of intelligent civilizations is short, meaning the value of L is small. Drake suggested that a large number of extraterrestrial civilizations would form, and he further speculated that the lack of evidence of such civilizations may be because technological civilizations tend to disappear rather quickly. Typical explanations include it is the nature of intelligent life to destroy itself, it is the nature of intelligent life to destroy others, they tend to experience a technological singularity, and others.

These lines of reasoning lead to the Great Filter hypothesis,[59] which states that since there are no observed extraterrestrial civilizations, despite the vast number of stars, then some step in the process must be acting as a filter to reduce the final value. According to this view, either it is very hard for intelligent life to arise, or the lifetime of such civilizations, or the period of time they reveal their existence, must be relatively short.

19.7   In fiction and popular culture

- Frederik Pohl's Hugo award-winning "Fermi and Frost", cites a paradox as evidence for the short lifetime of technical civilizations—that is, the possibility that once a civilization develops the power to destroy itself (perhaps by nuclear warfare), it does.

- Optimistic results of the equation along with unobserved extraterrestrials also serves as backdrop for humorous suggestions such as Terry Bisson's classic short story "They're Made Out of Meat," that there are many extraterrestrial civilizations but that they are deliberately ignoring humanity.[60]

- The equation was cited by Gene Roddenberry as supporting the multiplicity of inhabited planets shown in *Star Trek*, the television show he created. However, Roddenberry did not have the equation with him, and he was forced to "invent" it for his original proposal.[61] The invented equation created by Roddenberry is:

$$Ff^2(MgE) - C^1Ri^1 \cdot M = L/So$$

Drake has gently pointed out, however, that a number raised to the first power is merely the number itself. A poster with both versions of the equation was seen in the *Star Trek: Voyager* episode "Future's End."

- The equation is also cited in Michael Crichton's *Sphere*.[62]

- George Alec Effinger's short story "One" uses an expedition confident in the Drake equation as a backdrop to explore the psychological implications of a lone humanity.

- Alastair Reynolds' *Revelation Space* trilogy and short stories focus very much on the Drake equation and the Fermi paradox, using genocidal self-replicating machines as a great filter.

- Stephen Baxter's *Manifold Trilogy* explores the Drake equation and the Fermi paradox in three distinct perspectives.

- Ian R. MacLeod's 2001 novel "New Light On The Drake Equation" concerns a man who is obsessed by the Drake equation.

- The Ultimate Marvel comic book mini-series *Ultimate Secret* has Reed Richards examining the Drake equation and considering the Fermi paradox. He believes that advanced civilizations destroy themselves. In the story, it turns out that they are also destroyed by Gah Lak Tus.

- Eleanor Ann Arroway paraphrases the Drake equation several times in the film *Contact*, using the magnitude of N and its implications on the output value to justify the SETI program.

- The band Carbon Based Lifeforms mention the Drake equation in their song "Abiogenesis" in their 2006 album *World of Sleepers*.[63]

- Nick Warren's 2005 DJ mix compilation Global Underground 028: Shanghai features a track called "The Drake Equation" by Norwegian production duo Stian Klo & Thomas Nøkling under their "Seyton" alias.[64]

- The July 2013 issue of *Popular Science*, as a sidebar to an article about the Daleks of *Doctor Who*, includes an adaptation of the Drake equation, modified to include an additional factor dubbed the "Dalek Variable", rendering the equation thus:

$$N = R_* \cdot f_p \cdot n_e \cdot f_t \cdot f_i \cdot f_c \cdot L \cdot f_d$$

The added variable at the end is defined as the "fraction of those civilizations that can survive an alien attack." (Note: in the article, the first variable is presented with the asterisk as superscript.)[65]

19.8 See also

- Astrobiology

- Goldilocks principle

- Kardashev scale

- Planetary habitability

- Ufology

19.9   Notes

[1] The rendering of the equation here is slightly modified for clarity of presentation from the rendering in the cited source.[20]

19.10   References

[1] Burchell, M.J. (2006). "W(h)ither the Drake equation?".*International Journal of Astrobiology***5**(3): 243–250. Bibcode:20 doi:10.1017/S1473550406003107.

[2] http://arxiv.org/ftp/arxiv/papers/1112/1112.1506.pdf

[3] "Chapter 3 — Philosophy: "Solving the Drake Equation". SETI League. December 2002. Retrieved 2013-04-10.

[4] Cocconi, G.; Morisson, P. (1959). "Searching for Interstellar Communications"(PDF).*Nature***184**(4690): 844–846. doi:10.1038/184844a0. Retrieved 2013-04-10.

[5] Schilling, G.; MacRobert, A. M. (2013). "The Chance of Finding Aliens". *Sky & Telescope*. Retrieved 2013-04-10.

[6] newspaper, staff (8 November 1959). "Life On Other Planets?". *Sydney Morning Herald*. Retrieved 2015-10-02.

[7] "The Drake Equation Revisited: Part I". *Astrobiology Magazine*. 29 September 2003. Retrieved 2013-08-13.

[8] Zaun, H. (1 November 2011). "Es war wie eine 180-Grad-Wende von diesem peinlichen Geheimnis" [It was like a 180 turn from this embarrassing secret]. *Telepolis* (in German). Retrieved 2013-08-13.

[9] "Drake Equation Plaque". Retrieved 2013-08-13.

[10] Darling, D. J. "Green Bank conference (1961)". *The Encyclopedia of Science*. Retrieved 2013-08-13.

[11] Aguirre, L. (1 July 2008). "The Drake Equation". *Nova ScienceNow*. PBS. Retrieved 2010-03-07.

[12] "What do we need to know about to discover life in space?". SETI Institute. Retrieved 2013-04-16.

[13] Jones, D. S. (26 September 2001). "Beyond the Drake Equation". Retrieved 2013-04-17.

[14] "The Search For Life : The Drake Equation 2010 - Part I". BBC Four. 2010. Retrieved 2013-04-17.

[15] SETI: A celebration of the first 50 years. Keith Cooper. *Astronomy Now*. 2000

[16] Hetesi, Z.; Regaly, Z. (2006). "A new interpretation of Drake-equation"(PDF).*Journal of the British Interplanetary* 11–14. Bibcode:2006JBIS...59...11H.

[17] Maccone, C. (2010). "The Statistical Drake Equation".*Acta Astronautica***67**(11–12): 1366–1383. Bibcode:2010Ac. doi:10.1016/j.actaastro.2010.05.003.

[18] Brin, G. D. (1983). "The Great Silence – The Controversy Concerning Extraterrestrial Intelligent Life".*Quarter the Royal Astronomical Society* **24** (3): 283–309. Bibcode:1983QJRAS..24..283B.

[19] Zaitsev, A. (May 2005). "The Drake Equation: Adding a METI Factor". SETI League. Retrieved 2013-04-20.

[20] The Drake Equation Revisited: Interview with Planet Hunter Sara Seager Devin Powell.*Astrobiology Magazine* 4 2013.

[21] "A New Equation Reveals Our Exact Odds of Finding Alien Life". io9.com. External link in |publisher= (help)

[22] Drake, F.; Sobel, D. (1992). *Is Anyone Out There? The Scientific Search for Extraterrestrial Intelligence*. Delta. pp. 55–62. ISBN 0-385-31122-2.

[23] Glade, N.; Ballet, P.; Bastien, O. (2012). "A stochastic process approach of the drake equation parameters". *International Journal of Astrobiology* **11** (2): 103–108. arXiv:1112.1506. Bibcode:2012IJAsB..11..103G. doi:10.1017/S1473550411000413. Note: This reference has a table of 1961 values, claimed to be taken from Drake & Sobel, but these differ from the book.

[24] Wanjek, C. (5 January 2006). "Milky Way Churns Out Seven New Stars Per Year, Scientists Say". Goddard Space Flight Center. Retrieved 2008-05-08.

[25] Palmer, J. (11 January 2012). "Exoplanets are around every star, study suggests". BBC. Retrieved 2012-01-12.

[26] Cassan, A.; et al. (11 January 2012). "One or more bound planets per Milky Way star from microlensing observations". *Nature* **481** (7380): 167–169. arXiv:1202.0903. Bibcode:2012Natur.481..167C. doi:10.1038/nature10684. PMID 22237108.

[27] Overbye, Dennis (4 November 2013). "Far-Off Planets Like the Earth Dot the Galaxy". *New York Times*. Retrieved 5 November 2013.

[28] Petigura, Eric A.; Howard, Andrew W.; Marcy, Geoffrey W. (31 October 2013). "Prevalence of Earth-size planets orbiting Sun-like stars". *Proceedings of the National Academy of Sciences of the United States of America*. arXiv:1311.6806. Bibcode:2013PNAS..11019273P. doi:10.1073/pnas.1319909110. Retrieved 5 November 2013.

[29] Khan, Amina (4 November 2013). "Milky Way may host billions of Earth-size planets". *Los Angeles Times*. Retrieved 5 November 2013.

[30] Trimble, V. (1997). "Origin of the biologically important elements". *Origins of Life and Evolution of the Biosphere* **27** (1–3): 3–21. doi:10.1023/A:1006561811750. PMID 9150565.

[31] Lineweaver, C. H.; Fenner, Y.; Gibson, B. K. (2004). "The Galactic Habitable Zone and the Age Distribution of Complex Life in the Milky Way". *Science* **303** (5654): 59–62. arXiv:astro-ph/0401024. Bibcode:2004Sci...303...59L. doi:10.1126/science.1092322. PMID 14704421.

[32] "Red Dwarf Stars Could Leave Habitable Earth-Like Planets Vulnerable to Radiation". *SciTech Daily*. Retrieved 22 September 2015.

[33] Dressing, C. D.; Charbonneau, D. (2013). "The Occurrence Rate of Small Planets around Small Stars". *The Astrophysical Journal* **767**: 95. arXiv:1302.1647. Bibcode:2013ApJ...767...95D. doi:10.1088/0004-637X/767/1/95.

[34] Ward, Peter D.; Brownlee, Donald (2000). *Rare Earth: Why Complex Life is Uncommon in the Universe*. Copernicus Books (Springer Verlag). ISBN 0-387-98701-0.

[35] Davies, P. (2007). "Are Aliens Among Us?". *Scientific American* **297** (6): 62–69. doi:10.1038/scientificamerican1207-62.

[36] Crick, F. H. C.; Orgel, L. E. (1973). "Directed Panspermia" (PDF). *Icarus* **19** (3): 341–346. Bibcode:1973Icar...19..341C. doi:10.1016/0019-1035(73)90110-3.

[37] "Ernst Mayr on SETI". The Planetary Society.

[38] Rare Earth, p. xviii.: "We believe that life in the form of microbes or their equivalents is very common in the universe, perhaps more common than even Drake or Sagan envisioned. However, *complex* life - animals and higher plants = is likely to be far more rare than commonly assumed."

[39] Campbell, A. (13 March 2005). "Review of *Life's Solution* by Simon Conway Morris".

[40] Bonner, J. T. (1988). *The evolution of complexity by means of natural selection*. Princeton University Press. ISBN 0-691-08494-7.

[41] Forgan, D.; Elvis, M. (2011). "Extrasolar Asteroid Mining as Forensic Evidence for Extraterrestrial Intelligence". *International Journal of Astrobiology* **10** (4): 307. arXiv:1103.5369. Bibcode:2011IJAsB..10..307F. doi:10.1017/S1473550411000127.

[42] Shermer, M. (August 2002). "Why ET Hasn't Called". *Scientific American*: 21.

[43] Grinspoon, D. (2004). *Lonely Planets*.

[44] Goldsmith, D.; Owen, T. (1992). *The Search for Life in the Universe* (2nd ed.). Addison-Wesley. p. 415. ISBN 1-891389-16-5.

[45] "The value of N remains highly uncertain. Even if we had a perfect knowledge of the first two terms in the equation, there are still five remaining terms, each of which could be uncertain by factors of 1,000." from Wilson, TL (2001). "The search for extraterrestrial intelligence". *Nature* (Nature Publishing Group) **409** (6823): 1110—1114., or more informally, "The Drake Equation can have any value from "billions and billions" to zero", Michael Crichton, as quoted in Douglas A. Vakoch; et al. (2015). *The Drake equation - estimating the prevalence of extraterrestrial life through the ages.* Cambridge University Press. ISBN 978-1-10-707365-4., pp. 13

[46] Dean, T. (10 August 2009). "A review of the Drake Equation". *Cosmos Magazine.* Retrieved 2013-04-16.

[47] ""Beyond Drake's Equation" --New Insights into the Search for Extraterrestrial Civilizations". The Daily Galaxy. 10 December 2012. Retrieved 2013-04-15.

[48] Rare Earth, page 270: "When we take into account factors such as the abundance of planets and the location and lifetime of the habitable zone, the Drake Equation suggests that only between 1% and 0.001% of all stars might have planets with habitats similar to Earth. [...] If microbial life forms readily, then millions to hundreds of millions of planets in the galaxy have the *potential* for developing advanced life. (We expect that a much higher number will have microbial life.)"

[49] von Bloh, W.; Bounama, C.; Cuntz, M.; Franck, S. (2007). "The habitability of super-Earths in Gliese 581". *Astronomy & Astrophysics* **476** (3): 1365. arXiv:0705.3758. Bibcode:2007A&A...476.1365V. doi:10.1051/0004-6361:20077939.

[50] Selsis, F.; Kasting, J. F.; Levrard, B.; Paillet, J.; Ribas, I.; Delfosse, X. (2007). "Habitable planets around the star Gliese 581?". *Astronomy & Astrophysics* **476** (3): 1373. arXiv:0710.5294. Bibcode:2007A&A...476.1373S. doi:10.1051/0004-6361:20078091.

[51] Lineweaver, C. H.; Davis, T. M. (2002). "Does the rapid appearance of life on Earth suggest that life is common in the universe?". *Astrobiology* **2** (3): 293–304. arXiv:astro-ph/0205014. Bibcode:2002AsBio...2..293L. doi:10.1089/153110702762027871.PMID12530239.

[52] Forgan, D. (2009). "A numerical testbed for hypotheses of extraterrestrial life and intelligence". *International Journal of Astrobiology* **8** (2): 121–131. arXiv:0810.2222. Bibcode:2009IJAsB...8..121F. doi:10.1017/S1473550408004321.

[53] Dvorsky, G. (31 May 2007). "The Drake Equation is obsolete". *Sentient Developments.* Retrieved 2013-08-21.

[54] Tarter, J. (May–June 2006). "The Cosmic Haystack Is Large". *Skeptical Inquirer* **30** (3). Retrieved 2013-08-21.

[55] Alexander, A. "The Search for Extraterrestrial Intelligence: A Short History - Part 7: The Birth of the Drake Equation". The Planetary Society. Archived from the original on 2005-03-06.

[56] Jones, E. M. (1 March 1985). ""Where is everybody?" An account of Fermi's question" (PDF). Los Alamos National Laboratory. Retrieved 2013-08-21.

[57] Krauthammer, C. (29 December 2011). "Are we alone in the Universe?". *The Washington Post.* Retrieved 2013-08-21.

[58] Webb, S. (2002). *If the Universe Is Teeming with Aliens... Where Is Everybody?.* Praxis Publishing. ISBN 0-387-95501-1.

[59] Hanson, R. (15 September 1998). "The Great Filter — Are We Almost Past It?". Retrieved 2013-08-21.

[60] "They're made out of Meat, by Hugo and Nebula Winner Terry Bisson". Baetzler.de. Retrieved 7 March 2010.

[61] *The Making of Star Trek* by Stephen E. Whitfield and Gene Roddenberry, Ballantine Books, N. Y., 1968

[62] Crichton, Michael (2012). *Sphere.* Knopf Doubleday Publishing Group. p. &q=drake+equation 28. ISBN 978-0-307-81648-1.

[63] "Carbon Based Lifeforms – Abiogenesis". Last.fm. Retrieved 23 March 2012.

[64] "Seyton Discography at Discogs".

[65] Gregory Mone and Jim Rossiter, "The Dalek Variable", *Popular Science*, July 2013, p. 81.

19.11 Further reading

- Morton, Oliver (2002). "A Mirror in the Sky". In Graham Formelo. *It Must Be Beautiful*. Granta Books. ISBN 1-86207-555-7.

- Rood, Robert T.; James S. Trefil (1981). *Are We Alone? The Possibility of Extraterrestrial Civilizations*. New York: Scribner. ISBN 0684178427.

- Douglas A. Vakoch, et al.: *The Drake equation - estimating the prevalence of extraterrestrial life through the ages*. Cambridge University Press, Cambridge 2015, ISBN 978-1-10-707365-4.

19.12 External links

- Interactive Drake Equation Calculator

- "Only a matter of time, says Frank Drake". A Q&A with Frank Drake in February 2010.

- Frank Drake (December 2004). "The E.T. Equation, Recalculated". *Wired*.

- Macromedia Flash page allowing the user to modify Drake's values from PBS Nova

- The Drake Equation Astronomy Cast episode #23, includes full transcript.

- Animated simulation of the Drake equation.

- The Alien Equation 22 September 2010, BBC Radio program Discovery.

Chapter 20

Fermi paradox

This article is about the absence of evidence for extraterrestrial intelligence. For the type of estimation problem, see Fermi problem. For the television episode, see Where Is Everybody?.

 The **Fermi paradox** — or **Fermi's paradox** — is the apparent contradiction between high estimates of the probability of the existence of extraterrestrial civilizations, such as in the Drake equation, and the lack of evidence for such civilizations.[1] The basic points of the argument, made by physicists Enrico Fermi (1901–1954) and Michael H. Hart (born 1932), are:

- The Sun is a typical star, and there are billions of stars in the galaxy that are billions of years older.[2][3]

- With high probability, some of these stars will have Earth-like planets,[4][5] and if the earth is typical, some might develop intelligent life.

- Some of these civilizations might develop interstellar travel, a step the Earth is investigating now.

- Even at the slow pace of currently envisioned interstellar travel, the Milky Way galaxy could be completely traversed in about a million years.[6]

According to this line of thinking, the Earth should already have been visited by extraterrestrial aliens though Fermi saw no convincing evidence of this, nor any signs of alien intelligence anywhere in the observable universe, leading him to ask, "Where is everybody?"[7][8]

20.1   Overview

The age of the universe and its vast number of stars has led some to suggest that unless the rare Earth hypothesis holds true, extraterrestrial life should be common.[9] In an informal discussion in 1950, the physicist Enrico Fermi questioned why, if a multitude of advanced extraterrestrial civilizations exist in the Milky Way galaxy, evidence such as a flying saucer or Von Neumann probe have not yet been seen. Counterarguments suggest that intelligent extraterrestrial life does not exist or occurs so rarely or briefly that humans will never make contact with it.[10] Other common names for the *Fermi's question* ("Where are they?") include: the *Great Silence*,[11][12][13][14] and *silentium universi*[14][15] (Latin for "silence of the universe").

Michael H. Hart published in 1975 a detailed examination of the paradox,[6] which has since become a theoretical reference point for much of the research into what is now sometimes known as the **Fermi-Hart paradox**.[16] Interest in the paradox has spawned numerous scholarly works addressing it directly, while questions that relate to it have been addressed in fields as diverse as astronomy, biology, ecology, and philosophy.

20.2 Basis

The Fermi paradox is a conflict between arguments of scale and probability that seem to favor intelligent life being common in the universe, and a total lack of evidence of intelligent life having ever arisen anywhere other than on the Earth.

The first aspect of the Fermi paradox is a function of the scale or the large numbers involved: there are an estimated 200–400 billion stars in the Milky Way[17] ($2-4 \times 10^{11}$) and 70 sextillion (7×10^{22}) in the observable universe.[18] Even if intelligent life occurs on only a minuscule percentage of planets around these stars, there might still be a great number of extant civilizations, and if the percentage were high enough it would produce a significant number of extant civilizations in the Milky Way. This creates the assumption that Earth is merely a typical planet.

The second aspect of the Fermi paradox is the argument of probability: given intelligent life's ability to overcome scarcity, and its tendency to colonize new habitats, it seems possible that at least some civilizations would be technologically advanced, seek out new resources in space, and colonize their own star system and, subsequently, surrounding star systems. Since there is no conclusive evidence on Earth or elsewhere in the known universe of other intelligent life after 13.8 billion years of the universe's history, we have a conflict requiring a resolution. Some examples of possible resolutions are that intelligent life is rarer than we think, that our assumptions about the general development or behavior of intelligent species are flawed, or, more radically, that our current scientific understanding of the nature of the universe itself is quite incomplete.

The Fermi paradox can be asked in two ways. The first is, "Why are no aliens or their artifacts found here on Earth?" If interstellar travel is possible, even the "slow" kind nearly within the reach of Earth technology, then it would only take from 5 million to 50 million years to colonize the galaxy.[19] This is relatively brief on a geological scale, let alone a cosmological one. Since there are many stars older than the Sun, and since intelligent life might have evolved earlier elsewhere, the question then becomes why the galaxy has not been colonized already. Even if colonization is impractical or undesirable to all alien civilizations, large-scale *exploration* of the galaxy could be possible using various means of exploration. Travel times may well explain the lack of alien visits to Earth, but a sufficiently advanced civilization could potentially be observable over a significant fraction of the size of the observable universe.[20] Even if such civilizations are rare, the scale argument indicates they should exist somewhere at some point during the history of the universe, and since they could be detected from far away over a considerable period of time, many more potential sites for their origin are within range of our observation. It is unknown whether the paradox is stronger for our galaxy or for the universe as a whole.[21]

20.3 Name

In 1950, while working at Los Alamos National Laboratory, Fermi had a casual conversation while walking to lunch with colleagues Emil Konopinski, Edward Teller and Herbert York.[22] The men discussed a recent spate of UFO reports and an Alan Dunn cartoon[23] facetiously blaming the disappearance of municipal trashcans on marauding aliens. The conversation shifted to other subjects, until during lunch Fermi suddenly exclaimed, "Where are they?" (alternatively, *"Where is everybody?"*). Teller remembers, "The result of his question was general laughter because of the strange fact that in spite of Fermi's question coming from the clear blue, everybody around the table seemed to understand at once that he was talking about extraterrestrial life."[24] Herbert York recalls that Fermi followed up on his comment with a series of calculations on the probability of Earth-like planets, the probability of life, the likely rise and duration of high technology, etc., and concluded that we ought to have been visited long ago and many times over.

Although Fermi's name is most commonly associated with the paradox, he was not the first to ask the question. An earlier implicit mention was by Konstantin Tsiolkovsky in an unpublished manuscript from 1933.[25] He noted "people deny the presence of intelligent beings on the planets of the universe" because "(i) if such beings exist they would have visited Earth, and (ii) if such civilisations existed then they would have given us some sign of their existence." This was not a paradox for others, who took this to imply the absence of ETs, but it was for him, since he himself was a strong believer in extraterrestrial life and the possibility of space travel. Therefore, he speculated that mankind is not yet ready for higher beings to contact us.[26] That Tsiolkovsky himself may not have been the first to discover the paradox is suggested by his above-mentioned reference to other people's reasons for denying the existence of Extraterrestrial Civilisations (ETCs).

20.4   Drake equation

Main article: Drake equation

The theories and principles in the Drake equation are closely related to the Fermi paradox.[27] The equation was formulated by Frank Drake in 1961 in an attempt to find a systematic means to evaluate the numerous probabilities involved in the existence of alien life. The speculative equation considers the rate of star formation in the galaxy; the fraction of stars with planets and the number per star that are habitable; the fraction of those planets which develop life; the fraction that develop *intelligent* life; the fraction that have detectable, technological intelligent life; and finally the length of time such communicable civilizations are detectable. The fundamental problem is that the last four terms are completely unknown, rendering statistical estimates impossible.

The Drake equation has been used by both optimists and pessimists, with wildly differing results. Carl Sagan, using optimistic numbers, suggested as many as one million communicating civilizations in the Milky Way in 1966, though he later suggested that the actual number could be far smaller. Frank Tipler and John D. Barrow used pessimistic numbers and concluded that the average number of civilizations in a galaxy is much less than one.[28]

20.5   Empirical projects

Efforts to find signs of extraterrestrial intelligence have been made since 1960, and several are ongoing.[29] One challenge is the need to avoid an anthropocentric viewpoint.

20.5.1   Mainstream astronomy and SETI

Although astronomers do not usually search for extraterrestrials, they might observe some phenomenon that cannot be explained without positing an intelligent civilization as the source. This has been suspected several times. For example, pulsars, when first discovered, were called little green men (LGM) because of the precise repetition of their pulses.[30] In all cases, explanations with no need for intelligent life have been found for such observations,[31] but the possibility of discovery remains.[32] Proposed examples include asteroid mining that would change the appearance of debris disks around stars,[33] or spectral lines from nuclear waste disposal in stars.[34] An on-going example is the unusual transit light curves of star KIC 8462852, where natural interpretations are not fully convincing.[35] Although most likely a natural explanation will emerge, some scientists are investigating the remote possibility that it could be a sign of alien technology.[36][37][38]

20.5.2   Electromagnetic emissions

Further information: SETI, Project Ozma, Project Cyclops, Project Phoenix (SETI), SERENDIP and Allen Telescope Array
 Radio technology and the ability to construct a radio telescope are presumed to be a natural advance for technological species,[39] theoretically creating effects that might be detected over interstellar distances. The careful searching for non-natural radio emissions from space may lead to the detection of alien civilizations. Sensitive alien observers of the Solar System, for example, would note unusually intense radio waves for a G2 star due to Earth's television and telecommunication broadcasts. In the absence of an apparent natural cause, alien observers might infer the existence of a terrestrial civilization. It should be noted however that even the most sensitive radio telescopes currently available on Earth would not be able to detect non-directional radio signals even at a fraction of a light-year, so it is questionable whether any such signals could be detected by an extraterrestrial civilization. Such signals could be either "accidental" by-products of a civilization, or deliberate attempts to communicate, such as the Arecibo message. A number of astronomers and observatories have attempted and are attempting to detect such evidence, mostly through the SETI organization. Several decades of SETI analysis have not revealed any unusually bright or meaningfully repetitive radio emissions, although there have been several candidate signals.

In 2015, a study of galactic mid-infrared emissions came to the conclusion that "Kardashev Type-III civilisations are either very rare or do not exist in the local Universe".[40]

20.5.3 Electromagnetic emissions and information panspermia

The strategy of decoding of electromagnetic emission has to be influenced by the concept of *Information Panspermia* coined in [41] and which Webb (2015) attributed as Solution 23 of Fermi Paradox. It is shown that the human genome and hence the terrestrial life possess low Kolmogorov complexity and so the corresponding bit strings can be transmitted by Arecibo-type antenna to Galactic distances. Then the electromagnetic signals have to be analyzed on the subject of bit strings as traveling extraterrestrial life streams at von Neumann automata network.

20.5.4 Direct planetary observation

Exoplanet detection and classification is a very active sub-discipline in astronomy, and the first possibly terrestrial planet discovered within a star's habitable zone was found in 2007.[42] New refinements in exoplanet detection methods, and use of existing methods from space (such as the Kepler Mission, launched in 2009) are starting to detect and characterize Earth-size planets, and determine if they are within the habitable zones of their stars. Such observational refinements may allow us to better gauge how common potentially habitable worlds are.

Direct evidence for the existence of life on an exoplanet may eventually be observable, such as the detection of biotic signature gases (such as methane and oxygen) — or even the industrial air pollution of a technologically advanced civilization — in an exoplanet's atmosphere, by means of spectrographic analysis.[43]

Conjectures about interstellar probes

Further information: Von Neumann probe and Bracewell probe

Self-replicating probes could exhaustively explore a galaxy the size of the Milky Way in as little as half a million years. If even a single civilization in the Milky Way attempted this, such probes could spread throughout the entire galaxy. Another speculation for contact with an alien probe — one that would be trying to find human beings — is an alien Bracewell probe. Such hypothetical device would be an autonomous space probe whose purpose is to seek out and communicate with alien civilizations (as opposed to Von Neumann probes, which are usually described as purely exploratory). These were proposed as an alternative to carrying a slow speed-of-light dialogue between vastly distant neighbors. Rather than contending with the long delays a radio dialogue would suffer, a probe housing an artificial intelligence would seek out an alien civilization to carry on a close-range communication with the discovered civilization. The findings of such a probe would still have to be transmitted to the home civilization at light speed, but an information-gathering dialogue could be conducted in real time.[44]

Attempts to find alien probes

Direct exploration of the Solar System has yielded no evidence indicating a visit by aliens or their probes. Detailed exploration of areas of the Solar System where resources would be plentiful may yet produce evidence of alien exploration,[45][46] though the entirety of the Solar System is vast and difficult to investigate. Attempts to signal, attract, or activate hypothetical Bracewell probes in Earth's vicinity have not succeeded.[47]

Conjectures about stellar-scale artifacts

Further information: Dyson sphere, Kardashev scale, Alderson disk, Matrioshka brain, Stellar engine
In 1959, Freeman Dyson observed that every developing human civilization constantly increases its energy consumption, and, he conjectured, a civilization might try to harness a large part of the energy produced by a star. He proposed that a Dyson sphere could be a possible means: a shell or cloud of objects enclosing a star to absorb and utilize as much radiant energy as possible. Such a feat of astroengineering would drastically alter the observed spectrum of the star involved,

changing it at least partly from the normal emission lines of a natural stellar atmosphere to those of black body radiation, probably with a peak in the infrared. Dyson speculated that advanced alien civilizations might be detected by examining the spectra of stars and searching for such an altered spectrum.[48][49][50]

There have been some attempts to find evidence of the existence of Dyson spheres that would alter the spectra of their core stars.[51] These surveys have not yet located anything. Similarly, direct observation of thousands of galaxies has shown no explicit evidence of artificial construction or modifications.[49][50][52][53]

20.6 Hypothetical explanations for the paradox

20.6.1 No other civilizations have arisen

Those who think that extraterrestrial intelligent life does not exist argue that the conditions needed for the evolution of life — or at least the evolution of biological complexity — are rare or even unique to Earth. Under this assumption, called the rare Earth hypothesis, a rejection of the mediocrity principle, complex multicellular life is regarded as exceedingly unusual.[54] Similarly, it is possible that even if complex life is common, intelligence and technological civilizations are not.[55] To skeptics, the fact that in the history of life on the Earth only one species has developed a civilization to the point of being capable of spaceflight and radio technology, lends more credence to the idea that technologically advanced civilizations are rare in the universe.[56]

20.6.2 It is the nature of intelligent life to destroy itself

This is the argument that technological civilizations may usually or invariably destroy themselves before or shortly after developing radio or spaceflight technology. Possible means of annihilation are many,[57] including nuclear war, biological warfare, or accidental environmental contamination. This general theme is explored both in fiction and in scientific hypothesizing.[58] Indeed, there are probabilistic arguments which suggest that human extinction may occur sooner rather than later. In 1966, Sagan and Shklovskii speculated that technological civilizations will either tend to destroy themselves within a century of developing interstellar communicative capability or master their self-destructive tendencies and survive for billion-year timescales.[59] Self-annihilation may also be viewed in terms of thermodynamics: insofar as life is an ordered system that can sustain itself against the tendency to disorder, the "external transmission" or interstellar communicative phase may be the point at which the system becomes unstable and self-destructs.[60]

20.6.3 It is the nature of intelligent life to destroy others

See also: Technological singularity and Von Neumann probe

Another hypothesis is that an intelligent species beyond a certain point of technological capability will destroy other intelligent species as they appear. The idea that something, or someone, might be destroying intelligent life in the universe has been explored in the scientific literature.[11] A species might undertake such extermination out of expansionist motives, paranoia, or aggression. In 1981, cosmologist Edward Harrison argued that such behavior would be an act of prudence: an intelligent species that has overcome its own self-destructive tendencies might view any other species bent on galactic expansion as a threat.[61] It has also been suggested that a successful alien species would be a superpredator, as are humans.[62][63]

20.6.4 Life is periodically destroyed by naturally occurring events

On Earth, there have been numerous major extinction events that destroyed the majority of complex species alive at the time. The extinction of the dinosaurs is the best known example. These are thought to have been caused by events such as impact from a large meteorite, massive volcanic eruptions, or astronomical events such as gamma-ray bursts.[64] It may be the case that such extinction events are common throughout the universe and periodically destroy intelligent life, or at least its civilizations, before the species is able to develop the technology to communicate with other species.[65]

20.6.5   Inflation hypothesis and the youngness argument

Cosmologist Alan Guth proposed a multi-verse solution to the Fermi paradox. This hypothesis uses the synchronous gauge probability distribution, that young universes exceedingly outnumber older ones (by a factor of $e^{10^{37}}$ for every second of age). Therefore, averaged over all universes, universes with civilizations will almost always have just one, the first to develop. However, Guth notes "Perhaps this argument explains why SETI has not found any signals from alien civilizations, but I find it more plausible that it is merely a symptom that the synchronous gauge probability distribution is not the right one."[66]

20.6.6   Intelligent civilizations are too far apart in space or time

It may be that non-colonizing technologically capable alien civilizations exist, but that they are simply too far apart for meaningful two-way communication.[67] If two civilizations are separated by several thousand light-years, it is possible that one or both cultures may become extinct before meaningful dialogue can be established. Human searches may be able to detect their existence, but communication will remain impossible because of distance. It has been suggested that this problem might be ameliorated somewhat if contact/communication is made through a Bracewell probe. In this case at least one partner in the exchange may obtain meaningful information. Alternatively, a civilization may simply broadcast its knowledge, and leave it to the receiver to make what they may of it. This is similar to the transmission of information from ancient civilizations to the present,[68] and humanity has undertaken similar activities like the Arecibo message, which could transfer information about Earth's intelligent species, even if it never yields a response or does not yield a response in time for humanity to receive it. It is also possible that archaeological evidence of past civilizations may be detected through deep space observations — especially if they left behind large artifacts such as Dyson spheres.

A related speculation by Sagan and Newman suggests that if other civilizations exist, and are transmitting and exploring, their signals and probes simply have not arrived yet.[69] However, critics have noted that this is unlikely, since it requires that humanity's advancement has occurred at a very special point in time, while the Milky Way is in transition from empty to full. This is a tiny fraction of the lifespan of a galaxy under ordinary assumptions and calculations resulting from them, so the likelihood that we are in the midst of this transition is considered low in the paradox.[70]

20.6.7   It is too expensive to spread physically throughout the galaxy

See also: Project Daedalus, Project Orion (nuclear propulsion) and Project Longshot

Many speculations about the ability of an alien culture to colonize other star systems are based on the idea that interstellar travel is technologically feasible. While the current understanding of physics rules out the possibility of faster-than-light travel, it appears that there are no major theoretical barriers to the construction of "slow" interstellar ships, even though the engineering required is considerably beyond our present capabilities. This idea underlies the concept of the Von Neumann probe and the Bracewell probe as a potential evidence of extraterrestrial intelligence.

It is possible, however, that present scientific knowledge cannot properly gauge the feasibility and costs of such interstellar colonization. Theoretical barriers may not yet be understood, and the cost of materials and energy for such ventures may be so high as to make it unlikely that any civilization could afford to attempt it. Even if interstellar travel and colonization are possible, they may be difficult, leading to a colonization model based on percolation theory.[71] Colonization efforts may not occur as an unstoppable rush, but rather as an uneven tendency to "percolate" outwards, within an eventual slowing and termination of the effort given the enormous costs involved and the fact that colonies will inevitably develop a culture and civilization of their own. Colonization may thus occur in "clusters," with large areas remaining uncolonized at any one time.[71]

20.6.8   Human beings have not existed long enough

Humanity's ability to detect intelligent extraterrestrial life has existed for only a very brief period—from 1937 onwards, if the invention of the radio telescope is taken as the dividing line—and *Homo sapiens* is a geologically recent species. The whole period of modern human existence to date is a very brief period on a cosmological scale, and radio transmissions have only been propagated since 1895. Thus, it remains possible that human beings have neither existed long enough nor made themselves sufficiently detectable to be found by extraterrestrial intelligence.

20.6.9   Humans are not listening properly

There are some assumptions that underlie the SETI search programs that may cause searchers to miss signals that are present. Extraterrestrials might, for example, transmit signals that have a very high or low data rate, or employ unconventional (in our terms) data compression, frequencies, or modulations. Signals might be sent from non-main sequence star systems that we search with lower priority; current programs assume that most alien life will be orbiting Sun-like stars.[72]

The greatest challenge is the sheer size of the radio search needed to look for signals (effectively spanning the entire visible universe), the limited amount of resources committed to SETI, and the sensitivity of modern instruments. SETI estimates, for instance, that with a radio telescope as sensitive as the Arecibo Observatory, Earth's television and radio broadcasts would only be detectable at distances up to 0.3 light-years, less than 1/10 the distance to the nearest star. A signal is much easier to detect if the signal energy is limited to either a narrow range of frequencies, or directed at a specific part of the sky. Such signals could be detected at ranges of hundreds to tens of thousands of light-years distance. However, this means that detectors must be listening to an appropriate range of frequencies, and be in that region of space to which the beam is being sent. Many SETI searches, starting with the venerable Project Cyclops, go so far as to assume that extraterrestrial civilizations will be broadcasting a deliberate signal, like the Arecibo message, in order to be found.

Thus to detect alien civilizations through their radio emissions, Earth observers either need more sensitive instruments or must hope for fortunate circumstances: that the broadband radio emissions of alien radio technology are much stronger than our own; that one of SETI's programs is listening to the correct frequencies from the right regions of space; or that aliens are deliberately sending focused transmissions in our general direction.

20.6.10   Civilizations broadcast detectable radio signals only for a brief period of time

It may be that alien civilizations are detectable through their radio emissions for only a short time, reducing the likelihood of spotting them. There are two possibilities in this regard: civilizations outgrow radio through technological advance or, conversely, resource depletion cuts short the time in which a species broadcasts. These will potentially remain visible even after broadcast emission are replaced by less observable technology.[73]

More hypothetically, advanced alien civilizations evolve beyond broadcasting at all in the electromagnetic spectrum and communicate by principles of physics not yet understood by humans. Some scientists have hypothesized that advanced civilizations may send neutrino signals.[74] If such signals exist, they could be detectable by neutrino detectors that are now under construction for other goals.[75]

20.6.11   They tend to isolate themselves

It has been suggested that some advanced beings may divest themselves of physical form, create massive artificial virtual environments, transfer themselves into these environments through mind uploading, and exist totally within virtual worlds, ignoring the external physical universe.[76]

It may also be that intelligent alien life develop an "increasing disinterest" in their outside world.[77] Possibly any sufficiently advanced society will develop highly engaging media and entertainment well before the capacity for advanced space travel, and that the rate of appeal of these social contrivances is destined, because of their inherent reduced complexity, to overtake any desire for complex, expensive endeavors such as space exploration and communication. Once any sufficiently advanced civilization becomes able to master its environment, and most of its physical needs are met through technology, various "social and entertainment technologies", including virtual reality, are postulated to become the primary drivers and motivations of that civilization.[78]

20.6.12 They are too alien

Another possibility is that human theoreticians have underestimated how much alien life might differ from that on Earth. Aliens may be psychologically unwilling to attempt to communicate with human beings. Perhaps human mathematics is parochial to Earth and not shared by other life.[79] though others argue this can only apply to abstract math since the math associated with physics must be similar (in results, if not in methods.)[80]

Physiology might also cause a communication barrier. Carl Sagan speculated that an alien species might have a thought process orders of magnitude slower (or faster) than humans. Such a species could conceivably speak so slowly that it requires years to say even a simple phrase like "Hello". A message broadcast by that species might well seem like random background noise to humans, and therefore go undetected.

Another thought is that technological civilizations invariably experience a technological singularity and attain a post-biological character. Hypothetical civilizations of this sort may have advanced drastically enough to render communication impossible.[81][82]

20.6.13 They are non-technological

It may be that at least some civilizations of intelligent beings are not technological. Such civilizations would be very difficult for humans to detect.[83] While there are remote sensing techniques which could perhaps detect life-bearing planets without relying on the signs of technology,[84][85] none of them has any ability to tell if any detected life is intelligent. This is sometimes referred to as the "algae vs. alumnae" problem.[83]

20.6.14 Everyone is listening, no one is transmitting

Alien civilizations might be technically capable of contacting Earth, but are only listening instead of transmitting.[86] If all, or even most, civilizations act the same way, the galaxy could be full of civilizations eager for contact, but everyone is listening and no one is transmitting. This is the so-called *SETI Paradox*.[87]

The only civilization we know, the Earth, does not explicitly transmit, except for a few small efforts.[86] Even these efforts, and certainly any attempt to expand them, are controversial.[88] It is not even clear we would respond to a detected signal — the official policy within the SETI community[89] is that "[no] response to a signal or other evidence of extraterrestrial intelligence should be sent until appropriate international consultations have taken place." However, given the possible impact of any reply[90] it may be very difficult to obtain any consensus on "Who speaks for Earth?" and "What should we say?"

20.6.15 Earth is deliberately not contacted (zoo hypothesis)

The zoo hypothesis states that intelligent extraterrestrial life exists and does not contact life on Earth to allow for its natural evolution and development.[91] This hypothesis may break down under the uniformity of motive flaw: all it takes is a single culture or civilization to decide to act contrary to the imperative within our range of detection for it to be abrogated, and the probability of such a violation increases with the number of civilizations.[19]

Analysis of the inter-arrival times between civilizations in the galaxy based on common astrobiological assumptions suggests that since the initial civilization would have such a commanding lead over the later arrivals, it may have established what we call *zoo hypothesis* as a galactic/universal norm and the resultant "paradox" by a cultural founder effect with or without the continued activity of the founder.[92]

20.6.16   Earth is purposely isolated (planetarium hypothesis)

Main article: Planetarium hypothesis

A related idea to the zoo hypothesis is that, beyond a certain distance, the perceived universe is a simulated reality. The planetarium hypothesis[93] speculates that beings may have created this simulation so that the universe appears to be empty of other life.

20.6.17   It is dangerous to communicate

An alien civilization might feel it is too dangerous to communicate, either for us or for them. After all, when very different civilizations have met on Earth, the results have often been disastrous for one side or the other, and the same may well apply to interstellar contact. Even contact at a safe distance could lead to infection by computer code[94] or even ideas themselves.[95] Perhaps prudent civilizations actively hide not only from Earth but from everyone, out of fear of other civilizations.[96]

20.6.18   The Fermi paradox itself is what prevents communication

Perhaps the Fermi paradox itself — or the alien equivalent of it — is the reason for any civilization to avoid contact with other civilizations, even if no other obstacles existed. From any one civilization's point of view, it would be unlikely for them to be the first ones to make first contact. Therefore, according to this reasoning, it is likely that previous civilizations faced fatal problems with first contact and doing so should be avoided. So perhaps every civilization keeps quiet because of the possibility that there is a real reason for others to do so.[11]

20.6.19   They are here undetected

It is possible that a civilization advanced enough to travel between the stars could visit or observe our world while remaining undetected.[97]

20.6.20   They are here unacknowledged

Main article: UFO conspiracy theory

A significant fraction of the population believes that at least some UFOs (Unidentified Flying Objects) are spacecraft piloted by aliens.[98] While most of these are unrecognized or mistaken interpretations of mundane phenomena, there are those that remain puzzling even after investigation. The consensus scientific view is that although they may be unexplained, they do not rise to the level of convincing evidence.[99]

Similarly, it is theoretically possible that SETI groups are not reporting positive detections, or governments have been blocking signals or suppressing publication. This response might be attributed to security or economic interests from the potential use of advanced extraterrestrial technology. It has been suggested that the detection of an extraterrestrial radio signal or technology could well be the most highly secret information that exists.[100] Claims that this has already happened are common in the popular press,[101][102] but the scientists involved report the opposite experience — the press becomes informed and interested in a potential detection even before a signal can be confirmed.[103] Another problem with such a conspiracy theory is the diverse number of organisations and governments involved in science activities that might chance upon detections, of which SETI forms only a small part.

20.7 In science fiction and other media

Many, perhaps most, of the serious explanations for the Fermi Paradox have appeared in science fiction literature, along with many that are not so serious. Less commonly the Fermi Paradox appears in other media. Examples include:

20.7.1 Literature

- Books

  - *Revelation Space* (2000) by Alastair Reynolds and the other books in this series
  - *Manifold: Space* (2000) by Stephen Baxter and the other books in this series
  - *Existence* (2012) by David Brin[104]
  - *Quarantine* by Greg Egan.
  - *The Forge of God* (1987) by Greg Bear is a science fiction novel proposing a solution to the Fermi paradox in which civilizations that are detectable by electromagnetic radiation attract the attention of predatory alien technology designed to conduct a "search and destroy" function on behalf of its creators. Therefore, Earth has not detected other civilizations either because they are undetectable or because they have been destroyed.
  - *Rama Revealed* (1993) by Arthur C. Clarke: Civilizations are common but short-lived and too separated in space and time to become aware of others.
  - The Safehold series by David Weber postulates that the reason for the Fermi Paradox is a xenophobic race that declares war on any species that becomes sufficiently advanced.
  - The Three Body trilogy by Liu Cixin

- Short stories

  - "Fermi and Frost" (1985) by Frederik Pohl
  - "The New Cosmogony" (1971) by Stanisław Lem
  - "The Fermi Paradox Is Our Business Model" (August 11, 2010) by Charlie Jane Anders
  - "The Crystal Spheres" (1984) by David Brin

20.7.2 Film

- *The Fermi Paradox*[105]

20.7.3 Music

- *Fermi Paradox* (2002), album by Tub Ring

20.8 See also

- Abiogenesis
- Anthropic principle
- Astrobiology
- Extraterrestrial hypothesis
- Fermi problem
- Interstellar travel
- The Great Filter

20.9   Notes

20.10   References

[1] Krauthammer, C. (December 29, 2011). "Are we alone in the universe?". *The Washington Post*. Retrieved January 6, 2015.

[2] Chris Impe (2011). *The Living Cosmos: Our Search for Life in the Universe*. Cambridge University Press. ISBN 978-0-521-84780-3., page 282.

[3] Aguirre, V. Silva, G. R. Davies, S. Basu, J. Christensen-Dalsgaard, O. Creevey, T. S. Metcalfe, T. R. Bedding; et al. (2015). "Ages and fundamental properties of Kepler exoplanet host stars from asteroseismology.". *Monthly Notices of the Royal Astronomical Society* **452** (2): 2127. arXiv:1504.07992. Bibcode:2015MNRAS.452.2127S. doi:10.1093/mnras/stv1388. Accepted for publication in MNRAS. See Figure 15 in particular.

[4] Schilling, G. (June 13, 2012). "ScienceShot: Alien Earths Have Been Around for a While". *Science*. Retrieved January 6, 2015.

[5] Buchhave, L. A.; et al. (June 21, 2012). "An abundance of small exoplanets around stars with a wide range of metallicities". *Nature* **486**: 375. Bibcode:2012Natur.486..375B. doi:10.1038/nature11121.

[6] Hart, Michael H. (1975). "Explanation for the Absence of Extraterrestrials on Earth". *Quarterly Journal of the Royal Astronomical Society* **16**: 128–135. Bibcode:1975QJRAS..16..128H.

[7] Jones, E. M. (March 1, 1985). "'Where is everybody?' An account of Fermi's question" (PDF). Los Alamos National Laboratory. OSTI 785733. Retrieved January 12, 2013.

[8] Overbye, Dennis (August 3, 2015). "The Flip Side of Optimism About Life on Other Planets". *New York Times*. Retrieved October 29, 2015.

[9] Sagan, Carl *Cosmos*, Random House 2002 ISBN 0-375-50832-5

[10] Urban, Tim (June 17, 2014). "The Fermi Paradox". *Huffington Post*. Retrieved January 6, 2015.

[11] Brin, Glen David (1983). "The 'Great Silence': The Controversy Concerning Extraterrestrial Intelligent Life". *Quarterly Journal of the Royal Astronomical Society* **24**: 283–309. Bibcode:1983QJRAS..24..283B.

[12] James Annis (1999). "An Astrophysical Explanation for the Great Silence". arXiv:astro-ph/9901322.

[13] Bostrom, Nick (2007). "In Great Silence there is Great Hope" (PDF). Retrieved September 6, 2010.

[14] Milan M. Ćirković (2009). "Fermi's Paradox – The Last Challenge for Copernicanism?". *Serbian Astronomical Journal* **178** (178): 1–20. arXiv:0907.3432. Bibcode:2009SerAJ.178....1C. doi:10.2298/SAJ0978001C.

[15] Lem, Stanislaw (1983). *His Master's Voice*. Harvest Books. ISBN 0-15-640300-5.

[16] Wesson, Paul (1990). "Cosmology, extraterrestrial intelligence, and a resolution of the Fermi-Hart paradox". *Quarterly Journal of the Royal Astronomical Society* **31**: 161–170. Bibcode:1990QJRAS..31..161W.

[17] "NASA – Galaxy". Nasa.gov. November 29, 2007. Retrieved August 19, 2010.

[18] Craig, Andrew (July 22, 2003). "Astronomers count the stars". *BBC News* (BBC). Retrieved April 8, 2010.

[19] Crawford, I.A.. "Where are They? Maybe we are alone in the galaxy after all". *Scientific American*, July 2000, 38–43, (2000).

[20] Shklovskii & Sagan 1966, p. 364

[21] J. Richard Gott, III. "Chapter 19: Cosmological SETI Frequency Standards". In Zuckerman, Ben; Hart, Michael. *Extraterrestrials: Where Are They?*. Page 180.

[22] Shostak, Seth (October 25, 2001). "Our Galaxy Should Be Teeming With Civilizations, But Where Are They?". *Space.com*. Space.com. Archived from the original on April 15, 2006. Retrieved October 14, 2014.

[23] Dunne, Alan (1950). "Uncaptioned cartoon". New Yorker, 20 May 1950. Retrieved August 19, 2010.

[24] Jones, Eric "Where is everybody?". An account of Fermi's question". Los Alamos Technical report LA-10311-MS, March, 1985.

[25] Tsiolkovsky, K. 1933, *The Planets are Occupied by Living Beings*, Archives of the Tsiolkovsky State Museum of the History of Cosmonautics, Kaluga, Russia.

[26] Lytkin, V., Finney, B., & Alepko, L.; Finney; Alepko (Dec 1995). "Tsiolkovsky – Russian Cosmism and Extraterrestrial Intelligence". *Quarterly Journal of the Royal Astronomical Society* **36** (4): 369. Bibcode:1995QJRAS..36..369L.

[27] Gowdy, Robert H., VCU Department of Physics SETI: Search for ExtraTerrestrial Intelligence. The Interstellar Distance Problem, 2008

[28] Barrow, John D.; Tipler, Frank J. (1988). *The Anthropic Cosmological Principle*. Oxford University Press. p. 588. ISBN 978-0-19-282147-8. LCCN 87028148.

[29] See, for example, the SETI Institute, The Harvard SETI Home Page, or The Search for Extra Terrestrial Intelligence at Berkeley

[30] Wade, Nicholas (1975). "Discovery of pulsars: a graduate student's story". *Science* **189** (4200) (American Association for the Advancement of Science). pp. 358–364.

[31] Pulsars are now attributed to neutron stars, and Seyfert galaxies to an end-on view of the accretion onto the black holes.

[32] "NASA/CP2007-214567: Workshop Report on the Future of Intelligence In The Cosmos" (PDF). NASA.

[33] Duncan Forgan, Martin Elvis; Elvis (28 March 2011). "Extrasolar Asteroid Mining as Forensic Evidence for Extraterrestrial Intelligence". *International Journal of Astrobiology* **10** (4): 307–313. arXiv:1103.5369. Bibcode:2011IJAsB..10..307F. doi:10.1017/S1473550411000127.

[34] Whitmire, Daniel P., and David P. Wright. (1980). "Nuclear waste spectrum as evidence of technological extraterrestrial civilizations". *Icarus* **42.1**, pp. 149–156. doi:10.1016/0019-1035(80)90253-5.

[35] Boyajian, T. S.; LaCourse, D. M.; Rappaport, S. A.; Fabrycky, D.; Fischer, D. A.; Gandolfi, D.; Kennedy, G. M.; Liu, M. C.; Moor, A.; Olah, K.; Vida, K.; Wyatt, M. C.; Best, W. M. J.; Ciesla, F.; Csak, B.; Dupuy, T. J.; Handler, G.; Heng, K.; Korhonen, H.; Kovacs, J.; Kozakis, T.; Kriskovics, L.; Schmitt, J. R.; Szabo, Gy.; Szabo, R.; Wang, J.; Goodman, S.; Hoekstra, A.; Jek, K. J. (14 Sep 2015). "Planet Hunter X. KIC 8462852 - Where's the flux?". arXiv:1509.03622 [astro-ph.SR].

[36] Ross Anderson (13 Oct 2015). "The Most Mysterious Star in our Galaxy". The Atlantic.

[37] Ian O'Neill (14 Oct 2015). "Has Kepler Discovered an Alien Megastructure?". Discovery News.

[38] Wright, Jason T.; Cartier, Kimberly M. S.; Zhao, Ming; Jontof-Hutter, Daniel; Ford, Eric B. (2015). "The Ĝ Search for Extraterrestrial Civilizations with Large Energy Supplies. IV. The Signatures and Information Content of Transiting Megastructures". arXiv:1510.04606 [astro-ph.EP].

[39] Mullen, Leslie (2002). "Alien Intelligence Depends on Time Needed to Grow Brains". *Astrobiology Magazine*. Space.com. Retrieved April 21, 2006.

[40] Garrett, Michael (August 12, 2015). "The application of the Mid-IR radio correlation to the Ĝ sample and the search for advanced extraterrestrial civilisations" (PDF). *Astronomy & Astrophysics* **581**: L5. arXiv:1508.02624. Bibcode:2015A&A...581L...5G.doi:10.1051/0004-6361/201526687.

[41] Gurzadyan, V.G. (2005). "Kolmogorov complexity, string information, panspermia and the Fermi paradox". *Observatory* **125**: 352. arXiv:physics/0508010. Bibcode:2005Obs...125..352G.

[42] Udry, S.; Bonfils, X.; Delfosse, X.; Forveille, T.; Mayor, M.; Perrier, C.; Bouchy, F.; Lovis, C.; Pepe, F.; Queloz, D.; Bertaux, J.-L. (2007). "The HARPS search for southern extra-solar planets" (PDF). *Astronomy and Astrophysics* **469** (3): L43. arXiv:0704.3841. Bibcode:2007A&A...469L..43U. doi:10.1051/0004-6361:20077612.

[43] Matsos, Helen (June 16, 2009). "Habitable Planet Signposts". Astrobiology magazine. Retrieved August 19, 2010.

[44] Bracewell, R. N. (1960). "Communications from Superior Galactic Communities". *Nature* **186** (4726): 670–671. Bibcode:1960 doi:10.1038/186670a0.

[45] Papagiannis, M. D. (1978). "Are We all Alone, or could They be in the Asteroid Belt?". *Quarterly Journal of the Royal Astronomical Society* **19**: 277–281. Bibcode:1978QJRAS..19..277P.

[46] Robert A. Freitas Jr. (November 1983). "Extraterrestrial Intelligence in the Solar System: Resolving the Fermi Paradox". *Journal of the British Interplanetary Society* **36**. pp. 496–500.

[47] Freitas, Robert A. Jr; Valdes, F (1985). "The search for extraterrestrial artifacts (SETA)". *Acta Astronautica* **12** (12): 1027–1034. Bibcode:1985AcAau..12.1027F. doi:10.1016/0094-5765(85)90031-1. Retrieved August 19, 2010.

[48] Dyson, Freeman J. (1960). "Search for Artificial Stellar Sources of Infra-Red Radiation". *Science* **131** (3414): 1667–1668. Bibcode:1960Sci...131.1667D. doi:10.1126/science.131.3414.1667. PMID 17780673.

[49] Wright, J. T.; Mullan, B.; Sigurðsson, S.; Povich, M. S. (2014). "The Ĝ Infrared Search for Extraterrestrial Civilizations with Large Energy Supplies. I. Background and Justification". *The Astrophysical Journal* **792**: 26. arXiv:1408.1133. Bibcode:2014ApJ...792...26W. doi:10.1088/0004-637X/792/1/26.

[50] Wright, J. T.; Griffith, R.; Sigurðsson, S.; Povich, M. S.; Mullan, B. (2014). "The Ĝ Infrared Search for Extraterrestrial Civilizations with Large Energy Supplies. II. Framework, Strategy, and First Result". *The Astrophysical Journal* **792**: 27. arXiv:1408.1134. Bibcode:2014ApJ...792...27W. doi:10.1088/0004-637X/792/1/27.

[51] "Fermilab Dyson Sphere search program". Fermi National Accelerator Laboratory. Retrieved February 10, 2008.

[52] Wright, J. T.; Mullan, B; Sigurdsson, S; Povich, M. S (2014). "The Ĝ Infrared Search for Extraterrestrial Civilizations with Large Energy Supplies. III. The Reddest Extended Sources in WISE". *The Astrophysical Journal Supplement Series* **217** (2): 25. arXiv:1504.03418. Bibcode:2015ApJS..217...25G. doi:10.1088/0067-0049/217/2/25.

[53] "Alien Supercivilizations Absent from 100,000 Nearby Galaxies". Scientific American. April 17, 2015.

[54] Ward, Peter D.; Brownlee, Donald (January 14, 2000). *Rare Earth: Why Complex Life is Uncommon in the Universe* (1st ed.). Springer. p. 368. ISBN 978-0-387-98701-9.

[55] Lineweaver, Charles H (2008). *Paleontological tests: human-like intelligence is not a convergent feature of evolution.* From fossils to astrobiology. Springer. pp. 353–368. arXiv:0711.1751.

[56] "The Intelligent-Life Lottery". New York Times. Aug 18, 2014.

[57] Webb, 2nd edition, 2015, Chapters 36-39.

[58] Bostrom, Nick. "Existential Risks Analyzing Human Extinction Scenarios and Related Hazards". Retrieved October 4, 2009.

[59] Sagan, Carl. "Cosmic Search Vol. 1 No. 2". *Cosmic Search Magazine*. Retrieved 2015-07-21.

[60] Hawking, Stephen. "Life in the Universe". *Public Lectures*. University of Cambridge. Archived from the original on April 21, 2006. Retrieved May 11, 2006.

[61] Soter, Steven (2005). "SETI and the Cosmic Quarantine Hypothesis". *Astrobiology Magazine*. Space.com. Retrieved May 3, 2006.

[62] Archer, Michael (1989). "Slime Monsters Will Be Human Too". *Aust. Nat. Hist* **22**: 546–547.

[63] Webb 2002, p. 112

[64] Melott, A.L. and Lieberman, BS and Laird, CM and Martin, LD and Medvedev, MV and Thomas, BC and Cannizzo, JK and Gehrels, N. and Jackman, CH (2004). "Did a gamma-ray burst initiate the late Ordovician mass extinction?" (PDF). *International Journal of Astrobiology* (Cambridge Univ Press) **3** (1): 55–61. arXiv:astro-ph/0309415. Bibcode:2004IJAsB...3...55M. doi:10.1017/S1473550404001910.

[65] Nick Bostrom, Milan M. Ćirković. *Global catastrophic risks.* Section 12.5 – The Fermi Paradox and Mass Extinctions.

[66] Guth, Alan (2007). "Eternal Inflation and its Implications" (PDF). *Journal of Physics A: Mathematical and Theoretical* **40** (25): 6811–6826. arXiv:hep-th/0702178. Bibcode:2007JPhA...40.6811G. doi:10.1088/1751-8113/40/25/S25.

[67] Webb 2002, pp. 62–71

[68] Vakoch, Douglas (November 15, 2001). "Decoding E.T.: Ancient Tongues Point Way To Learning Alien Languages". SETI Institute. Retrieved August 19, 2010.

[69] Newman, W.T. and Sagan, C. (1981). "Galactic civilizations: Population dynamics and interstellar diffusion". *Icarus* **46** (3): 293–327. Bibcode:1981Icar...46..293N. doi:10.1016/0019-1035(81)90135-4.

[70] Brin, Glen David (1983). "The 'Great Silence': The Controversy Concerning Extraterrestrial Intelligent Life". *Quarterly Journal of the Royal Astronomical Society* **24**: 287, 298. Bibcode:1983QJRAS..24..283B.

[71] Landis, Geoffrey (1998). "The Fermi Paradox: An Approach Based on Percolation Theory". *Journal of the British Interplanetary Society* **51**: 163–166. Bibcode:1998JBIS...51..163L.

[72] Turnbull, Margaret C.; Tarter, Jill C. (2003). "Target Selection for SETI. I. A Catalog of Nearby Habitable Stellar Systems" (PDF). *The Astrophysical Journal Supplement Series* **145** (1): 181–198. arXiv:astro-ph/0210675. Bibcode:2003ApJS..145..181T. doi:10.1086/345779. Retrieved August 19, 2010.

[73] Stephenson, D. G (1984). "Solar Power Satellites as Interstellar Beacons". *Quarterly Journal of the Royal Astronomical Society* (Royal Astronomical Society) **25** (1): 80. Bibcode:1984QJRAS..25...80S.

[74] "Cosmic Search Vol. 1 No. 3". Bigear.org. September 21, 2004. Retrieved July 3, 2010.

[75] Learned, J; Pakvasa, S; Zee, A (2009). "Galactic neutrino communication". *Physics Letters B* **671** (1): 15–19. arXiv:0805.2429. Bibcode:2009PhLB..671...15L. doi:10.1016/j.physletb.2008.11.057.

[76] Bostrom, Nick (22 April 2008). "Where Are They?". MIT Technology Review. Retrieved 21 June 2015.

[77] Webb 2002, p. 86

[78] Webb, Chapter 15: "They Stay at Home and Surf the Web"

[79] Schombert, James. "Fermi's paradox (i.e. Where are they?)" *Cosmology Lectures*, University of Oregon.

[80] Hamming, RW (1998). "Mathematics on a distant planet". *The American mathematical monthly* **105** (7): 640–650. doi:10.2307/JSTOR 2589247.

[81] Long, K. F. (2011-11-25). *Deep Space Propulsion: A Roadmap to Interstellar Flight*. p. 114. ISBN 978-1-4614-0607-5. Retrieved 23 June 2015.

[82] Cook, Stephen P. "SETI: Assessing Imaginative Proposals". *Life on Earth and other Planetary Bodies*. p. 54. ISBN 978-94-007-4966-5.

[83] Tarter, Jill (2006). "What is SETI?a". *Annals of the New York Academy of Sciences* **950** (1): 269–75. Bibcode:2001NYASA.95 doi:10.1111/j.1749-6632.2001.tb02144.x. PMID 11797755.

[84] Steven V. W. Beckwith (2008). "Detecting Life-bearing Extrasolar Planets with Space Telescopes". *The Astrophysical Journal* (IOP Publishing) **684** (2,): 1404–1415. arXiv:0710.1444. Bibcode:2008ApJ...684.1404B. doi:10.1086/590466.

[85] Sparks, W.B. and Hough, J. and Germer, T.A. and Chen, F. and DasSarma, S. and DasSarma, P. and Robb, F.T. and Manset, N. and Kolokolova, L. and Reid, N.J.; et al. (2009). "{Detection of circular polarization in light scattered from photosynthetic microbes" (PDF). *Proceedings of the National Academy of Sciences* (National Acad Sciences) **106** (14–16): 7816. doi:10.1016/j.jqsrt.2009.02.028.

[86] Webb, Stephen (2015-05-18). *If the Universe Is Teeming with Aliens ... WHERE IS EVERYBODY?: Fifty Solutions to the Fermi Paradox and the Problem of Extraterrestrial Life*. ISBN 978-0-387-95501-8. Retrieved 21 June 2015.

[87] Alexander Zaitsev (2006). "The SETI paradox". arXiv:physics/0611283 [physics.gen-ph].

[88] "Should We Call the Cosmos Seeking ET? Or Is That Risky?". New York Times. Feb 13, 2015.

[89] "Declaration of Principles Concerning Activities Following the Detection of Extraterrestrial Intelligence".

[90] Michaud, M. (2003). "Ten decisions that could shake the world". *Space Policy* **19** (2): 131–950. doi:10.1016/S0265-9646(03)000 5.

[91] Ball, J (1973). "The zoo hypothesis". *Icarus* **19** (3): 347–349. Bibcode:1973Icar...19..347B. doi:10.1016/0019-1035(73)90111-5.

[92] Hair, Thomas W. (2011). "Temporal Dispersion of the Emergence of Intelligence: An Inter-arrival Time Analysis". *International Journal of Astrobiology* 10(2): 131–135 (2011). doi:10.1017/S1473550411000024

[93] Baxter, Stephen (2001). "The Planetarium Hypothesis: A Resolution of the Fermi Paradox". *Journal of the British Interplanetary Society* **54** (5/6): 210–216. Bibcode:2001JBIS...54..210B.

[94] Carrigan, Richard A. (2006). "Do potential SETI signals need to be decontaminated?". *Acta Astronautica* **58** (2): 112–117. Bibcode:2006AcAau..58..112C. doi:10.1016/j.actaastro.2005.05.004.

[95] Marsden, P. (1998). "Memetics and social contagion: Two sides of the same coin". *Journal of Memetics-Evolutionary Models of Information Transmission* **2** (2): 171–185.

[96] Beatriz Gato-Rivera (1970). "A Solution to the Fermi Paradox: The Solar System, Part of a Galactic Hypercivilization?". arXiv:physics/0512062 [physics.pop-ph].

[97] Tough, Allen (1986). "What Role Will Extraterrestrials Play in Humanity's Future?" (PDF). *Journal of the British Interplanetary Society* **39** (11): 492–498. Bibcode:1986JBIS...39..491T.

[98] Mogi, Ken (2014). "Free will and paranormal beliefs". *Frontiers in psychology* (Frontiers Media SA) **5**: 281. doi:10.3389/fpsyg. PMC 3980098. PMID 24765084.

[99] Shermer, Michael (2011). "UFOs, UAPs and CRAPs". *Scientific American* (Nature Publishing Group) **304** (4): 90–90. doi:10.1038/scientificamerican0411-90.

[100] A. Tough (1990). "A critical examination of factors that might encourage secrecy". *Acta Astronautica* **21** (21.): 97–102. Bibcode:1990AcAau..21...97T. doi:10.1016/0094-5765(90)90134-7.

[101] Ashley Vance (31 July 2006). "SETI urged to fess up over alien signals". The Guardian.

[102] "UFO Hunters Keep Pressing White House For Answers Through 'We The People' Petitions". Huffington Post. 6 Dec 2011.

[103] G. Seth Shostak (2009). *Confessions of an Alien Hunter: A Scientist's Search for Extraterrestrial Intelligence*. National Geographic. ISBN 978-1-4262-0392-3. Page 17.

[104] Brin, David (2012). *Existence*. Tor Books. ISBN 978-0-7653-0361-5.

[105] "The Fermi Paradox". IMDB.

20.11   Bibliography

- Crowe, Michael J. (2008). *The Extraterrestrial Life Debate, Antiquity to 1915*. University of Notre Dame Press. ISBN 978-0-268-02368-3.

- Shklovskii, Iosif; Sagan, Carl (1966). *Intelligent Life in the Universe*. San Francisco: Holden–Day. ISBN 1-892803-02-X.

- Webb, Stephen (2002). *If the Universe Is Teeming with Aliens... Where Is Everybody? Fifty solutions to the Fermi Paradox and the Problem of Extraterrestrial Life*. Copernicus Books. ISBN 0-387-95501-1.

- Webb, Stephen (2015). *If the Universe Is Teeming with Aliens... Where Is Everybody? Seventy five Solutions to the Fermi Paradox and the Problem of Extraterrestrial Life* (2nd ed.). Copernicus Books. ISBN 978-3-319-13235-8.

20.12   Further reading

- Ben Zuckerman and Michael H. Hart. Extraterrestrials: Where Are They? ISBN 0-521-44803-4 Amazon

- Michaud, Michael (2006). *Contact with Alien Civilizations: Our Hopes and Fears about Encountering Extraterrestrials*. Copernicus Books. ISBN 978-0-387-28598-6.

20.13 External links

- The dictionary definition of Fermi paradox at Wiktionary

- Interstellar Radio Messages

- So much space, so little time: why aliens haven't found us yet by Ian Sample, *The Guardian* January 18, 2007

- Fermi Paradox debate *Astrobiology Magazine* July 2002. Michael Meyer, Frank Drake, Christopher McKay, Donald Brownlee, & David Grinspoon.

- Translation of the documentary "Overcome the Great Silence!" provided by Aleksandr Leonidovich Zaitsev

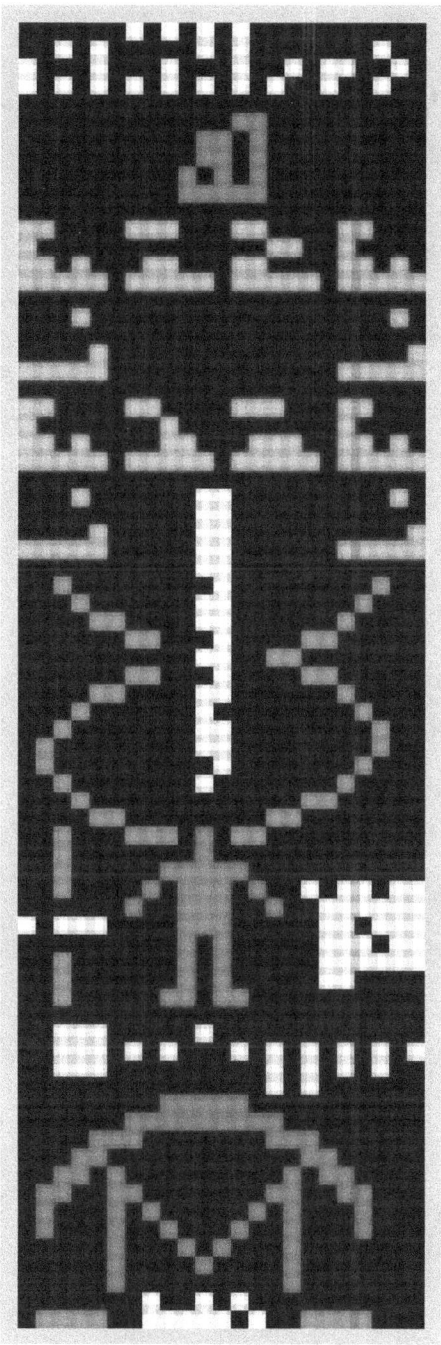

A graphical representation of the Arecibo message – Humanity's first attempt to use radio waves to actively communicate its existence to alien civilizations

Enrico Fermi (1901–1954)

Planet Earth

Los Alamos National Laboratory

An artist's depiction of the "little green man" described in the novel Martians, Go Home

Radio telescopes are often used by SETI projects

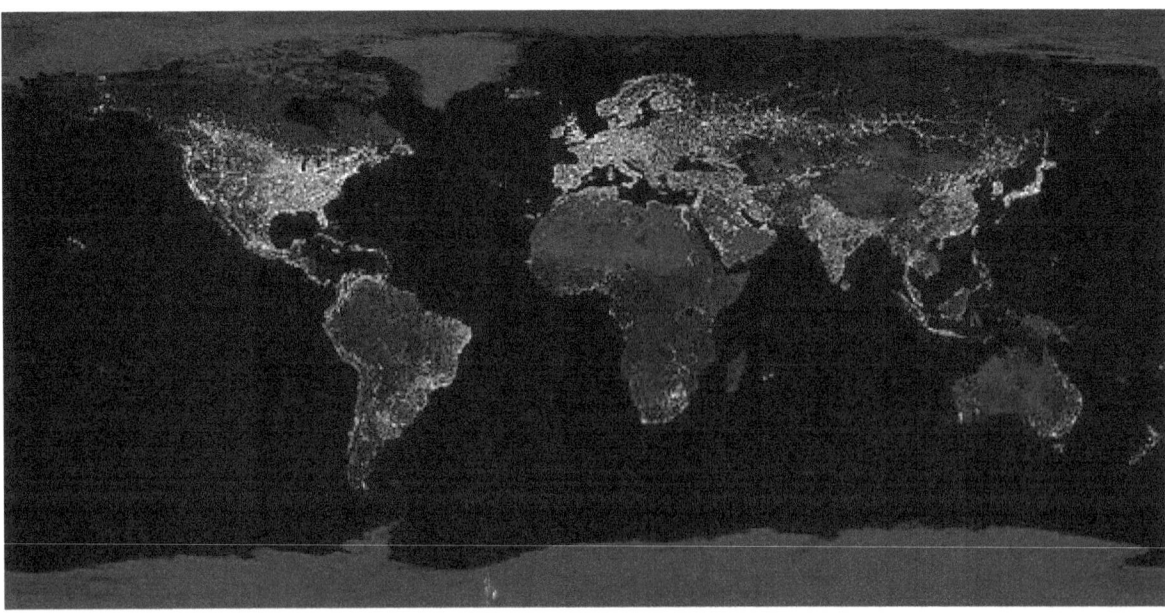

A composite picture of Earth at night, created with data from the Defense Meteorological Satellite Program (DMSP) Operational Linescan System (OLS). Large-scale artificial lighting produced by the human civilization is detectable from space.

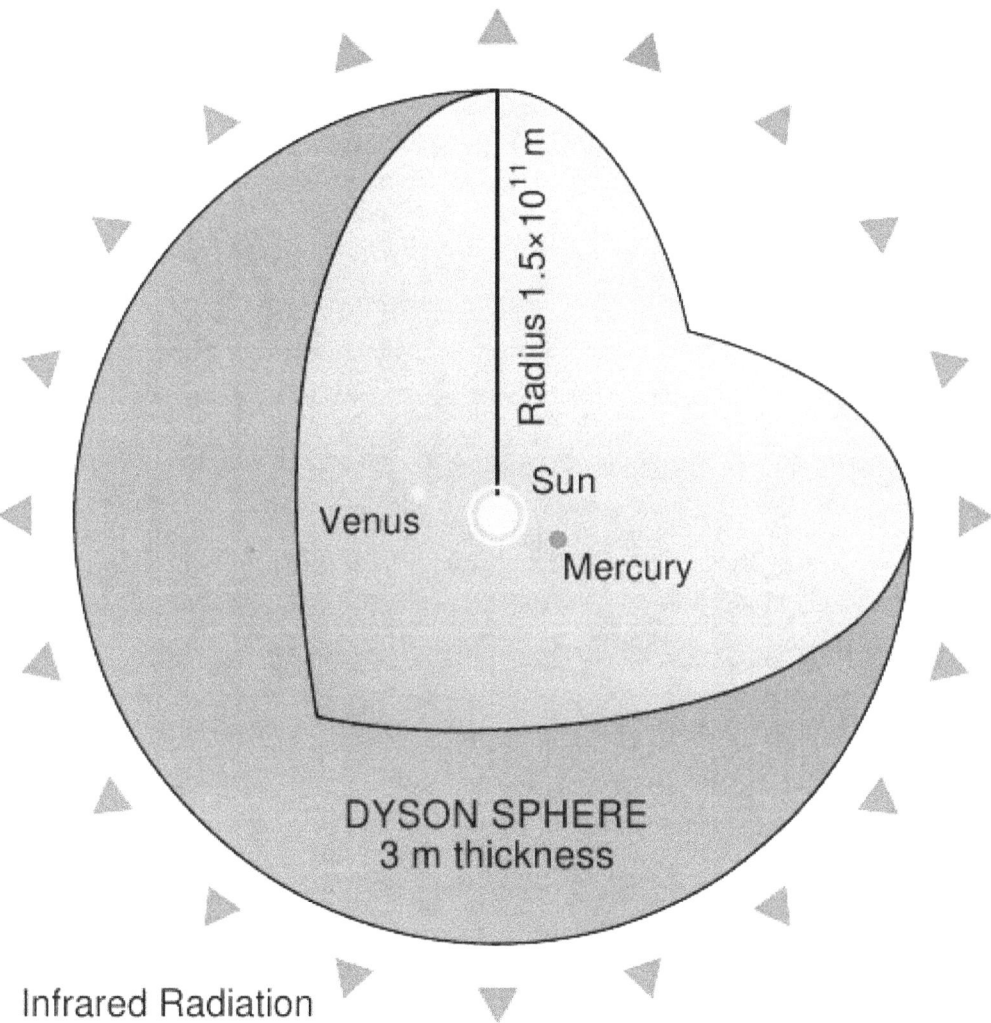

Radius 1.5×10^{11} m

Sun

Venus

Mercury

DYSON SPHERE
3 m thickness

Infrared Radiation

A variant of the speculative Dyson sphere. Such large scale artifacts would drastically alter the spectrum of a star.

A 23 kiloton tower shot called BADGER, fired as part of the Operation Upshot-Knothole nuclear test series.

NASA's conception of the Terrestrial Planet Finder

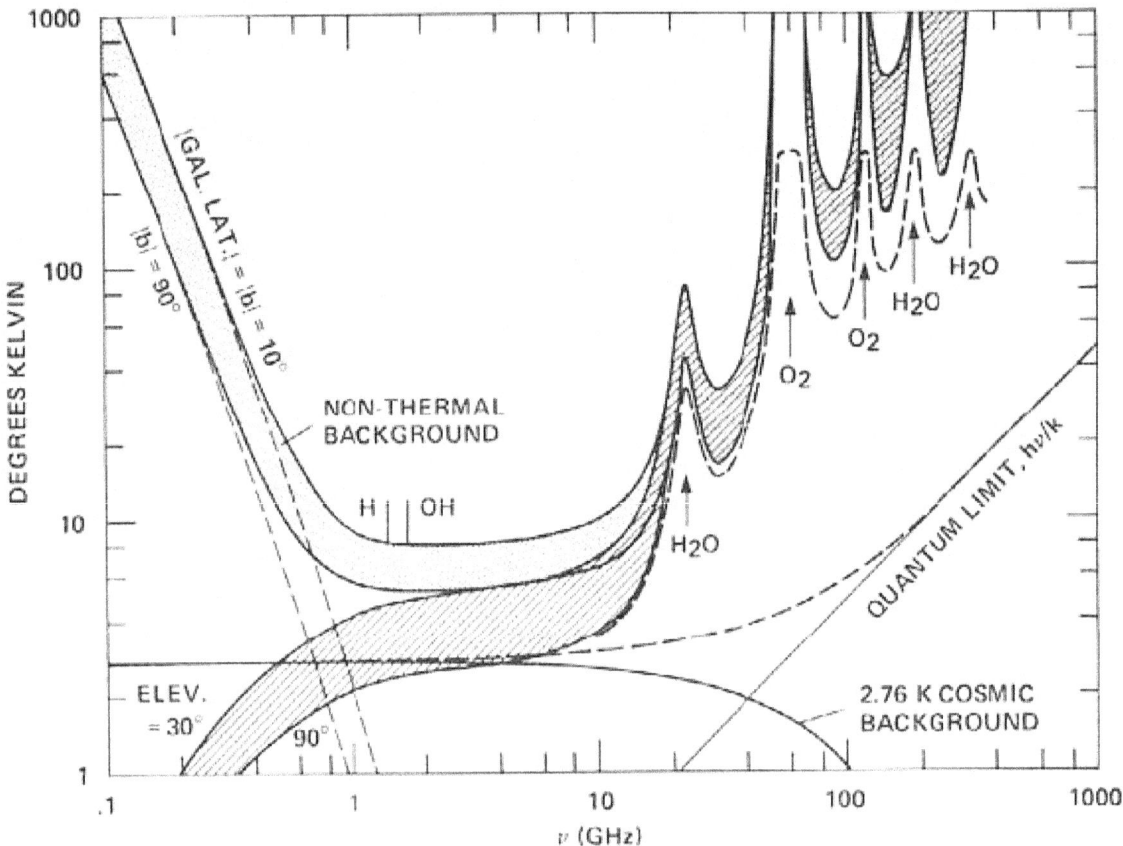

Microwave window as seen by a ground based system. From NASA report SP-419: SETI – the Search for Extraterrestrial Intelligence

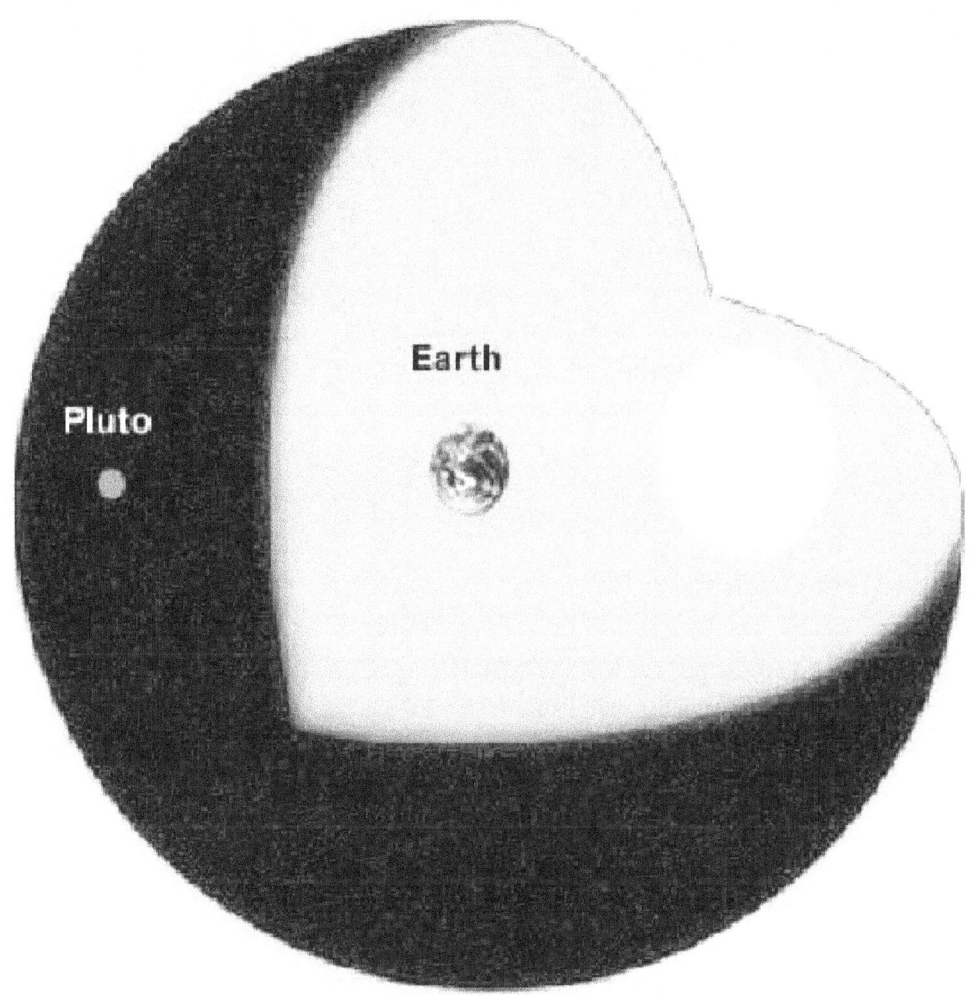

Schematic representation of a planetarium simulating the universe to humans. The "real" universe is outside the black sphere, the simulated one projected on/filtered through it.

Chapter 21

Cosmic pluralism

This article is about the concept of cosmic pluralism. For other uses of the term, see Pluralism (disambiguation).

Cosmic pluralism, the **plurality of worlds**, or simply **pluralism**, describes the philosophical belief in numerous "worlds" in addition to Earth (possibly an infinite number), which may harbour extraterrestrial life.

The debate over pluralism began as early as the time of Anaximander (c. 610 – c. 546 BC) as an abstract metaphysical argument,[1] long predating the scientific Copernican conception that the Earth is one of numerous planets. It has continued, in a variety of forms, until the modern era.

21.1   Ancient Greek debates

In Greek times, the debate was largely philosophical and did not conform to present notions of cosmology. Cosmic pluralism was a corollary to notions of infinity and the purported multitude of life-bearing worlds were more akin to parallel universes (either contemporaneously in space or infinitely recurring in time) than to different solar systems. After Anaximander opened the door to an infinite universe, a strong pluralist stance was adopted by the atomists, notably Leucippus, Democritus, and Epicurus. While these were prominent thinkers, their opponents—Plato and Aristotle—had greater effect. They argued that the Earth is unique and that there can be no other systems of worlds.[2][3] This stance neatly dovetailed with later Christian ideas[4] and pluralism was effectively suppressed for approximately a millennium.

21.2   Medieval thought

During the Middle Ages, cosmic pluralism was depicted in fictional Arabic literature. "The Adventures of Bulukiya", a tale from the *One Thousand and One Nights* (*Arabian Nights*), depicted a cosmos consisting of different worlds, some larger than Earth and each with their own inhabitants.[5] Fakhr ad-Din ar-Razi (1149–1209), in dealing with his conception of physics and the physical world in his Matalib, rejects the Aristotelian and Avicennian notion of the Earth's centrality within the universe, but instead argues that there are "a thousand thousand worlds (alfa alfi 'awalim) beyond this world such that each one of those worlds be bigger and more massive than this world as well as having the like of what this world has." To support his theological argument, he cites the Qur'anic verse, "All praise belongs to God, Lord of the Worlds," emphasizing the term "Worlds." Another traditional Islamic exegesis of this verse from Surah al-Fatiha, typified by Ibn Taymiyyah, interprets the "worlds" as being the heavenly and the earthly, or the angelic, the human, the animal, and the world of the djinn, similar to the traditional Christian exegesis of the "three heavens" [God's abode, stellar space, atmospheric space] of the Bible.

21.3   Scholastic thinkers

Eventually the Ptolemaic-Aristotelian system was challenged and pluralism reasserted, first tentatively by scholastics and then more seriously by followers of Copernicus. The telescope appeared to prove that a multitude of life was reasonable and an expression of God's creative omnipotence; still powerful theological opponents, meanwhile, continued to insist that although the Earth may have been displaced from the center of the cosmos, it was still the unique focus of God's creation. Thinkers such as Johannes Kepler were willing to admit the possibility of pluralism without truly supporting it.

21.4   Renaissance

Giordano Bruno introduced in his works the idea of multiple worlds instantiating the infinite possibilities of a pristine, indivisible One. Bruno (from the mouth of his character Philotheo) in his *De l'infinito universo et mondi* (1584) claims that "innumerable celestial bodies, stars, globes, suns and earths may be sensibly perceived therein by us and an infinite number of them may be inferred by our own reason."[6]

21.5   Enlightenment

During the Scientific Revolution and the later Enlightenment, cosmic pluralism became a mainstream possibility. Bernard le Bovier de Fontenelle's *Entretiens sur la pluralité des mondes* (*Conversations on the Plurality of Worlds*) of 1686 was an important work from this period, speculating on pluralism and describing the new Copernican cosmology.[7] Pluralism was also championed by philosophers such as John Locke, astronomers such as William Herschel and even politicians, including John Adams and Benjamin Franklin. As greater scientific skepticism and rigour were applied to the question it ceased to be simply a matter of philosophy and theology and was properly bounded by astronomy and biology.

The French astronomer Camille Flammarion was one of the chief proponents of cosmic pluralism during the latter half of the nineteenth century. His first book, *La pluralité des mondes habités* (1862) was a great popular success, going through 33 editions in its first twenty years. Flammarion was one of the first people to put forward the idea that extraterrestrial beings were genuinely alien, and not simply variations of earthly creatures.[8]

21.6   Modern thought

In the late nineteenth and twentieth centuries the term "cosmic pluralism" became largely archaic as knowledge diversified and the speculation on extraterrestrial life focused on particular bodies and observations. The historic debate continues to have modern parallels, however. Carl Sagan and Frank Drake, for instance, could well be considered "pluralists" while proponents of the Rare Earth hypothesis are modern skeptics.

21.7   See also

- Extraterrestrial life

- Extraterrestrial life in fiction

- Hindu cosmology

- Mediocrity principle

- Mormon cosmology

- Planetary habitability

21.8 References

[1] Simplicius, *Commentary on Aristotle's Physics*, 1121, 5–9

[2] Michael J. Crowe (1999). *The Extraterrestrial Life Debate, 1750–1900*. Courier Dover Publications. ISBN 0-486-40675-X.

[3] David Darling article

[4] Wiker, Benjamin D. (November 4, 2002). "Alien Ideas: Christianity and the Search for Extraterrestrial Life". *Crisis Magazine*. Archived from the original on February 10, 2003.

[5] Irwin, Robert (2003). *The Arabian Nights: A Companion*. Tauris Parke Paperbacks. p. 204 & 209. ISBN 1-86064-983-1.

[6] Hetherington, Norriss S., ed. (April 2014) [1993]. *Encyclopedia of Cosmology (Routledge Revivals): Historical, Philosophical, and Scientific Foundations of Modern Cosmology*. Routledge. p. 419. ISBN 9781317677666. Retrieved 29 March 2015.

[7] Conversations on the Plurality of Worlds— Bernard le Bovier de Fontenelle

[8] Flammarion, (Nicolas) Camille (1842–1925)— The Internet Encyclopedia of Science

21.9 Further reading

- Ernst Benz (1978). *Kosmische Bruderschaft. Die Pluralität der Welten. Zur Ideengeschichte des Ufo-Glaubens*. Aurum Verlag. ISBN 3-591-08061-6. *(later titled "Außerirdische Welten. Von Kopernikus zu den Ufos")*

Chapter 22

Search for extraterrestrial intelligence

"SETI" redirects here. For other uses, see SETI (disambiguation).

The **search for extraterrestrial intelligence (SETI)** is a collective term for the scientific search for intelligent extraterrestrial

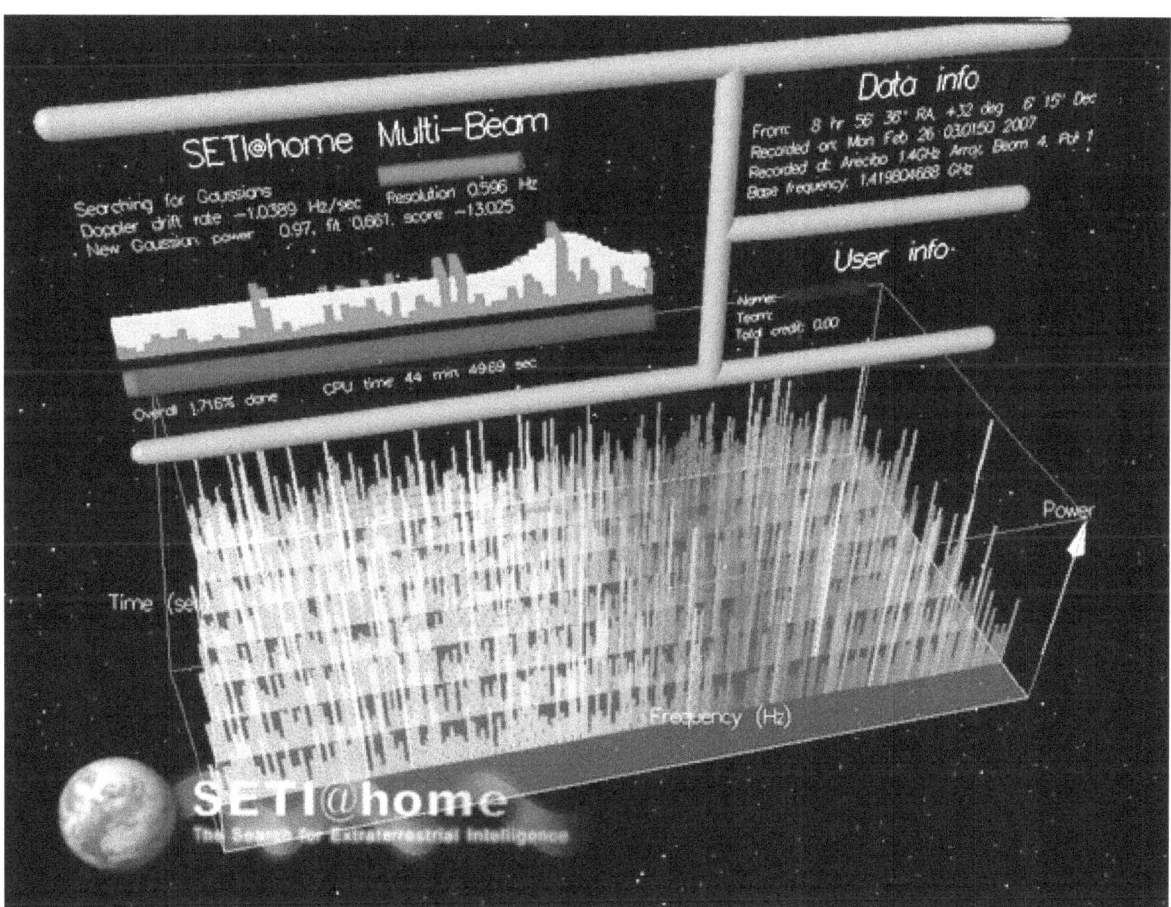

Screen shot of the screensaver for SETI@home, a distributed computing project in which volunteers donate idle computer power to analyze radio signals for signs of extraterrestrial intelligence

life. For example, monitoring electromagnetic radiation for signs of transmissions from civilizations on other worlds.[1][2]

There are great challenges in searching the universe for signs of intelligent life, including their identification and interpretation. As various SETI projects have progressed, some have criticized early claims by researchers as being too "euphoric".[3]

Scientific investigation of the potential phenomenon began shortly after the advent of radio in the early 1900s. Focused international efforts to answer a variety of scientific questions have been going on since the 1980s. More recently, Stephen Hawking, British physicist, and Yuri Milner, Russian billionaire announced a well-funded effort, called the Breakthrough Initiatives, to expand efforts to search for extraterrestrial life.[4]

22.1 History of SETI

22.1.1 Early work

As early as 1896, Nikola Tesla suggested that an extreme version of his wireless electrical transmission system could be used to contact beings on Mars.[5] In 1899 while conducting experiments at his Colorado Springs experimental station, he thought he had detected a signal from the planet since an odd repetitive static signal seemed to cut off when Mars set in the night sky. Analysis of Tesla's research has ranged from suggestions that Tesla detected nothing, he simply was misunderstanding the new technology he was working with,[6] to claims that Tesla may have been observing signals from Marconi's European radio experiments and even that he could have picked up naturally occurring Jovian plasma torus signals.[7] In the early 1900s, Guglielmo Marconi, Lord Kelvin, and David Peck Todd also stated their belief that radio could be used to contact Martians, with Marconi stating that his stations had also picked up potential Martian signals.[8]

On August 21–23, 1924, Mars entered an opposition closer to Earth than any time in a century before or the next 80 years.[9] In the United States, a "National Radio Silence Day" was promoted during a 36-hour period from August 21–23, with all radios quiet for five minutes on the hour, every hour. At the United States Naval Observatory, a radio receiver was lifted 3 kilometres (1.9 miles) above the ground in a dirigible tuned to a wavelength between 8 and 9 kilometres (approx. 6 mi), using a "radio-camera" developed by Amherst College and Charles Francis Jenkins. The program was led by David Peck Todd with the military assistance of Admiral Edward W. Eberle (Chief of Naval Operations), with William F. Friedman (chief cryptographer of the United States Army), assigned to translate any potential Martian messages.[10][11]

A 1959 paper by Philip Morrison and Giuseppe Cocconi first pointed out the possibility of searching the microwave spectrum, and proposed frequencies and a set of initial targets[12][13]

In 1960, Cornell University astronomer Frank Drake performed the first modern SETI experiment, named "Project Ozma", after the Queen of Oz in L. Frank Baum's fantasy books.[14] Drake used a radio telescope 26 metres (85 ft) in diameter at Green Bank, West Virginia, to examine the stars Tau Ceti and Epsilon Eridani near the 1.420-gigahertz marker frequency, a region of the radio spectrum dubbed the "water hole" due to its proximity to the hydrogen and hydroxyl radical spectral lines. A 400-kilohertz band was scanned around the marker frequency, using a single-channel receiver with a bandwidth of 100 hertz. He found nothing of interest.

TheSoviet scientists took a strong interest in SETI during the1960s and performed a number of searches with omnidirectio -nal antennas in the hope of picking up powerful radio signals.
Soviet astronomer Iosif Shklovsky wrote the pioneering book in the field, *Universe, Life, Intelligence* (1962), which was expanded upon by American astronomer Carl Sagan as the best-selling book *Intelligent Life in the Universe*(1966).[15]

In the March 1955 issue of *Scientific American*, John D. Kraus described a concept to scan the cosmos for natural radio signals using a flat-plane radio telescope equipped with a parabolic reflector. Within two years, his concept was approved for construction by Ohio State University. With US$71,000 total in grants from the National Science Foundation, construction began on a 20-acre (8.1 ha) plot in Delaware, Ohio. This Ohio State University Radio Observatory telescope was called "Big Ear". Later, it began the world's first continuous SETI program, called the Ohio State University SETI program.

In 1971, NASA funded a SETI study that involved Drake, Bernard M. Oliver of Hewlett-Packard Corporation, and others. The resulting report proposed the construction of an Earth-based radio telescope array with 1,500 dishes known as "Project Cyclops". The price tag for the Cyclops array was US$10 billion. Cyclops was not built, but the report[16] formed the basis of much SETI work that followed.

The OSU SETI program gained fame on August 15, 1977, when Jerry Ehman, a project volunteer, witnessed a startlingly strong signal received by the telescope. He quickly circled the indication on a printout and scribbled the exclamation

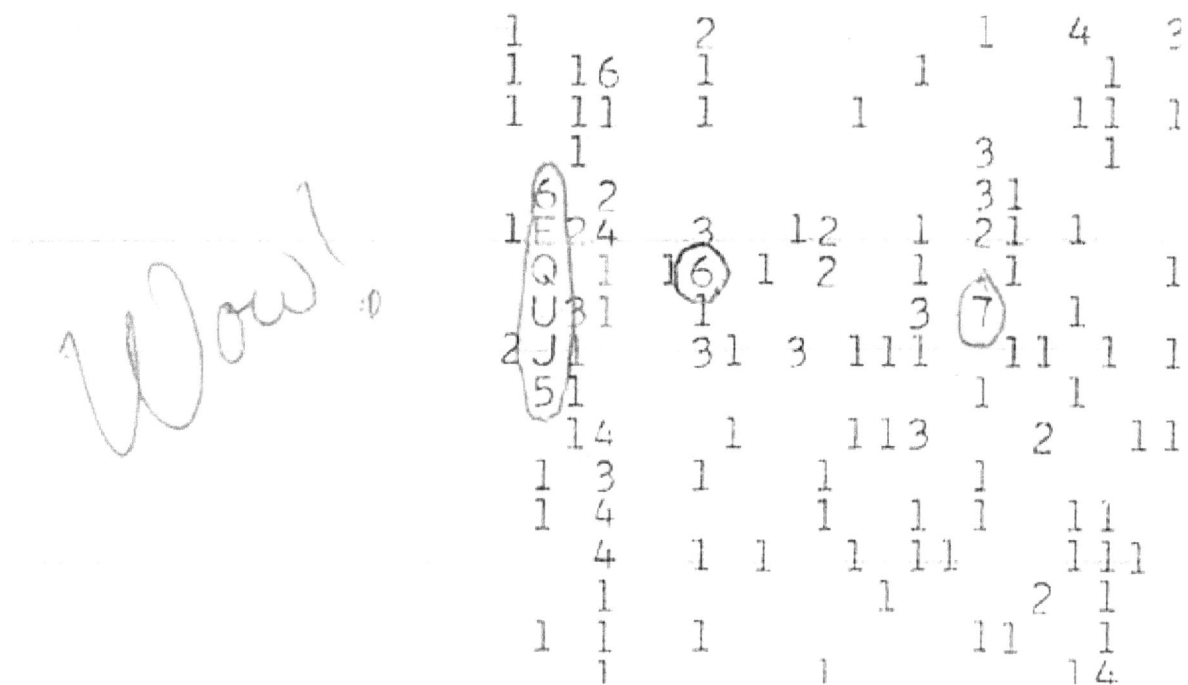

The WOW! Signal
Credit: The Ohio State University Radio Observatory and the North American AstroPhysical Observatory (NAAPO).

"Wow!" in the margin. Dubbed the *Wow! signal*, it is considered by some to be the best candidate for a radio signal from an artificial, extraterrestrial source ever discovered, but it has not been detected again in several additional searches.[17]

22.1.2   Sentinel, META, and BETA

In 1980, Carl Sagan, Bruce Murray, and Louis Friedman founded the U.S. Planetary Society, partly as a vehicle for SETI studies.

In the early 1980s, Harvard University physicist Paul Horowitz took the next step and proposed the design of a spectrum analyzer specifically intended to search for SETI transmissions. Traditional desktop spectrum analyzers were of little use for this job, as they sampled frequencies using banks of analog filters and so were restricted in the number of channels they could acquire. However, modern integrated-circuit digital signal processing (DSP) technology could be used to build autocorrelation receivers to check far more channels. This work led in 1981 to a portable spectrum analyzer named "Suitcase SETI" that had a capacity of 131,000 narrow band channels. After field tests that lasted into 1982, Suitcase SETI was put into use in 1983 with the 26-meter (85 ft) Harvard/Smithsonian radio telescope at Oak Ridge Observatory in Harvard, Massachusetts. This project was named "Sentinel" and continued into 1985.

Even 131,000 channels were not enough to search the sky in detail at a fast rate, so Suitcase SETI was followed in 1985 by Project "META", for "Megachannel Extra-Terrestrial Assay". The META spectrum analyzer had a capacity of 8.4 million channels and a channel resolution of 0.05 hertz. An important feature of META was its use of frequency Doppler shift to distinguish between signals of terrestrial and extraterrestrial origin. The project was led by Horowitz with the help of the Planetary Society, and was partly funded by movie maker Steven Spielberg. A second such effort, META II, was begun in Argentina in 1990, to search the southern sky. META II is still in operation, after an equipment upgrade in 1996.

The follow-on to META was named "BETA", for "Billion-channel Extraterrestrial Assay", and it commenced observation on October 30, 1995. The heart of BETA's processing capability consisted of 63 dedicated fast Fourier transform (FFT) engines, each capable of performing a 2^{22}-point complex FFTs in two seconds, and 21 general-purpose personal computers equipped with custom digital signal processing boards. This allowed BETA to receive 250 million simultaneous channels with a resolution of 0.5 hertz per channel. It scanned through the microwave spectrum from 1.400 to 1.720 gigahertz in eight hops, with two seconds of observation per hop. An important capability of the BETA search was rapid and automatic re-observation of candidate signals, achieved by observing the sky with two adjacent beams, one slightly to the east and the other slightly to the west. A successful candidate signal would first transit the east beam, and then the west beam and do so with a speed consistent with Earth's sidereal rotation rate. A third receiver observed the horizon to veto signals of obvious terrestrial origin. On March 23, 1999, the 26-meter radio telescope on which Sentinel, META and BETA were based was blown over by strong winds and seriously damaged.[18] This forced the BETA project to cease operation.

22.1.3   MOP and Project Phoenix

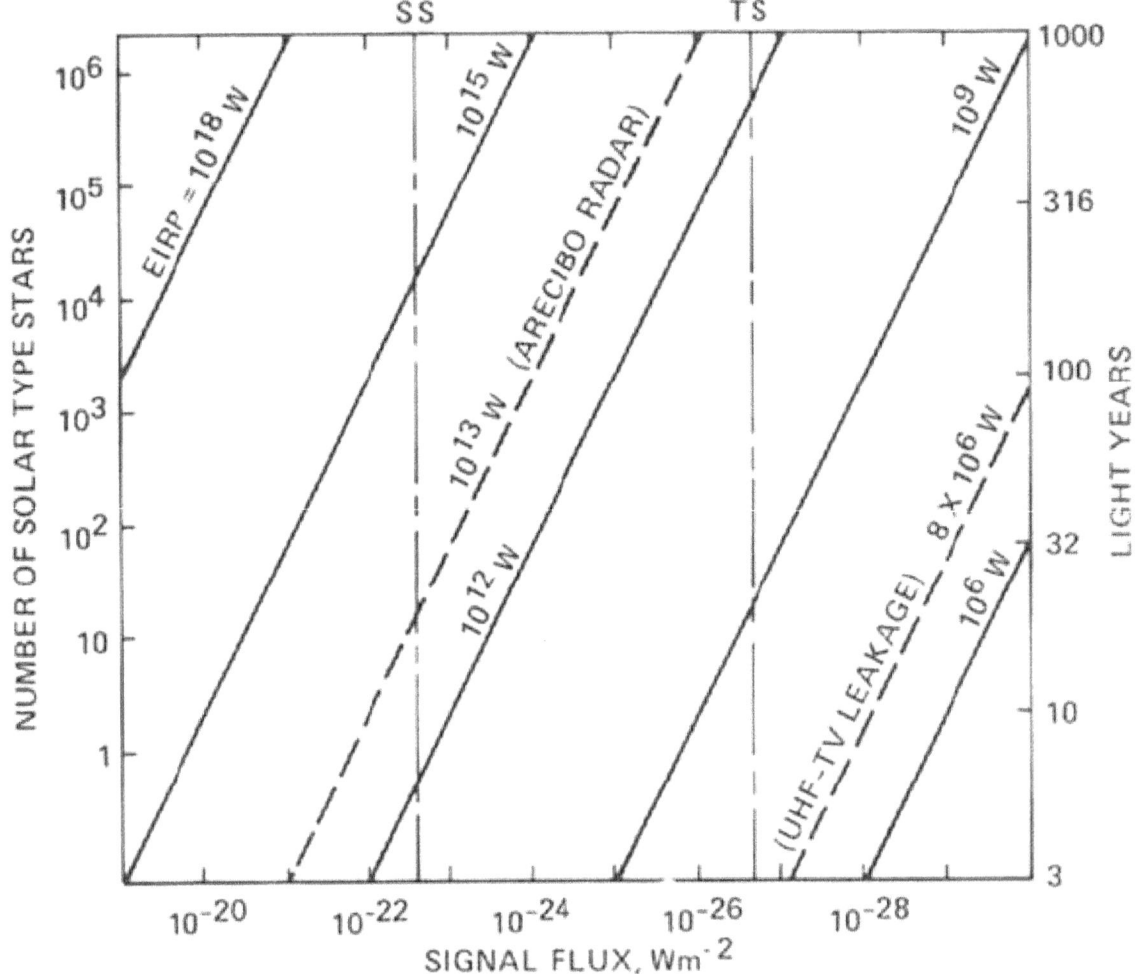

Sensitivity vs range for SETI radio searches. The diagonal lines show transmitters of different effective powers. The x-axis is the sensitivity of the search. The y-axis on the right is the range in light-years, and on the left is the number of Sun-like stars within this range. The vertical line labeled SS is the typical sensitivity achieved by a full sky search, such as BETA above. The vertical line labeled TS is the typical sensitivity achieved by a targeted search such as Phoenix.[19]

In 1978, the NASA SETI program had been heavily criticized by Senator William Proxmire, and funding for SETI research was removed from the NASA budget by Congress in 1981;[20] however, funding was restored in 1982, after Carl Sagan talked with Proxmire and convinced him of the program's value.[20] In 1992, the U.S. government funded an operational SETI program, in the form of the NASA Microwave Observing Program (MOP). MOP was planned as a long-term effort to conduct a general survey of the sky and also carry out targeted searches of 800 specific nearby stars. MOP was to be performed by radio antennas associated with the NASA Deep Space Network, as well as the 140-foot (43 m) radio telescope of the National Radio Astronomy Observatory at Green Bank, West Virginia and the 1,000-foot (300 m) radio telescope at the Arecibo Observatory in Puerto Rico. The signals were to be analyzed by spectrum analyzers, each with a capacity of 15 million channels. These spectrum analyzers could be grouped together to obtain greater capacity. Those used in the targeted search had a bandwidth of 1 hertz per channel, while those used in the sky survey had a bandwidth of 30 hertz per channel.

MOP drew the attention of the United States Congress, where the program was ridiculed[21] and canceled one year after its start.[20] SETI advocates continued without government funding, and in 1995 the nonprofit SETI Institute of Mountain View, California resurrected the MOP program under the name of Project "Phoenix", backed by private sources of funding. Project Phoenix, under the direction of Jill Tarter, is a continuation of the targeted search program from MOP and studies roughly 1,000 nearby Sun-like stars. From 1995 through March 2004, Phoenix conducted observations at the 64-meter (210 ft) Parkes radio telescope in Australia, the 140-foot (43 m) radio telescope of the National Radio Astronomy Observatory in Green Bank, West Virginia, and the 1,000-foot (300 m) radio telescope at the Arecibo Observatory in Puerto Rico. The project observed the equivalent of 800 stars over the available channels in the frequency range from 1200 to 3000 MHz. The search was sensitive enough to pick up transmitters with 1 GW EIRP to a distance of about 200 light-years. According to Prof. Tarter, in 2012 it costs around "$2 million per year to keep SETI research going at the SETI Institute" and approximately 10 times that to support "all kinds of SETI activity around the world."[22]

22.2   Ongoing radio searches

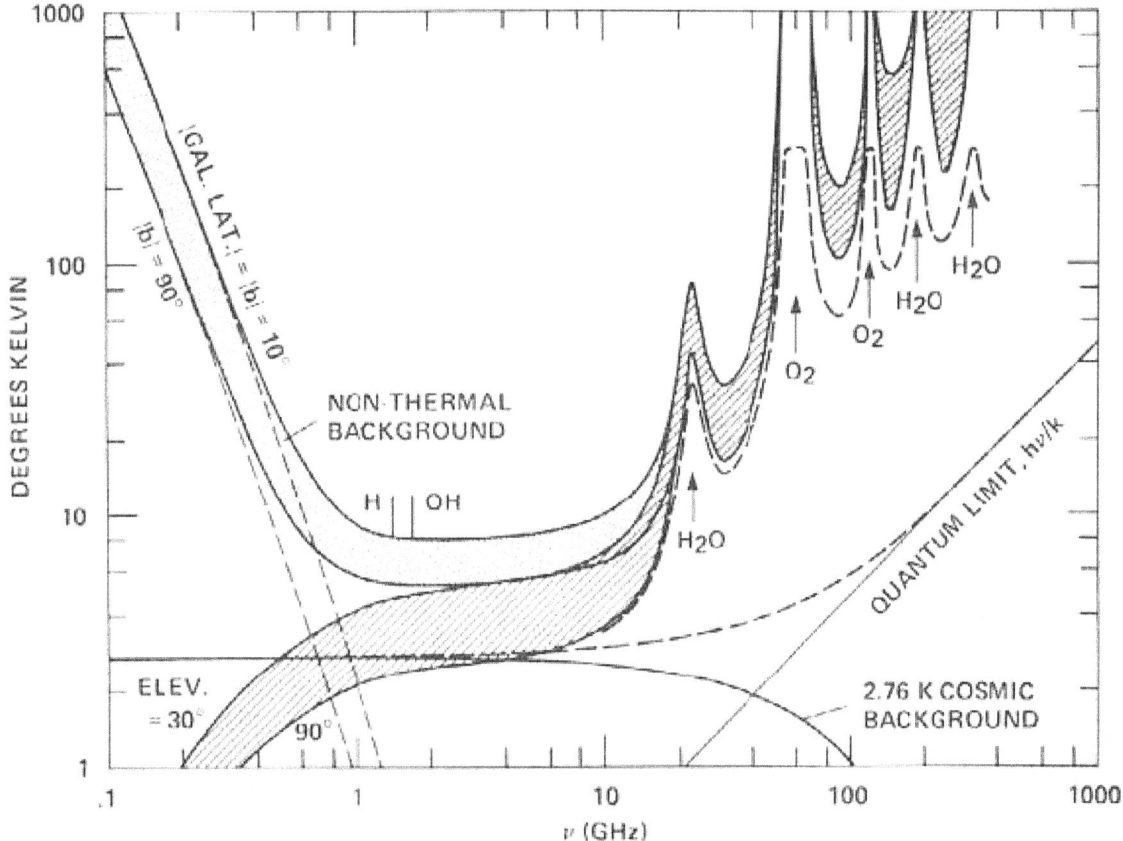

Microwave window as seen by a ground based system. From NASA report SP-419: SETI – the Search for Extraterrestrial Intelligence

Many radio frequencies penetrate Earth's atmosphere quite well, and this led to radio telescopes that investigate the cosmos using large radio antennas. Furthermore, human endeavors emit considerable electromagnetic radiation as a byproduct of communications such as television and radio. These signals would be easy to recognize as artificial due to their repetitive nature and narrow bandwidths. If this is typical, one way of discovering an extraterrestrial civilization might be to detect artificial radio emissions from a location outside the Solar System.

Many international radio telescopes are currently being used for radio SETI searches, including the Low Frequency Array (LOFAR) in Europe, the Murchison Widefield Array (MWA) in Australia, and the Lovell Telescope in the United Kingdom.[23]

22.2.1   Allen Telescope Array

Main article: Allen Telescope Array

The SETI Institute collaborated with the Radio Astronomy Laboratory at University of California, Berkeley to develop a specialized radio telescope array for SETI studies, something like a mini-cyclops array. Formerly known as the One Hectare Telescope (1HT), the concept was renamed the "Allen Telescope Array" (ATA) after the project's benefactor Paul Allen. Its sensitivity would be equivalent to a single large dish more than 100 meters in diameter if completed. Presently, the array under construction has 42 dishes at the Hat Creek Radio Observatory in rural northern California.[24][25]

The full array (ATA-350) is planned to consist of 350 or more offset-Gregorian radio dishes, each 6.1 meters (20 feet) in diameter. These dishes are the largest producible with commercially available satellite television dish technology. The ATA was planned for a 2007 completion date, at a very modest cost of US$25 million. The SETI Institute provided money for building the ATA while University of California, Berkeley designed the telescope and provided operational funding. The first portion of the array (ATA-42) became operational in October 2007 with 42 antennas. The DSP system planned for ATA-350 is extremely ambitious. Completion of the full 350 element array will depend on funding and the technical results from ATA-42.

ATA-42 (ATA) is designed to allow multiple observers simultaneous access to the interferometer output at the same time. Typically, the ATA snapshot imager (used for astronomical surveys and SETI) is run in parallel with the beam forming system (used primarily for SETI).[26] ATA also supports observations in multiple synthesized pencil beams at once, through a technique known as "multibeaming." Multibeaming provides an effective filter for identifying false positives in SETI, since a very distant transmitter must appear at only one point on the sky.[27][28][29]

SETI Institute's Center for SETI Research (CSR) uses ATA in the search for extraterrestrial intelligence, observing 12 hours a day, 7 days a week. From 2007-2015, ATA has identified hundreds of millions of technological signals. So far, all these signals have been assigned the status of noise or radio frequency interference because a) they appear to be generated by satellites or Earth-based transmitters, or b) they disappeared before the threshold time limit of ~1 hour.[30][31] Researchers in CSR are presently working on ways to reduce the threshold time limit, and to expand ATA's capabilities for detection of signals that may have embedded messages.[32]

Berkeley astronomers used the ATA to pursue several science topics, some of which might have turned up transient SETI signals,[33][34][35] until 2011, when the collaboration between the University of California and the SETI Institute was terminated. The DSP system planned for the ATA is extremely ambitious. The first portion of the array became operational in October 2007 with 42 antennas. Completion of the full 350 element array will depend on funding and the technical results from the 42-element sub-array.

CNET published an article and pictures about the Allen Telescope Array (ATA) on December 12, 2008.[36][37]

In April 2011, the ATA was forced to enter an 8-month "hibernation" due to funding shortfalls. Regular operation of the ATA was resumed on December 5, 2011.[38][39]

In 2012, new life was breathed into the ATA thanks to a $3.6M philanthropic donation by Franklin Antonio, Co-Founder and Chief Scientist of QUALCOMM Incorporated.[40] This gift supports upgrades of all the receivers on the ATA dishes to have dramatically (2x - 10x from 1–8 GHz) greater sensitivity than before and supporting sensitive observations over a wider frequency range from 1–18 GHz, though initially the radio frequency electronics go to only 12 GHz. As of July, 2013 the first of these receivers was installed and proven. Full installation on all 42 antennas is expected in June, 2014. ATA is especially well suited to the search for extraterrestrial intelligence SETI and to discovery of astronomical radio sources, such as heretofore unexplained non-repeating, possibly extragalactic, pulses known as fast radio bursts or FRBs.

22.2.2   SERENDIP

Main article: SERENDIP

SERENDIP (Search for Extraterrestrial Radio Emissions from Nearby Developed Intelligent Populations) is a SETI program launched in 1979 by the University of California, Berkeley.[41] SERENDIP takes advantage of ongoing "mainstream" radio telescope observations as a "piggy-back" or "commensal" program, using large radio telescopes including the NRAO 90m telescope at Green Bank and the Arecibo 305m telescope. Rather than having its own observation program, SERENDIP analyzes deep space radio telescope data that it obtains while other astronomers are using the telescopes.

The most recently deployed SERENDIP spectrometer, SERENDIP V.v, was installed at the Arecibo Observatory in June 2009 and is currently operational. The digital back-end instrument is an FPGA-based 128 million-channel digital spectrometer covering 200 MHz of bandwidth. It takes data commensally with the seven-beam Arecibo L-band Feed Array[42] (ALFA). The program has found around 400 suspicious signals, but there is not enough data to prove that they belong to extraterrestrial intelligence.[43]

22.2.3   Breakthrough Listen

Main article: Breakthrough Listen

Breakthrough Listen is a ten-year initiative with $100 million funding begun in July 2015 to actively search for intelligent extraterrestrial communications in the universe, in a substantially expanded way, using resources that had not previously been extensively used for the purpose.[44][45][46] It has been described as the most comprehensive search for alien communications to date.[45]

Announced in July 2015, the project will use thousands of hours every year on two major radiotelescopes, the Green Bank Observatory in West Virginia and the Parkes Observatory in Australia.[47] Previously, only about 24 to 36 hours of telescope per year was used in the search for alien life.[45] Furthermore, the Automated Planet Finder of Lick Observatory will search for optical signals coming from laser transmissions. For processing of the massive data, the experience of SETI and SETI@home will be used.[47] SETI founder Frank Drake is one of the project's scientists.[44][45]

22.3   Community SETI projects

22.3.1   SETI@home

Main article: SETI@home

SETI@home was conceived by David Gedye along with Craig Kasnoff and is a popular volunteer distributed computing project that was launched by the University of California, Berkeley, in May 1999. It was originally funded by The Planetary Society and Paramount Pictures, and later by the state of California. The project is run by director David P. Anderson and chief scientist Dan Werthimer. Any individual can become involved with SETI research by downloading the Berkeley Open Infrastructure for Network Computing (BOINC) software program, attaching to the SETI@home project, and allowing the program to run as a background process that uses idle computer power. The SETI@home program itself runs signal analysis on a "work unit" of data recorded from the central 2.5 MHz wide band of the SERENDIP IV instrument. After computation on the work unit is complete, the results are then automatically reported back to SETI@home servers at University of California, Berkeley. By June 28, 2009, the SETI@home project had over 180,000 active participants volunteering a total of over 290,000 computers. These computers give SETI@home an average computational power of 617 teraFLOPS.[48] In 2004 radio source SHGb02+14a set off speculation in the media that a signal had been detected but researchers noted the frequency drifted rapidly and the detection on three SETI@home computers fell within random chance.[49][50]

As of 2010, after 10 years of data collection, SETI@home has listened to that one frequency at every point of over 67 percent of the sky observable from Arecibo with at least three scans (out of the goal of nine scans), which covers about 20 percent of the full celestial sphere.[51]

22.3.2 SETI Net

SETI Net is a private search system created by a single individual. It is closely affiliated with the SETI League and is one of the project Argus stations (DM12jw).

The SETI Net station consists of off-the-shelf, consumer-grade electronics to minimize cost and to allow this design to be replicated as simply as possible. It has a 3-meter parabolic antenna that can be directed in azimuth and elevation, an LNA that covers the 1420 MHz spectrum, a receiver to reproduce the wideband audio, and a standard personal computer as the control device and for deploying the detection algorithms.

The antenna can be pointed and locked to one sky location, enabling the system to integrate on it for long periods. Currently the Wow! signal area is being monitored when it is above the horizon. All search data are collected and made available on the Internet archive.

SETI Net started operation in the early 1980s as a way to learn about the science of the search, and has developed several software packages for the amateur SETI community. It has provided an astronomical clock, a file manager to keep track of SETI data files, a spectrum analyzer optimized for amateur SETI, remote control of the station from the Internet, and other packages.

22.3.3 The SETI League and Project Argus

Founded in 1994 in response to the United States Congress cancellation of the NASA SETI program, The SETI League, Inc. is a membership-supported nonprofit organization with 1,500 members in 62 countries. This grass-roots alliance of amateur and professional radio astronomers is headed by executive director emeritus H. Paul Shuch, the engineer credited with developing the world's first commercial home satellite TV receiver. Many SETI League members are licensed radio amateurs and microwave experimenters. Others are digital signal processing experts and computer enthusiasts.

The SETI League pioneered the conversion of backyard satellite TV dishes 3 to 5 m (10–16 ft) in diameter into research-grade radio telescopes of modest sensitivity.[52] The organization concentrates on coordinating a global network of small, amateur-built radio telescopes under Project Argus, an all-sky survey seeking to achieve real-time coverage of the entire sky.[53] Project Argus was conceived as a continuation of the all-sky survey component of the late NASA SETI program (the targeted search having been continued by the SETI Institute's Project Phoenix). There are currently 143 Project Argus radio telescopes operating in 27 countries. Project Argus instruments typically exhibit sensitivity on the order of 10^{-23} Watts/square metre, or roughly equivalent to that achieved by the Ohio State University Big Ear radio telescope in 1977, when it detected the landmark "Wow!" candidate signal.

The name "Argus" derives from the mythical Greek guard-beast who had 100 eyes, and could see in all directions at once. In the SETI context, the name has been used for radio telescopes in fiction (Arthur C. Clarke, *"Imperial Earth"*; Carl Sagan, *"Contact"*), was the name initially used for the NASA study ultimately known as "Cyclops," and is the name given to an omnidirectional radio telescope design being developed at the Ohio State University.

22.4 Optical experiments

While most SETI sky searches have studied the radio spectrum, some SETI researchers have considered the possibility that alien civilizations might be using powerful lasers for interstellar communications at optical wavelengths. The idea was first suggested by R. N. Schwartz and Charles Hard Townes in a 1961 paper published in the journal *Nature* titled "Interstellar and Interplanetary Communication by Optical Masers". However, the 1971 Cyclops study discounted the possibility of optical SETI, reasoning that construction of a laser system that could outshine the bright central star of a remote star system would be too difficult. In 1983, Townes published a detailed study of the idea in the United States journal *Proceedings of the National Academy of Sciences*, which was met with widespread agreement by the SETI community.

There are two problems with optical SETI. The first problem is that lasers are highly "monochromatic", that is, they emit light only on one frequency, making it troublesome to figure out what frequency to look for. However, according to the uncertainty principle, emitting light in narrow pulses results in a broad spectrum of emission; the spread in frequency becomes higher as the pulse width becomes narrower, making it easier to detect an emission.

The other problem is that while radio transmissions can be broadcast in all directions, lasers are highly directional. This means that a laser beam could be easily blocked by clouds of interstellar dust, and Earth would have to cross its direct line of fire by chance to receive it.

Optical SETI supporters have conducted paper studies[54] of the effectiveness of using contemporary high-energy lasers and a ten-meter diameter mirror as an interstellar beacon. The analysis shows that an infrared pulse from a laser, focused into a narrow beam by such a mirror, would appear thousands of times brighter than the Sun to a distant civilization in the beam's line of fire. The Cyclops study proved incorrect in suggesting a laser beam would be inherently hard to see.

Such a system could be made to automatically steer itself through a target list, sending a pulse to each target at a constant rate. This would allow targeting of all Sun-like stars within a distance of 100 light-years. The studies have also described an automatic laser pulse detector system with a low-cost, two-meter mirror made of carbon composite materials, focusing on an array of light detectors. This automatic detector system could perform sky surveys to detect laser flashes from civilizations attempting contact.

Several optical SETI experiments are now in progress. A Harvard-Smithsonian group that includes Paul Horowitz designed a laser detector and mounted it on Harvard's 155 centimeters (61 inches) optical telescope. This telescope is currently being used for a more conventional star survey, and the optical SETI survey is "piggybacking" on that effort. Between October 1998 and November 1999, the survey inspected about 2,500 stars. Nothing that resembled an intentional laser signal was detected, but efforts continue. The Harvard-Smithsonian group is now working with Princeton University to mount a similar detector system on Princeton's 91-centimeter (36-inch) telescope. The Harvard and Princeton telescopes will be "ganged" to track the same targets at the same time, with the intent being to detect the same signal in both locations as a means of reducing errors from detector noise.

The Harvard-Smithsonian group is now building a dedicated all-sky optical survey system along the lines of that described above, featuring a 1.8-meter (72-inch) telescope. The new optical SETI survey telescope is being set up at the Oak Ridge Observatory in Harvard, Massachusetts.

The University of California, Berkeley, home of SERENDIP and SETI@home, is also conducting optical SETI searches. One is being directed by Geoffrey Marcy, an extrasolar planet hunter, and involves examination of records of spectra taken during extrasolar planet hunts for a continuous, rather than pulsed, laser signal. The other Berkeley optical SETI effort is more like that being pursued by the Harvard-Smithsonian group and is being directed by Dan Werthimer of Berkeley, who built the laser detector for the Harvard-Smithsonian group. The Berkeley survey uses a 76-centimeter (30-inch) automated telescope at Leuschner Observatory and an older laser detector built by Werthimer.

22.5   Gamma-ray bursts

Gamma-ray bursts (GRBs) are candidates for extraterrestrial communication. These high-energy bursts are observed about once per day and originate throughout the observable universe. SETI currently omits gamma ray frequencies in their monitoring and analysis because they are absorbed by the Earth's atmosphere and difficult to detect with ground-based receivers. In addition, the wide burst bandwidths pose a serious analysis challenge for modern digital signal processing systems. Still, the continued mysteries surrounding gamma-ray bursts have encouraged hypotheses invoking extraterrestrials. John A. Ball from the MIT Haystack Observatory suggests that an advanced civilization that has reached a technological singularity would be capable of transmitting a two-millisecond pulse encoding 1×10^{18} bits of information. This is "comparable to the estimated total information content of Earth's biosystem—genes and memes and including all libraries and computer media."[55]

22.6 Search for extraterrestrial artifacts

The possibility of using interstellar messenger probes in the search for extraterrestrial intelligence was first suggested by Ronald N. Bracewell in 1960 (see Bracewell probe), and the technical feasibility of this approach was demonstrated by the British Interplanetary Society's starship study Project Daedalus in 1978. Starting in 1979, Robert Freitas advanced arguments[56][57][58] for the proposition that physical space-probes are a superior mode of interstellar communication to radio signals. See Voyager Golden Record.

In recognition that any sufficiently advanced interstellar probe in the vicinity of Earth could easily monitor the terrestrial Internet, Invitation to ETI was established by Prof. Allen Tough in 1996, as a Web-based SETI experiment inviting such spacefaring probes to establish contact with humanity. The project's 100 Signatories includes prominent physical, biological, and social scientists, as well as artists, educators, entertainers, philosophers and futurists. Prof. H. Paul Shuch, executive director emeritus of The SETI League, serves as the project's Principal Investigator.

Inscribing a message in matter and transporting it to an interstellar destination can be enormously more energy efficient than communication using electromagnetic waves if delays larger than light transit time can be tolerated.[59] That said, for simple messages such as "hello," radio SETI could be far more efficient.[60] If energy requirement is used as a proxy for technical difficulty, then a solarcentric Search for Extraterrestrial Artifacts (SETA)[61] may be a useful supplement to traditional radio or optical searches.[62][63]

Much like the "preferred frequency" concept in SETI radio beacon theory, the Earth-Moon or Sun-Earth libration orbits[64] might therefore constitute the most universally convenient parking places for automated extraterrestrial spacecraft exploring arbitrary stellar systems. A viable long-term SETI program may be founded upon a search for these objects.

In 1979, Freitas and Valdes conducted a photographic search of the vicinity of the Earth-Moon triangular libration points L_4 and L_5, and of the solar-synchronized positions in the associated halo orbits, seeking possible orbiting extraterrestrial interstellar probes, but found nothing to a detection limit of about 14th magnitude.[64] The authors conducted a second, more comprehensive photographic search for probes in 1982[65] that examined the five Earth-Moon Lagrangian positions and included the solar-synchronized positions in the stable L4/L5 libration orbits, the potentially stable nonplanar orbits near L1/L2, Earth-Moon L_3, and also L_2 in the Sun-Earth system. Again no extraterrestrial probes were found to limiting magnitudes of 17–19th magnitude near L3/L4/L5, 10–18th magnitude for L_1/L_2, and 14–16th magnitude for Sun-Earth L_2.

In June 1983, Valdes and Freitas[66] used the 26 m radiotelescope at Hat Creek Radio Observatory to search for the tritium hyperfine line at 1516 MHz from 108 assorted astronomical objects, with emphasis on 53 nearby stars including all visible stars within a 20 light-year radius. The tritium frequency was deemed highly attractive for SETI work because (1) the isotope is cosmically rare, (2) the tritium hyperfine line is centered in the SETI waterhole region of the terrestrial microwave window, and (3) in addition to beacon signals, tritium hyperfine emission may occur as a byproduct of extensive nuclear fusion energy production by extraterrestrial civilizations. The wideband- and narrowband-channel observations achieved sensitivities of 5–14×10^{-21} W/m²/channel and 0.7-2×10^{-24} W/m²/channel, respectively, but no detections were made.

22.7 Technosignatures

See also: Technosignature and Megascale engineering

Technosignatures, including all signs of technology with the exception of the interstellar radio messages that define traditional SETI, are a recent avenue in the search for extraterrestrial intelligence. Technosignatures may originate from various sources, from megastructures such as Dyson spheres and space mirrors or space shaders[67] to the atmospheric contamination created by an industrial civilization,[68] or city lights on extrasolar planets, and may be detectable in the future with large hypertelescopes.[69]

Technosignatures can be divided into three broad categories: astroengineering projects, signals of planetary origin, and spacecraft within and outside the Solar System.

An astroengineering installation such as a Dyson sphere, designed to convert all of the incident radiation of its host star into energy, could be detected through the observation of an infrared excess from a solar analog star.[70] After examining some 100,000 nearby large galaxies, a team of researchers has concluded that none of them contain any obvious signs of highly advanced technological civilizations.[71][72][73] Another theoretical form of astroengineering, the Shkadov thruster, moves its host star by reflecting some of the star's light back on itself, and could be detected by observing if its transits across the star abruptly end with the thruster in front.[74] Asteroid mining within the Solar System is also a detectable technosignature of the first kind.[75]

Individual extrasolar planets can be analyzed for signs of technology. Avi Loeb of the Harvard-Smithsonian Center for Astrophysics has proposed that persistent light signals on the night side of an exoplanet can be an indication of the presence of cities and an advanced civilization.[76][77] In addition, the excess infrared radiation[69][78] and chemicals[79][80] produced by various industrial processes or terraforming efforts[81] may point to intelligence.

Clearly, light and heat detected from planets need to be distinguished from natural sources to conclusively prove the existence of civilization on a planet. However, as argued by the Colossus team,[82] a civilization heat signature should be within a "comfortable" temperature range, like terrestrial urban heat islands, i.e. only a few degrees warmer than the planet itself. In contrast, such natural sources as wild fires, volcanoes, etc. are significantly hotter, so they will be well distinguished by their maximum flux at a different wavelength.

Extraterrestrial craft are another target in the search for technosignatures. Magnetic sail interstellar spacecraft should be detectable over thousands of light-years of distance through the synchrotron radiation they would produce through interaction with the interstellar medium; other interstellar spacecraft designs may be detectable at more modest distances.[83] In addition, robotic probes within the Solar System are also being sought out with optical and radio searches.[84][85]

For a sufficiently advanced civilization, hyper energetic neutrinos from Planck scale accelerators should be detectable at a distance of many Mpc. [86]

22.8   Fermi paradox

Main article: Fermi paradox

Italian physicist Enrico Fermi suggested in the 1950s that if technologically advanced civilizations are common in the universe, then they should be detectable in one way or another. (According to those who were there,[87] Fermi either asked "Where are they?" or "Where is everybody?")

The Fermi paradox is commonly understood as asking why extraterrestrials have not visited Earth,[88] but the same reasoning applies to the question of why signals from extraterrestrials have not been heard. The SETI version of the question is sometimes referred to as "the Great Silence".

The Fermi paradox can be stated more completely as follows:

> The size and age of the universe incline us to believe that many technologically advanced civilizations must exist. However, this belief seems logically inconsistent with our lack of observational evidence to support it. Either (1) the initial assumption is incorrect and technologically advanced intelligent life is much rarer than we believe, or (2) our current observations are incomplete and we simply have not detected them yet, or (3) our search methodologies are flawed and we are not searching for the correct indicators.

There are multiple explanations proposed for the Fermi paradox,[89] ranging from analyses suggesting that intelligent life is rare (the "Rare Earth hypothesis"), to analyses suggesting that although extraterrestrial civilizations may be common, they would not communicate, or would not travel across interstellar distances.

Science writer Timothy Ferris has posited that since galactic societies are most likely only transitory, an obvious solution is an interstellar communications network, or a type of library consisting mostly of automated systems. They would store the cumulative knowledge of vanished civilizations and communicate that knowledge through the galaxy. Ferris calls this the "Interstellar Internet", with the various automated systems acting as network "servers". If such an Interstellar Internet exists, the hypothesis states, communications between servers are mostly through narrow-band, highly directional radio or laser links. Intercepting such signals is, as discussed earlier, very difficult. However, the network could maintain some broadcast nodes in hopes of making contact with new civilizations.

Although somewhat dated in terms of "information culture" arguments, not to mention the obvious technological problems of a system that could work effectively for billions of years and requires multiple lifeforms agreeing on certain basics of communications technologies, this hypothesis is actually testable (see below).

A significant problem is the vastness of space. Despite piggybacking on the world's most sensitive radio telescope, Charles Stuart Bowyer said, the instrument could not detect random radio noise emanating from a civilization like ours, which has been leaking radio and TV signals for less than 100 years. For SERENDIP and most other SETI projects to detect a signal from an extraterrestrial civilization, the civilization would have to be beaming a powerful signal directly at us. It also means that Earth civilization will only be detectable within a distance of 100 light-years.[90]

22.9 Post detection disclosure protocol

The International Academy of Astronautics (IAA) has a long-standing SETI Permanent Study Group (SPSG, formerly called the IAA SETI Committee), which addresses matters of SETI science, technology, and international policy. The SPSG meets in conjunction with the International Astronautical Congress (IAC) held annually at different locations around the world, and sponsors two SETI Symposia at each IAC. In 2005, the IAA established the **SETI: Post-Detection Science and Technology Taskgroup** (Chairman, Professor Paul Davies) "to act as a Standing Committee to be available to be called on at any time to advise and consult on questions stemming from the discovery of a putative signal of extraterrestrial intelligent (ETI) origin."

When awarded the 2009 TED Prize, SETI Institute's Jill Tarter outlined the organisation's "post detection protocol".[91] During NASA's funding of the project, an administrator would be first informed with the intention of informing the United States executive government. The current protocol for SETI Institute is to first internally investigate the signal, seeking independent verification and confirmation. During the process, the organisation's private financiers would be secretly informed. Once a signal has been verified, a telegram would be sent via the Central Bureau for Astronomical Telegrams. Following this process, Tarter says that the organisation will hold a press conference with the aim of broadcasting to the public. SETI Institute's Seth Shostak has claimed that knowledge of the discovery would likely leak as early as the verification process.[92]

However, the protocols mentioned apply only to radio SETI rather than for METI (Active SETI).[93] The intention for METI is covered under the SETI charter "Declaration of Principles Concerning Sending Communications with Extraterrestrial Intelligence".

The SETI Institute does not officially recognise the Wow! signal as of extraterrestrial origin (as it was unable to be verified). The SETI Institute has also publicly denied that the candidate signal Radio source SHGb02+14a is of extraterrestrial origin[94][95] though full details of the signal, such as its exact location have never been disclosed to the public. Although other volunteering projects such as Zooniverse credit users for discoveries, there is currently no crediting or early notification by SETI@Home following the discovery of a signal.

Some people, including Steven M. Greer,[96] have expressed cynicism that the general public might not be informed in the event of a genuine discovery of extraterrestrial intelligence due to significant vested interests. Some, such as Bruce Jakosky[97] have also argued that the official disclosure of extraterrestrial life may have far reaching and as yet undetermined implications for society, particularly for the world's religions.

22.10 Active SETI

Main article: Active SETI

Active SETI, also known as messaging to extraterrestrial intelligence (METI), consists of sending signals into space in the hope that they will be picked up by an alien intelligence.

22.10.1   Realized interstellar radio message projects

In November 1974, a largely symbolic attempt was made at the Arecibo Observatory to send a message to other worlds. Known as the Arecibo Message, it was sent towards the globular cluster M13, which is 25,000 light-years from Earth. Further IRMs Cosmic Call, Teen Age Message, Cosmic Call 2, and A Message From Earth were transmitted in 1999, 2001, 2003 and 2008 from the Evpatoria Planetary Radar.

22.10.2   Debate

Physicist Stephen Hawking, in his book *A Brief History of Time*, suggests that "alerting" extraterrestrial intelligences to our existence is foolhardy, citing mankind's history of treating his fellow man harshly in meetings of civilizations with a significant technology gap. He suggests, in view of this history, that we "lay low".

The concern over METI was raised by the science journal *Nature* in an editorial in October 2006, which commented on a recent meeting of the International Academy of Astronautics SETI study group. The editor said, "It is not obvious that all extraterrestrial civilizations will be benign, or that contact with even a benign one would not have serious repercussions" (Nature Vol 443 12 October 06 p 606). Astronomer and science fiction author David Brin has expressed similar concerns.[98]

Richard Carrigan, a particle physicist at the Fermi National Accelerator Laboratory near Chicago, Illinois, suggested that passive SETI could also be dangerous and that a signal released onto the Internet could act as a computer virus.[99] Computer security expert Bruce Schneier dismissed this possibility as a "bizarre movie-plot threat".[100]

To lend a quantitative basis to discussions of the risks of transmitting deliberate messages from Earth, the SETI Permanent Study Group of the International Academy of Astronautics adopted in 2007 a new analytical tool, the San Marino Scale.[101] Developed by Prof. Ivan Almar and Prof. H. Paul Shuch, the scale evaluates the significance of transmissions from Earth as a function of signal intensity and information content. Its adoption suggests that not all such transmissions are equal, and each must be evaluated separately before establishing blanket international policy regarding active SETI.

However, some scientists consider these fears about the dangers of METI as panic and irrational superstition; see, for example, Alexander L. Zaitsev's papers.[102][103] Biologist João Pedro de Magalhães also proposed in 2015 transmitting an invitation message to any extraterrestrial intelligences watching us already in the context of the Zoo Hypothesis and inviting them to respond, arguing this would not put us in any more danger than we are already if the Zoo Hypothesis is correct.[104]

On 13 February 2015, scientists (including Geoffrey Marcy, Seth Shostak, Frank Drake, Elon Musk and David Brin) at a convention of the American Association for the Advancement of Science, discussed Active SETI and whether transmitting a message to possible intelligent extraterrestrials in the Cosmos was a good idea;[105][106] one result was a statement, signed by many, that a "worldwide scientific, political and humanitarian discussion must occur before any message is sent".[107] On 28 March 2015, a related essay was written by Seth Shostak and published in *The New York Times*.[108]

22.10.3   Breakthrough Message

The Breakthrough Message program is an open competition announced in July 2015 to design a digital message that could be transmitted from Earth to an extraterrestrial civilization, with a US$1,000,000 prize pool. The message should be "representative of humanity and planet Earth". The program pledges "not to transmit any message until there has been a wide-ranging debate at high levels of science and politics on the risks and rewards of contacting advanced civilizations".[109]

22.11   Criticism

As various SETI projects have progressed, some have criticized early claims by researchers as being too "euphoric". For example, Peter Schenkel, while remaining a supporter of SETI projects, has written that

"[i]n light of new findings and insights, it seems appropriate to put excessive euphoria to rest and to take a more down-to-earth view ... We should quietly admit that the early estimates—that there may be a million, a hundred thousand, or ten thousand advanced extraterrestrial civilizations in our galaxy—may no longer be tenable."[1]

Clive Trotman presents some sobering but realistic calculations emphasizing the timeframe dimension.[110]

SETI has also occasionally been the target of criticism by those who suggest that it is a form of pseudoscience. In particular, critics allege that no observed phenomena suggest the existence of extraterrestrial intelligence, and furthermore that the assertion of the existence of extraterrestrial intelligence has no good Popperian criteria for falsifiability.[3]

In response, SETI advocates note, among other things, that the Drake Equation was never a hypothesis, and so never intended to be testable, nor to be "solved"; it was merely a clever representation of the agenda for the world's first scientific SETI meeting in 1961, and it serves as a tool in formulating testable hypotheses. Further, they note that the existence of intelligent life on Earth is a plausible reason to expect it elsewhere, and that individual SETI projects have clearly defined "stop" conditions.

22.12   See also

- Alien language

- Arecibo Message

- Astrobiology

- Astrobiology Science and Technology for Exploring Planets

- Breakthrough Listen

- Communication with extraterrestrial intelligence

- Cultural impact of extraterrestrial contact

- Darwin Mission

- First contact (science fiction)

- International Fellowship of Rotarian Amateur Astronomers

- Iosif Shklovsky

- Metalaw

- Nexus for Exoplanet System Science

- Ohio State University Radio Observatory (The Big Ear)

- Open SonATA

- SETI@home

- SETIcon

- SETI Institute

- setiQuest

- Wow! signal

- Xenoarchaeology

22.13   References

[1] Schenkel, Peter (May 2006). "SETI Requires a Skeptical Reappraisal". Skeptical Inquirer. Retrieved June 28, 2009.

[2] Moldwin, Mark (November 2004). "Why SETI is science and UFOlogy is not". Skeptical Inquirer.

[3] "SETI at 50". Nature 416 (7262): 316. 2009. Bibcode:2009Natur.461..316.. doi:10.1038/461316a.

[4] Katz, Gregory (July 20, 2015). "Searching for ET: Hawking to look for extraterrestrial life". AP News. Retrieved July 20, 2015.

[5] Seifer, Marc J. (1996). "Martian Fever (1895–1896)". Wizard : the life and times of Nikola Tesla: biography of a genius. Secaucus, New Jersey: Carol Pub. p. 157. ISBN 978-1-55972-329-9. OCLC 33865102.

[6] Spencer, John (1991). The UFO Encyclopedia. New York: Avon Books. ISBN 978-0-380-76887-5. OCLC 26211869.

[7] W. Bernard Carlson, Tesla: Inventor of the Electrical Age, Princeton University Press - 2013, pages 276-278.

[8] Corum, Kenneth L.; James F. Corum (1996). Nikola Tesla and the electrical signals of planetary origin (PDF). pp. 1, 6, 14. OCLC 68193760.

[9] Jacques Lasker. "A Primer on Mars Oppositions".

[10] Dick, Steven (1999). The Biological Universe: The Twentieth Century Extraterrestrial Life Debate. ISBN 0-521-34326-7.

[11] Prepare for Contact. Letters of Note (2009-11-06). Retrieved on 2011-10-14.

[12] Cocconi, Giuseppe & Philip Morrison (1959). "Searching for interstellar communications". Nature 184 (4690), pp. 844–846. doi:10.1038/184844a0.

[13] "Cosmic Search Vol. 1, No. 1". Retrieved 1 October 2014.

[14] "Science: Project Ozma." Time, April 18, 1960 (web version accessed 17 September 2010)

[15] Sagan, Carl; Iosif Shklovskii (1966). Intelligent Life in the Universe. ISBN 0-330-25125-2.

[16] "Project Cyclops: A Design Study of a System for Detecting Extraterrestrial Intelligent Life" (PDF). NASA. 1971. Retrieved October 12, 2014.

[17] Robert H. Gray. The Elusive WOW: Searching for Extraterrestrial Intelligence 2012. Palmer Square Press, Chicago. 242 pages. ISBN 978-0-9839584-4-4.

[18] Alan M. MacRobert. "SETI Searches Today". Sky and Telescope.

[19] Wolfe, JH; et al. (1979). "CP-2156, Chapter 5.5. SETI – The Search for Extraterrestrial Intelligence: Plans and Rationale". NASA. Retrieved July 1, 2009.

[20] Garber, Stephen J., Searching for Good Science: the Cancellation of NASA's SETI Program, Journal of the British Interplanetary Society, 52, pp. 3-12, 1999.

[21] "Ear to the Universe Is Plugged by Budget Cutters". The New York Times. October 7, 1993. Retrieved May 23, 2010.

[22] "Searching for Intelligent Aliens: Q&A with SETI Astronomer Jill Tarter". Space.com. May 22, 2012. Retrieved August 5, 2012.

[23] Siemion, Andrew (September 29, 2015). "Prepared Statement by Andrew Siemion - Hearing on Astrobiology Status Report - House Committee on Science, Space, and Technology". SpaceRef.com. Retrieved October 19, 2015. line feed character in |title= at position 63 (help)

[24] "Allen Telescope Array General Overview". SETI Institute. Archived from the original on 2006-04-28. Retrieved 2006-06-12.

[25] Welch, J.; Backer, D.; Blitz, L.; Bock, D.C.-J.; Bower, G.C.; Cheng, C.; Croft, S.; Dexter, M.; Engargiola, G.; Fields, E.; Forster, J.; Gutierrez-Kraybill, C.; Heiles, C.; Helfer, T.; Jorgensen, S.; Keating, G.; Lugten, J.; MacMahon, D.; Milgrome, O.; Thornton, D.; Urry, L.; van Leeuwen, J.; Werthimer, D.; Williams, P.H.; Wright, M.; Tarter, J.; Ackermann, R.; Atkinson, S.; Backus, P.; Barott, W.; Bradford, T.; Davis, M.; DeBoer, D.; Dreher, J.; Harp, G.; Jordan, J.; Kilsdonk, T.; Pierson, T.; Randall, K.; Ross, J.; Shostak, S.; Fleming, M.; Cork, C.; Vitouchkine, A.; Wadefalk, N.; Weinreb, S., "The Allen Telescope Array: The First Widefield, Panchromatic, Snapshot Radio Camera for Radio Astronomy and SETI." Proceedings of the IEEE , vol.97, no.8, pp.1438,1447, Aug. 2009 doi: 10.1109/JPROC.2009.2017103

[26] Gutierrez-Kraybill, Colby, <it>et al</it>, *Commensal observing with the Allen Telescope array: software command and control*. SPIE Astronomical Telescopes+ Instrumentation. International Society for Optics and Photonics, 2010, DOI: doi:10.1117/12. 857860, http://arxiv.org/pdf/1010.1567.pdf.

[27] Harp, G. R. "Customized beam forming at the Allen Telescope Array." ATA Memo Series 51 (2002), available at http://www. seti.org/sites/default/files/ATA-memo-series/memo51.pdf.

[28] Barott, William C., et al. "Real-time beamforming using high-speed FPGAs at the Allen Telescope Array." Radio Science 46.1 (2011).

[29] Harp, G. R. "Using Multiple Beams to Distinguish Radio Frequency Interference from SETI Signals." arXiv preprint arXiv: 1309.3826 (2013).

[30] Tarter, <it>et al</it>. *The first SETI observations with the Allen telescope array*. Acta Astronautica, **68** (3-4), 340–346 (2011).

[31] Backus, Peter R., and Allen Telescope Array Team. *The ATA Galactic Center Survey: SETI Observations in 2009*. Bulletin of the American Astronomical Society. Vol. 42. 2010

[32] Harp, Gerald R., et al. *A new class of SETI beacons that contain information*. Communication with Extraterrestrial Intelligence. State University of New York Press, 2011.

[33] Steve Croft <it>et al.</it> *The Allen Telescope Array Twenty-centimeter Survey—A 690 deg^2, 12 Epoch Radio Data Set. I. Catalog and Long-duration Transient Statistics*, 2010 ApJ 719 45

[34] Siemion, Andrew P. V. *et al.*, *The Allen Telescope Array Fly's Eye Survey for Fast Radio Transients*, 2012 ApJ 744 109

[35] Siemion, Andrew, *et al.* *Results from the Fly's Eye Fast Radio Transient Search at the Allen Telescope Array*. Bulletin of the American Astronomical Society. Vol. 43. 2011.

[36] Terdiman, Daniel. (2008-12-12) SETI's large-scale telescope scans the skies | Geek Gestalt – CNET News. News.cnet.com. Retrieved on 2011-10-14.

[37] Rendering of 350 image – Photos: Searching the heavens for life – CNET News. News.cnet.com (2008-12-12). Retrieved on 2011-10-14.

[38] The Great Beyond. Nature Blogs, ed. (25 April 2011). "SETI scope suspends search". Retrieved 26 April 2011.

[39] "SETI Search Resumes at Allen Telescope Array", SETI.org. December 5, 2011.

[40] Damon Arthur. "New Hat Creek receivers will let SETI delve deeper into space".

[41] "SERENDIP". UC Berkeley. Retrieved 2006-08-20.

[42] http://www.naic.edu/alfa/

[43] Л.М.Гиндилис. Радиопоиск: век двадцатый

[44] Feltman, Rachel (20 July 2015). "Stephen Hawking announces $100 million hunt for alien life". Washington Post. Retrieved 20 July 2015.

[45] Merali, Zeeya (20 July 2015). "Search for extraterrestrial intelligence gets a $100-million boost. Russian billionaire Yuri Milner announces most comprehensive hunt for alien life.". Nature News. Retrieved 20 July 2015.

[46] Rundle, Michael (20 July 2015). "$100m Breakthrough Listen is 'largest ever' search for alien civilisations". Wired. Retrieved 20 July 2015.

[47] Sample, Ian (20 July 2015). "Anybody out there? $100m radio wave project to scan far regions for alien life". The Guardian. Retrieved 20 July 2015.

[48] de Zutter, Willy. "SETI@home — Credit Overview". BOINCstats. Retrieved June 28, 2009.

[49] Whitehouse, David (2004-09-02). "Astronomers deny ET signal report". BBC News. Retrieved 24 April 2013.

[50] Alexander, Amir (2004-09-02). "SETI@home Leaders Deny Reports of Likely Extraterrestrial Signal". The Planetary Society. Archived from the original on 2011-07-26. Retrieved 2006-06-12.

[51] Alan M. MacRobert. "SETI Searches Today". Sky and Telescope (2010?).

[52] Chown, Marcus (April 1997). "The Alien Spotters". *New Scientist*: 28. Retrieved 2008-04-13.

[53] http://www.setileague.org/argus/

[54] Exers, Ronald; D. Cullers; J. Billingham; L. Scheffer, eds. (2003). *SETI 2020: A Roadmap for the Search for Extraterrestrial Intelligence*. SETI Press. ISBN 0-9666335-3-9.

[55] Ball, J.A. (1995). "Gamma-Ray Bursts: The ETI Hypothesis". *The Astrophysical Journal*.

[56] Freitas Jr., Robert A. (1980). "Interstellar probes — A new approach to SETI". Retrieved June 28, 2009.

[57] Freitas Jr., Robert A (1983). "Debunking the Myths of Interstellar Probes". Retrieved June 28, 2009.

[58] Freitas Jr., Robert A. (1983). "The Case for Interstellar Probes". Retrieved June 28, 2009.

[59] C. Rose and G. Wright (2 September 2004). "Inscribed matter as an energy efficient means of communication with an extraterrestrial civilization" (PDF). *Nature* **431** (7004): 47–9. Bibcode:2004Natur.431...47R. doi:10.1038/nature02884. PMID 15343327.

[60] Woodruff T. Sullivan (2 September 2004). "Message in a bottle". *Nature Magazine* **431** (7004): 27–28. doi:10.1038/431027a.

[61] Freitas Jr., Robert A (November 1983). "The Search for Extraterrestrial Artifacts (SETA)". Retrieved June 28, 2009.

[62] Editors (8 September 2004). "NY Times Editorial". *New York Times*.

[63] Rose, Christopher (September 2004). "Cosmic Communications". Retrieved August 1, 2010.

[64] Freitas Jr., Robert A; Valdes, Francisco (1980). "A Search for Natural or Artificial Objects Located at the Earth-Moon Libration Points". Retrieved June 28, 2009.

[65] Valdes, Francisco; Freitas Jr., Robert A (1983). "A Search for Objects near the Earth-Moon Lagrangian Points".

[66] Valdes, Francisco; Freitas Jr., Robert A (1986). "A Search for the Tritium Hyperfine Line from Nearby Stars".

[67] Korpela, Eric (2015). "Modeling Indications of Technology in Planetary Transit Light Curves -- Dark-side illumination". arXiv:1505.07399.

[68] Almár, Iván (2011). "SETI and astrobiology: The Rio Scale and the London Scale". *Acta Astronautica* **69** (9–10): 899–904. Bibcode:2011AcAau..69..899A. doi:10.1016/j.actaastro.2011.05.036.(subscription required)

[69] "Heat-Seeking, Alien-Hunting Telescope Could Be Ready In 5 Years". Space.com. 2013-06-07. Retrieved 2013-07-10.

[70] Freemann J. Dyson (1960). "Search for Artificial Stellar Sources of Infra-Red Radiation". *Science* **131** (3414): 1667–1668. Bibcode:1960Sci...131.1667D. doi:10.1126/science.131.3414.1667. PMID 17780673.

[71] www.scientificamerican.com 2015-04-17 Alien Supercivilizations Absent from 100,000 Nearby Galaxies

[72] Cornell University 2015-04-15 The Ĝ Infrared Search for Extraterrestrial Civilizations with Large Energy Supplies. III. The Reddest Extended Sources in WISE

[73] The Astrophysical Journal Supplement Series 2015-04-15 The Ĝ Infrared Search for Extraterrestrial Civilizations with Large Energy Supplies. III. The Reddest Extended Sources in WISE

[74] Villard, Ray (2013). "Alien 'Star Engine' Detectable in Exoplanet Data?". Retrieved 2013-07-08

[75] Duncan Forgan; Martin Elvis (2011). "Extrasolar Asteroid Mining as Forensic Evidence for Extraterrestrial Intelligence". arXiv:1103.5369 [astro-ph.EP].

[76] "SETI search urged to look for city lights". UPI.com. 2011-11-03. Retrieved 2013-07-10.

[77] Extrasolar Planets: Formation, Detection and Dynamics Rudolf Dvorak, page 14 John Wiley & Sons, 2007

[78] Povich, Matthew (11 August 2014). "Infrared Search for Extraterrestrial Civilizations with Large Energy Supplies". *astro-ph.GA* (Astrobiology Web). Retrieved 2014-08-19.

[79] "Satellite sniffs out chemical traces of atmospheric pollution / Observing the Earth / Our Activities / ESA". Esa.int. 2000-12-18. Retrieved 2013-07-10.

[80] "Haze on Saturn's Moon Titan Is Similar to Earth's Pollution". Space.com. 2013-06-06. Retrieved 2013-07-10.

[81] "Alien Hairspray May Help Us Find E.T.". Space.com. 2012-11-26. Retrieved 2013-07-10.

[82] "How to Find ET with Infrared Light". Astronomy.com. June 2013.

[83] Zubrin, Robert (1995). "Detection of Extraterrestrial Civilizations via the Spectral Signature of Advanced Interstellar Spacecraft". In Shostak, Seth. *Astronomical Society of the Pacific Conference Series*. Progress in the Search for Extraterrestrial Life. Astronomical Society of the Pacific. pp. 487–496. Bibcode:1995ASPC...74..487Z.

[84] Freitas, Robert (November 1983). "The Case for Interstellar Probes". *Journal of the British Interplanetary Society* **36**: 490–495. Bibcode:1983JBIS...36..490F.

[85] Tough, Allen (1998). "Small Smart Interstellar Probes" (PDF). *Journal of the British Interplanetary Society* **51**: 167–174.

[86] Lacki, B. C. (2015). SETI at Planck Energy: When Particle Physicists Become Cosmic Engineers. arXiv preprint arXiv: 1503.01509 .

[87] Jones, Eric (March 1985). ""Where is everybody?", An account of Fermi's question" (PDF). Los Alamos National Laboratory. Retrieved June 28, 2009.

[88] Ben Zuckerman and Michael H. Hart (editors), *Extraterrestrials: Where Are They?* Elsevier Science & Technology Books (1982), ISBN 9780080263427

[89] Stephen Webb, *Where is Everybody? Fifty Solutions to Fermi's Paradox*, Copernicus, 2002 edition, 978-0387955018

[90] "SETI Insensitive To Earth-like Signals". spacedaily.com. 1998. Retrieved February 8, 2013.

[91] Jill Tarter, 2009 TED Prize acceptance speech

[92] Coast to Coast AM interview February 24, 2000

[93] Pope, Nick What to do if we find extraterrestrial life? Who gets notified? Do we reply? Experts are already arguing msnbc.com 10/18/2010

[94] Alexander, Amir (2004-09-02). "SETI@home Leaders Deny Reports of Likely Extraterrestrial Signal". The Planetary Society. Retrieved 2006-06-12.

[95] Whitehouse, David (2004-09-02). "Astronomers deny ET signal report". BBC News. Retrieved 2006-06-12.

[96] Vance, Ashlee SETI urged to fess up over alien signals The Register 31 July 2006

[97] Siegel, Lee [The Meaning of Life http://nai.nasa.gov/news_stories/news_print.cfm?ID=138] NASA July 6, 2001

[98] Brin, David (June 2006). "Shouting at the Cosmos". Lifeboat Foundation. Retrieved June 28, 2009.

[99] Carrigan Jr., Robert A. (June 2006). "Do potential SETI signals need to be decontaminated?" (PDF). Fermi National Accelerator Laboratory.

[100] "A Science-Fiction Movie-Plot Threat". Retrieved March 13, 2011.

[101] Almár, Ivan. "The San Marino Scale". International Academy of Astronautics.

[102] Zaitsev, Alexander L. (September 2007). "Sending and searching for interstellar messages" (PDF). 58th International Astronautical Congress.

[103] Zaitsev, Alexander L. (April 2008). "Detection probability of terrestrial radio signals by a hostile super-civilization". *Journal of Radio Electronics* **5**.

[104] de Magalhaes, J. P. (2015). "A direct communication proposal to test the Zoo Hypothesis". arXiv:1509.03652.

[105] Borenstein, Seth (of AP News) (13 February 2015). "Should We Call the Cosmos Seeking ET? Or Is That Risky?". *The New York Times*. Retrieved 14 February 2015.

[106] Ghosh, Pallab (12 February 2015). "Scientist: 'Try to contact aliens'". *BBC News*. Retrieved 12 February 2015.

[107] Various (13 February 2015). "Statement - Regarding Messaging To Extraterrestrial Intelligence (METI) / Active Searches For Extraterrestrial Intelligence (Active SETI)". *University of California, Berkeley*. Retrieved 14 February 2015.

[108] Shostak, Seth (28 March 2015). "Should We Keep a Low Profile in Space?". *The New York Times*. Retrieved 29 March 2015.

[109] "Breakthrough Initiatives". *www.breakthroughinitiatives.org*. Retrieved 2015-07-24.

[110] Trotman, Clive (2004). *The Feathered Onion — Creation of Life in the Universe*. Wiley. ISBN 0-470-87187-3.

22.14   Further reading

- McConnell, Brian; Chuck Toporek (2001). *Beyond Contact: A Guide to SETI and Communicating with Alien Civilizations*. O'Reilly. ISBN 0-596-00037-5.

- Perelmuter, J.M. (2006). *The Sinusoidal Spaghetti*. iUniverse. ISBN 0-595-41713-2.

- MJ Carlotto (2007). "Detecting Patterns of a Technological Intelligence in Remotely Sensed Imagery" (PDF). *J British Interplanetary Society* **60**: 28–39. Bibcode:2007JBIS...60...28C.

- John B Campbell (2006). "Archaeology and direct imaging of exoplanets". In C. Aime & F. Vakili. *Proceedings of the International Astronomical Union* (PDF). Cambridge University Press. pp. 247 *ff*. ISBN 0-521-85607-8.

- Frank White: *The Seti Factor – How the Search for Extraterrestrial Intelligence Is Changing Our View of the Universe and Ourselves*. Walker & Company, New York 1990. ISBN 978-0-8027-1105-2

- David W. Swift: *Seti Pioneers — Scientists Talk about Their Search for Extraterrestrial Intelligence*. Univ. of Arizona Press, Tucson 1993. ISBN 0-8165-1119-5

- P.Morrison, J.Billingham, J.Wolfe:*The search for extraterrestrial intelligence-SETI.*NASA SP. Washington 1977.on

- Catran, Jack (1980). *Is There Intelligent Life on Earth?*. Lidiraven Books. ISBN 0-9361-6229-5.

22.15   External links

- Search for Extraterrestrial Intelligence at Home (SETI@Home)

- The SETI Institute

- Harvard University SETI Program

- University of California, Berkeley SETI Program

- The eerie silence Expanding the parameters of the search for technological and evolutionary footprints of extrasolar civilizations, beyond only radio signals. (Physics World). March 2, 2010.

- Konoplya, R. A.; Zhidenko, A. (2010). "Passage of radiation through wormholes of arbitrary shape". *Physical Review D* **81** (12). arXiv:1004.1284. Bibcode:2010PhRvD..8114036K. doi:10.1103/PhysRevD.81.124036.

- Project Dorothy

- Is it true that there could be intelligent life out there? physics.org page about SETI

- "SETI: Astronomy as a Contact Sport - A conversation with Jill Tarter", *Ideas Roadshow*, 2013

- The Rio Scale, a scale for rating SETI announcements.

- Stephen Hawking and Russian tycoon Yuri Milner kick off new search for E.T., $100 million funding to search star catalogue using SETI@home software, July 2015.

View of Arecibo Observatory in Puerto Rico with its 300 m (980 ft) dish- the world's largest. A small fraction of its observation time is devoted to SETI searches.

SETI@home logo

Chapter 23

Wow! signal

The Wow! signal

The **Wow! signal** was a strong narrowband radio signal detected by Jerry R. Ehman on August 15, 1977, while he was working on a SETI project at the Big Ear radio telescope of The Ohio State University, then located at Ohio Wesleyan University's Perkins Observatory in Delaware. Ohio. The signal bore the expected hallmarks of non-terrestrial and non-Solar System origin. The signal appears to have come from the northwest of the globular cluster of M55 in the constellation Sagittarius. near the Chi Sagittarii star group.

The entire signal sequence lasted for the full 72-second window that Big Ear was able to observe it, but has not been detected again. The signal has been the subject of significant media attention, and astronomers have tried many times in vain to find the signal again. Impressed by the relative resemblance of the expected signature of an interstellar signal in the antenna used, Ehman circled the signal on the computer printout and wrote the comment "Wow!" on its side, which became the name of the signal itself.

23.1   Background

The Wow! signal was detected by Jerry R. Ehman on August 15, 1977, while working on a SETI project at the now defunct Big Ear radio telescope, operated by Ohio State University,[1] which was then located at Ohio Wesleyan University's Perkins Observatory in Delaware, Ohio. Ehman spotted a surprising vertical column with the alphanumerical sequence "6EQUJ5" that had been recorded at 22:16 EST. With a red marker, he wrote "Wow!" in the margin of the printout and encircled the sequence.[2][3]

23.2   Interpretation of the paper chart

The circled alphanumeric code "6EQUJ5" describes the intensity variation of the signal. Each character represents 12 seconds (by 10 kHz). A space denotes an intensity between 0 and 1, the numbers 1 to 9 denote the correspondingly numbered intensities (from 1.0 to 9.9), and intensities of 10.0 and above are denoted by a letter ("A" corresponds to intensities between 10.0 and 11.0, "B" to 11.0 to 12.0, etc.). The value "U" (an intensity between 30.0 and 31.0) was the highest detected by the radio telescope; on a linear scale it was over 30 times louder than normal deep space.[1] The intensity in this case is the unitless signal-to-noise ratio, where noise was averaged for that band over the previous few minutes.[4]

Two different values for its frequency have been given: 1420.356 MHz (J. D. Kraus) and 1420.4556 MHz (J. R. Ehman). The frequency of the Wow! signal matches very closely with the hydrogen line, which is at 1420.40575177 MHz. The hydrogen line frequency is significant for SETI searchers because, it is reasoned, hydrogen is the most common element in the universe, and hydrogen resonates at about 1420.40575177 MHz, so extraterrestrials might use that frequency to transmit a strong signal.[1] The two different values given for the frequency of the Wow! signal (1420.356 MHz and 1420.4556 MHz) are the same distance apart from the hydrogen line—the first being about 0.0498 MHz (49.75177 kHz) less than the hydrogen line, and the second about 0.0498 MHz (49.84823 kHz) more. The bandwidth of the signal is less than 10 kHz (each column on the printout corresponds to a 10 kHz-wide channel; the signal is present in only one column).[5]

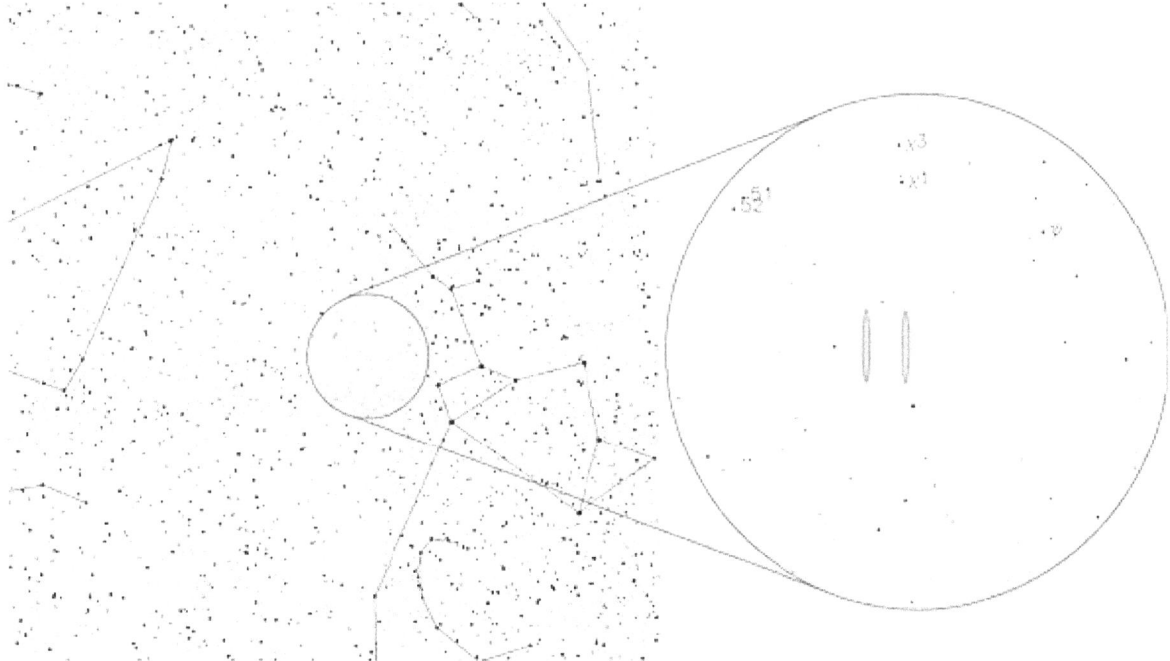

The location of the signal in the constellation Sagittarius, near the Chi Sagittarii star group. Because of the design of the experiment, the location may lie in either one of the two red bands, and there is also significant uncertainty in the declination (vertical axis). For clarity, the widths of the red bands are not drawn to scale; they should actually be narrower.

The original printout of the Wow! signal, complete with Jerry Ehman's famous exclamation, is preserved by the Ohio Historical Society.[6]

23.3 Location of the signal

Determining a precise location in the sky was complicated by the Big Ear telescope's use of two feed horns to search for signals, each pointing to a slightly different direction in the sky following Earth's rotation; the Wow! signal was detected in one of the horns but not in the other, and the data was processed in such a way that it is impossible to determine which of the two horns the signal entered.[7] There are, therefore, two possible right ascension values:

- $19^h22^m24.64^s \pm 5^s$ (positive horn)

- $19^h25^m17.01^s \pm 5^s$ (negative horn)

The declination was unambiguously determined to be $-27°03' \pm 20'$. The preceding values are all expressed in terms of the B1950.0 equinox.[8]

Converted into the J2000.0 equinox, the coordinates become RA= $19^h25^m31^s \pm 10^s$ or $19^h28^m22^s \pm 10^s$ and the declination becomes $-26°57' \pm 20'$. This region of the sky lies in the constellation Sagittarius, roughly 2.5 degrees south of the fifth-magnitude star group Chi Sagittarii, and about 3.5 degrees south of the plane of the ecliptic. Tau Sagittarii is the closest easily visible star.[9]

23.4 Time variation

The Big Ear telescope was fixed and used the rotation of the Earth to scan the sky. At the speed of the Earth's rotation, and given the width of the Big Ear's observation "window", the Big Ear could observe any given point for just 72 seconds.[10] A continuous extraterrestrial signal, therefore, would be expected to register for exactly 72 seconds, and the recorded intensity of that signal would show a gradual increase for the first 36 seconds—peaking when the signal reached the center of Big Ear's observation "window"— and then a gradual decrease.[11]

Therefore, both the length of the Wow! signal, 72 seconds, and the shape of the intensity graph may correspond to an extraterrestrial origin, as opposed to a stray terrestrial signal being picked up by the telescope.[12]

23.5 Searches for recurrence of the signal

Several attempts were made by Ehman as well as by other astronomers to detect and identify the signal again. The signal was expected to appear three minutes apart in each of the horns, but that did not happen.[12] Ehman unsuccessfully looked for recurrences using Big Ear in the months after the detection.[13]

In 1987 and 1989, Robert H. Gray searched for the event using the META array at Oak Ridge Observatory, but did not detect it.[13][14] In a July 1995 test of signal detection software to be used in its upcoming Project Argus search, SETI League executive director H. Paul Shuch made several drift-scan observations of the Wow! signal's coordinates with a 12-meter radio telescope at the National Radio Astronomy Observatory in Green Bank, West Virginia, also achieving a null result.

In 1995 and 1996, Gray again searched for the signal using the Very Large Array, which is significantly more sensitive than Big Ear.[13][14] Gray and Simon Ellingsen later searched for recurrences of the event in 1999 using the 26 m radio telescope at the University of Tasmania's Mount Pleasant Radio Observatory.[15] Six 14-hour observations were made at positions in the vicinity, but nothing like the Wow! signal was detected.[12][14]

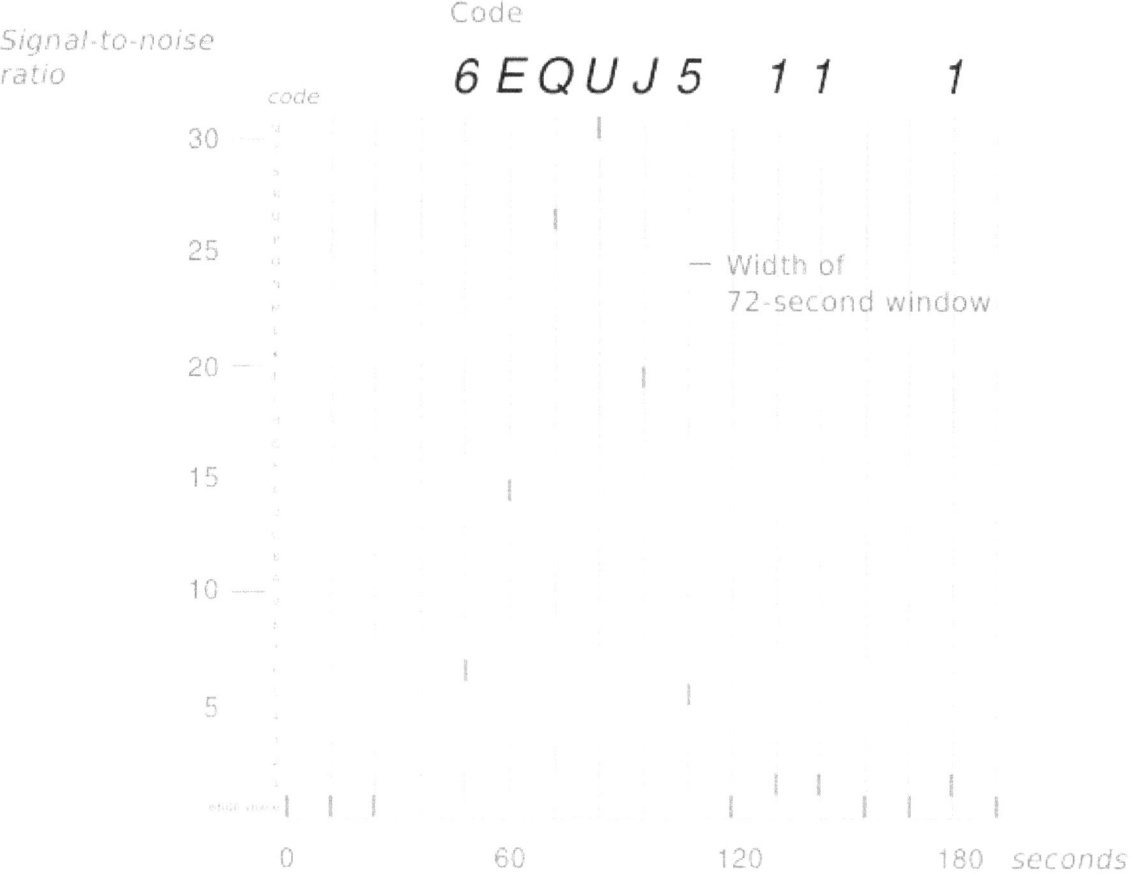

Plot of signal strength vs time

23.6   Speculation on the signal's origin

Interstellar scintillation of a weaker continuous signal—similar in effect to atmospheric twinkling—could be an explanation, but that would not exclude the possibility of the signal's being artificial in origin. But even the significantly more sensitive Very Large Array did not detect the signal, and the probability that a signal below the Very Large Array level could be detected by the Big Ear due to interstellar scintillation is low.[13] Other speculations include a rotating lighthouse-like source, a signal sweeping in frequency, or a one-time burst.[8]

Ehman has voiced doubts that the signal was of intelligent extraterrestrial origin: "We should have seen it again when we looked for it 50 times. Something suggests it was an Earth-sourced signal that simply got reflected off a piece of space debris."[16] He later recanted his skepticism somewhat, after further research showed an Earth-borne signal to be very unlikely, given the requirements of a space-borne reflector being bound to certain unrealistic requirements to sufficiently explain the signal.[17] Also, it is problematic to propose that the 1420 MHz signal originated from Earth since this is within the "protected spectrum": a bandwidth reserved for astronomical purposes in which terrestrial transmitters are forbidden to transmit.[18][19] In a 1997 paper, "The Big Ear Wow! Signal: What We Know and Don't Know About It After 20 Years", Ehman resists "drawing vast conclusions from half-vast data"—acknowledging the possibility that the source may have been military or otherwise a product of Earth-bound humans.[20]

The signal appears to have come from an area of the sky with no stars or planets, northwest of the globular cluster M55.[21]

23.7 Response

In 2012, on the 35th anniversary of the Wow! signal, Arecibo Observatory beamed a response from humanity, containing 10,000 Twitter messages, in the direction from which the signal originated.[22][23] Arecibo scientists have attempted to increase the chances of intelligent life receiving and decoding the celebrity videos and crowd-sourced tweets by attaching a repeating sequence header to each message that will let the recipient know that the messages are intentional and from another intelligent life form.[23]

23.8 See also

- Arecibo message, a 3 minute long message sent into space

- LGM-1, the first pulsar discovered, mistaken for an alien radio signal

23.9 References

[1] Krulwich, Robert (May 29, 2010). "Aliens Found In Ohio? The 'Wow!' Signal". National Public Radio. Retrieved 2013-07-01.

[2] "NGC, where's the wow". Retrieved May 24, 2015.

[3] Shuch, Dr. H. Paul. "SETI Sensitivity: Calibrating on a Wow! Signal". *SETI League*. Retrieved 22 May 2015.

[4] Ehman, Jerry. "Explanation of the Code "6EQUJ5" On the Wow! Computer Printout". Retrieved 2010-01-01.

[5] "bigearwow". Retrieved May 24, 2015.

[6] Wood, Lisa (July 3, 2010). "Wow!". Ohio Historical Society Collections Blog. Retrieved 2013-07-01.

[7] "Big Ear's Twin Feed Horns". Retrieved May 24, 2015.

[8] Gray, Robert; Kevin Marvel (2001). "A VLA Search for the Ohio State 'Wow'" (PDF). *The Astrophysical Journal* **546** (2): 1171–1177. Bibcode:2001ApJ...546.1171G. doi:10.1086/318272.

[9] "The Big Ear Wow! Signal". Retrieved May 24, 2015.

[10] "Skeptoid". Retrieved May 28, 2015.

[11] "EDN Moments". Retrieved May 28, 2015.

[12] Shostak, Seth (2002-12-05). "Interstellar Signal From the 70s Continues to Puzzle Researchers". Space.com.

[13] "The 'Wow!' Signal". The Discovery Channel network. Retrieved 2015-05-28.

[14] Gray, Robert H (2012). *The Elusive WOW: Searching for Extraterrestrial Intelligence*. Chicago: Palmer Square Press. ISBN 978-0-9839584-4-4.

[15] Gray, Robert; S. Ellingsen (2002). "A Search for Periodic Emissions at the Wow Locale". *The Astrophysical Journal* **578** (2): 967–971. Bibcode:2002ApJ...578..967G. doi:10.1086/342646.

[16] Kawa, Barry (1994-09-18). "The Wow! signal". Cleveland Plain Dealer. Retrieved 2006-06-12.

[17] Jerry R. Ehman (February 3, 1998). "The Big Ear Wow! Signal. What We Know and Don't Know About It After 20 Years". Retrieved February 27, 2010.

[18] "Frequencies Allocated to Radio Astronomy Used by the DSN". NASA. Retrieved November 2007.

[19] Committee on Radio Astronomy Frequencies Handbook for Radio Astronomy, European Science Foundation, 3rd edition, 2005, p. 101.

[20] Frank, Adam (July 10, 2012). "Talking To Aliens From Outer Space". NPR.

[21] Kiger, Patrick J. (June 21, 2012). "What is the Wow! Signal?". National Geographic Channel.

[22] Wolchover, Natalie (2012-06-27). "Possible Alien Message to Get Reply from Humanity". Discovery News.

[23] "Humanity Responds to 'Alien' Wow Signal, 35 Years Later". Space.com. 2012-08-12.

23.10   External links

Media related to Wow! signal at Wikimedia Commons

- Location on Google Sky

- Location on YourSky

- APOD NASA GOV NASA Signal 2011

- Dunning, Brian (December 25, 2012). "Was the Wow! Signal Alien?". Skeptoid.com. Retrieved 2013-07-01.

- Ehman, Jerry R. (May 28, 2010). "The Big Ear Wow! Signal (30th Anniversary Report)". North American AstroPhysical Observatory. Retrieved 2013-07-01.

Chapter 24

Communication with extraterrestrial intelligence

Communication with extraterrestrial intelligence (CETI) is a branch of the search for extraterrestrial intelligence that focuses on composing and deciphering messages that could theoretically be understood by another technological civilization. The best-known CETI experiment was the 1974 Arecibo message composed by Frank Drake and Carl Sagan. There are multiple independent organizations and individuals engaged in CETI research; the abbreviations CETI and SETI alone should not be taken as referring to any particular organization (such as the SETI Institute).

CETI research has focused on four broad areas: mathematical languages, pictorial systems such as the Arecibo message, algorithmic communication systems (ACETI) and computational approaches to detecting and deciphering "natural" language communication. There remain many undeciphered writing systems in human communication, such as Linear A, discovered by archeologists. Much of the research effort is directed at how to overcome similar problems of decipherment which arise in many scenarios of interplanetary communication.

On 13 February 2015, scientists (including David Grinspoon, Seth Shostak, and David Brin) at an annual meeting of the American Association for the Advancement of Science, discussed Active SETI and whether transmitting a message to possible intelligent extraterrestrials in the Cosmos was a good idea.[1][2] That same week, a statement was released, signed by many in the SETI community, that a "worldwide scientific, political and humanitarian discussion must occur before any message is sent".[3] On 28 March 2015, a related essay was written by Seth Shostak and published in the New York Times.[4]

24.1 History

In the nineteenth century there were many books and articles about the possible inhabitants of other planets. Many people believed that intelligent beings might live on the Moon, Mars, and Venus; but since travel to other planets was not yet possible, some people suggested ways to signal the extraterrestrials even before radio was discovered.

Carl Friedrich Gauss suggested that a giant triangle and three squares, the Pythagoras, could be drawn on the Siberian tundra. The outlines of the shapes would have been ten-mile-wide strips of pine forest, the interiors could be rye or wheat.

Joseph Johann Littrow proposed using the Sahara as a blackboard. Giant trenches several hundred yards wide could delineate twenty-mile-wide shapes. Then the trenches would be filled with water, and then enough kerosene could be poured on top of the water to burn for six hours. Using this method, a different signal could be sent every night.

Meanwhile, other astronomers were looking for signs of life on other planets. In 1822, Franz von Gruithuisen thought he saw a giant city and evidence of agriculture on the moon, but astronomers using more powerful instruments refuted his claims. Gruithuisen also believed he saw evidence of life on Venus. Ashen light had been observed on Venus, and he postulated that it was caused by a great fire festival put on by the inhabitants to celebrate their new emperor. Later he revised his position, stating that the Venusians could be burning their rainforest to make more farmland.[5]

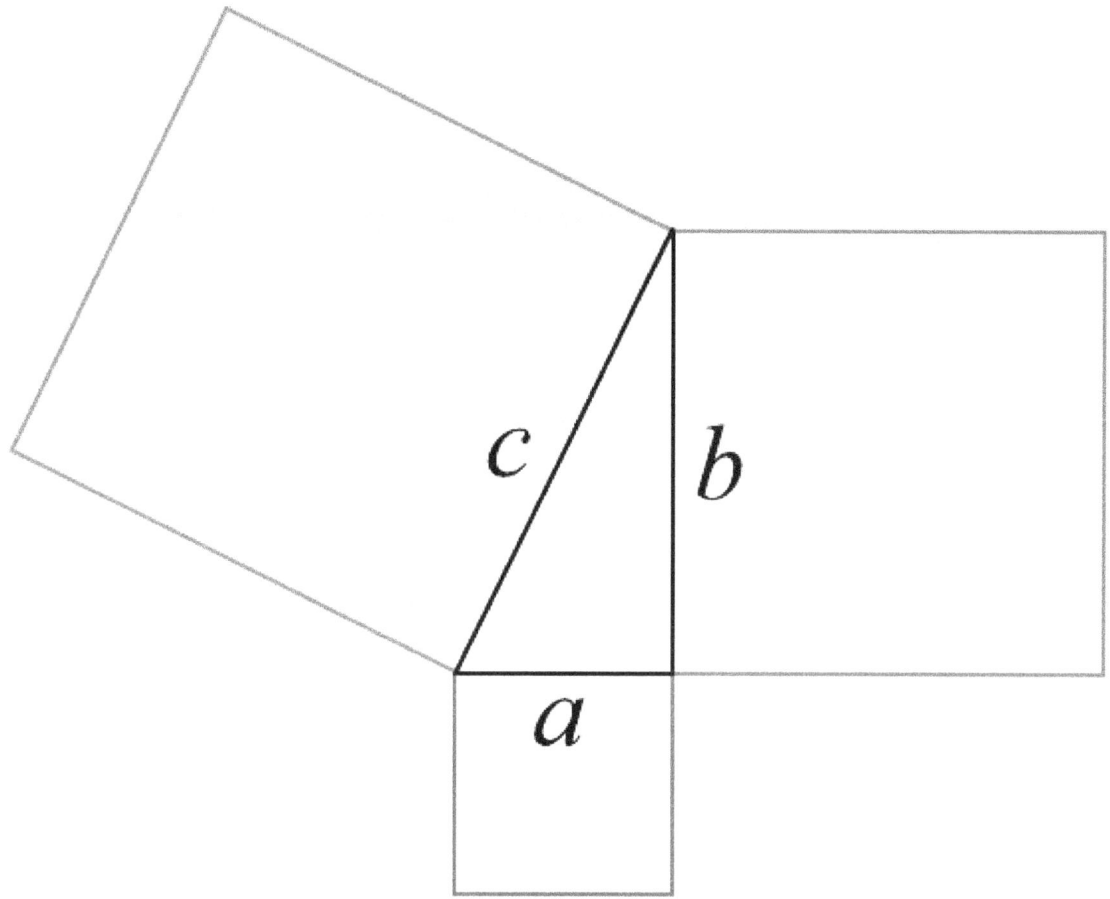

The Pythagoras.

By the late 1800s, the possibility of life on the moon was put to rest. Astronomers at that time believed in the Kant-Laplace hypothesis, which stated that the farthest planets from the sun are the oldest—therefore Mars was more likely to have advanced civilizations than Venus. It was evident that Venus was perpetually shrouded in clouds, so the Venusians probably would not be very good astronomers. Subsequent investigations focused on contacting Martians. In 1877 Giovanni Schiaparelli announced he had discovered "canali" ("channels" in Italian, which occur naturally, and mistranslated as "canals", which are artificial) on Mars—this was followed by thirty years of Mars enthusiasm.

The inventor Charles Cros was convinced that pinpoints of light observed on Mars and Venus were the lights of large cities. He spent years of his life trying to get funding for a giant mirror with which to signal the Martians. The mirror would be focused on the Martian desert, where the intense reflected sunlight could be used to burn figures into the Martian sand.

Inventor Nikola Tesla mentioned many times during his career that he thought his inventions such as his Tesla coil, used in the role of a "resonant receiver", could communicate with other planets[6][7] and even observed repetitive signals of what he believed were extraterrestrial radio communications coming from Venus or Mars in 1899. However, these "signals" turned out to be terrestrial radiation.

Around 1900, the Guzman Prize was created: the first person to establish interplanetary communication would be awarded 100,000 francs under one stipulation: Mars was excluded because Madame Guzman thought communicating with Mars would be too easy to deserve a prize.[8]

When the Martian canals proved illusory, it seemed that humans were alone in the solar system.

24.2 Mathematical and scientific languages

24.2.1 Lincos (Lingua cosmica)

Main article: Lincos (artificial language)

Published in 1960 by Hans Freudenthal, *Lincos: Design of a Language for Cosmic Intercourse*, expands upon Astraglossa to create a general-purpose language derived from basic mathematics and logic symbols.[9] Several researchers have further expanded upon Freudenthal's work. A Lincos-like dictionary was featured in the Carl Sagan novel *Contact* and film adaptation.

24.2.2 Astraglossa

Published in 1963 by Lancelot Hogben describes a system for combining numbers and operators in a series of short and long pulses. In Hogben's system, short pulses represent numbers, while trains of long pulses represent symbols for addition, subtraction, etc.[10]

24.2.3 Carl Sagan

In 1985, the science fiction novel *Contact* by Carl Sagan explored in some depth how a message might be constructed to allow communication with an alien civilization, using the prime numbers as a starting point, followed by various universal principles and facts of mathematics and science. Sagan also edited a non-fiction book on the subject,[11] which has been updated in 2011.[12]

24.2.4 A language based on the fundamental facts of science

Published in 1992 by Carl Devito and Richard Oehrle, is similar in syntax to Astraglossa and Lincos but builds its vocabulary around known physical properties.[13]

24.2.5 Busch general-purpose binary language used in Lone Signal transmissions

In 2010, Michael W. Busch created a general-purpose binary language[14] later used in the Lone Signal project[15] to transmit crowdsourced messages to extraterrestrial intelligence (METI). This was followed by an attempt to extend the syntax used in the Lone Signal hailing message to communicate in a way that, while neither mathematical nor strictly logical, was nonetheless understandable given the prior definition of terms and concepts in the Lone Signal hailing message.[16]

24.3 Pictorial messages

Pictorial communication systems seek to describe fundamental mathematical or physical concepts via simplified diagrams sent as bitmaps. These messages assume that the recipient has similar visual capabilities (weak assumption) and can understand basic mathematics and geometry (strong assumption because both are prerequisites for building the optimal shape for a radio or optical telescope). A common critique of these systems is that they assume a shared understanding of special shapes, which may not be the case with a species with substantially different vision, and therefore a different way of interpreting visual information. For instance, an arrow representing the movement of some object could be interpreted as a weapon firing.

24.3.1   Pioneer probes

The two Pioneer plaques were launched on Pioneer 10 and Pioneer 11 in 1972 and 1973, depicting the location of the Earth in the galaxy and the solar system, and the form of the human body.

24.3.2   Voyager probes

Launched in 1977, the Voyager probes carried two golden records that were inscribed with diagrams depicting the human form, our solar system and its location. Also included were recordings of pictures and sounds from Earth.

24.3.3   The Arecibo message

The Arecibo message, transmitted in 1974, was a 1679 pixel image with 73 rows and 23 columns. It shows the numbers one through ten, the atomic numbers of hydrogen, carbon, nitrogen, oxygen, and phosphorus, the formulas for the sugars and bases in the nucleotides of DNA, the number of nucleotides in DNA, the double helix structure of DNA, a figure of a human being and its height, the population of Earth, a diagram of our solar system, and an image of the Arecibo telescope with its diameter.

24.3.4   Cosmic Call messages

The *Cosmic Call* messages consisted of a few digital sections - "Rosetta Stone", copy of Arecibo Message, Bilingual Image Glossary, the Braastad message, as well as text, audio, video and other image files submitted for transmission by everyday people around the world. The "Rosetta Stone" was composed by Stephane Dumas and Yvan Dutil and represents a multi-page bitmap that builds a vocabulary of symbols representing numbers and mathematical operations. The message proceeds from basic mathematics to progressively more complex concepts, including physical processes and objects (such as a hydrogen atom). The message is designed with noise resistant format and characters, which make it resistant to alteration by noise. These messages were transmitted in 1999 and 2003 from Evpatoria Planetary Radar under scientific guidance of Alexander L. Zaitsev. Richard Braastad coordinated the overall project.

Stars to which messages were sent, are the following:[17]

24.4   Multi-modal messages

24.4.1   Teen-Age Message

Main article: Teen Age Message

The *Teen-Age Message*, composed by Russian scientists (Zaitsev, Gindilis, Pshenichner, Filippova) and teens, was transmitted from the 70-m dish of Yevpatoria Deep Space Center to six Sun-like stars on August 29 and September 3 and 4, 2001. The message consists of three parts:

Section 1 represents a coherent-sounding radio signal with slow Doppler wavelength tuning to imitate transmission from the Sun's center. This signal was transmitted in order to help extraterrestrials detect the TAM and diagnose the radio propagation effect of the interstellar medium.

Section 2 is analog information representing musical melodies performed on the theremin. This electric musical instrument produces a quasi-monochromatic signal, which is easily detectable across interstellar distances. There were seven musical compositions in the 1st Theremin Concert for Aliens. The 14 minute analog transmission of the theremin concert would take almost 50 hours by digital means; see The First Musical Interstellar Radio Message.

Section 3 represents a well-known Arecibo-like binary digital information: the logotype of the TAM, bilingual Russian and English greeting to aliens, and image glossary.

Stars to which the message was sent are the following:[17]

24.4.2 Cosmic Call 2 (Cosmic Call 2003) message

The Cosmic Call−2 message contained text, images, video, music, the Dutil/Dumas message, a copy of the 1974 Arecibo message, BIG = Bilingual Image Glossary, the AI program *Ella*, and the Braastad message.

24.5 Algorithmic messages

Algorithmic communication systems are a relatively new field within CETI. In these systems, which build upon early work on mathematical languages, the sender describes a small set of mathematics and logic symbols that form the basis for a rudimentary programming language that the recipient can run on a virtual machine. Algorithmic communication has a number of advantages over static pictorial and mathematical messages, including: localized communication (the recipient can probe and interact with the programs within a message, without transmitting a reply to the sender and then waiting years for a response), forward error correction (the message might contain algorithms that process data elsewhere in the message), and the ability to embed proxy agents within the message. In principle, a sophisticated program when run on a fast enough computing substrate, may exhibit complex behavior and perhaps intelligence.

24.5.1 CosmicOS

CosmicOS, designed by Paul Fitzpatrick at MIT, describes a virtual machine that is derived from lambda calculus.

24.5.2 Logic Gate Matrices

Logic Gate Matrices (a.k.a. LGM), developed by Brian McConnell, describes a universal virtual machine that is constructed by connecting coordinates in an n-dimensional space via mathematics and logic operations, for example: (1,0,0) <-- (OR (0,0,1) (0,0,2)). Using this method, one can describe an arbitrarily complex computing substrate as well as the instructions to be executed on it.

24.6 Natural language messages

This research focuses on the event that we receive a signal / message that is either not directed at us (eavesdropping) or one that is in its natural communicative form. To tackle this difficult but probable scenario, methods are being developed that will first detect if a signal has intelligent-like structure, categorize the type of structure detected and then decipher its content: from its physical level encoding and patterns to the parts-of-speech, which encode internal and external ontologies.[18][19]

Primarily, this structure modeling focuses on the search for generic human and inter-species language universals to devise computational methods by which language can be discriminated from non-language and core structural syntactic elements of unknown languages can be detected.[20] Aims of this research include: contributing to the understanding of language structure and the detection of intelligent language-like features in signals, to aid the search for extraterrestrial intelligence.[21][22]

The problem goal is therefore to separate language from non-language without dialogue, and learn something about the structure of language in the passing. The language may not be human (animals, aliens, computers...), the perceptual space can be unknown, and we cannot assume human language structure but must begin somewhere. We need to approach the language signal from a naive viewpoint, in effect, increasing our ignorance and assuming as little as possible.[23][24]

SETI scientist Laurance Doyle explains that the slope of a line that represents individual tokens in a stream of tokens can indicate whether the stream contains linguistic or other structured content. If the line angles at 45°, the stream contains such content. If the line is flat, it does not.[25][26]

24.7   SETI researchers

- Frank Drake (SETI Institute): SETI pioneer, composed the Arecibo message with Carl Sagan

- Dr John Elliott (SETI Research UK): research into developing strategies, which are based on receiving a 'natural' language message, that look at developing algorithms to detect if an ET signal has intelligent-like structure and if so, then how to decipher its content. Author of many papers in this area and a contributor to SETI's book on interstellar communication. Other contributions include message design and construction; member of: International Academy of Astronautics, SETI Permanent Study Group; International Task Group for the Post-detection identification of unknown radio signals.[18][19][20][21][22][23][24]

- Laurence Doyle (SETI Institute): studies animal communication, and has developed statistical measures of complexity in animal utterances as well as human language.

- Stephane Dumas: developed *Cosmic Call* messages, as well as a general technique for generating 2-D symbols that remain recognizable even if corrupted by noise.

- Yvan Dutil: developed *Cosmic Call* messages with Stephane Dumas.

- Paul Fitzpatrick (MIT): developed *CosmicOS* system based on lambda calculus

- Brian McConnell: developed framework for algorithmic communication systems (ACETI) from 2000-2002.

- Marvin Minsky (MIT AI researcher): Believes that aliens may think like humans because of shared constraints, permitting communication.[27] First proposed the idea of including algorithms within an interstellar message.

- Carl Sagan (deceased): co-authored the Arecibo message, and was heavily involved in SETI throughout his life.

- Douglas Vakoch (SETI Institute): studies CETI and has published numerous articles, as well as an upcoming book from MIT Press about interstellar communication.

- Alexander Zaitsev (IRE, Russia): composed *Teen Age Message* with Boris Pshenichner, Lev Gindilis, Lilia Filippova, et al., composed *Bilingual Image Glossary* for *Cosmic Call 2003 Message*. Scientific Manager of transmitting from Evpatoria Planetary Radar the Cosmic Call 1999, the Teen Age Message 2001, and the Cosmic Call 2003. Scientific consultant for A Message From Earth project.[28][29][30][31]

- Michael W. Busch: (Lone Signal) created the binary encoding system for the ongoing Lone Signal hailing message.

- Jacob Haqq Misra: (Lone Signal) is the Chief Science Officer for the ongoing Lone Signal active SETI project.

24.8   Connections with Interspecies Communication

John C. Lilly worked on teaching dolphins English (successful with rhythms, not with understandability, given their different mouth/blowhole shapes) and identifying whether extraterrestrial signals contain communication.

Laurance Doyle compares the complexity of cetacean and human languages to help determine if a specific signal from space is complex enough to represent a message that needs to be decoded.

Brenda McCowan studies signal complexity of humpback whales and extraterrestrial signals.

Robert Freitas has used density of brain processing (Sentience Quotient) to compare the difficulty of communicating with animals, including cetaceans, and extraterrestrials.

Self-explanatory languages like Lincos have been tried with radio waves to extraterrestrials, but not sound waves or other signals on earth. They assume recipients patient enough to analyze repetitive mathematical signals to understand the content, and may assume note-taking ability such as opposable thumbs.

24.9 See also

- Active SETI

- Nexus for Exoplanet System Science

- Pioneer plaque

- SETIcon

- Time capsule

- Waterhole (radio)

24.10 Notes

[1] Borenstein, Seth (of AP News) (13 February 2015). "Should We Call the Cosmos Seeking ET? Or Is That Risky?". *New York Times*. Retrieved 14 February 2015.

[2] Ghosh, Pallab (12 February 2015). "Scientist: 'Try to contact aliens'". *BBC News*. Retrieved 12 February 2015.

[3] Various (13 February 2015). "Statement - Regarding Messaging To Extraterrestrial Intelligence (METI) / Active Searches For Extraterrestrial Intelligence (Active SETI)". *University of California, Berkeley*. Retrieved 14 February 2015.

[4] Shostak, Seth (28 March 2015). "Should We Keep a Low Profile in Space?". *New York Times*. Retrieved 29 March 2015.

[5] Cattermole, P., & Moore, P. (1997). Atlas of Venus. Cambridge University Press.

[6] Seifer, Marc J. (1996). "Martian Fever (1895-1896)". *Wizard: the life and times of Nikola Tesla: biography of a genius*. Secaucus, New Jersey: Carol Pub. p. 157. ISBN 978-1-55972-329-9. OCLC 33865102.

[7] "Tesla at 75". *Time (magazine)* **18** (3). July 20, 1931. p. 3..

[8] Ley, Willy. Rockets, Missiles, and Space Travel (revised). New York: The Viking Press (1958)

[9] Freudenthal H, ed. (1960). *Lincos: Design of a Language for Cosmic Intercourse. Studies in Logic and the Foundations of Mathematics (Book 28)*. North-Holland, Amsterdam. ISBN 978-0-444-53393-7.

[10] Hogben, Lancelot (1963). *Science in Authority*. New York: W. W. Norton. ISBN 1245639935.

[11] Sagan, Carl. Communication with Extraterrestrial Intelligence. MIT Press. 1973. 428 pgs.ISBN 0262191067

[12] Vakoch, Douglas. Communication with Extraterrestrial Intelligence. SUNY Press. 2011. 500 pgs.

[13] Devito, C. and Oerle, R (1990). "A Language Based on the Fundamental Facts of Science". *Journal of the British Interplanetary Society* **43**: 561–568. PMID 11540499.

[14] Busch, Michael W.; Reddick, Rachel M. "Testing SETI Messages Design". *Astrobiology Science Conference 2010*. Archived from the original (PDF) on July 1, 2013.; Busch, Michael W.; Reddick, Rachel. M. "Testing SETI Messages (Extended version)". Archived from the original (PDF) on June 30, 2013.

[15] "Lone Signal - Encoding". Retrieved 7 July 2013.

[16] Chapman, Charles R. "Extending the syntax used by the Lone Signal Active SETI project". Archived from the original on August 13, 2013.

[17] Передача и поиски разумных сигналов во Вселенной

[18] Elliott, J. (2004). "Unsupervised Discovery of Language Structure in Audio Signals". *Proceedings of IASTED International Conference on Circuits, Signals and Systems, (CSS 2004), Clearwater Beach, Florida, USA.*

[19] Elliott, J. Atwell, E & Whyte, B. (2001). "First stage identification of syntactic elements, An extraterrestrial signal". *Proceedings of IAC 2001: the 52nd International Astronautical Congress*: AA–01–IAA.9.2.07.

[20] Elliott, J and Atwell, E. (2000). "Is anybody out there: the detection of intelligent and generic language-like features". *Journal of the British Interplanetary Society* **53**: 13–22. ISSN 0007-084X.

[21] Elliott, J. (2002a). "Detecting languageness". *Proceedings of 6th World Multi-Conference on Systemics, Cybernetics and Informatics (SCI 2002)* **IX**: 323–328.

[22] Elliott, J. (2002b). "The filtration of inter-galactic objets trouvés and the identification of the Lingua ex Machina hierarchy". *Proceedings of World Space Congress: The 53rd International Astronautical Congress*: IAA–02–IAA.9.2.10.

[23] Elliott, J, Atwell, E & Whyte, B. (2000). "Increasing our ignorance of language: identifying language structure in an unknown signal", in Daelemans, W (ed.) Proceedings of CoNLL-2000: International Conference on Computational Natural Language Learning", pp. 25–30 Association for Computational Linguistics.

[24] Elliott, J. (2007). "A Post-Detection Decipherment Matrix: Acta Astronautica. Journal of the International Academy of Astronautics Elsevier Science Ltd, England, AA2853".

[25] Freeman, David (March 5, 2012). "'Through The Wormhole' Host Morgan Freeman: 'We Can't Be' Alone In The Universe (VIDEO)". *Huffington Post*. Retrieved 25 May 2013.

[26] "Through the Wormhole: Information Theory : Video : Science Channel". Discovery Communications. Retrieved 25 May 2013.

[27] Minsky, Marvin (April 1985). "Communication with Alien Intelligence". *BYTE*. p. 127. Retrieved 27 October 2013.

[28] Zaitsev, Alexander (2002-03-18). "A Teen-Age Message to the Stars". Cplire.ru. Retrieved 2012-08-21.

[29] Zaitsev, A (5 Oct 2006). "Interstellar Radio Messages (http://www.cplire.ru/rus/ra&sr/index.html in Russian)".

[30] Zaitsev, A (August 29, 2001). "Messaging to Extraterrestrial Intelligence (METI)".

[31] Alexander Zaitsev (2011). "Clasificación de Mensajes de Radio Interestelares".

24.11   References

- Braastad, Richard, The Extraterrestrial Sermons, http://www.richardb.us/project.html

- Cattermole, P., & Moore, P. (1997). Atlas of Venus. Cambridge University Press.

- Dumas, Stepane. The 1999 and 2003 messages explained, http://www3.sympatico.ca/stephane_dumas/CETI/messages.pdf

- Dutil, Dumas, ACTIVE SETI PAGE, http://www3.sympatico.ca/stephane_dumas/CETI/default.htm

- EllaZ Systems, http://www.ellaz.com/AI/Default.aspx

- Martin, Martin C. 1991 SETI Puzzle, posted to sci.crypt, sci.astro, sci.space, rec.arts.sf-lovers and rec.puzzles. http://www.metahuman.org/martin/SETIPuzzle.html

- Communication with Alien Intelligence. 1985. Marvin Minsky, http://web.media.mit.edu/~{}minsky/papers/Alie.html

- Ley, Willy. Rockets, Missiles, and Space Travel (revised). New York: The Viking Press (1958)

- McConnell, Brian S. 2001 Beyond Contact: A Guide to SETI and Communicating with Alien Civilizations. O'Reilly, Cambridge, MA

- McConnell, Brian S. 2002 Algorithmic Communication with ETI & Mixed Media Message Composition

- McConnell, Brian S at al. 2006?, Between Worlds, SETI Institute/MIT Press

- Minsky, Marvin, talk given at Communication With Extraterrestrial Intelligence (CETI), Proceedings of a conference held at the Byurakan Astrophysical Observatory, Yerevan, USSR, 5–11 September 1971. Edited by Carl Sagan. Cambridge, MA: The Massachusetts Institute of Technology, 1973., p.ix

- Morrison, P. "Interstellar Communication." Bulletin of the Philosophical Society of Washington, 16, 78 (1962). Reprinted in A. G. W. Cameron, ed., Interstellar Communication.

- Team Encounter, Cosmic Call 2003,http://web.archive.org/web/20040406151735/www.teamencounter.com/mi message.asp

- Vakoch, D. A. (Ed.). (2011). Communication with Extraterrestrial Intelligence. New York: SUNY Press.

Chapter 25

Nexus for Exoplanet System Science

The **Nexus for Exoplanet System Science (NExSS)** initiative is a National Aeronautics and Space Administration (NASA) virtual institute designed to foster interdisciplinary collaboration in the search for life on exoplanets. Led by the Ames Research Center, the NASA Exoplanet Science Institute, and the Goddard Institute for Space Studies, NExSS will help organize the search for life on exoplanets from participating research teams and acquire new knowledge about exoplanets and extrasolar planetary systems.[2][3]

25.1   History

In 1995, astronomers using ground-based observatories discovered 51 Pegasi b, the first exoplanet orbiting a Sun-like star.[4] NASA launched the *Kepler* space telescope in 2009 to search for Earth-size exoplanets. By 2015, they had confirmed more than a thousand exoplanets,[note 1] while several thousand additional candidates awaited confirmation.[6]

To help coordinate efforts to sift through and understand the data, NASA needed a way for researchers to collaborate across disciplines. The success of the Virtual Planetary Laboratory research network at the University of Washington led Mary A. Voytek, director of the NASA Astrobiology Program, to model its structure and create the Nexus for Exoplanet System Science (NExSS) initiative.[1][7] Leaders from three NASA research centers will run the program: Natalie Batalha of NASA's Ames Research Center, Dawn Gelino of the NASA Exoplanet Science Institute, and Anthony Del Genio of NASA's Goddard Institute for Space Studies.[8]

25.2   Research

Functioning as a virtual institute, NExSS is currently composed of sixteen interdisciplinary science teams from ten universities, three NASA centers and two research institutes, who will work together to search for habitable explanets that can support life.[9] The US teams were initially selected from a total of about 200 proposals; however, the coalition is expected to expand nationally and internationally as the project gets underway.[10] Teams will also work with amateur citizen scientists who will have the ability to access the public Kepler data and search for exoplanets.[8]

NExSS will draw from scientific expertise in each of the four divisions of the Science Mission Directorate: Earth science, planetary science, heliophysics and astrophysics.[2] NExSS research will directly contribute to understanding and interpreting future exoplanet data from the upcoming launches of the Transiting Exoplanet Survey Satellite and James Webb Space Telescope, as well as the planned Wide Field Infrared Survey Telescope mission.[2]

Current NExSS research projects as of 2015:[2]

25.3   See also

- Astrobiology

- Astrochemistry

- Cosmochemistry

- Extraterrestrial atmospheres

- Extraterrestrial liquid water

- Planetary habitability

25.4   Notes

[1] There are 2030 confirmed exoplanets as of December 11, 2015.[5]

25.5   References

[1] Tollefson, Jeff (April 17, 2015). "Climate scientists join search for alien Earths." *Nature* 520, 420. doi:10.1038/520420a.

[2] Loff, Sarah (April 21, 2015). "NASA's NExSS Coalition to Lead Search for Life on Distant Worlds." NASA. Retrieved April 22, 2015.

[3] Gronstal, Aaron L. (April 2015). "NASA's Exoplanet Nexus — Part I: A History in Climate Studies, Part II: Looking to the Stars". NASA. GISS. Retrieved April 25, 2015.

[4] "NASA Taps UW Scientist in Search for Life Beyond the Solar System." *University of Wyoming News*. Retrieved April 25, 2015.

[5] Schneider, J. "Interactive Extra-solar Planets Catalog". *The Extrasolar Planets Encyclopedia*.

[6] Netburn, Deborah (April 24, 2015). "NASA gathers scientists to help find life beyond Earth". *Los Angeles Times*. Retrieved April 25, 2015.

[7] Kelley, Peter (April 22, 2015). "UW key player in new NASA coalition to search for life on distant worlds." *UW Today*. Retrieved May 3, 2015.

[8] Carreau, Mark (April 24, 2015). "NASA Widens Circle of Experts In Search for Life Beyond Earth." *Aviation Week*. Retrieved April 25, 2015.

[9] Sanders, Robert (April 22, 2015). "Astronomers join forces to speed discovery of habitable worlds." UC Berkeley News Center. Retrieved April 22, 2015.

[10] Cassidy, Chris (April 25, 2015). "NASA leaves local aerospace stars Harvard, MIT off 'unprecedented' mission to find extraterrestrial life." *Boston Herald*. Retrieved April 26, 2015.

[11] Carey, Bjorn (April 27, 2015). "Stanford and UC Berkeley partner on NASA's new effort to detect life on other planets." *Stanford Report*. Retrieved April 30, 2015.

[12] Goldberg, Logan (April 23, 2015). "UC Berkeley astronomers, Stanford physics professor lead NASA-funded project to discover habitable planets." *The Daily Californian*. Retrieved April 30, 2015.

[13] Sanders, Robert (April 22, 2015). "Astronomers join forces to speed discovery of habitable worlds." *UC Berkeley News Center*. Retrieved April 30, 2015.

[14] Beal, Tom (April 22, 2015). "UA, ASU teams to search for alien life." *Arizona Daily Star*. Retrieved April 30, 2015.

[15] Stolte, Daniel (April 20, 2015). "UA to Join 'A-Team' in Search for Earthlike Planets." *UA News*. Retrieved April 30, 2015.

[16] Cantillo, Laura; Barbara Kennedy (April 21, 2015). "New NASA coalition to lead search for life on distant worlds includes two leaders at Penn State." *Pennsylvania State University Science News*. Retrieved May 1, 2015.

[17] (April 21, 2015). A New Collaboration to Aid the Search for Life on Distant Worlds. SETI Institute. Retrieved April 25, 2015.

[18] Cassis, Nikki (April 21, 2015). "ASU team searching for signs of life in the stars." *ASU News*. Retrieved April 30, 2015.

[19] "NASA's NExSS Coalition to Lead Search for Life on Distant Worlds." Jet Propulsion Laboratory. California Institute of Technology. April 21, 2015. Retrieved May 3, 2015.

[20] Dietrich, Tamara (May 1, 2015). "Hampton University professor chosen to head NASA planet project." *Daily Press*. Retrieved May 3, 2015.

[21] "HU research team selected to lead NASA search for life on distant worlds." *HU News*. April 28, 2015. Retrieved May 3, 2015.

[22] Stephens, Tim (April 21, 2015). "UC Santa Cruz part of NASA coalition leading search for life on distant worlds." *UC Santa Cruz News Center*. Retrieved May 2, 2015.

[23] Victor, Jeff (May 1, 2015). "UW professor to head NASA team." *The Branding Iron*. Retrieved May 3, 2015.

[24] "NASA Taps UW Scientist in Search for Life Beyond the Solar System." *Sweetwater Now*. April 23, 2015. Retrieved May 3, 2015.

[25] Gottula, Todd (April 28, 2015). "UNK faculty Adam Jensen picked for NASA exoplanet project." *UNK News*. Retrieved May 1, 2015.

[26] Jensen, Kyle (April 29, 2015). "UW-led 'super group' boosts NASA's quest for extraterrestrial life." *Seattle Post-Intelligencer*. Retrieved May 3, 2015.

[27] Scharf, Caleb A. (April 24, 2015). NASA Goes Big and Bold for Exoplanet Science. *Scientific American*. Retrieved April 25, 2015.

[28] "NASA GISS to Help Lead Search For Habitable Exoplanets." *Astrobiology Magazine*. April 21, 2015. Retrieved May 1, 2015.

[29] Shelton, Jim (April 21, 2015). "Yale joins new NASA team searching for life outside the solar system." *Yale News*. Retrieved May 1, 2015.

Chapter 26

Extraterrestrial liquid water

Extraterrestrial liquid water (from the Latin words: *extra* ["outside of, beyond"] and *terrestris* ["of or belonging to Earth"]) is water in its liquid state that is found beyond Earth. It is a subject of wide interest because it is commonly thought to be one of the key prerequisites for extraterrestrial life.[1]

With oceanic water covering 71% of its surface, Earth is the only planet known to have stable bodies of liquid water on its surface,[2] and liquid water is essential to all known life forms on earth. The presence of water on the surface of Earth is a product of its atmospheric pressure and a stable orbit in the Sun's circumstellar habitable zone, though the origin of Earth's water remains unknown.

The main methods currently used for confirmation are absorption spectroscopy and geochemistry. These techniques have proven effective for atmospheric water vapour and ice. However, using current methods of astronomical spectroscopy it is substantially more difficult to detect liquid water on terrestrial planets, especially in the case of subsurface water. Due to this, astronomers, astrobiologists and planetary scientists use habitable zone, gravitational and tidal theory, models of planetary differentiation and radiometry to determine potential for liquid water. Water observed in volcanic activity can provide more compelling indirect evidence, as can fluvial features and the presence of antifreeze agents, such as salts or ammonia.

Using such methods, many scientists infer that liquid water once covered large areas of Venus and Mars. Water is thought to exist as liquid beneath the surface of planetary bodies, similar to groundwater on Earth. Water vapour is sometimes considered conclusive evidence for the presence of liquid water, although atmospheric water vapour may be found to exist in many places where liquid water does not. Similar indirect evidence, however, supports the existence of liquids below the surface of several moons and dwarf planets elsewhere in the Solar System.[1] Some are speculated to be large extraterrestrial "oceans".[1] Liquid water is thought to be common in other planetary systems, despite the lack of conclusive evidence, and there is a growing list of extrasolar candidates for liquid water.

26.1   Liquid water in the Solar System

26.1.1   Mars

Water on Mars exists today almost exclusively as ice, with a small amount present in the atmosphere as vapour. Some liquid water may occur transiently on the Martian surface today but only under certain conditions.[3] No large standing bodies of liquid water exist because the atmospheric pressure at the surface averages just 600 pascals (0.087 psi)—about 0.6% of Earth's mean sea level pressure—and because the global average temperature is far too low (210 K (−63 °C)), leading to either rapid evaporation or freezing. On 28 September, 2015, NASA announced that they found evidence that the recurring slope lineae are caused by flows of brine — hydrated salts.[4][5][6]

26.1.2 Europa

Scientists' consensus is that a layer of liquid water exists beneath Europa's surface, and that heat from tidal flexing allows the subsurface ocean to remain liquid.[7] It is estimated that the outer crust of solid ice is approximately 10–30 km (6–19 mi) thick, including a ductile "warm ice" layer, which could mean that the liquid ocean underneath may be about 100 km (60 mi) deep.[8] This leads to a volume of Europa's oceans of 3×10^{18} m^3, slightly more than two times the volume of Earth's oceans.

26.1.3 Enceladus

Enceladus, a moon of Saturn, has shown geysers of water, confirmed by the Cassini spacecraft in 2005 and analyzed more deeply in 2008. Gravimetric data in 2010-2011 confirmed a subsurface ocean. While previously believed to be localized, most likely in a portion of the southern hemisphere, evidence revealed in 2015 now suggests the subsurface ocean is global in nature.[9]

In addition to water, these geysers from vents near the south pole contained small amounts of salt, nitrogen, carbon dioxide, and volatile hydrocarbons. The melting of the ocean water and the geysers appear to be driven by tidal flux from Saturn.

26.1.4 Ganymede

A subsurface saline ocean is theorized to exist on Ganymede, a moon of Jupiter, following observation by the Hubble Space Telescope in 2015. Patterns in auroral belts and rocking of the magnetic field suggest the presence of an ocean. It is estimated to be 100 km deep with the surface lying below a crust of 150 km of ice.[10]

26.2 Methods of detection and confirmation

Most known extrasolar planetary systems appear to have very different compositions to the Solar System, though there is probably sample bias arising from the detection methods.

26.2.1 Spectroscopy

The most conclusive method for detection and confirmation of extraterrestrial liquid water is currently absorption spectroscopy. Liquid water has a distinct spectral signature to other states of water due to the state of its Hydrogen bonds. Despite the confirmation of extraterrestrial water vapor and ice, the spectral signature of liquid water is yet to be confirmed. The signatures of surface water on terrestrial planets may be undetectable through thick atmospheres across the vast distances of space using current technology.

Seasonal flows on warm Martian slopes, though strongly suggestive of briny liquid water, have yet to indicate this in spectroscopic analysis.

Water vapor has been confirmed in numerous objects via spectroscopy, though it does not by itself confirm the presence of liquid water. However, when combined with other observations, the possibility might be inferred. For example, the density of GJ 1214 b would suggest that a large fraction of its mass is water and follow-up detection by the Hubble telescope of the presence if water vapor strongly suggests that exotic materials like 'hot ice' or 'superfluid water' may be present.[11][12]

26.2.2 Geological indicators

Further information: Groundwater on Mars

Thomas Gold has posited that many Solar System bodies could potentially hold groundwater below the surface.[13]

It is thought that liquid water may exist in the Martian subsurface. Research suggests that in the past there was liquid water flowing on the surface,[14] creating large areas similar to Earth's oceans. However, the question remains as to where the water has gone.[15] There are a number[16] of direct and indirect proofs of water's presence either on or under the surface, e.g. stream beds, polar caps, spectroscopic measurement, eroded craters or minerals directly connected to the existence of liquid water (such as Goethite). In an article in the Journal of Geophysical Research, scientists studied Lake Vostok in Antarctica and discovered that it may have implications for liquid water still being on Mars. Through their research, scientists came to the conclusion that if Lake Vostok existed before the perennial glaciation began, that it is likely that the lake did not freeze all the way to the bottom. Due to this hypothesis, scientists say that if water had existed before the polar ice caps on Mars, it is likely that there is still liquid water below the ice caps that may even contain evidence of life.[17]

"Chaos terrain", a common feature on Europa's surface, is interpreted by some as regions where the subsurface ocean has melted through the icy crust.

26.2.3 Volcanic observation

Geysers have been found on Enceladus, a moon of Saturn, and Europa, moon of Jupiter.[18] These contain water vapour and could be indicators of liquid water deeper down.[19] It could also be just ice.[20] In June 2009, evidence was put forward for salty subterranean oceans on Enceladus.[21] On April 3, 2014, NASA reported that evidence for a large underground ocean of liquid water on Enceladus, moon of planet Saturn, had been found by the Cassini spacecraft. According to the scientists, evidence of an underground ocean suggests that Enceladus is one of the most likely places in the solar system to "host microbial life".[22][23]

26.2.4 Gravitational evidence

Scientists' consensus is that a layer of liquid water exists beneath Europa's surface, and that heat energy from tidal flexing allows the subsurface ocean to remain liquid.[24][25] The first hints of a subsurface ocean came from theoretical considerations of tidal heating (a consequence of Europa's slightly eccentric orbit and orbital resonance with the other Galilean moons).

Scientists used gravitational measurements from the Cassini spacecraft to confirm a water ocean under the crust of Enceladus.[22][23] Such tidal models have been used as theories for water layers in other Solar System moons.

26.2.5 Density calculation

Planetary scientists can use calculations of density to determine the composition of planets and their potential to possess liquid water, though the method is not highly accurate as the combination of many compounds and states can produce similar densities.

Scientists used low frequency radio signal from the Cassini probe to detect the existence of a layer of liquid water and ammonia beneath the surface of Saturn's moon Titan that are consistent with calculations of the moon's density.[26][27]

Initial analysis of 55 Cancri e's low density indicated that it consisted 30% supercritical fluid which Diana Valencia of the Massachusetts Institute of Technology proposed could be in the form of salty supercritical water,[28] though follow-up analysis of its transit failed to detect traces of either water or hydrogen.[29]

GJ 1214 b was the second exoplanet (after CoRoT-7b) to have an established mass and radius less than those of the giant Solar System planets. It is three times the size of Earth and about 6.5 times as massive. Its low density indicated that it is likely a mix of rock and water,[30] and follow-up observations using the Hubble telescope now seem to confirm that a large fraction of its mass is water, so it is a large waterworld. The high temperatures and pressures would form exotic materials like 'hot ice' or 'superfluid water'.[11][12]

26.2.6   Models of radioactive decay

Models of heat retention and heating via radioactive decay in smaller icy Solar System bodies suggest that Rhea, Titania, Oberon, Triton, Pluto, Eris, Sedna, and Orcus may have oceans underneath solid icy crusts approximately 100 km thick.[31] Of particular interest in these cases is the fact that the models indicate that the liquid layers are in direct contact with the rocky core, which allows efficient mixing of minerals and salts into the water. This is in contrast with the oceans that may be inside larger icy satellites like Ganymede, Callisto, or Titan, where layers of high-pressure phases of ice are thought to underlie the liquid water layer.[31]

Models of radioactive decay suggest that MOA-2007-BLG-192Lb, a small planet orbiting a small star could be as warm as the Earth and completely covered by a very deep ocean.[32]

26.2.7   Internal differentiation models

Models of Solar System objects indicate the presence of liquid water in their internal differentiation.

Some models of the dwarf planet Ceres, largest object in the asteroid belt indicate the possibility of a wet interior layer. Water vapor detected to be emitted by the dwarf planet[33][34] may be an indicator, thought sublimation of surface ice.

A global layer of liquid water thick enough to decouple the crust from the mantle is thought to be present on Titan, Europa and, with less certainty, Callisto, Ganymede[31] and Triton.[35][36] Other icy moons may also have internal oceans, or have once had internal oceans that have now frozen.[31]

26.2.8   Habitable zone

Main article: Habitable zone § Extrasolar discoveries
See also: Category:Exoplanets in the habitable zone
 A planet's orbit in the circumstellar habitable zone is a popular method used to predict its potential for surface water at its surface. Habitable zone theory has put forward several extrasolar candidates for liquid water, though they are highly speculative as a planet's orbit around a star alone does not guarantee that a planet it has liquid water. In addition to its orbit, a planetary mass object must have the potential for sufficient atmospheric pressure to support liquid water and a sufficient supply of hydrogen and oxygen at or near its surface.

The Gliese 581 system contains multiple planets that may be candidates for surface water, including Gliese 581 c,[37] Gliese 581 d might be warm enough for oceans if a greenhouse effect was operating,[38] Gliese 581 e.[39]

Gliese 667 C has three of them are in the habitable zone[40] including Gliese 667 Cc is estimated to have surface temperatures similar to Earth and a strong chance of liquid water.[41]

Kepler-22b one of the first 54 candidates found by the Kepler telescope and reported is 2.4 times the size of the Earth, with an estimated temperature of 22 °C. It is described as having the potential for surface water, though its composition is currently unknown.[42]

Among the 1,235 possible extrasolar planet candidates detected by NASA's planet-hunting Kepler space telescope during its first four months of operation, 54 are orbiting in the parent star's habitable 'Goldilocks' zone where liquid water could exist.[43] Five of these are near Earth-size.[44]

On 6 January 2015, NASA announced further observations conducted from May 2009 to April 2013 which included eight candidates between one to two times the size of Earth, orbiting in a habitable zone. Of these eight, six orbit stars that are similar to the Sun in size and temperature. Three of the newly confirmed exoplanets were found to orbit within habitable zones of stars similar to the Sun: two of the three, Kepler-438b and Kepler-442b, are near-Earth-size and likely rocky; the third, Kepler-440b, is a super-Earth.[45]

26.3   History

Lunar maria are vast basaltic plains on the Moon that were thought to be bodies of water by early astronomers, who referred to them as "seas". Galileo expressed some doubt about the lunar 'seas' in his *Dialogue Concerning the Two Chief World Systems*.[lower-alpha 1]

Before space probes were landed, the idea of oceans on Venus was credible science, but the planet was discovered to be much too hot.

Telescopic observations from the time of Galileo onward have shown that Mars has no features resembling watery oceans. Mars' dryness was long recognized, and gave credibility to the spurious Martian canals.

26.4 Evidence of past surface water

Assuming that the *Giant impact hypothesis* is correct, there were never real seas or oceans on the Moon, only perhaps a little moisture (liquid or ice) in some places, when the Moon had a thin atmosphere created by degassing of volcanoes or impacts of icy bodies.

The Dawn space probe found possible evidence of past water flow on the asteroid Vesta,[46] leading to speculation of underground reservoirs of water-ice.[47]

Astronomers speculate that Venus had liquid water and perhaps oceans in its very early history.[48] Given that Venus has been completely resurfaced by its own active geology, the idea of a primeval ocean is hard to test. Rock samples may one day give the answer.[49]

It was once thought that Mars might have dried up from something more Earth-like. The initial discovery of a cratered surface made this seem unlikely, but further evidence has changed this view. Liquid water may have existed on the surface of Mars in the distant past, and several basins on Mars have been proposed as dry sea beds.[50] The largest is Vastitas Borealis; others include Hellas Planitia and Argyre Planitia.

There is currently much debate over whether Mars once had an ocean of water in its northern hemisphere, and over what happened to it if it did. Recent findings by the Mars Exploration Rover mission indicate it had some long-term standing water in at least one location, but its extent is not known. The Opportunity Mars rover photographed bright veins of a mineral leading to conclusive confirmation of deposition by liquid water.[51]

On December 9, 2013, NASA reported that the planet Mars had a large freshwater lake (which could have been a hospitable environment for microbial life) based on evidence from the Curiosity rover studying Aeolis Palus near Mount Sharp in Gale Crater.[52][53]

Further information: Mars Ocean Hypothesis

26.5 Liquid water inside comets

Comets contain large proportions of water ice, but are generally thought to be completely frozen due to their small size and large distance from the Sun. However, studies on dust collected from comet Wild-2 show evidence for liquid water inside the comet at some point in the past.[54] It is yet unclear what source of heat may have caused melting of some of the comet's water ice.

Nevertheless, on 10 December 2014, scientists reported that the composition of water vapor from comet Churyumov–Gerasimenko, as determined by the *Rosetta* spacecraft, is substantially different from that found on Earth. That is, the ratio of deuterium to hydrogen in the water from the comet was determined to be three times that found for terrestrial water. This makes it very unlikely that water found on Earth came from comets such as comet Churyumov–Gerasimenko according to the scientists.[55][56]

26.6   Extrasolar habitable zone candidates for water

Most known extrasolar planetary systems appear to have very different compositions to the Solar System, though there is probably sample bias arising from the detection methods.

The goal of current searches is to find Earth-sized planets in the habitable zone of their planetary systems (also sometimes called the *Goldilocks zone*).[57] Planets with oceans could include Earth-sized moons of giant planets, though it remains speculative whether such 'moons' really exist. The Kepler telescope might be sensitive enough to detect them.[58] But there is evidence that rocky planets hosting water may be commonplace throughout the Milky Way.[59]

26.6.1   Gliese 667 C - three planets

Gliese 667 Cc was originally described as one of two 'super-Earth' planets around Gliese 667 C, a dim red star that is part of a triple star system. The stars of this system have a concentration of heavy elements only 25% that of our Sun's. Such elements are the building blocks of terrestrial planets so it was thought to be unusual for such star systems to have an abundance of low mass planets.[60] It seems that habitable planets can form in a greater variety of environments than previously thought.

Gliese 667 Cc, in a tight 28-day orbit of a dim red star, must receive 90% of the light that Earth receives, but most of its incoming light is in the infrared, a higher percentage of this incoming energy should be absorbed by the planet. The planet is expected to absorb about the same amount of energy from its star that Earth absorbs from the Sun, which would allow surface temperatures similar to Earth and perhaps liquid water.[41]

Further work published in June 2013 suggests that the system has six planets, and that three of them are in the habitable zone.[40]

26.6.2   GJ 1214 b

GJ 1214 b was the second exoplanet (after CoRoT-7b) to have an established mass and radius less than those of the giant Solar System planets. It is three times the size of Earth and about 6.5 times as massive. Its low density indicated that it is likely a mix of rock and water,[30] and follow-up observations using the Hubble telescope now seem to confirm that a large fraction of its mass is water, so it is a large waterworld. The high temperatures and pressures would form exotic materials like 'hot ice' or 'superfluid water'.[11][12]

26.6.3   HD 85512 b

HD 85512 b was discovered in August 2011. It is larger than Earth, but small enough to be probably a rocky world. It is on the borders of its star's habitable zone and might have liquid water, and is a potential candidate for a life-supporting world.[61][62]

26.6.4   Kapteyn b

Kapteyn b is a super-Earth orbiting within the habitable zone Kapteyn's Star, which is 13 light-years away and 11 billion years old.[63]

26.6.5   Kepler-22b

Kepler-22b is a planet 2.4 times the size of the Earth, with an estimated temperature of 22 °C. It was one of 54 candidates found by the Kepler telescope and reported in February as potentially habitable. It is the first of these to be formally confirmed using other telescopes. Its composition is currently unknown.[42] It is most likely to be an ocean planet with no dry land due to its large size and large mass that is halfway between being a terrestrial and Gas Giant.

26.6.6 Kepler-62e and Kepler-62f

The star Kepler-62 has two planets in the habitable zone.[64] Kepler-62f is only 40 percent larger than Earth, making it the exoplanet closest to the size of our planet known in the habitable zone of another star until the discovery of Kepler-186f.[65] Kepler-62e orbits on the inner edge of the habitable zone and is roughly 60 percent larger than Earth.[66]

26.6.7 Kepler-69c

This large rocky planet orbiting within the habitable zone of Kepler 69, which is similar to our sun and is 70% more massive than the Earth.[66]

26.6.8 Kepler-186f

Kepler-186f is only 10% larger than Earth, and orbits the red dwarf star Kepler-186 within the habitable zone. When announced on 17 April 2014, it was described as the most Earth-like sized planet so far discovered.[65]

26.7 See also

- Earth analog

- Extraterrestrial oceans

- Extraterrestrial water vapor

- List of nearest terrestrial exoplanet candidates

- Planetary habitability

- Super-Earth

- Terrestrial planet

- Water on terrestrial planets

26.8 References

Explanatory notes

[1] 'Salviati', who normally gives Galileo's own opinions, says:

> I say then that if there were in nature only one way for two surfaces to be illuminated by the sun so that one appears lighter than the other, and that this were by having one made of land and the other of water, it would be necessary to say that the moon's surface was partly terrene and partly aqueous. But because there are more ways known to us that could produce the same effect, and perhaps others that we do not know of, I shall not make bold to affirm one rather than another to exist on the moon... What is clearly seen in the moon is that the darker parts are all plains, with few rocks and ridges in them, though there are some. The brighter remainder is all full of rocks, mountains, round ridges, and other shapes, and in particular there are great ranges of mountains around the spots... I think that the material of the lunar globe is not land and water, and this alone is enough to prevent generations and alterations similar to ours.

Citations

[1] Dyches, Preston; Chou, Felcia (7 April 2015). "The Solar System and Beyond is Awash in Water". *NASA*. Retrieved 8 April 2015.

[2] "Earth". Nineplanets.org.

[3] "'NASA Mars Spacecraft Reveals a More Dynamic Red Planet".".

[4] Sample, Ian (28 September 2015). "Nasa scientists find evidence of flowing water on Mars". *The Guardian*. Retrieved 28 September 2015.

[5] Wall, Mike (28 September 2015). "Salty Water Flows on Mars Today, Boosting Odds for Life". *Space.com*. Retrieved 2015-09-28.

[6] Ojha, Lujendra; Wilhelm, Mary Beth; Murchie, Scott L.; McEwen, Alfred S.; et al. (28 September 2015). "Spectral evidence for hydrated salts in recurring slope lineae on Mars". *Nature Geoscience*. doi:10.1038/ngeo2546. Retrieved 2015-09-28.

[7] "Tidal Heating".

[8] "Water near surface of a Jupiter moon only temporary".

[9] Keith Wagstaff. "Saturn's Moon Enceladus Is Home to a Global Ocean". *NBC News*. Retrieved 3 October 2015.

[10] "'NASA's Hubble Observations Suggest Underground Ocean on Jupiter's Largest Moon".".

[11] "Distant 'water-world' confirmed". *BBC News*. Retrieved 3 October 2015.

[12] "Hubble Reveals a New Class of Extrasolar Planet". Retrieved 3 October 2015.

[13]

[14] "Science@NASA, The Case of the Missing Mars Water". Retrieved 2009-03-07.

[15] "Water on Mars: Where is it All?". Retrieved 2009-03-07.

[16] "Water at Martian south pole". 17 March 2004. Retrieved 29 September 2009.

[17] "A numerical model for an alternative origin of Lake Vostok and its exobiological implications for Mars". Retrieved 2009-04-08.

[18] Cook, Jia-Rui C.; Gutro, Rob; Brown, Dwayne; Harrington, J.D.; Fohn, Joe (12 December 2013). "Hubble Sees Evidence of Water Vapor at Jupiter Moon". *NASA*. Retrieved 12 December 2013.

[19] "Cassini Images of Enceladus Suggest Geysers Erupt Liquid Water at the Moon's South Pole". Ciclops.org. 2006-03-09. Retrieved 2012-01-22.

[20] "Saturn's Moon Enceladus Is Unlikely To Harbor Life". Sciencedaily.com. 2007-08-14. Retrieved 2012-01-22.

[21] "Possible salty ocean hidden in depths of Saturn moon". Astronomynow.com. 2009-06-25. Retrieved 2012-01-22.

[22] Platt, Jane; Bell, Brian (3 April 2014). "NASA Space Assets Detect Ocean inside Saturn Moon". *NASA*. Retrieved 3 April 2014.

[23] Iess, L.; Stevenson, D.J.; Parisi, M.; Hemingway, D.; Jacobson, R.A.; Lunine, J.I.; Nimmo, F.; Armstrong, J.w.; Asmar, S.w.; Ducci, M.; Tortora, P. (4 April 2014). "The Gravity Field and Interior Structure of Enceladus". *Science (journal)* **344** (6179): 78–80. Bibcode:2014Sci...344...78I. doi:10.1126/science.1250551. Retrieved 3 April 2014.

[24] "Tidal Heating". *geology.asu.edu*. Archived from the original on 2006-03-29.

[25] Greenberg, Richard (2005) *Europa: The Ocean Moon: Search for an Alien Biosphere*, Springer + Praxis Books, ISBN 978-3-540-27053-9.

[26] "Mysterious signal hints at subsurface ocean on Titan". Space.newscientist.com. Retrieved 2012-01-22.

[27] Briggs, Helen (2008-03-20). "Saturn moon may have hidden ocean". BBC News. Retrieved 2012-01-22.

[28] "Astrophile: Supercritical water world does somersaults". Newscientist.com. Retrieved 2012-01-22.

[29] D. Ehrenreich; et al. (October 2, 2012). "Hint of a transiting extended atmosphere on 55 Cancri b". *Astronomy & Astrophysics* **547**: A18. arXiv:1210.0531. Bibcode:2012A&A...547A..18E. doi:10.1051/0004-6361/201219981.

[30] "The small planet with a thick coat". Astronomynow.com. 2009-12-17. Retrieved 2012-01-22.

[31] Hussmann, Hauke; Sohl, Frank; Spohn, Tilman (November 2006). "Subsurface oceans and deep interiors of medium-sized outer planet satellites and large trans-neptunian objects" (PDF). *Icarus* **185** (1): 258–273. Bibcode:2006Icar..185..258H. doi:10.1016/j.icarus.2006.06.005.

[32] "Small Planet Discovered Orbiting Small Star". Sciencedaily.com. 2008-06-02. Retrieved 2012-01-22.

[33] Küppers, Michael; O'Rourke, Laurence; Bockelée-Morvan, Dominique; Zakharov, Vladimir; Lee, Seungwon; von Allmen, Paul; Carry, Benoit; Teyssier, David; Marston, Anthony; Müller, Thomas; Crovisier, Jacques; Barucci, M. Antonietta; Moreno, Raphael (2014). "Localized sources of water vapour on the dwarf planet (1) Ceres". *Nature* **505** (7484): 525–527. Bibcode:2014 Natur.505..525K. doi:10.1038/nature12918. ISSN 0028-0836. PMID24451541.

[34] Harrington, J.D. (22 January 2014). "Herschel Telescope Detects Water on Dwarf Planet - Release 14-021". *NASA*. Retrieved 22 January 2014.

[35] McKinnon, William B.; Kirk, Randolph L. (2007). "Triton". In Lucy Ann Adams McFadden, Lucy-Ann Adams, Paul Robert Weissman, Torrence V. Johnson. *Encyclopedia of the Solar System* (2nd ed.). Amsterdam; Boston: Academic Press. pp. 483–502. ISBN 978-0-12-088589-3.

[36] Javier Ruiz (December 2003). "Heat flow and depth to a possible internal ocean on Triton". *Icarus* **166** (2): 436–439. Bibcode:2003Icar..166..436R. doi:10.1016/j.icarus.2003.09.009.

[37] "New Planet Could Harbor Water and Life". Space.com. 2007-04-24. Retrieved 2012-01-22.

[38] "Scientists might have picked right star, wrong world for hosting life". MSNBC. 2007-06-18. Retrieved 2012-01-22.

[39] "Exoplanet near Gliese 581 star 'could host life'". BBC News. 2011-05-17. Retrieved 2012-01-22.

[40] "Three Planets in Habitable Zone of Nearby Star: Gliese 667c Reexamined". Retrieved 3 October 2015.

[41] "Super-Earth orbits in habitable zone of cool star". Retrieved 3 October 2015.

[42] "Kepler 22-b: Earth-like planet confirmed". BBC News. 2011-12-05. Retrieved 2012-01-22.

[43] "Kepler detects more than 1,200 possible planets". Spaceflightnow.com. Retrieved 2012-01-22.

[44] "NASA Finds Earth-Size Planet Candidates in Habitable Zone, Six Planet System". Sciencedaily.com. 2011-02-02. doi:10.1038 Retrieved 2012-01-22.

[45] Clavin, Whitney; Chou, Felicia; Johnson, Michele (6 January 2015). "NASA's Kepler Marks 1,000th Exoplanet Discovery, Uncovers More Small Worlds in Habitable Zones". *NASA*. Retrieved 6 January 2015.

[46] "Dawn probe spies possible water-cut gullies on Vesta". *BBC News*. Retrieved 3 October 2015.

[47] "Huge Asteroid Vesta May Be Packed With Water Ice". *Space.com*. Retrieved 3 October 2015.

[48] Owen, (2007), news.nationalgeographic.com/news/2007/11/071128-venus-earth_2.html

[49] *Did oceans on Venus harbour life?*, issue 2626 of *New Scientist* magazine.

[50] "Mars Probably Once Had A Huge Ocean". Sciencedaily.com. 2007-06-13. Retrieved 2012-01-22.

[51] Jpl.Nasa.Gov. "NASA Mars Rover Finds Mineral Vein Deposited by Water — NASA Jet Propulsion Laboratory". Jpl.nasa.gov. Retrieved 2012-01-22.

[52] Chang, Kenneth (December 9, 2013). "On Mars, an Ancient Lake and Perhaps Life". *New York Times*. Retrieved December 9, 2013.

[53] Various (December 9, 2013). "Science - Special Collection - Curiosity Rover on Mars". *Science*. Retrieved December 9, 2013.

[54] "Frozen comet's watery past: Discovery challenges paradigm of comets as 'dirty snowballs' frozen in time". Sciencedaily.com. 2011-04-05. doi:10.1016/j.gca.2011.03.026. Retrieved 2012-01-22.

[55] Agle, DC; Bauer, Markus (10 December 2014). "Rosetta Instrument Reignites Debate on Earth's Oceans". *NASA*. Retrieved 10 December 2014.

[56] Chang, Kenneth (10 December 2014). "Comet Data Clears Up Debate on Earth's Water". *New York Times*. Retrieved 10 December 2014.

[57] "Habitable planets may be common". Newscientist.com. Retrieved 2012-01-22.

[58] "The hunt for habitable exomoons". Astronomynow.com. 2009-09-04. Retrieved 2012-01-22.

[59] "Water, water everywhere". Astronomynow.com. Retrieved 2012-01-22.

[60] "New Super-Earth Detected Within the Habitable Zone of a Nearby Cool Star". Retrieved 3 October 2015.

[61] "Exoplanet Looks Potentially Lively". Scientificamerican.com. Retrieved 2012-01-22.

[62] "'Super-Earth,' 1 of 50 Newfound Alien Planets, Could Potentially Support Life". News.yahoo.com. 2011-09-12. Retrieved 2012-01-22.

[63] "Astronomers discover two new worlds orbiting ancient star next door: One may be warm enough to have liquid water". Retrieved 3 October 2015.

[64] "Kepler telescope spies 'most Earth-like' worlds to date". *BBC News*. Retrieved 3 October 2015.

[65] "'Most Earth-like planet yet' spotted by Kepler". *BBC News*. Retrieved 3 October 2015.

[66] "NASA's Kepler Discovers Its Smallest 'Habitable Zone' Planets to Date". *NASA*. Retrieved 3 October 2015.

26.9   External links

- The Extrasolar Planets Encyclopaedia

- Gliese 581: Extrasolar Planet Might Indeed Be Habitable

- Jupiter's Moon Europa: What Could Be Under The Ice?

- To Curious Aliens, Earth Would Stand Out As Living Planet

- Ocean-bearing Planets: Looking For Extraterrestrial Life In All The Right Places

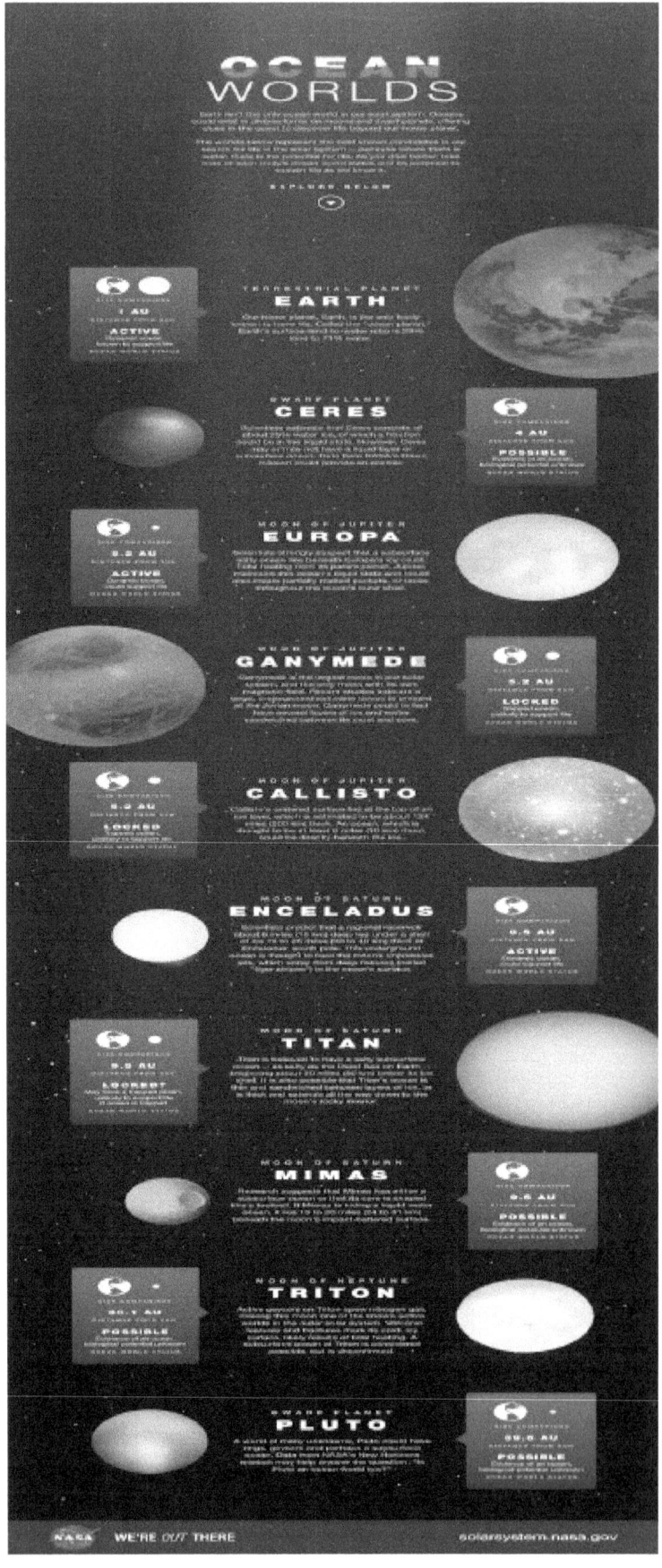

Warm season flows in Palikir Crater (inside Newton crater) on Mars. While there is intriguing but inconclusive evidence suggestive of extraterrestrial liquid water, it has so far eluded direct confirmation.

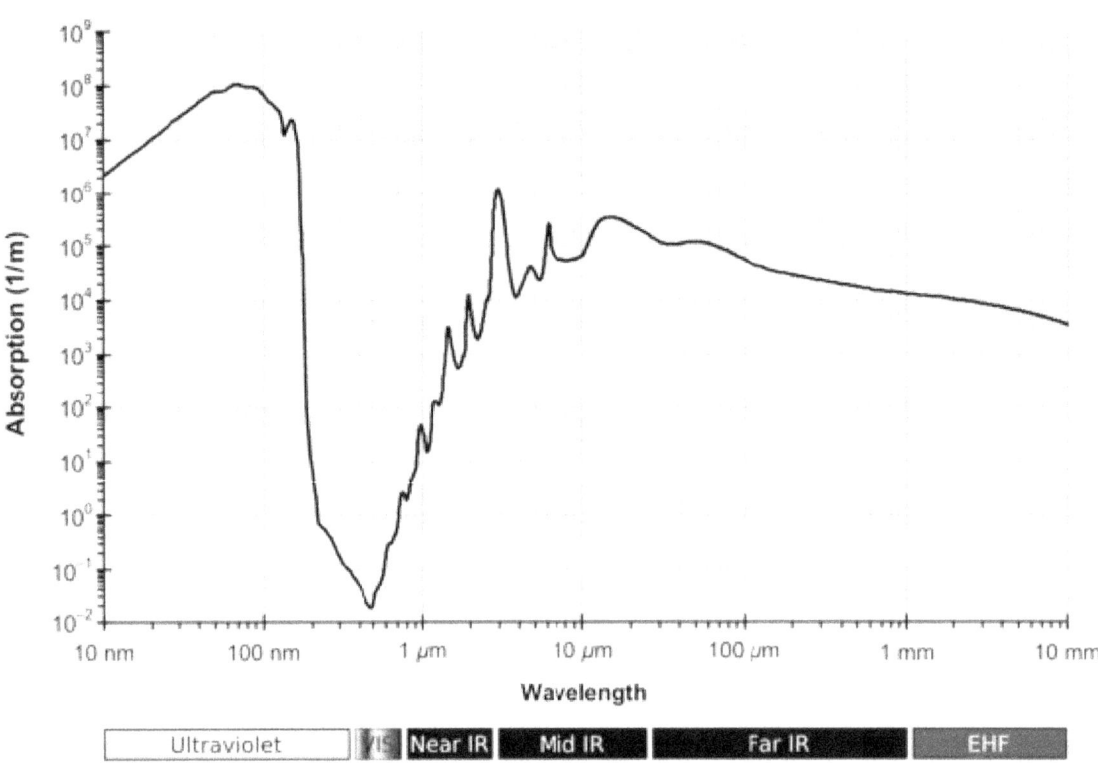

Absorption spectrum of liquid water

Liquid water has not been detected in spectroscopic analysis of suspected seasonal Martian flows.

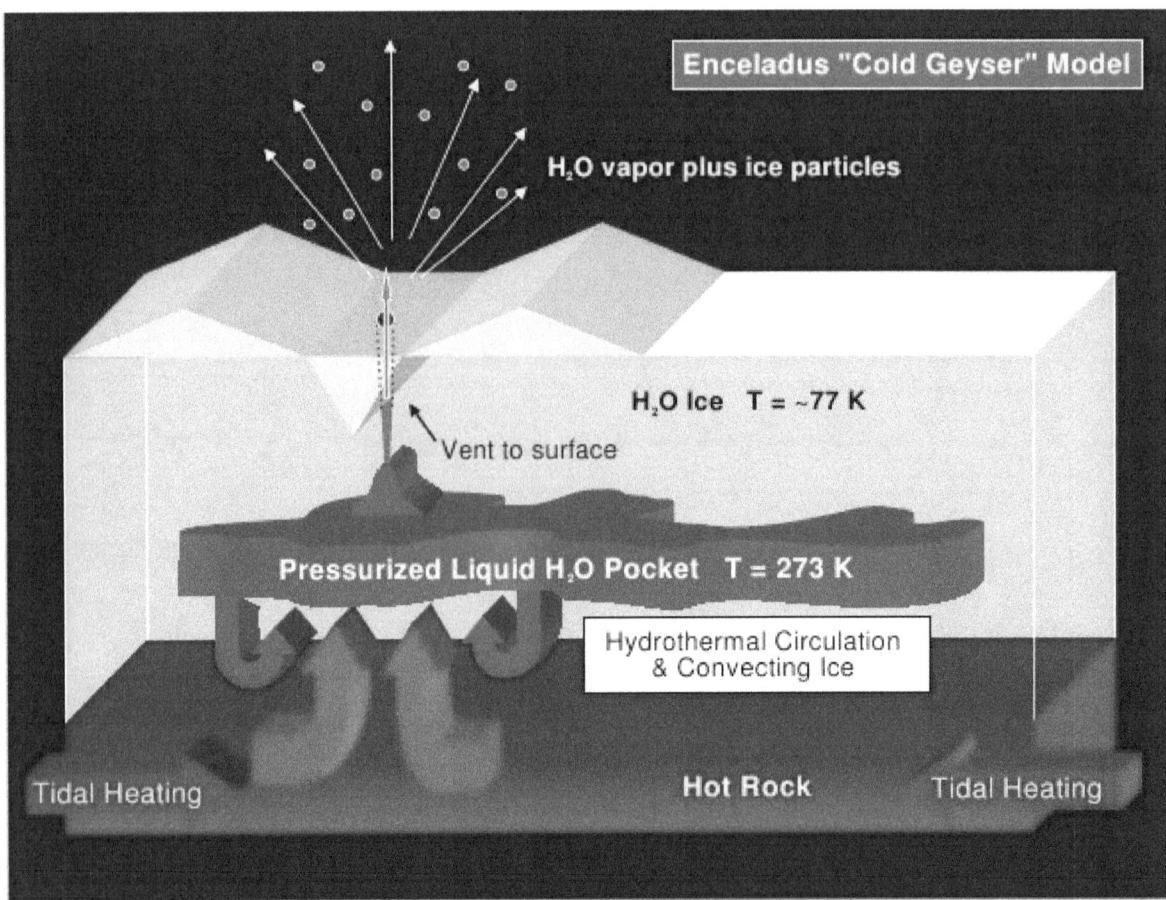

A possible mechanism for cryovolcanism on bodies like Enceladus

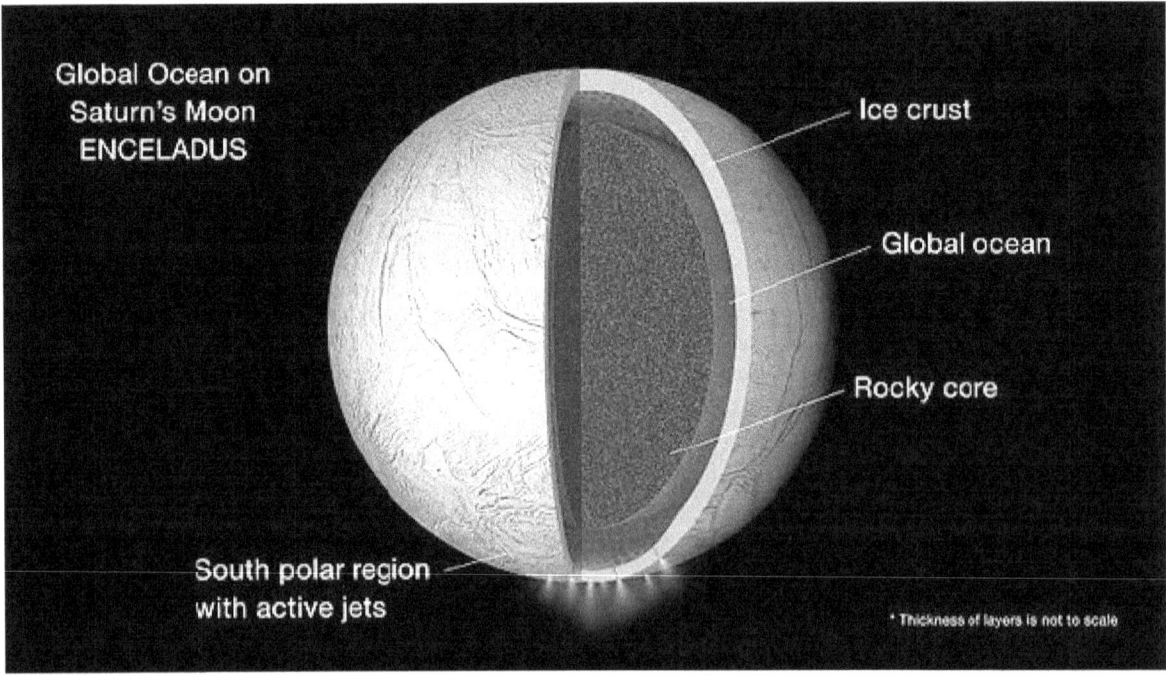

Artists conception of the subsurface water ocean confirmed on Enceladus in 2014 as calculated using gravitational measurements and density estimations.[22][23]

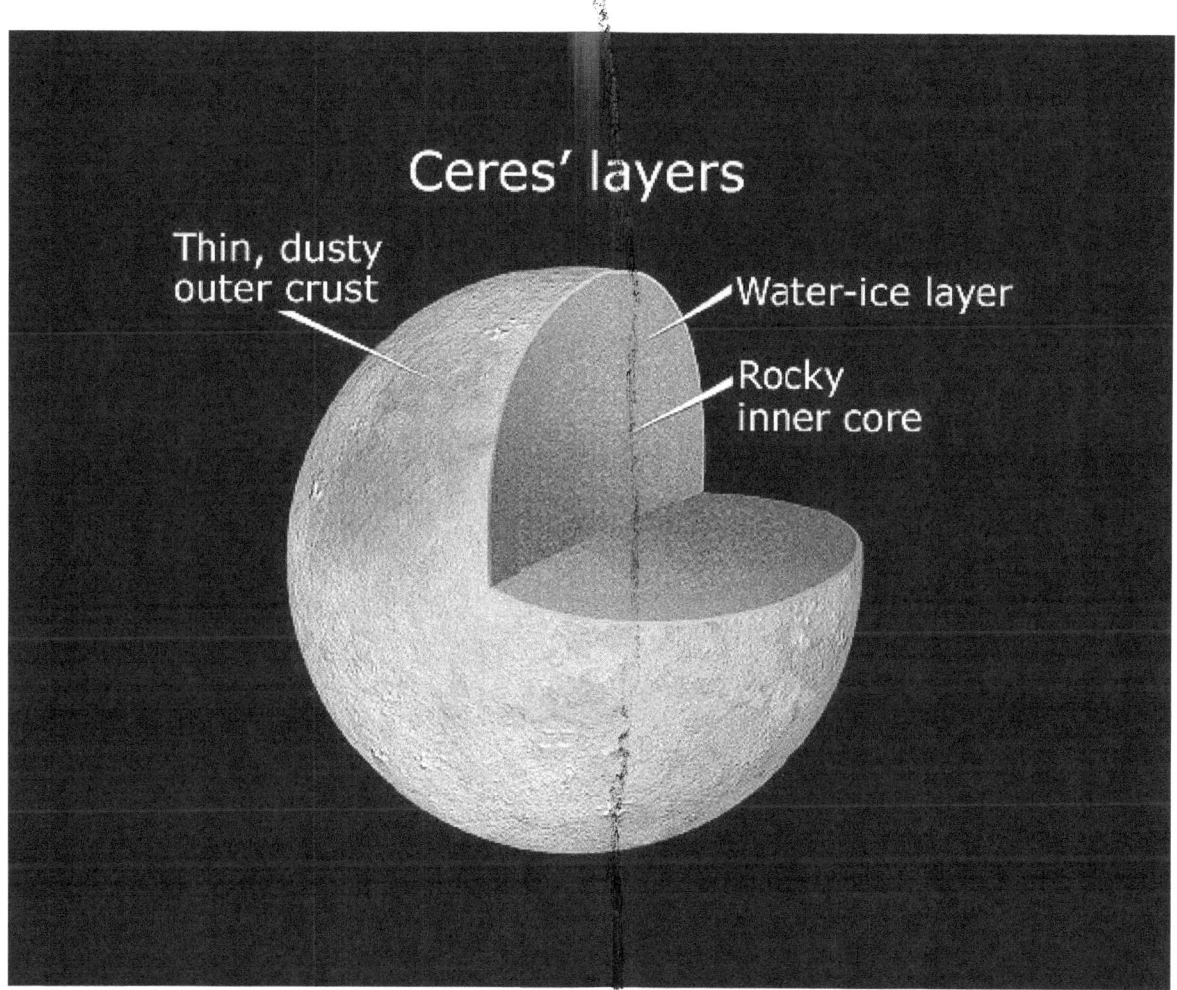

Diagram showing a possible internal structure of Ceres

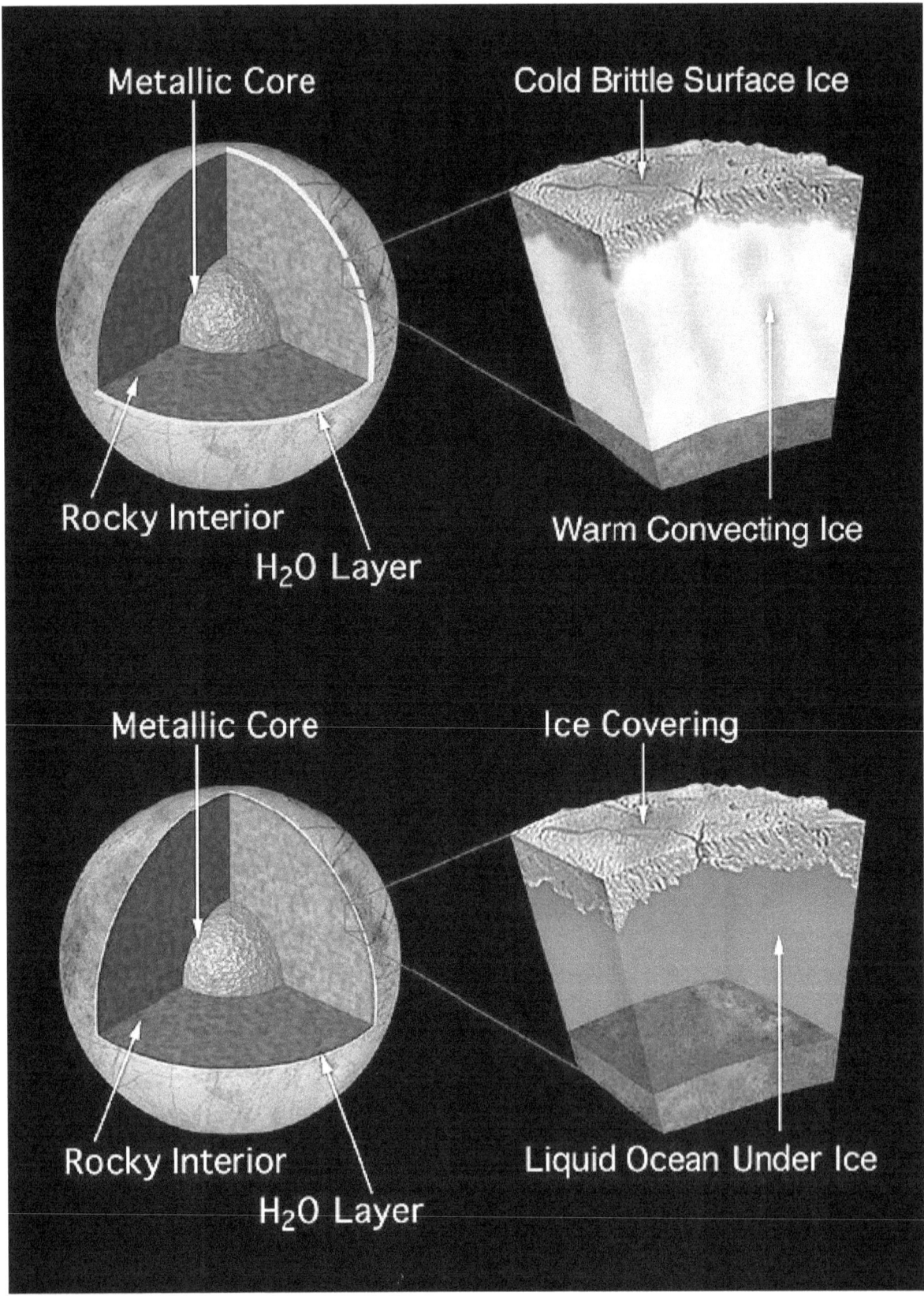

Two models for the composition of Europa suggest a large subsurface ocean of liquid water. Similar models have been proposed for other celestial bodies in the Solar System.

Artist's impression of Upsilon Andromedae d, portrayed as a class II planet with water vapor clouds, as seen from a hypothetical large moon with surface liquid water

An artist's impression of ancient Mars and its hypothesized oceans based on geological data

Chapter 27

Planetary protection

A Viking lander being prepared for dry heat sterilization - this remains the "Gold standard"[1] of present day planetary protection.

Planetary protection is a guiding principle in the design of an interplanetary mission, aiming to prevent biological contamination of both the target celestial body and the Earth. Planetary protection reflects both the unknown nature of the space environment and the desire of the scientific community to preserve the pristine nature of celestial bodies until they can be studied in detail.[2][3]

There are two types of interplanetary contamination. **Forward contamination** is the transfer of viable organisms from Earth to another celestial body. A major goal of planetary protection is to preserve the planetary record of natural processes by preventing introduction of Earth-originated life. **Back contamination** is the transfer of extraterrestrial organisms, if such exist, back to the Earth's biosphere.

27.1   History

The potential problem of lunar and planetary contamination was first raised at the International Astronautical Federation VIIth Congress in Rome in 1956.[4]

In 1958[5] the U.S. National Academy of Sciences (NAS) passed a resolution stating, "The National Academy of Sciences of the United States of America urges that scientists plan lunar and planetary studies with great care and deep concern so that initial operations do not compromise and make impossible forever after critical scientific experiments." This led to creation of the ad hoc Committee on Contamination by Extraterrestrial Exploration (CETEX), which met for a year and recommended that interplanetary spacecraft be sterilized, and stated, "The need for sterilization is only temporary. Mars and possibly Venus need to remain uncontaminated only until study by manned ships becomes possible"[6]

In 1959 planetary protection was transferred to the newly formed Committee on Space Research (COSPAR). COSPAR in 1964 issued Resolution 26

> affirms that the search for extraterrestrial life is an important objective of space research, that the planet of Mars may offer the only feasible opportunity to conduct this search during the foreseeable future, that contamination of this planet would make such a search far more difficult and possibly even prevent for all time an unequivocal result, that all practical steps should be taken to ensure that Mars be not biologically contaminated until such time as this search can have been satisfactorily carried out, and that cooperation in proper scheduling of experiments and use of adequate spacecraft sterilization techniques is required on the part of all deep space probe launching authorities to avoid such contamination.[7]

In 1967, the US, USSR, and UK ratified the United Nations Outer Space Treaty. The legal basis for planetary protection lies in Article IX of this treaty:

> "Article IX: ... States Parties to the Treaty shall pursue studies of outer space, including the Moon and other celestial bodies, and conduct exploration of them so as to avoid their harmful contamination and also adverse changes in the environment of the Earth resulting from the introduction of extraterrestrial matter and, where necessary, shall adopt appropriate measures for this purpose...[8][9]

This treaty has since been signed by almost all nation states, including all the current and aspiring space-faring nation states.

For forward contamination, the phrase to be interpreted is "harmful contamination". Two legal reviews came to differing interpretations of this clause (both reviews were unofficial). However the currently accepted interpretation is that "any contamination which would result in harm to a state's experiments or programs is to be avoided". NASA'S policy states explicitly that "the conduct of scientific investigations of possible extraterrestrial life forms, precursors, and remnants must not be jeopardized".[10]

27.2   COSPAR recommendations and categories

The Committee on Space Research (COSPAR) meets every two years, in a gathering of 2000 to 3000 scientists,[11] and one of its tasks is to develop recommendations for avoiding interplanetary contamination. Its legal basis is Article IX of the Outer Space Treaty [12] (see history below for details).

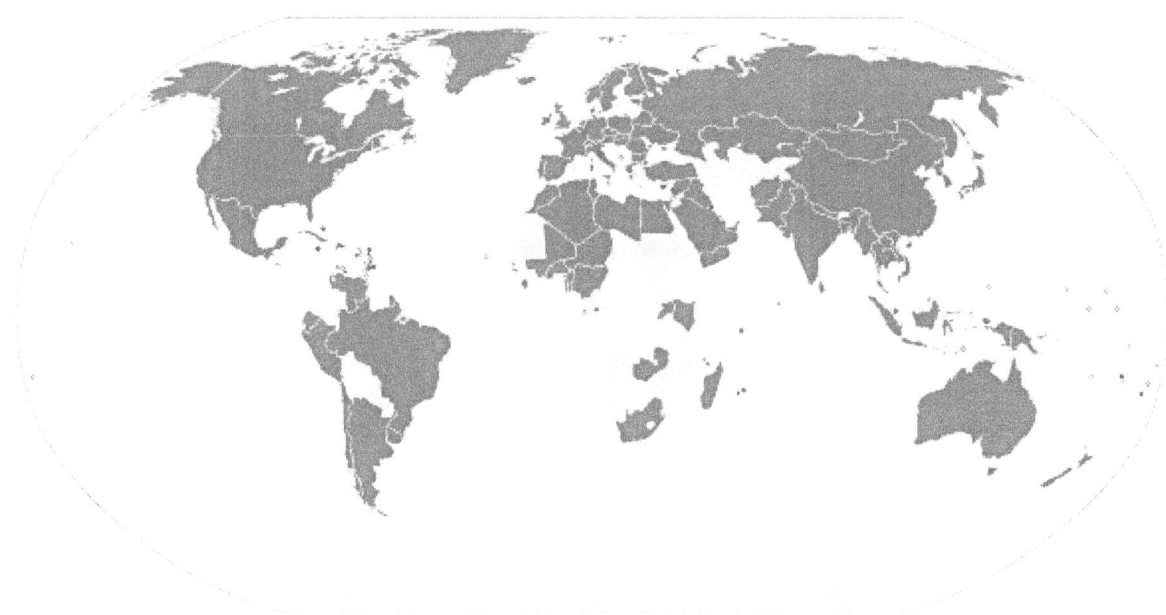

Signatories of the Outer Space Treaty - includes all current and aspiring space faring nation states. By signing the treaty, these nation states have all committed themselves to planetary protection.
Signed and ratified
Signed only
Not signed

Its recommendations depend on the type of space mission and the celestial body explored.[13] COSPAR categorizes the missions into 5 groups:

- *Category I:* Any mission to locations not of direct interest for chemical evolution or the origin of life, such as the Sun or Mercury. No planetary protection requirements.[14]

- *Category II:* Any mission to locations of significant interest for chemical evolution and the origin of life, but only a remote chance that spacecraft-borne contamination could compromise investigations. Examples include the Moon, Venus, and comets. Requires simple documentation only, primarily to outline intended or potential impact targets, and an end of mission report of any inadvertent impact site if such occurred.[14]

- *Category III:* Flyby and orbiter missions to locations of significant interest for chemical evolution and/or origin of life, and with a significant chance that contamination could compromise investigations e.g., Mars, Europa, Enceladus. Requires more involved documentation than Category II. Other requirements, depending on the mission, may include trajectory biasing, clean room assembly, possibly bioburden reduction, and if impact is a possibility, inventory of organics.[14]

- *Category IV:* Lander or probe missions to the same locations as Category III. Measures to be applied depend on the target body and the planned operations "Sterilization of the entire spacecraft may be required for landers and rovers with life detection experiments, and for those landing in or moving to a region where terrestrial microorganisms may survive and grow, or where indigenous life may be present. For other landers and rovers, the requirements would be for decontamination and partial sterilization of the landed hardware."[15]

  Missions to Mars in category IV are subclassified further:[13]

  - *Category IVa.* Landers that do not search for Martian life - uses the Viking lander pre-sterilization requirements, 300,000 spores per spacecraft and 300 spores per square meter.

- *Category IVb.* Landers that search for Martian life. Adds stringent extra requirements to prevent contamination of samples.
- *Category IVc.* Any component that accesses a Martian *special region* (see below) must be sterilized to at least to the Viking post-sterilization biological burden levels (30 spores total per spacecraft).

- *Category V:* This is further divided into unrestricted and restricted sample return.

  - Unrestricted Category V: samples from locations judged by scientific opinion to have no indigenous lifeforms. No special requirements
  - Restricted Category V: (where scientific opinion is unsure) the requirements include: absolute prohibition of destructive impact upon return, containment of all returned hardware which directly contacted the target body, and containment of any unsterilized sample returned to Earth.

For Category IV missions, after receiving the mission category a certain level of *biological burden* is allowed for the mission. In general this is expressed as a 'probability of contamination', required to be less than one chance in 10,000[16][17] of forward contamination per mission, but in the case of Mars Category IV missions (above) the requirement has been translated into a count of *Bacillus* spores per surface area, as an easy to use assay method.[14][18]

More extensive documentation is also required for Category IV. Other procedures required, depending on the mission, may include trajectory biasing, the use of clean rooms (Class 100,000 or better) during spacecraft assembly and testing, bioload reduction, partial sterilization of the hardware having direct contact with the target body, a bioshield for that hardware, and, in rare cases, complete sterilization of the entire spacecraft.[14]

For restricted Category V missions, the current recommendation[19] is that no uncontained samples should be returned unless sterilized. Since sterilization of the returned samples would destroy much of their science value, current proposals involve containment and quarantine procedures. For details, see Containment and quarantine below. Of course, Category V missions also have to fulfill the requirements of Category IV to protect the target body from forward contamination.

27.2.1 Mars special regions

A *special region* is a region classified by COSPAR within which terrestrial organisms could readily propagate, or one thought to have an elevated potential for existence of Martian life forms. This is understood to apply to any region on Mars where liquid water occurs, or can occasionally occur, based on the current understanding of requirements for life.

If a hard landing risks biological contamination of a special region, then the whole lander system must be sterilized to COSPAR category IVc.

27.3 Target categories

Some targets are easily categorized. Others are assigned provisional categories by COSPAR, pending future discoveries and research.

The 2009 COSPAR Workshop on Planetary Protection for Outer Planet Satellites and Small Solar System Bodies covered this in some detail. Most of these assessments are from that report, with some future refinements. This workshop also gave more precise definitions for some of the categories:[20][21]

27.3.1 Category I

"not of direct interest for understanding the process of chemical evolution or the origin of life." [22]

- Io, Sun, Mercury, Undifferentiated metamorphosed asteroids

27.3.2   Category II

... where there is only a remote chance that contamination carried by a spacecraft could jeopardize future exploration". In this case we define "remote chance" as "the absence of niches (places where terrestrial microorganisms could proliferate) and/or a very low likelihood of transfer to those places." [20][22]

- Callisto, Comets, Asteroids of category P, D, and C, Venus,[23] KBOs (< 1/2 size of Pluto)

27.3.3   Provisional Category II

- Ganymede, Titan, Triton, the Pluto-Charon system, and other large KBOs (> 1/2 size of Pluto),[24] Ceres.

Provisionally they assigned these objects to Category II. However, they state that more research is needed, because there is a remote possibility that the tidal interactions of Pluto and Charon could maintain some water reservoir below the surface. Similar considerations apply to the other larger KBOs

Triton they thought was insufficiently well understood at present to say it is definitely devoid of liquid water. The only close up observations to date are those of Voyager 2.

In a detailed discussion of Titan, scientists concluded that there was no danger of contamination of its surface, except short term adding of negligible amounts of organics, but Titan could have a below surface water reservoir that communicates with the surface, and if so this could be contaminated.

In the case of Ganymede, the question is, given that its surface shows pervasive signs of resurfacing, is there any communication with its subsurface ocean? They found no known mechanism by which this could happen, and Galileo found no evidence of cryovolcanism.

Initially they assigned it as Priority B minus, meaning that precursor missions are needed to assess its category before any surface missions. However after further discussion they provisionally assigned it to Category II, so no precursor missions required, depending on future research.

If there is cryovolcanism on Ganymede or Titan, the undersurface reservoir is thought to be 50 – 150 km below the surface. They were unable to find a process that could transfer the surface melted water back down through 50 km of ice to the under surface sea.[25] This is why they assigned both Ganymede and Titan a reasonably firm provisional Category II but pending results of future research.

They recommended more generally, that icy bodies that show signs of recent resurfacing need further discussion and might need to be assigned to a new category depending on future research.

This approach has been applied, for instance, to missions to Ceres. The planetary protection Category is subject for review during the mission of the Ceres orbiter depending on the results found.[26]

27.3.4   Category III / IV

"...where there is a significant chance that contamination carried by a spacecraft could jeopardize future exploration." We define "significant chance" as "the presence of niches (places where terrestrial microorganisms could proliferate) and the likelihood of transfer to those places." [20][22]

- Mars because of possible surface habitats

- Europa because of its subsurface ocean

- Enceladus because of evidence of water plumes.

27.3.5   Category V

Unrestricted Category V: "Earth-return missions from bodies deemed by scientific opinion to have no indigenous life forms."[22]

Restricted Category V: "Earth-return missions from bodies deemed by scientific opinion to be of significant interest to the process of chemical evolution and/or the origin of life."[22]

In the category V for sample return the conclusions so far are:[22]

- Restricted Category V: Mars, Europa, Enceladus
- Unrestricted Category V: Venus, Moon

with others to be decided.

27.3.6   Other objects

If there has been no activity for 3 billion years, it will not be possible to destroy the surface by terrestrial contamination, so can be treated as Category I. Otherwise, the category may need to be reassessed.

27.4   The Coleman-Sagan equation

The aim of the current regulations is to keep the number of microorganisms low enough so that the probability of contamination of Mars (and other targets) is acceptable. It is not an objective to make the probability of contamination zero.

The aim is to keep the probability of contamination of 1 chance in 10,000 of contamination per mission flown.[16] This figure is obtained typically by multiplying together the number of microorganisms on the spacecraft, the probability of growth on the target body, and a series of bioload reduction factors.

In detail the method used is the Coleman-Sagan equation.[27]

$$P_c = N_0 R P_S P_t P_R P_g \ .$$

where

N_0

R

P_S

P_t

P_R

P_g

Then the requirement is $P_c < 10^{-4}$

The 10^{-4} is a number chosen by Sagan et al., somewhat arbitrarily. Sagan and Coleman assumed that about 60 missions to the Mars surface would occur before the exobiology of Mars is thoroughly understood, 54 of those successful, and 30 flybys or orbiters, and the number was chosen to endure a probability to keep the planet free from contamination of at least 99.9% over the duration of the exploration period.[17]

27.4.1   Critiques

The Coleman Sagan equation has been criticised because the individual parameters are often not known to better than a magnitude or so. For example, the thickness of the surface ice of Europa is unknown, and may be thin in places, which can give rise to a high level of uncertainty in the equation.[28][29] It has also been criticised because of the inherent assumption made of an end to the protection period and future human exploration. In the case of Europa, this would only protect it with reasonable probability for the duration of the period of exploration.[28][29]

Greenberg has suggested an alternative, to use the natural contamination standard - that our missions to Europa should not have a higher chance of contaminating it than the chance of contamination by meteorites from Earth.[30][31]

> As long as the probability of people infecting other planets with terrestrial microbes is substantially smaller than the probability that such contamination happens naturally, exploration activities would, in our view, be doing no harm. We call this concept the natural contamination standard.

Another approach for Europa is the use of binary decision trees which is favoured by the *Committee on Planetary Protection Standards for Icy Bodies in the Outer Solar System* under the auspices of the Space Studies Board.[16] This goes through a series of seven steps, leading to a final decision on whether to go ahead with the mission or not.[32]

> Recommendation: Approaches to achieving planetary protection should not rely on the multiplication of bioload estimates and probabilities to calculate the likelihood of contaminating Solar System bodies with terrestrial organisms unless scientific data unequivocally define the values, statistical variation, and mutual independence of every factor used in the equation.
> Recommendation: Approaches to achieving planetary protection for missions to icy Solar System bodies should employ a series of binary decisions that consider one factor at a time to determine the appropriate level of planetary protection procedures to use.

27.5   Containment and quarantine for restricted Category V sample return

In the case of restricted Category V missions, Earth is protected through quarantine of sample and astronauts, and sample containment.

In the case of a Mars sample return, Missions would be designed so that no part of the capsule that encounters the Mars surface is exposed to the Earth environment.

One way to do that is to enclose the sample container within a larger outer container from Earth, in the vacuum of space. The integrity of any seals is essential and the system must also be monitored to check for the possibility of micro-meteorite damage during return to Earth.[33][34][35][36]

The recommendation of the ESF report is that [19]

> "No uncontained Mars materials, including space craft surfaces that have been exposed to the Mars environment should be returned to Earth unless sterilised"
> ..."For unsterilised samples returned to Earth, a programme of life detection and biohazard testing, or a proven sterilisation process, shall be undertaken as an absolute precondition for the controlled distribution of any portion of the sample."

No restricted category V returns have been carried out in recent times.

During the Apollo program the sample returns were regulated through the Extra-Terrestrial Exposure Law. This was rescinded in 1991, so new legislation would need to be enacted. The Apollo era quarantine procedures are of interest as the only attempt to date of a return to Earth of a sample that, at the time, was thought to have a remote possibility of including extraterrestrial life.

Samples and astronauts were quarantined in the Lunar Receiving Laboratory.[37] The methods used would be considered inadequate for containment by modern standards.[38] Also the lunar receiving laboratory would be judged a failure by its own design criteria as the sample return didn't contain the lunar material, with two failure points during the Apollo 11 return mission, at the splashdown and at the facility itself.

However the Lunar Receiving Laboratory was built quickly with only two years from start to finish, a time period now considered inadequate. Lessons learnt from it can help with design of any Mars sample return receiving facility.[39]

Design criteria for a proposed Mars Sample Return Facility, and for the return mission, have been developed by the American National Research Council,[40] and the European Space Foundation.[41] They concluded that it could be based on biohazard 4 containment but with more stringent requirements to contain unknown microorganisms possibly as small as or smaller than the smallest Earth microorganisms known, the ultramicrobacteria. The ESF study also recommended that it should be designed to contain the smaller gene transfer agents if possible, as these could potentially transfer DNA from martian microorganisms to terrestrial microorganisms if they have a shared evolutionary ancestry. It also needs to double as a clean room facility to protect the samples from terrestrial contamination that could confuse the sensitive life detection tests that would be used on the samples.

Before a sample return, new quarantine laws would be required. Environmental assessment would also be required, and various other domestic and international laws would need to be negotiated not present in the Apollo era.[42]

27.6 Decontamination procedures

For all spacecraft missions requiring decontamination, the starting point is clean room assembly in class 100 cleanrooms. These are rooms with less than 100 particles of size 0.5 μm or larger per cubic foot. Engineers wear cleanroom suits with only their eyes exposed. Components are sterilized individually before assembly, as far as possible, and they clean surfaces frequently with alcohol wipes during assembly.

For Category IVa missions (Mars landers that do not search for Martian life), the aim is to reduce the bioburden to 300,000 bacterial spores on any surface from which the spores could get into the Martian environment. Any heat tolerant components are heat sterilized to 114 °C. Sensitive electronics such as the core box of the rover including the computer, are sealed and vented through high-efficiency filters to keep any microbes inside.[43][44][45]

For more sensitive missions such as Category IVc (to Mars special regions), a far higher level of sterilization is required. This needs to be similar to implemented on the Viking landers, which were sterilized for a surface which, at the time, was thought to be potentially hospitable to life similar to special regions on Mars today.

In microbiology, it is usually impossible to prove that there are no microorganisms left viable, since many microorganisms are either not yet studied, or not cultivable. Instead, sterilization is done using a series of tenfold reductions of the numbers of microorganisms present. After a sufficient number of ten-fold reductions, the chance that there any microorganisms left will be extremely low.

The two Viking Mars landers were sterilized using dry heat sterilization. After preliminary cleaning to reduce the bioburden to similar levels to present day Category IVa spacecraft, the Viking spacecraft were heat-treated for 30 hours at 125 °C (five hours at 125 °C was considered enough to reduce the population tenfold even for enclosed parts of the spacecraft, so this was enough for a million-fold reduction of the originally low population).[46]

Modern materials however are often not designed to handle such temperatures, especially since modern spacecraft often use "commercial off the shelf" components. Problems encountered include nanoscale features only a few atoms thick, plastic packaging, and conductive epoxy attachment methods. Also many instrument sensors cannot be exposed to high temperature, and high temperature can interfere with critical alignments of instruments.[46]

As a result, new methods are needed to sterilize a modern spacecraft to the higher categories such as Category IVc for Mars, similar to Viking.[46] Methods under evaluation, or already approved, include:

- Vapour phase hydrogen peroxide - effective, but can affect finishes, lubricants and materials that use aromatic rings and sulfur bonds. This has been established, reviewed, and a NASA/ESA specification for use of VHP has been approved by the Planetary Protection Officer, but it has not yet been formally published.[47]

- Ethylene oxide - this is widely used in the medical industry, and can be used for materials not compatible with hydrogen peroxide. Is under consideration for missions such as ExoMars.

- Gamma radiation and electron beams have been suggested as a method of sterilization, as they are used extensively in the medical industry. They need to be tested for compatibility with spacecraft materials and hardware geometries, and is not yet ready for review.

Some other methods are of interest as they can sterilize the spacecraft after arrival on the planet.

- Supercritical carbon dioxide snow (Mars) - is most effective against traces of organic compounds rather than whole microorganisms. Has the advantage though that it eliminates the organic traces - while other methods kill the microorganisms, they leave organic traces that can confuse life detection instruments. Is under study by JPL and ESA.

- Passive sterilization through UV radiation (Mars). Highly effective against many microorganisms, but not all, as a *Bacillus* strain found in spacecraft assembly facilities is particularly resistant to UV radiation. Is also complicated by possible shadowing by dust and spacecraft hardware.

- Passive sterilization through particle fluxes (Europa). Plans for missions to Europa take credit for reductions due to this.

27.6.1   Bioburden detection and assessment

The spore count is used as an indirect measure of the number of microorganisms present. Typically 99% of microorganisms by species will be non-spore forming and able to survive in dormant states, and so the actual number of viable dormant microorganisms remaining on the sterilized spacecraft is expected to be many times the number of spore-forming microorganisms.

One new spore method approved is the "Rapid Spore Assay". This is based on commercial rapid assay systems, detects spores directly and not just viable microorganisms and gives results in 5 hours instead of 72 hours.[46]

27.6.2   Challenges

It is also long been recognized that spacecraft cleaning rooms harbour polyextremophiles as the only microbes able to survive in them.[48][49][50][51] For example, in a recent study, microbes from swabs of *Curiosity* rover were subjected to desiccation, UV exposure, cold and pH extremes. Nearly 11% of the 377 strains survived more than one of these severe conditions.[51]

This does not mean that these microbes have contaminated Mars. This is just the first stage of the process of bioburden reduction. To contaminate Mars they also have to survive the low temperature, vacuum, UV and ionizing radiation during the months long journey to Mars, and then have to encounter a habitat on Mars and start reproducing there. Whether this has happened or not is a matter of probability. The aim of planetary protection is to make this probability as low as possible. The currently accepted target probability of contamination per mission is to reduce it to less than 0.01%, though in the special case of Mars, scientists also rely on the hostile conditions on Mars to take the place of the final stage of heat treatment decimal reduction used for Viking. But with current technology scientists cannot reduce probabilities to zero.

27.6.3   New methods

Two recent molecular methods have been approved[46] for assessment of microbial contamination on spacecraft surfaces.[44][52]

- Adenosine triphosphate (ATP) detection - this is a key element in cellular metabolism. This method is able to detect non cultivable organisms. It can also be triggered by non viable biological material so can give a "false positive".

- Limulus Amebocyte Lysate assay - detects lipopolysaccharides (LPS). This compound is only present in Gram-negative bacteria. The standard assay analyses spores from microbes that are primarily Gram-positive, making it difficult to relate the two methods.

27.7 Impact prevention

This particularly applies to orbital missions, Category III, as they are sterilized to a lower standard than missions to the surface. It is also relevant to landers, as an impact gives more opportunity for forward contamination, and impact could be on an unplanned target, such as a special region on Mars.

The requirement for an orbital mission is that it needs to remain in orbit for at least 20 years after arrival at Mars with probability of at least 99% and for 50 years with probability at least 95%. This requirement can be dropped if the mission is sterilized to Viking sterilization standard.[53]

In the Viking era (1970s), the requirement was given as a single figure, that any orbital mission should have a probability of less than 0.003% probability of impact during the current exploratory phase of exploration of Mars.[54]

For both landers and orbiters, the technique of trajectory biasing is used during approach to the target. The spacecraft trajectory is designed so that if communications are lost, it will miss the target.

27.7.1 Issues with impact prevention

Despite these measures, there has been one notable failure of this element of the planetary protection policy, the Mars Climate Orbiter which was sterilized only to Category III, and crashed on Mars due to a mix-up of imperial and metric units. The office of planetary protection stated that it is likely that it burnt up in the atmosphere, but if it survived to the ground, then it could cause forward contamination.[55]

Mars Observer is another Category III mission with potential planetary contamination. Communications were lost three days before its orbital insertion maneuver. It seems most likely it did not succeed in entering into orbit around Mars, and simply continued past on a heliocentric orbit. If it did succeed in following its automatic programming, and attempted the manoeuvre, however, there is a chance it crashed on Mars.

Three landers had hard landings on Mars, which is potentially an issue for planetary protection. These are *Beagle 2*, the *Mars Polar Lander*, and *Deep Space 2*.

27.8 Controversies

27.8.1 Meteorite argument

Alberto G. Fairén and Dirk Schulze-Makuch published an article in *Nature* recommending that planetary protection measures need to be scaled down. They gave as their main reason for this, that exchange of meteorites between Earth and Mars means that any life on Earth that could survive on Mars has already got there and vice versa.[56]

Zubrin used similar arguments in favour of his view that the back contamination risk has no scientific validity.[57][58]

27.8.2 Rebuttal by NRC

The meteorite argument was examined by the NRC in the context of back contamination. It is thought that all the Martian meteorites originate in relatively few impacts every few million years on Mars. The impactors would be kilometers in diameter and the craters they form on Mars tens of kilometers in diameter. Models of impacts on Mars are consistent with these findings.[59][60][61]

Earth receives a steady stream of meteorites from Mars, but they come from relatively few original impactors, and transfer was more likely in the early Solar System. Also some life forms viable on both Mars and on Earth might be unable to survive transfer on a meteorite, and there is so far no direct evidence of any transfer of life from Mars to Earth in this way.

The NRC concluded that though transfer is possible, the evidence from meteorite exchange does not eliminate the need for back contamination protection methods.[62]

Impacts on Earth able to send microorganisms to Mars are also infrequent. Impactors of 10 km across or larger can send debris to Mars through the Earth's atmosphere but these occur rarely, and were more common in the early Solar System.

27.8.3 Proposal to end planetary protection for Mars

In a paper "The Over Protection of Mars" in astrobiology magazine, in 2013, Alberto Fairén and Dirk Schulze-Makuch suggested that we no longer need to protect Mars, using Zubrin's meteorite transfer argument essentially.[63] This was rebutted in a follow up article "Appropriate Protection of Mars", in nature by the current and previous planetary protection officers Catherine Conley and John Rummel.[64][65]

27.8.4 Critique of Category V containment measures

The scientific consensus is that the potential for large-scale effects, either through pathogenesis or ecological disruption, is extremely small.[40][66][67][68][69] Nevertheless, returned samples from Mars will be treated as potentially biohazardous until scientists can determine that the returned samples are safe. The goal is to reduce the probability of release of a Mars particle to less than one in a million.[67]

The International Committee Against Mars Sample Return[70] agree with the assessment of low probability of large-scale effects, but consider the proposed containment measures insufficient, given the possible severity of the worst-case scenario. They come to this conclusion partly as a result of considerations of human error and the novelty of the mission proposal. Consequently, they advocate much more in situ research before undertaking a Mars Sample Return (MSR).[70][71]

27.9 Proposal for extension of protection to non-biological considerations

A COSPAR workshop in 2010, looked issues to do with protecting areas from non biological contamination.[72][73] They recommended that COSPAR expand its remit to include such issues.

Recommendations of the workshop include:

> **Recommendation 3** COSPAR should add a separate and parallel policy to
> provide guidance on requirements/best practices for protection of non-living/nonlife-related aspects of
> Outer Space and celestial bodies

Ideas for doing this suggested since the workshop include protected special regions, or "Planetary Parks"[74] to keep regions of the Solar System pristine for future scientific investigation, and also for ethical reasons.

27.10 Proposal to extend planetary protection

Astrobiologist Christopher McKay has argued that until we have better understanding of Mars, our explorations should be biologically reversible.[75][76] For instance if all the microorganisms introduced to Mars so far remain dormant within the spacecraft, they could in principle be removed in the future, leaving Mars completely free of contamination from modern Earth lifeforms.

In the 2010 workshop one of the recommendations for future consideration was to extend the period for contamination prevention to the maximum viable lifetime of dormant microorganisms introduced to the planet.

"'**Recommendation 4.**' COSPAR should consider that the appropriate protection of potential indigenous extraterrestrial life shall include avoiding the harmful contamination of any habitable environment — whether extant or foreseeable— within the maximum potential time of viability of any terrestrial organisms (including microbial spores) that may be introduced into that environment by human or robotic activity."[73]

In the case of Europa, a similar idea has been suggested, that it is not enough to keep it free from contamination during our current exploration period. It might be that Europa is of sufficient scientific interest that we have a duty to keep it pristine for future generations to study as well. This was the majority view of the 2000 task force examining Europa, though there was a minority view of the same task force that such strong protection measures are not required.

"One consequence of this view is that Europa must be protected from contamination for an open-ended period, until it can be demonstrated that no ocean exists or that no organisms are present. Thus, we need to be concerned that over a time scale on the order of 10 million to 100 million years (an approximate age for the surface of Europa), any contaminating material is likely to be carried into the deep ice crust or into the underlying ocean."[77]

27.11   See also

- Astrobiology

- ExoMars

- List of microorganisms tested in outer space

- Mars 2020 rover mission

- Panspermia

27.12   References

[1] Assessment of Planetary Protection and Contamination Control Technologies for Future Planetary Science Missions, Jet Propulsion Laboratory, January 24, 2011
3.1.1 Microbial Reduction Methodologies.

   "This protocol was defined in concert with Viking, the first mission to face the most stringent planetary protection requirements; its implementation remains the gold standard today."

[2] Tánczer, John D. Rummel; Ketskeméty, L.; Lévai, G. (1989). "Planetary protection policy overview and application to future missions". *Advances in Space Research* **9** (6): 181–184. Bibcode:1989AdSpR...9..181T. doi:10.1016/0273-1177(89)90161-0. PMID 11537370. Retrieved 2012-09-11.

[3] Portree, David S.F. (2 October 2013). "Spraying Bugs on Mars (1964)". *Wired (magazine)*. Retrieved 3 October 2013.

[4] NASA Office of Planetary Protection. "Planetary Protection History". Retrieved 2013-07-13.

[5] Preventing the Forward Contamination of Mars (2006) - Page 12

[6] Preventing the Forward Contamination of Mars

[7] Preventing the Forward Contamination of Mars - p12 quotes from COSPAR 1964 Resolution 26

[8] Full text of the Outer Space Treaty Treaty on Principles Governing the Activities of States in the Exploration and Use of Outer Space, including the Moon and Other Celestial Bodies - *See Article IX*

[9] Centre National d'Etudes Spatiales (CNES) (2008). "Planetary protection treaties and recommendations". Retrieved 2012-09-11.

[10] Preventing the Forward Contamination of Mars, page 13 Summarizes this para in the book:

> A policy review of the Outer Space Treaty concluded that, while Article IX "imposed international obligations on all state parties to protect and preserve the environmental integrity of outer space and celestial bodies such as Mars," there is no definition as to what constitutes harmful contamination, nor does the treaty specify under what circumstances it would be necessary to "adopt appropriate measures" or which measures would in fact be "appropriate"
>
> An earlier legal review, however, argued that "if the assumption is made that the parties to the treaty were not merely being verbose" and "harmful contamination" is not simply redundant, "harmful" should be interpreted as "harmful to the interests of other states," and since "states have an interest in protecting their ongoing space programs," Article IX must mean that "any contamination which would result in harm to a state's experiments or programs is to be avoided"
>
> Current NASA policy states that the goal of NASA's forward contamination planetary protection policy is the protection of scientific investigations, declaring explicitly that "the conduct of scientific investigations of possible extraterrestrial life forms, precursors, and remnants must not be jeopardized"

[11] COSPAR scientific assemblies

[12] Preventing the Forward Contamination of Mars (2006) - Page 13

[13] COSPAR PLANETARY PROTECTION POLICY (20 October 2002; As Amended to 24 March 2011)

[14] Office of Planetary Protection - About The Categories

[15] "Mission Design And Requirements". *Office of Planetary Protection*.

[16] Planetary Protection Standards for Icy Bodies in the Outer Solar System - about the *Committee on Planetary Protection Standards for Icy Bodies in the Outer Solar System*

[17] Carl Sagan and Sidney Coleman Decontamination Standards for Martian Exploration Programs, Chapter 28 from Biology and the Exploration of Mars: Report of a Study edited by Colin Stephenson Pittendrigh, Wolf Vishniac, J. P. T. Pearman, National Academies, 1966 - Life on other planets

[18] *Keeping it clean*

[19] Mars Sample Return backward contamination – Strategic advice and requirements- foreword and section 1.2

[20] COSPAR Workshop on Planetary Protection for Outer Planet Satellites and Small Solar System Bodies European Space Policy Institute (ESPI), 15–17 April 2009

[21] COSPAR power point type presentation, gives good overview of the detailed category decisions

[22] "Mission Categories". *Office of Planetary Protection*.

[23] Assessment of Planetary Protection Requirements for Venus Missions -- Letter Report

[24] COSPAR Final

[25] COSPAR Workshop on Planetary Protection for Titan and Ganymede

[26] Catharine Conley Planetary Protection for the Dawn Mission, NASA HQ, Jan 2013

[27] edited by Muriel Gargaud, Ricardo Amils, Henderson James Cleaves, Michel Viso, Daniele Pinti Encyclopedia of Astrobiology, Volume 1 page 325

[28] Richard Greenberg, Richard J. Greenberg Unmasking Europa: the search for life on Jupiter's ocean moon ISBN 0387479368

[29] Paul Gilster Europa: Thin Ice and Contamination

[30] B. Randall Tufts, Richard Greenberg Infecting Other World, American Scientist, July 2001

[31] Europa the Ocean Moon, Search for an Alien Biosphere, chapter 21.5.2 Standards and Risks

[32] Committee on Planetary Protection Standards for Icy Bodies in the Outer Solar System; Space Studies Board; Division on Engineering and Physical Sciences; National Research Council Assessment of Planetary Protection Requirements for Spacecraft Missions to Icy Solar System Bodies (2012) / 2 Binary Decision Trees

[33] Designing a Box to Return Samples From Mars Astrobio.net. 2013

[34] Office of Planetary Protection: Mars Sample Quarantine Protocol Workshop

[35] Mars sample return mission concept study (for decadal review 2010)

[36] Proof of concept of a Bio-Containment System for Mars Sample Return Mission

[37] Richard S. Johnston, John A. Mason, Bennie C. Wooley, Gary W. McCollum, Bernard J. Mieszkuc BIOMEDICAL RESULTS OF APOLLO, SECTION V, CHAPTER 1, THE LUNAR QUARANTINE PROGRAM

[38] Nancy Atkinson How to Handle Moon Rocks and Lunar Bugs: A Personal History of Apollo's Lunar Receiving Lab, Universe Today, July 2009. See quote from: McLane who lead the group that designed and built the Lunar Receiving Facility:

> "The best that I hear now is that the techniques of isolation we used wouldn't be adequate for a sample coming back from Mars, so somebody else has a big job on their hands."

[39] The Quarantine and Certification of Martian Samples - Chapter 7: Lessons Learned from the Quarantine of Apollo Lunar Samples, Committee on Planetary and Lunar Exploration, Space Studies Board

[40] Assessment of Planetary Protection Requirements for Mars Sample Return Missions (Report), National Research Council, 2009.

[41] European Science Foundation - Mars Sample Return backward contamination - strategic advice July, 2012, ISBN 978-2-918428-67-1

[42] M. S. Race Planetary Protection, Legal Ambiguity, and the Decision Making Process for Mars Sample Return Adv. Space Res. vol 18 no 1/2 pp (1/2)345-(1/2)350 1996

[43] In-situ Exploration and Sample Return: Planetary Protection Technologies JPL - Mars Exploration Rovers

[44] Office of Planetary Protection (August 28, 2012). "Office of Planetary Protection - Methods and Implementation". *NASA.* Retrieved 2012-09-11.

[45] Benton C. Clark (2004). "Temperature–time issues in bioburden control for planetary protection". *Advances in Space Research* **34** (11): 2314–2319. Bibcode:2004AdSpR..34.2314C. doi:10.1016/j.asr.2003.06.037.

[46] Assessment of Planetary Protection and Contamination Control Technologies for Future Planetary Science Missions see Section 3.1.2 Bio-burden Detection and Assessment. January 24, JPL, 2011

[47] Fei Chen, Terri Mckay, James Andy Spry, Anthony Colozza, Salvador Distefano, Robert Cataldo Planetary Protection Concerns During Pre-Launch Radioisotope Power System Final Integration Activities - includes the draft specification of VHP sterilization and details of how it would be implemented. Proceedings of Nuclear and Emerging Technologies for Space 2013, Albuquerque, NM, February 25–28, 2013 Paper 6766

[48] Microbial characterization of the Mars Odyssey spacecraft and its encapsulation facility, Environ Microbiol. 2003 Oct;5(10):977-85., "Several spore-forming isolates were resistant to gamma-radiation, UV, H2O2 and desiccation, and one Acinetobacter radioresistens isolate and several Aureobasidium, isolated directly from the spacecraft, survived various conditions."

[49] Recurrent isolation of extremotolerant bacteria from the clean room where Phoenix spacecraft components were assembled Astrobiology. 2010 Apr;10(3):325-35. doi: 10.1089/ast.2009.0396 "Extremotolerant bacteria that could potentially survive conditions experienced en route to Mars or on the planet's surface were isolated with a series of cultivation-based assays that promoted the growth of a variety of organisms, including spore formers, mesophilic heterotrophs, anaerobes, thermophiles, psychrophiles, alkaliphiles, and bacteria resistant to UVC radiation and hydrogen peroxide exposure".

[50] Webster, Guy (6 November 2013). "Rare New Microbe Found in Two Distant Clean Rooms". *NASA.* Retrieved 6 November 2013.

[51] Madhusoodanan, Jyoti (19 May 2014). "Microbial stowaways to Mars identified". *Nature (journal)*. doi:10.1038/nature.2014.1 Retrieved 23 May 2014.

[52] A. Debus (2004). "Estimation and assessment of Mars contamination". *Advances in Space Research* **35** (9): 1648–1653. Bibcode:2005AdSpR..35.1648D. doi:10.1016/j.asr.2005.04.084. PMID 16175730.

[53] Preventing the Forward Contamination of Mars (2006) Page 27 (footnote to page 26) of chapter 2 Policies and Practices in Planetary Protection

[54] Preventing the Forward Contamination of Mars (2006) Page 22 of chapter 2 Policies and Practices in Planetary Protection

[55] Mars Climate Orbiter page at

[56] Alberto G. Fairén & Dirk Schulze-Makuch The Over Protection of Mars Nature Geoscience 6, 510–511 (2013) doi:10.1038/n6

[57] Robert Zubrin "Contamination From Mars: No Threat", The Planetary Report July/Aug. 2000, P.4–5

[58] transcription of a tele-conference interview with ROBERT ZUBRIN conducted on March 30, 2001 by the class members of STS497 I, "Space Colonization"; Instructor: Dr. Chris Churchill

[59] O. Eugster, G. F. Herzog, K. Marti, M. W. Caffee Irradiation Records, Cosmic-Ray Exposure Ages, and Transfer Times of Meteorites, see section 4.5 Martian Meteorites LPI, 2006

[60] L.E. NYQUIST1, D.D. BOGARD1, C.-Y. SHIH2, A. GRESHAKE3, D. STÖFFLER AGES AND GEOLOGIC HISTORIES OF MARTIAN METEORITES 2001

[61] Tony Irving Martian Meteorites - has graphs of ejection ages - site maintained by Tony Irving for up to date information on Martian meteorites

[62] "5: The Potential for Large-Scale Effects"". Assessment of Planetary Protection Requirements for Mars Sample Return Missions (Report). National Research Council. 2009. p. 48. *Despite suggestions to the contrary, it is simply not possible, on the basis of current knowledge, to determine whether viable martian life forms have already been delivered to the Earth. Certainly in the modern era there is no evidence for large-scale or other negative effects that are attributable to the frequent deliveries to Earth of essentially unaltered Martian rocks. However the possibility that such effects occurred in the distant past cannot be discounted. Thus it is not appropriate to argue that the existence of martian microbes on Earth negates the need to treat as potentially hazardous any samples returned from Mars via robotic spacecraft.*

[63] The overprotection of Mars

[64] Appropriate protection of Mars. Nature. Catherine Conley and John Rummel

[65] The Overprotection of Mars?, astrobio.net. Andrew Williams - Nov 18, 2013 - summarizes both papers on the subject, with links to originals

[66] http://mepag.nasa.gov/reports/iMARS_FinalReport.pdf Preliminary Planning for an International Mars Sample Return Mission Report of the International Mars Architecture for the Return of Samples (iMARS) Working Group June 1, 2008

[67] European Science Foundation - Mars Sample Return backward contamination - Strategic advice and requirements July, 2012. ISBN 978-2-918428-67-1 - see Back Planetary Protection section. (for more details of the document see abstract)

[68] Joshua Lederberg Parasites Face a Perpetual Dilemma Volume 65, Number 2, 1999 / American Society for Microbiology News 77.

[69] http://planetaryprotection.nasa.gov/summary/msr Mars Sample Return: Issues and Recommendations. Task Group on Issues in Sample Return. National Academies Press, Washington, DC (1997).

[70] International Committee Against Mars Sample Return

[71] Barry E. DiGregorio The dilemma of Mars sample return *August 2001 Vol. 31, No. 8, pp 18–27.*

[72] Rummel, J., Race, M., and Horneck, G. eds. 2011. COSPAR Workshop on Ethical Considerations for Planetary Protection in Space Exploration COSPAR, Paris, 51 pp.

[73] Ethical Considerations for Planetary Protection in Space Exploration: A Workshop. Astrobiology. 2012 November; 12(11): 1017–1023.

[74] 'Planetary Parks' Could Protect Space Wilderness by Leonard David, SPACE.com's Space Insider Columnist, January 17, 2013

[75] Christopher P. McKay Planetary Ecosynthesis on Mars: Restoration Ecology and Environmental Ethics NASA Ames Research Center

[76] Christopher P. McKay Biologically Reversible Exploration Science 6 February 2009: Vol. 323 no. 5915 p. 718 doi:10.1126/

[77] Preventing the forward contamination of Europa - Executive Summary page 2 National Academies Press

27.13   General references

- Sagan. C.; Coleman. S. (1965). "Spacecraft sterilization standards and contamination of Mars". *Journal of Astronautics and Aeronautics* **3** (5): 22–27.

- L. I. Tennen (2006). "Evolution of the planetary protection policy: conflict of science and jurisprudence". *Advances in Space Research* **34** (11): 2354–2362. Bibcode:2004AdSpR..34.2354T. doi:10.1016/j.asr.2004.01.018.

- L. Perek (2006). "Planetary protection: lessons learned". *Advances in Space Research* **34** (11): 2354–2362. Bibcode:2004AdSpR..34.2368P. doi:10.1016/j.asr.2003.02.066.

- J. D. Rummel; P. D. Stabekis; D. L. Devincenzi; J. B. Barengoltz (2002). "COSPAR's planetary protection policy: A consolidated draft". *Advances in Space Research* **30** (6): 1567–1571. Bibcode:2002AdSpR..30.1567R. doi:10.1016/S0273-1177(02)00479-9.

- D. L. DeVincenzi; P. Stabekis & J. Barengoltz (1996). "Refinement of planetary protection policy for Mars missions". *Advances in Space Research* **18** (1–2): 311–316. Bibcode:1996AdSpR..18..311D. doi:10.1016/0273-1177(95)00821-U. PMID 11538978.

- J. Barengoltz & P. D. Stabekis (1983). "U.S. planetary protection program: Implementation highlights". *Advances in Space Research* **3** (8): 5–12. Bibcode:1983AdSpR...3....5B. doi:10.1016/0273-1177(83)90166-7.

- L. P. Daspit, Stern. Cortright (1975). "Viking heat sterilization—Progress and problems". *Acta Astronautica* **2** (7–8): 649–666. doi:10.1016/0094-5765(75)90007-7.

27.14   External links

- No bugs please. this is a clean planet! (ESA article)

- COSPAR planetary protection policy, July 2008 (COSPAR article)

- NASA Planetary Protection Website

- JPL Develops High-Speed Test to Improve Pathogen Decontamination at JPL.

- Geoethics in Planetary and Space Exploration

- Catharine Conley: NASA & international planetary protection policy, methodology & applications, The Space Show, October 2012

Chapter 28

Potential cultural impact of extraterrestrial contact

This article is about consideration of the possible effects on humanity of potential future extraterrestrial contact. For the fictional treatment of the subject, see First contact (science fiction). For the search for intelligent life beyond Earth, see Search for extraterrestrial intelligence.

The **cultural impact of extraterrestrial contact** is the corpus of changes to terrestrial science, technology, religion, politics, and ecosystems resulting from contact with an extraterrestrial civilization. Although closely related to it, the study of the cultural impact of extraterrestrial contact is distinct from the search for extraterrestrial intelligence (SETI), which attempts to locate intelligent life as opposed to analyzing the implications of contact with that life.

The potential changes from extraterrestrial contact could vary greatly in magnitude and type, based on the extraterrestrial civilization's level of technological advancement, degree of benevolence or malevolence, and level of mutual comprehension between itself and humanity.[1] The medium through which humanity is contacted, be it electromagnetic radiation, direct physical interaction, extraterrestrial artefact, or otherwise, may also influence the results of contact. Incorporating these factors, various systems have been created to assess the implications of extraterrestrial contact.

The implications of extraterrestrial contact, particularly with a technologically superior civilization, have often been likened to the meeting of two vastly different human cultures on Earth, an historical precedent being the Columbian Exchange. Such meetings have generally led to the destruction of the civilization receiving contact (as opposed to the "contactor", which initiates contact), and therefore destruction of human civilization is a possible outcome.[2] However, the absence of any such contact to date means such conjecture is largely speculative.

28.1 Background

28.1.1 Search for extraterrestrial intelligence

Main article: Search for extraterrestrial intelligence

To detect extraterrestrial civilizations with radio telescopes, one must identify an artificial, coherent signal against a background of various natural phenomena that also produce radio waves. Telescopes capable of this include the Arecibo Observatory in Puerto Rico and the newer[4] Allen Telescope Array in Hat Creek, California. Various programs to detect extraterrestrial intelligence have had government funding in the past. Project Cyclops was commissioned by NASA in the 1970s to investigate the most effective way to search for signals from intelligent extraterrestrial sources,[3] but the report's recommendations were set aside in favor of the much more modest approach of Messaging to Extra-Terrestrial Intelligence (METI), the sending of messages that intelligent extraterrestrial beings might intercept. NASA then drastically reduced funding for SETI programs, which have since turned to private donations to continue their search.[5]

With the discovery in the late 20th and early 21st centuries of numerous extrasolar planets, some of which may be habitable, governments have once more become interested in funding new programs. In 2006 the European Space Agency

The Arecibo Observatory, one of the radio telescopes used in the search for extraterrestrial intelligence (SETI)

launched COROT, the first spacecraft dedicated to the search for exoplanets,[6] and in 2009 NASA launched the *Kepler* space observatory for the same purpose.[7] By February 2013 *Kepler* had detected 105[8] of the 2030 confirmed exoplanets, and one of them, Kepler-22b, is potentially habitable.[9] After it was discovered, the SETI Institute resumed the search for an intelligent extraterrestrial civilization, focusing on *Kepler*'s candidate planets,[10] with funding from the United States Air Force.[11]

Newly discovered planets, particularly ones that are potentially habitable, have enabled SETI and METI programs to refocus projects for communication with extraterrestrial intelligence. In 2009 A Message From Earth (AMFE) was sent toward the Gliese 581 system, which contains two potentially habitable planets, the confirmed Gliese 581 d and the more habitable but unconfirmed Gliese 581 g.[12] In the SETILive project, which began in 2012, human volunteers analyze data from the Allen Telescope Array to search for possible alien signals that computers might miss because of terrestrial radio interference.[13] The data for the study is obtained by observing *Kepler* target stars with the radio telescope.[10]

In addition to radio-based methods, some projects, such as SEVENDIP (Search for Extraterrestrial Visible Emissions from Nearby Developed Intelligent Populations) at the University of California, Berkeley, are using other regions of the electromagnetic spectrum to search for extraterrestrial signals.[14] Various other projects are not searching for coherent signals, but want to rather use electromagnetic radiation to find other evidence of extraterrestrial intelligence, such as megascale astroengineering projects.[15]

Several signals, such as the Wow! signal, have been detected in the history of the search for extraterrestrial intelligence, but none have yet been confirmed as being of intelligent origin.[16]

28.1.2   Impact assessment

The implications of extraterrestrial contact depend on the method of discovery, the nature of the extraterrestrial beings, and their location relative to the Earth.[17] Considering these factors, the Rio Scale has been devised in order to provide a more quantitative picture of the results of extraterrestrial contact.[17] More specifically, the scale gauges whether communication was conducted through radio, the information content of any messages, and whether discovery arose from a deliberately beamed message (and if so, whether the detection was the result of a specialized SETI effort or through general astronomical observations) or by the detection of occurrences such as radiation leakage from astroengineering installations.[18] The question of whether or not a purported extraterrestrial signal has been confirmed as authentic, and with what degree of confidence, will also influence the impact of the contact.[18] The Rio Scale was modified in 2011 to include a consideration of whether contact was achieved through an interstellar message or through a physical extraterrestrial artifact, with a suggestion that the definition of *artifact* be expanded to include "technosignatures", including all indications of intelligent extraterrestrial life other than the interstellar radio messages sought by traditional SETI programs.[19]

A study by astronomer Steven J. Dick at the United States Naval Observatory considered the cultural impact of extraterrestrial contact by analyzing events of similar significance in the history of science.[20] The study argues that the impact would be most strongly influenced by the information content of the message received, if any.[20] It distinguishes short-term and long-term impact.[20] Seeing radio-based contact as a more plausible scenario than a visit from extraterrestrial spacecraft, the study rejects the commonly stated analogy of European colonization of the Americas as an accurate model for information-only contact, preferring events of profound scientific significance, such as the Copernican and Darwinian revolutions, as more predictive of how humanity might be impacted by extraterrestrial contact.[20]

The physical distance between the two civilizations has also been used to assess the cultural impact of extraterrestrial contact. Historical examples show that the greater the distance, the less the contacted civilization perceives a threat to itself and its culture.[21] Therefore, contact occurring within the Solar System, and especially in the immediate vicinity of Earth, is likely to be the most disruptive and negative for humanity.[21] On a smaller scale, people close to the epicenter of contact would experience a greater effect than would those living farther away, and a contact having multiple epicenters would cause a greater shock than one with a single epicenter.[21] Space scientists Martin Dominik and John Zarnecki state that in the absence of any data on the nature of extraterrestrial intelligence, one must predict the cultural impact of extraterrestrial contact on the basis of generalizations encompassing all life and of analogies with history.[22]

The beliefs of the general public about the effect of extraterrestrial contact have also been studied. A poll of United States and Chinese university students in 2000 provides factor analysis of responses to questions about, *inter alia*, the participants' belief that extraterrestrial life exists in the Universe, that such life may be intelligent, and that humans will eventually make contact with it.[23] The study shows significant weighted correlations between participants' belief that extraterrestrial contact may either conflict with or enrich their personal religious beliefs and how conservative such religious beliefs are. The more conservative the respondents, the more harmful they considered extraterrestrial contact to be. Other significant correlation patterns indicate that participants took the view that the search for extraterrestrial intelligence may be futile or even harmful.[23]

28.1.3   Post-detection protocols

Main article: Post-detection protocol

Various protocols have been drawn up detailing a course of action for scientists and governments after extraterrestrial contact. Post-detection protocols must address three issues: what to do in the first weeks after receiving a message from an extraterrestrial source; whether or not to send a reply; and analyzing the long-term consequences of the message received.[24] No post-detection protocol, however, is binding under national or international law,[22] and Dominik and Zarnecki consider the protocols likely to be ignored if contact occurs.[22]

One of the first post-detection protocols, the "Declaration of Principles for Activities Following the Detection of Extraterrestrial Intelligence", was created by the SETI Permanent Committee of the International Academy of Astronautics (IAA).[24] It was later approved by the Board of Trustees of the IAA and by the International Institute of Space Law,[24] and still later by the International Astronomical Union (IAU), the Committee on Space Research, the International Union of Radio Science, and others.[24] It was subsequently endorsed by most researchers involved in the search for extraterrestrial intelligence,[25] including the SETI Institute.[26]

The Declaration of Principles contains the following broad provisions:[27]

1. Any person or organization detecting a signal should try to verify that it is likely to be of intelligent origin before announcing it.

2. The discoverer of a signal should, for the purposes of independent verification, communicate with other signatories of the Declaration before making a public announcement, and should also inform their national authorities.

3. Once a given astronomical observation has been determined to be a credible extraterrestrial signal, the astronomical community should be informed through the Central Bureau for Astronomical Telegrams of the IAU. The Secretary-General of the United Nations and various other global scientific unions should also be informed.

4. Following confirmation of an observation's extraterrestrial origin, news of the discovery should be made public. The discoverer has the right to make the first public announcement.

5. All data confirming the discovery should be published to the international scientific community and stored in an accessible form as permanently as possible.

6. Should evidence for extraterrestrial intelligence take the form of electromagnetic signals, the Secretary-General of the International Telecommunications Union (ITU) should be contacted, and may request in the next ITU Weekly Circular to minimize terrestrial use of the electromagnetic frequency bands in which the signal was detected.

7. Neither the discoverer nor anyone else should respond to an observed extraterrestrial intelligence; doing so requires international agreement under separate procedures.

8. The SETI Permanent Committee of the IAA and Commission 51 of the IAU should continually review procedures regarding detection of extraterrestrial intelligence and management of data related to such discoveries. A committee comprising members from various international scientific unions, and other bodies designated by the committee, should regulate continued SETI research.

A separate "Proposed Agreement on the Sending of Communications to Extraterrestrial Intelligence" was subsequently created.[28] It proposes an international commission, membership of which would be open to all interested nations, to be constituted on detection of extraterrestrial intelligence.[28] This commission would decide whether to send a message to the extraterrestrial intelligence, and if so, would determine the contents of the message on the basis of principles such as justice, respect for cultural diversity, honesty, and respect for property and territory.[28] The draft proposes to forbid the sending of any message by an individual nation or organization without the permission of the commission, and suggests that, if the detected intelligence poses a danger to human civilization, the United Nations Security Council should authorize any message to extraterrestrial intelligence.[28] However, this proposal, like all others, has not been incorporated into national or international law.[28]

Paul Davies, a member of the SETI Post-Detection Taskgroup, has stated that post-detection protocols, calling for international consultation before taking any major steps regarding the detection, are unlikely to be followed by astronomers, who would put the advancement of their careers over the word of a protocol that is not part of national or international law.[29]

28.2 Contact scenarios and considerations

Scientific literature and science fiction put forward various models of the ways in which extraterrestrial and human civilizations might interact. Their predictions range widely, from sophisticated civilizations that could advance human civilization in many areas to imperial powers that might draw upon the forces necessary to subjugate humanity.[1] Some theories suggest that an extraterrestrial civilization could be advanced enough to dispense with biology, living instead inside of advanced computers.[1]

The implications of discovery depend very much on the level of aggressiveness of the civilization interacting with humanity,[130] its ethics,[131] and how much human and extraterrestrial biologies have in common.[132] These factors will govern the quantity and type of dialogue that can take place.[132] The question of whether contact is physical or through electromagnetic signals will also govern the magnitude of the long-term implications of contact.[133] In the case of communication using electromagnetic signals, the long silence between the reception of one message and another would mean that the content of any message would particularly affect the consequences of contact,[134] as would the extent of mutual comprehension.[135]

28.2.1   Friendly civilizations

Many writers have speculated on the ways in which a friendly civilization might interact with humankind. Albert Harrison, a professor emeritus of psychology at the University of California, Davis,[136] thought that a highly advanced civilization might teach humanity such things as a physical theory of everything, how to use zero-point energy, or how to travel faster than light.[137] They suggest that collaboration with such a civilization could initially be in the arts and humanities before moving to the hard sciences, and even that artists may spearhead collaboration.[138] Seth D. Baum, of the Global Catastrophic Risk Institute, and others consider that the greater longevity of cooperative civilizations in comparison to uncooperative and aggressive ones might render extraterrestrial civilizations in general more likely to aid humanity.[139] In contrast to these views, however, Paolo Musso, a member of the SETI Permanent Study Group of the International Academy of Astronautics (IAA) and the Pontifical Academy of Sciences, took the view that extraterrestrial civilizations possess, like humans, a morality driven not entirely by altruism but for individual benefit as well, thus leaving open the possibility that at least *some* extraterrestrial civilizations are hostile.[140]

Futurist Allen Tough suggests that an extremely advanced extraterrestrial civilization, recalling its own past of war and plunder and knowing that it possesses superweapons that could destroy it, would be likely to try to help humans rather than to destroy them.[141] He identifies three approaches that a friendly civilization might take to help humanity:[141]

- Intervention only to avert catastrophe: this would involve occasional limited intervention to stop events that could destroy human civilization completely, such as nuclear war or asteroid impact.[141]

- Advice and action with consent: under this approach, the extraterrestrials would be more closely involved in terrestrial affairs, advising world leaders and acting with their consent to protect against danger.[141]

- Forcible corrective action: the extraterrestrials could require humanity to reduce major risks against its will, intending to help humans advance to the next stage of civilization.[141]

Tough considers advising and acting only with consent to be a more likely choice than the forceful option. While coercive aid may be possible, and advanced extraterrestrials would recognize their own practices as superior to those of humanity, it may be unlikely that this method would be used in cultural cooperation.[141] Lemarchand suggests that instruction of a civilization in its "technological adolescence", such as humanity, would probably focus on morality and ethics rather than on science and technology, to ensure that the civilization did not destroy itself with technology it was not yet ready to use.[142]

According to Tough, it is unlikely that the avoidance of immediate dangers and prevention of future catastrophes would be conducted through radio, as these tasks would demand constant surveillance and quick action.[141] However, cultural cooperation might take place through radio or a space probe in the Solar System, as radio waves could be used to communicate information about advanced technologies and cultures to humanity.[141]

Even if an ancient and advanced extraterrestrial civilization wished to help humanity, humans could suffer from a loss of identity and confidence due to the technological and cultural prowess of the extraterrestrial civilization.[143] However, a friendly civilization may calibrate its contact with humanity in such a way as to minimize unintended consequences.[130] Michael A. G. Michaud suggests that a friendly and advanced extraterrestrial civilization may even avoid all contact with an emerging intelligent species like humanity, to ensure that the less advanced civilization can develop naturally at its own pace.[144]

28.2.2   Hostile civilizations

Science fiction films often depict humans successfully repelling alien invasions, but scientists more often take the view that an extraterrestrial civilization with sufficient power to reach the Earth would be able to destroy human civilization with minimal effort.[3][45] Operations that are enormous on a human scale, such as destroying all major population centers on a planet, bombarding a planet with deadly neutrons, or even traveling to another planetary system in order to lay waste to it, may be important tools for a hostile and totalitarian civilization.[46]

Deardorff speculates that a small proportion of the intelligent life forms in the galaxy may be aggressive, but the actual aggressiveness or benevolence of the civilizations would cover a wide spectrum, with some civilizations "policing" others.[30] According to Harrison and Dick, hostile extraterrestrial life may indeed be rare in the Universe, just as belligerent and autocratic nations on Earth have been the ones that lasted for the shortest periods of time, and humanity is seeing a shift away from these characteristics in its own sociopolitical systems.[37] In addition, the causes of war may be diminished greatly for a civilization with access to the galaxy, as there are prodigious quantities of natural resources in space accessible without resort to violence.[3][47]

SETI researcher Carl Sagan believed that a civilization with the technological prowess needed to reach the stars and come to Earth must have transcended war to be able to avoid self-destruction. Representatives of such a civilization would treat humanity with dignity and respect, and humanity, with its relatively backward technology, would have no choice but to reciprocate.[48] Seth Shostak, an astronomer at the SETI Institute, disagrees, stating that the finite quantity of resources in the galaxy would cultivate aggression in any intelligent species, and that an explorer civilization that would want to contact humanity would be aggressive.[49] Similarly, Ragbir Bhathal claims that since the laws of evolution would be the same on another habitable planet as they are on Earth, an extremely advanced extraterrestrial civilization may have the motivation to colonize humanity, much as British colonizers did to the aboriginal peoples of Australia.[50]

Disputing these analyses, David Brin states that while an extraterrestrial civilization may have an imperative to act for no benefit to itself, it would be naïve to suggest that such a trait would be prevalent throughout the galaxy.[51] Brin points to the fact that in many moral systems on Earth, such as the Aztec or Carthaginian one, non-military killing has been accepted and even "exalted" by society, and further mentions that such acts are not confined to humans but can be found throughout the animal kingdom.[51]

Baum *et al.* speculate that highly advanced civilizations are unlikely to come to Earth to enslave humans, as the achievement of their level of advancement would have required them to solve the problems of labor and resources by other means, such as creating a sustainable environment and using mechanized labor.[39] Moreover, humans may be an unsuitable food source for extraterrestrials because of marked differences in biochemistry.[3] For example, the chirality of molecules used by terrestrial biota may differ from those used by extraterrestrial beings.[39]

Politicians have also commented on the likely human reaction to contact with hostile species. In his 1987 speech to the United Nations General Assembly, Ronald Reagan said, "I occasionally think how quickly our differences worldwide would vanish if we were facing an alien threat from outside this world."[52]

28.2.3   Equally advanced and more advanced civilizations

Robert Freitas speculated in 1978 that the technological advancement and energy usage of a civilization, measured either relative to another civilization or in absolute terms by its rating on the Kardashev scale, may play an important role in the result of extraterrestrial contact.[53] Given the infeasibility of interstellar space flight for civilizations at a technological level similar to that of humanity, interactions between such civilizations would have to take place by radio. Because of the long transit times of radio waves between stars, such interactions would not lead to the establishment of diplomatic relations, nor any significant future interaction at all, between the two civilizations.[53]

According to Freitas, direct contact with civilizations significantly more advanced than humanity would have to take place within the Solar System, as only the more advanced society would have the resources and technology to cross interstellar space.[54] Consequently, such contact could only be with civilizations rated as Type II or higher on the Kardashev scale, as Type I civilizations would be incapable of regular interstellar travel.[54] Freitas expected that such interactions would be carefully planned by the more advanced civilization to avoid mass societal shock for humanity.[54]

However much planning an extraterrestrial civilization may do before contacting humanity, the humans may experience great shock and terror on their arrival, especially as they would lack any understanding of the contacting civilization. Ben Finney compares the situation to that of the tribespeople of New Guinea, an island that was settled fifty thousand years ago during the last glacial period but saw little contact with the outside world until the arrival of European colonial powers in the late 19th and early 20th centuries. The huge difference between the indigenous stone-age society and the Europeans' technical civilization caused unexpected behaviors among the native populations known as cargo cults: to coax the gods into bringing them the technology that the Europeans possessed, the natives created wooden "radio stations" and "airstrips" as a form of sympathetic magic. Finney argues that humanity may misunderstand the true meaning of an extraterrestrial transmission to Earth, much as the people of New Guinea could not understand the source of modern goods and technologies. He concludes that the results of extraterrestrial contact will become known over the long term with rigorous study, rather than as fast, sharp events briefly making newspaper headlines.[35]

Billingham has suggested that a civilization which is far more technologically advanced than humanity is also likely to be culturally and ethically advanced, and would therefore be unlikely to conduct astroengineering projects that would harm human civilization. Such projects could include Dyson spheres, which completely enclose stars and capture all energy coming from them. Even if well within the capability of an advanced civilization and providing an enormous amount of energy, such a project would not be undertaken.[55] For similar reasons, such civilizations would not readily give humanity the knowledge required to build such devices.[55] Nevertheless, the existence of such capabilities would at least show that civilizations have survived "technological adolescence".[55] Despite the caution that such an advanced civilization would exercise in dealing with the less mature human civilization, Sagan imagined that an advanced civilization might send those on Earth an *Encyclopædia Galactica* describing the sciences and cultures of many extraterrestrial societies.[56]

Whether an advanced extraterrestrial civilization would send humanity a decipherable message is a matter of debate in itself. Sagan argued that a highly advanced extraterrestrial civilization would bear in mind that they were communicating with a relatively primitive one and therefore would try to ensure that the receiving civilization would be able to understand the message.[57] Arguing against this view, astronomer Guillermo Lemarchand stated that an advanced civilization would probably encrypt a message with high information content, such as an *Encyclopædia Galactica*, in order to ensure that only other ethically advanced civilizations would be able to understand it.[57]

28.2.4   Interstellar groups of civilizations

Given the age of the galaxy, Harrison surmises that there exist several "galactic clubs", groupings of multiple civilizations from across the galaxy.[47] Such clubs could begin as loose confederations or alliances, eventually developing into powerful unions of many civilizations.[47] If humanity could enter into a dialogue with one extraterrestrial civilization, it might be able to join such a galactic club. As more extraterrestrial civilizations, or unions thereof, are found, these could also become assimilated into such a club.[47] Sebastian von Hoerner has suggested that entry into a galactic club may be a way for humanity to handle the culture shock arising from contact with an advanced extraterrestrial civilization.[58]

Whether a broad spectrum of civilizations from many places in the galaxy would even be able to cooperate is disputed by Michaud, who states that civilizations with huge differences in the technologies and resources at their command "may not consider themselves even remotely equal".[59] It is unlikely that humanity would meet the basic requirements for membership at its current low level of technological advancement.[39] A galactic club may, William Hamilton speculates, set extremely high entrance requirements that are unlikely to be met by less advanced civilizations.[59]

Michaud suggests that an interstellar grouping of civilizations might take the form of an empire, which need not necessarily be a force for evil, but may provide for peace and security throughout its jurisdiction.[60] Owing to the distances between the stars, such an empire would not necessarily maintain control solely by military force, but may rather tolerate local cultures and institutions to the extent that these would not pose a threat to the central imperial authority.[60] Such tolerance may, as has happened historically on Earth, extend to allowing nominal self-rule of specific regions by existing institutions, while maintaining that area as a puppet or client state to accomplish the aims of the imperial power.[60] However, particularly advanced powers may use methods, including faster-than-light travel, to make centralized administration more effective.[60]

In contrast to the belief that an extraterrestrial civilization would want to establish an empire, Ćirković proposes that an extraterrestrial civilization would maintain equilibrium rather than expand outward.[61] In such an equilibrium, a civilization would only colonize a small number of stars, aiming to maximize efficiency rather than to expand massive and unsustainable imperial structures.[61] This contrasts with the classic Kardashev Type III civilization, which has access to the energy output of an entire galaxy and is not subject to any limits on its future expansion.[61] According to this view, advanced civilizations may not resemble the classic examples in science fiction, but might more closely reflect the small, independent Greek city-states, with an emphasis on cultural rather than territorial growth.[61]

28.2.5 Extraterrestrial artifacts

See also: Bracewell probe

An extraterrestrial civilization may choose to communicate with humanity by means of artifacts or probes rather than by radio, for various reasons. While probes may take a long time to reach the Solar System, once there they would be able to hold a sustained dialogue that would be impossible using radio from hundreds or thousands of light-years away.[62] Radio would be completely unsuitable for surveillance and continued monitoring of a civilization, and should an extraterrestrial civilization wish to perform these activities on humanity, artifacts may be the only option other than to send large, crewed spacecraft to the Solar System.[62]

Although faster-than-light travel has been seriously considered by physicists such as Miguel Alcubierre,[63] Tough speculates that the enormous amount of energy required to achieve such speeds under currently proposed mechanisms means that robotic probes traveling at conventional speeds will still have an advantage for various applications.[62] 2013 research at NASA's Johnson Space Center, however, shows that faster-than-light travel with the Alcubierre drive requires dramatically less energy than previously thought,[64] needing only about 1 metric ton of exotic mass-energy[65] to move a spacecraft at 10 times the speed of light, in contrast to previous estimates that stated that only a Jupiter-mass object would contain sufficient energy to power a faster-than-light spacecraft.[note 1]

According to Tough, an extraterrestrial civilization might want to send various types of information to humanity by means of artifacts, such as an *Encyclopædia Galactica*, containing the wisdom of countless extraterrestrial cultures, or perhaps an invitation to engage in diplomacy with them.[62] A civilization that sees itself on the brink of decline might use the abilities it still possesses to send probes throughout the galaxy, with its cultures, values, religions, sciences, technologies, and laws, so that they may not die along with their civilization.[62]

Freitas finds numerous reasons why interstellar probes may be a preferred method of communication among extraterrestrial civilizations wishing to make contact with Earth. A civilization aiming to learn more about the distribution of life within the galaxy might, he speculates, send probes to a large number of star systems, rather than using radio, as one cannot ensure a response by radio but can (he says) ensure that probes will return to their sender with data on the star systems they survey.[66] Furthermore, probes would enable the surveying of non-intelligent populations, or those not yet capable of space navigation (like humans before the 20th century), as well as intelligent populations that might not wish to provide information about themselves and their planets to extraterrestrial civilizations.[66] In addition, the greater energy required to send living beings rather than a robotic probe would, according to Michaud, be only used for purposes such as a one-way migration.[67]

Freitas points out that probes, unlike the interstellar radio waves commonly targeted by SETI searches, could store information for long, perhaps geological, timescales,[66] and could emit strong radio signals unambiguously recognizable as being of intelligent origin, rather than being dismissed as a UFO or a natural phenomenon.[66] Probes could also modify any signal they send to suit the system they were in, which would be impossible for a radio transmission originating from outside the target star system.[66] Moreover, the use of small robotic probes with widely distributed beacons in individual systems, rather than a small number of powerful, centralized beacons, would provide a security advantage to the civilization using them.[66] Rather than revealing the location of a radio beacon powerful enough to signal the whole galaxy and risk such a powerful device being compromised, decentralized beacons installed on robotic probes need not reveal any information that an extraterrestrial civilization prefers others not to have.[66]

Given the age of the Milky Way galaxy, an ancient extraterrestrial civilization may have existed and sent probes to the Solar System millions or even billions of years before the evolution of *Homo sapiens*.[67] Thus, a probe sent may have been nonfunctional for millions of years before humans learn of its existence.[67] Such a "dead" probe would not pose an imminent threat to humanity, but would prove that interstellar flight is possible.[67] However, if an active probe were to be discovered, humans would react much more strongly than they would to the discovery of a probe that has long since ceased to function.[67]

28.3   Further implications of contact

28.3.1   Theological

See also: Exotheology

The confirmation of extraterrestrial intelligence could have a profound impact on religious doctrines, potentially causing theologians to reinterpret scriptures to accommodate the new discoveries.[68] However, a survey of people with many different religious beliefs indicated that their faith would not be affected by the discovery of extraterrestrial intelligence,[68] and another study, conducted by Ted Peters of the Pacific Lutheran Theological Seminary, shows that most people would not consider their religious beliefs superseded by it.[69] Surveys of religious leaders indicate that only a small percentage are concerned that the existence of extraterrestrial intelligence might fundamentally contradict the views of the adherents of their religion.[70] Gabriel Funes, the chief astronomer of the Vatican Observatory and a papal adviser on science, has stated that the Catholic Church would be likely to welcome extraterrestrial visitors warmly.[71]

Contact with extraterrestrial intelligence would not be completely inconsequential for religion. The Peters study showed that most non-religious people, and a significant majority of religious people, believe that the world could face a religious crisis, even if their own beliefs were unaffected.[69] Contact with extraterrestrial intelligence would be most likely to cause a problem for western religions, in particular traditionalist Christianity, because of the geocentric nature of western faiths.[72] The discovery of extraterrestrial life would not contradict basic conceptions of God, however, and seeing that science has challenged established dogma in the past, for example with the theory of evolution and the teachings of Giordano Bruno, it is likely that existing religions will adapt similarly to the new circumstances.[73] In the view of Musso, however, a global religious crisis would be unlikely even for Abrahamic faiths, as the studies of himself and others on Christianity, the most "anthropocentric" religion, see no conflict between that religion and the existence of extraterrestrial intelligence.[40] In addition, the cultural and religious values of extraterrestrial species would likely be shared over centuries if contact is to occur by radio, meaning that rather than causing a huge shock to humanity, such information would be viewed much as archaeologists and historians view ancient artifacts and texts.[40]

Funes speculates that a decipherable message from extraterrestrial intelligence could initiate an interstellar exchange of knowledge in various disciplines, including whatever religions an extraterrestrial civilization may host.[74] Billingham further suggests that an extremely advanced and friendly extraterrestrial civilization might put an end to present-day religious conflicts and lead to greater religious toleration worldwide.[75] On the other hand, Jill Tarter puts forward the view that contact with extraterrestrial intelligence might eliminate religion as we know it and introduce humanity to an all-encompassing faith.[2]

28.3.2   Political

Tim Folger speculates that news of radio contact with an extraterrestrial civilization would prove impossible to suppress and would travel rapidly.[56] though Cold War scientific literature on the subject contradicts this.[36] Media coverage of the discovery would probably die down quickly, though, as scientists began to decipher the message and learn its true impact.[56] Different branches of government (for example legislative, executive, and judiciary) may pursue their own policies, potentially giving rise to power struggles.[76] Even in the event of a single contact with no follow-up, radio contact may prompt fierce disagreements as to which bodies have the authority to represent humanity as a whole.[39] Michaud hypothesizes that the fear arising from direct contact may cause nation-states to put aside their conflicts and work together for the common defense of humanity.[77]

Apart from the question of who would represent the Earth as a whole, contact could create other international problems, such as the degree of involvement of governments foreign to the one whose radio astronomers received the signal.[78] The United Nations discussed various issues of foreign relations immediately before the launch of the Voyager probes,[79] which in 2012 left the Solar System carrying a golden record in case they are found by extraterrestrial intelligence.[80] Among the issues discussed were what messages would best represent humanity, what format they should take, how to convey the cultural history of the Earth, and what international groups should be formed to study extraterrestrial intelligence in greater detail.[79]

According to Luca Codignola of the University of Genoa, contact with a powerful extraterrestrial civilization is comparable to occasions where one powerful civilization destroyed another, such as the arrival of Christopher Columbus and Hernán Cortés into the Americas and the subsequent destruction of the indigenous civilizations and their ways of life.[2] However, the applicability of such a model to contact with extraterrestrial civilizations, and that specific interpretation of the arrival of the European colonists to the Americas, have been disputed.[81] Even so, any large difference between the power of an extraterrestrial civilization and our own could be demoralizing and potentially cause or accelerate the collapse of human society.[39] Being discovered by a "superior" extraterrestrial civilization, and continued contact with it, might have psychological effects that could destroy a civilization, as is claimed to have happened in the past on Earth.[21]

Even in the absence of close contact between humanity and extraterrestrials, high-information messages from an extraterrestrial civilization to humanity have the potential to cause a great cultural shock.[58] Sociologist Donald Tarter has conjectured that knowledge of extraterrestrial culture and theology has the potential to compromise human allegiance to existing organizational structures and institutions.[58] The cultural shock of meeting an extraterrestrial civilization may be spread over decades or even centuries if an extraterrestrial message to humanity is extremely difficult to decipher.[58]

28.3.3 Legal

See also: Metalaw

Contact with extraterrestrial civilizations would raise legal questions, such as the rights of the extraterrestrial beings. An extraterrestrial arriving on Earth would only have the protection of animal cruelty statutes.[82] Much as various classes of human being, such as women, children, and indigenous people, were initially denied human rights, so might extraterrestrial beings, who could therefore be legally owned and killed.[83] If such a species were not to be treated as a legal animal, there would arise the challenge of defining the boundary between a legal person and a legal animal, considering the numerous factors that constitute intelligence.[84]

Freitas considers that even if an extraterrestrial being were to be afforded legal personhood, problems of nationality and immigration would arise. An extraterrestrial being would not have a legally recognized earthly citizenship, and drastic legal measures might be required in order to account for the technically illegal immigration of extraterrestrial individuals.[85]

If contact were to take place through electromagnetic signals, these issues would not arise. Rather, issues relating to patent and copyright law regarding who, if anyone, has rights to the information from the extraterrestrial civilization would be the primary legal problem.[82]

28.3.4 Scientific and technological

The scientific and technological impact of extraterrestrial contact through electromagnetic waves would probably be quite small, especially at first.[86] However, if the message contains a large amount of information, deciphering it could give humans access to a galactic heritage perhaps predating the human species itself, which may greatly advance our technology and science.[86] A possible negative effect could be to demoralize research scientists as they come to know that what they are researching may already be known to another civilization.[86]

On the other hand, extraterrestrial civilizations with malicious intent could send information that could enable human civilization to destroy itself,[86] such as powerful computer viruses or information on how to make extremely potent weapons that humans would not yet be able to use responsibly.[39] While the motives for such an action are unknown, it would require minimal energy use on the part of the extraterrestrials.[86] According to Musso, however, computer viruses in particular will be nearly impossible unless extraterrestrials possess detailed knowledge of human computer architectures, which would only happen if a human message sent to the stars were protected with little thought to security.[40] Even a virtual machine on which extraterrestrials could run computer programs could be designed specifically for the purpose, bearing little relation to computer systems commonly used on Earth.[40] In addition, humans could send messages to extraterrestrials detailing that they do not want access to the *Encyclopædia Galactica* until they have reached a suitable level of technological advancement, thus mitigating harmful impacts of extraterrestrial technology.[40]

Extraterrestrial technology could have profound impacts on the nature of human culture and civilization. Just as television provided a new outlet for a wide variety of political, religious, and social groups, and as the printing press made the Bible available to the common people of Europe, allowing them to interpret it for themselves, so an extraterrestrial technology might change humanity in ways not immediately apparent.[87] Harrison speculates that a knowledge of extraterrestrial technologies could increase the gap between scientific and cultural progress, leading to societal shock and an inability to compensate for negative effects of technology.[87] He gives the example of improvements in agricultural technology during the Industrial Revolution, which displaced thousands of farm laborers until society could retrain them for jobs suited to the new social order.[87] Contact with an extraterrestrial civilization far more advanced than humanity could cause a much greater shock than the Industrial Revolution, or anything previously experienced by humanity.[87]

Michaud suggests that humanity could be impacted by an influx of extraterrestrial science and technology in the same way that medieval European scholars were impacted by the knowledge of Arab scientists.[88] Humanity might at first revere the knowledge as having the potential to advance the human species, and might even feel inferior to the extraterrestrial species, but would gradually grow in arrogance as it gained more and more intimate knowledge of the science, technology, and other cultural developments of an advanced extraterrestrial civilization.[88]

The discovery of extraterrestrial intelligence would have various impacts on biology and astrobiology. The discovery of extraterrestrial life in any form, intelligent or non-intelligent, would give humanity greater insight into the nature of life on Earth and would improve the conception of how the tree of life is organized.[89] Human biologists could learn about extraterrestrial biochemistry and observe how it differs from that found on Earth.[89] This knowledge could help human civilization to learn which aspects of life are common throughout the universe and which are specific to Earth.[89]

28.3.5   Ecological and biological-warfare impacts

See also: Planetary protection

An extraterrestrial civilization might bring to Earth pathogens or invasive life forms that do not harm its own biosphere.[39] Alien pathogens could decimate the human population, which would have no immunity to them, or they might use terrestrial livestock or plants as hosts, causing indirect harm to humans.[39] Invasive organisms brought by extraterrestrial civilizations could cause great ecological harm because of the terrestrial biosphere's lack of defenses against them.[39]

On the other hand, pathogens and invasive species of extraterrestrial origin might differ enough from terrestrial organisms in their biology to have no adverse effects.[39] Furthermore, pathogens and parasites on Earth are generally suited to only a small and exclusive set of environments,[90] to which extraterrestrial pathogens would have had no opportunity to adapt.

If an extraterrestrial civilization bearing malice towards humanity gained sufficient knowledge of terrestrial biology and weaknesses in the immune systems of terrestrial biota, it might be able to create extremely potent biological weapons.[39] Even a civilization without malicious intent could inadvertently cause harm to humanity by not taking account of all the risks of their actions.[39]

According to Baum, even if an extraterrestrial civilization were to communicate using electromagnetic signals alone, it could send humanity information with which humans themselves could create lethal biological weapons.[39]

28.4   References

[1] Harrison, A. A. (2011). "Fear, pandemonium, equanimity and delight: Human responses to extra-terrestrial life". *Philosophical Transactions of the Royal Society A: Mathematical, Physical and Engineering Sciences* **369** (1936): 656. Bibcode:2011RSPTA. 369..656H. doi:10.1098/rsta.2010.0229. Archived from the original(PDF)on 24 December 2012. Retrieved 5 April 2012.

[2] Kazan, Casey (1 August 2008). "The Impact of ET Contact: Europe's Scientists Discuss The Future of Humans in Space". Daily Galaxy. Retrieved 21 April 2012.

[3] Kaku, Michio (2009). "Extraterrestrials and UFOs". *Physics of the Impossible: A Scientific Exploration into the World of Phasers, Force Fields, Teleportation, and Time Travel.* Knopf Doubleday Publishing Group. pp. 126–153. ISBN 0-307-27882-4.

[4] Terdiman, Daniel (12 December 2008). "SETI's large-scale telescope scans the skies". *CNET News.* Archived from the original on 24 December 2012. Retrieved 27 March 2012.

[5] "Center for SETI Research". *SETI Institute website.* SETI Institute. Archived from the original on 24 December 2012. Retrieved 31 May 2012.

[6] "Europe goes searching for rocky planets" (Press release). ESA. 26 October 2006. Archived from the original on 2006. Retrieved 26 March 2012.

[7] BBC Staff (7 March 2009). "Nasa launches Earth hunter probe". *BBC News*. Archived from the original on 24 December 2012. Retrieved 27 March 2012.

[8] "Kepler: A Search for Habitable Planets". *kepler.nasa.gov*. Archived from the original on 24 December 2012. Retrieved 7 July 2012.

[9] Klotz, Irene (5 December 2011). "Alien Planet Could Host Life". *Discovery News*. Archived from the original on 24 December 2012. Retrieved 27 March 2012.

[10] Ian O'Neill (5 December 2011). "SETI to Hunt for Aliens on Kepler's Worlds". Discovery News. Archived from the original on 24 December 2012.

[11] Mack, Eric (7 December 2011). "Kepler 22-b a top target in restarted SETI alien search". *CNET News Crave*. CNET. Archived from the original on 24 December 2012. Retrieved 27 March 2012.

[12] Cooper, Keith (3 May 2010). "SETI: Cosmic Call". *Astronomy Now*. Archived from the original on 24 December 2012. Retrieved 27 March 2012.

[13] Moskowitz, Clara (29 February 2012). "New Site Lets you Search for Extraterrestrial Life". *Space.com*. Space.com. Archived from the original on 24 December 2012. Retrieved 27 March 2012.

[14] "The Search for Extra Terrestrial Intelligence at Berkeley". University of California at Berkeley. Archived from the original on 24 December 2012. Retrieved 5 April 2012.

[15] "DYSON/IR Excess". *The Search for Extra Terrestrial Intelligence at UC Berkeley*. University of California, Berkeley. Archived from the original on 24 December 2012. Retrieved 7 July 2012.

[16] Krulwich, Robert (28 May 2010). "Aliens Found In Ohio! The 'Wow!' Signal". *Krulwich Wonders*. National Public Radio. Archived from the original on 24 December 2012. Retrieved 31 May 2012.

[17] Almár, Iván; Tarter, Jill (2011). "The discovery of ETI as a high-consequence, low-probability event". *Acta Astronautica* **68** (3–4): 358. Bibcode:2011AcAau..68..358A. doi:10.1016/j.actaastro.2009.07.007. Archived from the original on 24 December 2012.(subscription required)

[18] Almár, Iván (1995) [1993]. Seth Shostak, ed. *The Consequences of a Discovery: Different Scenarios. Progress in the Search for Extraterrestrial Life*. Astronomical Society of the Pacific Conference Series. Astronomical Society of the Pacific. Bibcode:1995ASPC...74..499A.ISBN0-937707-93-7. Archived fromthe original on24December2012.

[19] Almár, Iván (2011). "SETI and astrobiology: The Rio Scale and the London Scale". *Acta Astronautica* **69** (9–10): 899–904. Bibcode:2011AcAau..69..899A. doi:10.1016/j.actaastro.2011.05.036.(subscription required)

[20] Dick, S. (1995). *Consequences of Success in SETI: Lessons from the History of Science*. A New Era in Bioastronomy. Astronomical Society of the Pacific Conference Series. pp. 521–532. Bibcode:1995ASPC...74..521D.

[21] Schetsche, Michael (1 July 2005) [7 January 2005]. "SETI (Search for Extraterrestrial Intelligence) and the Consequences: Futurological Reflections on the Confrontation of Mankind with an Extraterrestrial Civilization" (PDF). Astrosociology.com. Retrieved 20 May 2012.

[22] Dominik, Martin & John C. Zarnecki (2011). "The detection of extra-terrestrial life and the consequences for science and society". *Philosophical Transactions of the Royal Society A* **369** (1936): 499–507. Bibcode:2011RSPTA.369..499D. doi:10.1098/rsta.2010.0236.(audio supplement)

[23] Vakoch, D.A & Y. S. Lee (2000). "Reactions to Receipt of a Message from Extraterrestrial Intelligence: A Cross-Cultural Empirical Study". *Acta Astronautica* **46** (10–12): 737–744. Bibcode:2000AcAau..46..737V. doi:10.1016/S0094-5765(00)00041-2.(subscription required)

[24] Billingham, John (August 1991). "SETI Post-Detection Protocols: What Do You Do After Detecting a Signal?". In Shostak, Seth. *ASP Conference Series*. Third Decennial US-USSR Conference on SETI. University of California, Santa Cruz: Astronomical Society of the Pacific. pp. 417–426. Archived from the original on 24 December 2012. Retrieved 30 June 2012.

[25] Norris, Ray (2002). "Bioastronomy 2002: Life Among the Stars". In Norris, R; F. Stoolman. *Proceedings of the IAU*. Bioastronomy 2002: Life Among the Stars. International Astronomical Union. Bibcode:2004IAUS..213..493N. Archived from the original on 24 December 2012. Retrieved 2 July 2012.

[26] SETI Permanent Committee. International Academy of Astronautics. "Protocols for an ETI Signal Detection". *www.seti.org*. SETI Institute. Retrieved 2 July 2012.

[27] Permanent SETI Committee. International Academy of Astronautics (17 August 1997). "Declaration of Principles for Activities Following the Detection of Extraterrestrial Intelligence". *setileague.org*. The SETI League, Inc. Archived from the original on 24 December 2012. Retrieved 2 July 2012.

[28] Michaud, Michael A. G. (March–April 1992). "An international agreement concerning the detection of extraterrestrial intelligence". *Acta Astronautica* **26** (3–4): 291–294. Bibcode:1992AcAau..26..291M. doi:10.1016/0094-5765(92)90114-X. Archived from the original on 24 December 2012.(subscription required)

[29] Zasky, Jason. "If ET Calls, Who Answers?". *Failure Magazine*. Failure Magazine LLC. Archived from the original on 24 December 2012. Retrieved 2 July 2012.

[30] Deardorff, James W. (1986). "Possible Extraterrestrial Strategy for Earth". *Quarterly Journal of the Royal Astronomical Society* **27**: 94. Bibcode:1986QJRAS..27...94D.

[31] Baum,S.D. (2010). "Universalist ethics in extraterrestrial encounter".*Acta Astronautica***66**(3–4): 617–201. doi:10.1016/j.ac required)

[32] Dick, Steven (2000). "Extraterrestrials and Objective Knowledge". In Tough, Allen. *When SETI Succeeds: The Impact of High-Information Contact*. pp. 47–48. Archived from the original (PDF) on 24 December 2012.

[33] Chaisson, Eric J. (2000). "Null or Negative Effects of ETI Contact in the Next Millennium". In Tough, Allen. *When SETI Succeeds: The Impact of High-Information Contact*. p. 59. Archived from the original (PDF) on 24 December 2012.

[34] Michael, Donald N.; et al. "Proposed Studies on the Implications of Peaceful Space Activities for Human Affairs". pp. 182–184. Archived from the original (PDF) on 24 December 2012. Retrieved 19 May 2012.

[35] Finney, Ben (1990). "The impact of contact". *Acta Astronautica* **21** (2): 117. Bibcode:1990AcAau..21..117F. doi:10.1016/0094-5765(90)90137-A.(subscription required)

[36] "UC Davis Psychology, Albert Harrison". Archived from the original on 24 December 2012. Retrieved 13 July 2012.

[37] Harrison, Albert & Steven Dick (July 2000). "Contact: Long-Term Implications for Humanity". In Tough, Allen. *When SETI Succeeds: The Impact of High-Information Contact*. pp. 7–29. Archived from the original (PDF) on 24 December 2012.

[38] Hines, David (July 2000). "The Role of Artists in Post-Contact Self-Identity". In Tough, Allen. *When SETI Succeeds: The Impact of High-Information Contact*. pp. 55–56. Archived from the original (PDF) on 24 December 2012.

[39] Baum, Seth D.; Haqq-Misra, Jacob D.; Domagal-Goldman, Shawn D. (2011). "Would contact with extraterrestrials benefit or harm humanity? A scenario analysis". *Acta Astronautica* **68** (11–12): 2114–2129. arXiv:1104.4462v2. Bibcode:2011AcAau..68.2114B.doi:10.1016/j.actaastro.2010.10.012.(subscription required)

[40] Musso, Paolo (September–October 2012). "The problem of active SETI: An overview".*Acta Astronautica***78**: 43–54. Bibcode: doi:10.1016/j.actaastro.2011.12.019. Archived from the original on 24 December 2012. Retrieved 8 December 2012.(subscription required)

[41] Tough, Allen (1986). "What Role will Extraterrestrials Play in Humanity's Future?". *Journal of the British Interplanetary Society* **39**: 491–498. Bibcode:1986JBIS...39..491T. Archived from the original (PDF) on 24 December 2012.

[42] Lemarchand, Guillermo A. (2008). "Counting on Beauty: The role of aesthetic, ethical, and physical universal principles for interstellar communication" **0807**: 4518. arXiv:0807.4518. Bibcode:2008arXiv0807.4518L.

[43] Tough, Allen (July 2000). "An Extraordinary Event". In Tough, Allen. *When SETI Succeeds: The Impact of High-Information Contact*. pp. 1–6. Archived from the original (PDF) on 24 December 2012.

[44] Michaud, Michael A. G. (2007). "Reformulating the Problem: Explanations Common to Both". *Contact with Alien Civilizations: Our Hopes and Fears about Encountering Extraterrestrials*. New York, New York, United States: Copernicus Books. pp. 181–184. ISBN 978-0-387-28598-6. Archived from the original (PDF) on 24 December 2012.

[45] Boucher, Geoff (13 March 2012). "'Alien Encounters': A few sage (and Sagan) thoughts on invasion". *Los Angeles Times Hero Complex*. Los Angeles Times. Archived from the original on 24 December 2012. Retrieved 28 March 2012.

[46] Freitas, Robert (1978). "Interstellar War". *Xenology: An Introduction to the Scientific Study of Extraterrestrial Life, Intelligence, and Civilization*. Xenology Research Institute. Archived from the original on 24 December 2012.

[47] Harrison, Albert (July 2000). "Networking with Our Galactic Neighbors". In Tough, Allen. *When SETI Succeeds: The Impact of High-Information Contact*. pp. 107–114. Archived from the original (PDF) on 24 December 2012.

[48] "Space Alien Encounter Scenario Has Scientists Saying How We Will React". *HuffPost Science*. Huffington Post. 30 March 2012. Archived from the original on 24 December 2012. Retrieved 30 March 2012.

[49] Chow, Denise (17 May 2012). "When Aliens Attack: 'Battleship' Strategy with SETI Astronomer Seth Shostak". *Search for Life*. Space.com. Archived from the original on 24 December 2012. Retrieved 19 May 2012.

[50] Bhathal, Ragbir (July 2000). "Human Analogues May Portend ET Conduct Toward Humanity". In Tough, Allen. *When SETI Succeeds: The Impact of High-Information Contact*. p. 57. Archived from the original (PDF) on 24 December 2012.

[51] Brin, David (2009). "The Dangers of First Contact: The Moral Nature of Extraterrestrial Intelligence and a Contrarian Perspective on Altruism" (PDF). *Skeptic Magazine* **15** (3): 2–9.

[52] Hoberman, J. (2 November 2008). "The Cold War Sci-Fi Parable That Fell to Earth". *New York Times Movies*. The New York Times. Archived from the original on 24 December 2012. Retrieved 28 March 2012.

[53] Freitas Jr., Robert A. (2008) [1975–1979]. "Encounters Between Equals: The 0/0 Contact". *Xenology: An Introduction to the Scientific Study of Extraterrestrial Life, Intelligence, and Civilization*. Sacramento, California, United States: Xenology Research Institute. Archived from the original on 24 December 2012.

[54] Freitas Jr., Robert A. (2008) [1975–1979]. "Gods and Primitives: The 11/0 Contact". *Xenology: An Introduction to the Scientific Study of Extraterrestrial Life, Intelligence, and Civilization*. Sacramento, California, United States: Xenology Research Institute. Archived from the original on 24 December 2012.

[55] Billingham, John (2000). "Astronomical Society of the Pacific Conference Series". In Lemarchand, G.; Meech, K. *Summary of Results of the Seminar on the Cultural Impact of Extraterrestrial Contact*. A New Era in Bioastronomy. Astronomical Society of the Pacific. pp. 667–678. Bibcode:2000ASPC..213..667B. Retrieved 6 May 2012.

[56] Folger, Tim (3 January 2011). "Contact: The Day After". *Scientific American*. Nature Publishing Group. pp. 40–45. Archived from the original (PDF) on 24 December 2012. Retrieved 6 May 2012.

[57] Michaud, Michael A. G. (2007). "Assumptions: After Contact: The Message Will Be Comprehensible". *Contact with Alien Civilizations: Our Hopes and Fears about Encountering Extraterrestrials*. New York, New York, United States: Copernicus Books. pp. 279–282. ISBN 978-0-387-28598-6. Archived from the original (PDF) on 24 December 2012.

[58] Michaud, Michael A. G. (2007). "Fears: Cultural Shock". *Contact with Alien Civilizations: Our Hopes and Fears about Encountering Extraterrestrials*. New York, New York, United States: Copernicus Books. pp. 233–238. ISBN 978-0-387-28598-6. Archived from the original (PDF) on 24 December 2012.

[59] Michaud, Michael A. G. (2007). "Assumptions: After Contact: The Galactic Club Exists". *Contact with Alien Civilizations: Our Hopes and Fears about Encountering Extraterrestrials*. New York, New York, United States: Copernicus Books. p. 316. ISBN 978-0-387-28598-6. Archived from the original (PDF) on 24 December 2012.

[60] Michaud, Michael A. G. (2007). "Assumptions: After Contact: Interstellar Empires Do Not Exist". *Contact with Alien Civilizations: Our Hopes and Fears about Encountering Extraterrestrials*. New York, New York, United States: Copernicus Books. pp. 317–322. ISBN 978-0-387-28598-6. Archived from the original (PDF) on 24 December 2012.

[61] Ćirković, Milan M. (2008). "Against the Empire". *Journal of the British Interplanetary Society* **61**: 246–254. arXiv:0805.1821. Bibcode:2008JBIS...61..246C.

[62] Tough, Allen (1998). "Small Smart Interstellar Probes" (PDF). *Journal of the British Interplanetary Society* **51**: 167–174.

[63] Alcubierre, Miguel (1994). "The warp drive: hyper-fast travel within general relativity". *Classical and Quantum Gravity* **11** (5): L73–L77. arXiv:gr-qc/0009013. Bibcode:1994CQGra..11L..73A. doi:10.1088/0264-9381/11/5/001.

[64] Moskowitz, Clara (17 September 2012). "Warp Drive May Be More Feasible Than Thought, Scientists Say". *Space.com*. Space.com. Retrieved October 27, 2012.

[65] Jenner, Lynn (23 February 2008). "NASA - Voyager Facts". National Aeronautics and Space Administration. Retrieved October 27, 2012.

[66] Freitas, Robert (November 1983). "The Case for Interstellar Probes". *Journal of the British Interplanetary Society* **36**: 490–495. Bibcode:1983JBIS...36..490F.

[67] Michaud, Michael A. G. (2007). "The Consequences of Contact: Scenarios of Contact: Close to Home". *Contact with Alien Civilizations: Our Hopes and Fears about Encountering Extraterrestrials* (PDF). New York, New York, United States: Copernicus Books. pp. 211–212. ISBN 978-0-387-28598-6.

[68] Choi, Charles Q. (24 January 2011). "Could Extraterrestrial Intelligence Sway Religious Beliefs?". *Space.com*. Space.com. Retrieved 30 March 2012.

[69] Peters, T. (2011). "The implications of the discovery of extra-terrestrial life for religion". *Philosophical Transactions of the Royal Society A: Mathematical, Physical and Engineering Sciences* **369** (1936): 644. Bibcode:2011RSPTA.369..644P. doi:10.1098/rsta.2010.0234.

[70] McAdamis, E.M. (2011). "Astrosociology and the Capacity of Major World Religions to Contextualize the Possibility of Life Beyond Earth". *Physics Procedia* **20**: 338. Bibcode:2011PhPro..20..338M. doi:10.1016/j.phpro.2011.08.031.

[71] Keim, Brandon (13 June 2008). "Christian Theologians Prepare for Extraterrestrial Life". *Wired*. Condé Nast. Retrieved 20 May 2012.

[72] Kaufman, Marc (2012). *First Contact: Scientific Breakthroughs in the Hunt for Life Beyond Earth*. Simon and Schuster. ISBN 978-1-4391-0901-4. External link in |title= (help)

[73] Freitas Jr., Robert A. (2008) [1975–1979]. "The Religiou Response Contact". *Xenology: An Introduction to the Scientific Study of Extraterrestrial Life, Intelligence, and Civilization*. Sacramento, California, United States: Xenology Research Institute.

[74] Lemarchand, Guillermo A. (2000). "Speculations on the First Contact: Encyclopedia Galactica or the Music of the Spheres?" (PDF). In Tough, Allen. *When SETI Succeeds: The Impact of High-Information Contact*. pp. 153–163.

[75] Billingham, John (2000). "Who Said What: A Summary and Eleven Conclusions" (PDF). In Tough, Allen. *When SETI Succeeds: The Impact of High-InformationContact*. pp. 33–39.

[76] Michaud, Michael A. G. (2007). "Annex: Preparing: Preparing Governments for Contact". *Contact with Alien Civilizations: Our Hopes and Fears about Encountering Extraterrestrials* (PDF). New York, New York, United States: Copernicus Books. pp. 366–368. ISBN 978-0-387-28598-6.

[77] Michaud, Michael A. G. (2007). "Assumptions: After Contact: Contact Will Unify Humankind". *Contact with Alien Civilizations: Our Hopes and Fears about Encountering Extraterrestrials* (PDF). New York, New York, United States: Copernicus Books. pp. 292–293. ISBN 978-0-387-28598-6.

[78] Michaud, Michael A. G. (2007). "Mixed Emotions: Political Reactions". *Contact with Alien Civilizations: Our Hopes and Fears about Encountering Extraterrestrials* (PDF). New York, New York, United States: Copernicus Books. p. 253. ISBN 978-0-387-28598-6.

[79] Othman, M. (2011). "Supra-Earth affairs". *Philosophical Transactions of the Royal Society A: Mathematical, Physical and Engineering Sciences* **369** (1936): 693. Bibcode:2011RSPTA.369..693O. doi:10.1098/rsta.2010.0311.

[80] NASA Jet Propulsion Laboratory. "Voyager – The Interstellar Mission". Retrieved 12 May 2012.

[81] Mann, Adam (4 April 2012). "Q&A: The Anthropology of Searching for Aliens". *Wired Science*. Wired. Retrieved 21 April 2012.

[82] Freitas, Robert (1978). "Legal Issues of First Contact". *Xenology: An Introduction to the Scientific Study of Extraterrestrial Life, Intelligence, and Civilization*. Xenology Research Institute.

[83] Freitas, Robert (1978). "Alien Animals". *Xenology: An Introduction to the Scientific Study of Extraterrestrial Life, Intelligence, and Civilization*. Xenology Research Institute.

[84] Freitas, Robert (1978). "Legal Standards of Personhood". *Xenology: An Introduction to the Scientific Study of Extraterrestrial Life, Intelligence, and Civilization*. Xenology Research Institute.

[85] Freitas, Robert (1978). "Extraterrestrial Persons". *Xenology: An Introduction to the Scientific Study of Extraterrestrial Life, Intelligence, and Civilization*. Xenology Research Institute.

[86] Freitas, Robert (1978). "Impact on Science and Technology". *Xenology: An Introduction to the Scientific Study of Extraterrestrial Life, Intelligence, and Civilization*. Xenology Research Institute.

[87] Harrison, Albert A. (2002). *After Contact: The Human Response To Extraterrestrial Life*. Basic Books. ISBN 978-0-7382-0846-6.

[88] Michaud, Michael A. G. (2007). "Fears: The End of Hubris". *Contact with Alien Civilizations: Our Hopes and Fears about Encountering Extraterrestrials* (PDF). New York, New York, United States: Copernicus Books. pp. 232–233. ISBN 978-0-387-28598-6.

[89] McKay, C. P. (2011). "The search for life in our Solar System and the implications for science and society". *Philosophical Transactions of the Royal Society A: Mathematical, Physical and Engineering Sciences* **369**(1936): 594. Bibcode:2011RSPTA.369..594M. doi:10.1098/rsta.2010.0247.

[90] Brant, Sara V.; Loker, Eric S. (2005). "Can Specialized Pathogens Colonize Distantly Related Hosts? Schistosome Evolution as a Case Study". *PLoS Pathogens* **1** (3): 28–31. doi:10.1371/journal.ppat.0010038. PMC 1291355. PMID 16322771.

28.5 Notes

[1] The original article stated that the mass-energy required would be roughly equal to that of the Voyager 1 spacecraft, and the mass of that spacecraft itself is, according to NASA, currently about 733 kg. Therefore, "about one metric ton" was used in the text.

28.6 Further reading

- Steven J. Dick: *The impact of discovering life beyond earth*. Cambridge University Press, Cambridge 2015, ISBN 978-1-107-10998-8.

28.7 External links

- SETI Institute

- Cultural Aspects of SETI

- Introduction to ExtraTerrestrial Intelligence

An advanced, friendly extraterrestrial civilization might force humanity to eliminate risks that could destroy its fledgling civilization.

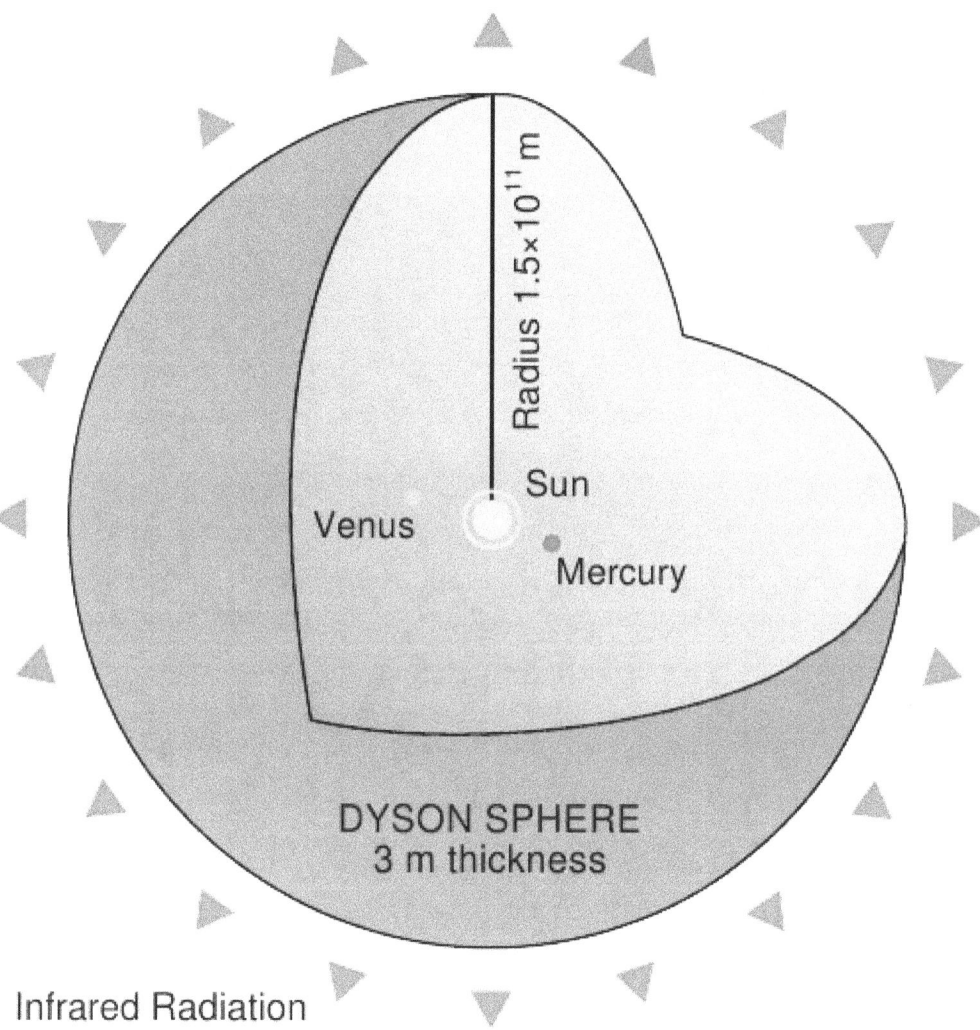

It is suggested that technologically advanced extraterrestrial civilization would probably be ethically advanced as well and would not attempt projects with severe ecological implications for other species, like the construction of a Dyson sphere.

Robotic probes may be preferable to radio waves or microwaves as a means of interstellar communication.

Chapter 29

Aurelia and Blue Moon

Aurelia and **Blue Moon** are hypothetical examples of a planet and a moon on which extraterrestrial life could evolve. They are the outcome of a collaboration between television company Blue Wave Productions Ltd. and a group of American and British scientists who were collectively commissioned by National Geographic. The team used a combination of accretion theory, climatology, and xenobiology to imagine the most likely locations for extraterrestrial life and most probable evolutionary path such life would take.

The beginning concepts appeared in a two-part television broadcast called *Alien Worlds*, aired in 2005 in the UK by Channel 4. Channel 4 has also released a DVD of the program. The show was also aired on the National Geographic Channel as *Extraterrestrial* on Monday, May 30, 2005[1] and focuses more on the alien life on the two worlds.

The first program in the series focused on Aurelia, a hypothetical Earth-sized extrasolar planet orbiting a red dwarf star in our local area of the Milky Way. The second focuses on a moon called Blue Moon, which orbits an enormous gas giant that is itself orbiting a binary star system.

Aurelia and Blue Moon both featured in the touring exhibition The Science of Aliens.

29.1   Reasons for theorizing

Discoveries regarding extrasolar planets were first published in 1989 raising the prospect of whether life (as we know it or imagine it) could be supported on other planets. It is currently believed that for this to happen a planet must orbit in a relatively narrow band around its parent star, where temperatures are suitable for water to exist as a liquid. This region is called the habitable zone.

The most Earth-like exoplanets yet found, Gliese 667 Cc and Gliese 581 g (disputed), have masses larger than Earth's and orbit red dwarf stars in the habitable zone.

The sensitivity of current detection methods makes it difficult for scientists to search for terrestrial planets smaller than this. To allow smaller bodies to be detected, NASA was studying a project called the Terrestrial Planet Finder (TPF), a two-telescope concept slated to begin launching around 2014. However, Congressional spending limits under House Resolution 20 passed on January 31, 2007 by the U.S. House of Representatives and February 14 by the U.S. Senate have all but canceled the program.

Prior to TPF's cancellation, astrophysicists had begun speculating about the best places to point the telescope in order to find Earth-like planets. Whereas life on Earth has formed around a stable yellow dwarf, solar twins are not as common in the galaxy as red dwarf stars (which have a mass of less than one-half that of the Sun and consequently emit less heat), or bigger, brighter blue giants. In addition, it is estimated that more than a quarter of all stars are at least binary systems, with as many as 10% of these systems containing more than two stars (trinary etc.)—unlike our own sun, which has no companion. Therefore, it may be prudent to consider how life might evolve in such environments. Such speculation may still be of use should a future planet-finding telescope be launched, and possibly for NASA's Kepler mission.

29.2 Aurelia

The scientists on the project theorized that aiming the TPF at a red dwarf star might yield the best opportunities for seeing smaller planets. Due to the slow rate at which they burn hydrogen, red dwarfs have an enormous estimated lifespan; allowing plenty of time for life to evolve on surrounding planets. Also, red dwarves are very common in the universe. Therefore, if they support habitable planets, it substantially increases the chances of finding life in the universe. However, being much dimmer than other stars, it will be harder to detect planetary systems around them. In addition, lower gravity would limit the potential size of a system. The discovery of Gliese 581 g raises hopes of finding more red dwarf systems, including potentially habitable ones.

29.2.1 Tidal lock

However, the dwarf's smaller nature and fainter heat/light output would mean that such a planet would need to be particularly close to the star's surface. The cost of such an orbit would be that an Earth-sized body would become tidally locked. When this happens, the object presents the same face to its parent at all times as it orbits, just as the Moon does with the Earth (more technically, one sidereal day is exactly equal to one year for the orbiting body).

29.2.2 Traditional theories

Traditional scientific theories proposed that such a tidally locked planet might be incapable of holding on to an atmosphere. Having such a slow rotation would weaken the magnetic effect that protects the atmosphere from being blown away by solar wind (see Rare Earth hypothesis).

29.2.3 Traditional assumptions tested

Nonetheless, the scientists employed by the programme decided to test the traditional assumptions for such a planet and start a model out for it from a protoplanetary disk through to its eventual death. Their estimations suggested such a planet could indeed hold on to its atmosphere, although with freakishly unusual results by Earth standards. Aurelia would be gravitationally locked to its star (a red dwarf). Due to this, Aurelia would not have seasons or a day/night cycle,[1] as half of Aurelia would be in perpetual darkness and would be in a permanent ice age. The other half would contain a giant, unending hurricane with permanent torrential rain at the point directly opposite the local star. In between these two zones would be a place suitable for life.

This hurricane could perhaps generate enormous waves in the ocean and the waves would migrate outwards. Oceanographers should test how high these waves would be in the postulated nearby swamps and delta area. They would be wind driven waves and would not reach from the top of an ocean to the bottom like a tsunami. Nonetheless, waves as big and as devastating as those that humans call freak waves might be regular.
Simple bacterial and algal life would not be threatened.

29.2.4 Continued theories

The theorizations continued, and assuming that there was land in this habitable zone, it would be likely to form large networks of river deltas and swampland, due to rain runoff from the nearby storm.

At the far end of assumptions about Aurelia were attempts to construct lifeforms based on Earthly evolutionary models and how ecosystems might develop. The scientists' assumptions included the idea that the long life of a red dwarf allows for evolution to fine-tune any ecosystem on the planet. The scientists involved in the project hypothesized that the vast majority, if not all, of extra-solar biology will be carbon-based.

This assumption is often referred to by critics as carbon chauvinism, as it may be possible for life to form that is not based on carbon.

From this carbon-based hypothesis the scientific team assumed some form of staple photosynthesizing animal/plant combination would be the principal autotroph. They decided upon a plant-like creature called a Stinger Fan. It has five hearts and limited mobility. Its fan-like leaves trap the red dwarf star's energy to produce sugars. Its hearts pump them around its body.[2]

Feeding upon the Stinger Fans are six-legged semi-amphibious beaver like creatures called Mudpods. They use their long, continually growing thumb claws to cut down a Stinger Fan and dam the river systems, creating artificial lagoons and swamps which provide safety from predators.[3] Upon that animal, a large emu-like animal, the Gulphog, is the main predator. These 2 meter tall carnivores live socially in packs, and display promising signs of intelligence. Finally, there is a second semi-amphibious creature called the Hysteria - a cross between a plague of tadpoles and piranha. These tiny, orange creatures can collect together (in a manner similar to slime molds) and form one huge super-organism, moving together up banks to paralyze and consume other animals. Scabian Slugs that live by the water can fall victim to the Hysteria, but it can take something as large as a Gulphog to satisfy them.

The planet's ecosystem suffers from a number of particular peculiarities, most notably evolutionary quirks to allow all living organisms to detect and avoid solar flares. Red dwarf stars are unstable and eject frequent solar flares. Such intense ultraviolet radiation is deadly to all carbon-based life forms as it breaks down the atomic bonds formed by organic compounds. The Gulphogs have adapted by having an ultraviolet light sensitive eye on top of their heads. Stinger Fans fold up to protect themselves. Mudpods have sensitive backs that can sense the ultraviolet rays. The Hysteria's adaption is unknown. However, the flare stage might only be when the red dwarfs are relatively young.

See also

- Habitability of red dwarf systems

- Gliese 581

- HD 85512

- Kepler-22

- Gliese 667C

29.3   Blue Moon

Blue Moon is covered in life-giving water and an atmosphere so dense that enormous creatures hypothetically can take flight. The Blue Moon orbits a Water Cloud Jovian planet (a Jupiter-like planet that is cool enough to have visible rain clouds in its atmosphere) orbiting a close binary star system. The Blue Moon itself is roughly an earth mass but has an air pressure around three times that of Earth's at sea level.

A distinguishing feature of Blue Moon is that it has no polar ice caps: the thick atmosphere keeps temperatures constant across the moon's surface. There is also a greenish haze over the moon from large carpets of floating moss and algae.

The denser atmosphere allows more massive creatures to remain airborne than on Earth. Skywhales, gargantuan whale-like animals which evolved away from the ocean into the air, fill the ecological niche this creates. Because of the increased muscle power from excess atmospheric oxygen, these creatures can have wingspans of ten meters and remain airborne their entire lives. They feed on the previously mentioned Air Moss. They evolved from seagoing animals into flying ones in one evolutionary leap.

High levels of oxygen (30% of the atmosphere) push the atmosphere to the brink of spontaneous combustion during lightning storms. Carbon dioxide levels are thirty times higher than on Earth making the air clammy and warm. Like our moon, Blue Moon is tidally locked, meaning it keeps the same side of the moon faced towards its planet.

With an orbital time of roughly ten days, that means five days of continuous night and five days of continuous daytime. The long days and nights also create strong cross-hemisphere winds that help keep the Skywhales afloat, in addition to the density of the atmosphere and its increased oxygen concentration compared to Earth.

Skywhales are prey to the insect-like caped Stalkers, colony-living predators that have several different tasks. Scouts find skywhales and mark them with a special scent, then return to the nest to spread the word. Workers then swarm out in huge numbers, detecting the whale and working together to bring them down from the sky and kill them. Finally, there is

A hypothetical rendition of The Blue Moon orbiting high above the plane of its parent gas giant planet.

a queen, who stays in the nest and constantly lays eggs that become new stalkers. This lifestyle is based on earth's hornets. The Stalkers are also prey, for the Pagoda branches are draped with the lethal webs of the plant-like ghost traps. Once a Stalker is caught in a ghost trap web, the carnivore uses its tentacles to lift its catch up into its mouth, to be digested by the acid in a primitive stomach.

As well as Skywhales, giant Kites also fly above the forest canopy. These parasol-like grazers can grow up to 5 m (16 ft) in diameter and still stay airborne. Their tethers help control their floating, while their jellyfish-like tentacles snatch Helibug larvæ from the water-filled skyponds. Helibugs have a trilaterally symmetrical body plan, with three eyes, three wings, three legs, three mouth parts and three tongues.

70% of Blue Moon's land mass is coated in two main plant types, pagoda trees and balloon plants. Pagoda trees interconnect with each other to allow them to grow 700 ft (210 m) tall. Their hollow leaves collect rainwater, since the trees are too tall to draw it from the ground. Balloon plants release their seeds by filling them with hydrogen to float in the dense atmosphere, in a way similar to kelp on Earth.

The Blue Moon is threatened by mass wildfires that can wipe out entire pagoda forests. Balloon plants grow in the gaps resulting. The floating balloons released by the plants are full of explosive hydrogen, and when a fire hits, they explode like bombs, releasing seeds flying through the air. Skywhales and Kites will gain altitude until the fire ends. The Stalkers' escape strategy is unknown.

29.3.1   See also

- 55 Cancri

- HD 28185

- HD 10180

29.4   See also

General

- *Alien Planet* (docufiction)

29.5   References

[1] Lovgran, Stefan (3 June 2005). "Flying Whales, Other Aliens Theorized by Scientists". *National Geographic News*. Retrieved 16 October 2011.

[2] Bell, Richard (25 November 2005). "Stinger Fans on Alien Worlds". *Richard Bell's Wild West Yorkshire Nature Diary*. Archived from the original on 19 May 2007. Retrieved 29 April 2007.

[3] Cooper, Sean (February 2006). "Alien Animal Planet". *Wired* **14** (2). Retrieved 30 March 2012.

29.6   External links

- "Extraterrestrial" from National Geographic Channel

- "Extraterrestrial" on YouTube (full program, official upload)

Chapter 30

Kardashev scale

This article is about a measuring method. For the album by Greydon Square, see The Kardashev Scale (album).

The **Kardashev scale** is a method of measuring a civilization's level of technological advancement, based on the amount of energy a civilization is able to utilize.[1] The scale has three designated categories called *Type I, II*, and *III*. A Type I civilization uses only resources available on its home planet, Type II harnesses all needed energy from its local star, and Type III of its galaxy.[2] The scale is hypothetical, and regards energy consumption on a cosmic scale. It was first proposed in 1964 by the Soviet astronomer Nikolai Kardashev. Various extensions of the scale have been proposed since, from a wider range of power levels (types 0, IV and V) to the use of metrics other than pure power.

30.1 Definition

In 1964, Kardashev defined three levels of civilizations, based on the order of magnitude of power available to them:

Type I

"Technological level close to the level presently attained on earth, with energy consumption at $\approx 4 \times 10^{19}$ erg/sec (4×10^{12} watts)."[1] Guillermo A. Lemarchand stated this as "A level near contemporary terrestrial civilization with an energy capability equivalent to the solar insolation on Earth, between 10^{16} and 10^{17} watts."[3]

Type II

"A civilization capable of harnessing the energy radiated by its own star (for example, the stage of successful construction of a Dyson sphere), "with energy consumption at $\approx 4 \times 10^{33}$ erg/sec."[1] Lemarchand stated this as "A civilization capable of utilizing and channeling the entire radiation output of its star. The energy utilization would then be comparable to the luminosity of our Sun, about 4×10^{33} erg/sec (4×10^{26} watts)."[3]

Type III

"A civilization in possession of energy on the scale of its own galaxy, with energy consumption at $\approx 4 \times 10^{44}$ erg/sec."[1] Lemarchand stated this as "A civilization with access to the power comparable to the luminosity of the entire Milky Way galaxy, about 4×10^{44} erg/sec (4×10^{47} watts)."[3]

30.2   Current status of human civilization

Further information: World energy resources and consumption

Michio Kaku suggested that humans may attain Type I status in 100–200* years, Type II status in a few thousand years, and Type III status in 100,000 to a million years.[4]

Carl Sagan suggested defining intermediate values (not considered in Kardashev's original scale) by interpolating and extrapolating the values given above for types I (10^{16} W), II (10^{26} W) and III (10^{36} W), which would produce the formula

$$K = \frac{\log_{10} P - 6}{10}$$

where value K is a civilization's Kardashev rating and P is the power it uses, in watts. Using this extrapolation, a "Type 0" civilization, not defined by Kardashev, would control about 1 MW of power, and humanity's civilization type as of 1973 was about 0.7 (apparently using 10 terawatt (TW) as the value for 1970s humanity).[5]

In 2012, total world energy consumption was 553 exajoules (553×10^{18} J=153,611 TWh),[6] equivalent to an average power consumption of 17.54 TW (or 0.724 on Sagan's Kardashev scale).

30.3   Observational evidence

In 2015, a study of galactic mid-infrared emissions came to the conclusion that "Kardashev Type-III civilizations are either very rare or do not exist in the local Universe".[7] On October 14, 2015, the realization of a strange pattern of light surrounding star KIC 8462852 has raised speculation that a Dyson Sphere (Type II civilization) may have been discovered.[8][9][10][11][12]

30.4   Energy development

See also: Energy development

30.4.1   Type I civilization methods

- Large-scale application of fusion power. According to mass-energy equivalence, Type I implies the conversion of about 2 kg of matter to energy per second. An equivalent energy release could theoretically be achieved by fusing approximately 280 kg of hydrogen into helium per second,[13] a rate roughly equivalent to 8.9×10^9 kg/year. A cubic km of water contains about 10^{11} kg of hydrogen, and the Earth's oceans contain about 1.3×10^9 cubic km of water, meaning that humans on Earth could sustain this rate of consumption over geological time-scales.

- Antimatter in large quantities would have a mechanism to produce power on a scale several magnitudes above our current level of technology. In antimatter-matter collisions, the entire rest mass of the particles is converted to radiant energy. Their energy density (energy released per mass) is about four orders of magnitude greater than that from using nuclear fission, and about two orders of magnitude greater than the best possible yield from fusion.[14] The reaction of 1 kg of anti-matter with 1 kg of matter would produce 1.8×10^{17} J (180 petajoules) of energy.[15] Although antimatter is sometimes proposed as a source of energy, this is not feasible. Artificially producing antimatter – according to current understanding of the laws of physics – involves first converting energy into mass, so no net gain results. Artificially created antimatter is only usable as a medium of energy storage, not as an energy source, unless future technological developments (contrary to the conservation of the baryon number, such as a CP violation in favour of antimatter) allow the conversion of ordinary matter into anti-matter.

Theoretically, humans may in the future have the capability to cultivate and harvest a number of naturally occurring sources of antimatter.[16][17][18]

- Renewable energy through converting sunlight into electricity — either by using solar cells and concentrating solar power or indirectly through wind and hydroelectric power. There is no known way for human civilization to use the equivalent of the Earth's total absorbed solar energy without completely coating the surface with human-made structures, which is not feasible with current technology. However, if a civilization constructed very large space-based solar power satellites, Type I power levels might become achievable.

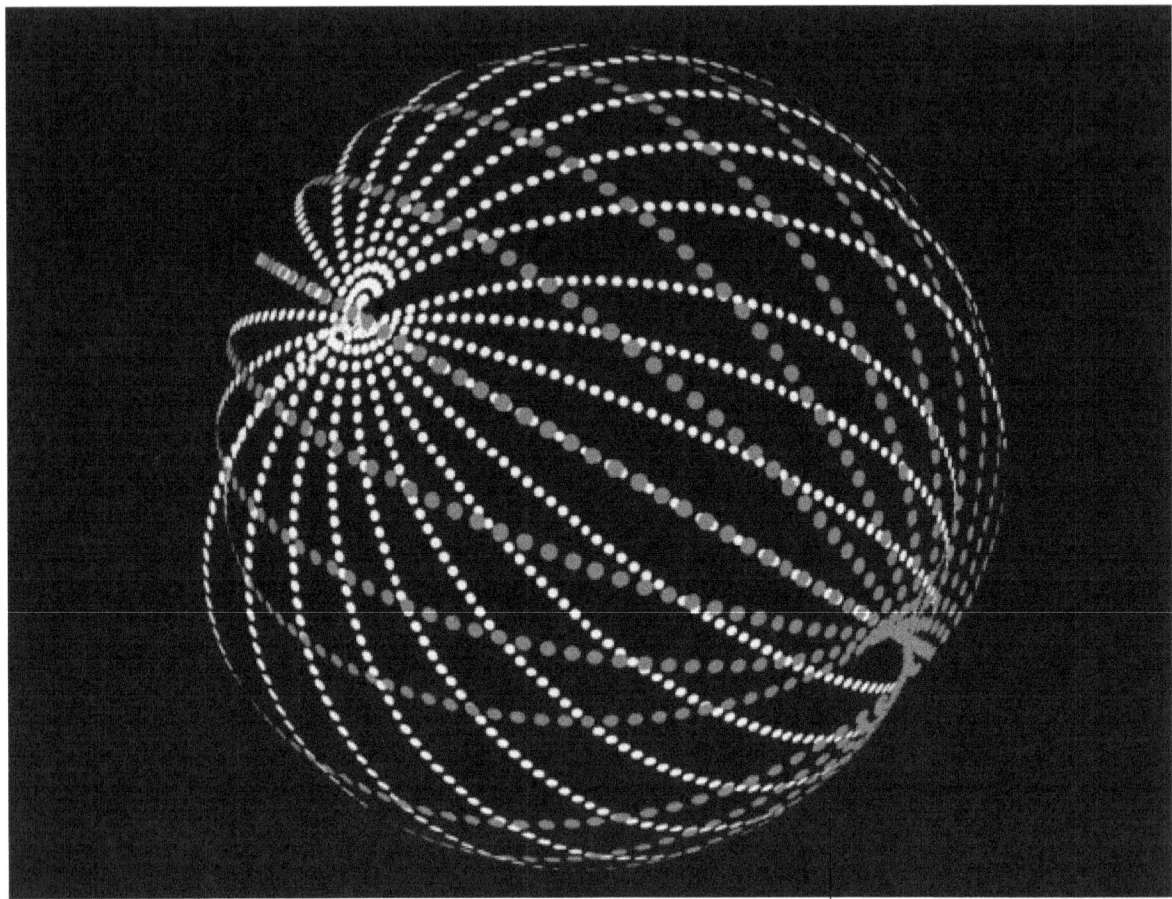

Figure of a Dyson swarm surrounding a star

30.4.2 Type II civilization methods

- Type II civilizations might use the same techniques employed by a Type I civilization, but applied to a large number of planets in a large number of solar systems.

- A Dyson sphere or Dyson swarm and similar constructs are hypothetical megastructures originally described by Freeman Dyson as a system of orbiting solar power satellites meant to enclose a star completely and capture most or all of its energy output.[19]

- Perhaps a more exotic means to generate usable energy would be to feed a stellar mass into a black hole, and collect photons emitted by the accretion disc.[20][21] Less exotic would be simply to capture photons already escaping from the accretion disc, reducing a black hole's angular momentum; known as the Penrose process.

- Star lifting is a process where an advanced civilization could remove a substantial portion of a star's matter in a controlled manner for other uses.

- Antimatter is likely to be produced as an industrial byproduct of a number of megascale engineering processes (such as the aforementioned star lifting) and therefore could be recycled.

- In multiple-star systems of a sufficiently large number of stars, absorbing a small but significant fraction of the output of each individual star.

30.4.3 Type III civilization methods

- Type III civilizations might use the same techniques employed by a Type II civilization, but applied to all possible stars of one or more galaxies individually.[2]

- They may also be able to tap into the energy released from the supermassive black holes which are believed to exist at the center of most galaxies.

- White holes, if they exist, theoretically could provide large amounts of energy from collecting the matter propelling outwards.

- Capturing the energy of gamma-ray bursts is another theoretically possible power source for a highly advanced civilization.

- The emissions from quasars can be readily compared to those of small active galaxies and could provide a massive power source if collectable.

30.5 Civilization implications

There are many historical examples of human civilization undergoing large-scale transitions, such as the Industrial Revolution. The transition between Kardashev scale levels could potentially represent similarly dramatic periods of social upheaval, since they entail surpassing the hard limits of the resources available in a civilization's existing territory. A common speculation[22] suggests that the transition from Type 0 to Type I might carry a strong risk of self-destruction since, in some scenarios, there would no longer be room for further expansion on the civilization's home planet, as in a Malthusian catastrophe. Excessive use of energy without adequate disposal of heat, for example, could plausibly make the planet of a civilization approaching Type I unsuitable to the biology of the dominant life-forms and their food sources. If Earth is an example, then sea temperatures in excess of 35 C would jeopardize marine life and make the cooling of mammals to temperatures suitable for their metabolism difficult if not impossible. Of course, these theoretical speculations may not become problems in reality thanks to evolution or the application of future engineering and technology. Also, by the time a civilization reaches Type I it may have colonized other planets or created O'Neill-type colonies, so that waste heat could be distributed throughout the solar system.

30.6 Extensions to the original scale

Many extensions and modifications to the Kardashev scale have been proposed.

- **Type 4 Kardashev Rating:** The most straightforward extend the scale to even more hypothetical Type IV beings who can control or use the entire universe or Type V who control collections of universes. The power output of the visible universe is within a few orders of magnitude of 10^{45} W. Such a civilization approaches or surpasses the limits of speculation based on current scientific understanding, and may not be possible.

  - Zoltán Galántai has argued that such a civilization could not be detected, as its activities would be indistinguishable from the workings of nature (there being nothing to compare them to).[23]

- **Type 4 Kardashev-Kaku Ratings (Michio Kaku):** In his book *Parallel Worlds*, Michio Kaku has discussed a Type IV civilization that could harness "extragalactic" energy sources such as dark energy.[24]

- **Kardashev Alternate Rating Characteristics:** Other proposed changes to the scale use different metrics such as 'mastery' of systems, amount of information used, or progress in control of the very small as opposed to the very large.

- **Planet Mastery (Robert Zubrin):** Metrics other than pure power usage have also been proposed. One is 'mastery' of a planet, system or galaxy rather than considering energy alone.[25]

- **Information Mastery (Carl Sagan):** Alternatively, Carl Sagan suggested adding another dimension in addition to pure energy usage: the information available to the civilization.

  - He assigned the letter A to represent 10^6 unique bits of information (less than any recorded human culture) and each successive letter to represent an order of magnitude increase, so that a level Z civilization would have 10^{31} bits.

  - In this classification, 1973 Earth is a 0.7 H civilization, with access to 10^{13} bits of information.

  - Sagan believed that no civilization has yet reached level Z, conjecturing that so much unique information would exceed that of all the intelligent species in a galactic supercluster and observing that the universe is not old enough to exchange information effectively over larger distances.

  - The information and energy axes are not strictly interdependent, so that even a level Z civilization would not need to be Kardashev Type III.[5]

- **Microdimensional Mastery (John Barrow):** John D. Barrow, going by the fact that humans have found it more cost-effective to extend any abilities to manipulate their environment over increasingly smaller dimensions rather than increasingly larger ones, reverses the classification downward from Type I-minus to Type Omega-minus:

  - **Type I-minus** is capable of manipulating objects over the scale of themselves: building structures, mining, joining and breaking solids;

  - **Type II-minus** is capable of manipulating genes and altering the development of living things, transplanting or replacing parts of themselves, reading and engineering their genetic code;

  - **Type III-minus** is capable of manipulating molecules and molecular bonds, creating new materials;

  - **Type IV-minus** is capable of manipulating individual atoms, creating nanotechnologies on the atomic scale and creating complex forms of artificial life;

  - **Type V-minus** is capable of manipulating the atomic nucleus and engineering the nucleons that compose it;

  - **Type VI-minus** is capable of manipulating the most elementary particles of matter (quarks and leptons) to create organized complexity among populations of elementary particles; culminating in,

  - **Type Omega-minus** is capable of manipulating the basic structure of space and time.[26]

According to this scale, human civilization is between III- and IV-minus.

- **Civilizational Range (Robert Zubrin):** Robert Zubrin adapts the Kardashev scale to refer to how widespread a civilization is in space, rather than to its energy use.

  - In his definition, a Type I civilization has spread across its planet.

  - A Type II has extensive colonies in its respective stellar system, and

  - A Type III has colonized its galaxy.[25]

30.7 Criticism

It has been argued that, because we cannot understand advanced civilizations, we cannot predict their behavior. Thus, the Kardashev scale may not be relevant or useful for classifying extraterrestrial civilizations. This central argument is found in the book *Evolving the Alien: The Science of Extraterrestrial Life*.[27]

30.8   See also

- Astroengineering

- Clarke's three laws

- Drake equation

- KIC 8462852

- Orders of magnitude (power)

- Orders of magnitude (energy)

- Terraforming

- White's Law

- World energy resources and consumption

30.9   References

[1] Kardashev, Nikolai (1964). "Transmission of Information by Extraterrestrial Civilizations". *Soviet Astronomy* **8**: 217. Bibcode:

[2] Kardashev, Nikolai. "On the Inevitability and the Possible Structures of Supercivilizations". The search for extraterrestrial life: Recent developments; Proceedings of the Symposium, Boston, MA, June 18–21, 1984 (A86-38126 17-88). Dordrecht, D. Reidel Publishing Co., 1985, p. 497–504.

[3] Lemarchand, Guillermo A. "Detectability of Extraterrestrial Technological Activities". Coseti..

[4] Kaku, Michio (2010). "The Physics of Interstellar Travel: To one day, reach the stars.". Retrieved 2010-08-29.

[5] Sagan, Carl (October 2000) [1973]. Jerome Agel, ed. *Cosmic Connection: An Extraterrestrial Perspective*. Freeman J. Dyson, David Morrison. Cambridge Press. ISBN 0-521-78303-8. Retrieved 2008-01-01. I would suggest Type 1.0 as a civilization using 10^{16} watts for interstellar communication; Type 1.1, 10^{17} watts; Type 1.2, 10^{18} watts, and so on. Our present civilization would be classed as something like Type 0.7.

[6] "Total Primary Energy Consumption 2008-2012" (cfm). *Statistical Review of World Energy 2008-2012*. U.S. Energy Information Administration.

[7] Garrett, Michael (2015). "The application of the Mid-IR radio correlation to the S\hat{G}S sample and the search for advanced extraterrestrial civilizations". *Astronomy & Astrophysics* **581**: L5. arXiv:1508.02624. doi:10.1051/0004-6361/201526687. line feed character in |title= at position 72 (help)

[8] "The Most Mysterious Star in Our Galaxy". *The Atlantic*. https://plus.google.com/109258622984321091629. Retrieved 2015-10-15. External link in |publisher= (help)

[9] Kaplan, Sarah (2015-10-15). "The strange star that has serious scientists talking about an alien megastructure". *The Washington Post*. ISSN 0190-8286. Retrieved 2015-10-15.

[10] "Citizen scientists catch cloud of comets orbiting distant star". *New Scientist*. Retrieved 2015-10-15.

[11] Plait, Phil (2015-10-14). "Did Astronomers Find Evidence of an Alien Civilization? (Probably Not. But Still Cool.)". *Slate*. ISSN 1091-2339. Retrieved 2015-10-15.

[12] "Astronomers think they have found an alien megastructure". *The Independent*. Retrieved 2015-10-15.

[13] Souers, P. C. (1986). *Hydrogen properties for fusion energy*. University of California Press. p. 4. ISBN 978-0-520-05500-1.

[14] Borowski, Steve K. (1987-07-29). "Comparison of Fusion/Anti-matter Propulsion Systems for Interplanetary Travel" (PDF). *Technical Memorandum 107030*. San Diego, California, USA: National Aeronautics and Space Administration. pp. 1–3. Retrieved 2008-01-28.

[15] By the mass-energy equivalence formula $E = mc^2$. See anti-matter as a fuel source for the energy comparisons.

[16] Than, Ker (August 10, 2011). "Antimatter Found Orbiting Earth—A First". *National Geographic News*.

[17] Adriani; Barbarino; Bazilevskaya; Bellotti; Boezio; Bogomolov; Bongi; Bonvicini; Borisov (2011). "The discovery of geomagnetically trapped cosmic ray antiprotons". *The Astrophysical Journal* **736** (29): L1. arXiv:1107.4882. Bibcode:2011ApJ...736 L....1H.doi:10.1088/2041-8205/736/1/L1.

[18] "Antimatter caught streaming from thunderstorms on Earth". *BBC News*. 2011-01-11.

[19] Dyson, Freeman J. (1966). Marshak, R. E., ed. "The Search for Extraterrestrial Technology". *Perspectives in Modern Physics* (New York: John Wiley & Sons).

[20] Newman, Phil (2001-10-22). "New Energy Source "Wrings" Power from Black Hole Spin". NASA. Archived from the original on 2008-02-09. Retrieved 2008-02-19.

[21] Schutz, Bernard F. (1985). *A First Course in General Relativity*. New York: Cambridge University Press. pp. 304, 305. ISBN 0-521-27703-5.

[22] Dyson, Freeman (1960-06-03). "Search for Artificial Stellar Sources of Infrared Radiation". *Science* (New York: W. A. Benjamin, Inc) **131** (3414): 1667–1668. Bibcode:1960Sci...131.1667D. doi:10.1126/science.131.3414.1667. PMID 17780673. Retrieved 2008-01-30.

[23] Galántai, Zoltán (September 7, 2003). "Long Futures and Type IV Civilizations" (PDF). Retrieved 2014-11-03.

[24] Kaku, Michio (2005). *Parallel Worlds: The Science of Alternative Universes and Our Future in the Cosmos*. New York: Doubleday. p. 317. ISBN 0-7139-9728-1.

[25] Zubrin, Robert (1999). *Entering Space: Creating a Spacefaring Civilization*. ISBN 978-1585420360.

[26] Barrow, John (1998). *Impossibility: Limits of Science and the Science of Limits*. Oxford: Oxford University Press. p. 133. ISBN 978-0198518907.

[27] Jack Cohen and Ian Stewart: *Evolving the Alien: The Science of Extraterrestrial Life*, Ebury Press, 2002. ISBN 0-09-187927-2

30.10   Further reading

- Dyson, Freeman J. *Energy in the Universe* Article in September 1971 *Scientific American* magazine (Special September Issue on *Energy*)

- Rusinek, Marvin (1998). "Energy Consumption of Europe". *The Physics Factbook*.

- Wind Powering America

- Clean Energy for Planetary Survival: International Development Research Centre

- LBL Scientists Research Global Warming

- E^3 Handbook

- Clarke H2 energy systems

- Holdren, John P.; Carl Kaysen (2003). "Environmental Change and the Human Condition" (PDF). *Bulletin Fall*. pp. 24–31. Retrieved 2006-08-10.

- Dordrecht, D. (1985). "Exponential Expansion. Galactic Destiny or Technological Hubris?". In B. R. Finney, M. D. Papagiannis. *The Search for Extraterrestrial Life: Recent Developments*. Reidel Publ. Co. pp. 465–463.

- Shkadov Thruster

- Korotayev, A.; Malkov, A.; Khaltourina, D. (2006). *Introduction to Social Macrodynamics: Compact Macromodels of the World System Growth*. Moscow: URSS. ISBN 5-484-00414-4.

- Kardashev, Nikolai (March 1997). "Cosmology and Civilizations". *Astrophysics and Space Science* **252**: 25. doi:10.1023/A:1000837427320.

- *Supercivilizations as Possible Products of the Progressive Evolution of Matter*: also by Kardashev

- *Search for Artificial Stellar Sources of Infrared Radiation*, by Freeman J. Dyson

- *The Radio Search For Intelligent Extraterrestral Life*, by Frank Drake

- Freitas Jr., Robert A. *Energy and Culture (chapter 15)*.

- Griffin, John. *Operation TOGA: Type One Go Ahead*. ISBN 1-4502-0702-2.

30.11 External links

- Kardashev civilizations

- Astrobiology: The Living Universe

- Detectability of Extraterrestrial Technological Activities

- Flash Animation on Civilizations

- After Kardashev: Farewell to Super Civilizations

- Exotic Civilizations: Beyond Kardashev

- Description of civilization types from Dr. Michio Kaku

- Search for Type III civilizations

Chapter 31

Metalaw

Metalaw is a concept of space law closely related to the scientific Search for Extraterrestrial Intelligence (SETI).[1]

Immanuel Kant and his Categorical imperative: "Act only according to that maxim whereby you can at the same time will that it should become a **universal law**" is the forerunner of the Metalaw, because if we wish to detect the intelligent signals from the Universe, we have to act in a similar manner, that is, we must also transmit intelligent signals into the Universe.

31.1 Andrew Haley and the Origin of Metalaw

First articulated by attorney Andrew G. Haley in 1956, Metalaw was the term Haley coined to refer to his hypothesis regarding the proposed existence of fundamental legal precepts of theoretically universal application to all intelligences, both human and hypothesized intelligent extraterrestrial life. Writer Frank G. Anderson proposed that the definition be expanded to cover all intelligent species, extraterrestrial and terrestrial - which would include any/all intelligent animal life.

In 1956, Haley first published an article entitled "Space Law and Metalaw – A Synoptic View,"[2] in which Haley first proposed what he called an "Interstellar Golden Rule": Do unto others as they would have you do unto them. According to Haley, humans can project only one principle of human law onto our possible future relations with extraterrestrial intelligence: "the stark concept of absolute equity." Haley developed his formulation of Metalaw somewhat further in various papers and a 1963 book.[3]

31.2 Elaboration by Ernst Fasan

Significant elaboration of Haley's ideas did not take place until the publication in 1970 of *Relations with Alien Intelligences: The Scientific Basis of Metalaw*,[4] written by Dr. Ernst Fasan.

In *Relations with Alien Intelligences*, Fasan proposed Metalaw is "the entire sum of legal rules regulating relationships between different races in the universe." Metalaw is the "first and basic 'law' between races" providing the "ground rules" for a relationship if and when humans establish communication with or encounter an intelligent extraterrestrial race elsewhere in the universe. Fasan asserted that these rules would govern both human conduct and that of extraterrestrial races so as to avoid mutually harmful activities.

In later papers[5][6] published in the 1990s that more directly related Metalaw to SETI, Fasan proposed a simple 3-prong formula of metalegal principles. That formula involves:

1. A prohibition on damaging the other race.

2. The right of a race to self-defense.

3. The right to adequate living space.

31.3   Criticism of Metalaw

Several authors have criticized the metalegal principles proposed by Haley and Fasan for their reliance on Immanuel Kant's Categorical Imperative and on an approach to legal science and jurisprudence known as natural law theory. In jurisprudence, natural law theory refers generally to the view that links law to morality and proposes that just laws are immanent in nature and independent of the lawgiver, waiting to be discovered or found (as opposed to created by humans), usually by means of reason alone.[7]

Other commentators have noted that Haley's formulation of Metalaw depends heavily upon subjective or relative (and therefore inadequate) concepts of "good" and "bad."[8] Critics have noted that there is no guarantee that other civilizations would abide by Haley's assertions regarding equity among intelligent races in the universe.[9] Haley's failure to acknowledge the obvious anthropocentric limits of natural law theory has led some to note that the cultural concept of rules or law is itself anthropocentric.[10]

31.4   Metalaw in Popular Culture

In *Have Space Suit -- Will Travel*, a 1958 story by science fiction author Robert A. Heinlein and published two years after Haley's 1956 paper, one of the characters mentions "space law and meta-law."

G. Harry Stine (under his pen-name Lee Correy) wrote a short novel on the topic - "A Matter of Metalaw" (DAW_Books, October 1986).

31.5   References

[1] Adam Chase Korbitz, The Limits of Metalaw and the Need for Further Elaboration, Paper IAC-10-A4.2.10, presented at the 39th Symposium on the Search for Extraterrestrial Intelligence, 61st International Astronautical Congress, 2010, Prague, Czech Republic

[2] Andrew G. Haley, Space law and Metalaw – A Synoptic View, Harvard Law Record 23 (November 8, 1956)

[3] Andrew G. Haley, Space Law and Government, Appleton Century Crofts, New York, 1963

[4] Ernst Fasan, Relations with Alien Intelligences: The Scientific Basis of Metalaw, Berlin Verlag, Berlin, 1970

[5] Ernst Fasan, Discovery of ETI: Terrestrial and Extraterrestrial Legal Implications, Acta Astronautica 21 (2) (1990) 131-135

[6] Ernst Fasan, Legal Consequences of a SETI Detection, Acta Astronautica 42 (10-12) (1998) 677-679

[7] G.S. Robinson, Ecological foundations of Haley"s Metalaw, Journal of the British Interplanetary Society 22 (1969) 266-274

[8] F. Lyall, P.B. Larsen, Space Law: A Treatise, Ashgate Publishing Company, Burlington VT, 2009

[9] G.H. Reynolds, International space law: Into the Twenty-First Century, Vanderbilt Journal of Transnational Law 25 (1992) 225-255

[10] G.S. Robinson, note 7, supra

31.6 External links

- A Brief Introduction to Metalaw, by Adam Chase Korbitz

- Metalaw and SETI

- Metalaw and Interstellar Relations, by R.A. Freitas

Chapter 32

Rare Earth hypothesis

In planetary astronomy and astrobiology, the **Rare Earth Hypothesis** argues that the origin of life and the evolution of biological complexity such as sexually reproducing, multicellular organisms on Earth (and, subsequently, human intelligence) required an improbable combination of astrophysical and geological events and circumstances. The hypothesis argues that complex extraterrestrial life is a very improbable phenomenon and likely to be extremely rare. The term "Rare Earth" originates from *Rare Earth: Why Complex Life Is Uncommon in the Universe* (2000), a book by Peter Ward, a geologist and paleontologist, and Donald E. Brownlee, an astronomer and astrobiologist, both faculty members at the University of Washington.

An alternative view point was argued by Carl Sagan and Frank Drake, among others. It holds that Earth is a typical rocky planet in a typical planetary system, located in a non-exceptional region of a common barred-spiral galaxy. Given the principle of mediocrity (in the same vein as the Copernican principle), it is probable that the universe teems with complex life. Ward and Brownlee argue to the contrary: that planets, planetary systems, and galactic regions that are as friendly to complex life as are the Earth, the Solar System, and our region of the Milky Way are very rare.

32.1 Rare Earth's requirements for complex life

The Rare Earth hypothesis argues that the evolution of biological complexity requires a host of fortuitous circumstances, such as a galactic habitable zone, a central star and planetary system having the requisite character, the circumstellar habitable zone, a right sized terrestrial planet, the advantage of a gas giant guardian and large natural satellite, conditions needed to ensure the planet has a magnetosphere and plate tectonics, the chemistry of the lithosphere, atmosphere, and oceans, the role of "evolutionary pumps" such as massive glaciation and rare bolide impacts, and whatever led to the appearance of the eukaryote cell, sexual reproduction and the Cambrian explosion of animal, plant, and fungi phyla. The evolution of human intelligence may have required yet further events, which are extremely unlikely to have happened were it not for the Cretaceous–Paleogene extinction event 66 million years ago which saw the decline of dinosaurs as the dominant terrestrial vertebrates.

In order for a small rocky planet to support complex life, Ward and Brownlee argue, the values of several variables must fall within narrow ranges. The universe is so vast that it could contain many Earth-like planets. But if such planets exist, they are likely to be separated from each other by many thousands of light years. Such distances may preclude communication among any intelligent species evolving on such planets, which would solve the Fermi paradox: "If extraterrestrial aliens are common, why aren't they obvious?"[1]

32.1.1 The right location in the right kind of galaxy

Rare Earth suggests that much of the known universe, including large parts of our galaxy, cannot support complex life; Ward and Brownlee refer to such regions as "dead zones". Those parts of a galaxy where complex life is possible make up the galactic habitable zone. This zone is primarily a function of distance from the galactic center. As that distance

452

The Rare Earth Hypothesis argues that planets with complex life, like Earth, are exceptionally rare

increases:

1. Star metallicity declines. Metals (which in astronomy means all elements other than hydrogen and helium) are necessary to the formation of terrestrial planets.

2. The X-ray and gamma ray radiation from the black hole at the galactic center, and from nearby neutron stars, becomes less intense. Radiation of this nature is considered dangerous to complex life, hence the Rare Earth hypothesis predicts that the early universe, and galactic regions where stellar density is high and supernovae are common, will be unfit for the development of complex life.[3]

3. Gravitational perturbation of planets and planetesimals by nearby stars becomes less likely as the density of stars decreases. Hence the further a planet lies from the galactic center or a spiral arm, the less likely it is to be struck by a large bolide. A sufficiently large impact may extinguish all complex life on a planet.

The dense centre of galaxies such as NGC 7331 (often referred to as a "twin" of the Milky Way[?]) have high levels of radiation which are dangerous to complex life

Item #1 rules out the outer reaches of a galaxy; #2 and #3 rule out galactic inner regions. As one moves from the center of a galaxy to its furthest extremity, the ability to support life rises then falls. Hence the galactic habitable zone may be ring-shaped, sandwiched between its uninhabitable center and outer reaches.

While a planetary system may enjoy a location favorable to complex life, it must also maintain that location for a span of time sufficiently long for complex life to evolve. Hence a central star with a galactic orbit that steers clear of galactic regions where radiation levels are high, such as the galactic center and the spiral arms, would appear most favourable. If the central star's galactic orbit is eccentric (elliptic or hyperbolic), it will pass through some spiral arms, but if the orbit is a near perfect circle and the orbital velocity equals the "rotational" velocity of the spiral arms, the star will drift into a spiral arm region only gradually—if at all. Therefore, *Rare Earth* proponents conclude that a life-bearing star must have a galactic orbit that is nearly circular about the center of its galaxy. The required synchronization of the orbital velocity of a central star with the wave velocity of the spiral arms can occur only within a fairly narrow range of distances from the galactic center. This region is termed the "galactic habitable zone". Lineweaver et al.[4] calculate that the galactic habitable zone is a ring 7 to 9 kiloparsecs in diameter, that includes no more than 10% of the stars in the Milky Way.[5] Based on conservative estimates of the total number of stars in the galaxy, this could represent something like 20 to 40 billion stars. Gonzalez, *et al.*[6] would halve these numbers; he estimates that at most 5% of stars in the Milky Way fall in the galactic habitable zone.

Approximately 77% of observed galaxies are spiral galaxies[7] and two-thirds of all spiral galaxies, are barred and more than half, like the Milky Way, exhibit multiple arms.[8] What makes our galaxy different, according to Rare Earth, is that it is unusually quiet and dim (see argument below), representing just 7% of its kind.[9] Even so, this would still represent more than 200 billion galaxies in the known universe.

A reason that our galaxy is considered rare by Rare Earth is because it appears to have suffered fewer collisions with other galaxies over the last 10 billion years, and its peaceful history may have made it more hospitable to complex life than galaxies which have suffered more collisions, and consequently more supernovae and other disturbances.[10] The level of activity of the black hole at the centre of the Milky Way may also be important: too much or too little and the conditions for life may be even rarer. The Milky Way black hole appears to be just right.[11] The orbit of the Sun around the center of the Milky Way is indeed almost perfectly circular, with a period of 226 Ma (1 Ma = 1 million years), one closely matching the rotational period of the galaxy. However, the majority of stars in barred spiral galaxies populate the spiral arms rather than the halo and tend to move in gravitationally aligned orbits, so there is little that is unusual about the Sun's orbit. While the Rare Earth hypothesis predicts that the Sun should rarely, if ever, have passed through a spiral arm

According to Rare Earth, globular clusters are unlikely to support life.

since its formation, astronomer Karen Masters has calculated that the orbit of the Sun takes it through a major spiral arm approximately every 100 million years.[12] Some researchers have suggested that several mass extinctions do correspond with previous crossings of the spiral arms.[13]

32.1.2   Orbiting at the right distance from the right type of star

The terrestrial example suggests that complex life requires water in the liquid state, and a central star's planet must therefore be at an appropriate distance. This is the core of the notion of the habitable zone or Goldilocks Principle.[14] The habitable zone forms a ring around the central star. If a planet orbits its sun too closely or too far away, the surface temperature is incompatible with water being in liquid form.

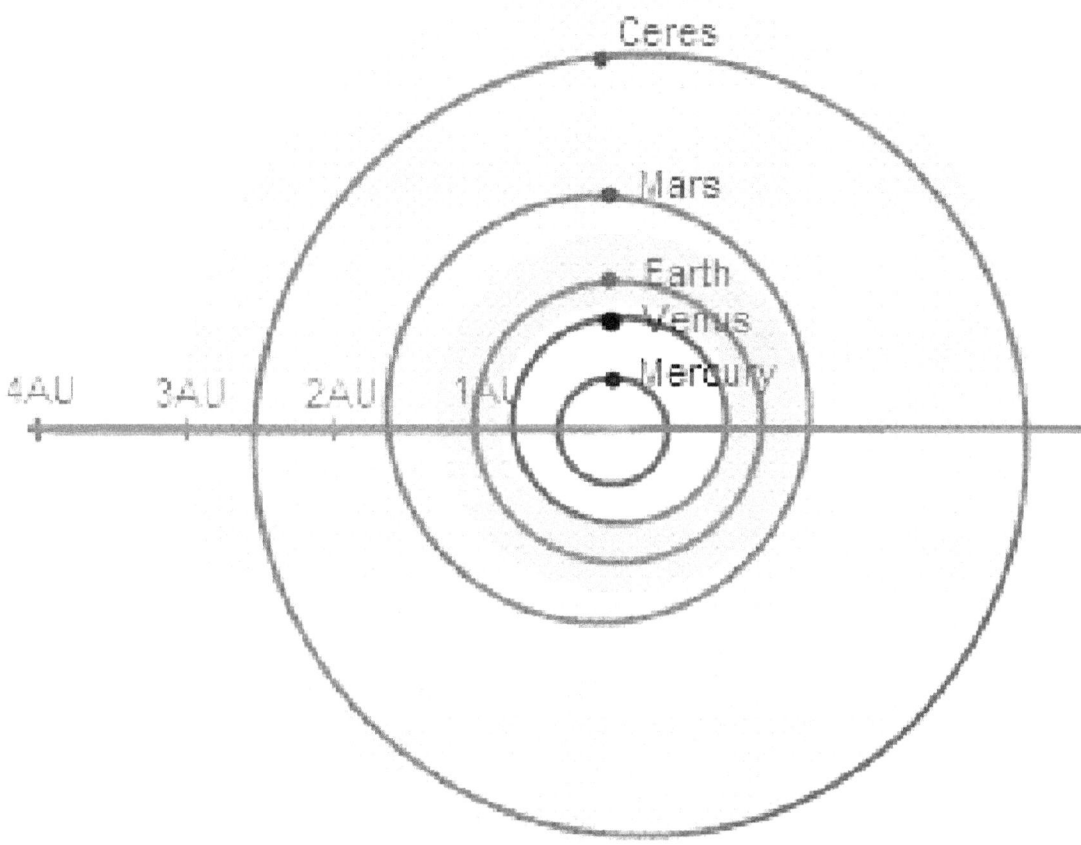

According to the hypothesis, Earth has an improbable orbit in the very narrow habitable zone (dark green) around the Sun.

The habitable zone varies with the type and age of the central star. For advanced life the star must have a high degree of stability. The habitable zone for a main sequence star very gradually moves out over time until the star becomes a white dwarf, at which time the habitable zone vanishes. The habitable zone is closely connected to the greenhouse warming afforded by atmospheric water vapor (H
2O), carbon dioxide (CO_2), and/or other greenhouse gases. Even though the Earth's atmosphere contains a water vapor concentration from 0% (in arid regions) to 4% (in rain forest and ocean regions) and -as of June 2013- only 400 parts per million of CO_2, these small amounts suffice to raise the average surface temperature of the Earth by about 40 °C from what it would otherwise be,[15] with the dominant contribution being due to water vapor, which together with clouds makes up between 66% and 85% of Earth's greenhouse effect, with CO_2 contributing between 9% and 26% of the effect.[16]

Rocky planets must orbit within the habitable zone for life to form. Although the habitable zone of such hot stars as Sirius or Vega is wide:

1. Rocky planets that form too close to the star to lie within the habitable zone cannot sustain life; however, life could arise on a moon of a gas giant. Hot stars also emit much more ultraviolet radiation that ionizes any planetary atmosphere.

2. Hot stars, as mentioned above, may become red giants before advanced life evolves on their planets.

These considerations rule out the massive and powerful stars of type F6 to O (see stellar classification) as homes to evolved metazoan life.

Small red dwarf stars conversely have small habitable zones wherein planets are in tidal lock—one side always faces the star and becomes very hot and the other always faces away and becomes very cold—and are also at increased risk of solar flares (see Aurelia) that would tend to ionize the atmosphere and be otherwise inimical to complex life. *Rare Earth* proponents argue that life therefore cannot arise in such systems and that only central stars that range from F7 to K1 stars are hospitable. Such stars are rare: G type stars such as the Sun (between the hotter F and cooler K) comprise only 9%[17] of the hydrogen-burning stars in the Milky Way.

Such aged stars as red giants and white dwarfs are also unlikely to support life. Red giants are common in globular clusters and elliptical galaxies. White dwarfs are mostly dying stars that have already completed their red giant phase. Stars that become red giants expand into or overheat the habitable zones of their youth and middle age (though theoretically planets at a much greater distance may become habitable).

An energy output that varies with the lifetime of the star will very likely prevent life (e.g., as Cepheid variables). A sudden decrease, even if brief, may freeze the water of orbiting planets, and a significant increase may evaporate them and cause a greenhouse effect that may prevent the oceans from reforming.

Life without complex chemistry is unknown. Such chemistry requires metals, namely elements other than hydrogen or helium and thereby suggests that a planetary system rich in metals is a necessity for life. The only known mechanism for creating and dispersing metals is a supernova explosion. The absorption spectrum of a star reveals the presence of metals within, and studies of stellar spectra reveal that many, perhaps most, stars are poor in metals. Low metallicity characterizes the early universe: globular clusters and other stars that formed when the universe was young, stars in most galaxies other than large spirals, and stars in the outer regions of all galaxies. Metal-rich central stars capable of supporting complex life are therefore believed to be most common in the quiet suburbs of the larger spiral galaxies—where radiation also happens to be weak.[18]

32.1.3 With the right arrangement of planets

Depiction of the Sun and planets of the Solar System and the sequence of planets. Rare Earth argues that without such an arrangement, in particular the presence of the massive gas giant Jupiter (fifth planet from the Sun and the largest), complex life on Earth would not have arisen.

Rare Earth proponents argue that a planetary system capable of sustaining complex life must be structured more or less like the Solar System, with small and rocky inner planets and outer gas giants.[19] Without the protection of 'celestial vacuum cleaner' planets with strong gravitational pull, the number of asteroid collisions may have been larger, and a greater number of mass extinction events may have occurred.

In addition, the arrangement of the Solar System is not only rare but optimal as the large mass and gravitational attraction of the gas giants provide protection for the inner rocky planets from Small Solar System body impacts and asteroid bombardment.

32.1.4   A continuously stable orbit

Rare Earth argues that a gas giant must not be too close to a body upon which life is developing, unless that body is one of its moons. Close placement of gas giant(s) could disrupt the orbit of a potential life-bearing planet, either directly or by drifting into the habitable zone.

Newtonian dynamics can produce chaotic planetary orbits, especially in a system having large planets at high orbital eccentricity.[20]

The need for stable orbits rules out stars with systems of planets that contain large planets with orbits close to the host star (called "hot Jupiters"). It is believed that hot Jupiters formed much further from their parent stars than they are now (see planetary migration), and have migrated inwards to their current orbits. In the process, they would have catastrophically disrupted the orbits of any planets in the habitable zone.[21]

32.1.5   A terrestrial planet of the right size

Planets of the Solar System to scale. Rare Earth argues that complex life cannot exist on large gaseous planets like Jupiter and Saturn (top row) a Uranus and Neptune (top middle) or smaller planets such as Mars and Mercury

It is argued that life requires terrestrial planets like Earth and as gas giants lack such a surface, that complex life cannot arise there.[22]

A planet that is too small cannot hold much of an atmosphere. Hence the surface temperature becomes more variable and the average temperature drops. Substantial and long-lasting oceans become impossible. A small planet will also tend to have a rough surface, with large mountains and deep canyons. The core will cool faster, and plate tectonics will either not last as long as they would on a larger planet or may not occur at all. A planet that is too large will retain too much of its atmosphere and will be like Venus. Venus is similar in size and mass to Earth, but has a surface atmosphere pressure that is 92 times that of Earth's. Venus mean surface temperature is 735 K (462 °C; 863 °F) making Venus the hottest planet in the Solar System. Earth had a similar early atmosphere to Venus, but lost it in the giant impact event.[23]

32.1.6 With plate tectonics

Rare Earth proponents argue that plate tectonics and a large magnetic field are essential for the emergence and sustenance of complex life.[24] Ward & Brownlee assert that biodiversity, global temperature regulation, the carbon cycle, and the magnetic field of the Earth that make it habitable for complex terrestrial life all depend on plate tectonics.[25]

Ward & Brownlee contend that the lack of mountain chains elsewhere in the Solar System is direct evidence that Earth is the only body with plate tectonics and as such the only body capable of supporting life.[26]

Plate tectonics is dependent on chemical composition and a long-lasting source of heat in the form of radioactive decay occurring deep in the planet's interior. Continents must also be made up of less dense felsic rocks that "float" on underlying denser mafic rock. Taylor[27] emphasizes that subduction zones (an essential part of plate tectonics) require the lubricating action of ample water; on Earth, such zones exist only at the bottom of oceans.

Ward & Brownlee and others such as Tilman Spohn of the German Space Research Centre Institute of Planetary Research[28] argue that plate tectonics provides a means of biochemical cycling which promotes complex life on Earth and that water is required to lubricate planetary plates.

Plate tectonics and as a result continental drift and the creation of separate land masses would create diversified ecosystems which is thought to have promoted the diversification of species, and that diversity is one of the strongest defences against extinction.[29]

An example of species diversification and later competition on Earth's continents is the Great American Interchange. This was the result of the tectonically induced connection between North & Middle America with the South American continent, at around 3.5 to 3 Ma. The previously undisturbed fauna of South America could evolve in their own way for about 30 million years, since Antarctica separated. Many species were subsequently wiped out in mainly South America by competing Northern American animals.

32.1.7 A large moon

The Moon is unusual because the other rocky planets in the Solar System either have no satellites (Mercury and Venus), or have tiny satellites that are probably captured asteroids (Mars).

The giant impact theory hypothesizes that the Moon resulted from the impact of a Mars-sized body, Theia, with the very young Earth. This giant impact also gave the Earth its axial tilt and velocity of rotation.[27] Rapid rotation reduces the daily variation in temperature and makes photosynthesis viable.[30] The *Rare Earth* hypothesis further argues that the axial tilt cannot be too large or too small (relative to the orbital plane). A planet with a large tilt (inclination) will experience extreme seasonal variations in climate, unfriendly to complex life. A planet with little or no tilt will lack the stimulus to evolution that climate variation provides. In this view, the Earth's tilt is "just right". The gravity of a large satellite also stabilizes the planet's tilt; without this effect the variation in tilt would be chaotic, probably making complex life forms on land impossible.[31]

If the Earth had no Moon, the ocean tides resulting solely from the Sun's gravity would be only half that of the lunar tides. A large satellite gives rise to tidal pools, which may be essential for the formation of complex life, though this is far from certain.[32]

A large satellite also increases the likelihood of plate tectonics through the effect of tidal forces on the planet's crust. The impact that formed the Moon may also have initiated plate tectonics, without which the continental crust would cover the entire planet, leaving no room for oceanic crust. It is possible that the large scale mantle convection needed to drive plate tectonics could not have emerged in the absence of crustal inhomogeneity.

If a giant impact is the only way for a rocky inner planet to acquire a large satellite, any planet in the circumstellar habitable zone will need to form as a double planet in order that there be an impacting object sufficiently massive to give rise in due

The Great American Interchange on Earth, around ~ 3.5 to 3 Ma, an example of species competition, resulting from continental plate interaction

course to a large satellite. An impacting object of this nature is not necessarily improbable.

Tide pools resulting from tidal interaction of the Moon are said to have promoted the evolution complex life.

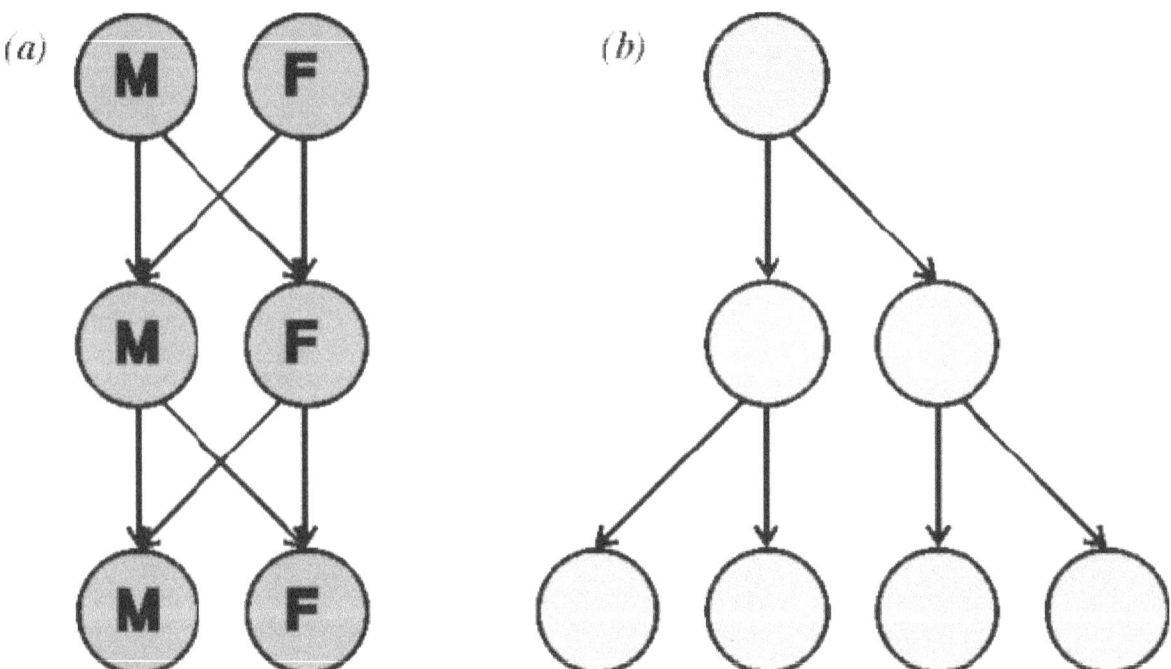

This diagram illustrates the twofold cost of sex. If each individual were to contribute to the same number of offspring (two), (a) the sexual population remains the same size each generation, where the (b) asexual population doubles in size each generation

32.1.8   One or more evolutionary triggers for complex life

Regardless of whether planets with similar physical attributes to the Earth are rare or not, some argue that life usually remains simple bacteria. Biochemist Nick Lane argues that simple cells (prokaryotes) emerged soon after Earth's formation, but almost half the planet's life had passed before they evolved into complex ones (eukaryotes) all of whom share a common ancestor, this event can only have happened once. In some views, prokaryotes lack the cellular architecture to evolve into eukaryotes because a bacterium expanded up to eukaryotic proportions would have tens of thousands of times less energy available; two billion years ago, one simple cell incorporated itself into another, multiplied, and evolved into mitochondria that supplied the vast increase in available energy that enabled the evolution of complex life. If this incorporation occurred only once in four billion years or is otherwise unlikely, then life on most planets remains simple.[33] An alternative view that mitochondria evolution was environmentally triggered, and that mitochondria containing organisms appear very soon after first traces of oxygen appear in Earth's atmosphere.[34]

The evolution of sexual reproduction as well as its maintenance, is another mystery in biology. The purpose of sexual reproduction is unclear, as in many organisms it has a 50% cost (fitness disadvantage) in relation to asexual reproduction.[35] Mating types (types of gametes, according to their compatibility) may have arisen as a result of anisogamy (gamete dimorphism), or the male and female genders may have evolved before anisogamy.[36][37] It is also unknown why most sexual organisms use a binary mating system,[38] and why some organisms have gamete dimorphism. Charles Darwin was the first to suggest that sexual selection drives speciation (the formation of species); without sexual reproduction it is unlikely that complex life would have evolved.

32.1.9   The right time in evolution

While life on Earth is regarded to have spawned relatively early in the planet's history, the evolution to complex organisms took around 800 million years[39] Civilizations on Earth have existed for ~10,000 years and radio communication with space is not older than 80 years. Relative to the age of our solar system (~4.57 Ga) this is a tiny age span, an age span in which extreme climatic variations, super volcanoes or large meteorite impacts were absent. These events would severely harm intelligent life, as well as life in general. For example, the Permian-Triassic mass extinction, caused by widespread and continuous volcanic eruptions in an area the size of Western Europe, led to the extinction of 95% of known species around 251.2 Ma ago. About 65 million years ago, the Chicxulub impact at the Cretaceous–Paleogene boundary (~65.5 Ma) on the Yucatán peninsula in Mexico led to a mass extinction of the most advanced species at that time.

If intelligent extraterrestrial civilizations did exist and with such an intelligence level that they could make contact with distant Earth, they would have to live in the same time span in evolution. The nearest Earth-like planets are around 11.9 light years away; probable planets as Tau Ceti e and f around the star Tau Ceti in the constellation of Cetus, a star considered to be 5.8 Ga; 1.23 billion years older than the Sun.

Under the assumption that both the explosion of life and the development of civilization were to be relative to the planet's age, they would have spawned 723 Ma and 12.691 ka, respectively. The time between the life explosion if that had existed on an exoplanet and the dawn of civilizations is thus very large and the time between civilization and radio signals evenly so.

The risk of intelligent-life destruction is not a Drake equation factor: in the 33 million years since the Eocene-Oligocene extinction event there have been no major mass extinctions.

The chance of bigger impacts in the time span of evolution to intelligent life depends on the amount of shielding by larger bodies, such as our system's Jupiter or the Moon. The chance of a large impact and resulting mass extinction happening in a multi-planetary "protected" system is, however, impossible to predict.

32.2   Rare Earth equation

The following discussion is adapted from Cramer.[40] The Rare Earth equation is Ward and Brownlee's riposte to the Drake equation. It calculates N, the number of Earth-like planets in the Milky Way having complex life forms, as:

$$N = N^* \cdot n_e \cdot f_g \cdot f_p \cdot f_{pm} \cdot f_i \cdot f_c \cdot f_l \cdot f_m \cdot f_j \cdot f_{me} \text{ [41]}$$

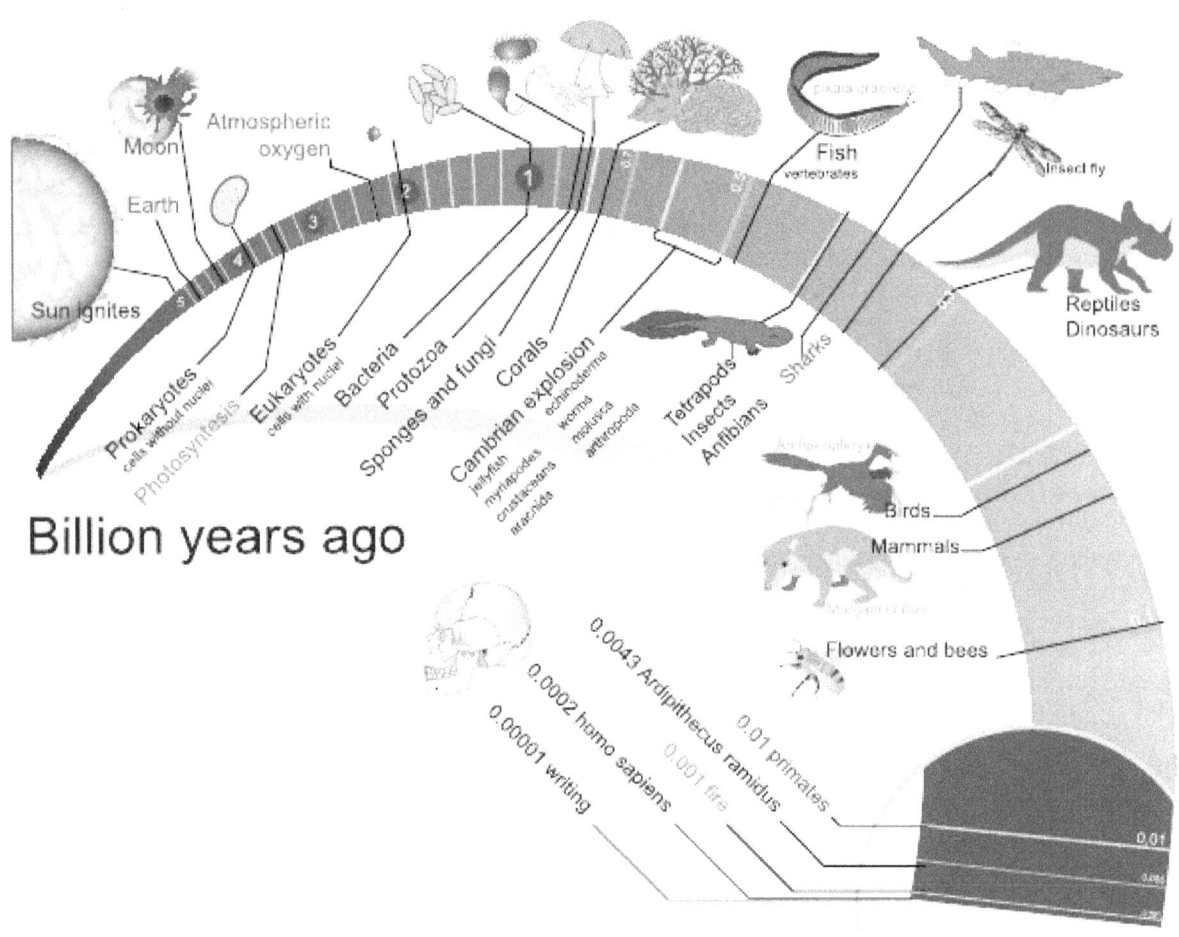

Billion years ago

Timeline of evolution: human writings exists for only 0.000218% of Earth's history.

where:

- N^* is the number of stars in the Milky Way. This number is not well-estimated, because the Milky Way's mass is not well estimated. Moreover, there is little information about the number of very small stars. N^* is at least 100 billion, and may be as high as 500 billion, if there are many low visibility stars.

- n_e is the average number of planets in a star's habitable zone. This zone is fairly narrow, because constrained by the requirement that the average planetary temperature be consistent with water remaining liquid throughout the time required for complex life to evolve. Thus $n_e = 1$ is a likely upper bound.

We assume $N^* \cdot n_e = 5 \cdot 10^{11}$. The Rare Earth hypothesis can then be viewed as asserting that the product of the other nine Rare Earth equation factors listed below, which are all fractions, is no greater than 10^{-10} and could plausibly be as small as 10^{-12}. In the latter case, N could be as small as 0 or 1. Ward and Brownlee do not actually calculate the value of N, because the numerical values of quite a few of the factors below can only be conjectured. They cannot be estimated simply because we have but one data point: the Earth, a rocky planet orbiting a G2 star in a quiet suburb of a large barred spiral galaxy, and the home of the only intelligent species we know, namely ourselves.

- f_g is the fraction of stars in the galactic habitable zone (Ward, Brownlee, and Gonzalez estimate this factor as $0.1^{[6]}$).

- f_p is the fraction of stars in the Milky Way with planets.

- f_{pm} is the fraction of planets that are rocky ("metallic") rather than gaseous.

- f_l is the fraction of habitable planets where microbial life arises. Ward and Brownlee believe this fraction is unlikely to be small.

- f_c is the fraction of planets where complex life evolves. For 80% of the time since microbial life first appeared on the Earth, there was only bacterial life. Hence Ward and Brownlee argue that this fraction may be very small.

- f_l is the fraction of the total lifespan of a planet during which complex life is present. Complex life cannot endure indefinitely, because the energy put out by the sort of star that allows complex life to emerge gradually rises, and the central star eventually becomes a red giant, engulfing all planets in the planetary habitable zone. Also, given enough time, a catastrophic extinction of all complex life becomes ever more likely.

- f_m is the fraction of habitable planets with a large moon. If the giant impact theory of the Moon's origin is correct, this fraction is small.

- f_j is the fraction of planetary systems with large Jovian planets. This fraction could be large.

- f_{me} is the fraction of planets with a sufficiently low number of extinction events. Ward and Brownlee argue that the low number of such events the Earth has experienced since the Cambrian explosion may be unusual, in which case this fraction would be small.

The Rare Earth equation, unlike the Drake equation, does not factor the probability that complex life evolves into intelligent life that discovers technology (Ward and Brownlee are not evolutionary biologists). Barrow and Tipler[42] review the consensus among such biologists that the evolutionary path from primitive Cambrian chordates, e.g., *Pikaia* to *Homo sapiens*, was a highly improbable event. For example, the large brains of humans have marked adaptive disadvantages, requiring as they do an expensive metabolism, a long gestation period, and a childhood lasting more than 25% of the average total life span. Other improbable features of humans include:

- Being one of a handful of extant bipedal land (non-avian) vertebrate. Combined with an unusual eye–hand coordination, this permits dextrous manipulations of the physical environment with the hands;

- A vocal apparatus far more expressive than that of any other mammal, enabling speech. Speech makes it possible for humans to interact cooperatively, to share knowledge, and to acquire a culture;

- The capability of formulating abstractions to a degree permitting the invention of mathematics, and the discovery of science and technology. Only recently did humans acquire anything like their current scientific and technological sophistication.

32.3   Advocates

Authors that advocate the Rare Earth hypothesis:

- Stuart Ross Taylor,[27] a specialist on the solar system, firmly believes in the hypothesis. Taylor concludes that the solar system is probably very unusual, because it resulted from so many chance factors and events.

- Stephen Webb,[1] a physicist, mainly presents and rejects candidate solutions for the Fermi paradox. The Rare Earth hypothesis emerges as one of the few solutions left standing by the end of the book.

- Simon Conway Morris, a paleontologist, endorses the Rare Earth hypothesis in chapter 5 of his *Life's Solution: Inevitable Humans in a Lonely Universe*,[43] and cites Ward and Brownlee's book with approval.[44]

- John D. Barrow and Frank J. Tipler (1986, 3.2, 8.7, 9), cosmologists, vigorously defend the hypothesis that humans are likely to be the only intelligent life in the Milky Way, and perhaps the entire universe. But this hypothesis is not central to their book *The Anthropic Cosmological Principle*, a very thorough study of the anthropic principle, and of how the laws of physics are peculiarly suited to enable the emergence of complexity in nature.

- Ray Kurzweil, a computer pioneer and self-proclaimed Singularitarian, argues in *The Singularity Is Near* that the coming Singularity requires that Earth be the first planet on which sentient, technology-using life evolved. Although other Earth-like planets could exist, Earth must be the most evolutionarily advanced, because otherwise we would have seen evidence that another culture had experienced the Singularity and expanded to harness the full computational capacity of the physical universe.

- John Gribbin, a prolific science writer, defends the hypothesis in a book devoted to it called *Alone in the Universe: Why our planet is unique*.[45]

- Guillermo Gonzalez, astrophysicist who coined the term Galactic Habitable Zone uses the hypothesis in his book *The Privileged Planet* to promote the concept of intelligent design.[46]

- Michael H. Hart, astrophysicist who proposed a very narrow habitable zone based on climate studies edited the influential book "Extraterrestrials: Where are They" and authored "Atmospheric Evolution, the Drake Equation and DNA: Sparse Life in an Infinite Universe"[47]

- Howard Alan Smith, PhD., astrophysicist and author of 'Let there be light: modern cosmology and Kabbalah : a new conversation between science and religion'[48]

32.4 Criticism

Cases against the Rare Earth Hypothesis take various forms.

32.4.1 Anthropic reasoning

The hypothesis concludes, more or less, that complex life is rare because it can evolve only on the surface of an Earth-like planet or on a suitable satellite of a planet. Some biologists, such as Jack Cohen, believe this assumption too restrictive and unimaginative; they see it as a form of circular reasoning.

According to David Darling, the Rare Earth hypothesis is neither hypothesis nor prediction, but merely a description of how life arose on Earth.[49] In his view Ward and Brownlee have done nothing more than select the factors that best suit their case.

> What matters is not whether there's anything unusual about the Earth; there's going to be something idiosyncratic about every planet in space. What matters is whether any of Earth's circumstances are not only unusual but also essential for complex life. So far we've seen nothing to suggest there is.[50]

Critics also argue that there is a link between the Rare Earth Hypothesis and the creationist ideas of intelligent design.[51]

32.4.2 Exoplanets around main sequence stars are being discovered in large numbers

See also: Estimated frequency of Earth-like planets

An increasing number of extrasolar planet discoveries are being made with 2030 planets in 1288 planetary systems known. Rare Earth proponents argue life cannot arise outside Sun-like systems. However, some exobiologists have suggested that stars outside this range may give rise to life under the right circumstances; this possibility is a central point of contention to the theory because these late-K and M category stars make up about 82% of all hydrogen-burning stars.[17]

Current technology limits the testing of important Rare Earth Criteria: surface water, tectonic plates, a large moon and biosignatures are currently undetectable. Though planets the size of Earth are difficult to detect and classify, scientists now conclude that rocky planets are common around Sun-like stars.[52] The Earth Similarity Index (ESI) of mass, radius and temperature provides a means of measurement, but falls short of the full Rare Earth criteria.[53][54]

32.4.3   Rocky planets orbiting within habitable zones may not be rare

Some argue that Rare Earth's estimates of rocky planets in habitable zones (n_e in the Rare Earth equation) are too restrictive. James Kasting cites the Titius-Bode law to contend that it is a misnomer to describe habitable zones as narrow when there is a 50% chance of at least one planet orbiting within one.[56] In 2013 a study that was published in the journal Proceedings of the National Academy of Sciences calculated that about "one in five" of all sun-like stars are expected to have earthlike planets "within the habitable zones of their stars"; 8.8 billion of them therefore exist in the Milky Way galaxy alone.[57] On 4 November 2013, astronomers reported, based on *Kepler* space mission data, that there could be as many as 40 billion Earth-sized planets orbiting in the habitable zones of sun-like stars and red dwarf stars within the Milky Way Galaxy.[58][59] 11 billion of these estimated planets may be orbiting sun-like stars.[60]

32.4.4   Uncertainty over Jupiter's role

The requirement for a system to have a Jovian planet as protector (Rare Earth equation factor f_j) has been challenged and this has a bearing on the number of proposed extinction events (Rare Earth equation factor f_{me}). Kasting's 2001 review of Rare Earth questions whether a Jupiter protector has any bearing on the frequency of complex life.[61] Computer modelling including the 2005 Nice model and 2007 Nice 2 model yield inconclusive results in relation to Jupiter's gravitational influence and impacts on the inner planets.[62] A study by Horner & Jones (2008) using computer simulation found that while the total effect on all orbital bodies within the Solar System is unclear, Jupiter has caused more impacts on Earth than it has prevented.[63] Lexell's Comet, a 1770 near miss that passed closer to Earth than any other comet in recorded history, was known to be caused by the gravitational influence of Jupiter.[64]

32.4.5   Plate tectonics may not be unique to Earth

Ward & Brownlee argue that tectonics is necessary to support biogeochemical cycles required for complex life to arise and predicted that such geological features would not be found outside of Earth, pointing to a lack of observable orogenic evidence, specifically in the form of mountain ranges and subduction zones.[66] However, recent evidence points to similar activity either having occurred or continuing to occur elsewhere. The geology of Pluto, for example, described by Ward & Brownlee as "without mountains or volcanoes ... devoid of volcanic activity",[18] has since been found to be quite the contrary, with a geologically active surface possessing organic molecules[67] and mountain ranges[68] like Norgay Montes and Hillary Montes comparable in relative size to those of Earth, and observations suggest the involvement of endogenic processes.[69] Plate tectonics has been suggested as a hypothesis for the Martian dichotomy and in 2012 Geologist An Yin put forward evidence for active plate tectonics on Mars.[70] Europa has long suspected to have plate tectonics[71] and in 2014 that NASA announced evidence of active subduction.[72] Kasting suggests that there is nothing unusual about the occurrence of plate tectonics in large rocky planets and liquid water on the surface as most should generate internal heat even without the assistance of radioactive elements.[61] Studies by Valencia[73] and Cowan[74] suggest that plate tectonics may be inevitable for terrestrial planets Earth sized or larger, that is, Super-Earths, which are now known to be more common in planetary systems.[75]

Free oxygen may neither be rare nor a prerequisite for multicellular life

The hypothesis that molecular oxygen, necessary for animal life to exist is rare and that a Great Oxygenation Event (a condition for Rare Earth equation factor f_c), could only have been triggered and sustained by tectonics as occurred on Earth, appears to have been invalidated by more recent discoveries.

Ward & Brownlee ask "whether oxygenation, and hence the rise of animals, would ever have occurred on a world where there were no continents to erode".[76] Extraterrestrial free oxygen has recently been detected around other solid objects, including Mercury,[77] Venus,[78] Mars[79] Jupiter's four Galilean moons,[80] Saturn's moons Enceladus,[81] Dione[82][83] and Rhea[84] and even the atmosphere of a comet.[85] This has led scientists to speculate whether processes other than photosynthesis could be capable of generating an environment rich in free oxygen. Wordsworth (2014), concludes that oxygen generated through photodissociation may be not only be likely on Earth-like exoplanets, but could actually lead to false positive detections of life.[86] Narita (2015) suggests Titania as a geochemical mechanism for producing oxygen atmospheres.[87]

Since Ward & Brownlee's assertion that "there is irrefutable evidence that oxygen is a necessary ingredient for animal life",[76] anaerobic metazoa have been found that indeed do metabolise without oxygen. Spinoloricus nov. sp., for example, a species discovered in the hypersaline anoxic L'Atalante basin at the bottom of the Mediterranean Sea in 2010, appears to metabolise with hydrogen, lacking mitochondria and instead using hydrogenosomes.[88][89] Stevenson (2015) has proposed other membrane alternatives for complex life in worlds without oxygen.[90] Independent studies by Schirrmeister and by Mills concluded that Earth's multicellular life existed prior to the Great Oxygenation Event, not as a consequence of it.[91][92]

NASA scientists Hartman and McKay argue that plate tectonics may in fact slow the rise of oxygenation (and thus stymie complex life rather than promote it).[93] Computer modelling by Tilman Spohn in 2014 found that plate tectonics on Earth may have arisen from the effects of complex life's emergence, rather than the other way around as the Rare Earth might suggest. The action of lichens on rock may have contributed to the formation of subduction zones in the presence of water.[94] Kasting argues that if oxygenation caused the Cambrian explosion than any planet with oxygen producing photosynthesis should have complex life.[95]

A magnetic field may not be a requirement

The importance of Earth's magnetic field to the development of complex life has been disputed. Kasting argues that the atmosphere provides sufficient protection against cosmic rays even during times of magnetic pole reversal and atmosphere loss by sputtering.[96] Kasting also dismisses the role of the magnetic field in the evolution of eukaryotes citing the age of the oldest known magnetofossils.[97]

32.4.6 A large moon may neither be rare nor necessary

The requirement of a large moon (Rare Earth equation factor f_m) has also been challenged. Though even if it were required, such an occurrence may not be as unique as predicted by the Rare Earth Hypothesis. Recent work by Edward Belbruno and J. Richard Gott of Princeton University suggests that giant impacts such as those that may have formed the Moon can indeed form in planetary trojan points (L_4 or L_5 Lagrangian point) which means that similar circumstances may occur in other planetary systems.[98]

Rare Earth's assertion that the Moon's stabilization of Earth's obliquity and spin is a requirement for complex life has been questioned. Kasting argues that a moonless Earth would still possess habitats with climates suitable for complex life and questions whether the spin rate of a moonless Earth can be predicted.[61] Although the giant impact theory posits that the impact forming the Moon increased Earth's rotational speed to make a day about 5 hours long, the Moon has slowly "stolen" much of this speed to reduce Earth's solar day since then to about 24 hours and continues to do so: in 100 million years Earth's solar day will be roughly 24 hours 38 minutes (the same as Mars's solar day); in 1 billion years, 30 hours 23 minutes. Larger secondary bodies would exert proportionally larger tidal forces that would in turn decelerate their primaries faster and potentially increase the solar day of a planet in all other respects like earth to over 120 hours within a few billion years. This long solar day would make effective heat dissipation for organisms in the tropics and subtropics extremely difficult in a similar manner to tidal locking to a red dwarf star. Short days (high rotation speed) causes high wind speeds at ground level. Long days (slow rotation speed) causes the day/night temperatures to be too extreme.[99]

Many Rare Earth proponents argue that the Earth's plate tectonics would probably not exist if not for the tidal forces of the Moon.[100][101] The hypothesis that the Moon's tidal influence initiated or sustained Earth's plate tectonics remains unproven, though at least one study implies a temporal correlation to the formation of the Moon.[102] Evidence for the past existence of plate tectonics on planets like Mars[103] which may never have had a large moon would counter this argument. Kasting argues that a large moon is not required to initiate plate tectonics.[61]

32.4.7 Complex life may arise in alternative habitats

See also: Alternative biochemistry

Rare Earth proponents argue that simple life may be common, though complex life requires specific environmental conditions to arise. Dirk Schulze-Makuch argues that there is no evidence to support this conclusion, hypothesizing alternative biochemistries as a method for complex life to arise in completely alien conditions.[104] While Rare Earth proponents argue that only microbial extremophiles could exist in subsurface habitats beyond Earth, some argue that complex life can also arise in these environments. Critics cite examples of extremophile animals such as the Hesiocaeca methanicola, an animal that inhabits ocean floor methane clathrates substances more commonly found in the outer Solar System or the Tardigrade which can survive in the vacuum of space[105] as complex life that thrive in "alien" environments. Jill Tarter counters the classic counterargument that these species adapted to these environments rather than arose in them, by suggesting that we cannot assume conditions for life to emerge which are not actually known.[106] There are suggestions that complex life could arise in sub-surface conditions which may be similar to those where life may have arisen on Earth, such as the tidally heated subsurfaces of Europa or Enceladus.[107][108] Ancient circumvental ecosystems such as these support complex life on Earth such as Riftia pachyptila that exist completely independent of the surface biosphere.[109]

32.5   Notes

[1] Webb 2002

[2] 1 Morphology of Our Galaxy's 'Twin' Spitzer Space Telescope. Jet Propulsion Laboratory. NASA.

[3] Ward & Brownlee 2000, pp. 27–29

[4] Lineweaver, Charles H.; Fenner, Yeshe; Gibson, Brad K. (2004). "The Galactic Habitable Zone and the Age Distribution of Complex Life in the Milky Way" (PDF). Science 303 (5654): 59–62. arXiv:astro-ph/0401024. Bibcode:2004Sci...303...59L. doi:10.1126/science.1092322. PMID 14704421.

[5] Ward & Brownlee 2000, p. 32

[6] Gonzalez, Brownlee & Ward 2001

[7] Loveday, J. (February 1996). "The APM Bright Galaxy Catalogue". Monthly Notices of the Royal Astronomical Society 278 (4): 1025–1048. arXiv:astro-ph/9603040. Bibcode:1996MNRAS.278.1025L. doi:10.1093/mnras/278.4.1025.

[8] D. Mihalas (1968). Galactic Astronomy. W. H. Freeman. ISBN 978-0-7167-0326-6.

[9] Hammer, F.; Puech, M.; Chemin, L.; Flores, H.; Lehnert, M. D. (2007). "The Milky Way, an Exceptionally Quiet Galaxy: Implications for the Formation of Spiral Galaxies". The Astrophysical Journal 662 (1): 322–334. doi:10.1086/516727. ISSN 0004-637X.

[10] "Sibling Rivalry". New Scientist. 31 March 2012.

[11] Scharf. 2012

[12] How often does the Sun pass through a spiral arm in the Milky Way?, Karen Masters, Curious About Astronomy

[13] Dartnell 2007, p. 75

[14] Hart, M.H. (January 1979). "Habitable Zones Around Main Sequence Stars". Icarus 37 (1): 351–7. Bibcode:1979Icar...37..351H. doi:10.1016/0019-1035(79)90141-6.

[15] Ward & Brownlee 2000, p. 18

[16] Schmidt, Gavin (6 April 2005). "Water vapour: feedback or forcing?". RealClimate.

[17] The One Hundred Nearest Star Systems. Research Consortium on Nearby Stars.

[18] Ward & Brownlee 2000, pp. 15–33

[19] Minard, Anne (27 August 2007). "Jupiter Both an Impact Source and Shield for Earth". Retrieved 2014-01-14. without the long, peaceful periods offered by Jupiter's shield, intelligent life on Earth would never have been able to take hold.

[20] Hinse, T.C. "Chaos and Planet-Particle Dynamics within the Habitable Zone of Extrasolar Planetary Systems (A qualitative numerical stability study)" (PDF). Niels Bohr Institute. Retrieved 2007-10-31. Main simulation results observed: [1] The presence of high-order mean-motion resonances for large values of giant planet eccentricity [2] Chaos dominated dynamics within the habitable zone(s) at large values of giant planet mass.

[21] "Once you realize that most of the known extrasolar planets have highly eccentric orbits (like the planets in Upsilon Andromedae), you begin to wonder if there might be something special about our solar system" (UCBerkeleyNews quoting Extra solar planetary researcher Eric Ford.) Sanders, Robert (13 April 2005). "Wayward planet knocks extrasolar planets for a loop". Retrieved 2007-10-31.

[22] pg 220 Ward & Brownlee

[23] Lissauer 1999, as summarized by Conway Morris 2003, p. 92; also see Comins 1993

[24] Ward & Brownlee 2000, p. 191

[25] Ward & Brownlee 2000, p. 194

[26] Ward & Brownlee 2000, p. 200

[27] Taylor 1998

[28] http://www.space.com/4076-plate-tectonics-essential-alien-life.html

[29] Ward, R. D. & Brownlee, D. 2000. *Plate tectonics essential for complex evolution* - Rare Earth - Copernicus Books

[30] scientificamerican.com, Fact or Fiction: The Days (and Nights) Are Getting Longer, By Adam Hadhazy, June 14, 2010

[31] Dartnell 2007, pp. 69–70

[32] A formal description of the hypothesis is given in: Lathe, Richard (March 2004). "Fast tidal cycling and the origin of life". *Icarus* **168** (1): 18–22. Bibcode:2004Icar..168...18L. doi:10.1016/j.icarus.2003.10.018. tidal cycling, resembling the polymerase chain reaction (PCR) mechanism, could only replicate and amplify DNA-like polymers. This mechanism suggests constraints on the evolution of extra-terrestrial life. It is taught less formally here: Schombert, James. "Origin of Life". University of Oregon. Retrieved 2007-10-31. with the vastness of the Earth's oceans it is statistically very improbable that these early proteins would ever link up. The solution is that the huge tides from the Moon produced inland tidal pools, which would fill and evaporate on a regular basis to produce high concentrations of amino acids.

[33] Lane, 2012

[34] Origin of Mitochondria

[35] Ridley M (2004) Evolution, 3rd edition. Blackwell Publishing, p. 314.

[36] T. Togashi, P. Cox (Eds.) *The Evolution of Anisogamy*. Cambridge University Press, Cambridge; 2011, p. 22-29.

[37] Beukeboom, L. & Perrin, N. (2014). *The Evolution of Sex Determination*. Oxford University Press, p. 25 . Online resources, .

[38] Czárán, T.L.; Hoekstra, R.F. (2006). "Evolution of sexual asymmetry". *BMC Evolutionary Biology* **4**: 34–46. doi:10.1186/1471-2148-4-34.

[39] (English) 800 million years for complex organ evolution - Heidelberg University

[40] Cramer 2000

[41] Ward & Brownlee 2000, pp. 271–5

[42] Barrow, John D.; Tipler, Frank J. (1988). *The Anthropic Cosmological Principle*. Oxford University Press. ISBN 978-0-19-282147-8. LCCN 87028148. Section 3.2

[43] Conway Morris 2003, Ch. 5

[44] Conway Morris, 2003, p. 344, n. 1

[45] Gribbin 2011

[46] arxiv.org, Iowa State University, Guillermo Gonzalez Galactic Habitable Zone

[47] Extraterrestrials: Where are They? 2nd ed., Eds. Ben Zuckerman and Michael H. Hart (Cambridge: Press Syndicate of the University of Cambridge, 1995), 153.

[48] http://winteryknight.com/2011/01/25/harvard-astrophysicist-backs-the-rare-earth-hypothesis/

[49] Darling, David (2001). *Life Everywhere: The Maverick Science of Astrobiology*. Basic Books/Perseus. ISBN 0-585-41822-5.

[50] Darling 2001, p. 103

[51] Frazier, Kendrick. 'Was the 'Rare Earth' Hypothesis Influenced by a Creationist?' The Skeptical Inquirer. November 1, 2001

[52] Howard, Andrew W.; Sanchis-Ojeda, Roberto; Marcy, Geoffrey W.; Johnson, John Asher; Winn, Joshua N.; Isaacson, Howard; Fischer, Debra A.; Fulton, Benjamin J.; Sinukoff, Evan; Fortney, Jonathan J. (2013). "A rocky composition for an Earth-sized exoplanet". *Nature* **503** (7476): 381–384. doi:10.1038/nature12767. ISSN 0028-0836.

[53] http://www.wired.co.uk/news/archive/2011-11/21/exoplanet-indices

[54] Stuart Gary New approach in search for alien life ABC Online. November 22, 2011

[55] Clavin, Whitney; Chou, Felicia; Johnson, Michele (6 January 2015). "NASA's Kepler Marks 1,000th Exoplanet Discovery, Uncovers More Small Worlds in Habitable Zones". *NASA*. Retrieved 6 January 2015.

[56] Kasting 2001, pp. 123

[57] Borenstein, Seth (4 November 2013). "8.8 billion habitable Earth-size planets exist in Milky Way alone". nbcnews.com/. Retrieved 2013-11-05.

[58] Overbye, Dennis (4 November 2013). "Far-Off Planets Like the Earth Dot the Galaxy". *New York Times*. Retrieved 5 November 2013.

[59] Petigura, Eric A.; Howard, Andrew W.; Marcy, Geoffrey W. (31 October 2013). "Prevalence of Earth-size planets orbiting Sun-like stars". *Proceedings of the National Academy of Sciences of the United States of America*. arXiv:1311.6806. Bibcode:2013PNAS..11019273P. doi:10.1073/pnas.1319909110. Retrieved 5 November 2013.

[60] Khan, Amina (November 4, 2013). "Milky Way may host billions of Earth-size planets". *Los Angeles Times*. Retrieved November 5, 2013.

[61] Kasting, James (2001). "Peter Ward and Donald Brownlee's "Rare Earth"". *Perspectives in Biology and Medicine* **44** (1): 118–120.

[62] Brumhel, Geoff (2007). "Jupiter's protective pull questioned". *news@nature*. doi:10.1038/news070820-11. ISSN 1744-7933.

[63] Horner, J.; Jones, B.W. (2008). "Jupiter – friend or foe? I: the asteroids" (PDF). *International Journal of Astrobiology* **7** (3&4): 251–261. arXiv:0806.2795. Bibcode:2008IJAsB...7..251H. doi:10.1017/S1473550408004187.

[64] Cooper, Keith (2012-03-12). "Villain in disguise: Jupiter's role in impacts on Earth". Retrieved 2015-09-02.

[65] Gipson, Lillian (24 July 2015). "New Horizons Discovers Flowing Ices on Pluto". *NASA*. Retrieved 24 July 2015.

[66] Ward & Brownlee 2000, pp. 191–193

[67] Stern, S. A.; Cunningham, N. J.; Hain, M. J.; Spencer, J. R.; Shinn, A. (2012). "FIRST ULTRAVIOLET REFLECTANCE SPECTRA OF PLUTO AND CHARON BY THEHUBBLE SPACE TELESCOPECOSMIC ORIGINS SPECTROGRAPH: DETECTION OF ABSORPTION FEATURES AND EVIDENCE FOR TEMPORAL CHANGE". *The Astronomical Journal* **143** (1): 22. doi:10.1088/0004-6256/143/1/22. ISSN 0004-6256.

[68] Hand, Eric (2015). "UPDATED: Pluto's icy face revealed, spacecraft 'phones home'". *Science*. doi:10.1126/science.aac8847. ISSN 0036-8075.

[69] Barr, Amy C.; Collins, Geoffrey C. (2015). "Tectonic activity on Pluto after the Charon-forming impact". *Icarus* **246**: 146–155. doi:10.1016/j.icarus.2014.03.042. ISSN 0019-1035.

[70] Yin, A. (2012). "Structural analysis of the Valles Marineris fault zone: Possible evidence for large-scale strike-slip faulting on Mars". *Lithosphere* **4** (4): 286–330. doi:10.1130/L192.1. ISSN 1941-8264.

[71] Greenberg, Richard; Geissler, Paul; Tufts, B. Randall; Hoppa, Gregory V. (2000). "Habitability of Europa's crust: The role of tidal-tectonic processes". *Journal of Geophysical Research* **105** (E7): 17551. Bibcode:2000JGR...10517551G. doi:10.1029/1999JE001147.ISSN 0148-0227.

[72] "Scientists Find Evidence of 'Diving' Tectonic Plates on Europa". http://www.jpl.nasa.gov/. NASA. September 8, 2014. Retrieved 30 August 2015. External link in |website= (help)

[73] Valencia, Diana; O'Connell, Richard J.; Sasselov, Dimitar D (November 2007). "Inevitability of Plate Tectonics on Super-Earths". *Astrophysical Journal Letters* **670**(1): L45–L48. arXiv:0710.0699. Bibcode:2007ApJ...670L..45V.doi:10.1086/5240.

[74] Cowan, Nicolas B.; Abbot, Dorian S. (2014). "WATER CYCLING BETWEEN OCEAN AND MANTLE: SUPER-EARTHS NEED NOT BE WATERWORLDS". *The Astrophysical Journal* **781** (1): 27. doi:10.1088/0004-637X/781/1/27. ISSN 0004-637X.

[75] Mayor, M.; Udry, S.; Pepe, F.; Lovis, C. (2011). "Exoplanets: the quest for Earth twins". *Philosophical Transactions of the Royal Society A: Mathematical, Physical and Engineering Sciences* **369** (1936): 574. doi:10.1098/rsta.2010.0245. ISSN 1364-503X.

[76] Ward & Brownlee 2000, p. 217

[77] Killen, Rosemary; Cremonese, Gabrielle; Lammer, Helmut; et al. (2007). "Processes that Promote and Deplete the Exosphere of Mercury". *Space Science Reviews* **132** (2-4): 433–509. Bibcode:2007SSRv..132..433K. doi:10.1007/s11214-007-9232-0.

[78] Gröller, H.; Shematovich, V. I.; Lichtenegger, H. I. M.; Lammer, H.; Pfleger, M.; Kulikov, Yu. N.; Macher, W.; Amerstorfer, U. V.; Biernat, H. K. (2010). "Venus' atomic hot oxygen environment". *Journal of Geophysical Research* **115** (E12). doi:10.1029/2010JE003697. ISSN 0148-0227.

[79] Mahaffy, P. R.; Webster, C. R.; Atreya, S. K.; Franz, H.; Wong, M.; Conrad, P. G.; Harpold, D.; Jones, J. J.; Leshin, L. A.; Manning, H.; Owen, T.; Pepin, R. O.; Squyres, S.; Trainer, M.; Kemppinen, O.; Bridges, N.; Johnson, J. R.; Minitti, M.; Cremers, D.; Bell, J. F.; Edgar, L.; Farmer, J.; Godber, A.; Wadhwa, M.; Wellington, D.; McEwan, I.; Newman, C.; Richardson, M.; Charpentier, A.; Peret, L.; King, P.; Blank, J.; Weigle, G.; Schmidt, M.; Li, S.; Milliken, R.; Robertson, K.; Sun, V.; Baker, M.; Edwards, C.; Ehlmann, B.; Farley, K.; Griffes, J.; Grotzinger, J.; Miller, H.; Newcombe, M.; Pilorget, C.; Rice, M.; Siebach, K.; Stack, K.; Stolper, E.; Brunet, C.; Hipkin, V.; Leveille, R.; Marchand, G.; Sanchez, P. S.; Favot, L.; Cody, G.; Steele, A.; Fluckiger, L.; Lees, D.; Nefian, A.; Martin, M.; Gailhanou, M.; Westall, F.; Israel, G.; Agard, C.; Baroukh, J.; Donny, C.; Gaboriaud, A.; Guillemot, P.; Lafaille, V.; Lorigny, E.; Paillet, A.; Perez, R.; Saccoccio, M.; Yana, C.; Armiens-Aparicio, C.; Rodriguez, J. C.; Blazquez, I. C.; Gomez, F. G.; Gomez-Elvira, J.; Hettrich, S.; Malvitte, A. L.; Jimenez, M. M.; Martinez-Frias, J.; Martin-Soler, J.; Martin-Torres, F. J.; Jurado, A. M.; Mora-Sotomayor, L.; Caro, G. M.; Lopez, S. N.; Peinado-Gonzalez, V.; Pla-Garcia, J.; Manfredi, J. A. R.; Romeral-Planello, J. J.; Fuentes, S. A. S.; Martinez, E. S.; Redondo, J. T.; Urqui-O'Callaghan, R.; Mier, M.-P. Z.; Chipera, S.; Lacour, J.-L.; Mauchien, P.; Sirven, J.-B.; Fairen, A.; Hayes, A.; Joseph, J.; Sullivan, R.; Thomas, P.; Dupont, A.; Lundberg, A.; Melikechi, N.; Mezzacappa, A.; DeMarines, J.; Grinspoon, D.; Reitz, G.; Prats, B.; Atlaskin, E.; Genzer, M.; Harri, A.-M.; Haukka, H.; Kahanpaa, H.; Kauhanen, J.; Kemppinen, O.; Paton, M.; Polkko, J.; Schmidt, W.; Siili, T.; Fabre, C.; Wray, J.; Wilhelm, M. B.; Poitrasson, F.; Patel, K.; Gorevan, S.; Indyk, S.; Paulsen, G.; Gupta, S.; Bish, D.; Schieber, J.; Gondet, B.; Langevin, Y.; Geffroy, C.; Baratoux, D.; Berger, G.; Cros, A.; d'Uston, C.; Forni, O.; Gasnault, O.; Lasue, J.; Lee, Q.-M.; Maurice, S.; Meslin, P.-Y.; Pallier, E.; Parot, Y.; Pinet, P.; Schroder, S.; Toplis, M.; Lewin, E.; Brunner, W.; Heydari, E.; Achilles, C.; Oehler, D.; Sutter, B.; Cabane, M.; Coscia, D.; Israel, G.; Szopa, C.; Dromart, G.; Robert, F.; Sautter, V.; Le Mouelic, S.; Mangold, N.; Nachon, M.; Buch, A.; Stalport, F.; Coll, P.; Francois, P.; Raulin, F.; Teinturier, S.; Cameron, J.; Clegg, S.; Cousin, A.; DeLapp, D.; Dingler, R.; Jackson, R. S.; Johnstone, S.; Lanza, N.; Little, C.; Nelson, T.; Wiens, R. C.; Williams, R. B.; Jones, A.; Kirkland, L.; Treiman, A.; Baker, B.; Cantor, B.; Caplinger, M.; Davis, S.; Duston, B.; Edgett, K.; Fay, D.; Hardgrove, C.; Harker, D.; Herrera, P.; Jensen, E.; Kennedy, M. R.; Krezoski, G.; Krysak, D.; Lipkaman, L.; Malin, M.; McCartney, E.; McNair, S.; Nixon, B.; Posiolova, L.; Ravine, M.; Salamon, A.; Saper, L.; Stoiber, K.; Supulver, K.; Van Beek, J.; Van Beek, T.; Zimdar, R.; French, K. L.; Iagnemma, K.; Miller, K.; Summons, R.; Goesmann, F.; Goetz, W.; Hviid, S.; Johnson, M.; Lefavor, M.; Lyness, E.; Breves, E.; Dyar, M. D.; Fassett, C.; Blake, D. F.; Bristow, T.; DesMarais, D.; Edwards, L.; Haberle, R.; Hoehler, T.; Hollingsworth, J.; Kahre, M.; Keely, L.; McKay, C.; Wilhelm, M. B.; Bleacher, L.; Brinckerhoff, W.; Choi, D.; Dworkin, J. P.; Eigenbrode, J.; Floyd, M.; Freissinet, C.; Garvin, J.; Glavin, D.; Jones, A.; Martin, D. K.; McAdam, A.; Pavlov, A.; Raaen, E.; Smith, M. D.; Stern, J.; Tan, F.; Meyer, M.; Posner, A.; Voytek, M.; Anderson, R. C.; Aubrey, A.; Beegle, L. W.; Behar, A.; Blaney, D.; Brinza, D.; Calef, F.; Christensen, L.; Crisp, J. A.; DeFlores, L.; Ehlmann, B.; Feldman, J.; Feldman, S.; Flesch, G.; Hurowitz, J.; Jun, I.; Keymeulen, D.; Maki, J.; Mischna, M.; Morookian, J. M.; Parker, T.; Pavri, B.; Schoppers, M.; Sengstacken, A.; Simmonds, J. J.; Spanovich, N.; Juarez, M. d. l. T.; Vasavada, A. R.; Yen, A.; Archer, P. D.; Cucinotta, F.; Ming, D.; Morris, R. V.; Niles, P.; Rampe, E.; Nolan, T.; Fisk, M.; Radziemski, L.; Barraclough, B.; Bender, S.; Berman, D.; Dobrea, E. N.; Tokar, R.; Vaniman, D.; Williams, R. M. E.; Yingst, A.; Lewis, K.; Cleghorn, T.; Huntress, W.; Manhes, G.; Hudgins, J.; Olson, T.; Stewart, N.; Sarrazin, P.; Grant, J.; Vicenzi, E.; Wilson, S. A.; Bullock, M.; Ehresmann, B.; Hamilton, V.; Hassler, D.; Peterson, J.; Rafkin, S.; Zeitlin, C.; Fedosov, F.; Golovin, D.; Karpushkina, N.; Kozyrev, A.; Litvak, M.; Malakhov, A.; Mitrofanov, I.; Mokrousov, M.; Nikiforov, S.; Prokhorov, V.; Sanin, A.; Tretyakov, V.; Varenikov, A.; Vostrukhin, A.; Kuzmin, R.; Clark, B.; Wolff, M.; McLennan, S.; Botta, O.; Drake, D.; Bean, K.; Lemmon, M.; Schwenzer, S. P.; Anderson, R. B.; Herkenhoff, K.; Lee, E. M.; Sucharski, R.; Hernandez, M. A. d. P.; Avalos, J. J. B.; Ramos, M.; Kim, M.-H.; Malespin, C.; Plante, I.; Muller, J.-P.; Navarro-Gonzalez, R.; Ewing, R.; Boynton, W.; Downs, R.; Fitzgibbon, M.; Harshman, K.; Morrison, S.; Dietrich, W.; Kortmann, O.; Palucis, M.; Sumner, D. Y.; Williams, A.; Lugmair, G.; Wilson, M. A.; Rubin, D.; Jakosky, B.; Balic-Zunic, T.; Frydenvang,

J.; Jensen, J. K.; Kinch, K.; Koefoed, A.; Madsen, M. B.; Stipp, S. L. S.; Boyd, N.; Campbell, J. L.; Gellert, R.; Perrett, G.; Pradler, I.; VanBommel, S.; Jacob, S.; Rowland, S.; Atlaskin, E.; Savijarvi, H.; Boehm, E.; Bottcher, S.; Burmeister, S.; Guo, J.; Kohler, J.; Garcia, C. M.; Mueller-Mellin, R.; Wimmer-Schweingruber, R.; Bridges, J. C.; McConnochie, T.; Benna, M.; Bower, H.; Brunner, A.; Blau, H.; Boucher, T.; Carmosino, M.; Elliott, H.; Halleaux, D.; Renno, N.; Elliott, B.; Spray, J.; Thompson, L.; Gordon, S.; Newsom, H.; Ollila, A.; Williams, J.; Vasconcelos, P.; Bentz, J.; Nealson, K.; Popa, R.; Kah, L. C.; Moersch, J.; Tate, C.; Day, M.; Kocurek, G.; Hallet, B.; Sletten, R.; Francis, R.; McCullough, E.; Cloutis, E.; ten Kate, I. L.; Kuzmin, R.; Arvidson, R.; Fraeman, A.; Scholes, D.; Slavney, S.; Stein, T.; Ward, J.; Berger, J.; Moores, J. E. (2013). "Abundance and Isotopic Composition of Gases in the Martian Atmosphere from the Curiosity Rover". *Science* **341** (6143): 263–266. doi:10.1126/science.1237966. ISSN 0036-8075.

[80] Spencer, John R.; Calvin, Wendy M.; Person, Michael J. (1995). "Charge-coupled device spectra of the Galilean satellites: Molecular oxygen on Ganymede". *Journal of Geophysical Research* **100** (E9): 19049. doi:10.1029/95JE01503. ISSN 0148-0227.

[81] Esposito, Larry W.; Barth, Charles A.; Colwell, Joshua E.; Lawrence, George M.; McClintock, William E.; Stewart, A. Ian F.; Keller, H. Uwe; Korth, Axel; Lauche, Hans; Festou, Michel C.; Lane, Arthur L.; Hansen, Candice J.; Maki, Justin N.; West, Robert A.; Jahn, Herbert; Reulke, Ralf; Warlich, Kerstin; Shemansky, Donald E.; Yung, Yuk L. (2004). "The Cassini Ultraviolet Imaging Spectrograph Investigation". *Space Science Reviews* **115** (1-4): 299–361. doi:10.1007/s11214-004-1455-8. ISSN 0038-6308.

[82] Tokar, R. L.; Johnson, R. E.; Thomsen, M. F.; Sittler, E. C.; Coates, A. J.; Wilson, R. J.; Crary, F. J.; Young, D. T.; Jones, G. H. (2012). "Detection of exospheric O2+at Saturn's moon Dione". *Geophysical Research Letters* **39** (3): n/a–n/a. doi:10.1029/2011GL050452. ISSN 0094-8276.

[83] Glein, Christopher R.; Baross, John A.; Waite, J. Hunter (2015). "The pH of Enceladus' ocean". *Geochimica et Cosmochimica Acta* **162**: 202–219. doi:10.1016/j.gca.2015.04.017. ISSN 0016-7037.

[84] Teolis, B. D.; Jones, G. H.; Miles, P. F.; Tokar, R. L.; Magee, B. A.; Waite, J. H.; Roussos, E.; Young, D. T.; Crary, F. J.; Coates, A. J.; Johnson, R. E.; Tseng, W.- L.; Baragiola, R. A. (2010). "Cassini Finds an Oxygen-Carbon Dioxide Atmosphere at Saturn's Icy Moon Rhea". *Science* **330** (6012): 1813–1815. doi:10.1126/science.1198366. ISSN 0036-8075.

[85] http://gizmodo.com/theres-primordial-oxygen-leaking-from-rosettas-comet-1739333271

[86] Hall, D. T.; Strobel, D. F.; Feldman, P. D.; McGrath, M. A.; Weaver, H. A. (1995). "Detection of an oxygen atmosphere on Jupiter's moon Europa". *Nature* **373** (6516): 677–679. doi:10.1038/373677a0. ISSN 0028-0836.

[87] Narita, Norio; Enomoto, Takafumi; Masaoka, Shigeyuki; Kusakabe, Nobuhiko (2015). "Titania may produce abiotic oxygen atmospheres on habitable exoplanets". *Scientific Reports* **5**: 13977. doi:10.1038/srep13977. ISSN 2045-2322.

[88] Oxygen-Free Animals Discovered-A First. National Geographic news

[89] Danovaro R; Dell'anno A; Pusceddu A; Gambi C; et al. (April 2010). "The first metazoa living in permanently anoxic conditions". *BMC Biology* **8** (1): 30. doi:10.1186/1741-7007-8-30. PMC 2907586. PMID 20370908.

[90] Stevenson, J.; Lunine, J.; Clancy, P. (2015). "Membrane alternatives in worlds without oxygen: Creation of an azotosome". *Science Advances* **1** (1): e1400067–e1400067. doi:10.1126/sciadv.1400067. ISSN 2375-2548.

[91] Schirrmeister, B. E.; de Vos, J. M.; Antonelli, A.; Bagheri, H. C. (2013). "Evolution of multicellularity coincided with increased diversification of cyanobacteria and the Great Oxidation Event". *Proceedings of the National Academy of Sciences* **110** (5): 1791–1796. doi:10.1073/pnas.1209927110. ISSN 0027-8424.

[92] Mills, D. B.; Ward, L. M.; Jones, C.; Sweeten, B.; Forth, M.; Treusch, A. H.; Canfield, D. E. (2014). "Oxygen requirements of the earliest animals". *Proceedings of the National Academy of Sciences* **111** (11): 4168–4172. doi:10.1073/pnas.1400547111. ISSN 0027-8424.

[93] Hartman H, McKay CP "Oxygenic photosynthesis and the oxidation state of Mars." Planet Space Sci. 1995 Jan-Feb;43(1-2):123-8.

[94] Choi, Charles Q. (2014). "Does a Planet Need Life to Create Continents?". *Astrobiology Magazine*. retrieved 2014-01-06

[95] Kasting 2001, pp. 130

[96] Kasting 2001, pp. 118–120

[97] Kasting 2001, pp. 128–129

[98] Belbruno, E.; J. Richard Gott III (2005). "Where Did The Moon Come From?". *The Astronomical Journal* **129** (3): 1724–45. arXiv:astro-ph/0405372. Bibcode:2005AJ....129.1724B. doi:10.1086/427539.

[99] discovery.com What If Earth Became Tidally Locked? Feb 2, 2013

[100] Ward & Brownlee 2000, p. 233

[101] Nick, Hoffman (2001-06-11). "The Moon And Plate Tectonics: Why We Are Alone". *Space Daily*. Retrieved 2015-08-08.

[102] Turner, S.; Rushmer, T.; Reagan, M.; Moyen, J.-F. (2014). "Heading down early on? Start of subduction on Earth". *Geology* **42** (2): 139–142. doi:10.1130/G34886.1. ISSN 0091-7613.

[103] UCLA scientist discovers plate tectonics on Mars By Stuart Wolpert August 09, 2012

[104] Dirk Schulze-Makuch; Louis Neal Irwin (2 October 2008). *Life in the Universe: Expectations and Constraints*. Springer Science & Business Media. p. 162. ISBN 978-3-540-76816-6.

[105] Dean, Cornelia (September 7, 2015). "The Tardigrade: Practically Invisible, Indestructible 'Water Bears'". *New York Times*. Retrieved September 7, 2015.

[106] Tarter, Jill. "Exoplanets, Extremophiles, and the Search for Extraterrestrial Intelligence" (PDF). State University of New York Press. Retrieved 2015-09-11.

[107] Reynolds, R.T.; McKay, C.P.; Kasting, J.F. (1987). "Europa, Tidally Heated Oceans, and Habitable Zones Around Giant Planets". *Advances in Space Research* **7** (5): 125–132. Bibcode:1987AdSpR...7..125R. doi:10.1016/0273-1177(87)90364-4.

[108] For a detailed critique of the Rare Earth hypothesis along these lines, see Cohen & Stewart 2002.

[109] Vaclav Smil (2003). *The Earth's Biosphere: Evolution, Dynamics, and Change*. MIT Press. p. 166. ISBN 978-0-262-69298-4.

32.6   References

- 'Hundreds of worlds' in Milky Way

- Barrow, John D.; Tipler, Frank J. (1988). *The Anthropic Cosmological Principle*. Oxford University Press. ISBN 978-0-19-282147-8. LCCN 87028148.

- Cirkovic, Milan M., and Bradbury, Robert J., 2006, "Galactic Gradients, Postbiological Evolution, and the Apparent Failure of SETI." New Astronomy, vol. 11, pp. 628–639.

- Comins, Neil F. (1993). *What If the Moon Didn't Exist? Voyages to Earths that might have been*. HarperCollins.

- Conway Morris, Simon (2003). *Life's Solution: Inevitable Humans in a Lonely Universe*. Cambridge University Press. ISBN 0 521 82704 3.

- Cohen, Jack; Stewart, Ian (2002). *What Does a Martian Look Like: The Science of Extraterrestrial Life*. Ebury Press. ISBN 0-09-187927-2.

- Cramer, John G. (September 2000). "The 'Rare Earth' Hypothesis". *Analog Science Fiction & Fact Magazine*.

- Dartnell, Lewis (2007). *Life in the Universe, a Beginner's Guide*. Oxford: One World.

- Gonzalez, Guillermo; Brownlee, Donald; Ward, Peter (July 2001). "The Galactic Habitable Zone: Galactic Chemical Evolution".*Icarus***152**(1): 185–200. arXiv:astro-ph/0103165. Bibcode:2001Icar..152..185G.doi:10.1006/ica

- Gribbin, John (2011). *Alone in the Universe: Why our planet is unique*. Wiley.

- Kasting, James; Whitmire, D. P.; Reynolds, R. T. (1993). "Habitable zones around main sequence stars". *Icarus* **101** (1): 108–28. Bibcode:1993Icar..101..108K. doi:10.1006/icar.1993.1010. PMID 11536936.

- Kasting, James (2001). "Peter Ward and Donald Brownlee's "Rare Earth"". *Perspectives in Biology and Medicine* **44** (1): 118–120.

- Kirschvink, Joseph L.; Ripperdan, Robert L.; Evans, David A. (1997). "Evidence for a Large-Scale Reorganization of Early Cambrian Continental Masses by Inertial Interchange True Polar Wander". *Science* **277** (5325): 541–45. doi:10.1126/science.277.5325.541.

- Knoll, Andrew H (2003). *Life on a Young Planet: The First Three Billion Years of Evolution on Earth*. Princeton University Press.

- Lane, Nick (28 June 2012). "Life: is it inevitable or just a fluke?". *New Scientist* (2870). Retrieved 1 July 2012.

- Lineweaver, Charles H.; Fenner, Yeshe; Gibson, Brad K. (2004). "The Galactic Habitable Zone and the Age Distribution of Complex Life in the Milky Way" (PDF). *Science* **303** (5654): 59–62. arXiv:astro-ph/0401024. Bibcode:2004Sci...303...59L. doi:10.1126/science.1092322. PMID 14704421.

- Lissauer, J.J. (December 1999). "How common are habitable planets?". *Nature* **402** (6761 Suppl): C11–4. doi:10.1038/35011503. PMID 10591221.

- Prantzos, Nikos (March 2008). Bada, J.; et al., eds. "On the Galactic Habitable Zone". *Space Science Reviews* **135** (1-4): 313–322. arXiv:astro-ph/0612316. Bibcode:2008SSRv..135..313P. doi:10.1007/s11214-007-9236-9.

- Ross, Hugh (1993). "Some of the parameters of the galaxy-sun-earth-moon system necessary for advanced life". *The Creator and the Cosmos* (2nd ed.). Colorado Springs CO: NavPress.

- Scharf, Caleb (17 July 2012). "How Black Holes Shape the Galaxies, Stars and Planets around Them". *Scientific American*.

- Taylor, Stuart Ross (1998). *Destiny or Chance: Our Solar System and Its Place in the Cosmos*. Cambridge University Press.

- Tipler FJ(2003). "Intelligent Life in Cosmology".*International Journal of Astrobiology* **2**(2): 141–8. arXiv:0704. Bibcode:2003IJAsB...2..141T. doi:10.1017/S1473550403001526.

- Ward, Peter D.; Brownlee, Donald (2000). *Rare Earth: Why Complex Life is Uncommon in the Universe*. Copernicus Books (Springer Verlag). ISBN 0-387-98701-0.

- Webb, Stephen (2002). *Where is Everybody? (If the universe is teeming with aliens, Where is Everybody?: Fifty solutions to the Fermi paradox and the problem of extraterrestrial life)*. Copernicus Books (Springer Verlag).

- Stenger V (1999). "The Anthropic Coincidences: A Natural Explanation". *The Skeptical Intelligencer* **3**: 3.

- Waltham, David (2013). *Lucky Planet*. Basic Books. A recent defense of the Rare Earth Hypothesis by a UK geologist.

32.7   External links

- Home page of *Rare Earth* (archival)

- Reviews of *Rare Earth*:

  - Athena Andreadis, PhD in molecular biology.

  - Kendrick Frazier, editor, *Skeptical Inquirer*.

- "Galactic Habitable Zone." *Astrobiology Magazine*. May 18, 2001.

- Gregg Easterbrook, "Are We Alone?" *The Atlantic Monthly*, August 1988. Article that anticipates REH in some respects.

- Solstation.com: "Stars and Habitable Planets."

- Recer, Paul (June 1, 1999). "Radio astronomers measure sun's orbit around Milky Way". *Houston Chronicle.* Associated Press. Archived from the original on 1999-10-11.

- Falcon-Lang, Howard (9 December 2011). "Life on Earth: Is our planet special?". *BBC News.*

- Morison, Ian (24 September 2014). "Are We Alone? The search for life beyond the Earth". Gresham College.

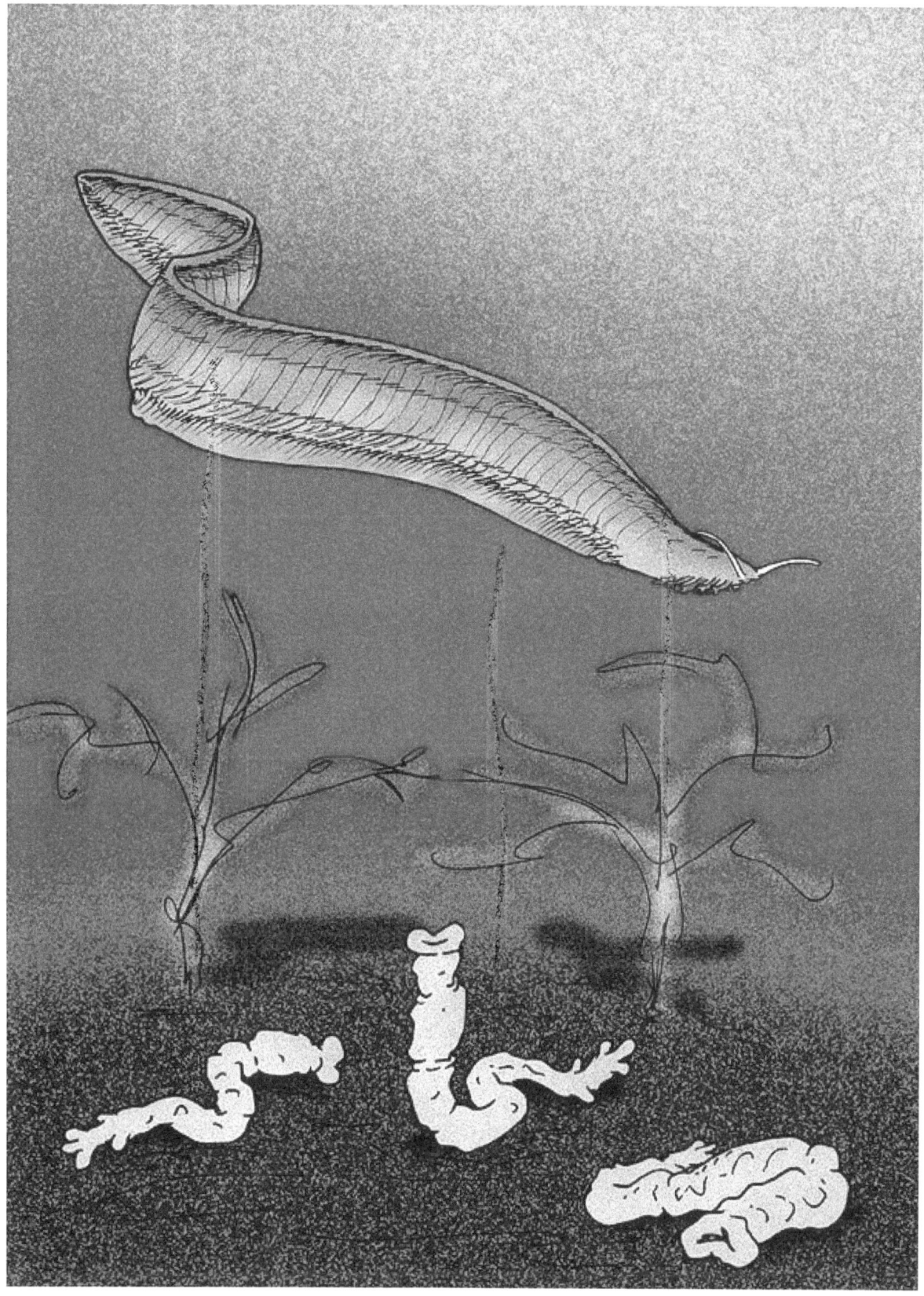

According to Rare Earth, the Cambrian explosion that saw extreme diversification of chordata from simple forms like Pikaia (pictured) was an improbable event

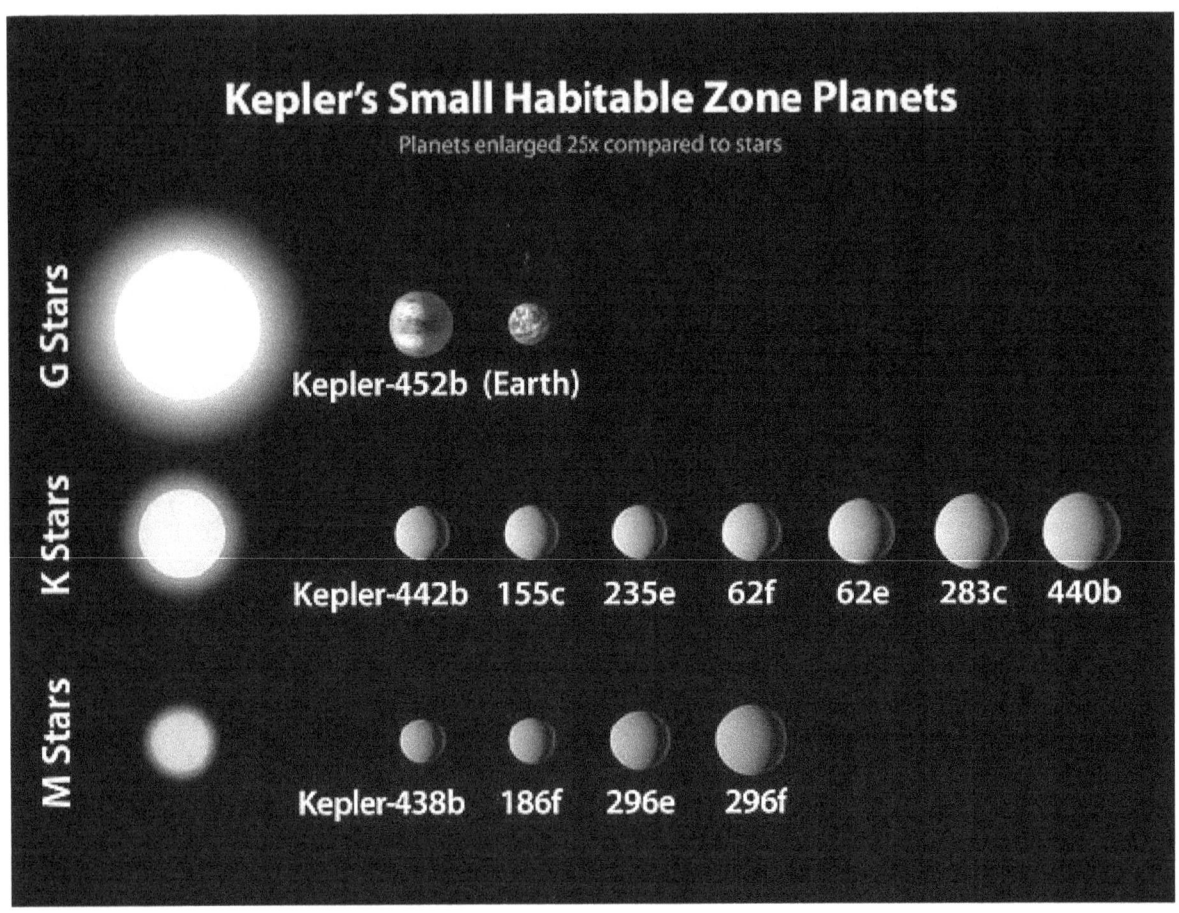

Planets similar to Earth in size are being found in relatively large number in the habitable zones of similar stars. The 2015 infographic depicts Kepler-62e, Kepler-62f, Kepler-186f, Kepler-296e, Kepler-296f, Kepler-438b, Kepler-440b, Kepler-442b.[55]

*Geological discoveries like the active features of **Pluto**'s Tombaugh Regio appear to contradict the argument that geologically active worlds like Earth are rare.*[651]

Animals like Spinoloricus nov. sp. appear to defy the premise that animal life would not exist without oxygen

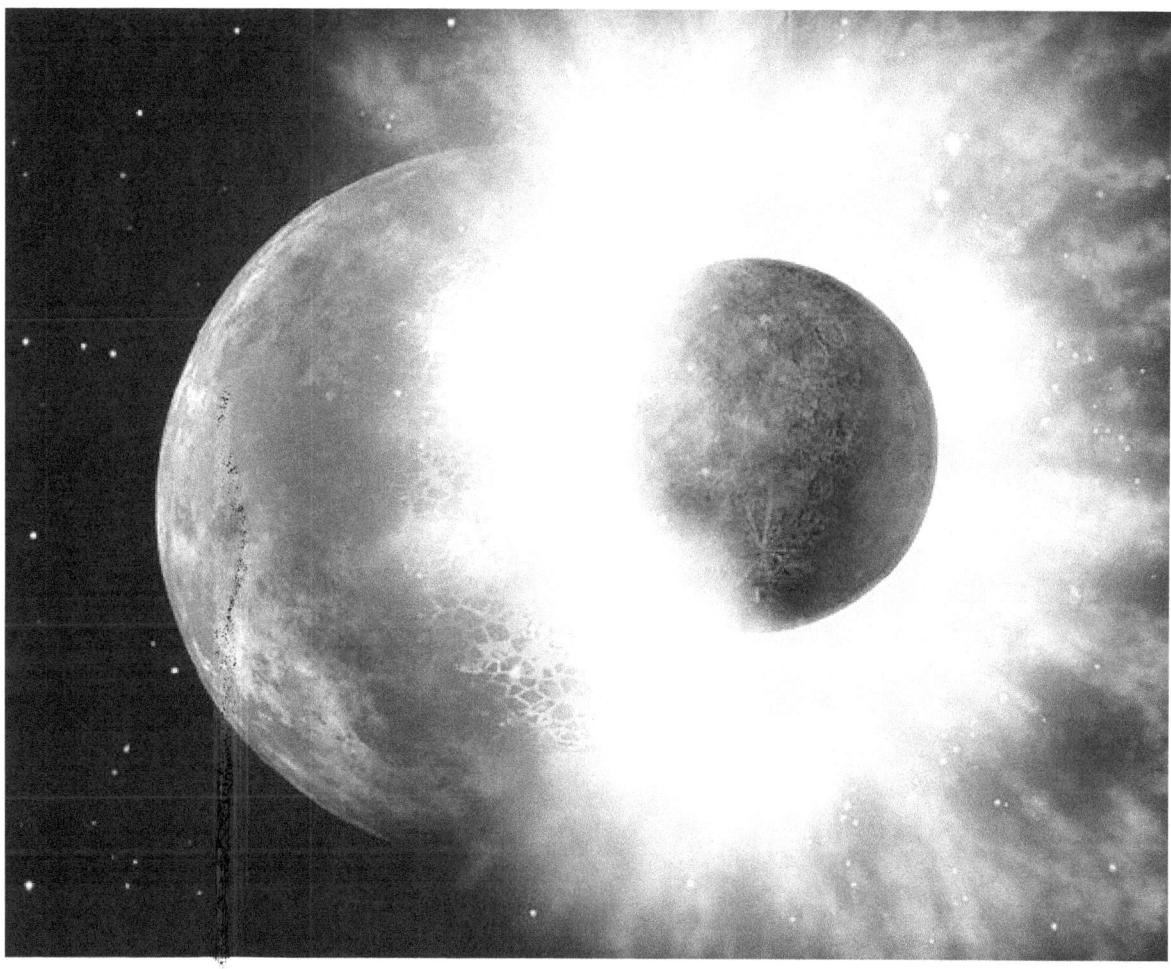

Artist's depiction of a collision between two planetary bodies. Such an impact may not be necessary to generate rotational speed, rotational axis, plate tectonics or magnetic field.

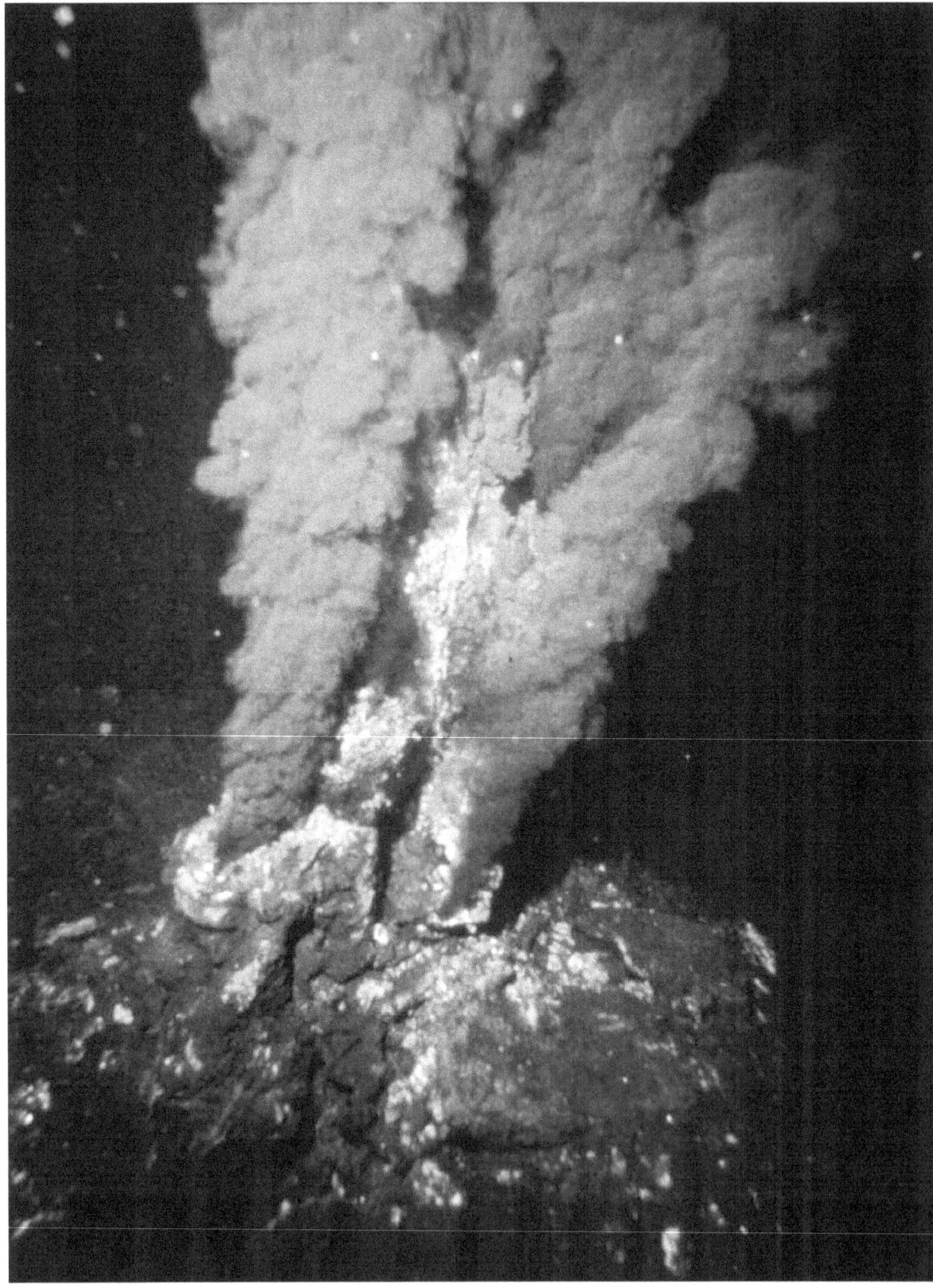

Complex life may exist elsewhere in environments similar to those found around black smokers on Earth.

Chapter 33

Sentience quotient

The **sentience quotient** concept was introduced by Robert A. Freitas Jr. in the late 1970s.[1] It defines sentience as the relationship between the information processing rate (bit/s) of each individual processing unit (neuron), the weight/size of a single unit and the total number of processing units (expressed as mass).

This is a non-standard usage of the word "sentience" which in standard usage relates to an individual organism's capacity to perceive the world subjectively (The word "sentience" is derived from the Latin "sentire" meaning "to feel" and is closely related to the word "sentiment." Intelligence or cognitive capacity is better denoted by the word "sapience" and not "sentience.")

The potential and total processing capacity of a brain, based on the amount of neurons and the processing rate and mass of a single one, combined with its design (myelin coating and specialized areas and so on) and programming, lays the foundations of the brain level of the individual. Not just in humans, but in all organisms, even artificial ones such as computers (although their "brain" is not based on neurons).

33.1 Definition

The sentience quotient (SQ) of an individual is a measure of the efficiency of an individual brain, not its relative intelligence, and is defined as:

$$SQ = \log_{10}\left(\frac{I}{M}\right)$$

where I is the information processing rate (bits/s) and M is the mass of the brain (kg). The lower limit of SQ is approximately -70, while the upper (quantum) limit is about 50.[1]

33.2 SQ's of various entities

According to this equation, humans have an SQ of +13. A human neuron has an average mass of about 10^{-10} kg and one neuron can process 1000-3000 bit/s, giving us an SQ rating of +13. All other animals with a nervous system (or all "neuronal sentience") from insects to mammals, cluster within several points of the human value. Plants cluster around an SQ of -2. Carnivorous plants have an SQ of +1, while the Cray-1 had an SQ of +9. IBM Watson, which achieves 80 TFLOPS[2] (using 64-bit words) and consists of 90 IBM Power 750 weighing approximately 100 kilograms (220 lb) each,[3] has an SQ in the range of +11—+12.

The theoretical superconducting Josephson junction electronic gates could weigh 10^{-12} kg and process 10^{11} bits/s, giving an "electronic sentience" made of these components an SQ of +23.

33.3 SQ spectrum

SQ is not limited to sentient beings so far encountered by humans, but extends to all possible sentiences, defining an expected range.

The lowest SQ possible would have just one neuron with the mass of the whole universe (10^{52} kg) and require a time equal to the age of the universe (10^{18} seconds) to process just one bit, giving a minimum SQ of −70.[1] It has been argued that under multiverse theory, an infinitely low SQ is theoretically possible, though Freitas is not known to have commented on this possibility himself.

The fundamental upper limit to brain efficiency is imposed by the laws of quantum mechanics: all information, to be acted upon, must be represented physically and be carried by matter-energy "markers." According to the Uncertainty Principle in quantum mechanics, the lower limit for the accuracy with which energy can be measured—the minimum measurable energy level for a marker carrying one bit–is given by Planck's constant h divided by T, the duration of the measurement. If one energy level is used to represent one bit, then the maximum bit rate of a brain is equal to the total energy available E ($= mc_0{}^2$) for representing information, divided by the minimum measurable energy per bit (h/T) divided by the minimum time required for readout (T): $mc_0{}^2/h = 10^{50}$ (bit/s)/kg. Hence the maximum possible SQ is +50.[1]

33.4 Implications for interspecies communication

According to Freitas, an alien civilization having their consciousness running on non-biological hardware (such as quantum-mechanical circuits) could have an SQ of 23+, 10 orders of magnitude more than the human SQ. Freitas states that such a gap in SQ "may affect our ability, and the desirability, of communicating with extraterrestrial beings...It may be that there is a minimum SQ "communication gap," an intellectual distance beyond which no two entities can meaningfully converse." [1] For example, an alien civilization may form a Matrioshka brain or a black hole and communicate using neutrinos or gamma-ray bursts at bandwidths that exceed our receiving capabilities.

> At present, human scientists are attempting to communicate outside our species to primates and cetaceans, and in a limited way to a few other vertebrates. This is inordinately difficult, and yet it represents a gap of at most a few SQ points. The farthest we can reach in our "communication" with vegetation is when we plant, water, or fertilize it, but it is evident that messages transmitted across an SQ gap of 10 points or more cannot be very meaningful. What, then, could an SQ +50 Superbeing possibly have to say to us?
> — Robert A. Freitas Jr

33.5 Analysis, misconceptions and criticism

The sentience quotient (SQ) is misleading and is not a measure of the *efficiency* of an *individual brain* as in from person to person. Efficiency here means the elemental *effectivity* of utilizing matter in implementing or constructing a certain "neuron design". As there is a constant firing rate for a given neuron type, information processing power I is proportional to brain mass M. This removes any assignable individuality between persons. The SQ value for humans would also apply to all mammals. The Drosophila fly has a similar firing rate, this implies the same SQ value is universal for most neuron equipped organisms on Earth.[4] The idea of this concept is to use, in a similar fashion as specific weight, a measure for comparison. The SQ equation serves to show the basic underlying potential when comparing two fundamental different "neuron designs". This is due to the logarithm being used as a simplified scale which only show significant differences in the order of one order of magnitude apart for $\frac{I}{M}$.

33.6 See also

- AI

- Artificial consciousness

- Communalness

- Consciousness

- Fermi Paradox

- Encephalization quotient

- Hormonal sentience

- Intelligence quotient

- Sentience

- Transhumanism

- Xenolinguistics

33.7   Notes

[1] Dr. Freitas, Robert A. Jr., *Xenopsychology*, Analog Science Fiction/Science Fact, Vol. 104, April 1984, pp 41-53

[2] Tony Pearson, IBM Watson - How to build your own "Watson Jr." in your basement, Inside System Storage

[3] IBM, 8233-E8B (IBM Power 750 Express)

[4] *Refractory Sampling Links Efficiency and Costs of Sensory Encoding to Stimulus Statistics* Zhuoyi Song, Mikko Juusola, The Journal of Neuroscience, May 21, 2014

33.8   References

Moravec, Hans. "When will computer hardware match the human brain?" *Journal of Evolution and Technology* 1998, Vol. 1. Last accessed 11 April 2008.

33.9   References to SQ

- Hypnotic experiments with 'time sense', Dr. Bernard Aaronson, Bureau of Research in Neuropsychology & Psychiatry, described by Freitas in "Timeless Minds," Omni 6 (February 1984):38.

- Sentience Quotient is referenced in the Artificial Intelligence issue of Daedalus, Proceedings of the American Academy of Arts and Sciences. [Probably Volume 117, Winter 1988.]

- "Sentience Quotient" shows up in Star Trek Voyager episode "Eye of the Needle", broadcast Monday, February 20, 1995, in the review by Tim Wright.

- "Sentience Quotient" appears in the Transhuman Terminology Page by Anders Sandberg, in the Lextropicon collected by Max More, and in the Nanotechnology Glossary by 7th Wave Inc.

- "Sentience Quotient" as defined by Freitas is mentioned in: Linda MacDonald Glenn, "Biotechnology at the Margins of Personhood: An Evolving Legal Paradigm," Journal of Evolution and Technology, Vol. 13 (2002).

- "Sentience Quotient" is described and referenced in Ray Kurzweil, The Singularity is Near, Viking, 2005, p. 536; and in Ray Kurzweil, How to Create a Mind, Viking, 2012, p. 316.

Chapter 34

Zoo hypothesis

The **zoo hypothesis** is one of many theoretical explanations for the Fermi paradox. The hypothesis speculates as to the assumed behavior and existence of technically advanced extraterrestrial life and the reasons they refrain from contacting Earth. One interpretation of the hypothesis argues that intelligent alien life intentionally ignores Earth to allow for natural evolution and sociocultural development. The hypothesis seeks to explain the apparent absence of extraterrestrial life despite its generally accepted plausibility and hence the reasonable expectation of its existence.[1]

Aliens might, for example, choose to allow contact once the human species has passed certain technological, political, or ethical standards. They might withhold contact until humans force contact upon them, possibly by sending a spacecraft to planets they inhabit. Alternatively, a reluctance to initiate contact could reflect a sensible desire to minimize risk. An alien society with advanced remote-sensing technologies may conclude that direct contact with neighbors confers added risks to oneself without an added benefit.

34.1 Assumptions

The zoo hypothesis assumes first that a large number of alien cultures exist, and second that these aliens have great reverence for independent, natural evolution and development. In particular, assuming that intelligence is a physical process that acts to maximize the diversity of a system's accessible futures,[2] a fundamental motivation for the zoo hypothesis would be that premature contact would "unintelligently" reduce the overall diversity of paths the universe itself could take.

These ideas are perhaps most plausible if there is a relatively universal cultural or legal policy among a plurality of extraterrestrial civilizations necessitating isolation with respect to civilizations at Earth-like stages of development. In a Universe without a hegemonic power, random single civilizations with independent principles would make contact. This makes a crowded Universe with clearly defined rules seem more plausible.[3]

If there is a plurality of alien cultures, however, this theory may break down under the uniformity of motive concept because it would take just a single extraterrestrial civilization to decide to act contrary to the imperative within our range of detection for it to be abrogated, and the probability of such a violation increases with the number of civilizations.[4] This idea, however, becomes more plausible if all civilizations tend to evolve similar cultural standards and values with regard to contact much like convergent evolution on Earth has independently evolved eyes on numerous occasions,[5] or all civilizations follow the lead of some particularly distinguished civilization . . . the first civilization.[6]

34.2 Fermi paradox

With this in mind, a modified Zoo Hypothesis becomes a more appealing answer to the Fermi paradox. The time between the emergence of the first civilization within the Milky Way and all subsequent civilizations could be enormous. Monte Carlo simulation shows the first few inter-arrival times between emergent civilizations would be similar in length to geologic epochs on Earth. Just what could a civilization do with a ten-million, one-hundred-million, or half-billion-year head start?[7]

Even if this first grand civilization is long gone, their initial legacy could live on in the form of a passed-down tradition, or perhaps an artificial life form dedicated to such a goal without the risk of death. Beyond this, it does not even have to be the first civilization, but simply the first to spread its doctrine and control over a large volume of the galaxy. If just one civilization gained this hegemony in the distant past, it could form an unbroken chain of taboo against rapacious colonization in favour of non-interference in those civilizations that follow. The uniformity of motive concept previously mentioned would become moot in such a situation.

If the oldest civilization still present in the Milky Way has, for example, a 100-million-year time advantage over the next oldest civilization, then it is conceivable that they could be in the singular position of being able to control, monitor, influence or isolate the emergence of every civilization that follows within their sphere of influence. This is analogous to what happens on Earth within our own civilization on a daily basis, in that everyone born on this planet is born into a pre-existing system of familial associations, customs, traditions and laws that were already long established before our birth and which we have little or no control over.[8]

34.3   Appearance in fiction

- In Olaf Stapledon's 1937 novel *Star Maker*, great care is taken by the Symbiont race to keep its existence hidden from "pre-utopian" primitives, "lest they should lose their independence of mind". It is only when such worlds become utopian-level space travellers that the Symbionts make contact and bring the young utopia to an equal footing.

- Arthur C. Clarke's *The Sentinel* (first published in 1951) and its later novel adaptation *2001: A Space Odyssey* (1968) feature a beacon which is activated when the human race discovers it on the moon. An alien race has apparently visited us in the distant past.

- In *Childhood's End*, a novel by Arthur C. Clarke published in 1953, the alien cultures had been observing and registering the Earth's evolution and human history for thousands (perhaps millions) of years. At the beginning of the book, when mankind is about to achieve spaceflight, the aliens reveal their existence and quickly end the arms race, colonialism, racial segregation and the Cold War.

- In *Star Trek*, the Federation (including humans) has a strict Prime Directive policy of nonintervention with less technologically advanced cultures which the Federation encounters. The threshold of inclusion is the independent technological development of faster-than-light propulsion. In the show's canon the Vulcan race limited their encounters to observation until humans made their first warp flight, after which they initiated first contact, indicating the practice predated the Human race's advance of this threshold. Additionally, in the episode The Chase (TNG), a message from a first (or early) civilization is discovered, hidden in DNA of life spread across many worlds, something that could only have been fully discovered after a race had become sufficiently advanced.

- In Julian May's 1987 novel *Intervention*, the five alien races of the Galactic Milieu keep the Earth under surveillance, but do not intervene until humans demonstrate mental and ethical maturity through a paranormal prayer of peace.

- In Robert J. Sawyer's SF novel *Calculating God* (2000), Hollus, a scientist from an advanced alien civilization, denies that her government is operating under the prime directive.

- In *Hard to Be a God* by Arkady and Boris Strugatsky, the (unnamed) medieval-esque planet where the novel takes action is protected by the advanced civilization of Earth, and the observers from Earth present on the planet are forbidden to intervene and make overt contact. One of the major themes of the novel is the ethical dilemma presented by such a stance to the observers.

- In *Speaker for the Dead* by Orson Scott Card, the human xenobiologists and xenologers, biologists and anthropologists observing alien life, are forbidden from giving the native species, the Pequeninos, any technology or information. When one of the xenobiologists is killed in an alien ceremony, they are forbidden to mention it. This happens again until Ender Wiggin, the main character of Ender's Game, explains to the Pequeninos that humans cannot partake in the ceremony because it kills them. While this is not exactly an example of the zoo hypothesis, since humanity makes contact, it is very similar and the humans seek to keep the Pequeninos ignorant of technology.

- In *South Park*'s inaugural episode of season seven, "Cancelled," aliens refrain from contacting Earth because the planet is the subject and setting of a reality television show. Unlike most variations of the zoo hypothesis where contact is not initiated in order to allow organic socioeconomic, cultural, and technological development, the aliens in this episode refrain from contact for the sole purpose of entertainment. In essence, the aliens treat all of Earth like the titular character in *The Truman Show* in order to maintain the show's integrity.

34.4 References

[1] Ball, John A. (Jul 1973). "The Zoo Hypothesis". *Icarus* **19** (3): 347–349. Bibcode:1973Icar...19..347B. doi:10.1016/0019-1035(73)90111-5.

[2] A. D. Wissner-Gross, "Causal entropic forces", *Physical Review Letters* 110, 168702 (2013).

[3] Soter, S. (2005). Astrobiol. Mag. 17 Oct "SETI and the Cosmic Quarantine Hypothesis"

[4] Crawford, I.A., "Where are They? Maybe we are alone in the galaxy after all", Scientific American, July 2000, 38–43, (2000).

[5] Kozmik, Z.; Ruzickova, J.; Jonasova, K.; Matsumoto, Y.; Vopalensky, P.; Kozmikova, I.; Strnad, H.; Kawamura, S. et al. (Jul 2008). "Assembly of the cnidarian camera-type eye from vertebrate-like components". *Proceedings of the National Academy of Sciences of the United States of America* 105 (26): 8989–8993.

[6] Bracewell, R. (1982). Pre-emption of the Galaxy by the First AdvancedCivilization, Pergmon Press, Oxford.

[7] Kardashev scale Kardashev, N.S. (1964). *Soviet Astronomy*. 8, 217

[8] Hair, T. W. (2011). "Temporal dispersion of the emergence of intelligence: An inter-arrival time analysis". *International Journal of Astrobiology* **10** (2): 131. Bibcode:2011IJAsB..10..131H. doi:10.1017/S1473550411000024.

{{int:Coll-attribution-page}}

- **Extraterrestrial life** *Source:* https://en.wikipedia.org/wiki/Extraterrestrial_life?oldid=694836586 *Contributors:* Damian Yerrick, Dreamyshade ,Eloquence, Mav, Bryan Derksen, Robert Merkel, Zundark, The Anome, Ed Poor, TomCerul, Graft, Edward, Patrick, Infrogmation, Bewilde-beast, Lexor, Kku, Ixfd64, Gdvorsky, Skysmith, Paul A. Minesweeper, Alfio, Kosebamse, Tregoweth, Ahoerstemeier, Cyp, JWSchmidt, Big-FatBuddha, Darkwind, Julesd, Glenn, Andres, Corixidae, Mxn, Guaka, Timwi, RickK, WhisperToMe, Steinsky, Nv8200pa, Martinphi, Thue,Earthsound, Wetman, Proteus, Johnleemk, PuzzletChung, Riddley, Phil Boswell, Nufy8, Moriori, Tlogmer, Goethean, Romann, Modulatum,Lowellian, PedroPVZ, Rursus, SchmuckyTheCat, Ojigiri~enwiki, Jondel, Rasmus Faber, Hadal, Mushroom, Lupo, Dina, Tobias Bergemann,David Gerard, Giftlite, Dbenbenn, MPF, Lethe, Tom harrison, Ich, Peruvianllama, TomViza, Maha ts, Gracefool, Eequor, Glengarry, Simul-era, Gadfium, Utcursch, J~enwiki, Sonjaaa, Antandrus, Beland, OverlordQ, Robert Brockway, Mark5677, Jossi, Redroach, Latitude0116,Balcer, Kevin B12, Mysidia, Phil1988, Icairns, Sam Hocevar, Arcturus, KeithTyler, Joyous !, Damieng, Adashiel, Valmi, Jimaginator, Eep², Mike Rosoft, Natrij, Freakofnurture, JTN, Gimmick Account, Discospinster, Rich Farmbrough, Rhobite, Cacycle, Vsmith, ArnoldReinhold, Dave souza, LindsayH, Antaeus Feldspar, Dbachmann, Manil, Grutter, SpookyMulder, Jackqu7, Bender235, MisterBadIdea, Cyclopia,Kaisershatner, Ben Standeven, Pedant, Brian0918, RJHall, CanisRufus, Jpittman, MBisanz, El C, Shrike, Lycurgus, Nonpareility, Kwamik-agami, RoyBoy, EurekaLott, Dustinasby, Bastique, Bobo192, Dystopos, Enric Naval, Viriditas, GTubio, Rbj, Angie Y., Mytildebang, Jag123,Forteanajones, La goutte de pluie, Aejelen, Pierre2012 , John Fader, Polylerus, Gsklee, Marcino, OGoncho, Storm Rider, Zachlipton, Alan-sohn, Coma28, Transfinite, Blahma, Neitram, 119, Arthena, Visviva, Keolah, Andrewpmk, Riana, AzaToth, Hinotori, James.england, Fritzpoll,Daniel.inform, Alex '05, Cjnm, Mysdaao, Mrestko, DreamGuy, Wtmitchell, Velella, Max rspct, Docboat, Sudachi, Jesvane, LFaraone, Bsad-owski1, Sfacets, Computerjoe, Kusma, Arthur Warrington Thomas, LukeSurl, Kazvorpal, Ceyockey, Centauri, BerserkerBen, Duke33, Njk,Alex.g, WilliamKF, MickWest, Angr, Linespoacher, Roboshed, OwenX, Woohookitty, TigerShark, Etacar11, Molloy, Pinball22, Mathmo,Jersyko, StradivariusTV, Urod, WadeSimMiser, Jeff3000, MONGO, Moormand, Oreckel, Dmol, Tomhllis, Grika, Optichan, KFan II, Norro, Hollyl, CharlesC, Waldir, Zzyzx11, Wayward, Dromedary, Essjay, MarcoTolo, Mandarax, Nivedh, Ashmoo, Graham87, Marskell, WBardwin, Magister Mathematicae, V8rik, BD2412, MikeDockery, OGRastamon, FreplySpang, Quantum bird, Kane5187, Ciroa, Kafuffle, Ketiltrout,Drbogdan, Sjakkalle, Rjwilmsi, Bremen, Jake Wartenberg, Panoptical, Hiberniantears, Linuxbeak, JHMM13, Adamacious, Stardust8212,Seraphimblade, Mred64, Ligulem, Bubba73, Ivan.Romero~enwiki, Bhadani, Yamamoto Ichiro, Scorpionman, FayssalF, Falphin, Lostsocks,Ian Pitchford, SchuminWeb, Opeth, Tom-b, Petruchi11, Doc glasgow, Nihiltres, JdforresterBot, Crazycomputers, Harmil, Bitoffish, MacRus-gail, Rune.welsh, RexNL, Wctaiwan, Gurch, Str1977, KFP, Alphachimp, Malhonen, Bmicomp, LeCire-enwiki, Srleffler, Wrightbus, Kri.Spencerk, Startaq, Butros, Nicholasink, Lord Patrick, Deyyaz, Visor, HKT, GangofOne, Jared Preston, DVdm, Mhking, Antiuser, Bgwhite,Sheean, Hall Monitor, Bomb319, Gwernol, Elfguy, UkPaolo, Theymos, Neatherday, Sceptre, Blightsoot, Stan2525, Rtkat3, Pip2andahalf,Phantomsteve, RussBot, Peoplesunionpro, WritersCramp, Muchness, WAvegetarian, Witan, Zafiroblue05, Chris Capoccia, SpuriousQ, RaquelBaranow, Devahn58, BillMasen, Varenius, Stephenb, Cate, Gaius Cornelius, Bovineone, Varnav, Wimt, Cumado19, Shanel, NawlinWiki,Vyran, Wiki alf, Erielhonan, Jaxl, Djadek, Kvn8907, Alisha Gergett, Mersenne, Introgressive, RazorICE, Bmdavll, Lykaestria, Anetode,Banes, Moe Epsilon, Truthdowser, MSJapan, Phaleux, Rdl381, Db firs, Adreamsoul, Mysid, Kortoso, Barnabypage, Csobankai Aladar, Elk-man, Aaronb1215, Ignitus, Alpha 4615, Bantosh, Nick123, Wknight94, Elysianfields, DocXango, Jangam, Poleary, 2over0, Rudrasharman,Encephalon, Ageekgal, Knotnic, Theda, Closedmouth, Rpvdk, Arthur Rubin, Pb30, Josh3580, Reyk, Sultan of smoov, JQF, Saudade7, PetriKrohn, JoanneB, Peyna, Sitenl, Bagheera, Urocyon, Tyrenius, Nixer, Ilmari Karonen, Neoaeolian, Bluezy, Junglecat, Tzepish, NeilN, Crni-Bombarder!!!, DVD R W, WesleyDodds, Dragon of the Pants, Hiddekel, Sardanaphalus, Vanka5, Attilios, Tttrung, SmackBot, Aim Here,Brammers, Dubbin, Unschool, Haymaker, Vati115, Zazaban, KnowledgeOfSelf, TestPilot, Hydrogen Iodide, Although, TBH, C.Fred, Vald,Jacek Kendysz, Jagged 85, Watercolour, Davewild, Dims, Jrockley, Delldot, Michaelll, Jab843, Timeshifter, Edgar181, HalfShadow, Alex earlier account, Septegram, Xaosflux, Yamaguchi先生, Cuddlyopedia, Linam97, Gilliam, Portillo, Ohnoitsjamie, Hmains, Skizzik, Martial Law, Weirdoactor, Rmosler2100, Chris the speller, Kurykh, Persian Poet Gal, Cattus, Fluri, Timneu22, Hibernian, Dlohcierekim's sock, J. Spencer,Eff er, Ai.kefu, Baa, CMacMillan, MaxSem, Salmar, Zsinj, Can't sleep, clown will eat me, RyanEberhart, Jeffire, Abyssal, Kelvin Case,OrphanBot, Sephiroth BCR, Nima Baghaei, TKD, Addshore, RedHillian, Runefurb, Khoikhoi, Soosed, Aldaron, Krich, Ianmacm, CanDo,Downtown dan seattle, Khukri, Decltype, Bowlhover, Nakon, TedE, James McNally, John D. Croft, Richard001, Gauntlet~enwiki, Lpgeffen,Illlaaa, 44Dume, Polonium, Junyor, WoodyWerm, Sigma 7, ElizabethFong, Bejnar, Jagg2499, Ace ETP, Thor Dockweiler, CIS, Cor anglais16, Cast, The undertow, SashatoBot, Vildricianus, SingCal, Producercunningham, Swatjester, Aximilli Isthill, Xeroes, Kuru, John, J 1982,Heimstern, Katstevens, Almkglor, Perfectblue97, JorisvS, JayMan, Tlesher, JohnWittle, IronGargoyle, Putnamehere3145, 041744, Ckatz,RandomCritic, MarkSutton, Benjaminlobato, Slakr, Hvn0413, Booksworm, Nyarkni, NJMauthor, George The Dragon, Dicklyon, Nephalim,Mets501, Rtkw, Dhp1080, AdultSwim, Anoyee, Midnightblueowl, Ryulong, Purplekitty, Danilot, Dradious, H, Suthaar, Nicolharper, Shak-ingSpirit, SimonD, Iridescent, Michaelbusch, Jason.grossman, Joseph Solis in Australia, Debeo Morium, Newone, StephenBuxton, Twas Now,Nubzor, TheBreeze, Gerfinch, LordRahl, Beno1000, CapitalR, DavidOaks, Courcelles, Cheesemonger, Matlefebvre20, PaddyM, Tawker-bot2, The Letter J, Gveret Tered, ChrisCork, TheFloydman, JForget, MrPeabody, Brainbark, JF Mephisto, CmdrObot, Jargon, Ben groulx,Tobes00, Dycedarg, Calibanu, Falconfly, Makeemlighter, Ruslik0, Rasd, KnightLago, Benwildeboer, Evan7257, Dgw, N2e, AshLin, Chmee2,Moreschi, Elsand bag57, QuinnJL, Rgonsalv , Dan Fuhry, MrFish, Ebonize christian, Alexignatiou~enwiki, FatBaka, CMG, Xaman, Shanoman,Jeek~enwiki, Slazenger, Cydebot, Shitmaster 5, Ik the Toaster, Bluecurio, Steel, Michaelas10, Gogo Dodo, Jkokavec, Crowish, Flowerpotman,Daniel J. Leivick, Mycroft.Holmes, Gilabarak, Karafias, Tawkerbot4, FDV, Itsfun, Robertinventor, Vanwiek, Dinnerbone, Garik, Har057,Protious, SpK, Omicronpersei8, Shashankgupta, JodyB, Arb, UberScienceNerd, Nol888, Nadirali, Casliber, Epbr123, Casual Moose, Pajz,Marbie, Pstanton, Aaron Pelzer, Indef blocked user 001, PhilMacdonald, Ucanlookitup, Andyjsmith, 24fan24, Headbomb, Simeon H, Ru-gadh, Marek69, Sean7phil, Timebender13, John254, A3RO, SGGH, Jakerake, Pmrobert49, Kaishinjou, RickKJr, Catsmoke, Eljamoquio,Mikeeg555, Whoda, Cooljuno411, FreeKresge, Eriemachmer, FreshFruitsRule, Keyvez, Noclevername, Mmortal03, Mentifisto, Hmrox, Cy-clonenim , AntiVandalBot, Majorly, Gioto, Luna Santin, Seaphoto, Opelio, Paul from Michigan, QuiteUnusual, Prolog, TrulyUnusual.com, Pwhitwor, TimVickers, Esteray, Electromagnet, Kevin Nelson, Robsmyth40, Skynet1216, ARTEST4ECHO, AubreyEllenShomo, RobertA, Mitchell, Myanw, Mad Pierrot, Deadbeef, GWhitewood, JackSparrow Ninja, Peter Harriman, JAnDbot, Xhienne, 24630, Husond, Barek,MER-C, Reduxx, Inks.LWC, Ericoides, Hodgetts, Smiddle, Plm209, Andonic, Kerotan, Jarkeld, Y2kcrazyjoker4, Connormah, Bakilas, Pedro,Bongwarrior, VoABot II, BruceDude, James166, StudierMalMarburg, AuburnPilot, JNW, JamesBWatson, TL789, Azznrivera, Swpb, Mutabl-eye, BobTheMad, Violentbob, Jim Douglas, Truancy07, Tonyfaull, Avicennasis, BrianGV, Catgut, Indon, ClovisPt, C.lettinga, Walkerlamond,Sgr927, IkonicDeath, Ali'i, Dinohunter, BatteryIncluded, Beetfarm Louie, Tenjikuronin, Vssun, JoergenB, TehBrandon, Nikolaj Christensen,DerHexer, Hi hi puka puka, Edward321, Saxophlute, NatureA16, Stephenchou0722, Zawarq, Kamikaze, MartinBot, BetBot ~enwiki, Arjun01,

Zogundar, Sm8900, Mitchhe 91, Thomasshaw, Deepsupport, CommonsDelinker, AlexiusHoratius, Damodar87, LittleOldMe old, Ash, Alex-Parky, Ulisse0, Master of Tofu, J.delanoy, Sasajid, Pharaoh of the Wizards, Mattlewisthepimp, CFCF, Trusilver, ProfButler, Ville V. Kokko, Bogey97, Hans Dunkelberg, Rhinestone K, Uncle Dick, Maurice Carbonaro, All Is One, Jesant13, Harold56, Dada124C41+, NerdyNSK, Ian.thomson, Darth Mike, Gzkn, Acalamari, M C Y 1008, PeterH2, TheChrisD, BrokenSphere, Pegasus1457, Katalaveno, Magiwand, Dark-Falls, Firedraikke, McSly, SpigotMap, Ryan Postlethwaite, Enricorpg, BeZet, Vanished User 4517, Belovedfreak, NewEnglandYankee, Rominandreu, Rwessel, SJP, G.Batty, AliensPhD, Malerin, Biglovinb, Workofthedevil, Umair82, West109, MetsFan76, Quogud, Donmarkdixon, Chppxbx, Bettyboop2336, Cometstyles, Kenneth M Burke, Myrealana, Nrorres, Vanished user 39948282, Treisijs, Rising*From*Ashes, Mirage GSM, Pdcook, Ja 62, Mwmillar, Trunkalunk, Suuperturtle, CardinalDan, Idioma-bot, Tobynsaunders, Gueedo, Muoi, Wikieditor06, Light of Shadow, Lights, Caribbean H.Q., Char555, Deor, McNoddy–enwiki, TreasuryTag, A.Ou, TallNapoleon, Hersfold, Jeff G., Orthologist, AlnoktaBOT, Maghnus, Newyowker, Boooobzzz, QuackGuru, Mwamanator, Philip Trueman, DoorsAjar, TXiKiBoT, Oshwah, Cosmic Latte, Convolution223, Mrkwtrs, Allbee Honnête, Jkstark, Jacob Lundberg, Alan Rockefeller, Technopat, Malljaja, Ann Stouter, Huwe2, Drestros power, Arnon Chaffin, Someguy1221, Cosmium, Warrush, Supersmarty, Oxfordwang, Anna Lincoln, Cherhillsnow, Wiikipedian, Martin451, BwDraco, LeaveSleaves, Cremepuff222, Mazarin07, Hippy-dippy, Madhero88, Discgolfrules, Tcgriffin, Jdman24, Comrade Tux, Strangerer, Falcon8765, Sardonicone, Scubasteve111, !dea4u, Sesshomaru, SchumiChamp, Insanity Incarnate, Calvin2011, Brianga, Balderunner, Atmgnef, Uranometria, Zungaphile, Logan, Closenplay, Planet-man828, Megasquid500, Daveh4h, Dinofang, NHRHS2010, Gaelen S., Ponyo, Xzazerx, Randommelon, Dusti, Ellio-3.14, Sonicology, Wibubba48, Ziplock74, Zephyrus67, Lemonflash, Dawn Bard, METIfan, Charsniper, JJ Da Kid, Indelible sinn, Yintan, Tataryn, August Dominus, Crash Underride, Dinokid, AlbertHall, Raekshea, Super Sonic Sloth, Flyer22 Reborn, Radon210, Alexbrn, RoGGe, Undead Herle King, Oxymoron83, Faradayplank, Smilesfozwood, AngelOfSadness, Lisatwo, Steven Crossin, Lightmouse, Techman224, Lumentec, Nancy, Spartan-James, StaticGull, Hamiltondaniel, Dust Filter, Dabomb87, Hawkesy12345678, Neo., Kalidasa 777, Lloydpick, Goodergrammar, Gantuya eng, Revelian, Stillwaterising, Kanonkas, Tolerancebelowzero, Sabbath73, Yaanu4, Runn–enwiki, Hpforlife, Loren.wilton, Martarius, Sfan00 IMG, Elassint, ClueBot, Cmmmm, BlazeMaster4R3AL, Foxj, The Thing That Should Not Be, Vikasatkin, Diazenefz, Gaia Octavia Agrippa, R000t, Shade11sayshello, Warder Alduren, Spandrawn, Ben5ive, SuperHamster, JTBX, Boing! said Zebedee, FractalFusion, CounterVandalismBot, Niceguyedc, Blanchardb, Fallenfromthesky, Cavalryman101, Facemuncher, Btroxmysox, Ernstblumberg, Amirreza, Supergodzilla2090, Excirial, -Midorihana-, Jusdafax, Unrealpotboy3154, Panyd, Vimalkdwivedi, NewBeat, SpikeToronto, Brnabepo2ndof4, Fishing ko, Cenarium, Peter.C, TheRedPenOfDoom, Tnxman307, M.O.X, Istok, Razorflame, Redthoreau, Hoth194, Mikaey, Froogle62, Tired time, Thingg, 7, King of the World 5, Footballfan190, Mezack26, WxHalo, Versus22, Porchcorpter, PCHS-NJROTC, Berean Hunter, Docsavage20, SoxBot III, Hattiel, Vanished user uih38riiw4hjlsd, Youhatemeandidontcare, DumZiBoT, Cowardly Lion, BarretB, XLinkBot, Bjrcoolguy210, Tarheel95, Fastily, Pichpich, Gnowor, Wakawaka1, CGaldi, Andrzej Kmicic, 68Kustom, Fullsick, Badgernet, Truthnlove, SelfQ, Herwest299, Thatguyflint, D.M. from Ukraine, Olyus, Coltonmj11, Haqqmisra, Addbot, Xp54321, Cxz111, Basilicofresco, Manwithahat, Gravitophoton, Tenev, Unforgiven666, Morriswa, Zellfaze, Joegriff4, Ronhjones, Hollando, CanadianLinuxUser, Bnaur, Cst17, Hugh wiki01, MrOllie, Proxima Centauri, Woo united11, Morning277, Chamal N, Glane23, Juneretos, Lollimaroll, AnnaFrance, Quietmare, Theking9794, 5 albert square, Avs dps, Flump1, Jaguar65, SmartBagel, Tassedethe, DubaiTerminator, Changin history n00bs, Tide rolls, Bfigura's puppy, Lightbot, Tototrotroll, Gail, BazzookaJoe, Phantom in ca, Quantumobserver, GiantPea, Bartledan, BlackMarlin, Megaman en m, Alfie66, Twiggfulysmart, Wellminesthesun, Zobango, TheSuave, Yobot, Ultrasquad, Anomalieshunters, Senator Palpatine, Fraggle81, AliDinegor, Cflm001, Legobot II, Akalvi, AdjutantRefluxSoliton, VertigoOne, Texas™, Drcocapepper13, Ningauble, Black vuh jj, Againme, Knoital, MacTire02, Magog the Ogre, AnomieBOT, Angry bee, DoctorJoeE, Jim1138, IRP, Neko85, Zuko117, Piano non troppo, Llcbwoc, Kingpin13, Yachtsman1, Flewis, Saidaziz eng, Materialscientist, Cavemanug1001, The High Fin Sperm Whale, Bestie4, Citation bot, UFOForum, E2eamon, Vuerqex, Raphaelzola, Keli2189, Neurolysis, LilHelpa, The Dean Man, Coolboyno1, Vinego, Random astronomer, Sionus, Capricorn42, Newzebras, Bihco, Pontificalibus, ChildofMidnight, Kmcdm, Darkestdeath1, Grim23, Dkelly1966, Tyrol5, Mlpearc, Singster, Gap9551, Bigaireatscheese, Jamesslot3, J04n, GrouchoBot, Ataleh, MrVoodooArmy, Skraddarbacken, Shirik, Italomex, Hoifon, Mathonius, Amaury, Voltorb, Falcon189, Cailex, Mnance1975, Heavenly Chihuahua, Lexy-lou, Shadowjams, Vanished user giiw8u4ikmfw823foine2, Glump1, Aaron Kauppi, Jacobkailynzeeba, Erik9, Islas93, Bwham, Griffinofwales, Minkala, BoomerAB, MrAwesomeDudePerson, The Skull Fiend, Captain Weirdo the Great, D10hitman, Rock4, Bobwrits, Johneamanley, Hobbit119, Raj6, Bendover61, Exchangor, Nickucla, Banshee lover, Sanpitch, ReneVenegas95, Knubbe kub, Bobri28, Cargoking, ZOLTAN THE gOd, Mj12cz, HJ Mitchell, Jamsam70, The One & Only Fools and Horses–enwiki, Citation bot 1, Pizik, Dogshitface, GaussianCopula, Amplitude101, Redrose64, Cubs197, JKDw, Pinethicket, I dream of horses, Vicenarian, Flaneurshields, Usukasshkeurmom, Juliaashford, Tom.Reding, Supreme Deliciousness, Calmer Waters, Yahia.barie, Geogene, A8UDI, Jschnur, Jakuplatzen, Xfact, Serols, Militaryman433, BobaTyroneFett, Ooooga Booga 50039, Hairy Tom, Milzmilzmilz, SkyMachine, Jokerapid, AATroop, Chachap, Cmdahler, Trappist the monk, Ksanexx, Fuad Thahir, Tycopup, Marvinandnilo, Vrenator, Suomi Finland 2009, MrX, Vancouver Outlaw, Begoon, Fmndu9aegh79q3, Lynn Wilbur, Reaper Eternal, --(Add Name Here)--, Perrielover123, Urface24, Boucher-craig, Tbhotch, Stroppolo, TheMesquito, K602, Marie Poise, Manuron, DARTH SIDIOUS 2, Firemanic9, Svr21121, Davidjess, Onel5969, Mean as custard, The Fudginator, Maryjaneadams, RjwilmsiBot, Dudemister67, SpartanGreg09, Jake576, Ilikecheeseandcereal, MyFellowAmericans234, Vermastr, Latinlover2010, DASHBot, Bob512, ImprovingWiki, John of Reading, Orphan Wiki, Aeather96, Ghostofnemo, Immunize, Gerow, Heracles31, ScottyBerg, 333beth333, Notanoption, GoingBatty, RA0808, JustinRussoLove, Minimac's Clone, Icantthinkofasuitableusername, Infringement153, The Mysterious El Willstro, Nailer111, Mr Hat 971, Wikipelli, Sinalese, K6ka, WaightZter, Claytonmaner, Solomonfromfinland, Thecheesykid, Sundance218, Qwertyrandom, John Cline, John Mentor, Fæ, A2soup, DalitDynamite, Mattimole, The next dimensions, Carva01, ElationAviation, Cami37, Reve316, AnArdentReader, Aga1234567, PhreshKiid, Tylerball22, Force92i, Logxep, QEDK, Wayne Slam, David J Johnson, Litjohn, Cit helper, 567895th, JoeSperrazza, TyA, Yotvata, Brandmeister, Coasterlover1994, Angus hendrick, L Kensington, Societies, Senjuto, Hell of howling, Donner60, I am beast 97, Gorvindo, EvenGreenerFish, Ego White Tray, Bill william compton, Status, Whydoyoucare555, Hman99, ChiZeroOne, Brotherbro, Maxigorilla, Iketsi, Forever Dusk, Haloreach89, TYelliot, Oursana, Rocketrod1960, E.f.harper, Gary Dee, Omni 98, Petrb, Grapple X, ClueBot NG, Kimubone, Thilot, Namelesslamnot, Lalalaa12, Giordanobrunoburning, MelbourneStar, This lousy T-shirt, Prathfig, Aarishshaheen, Gilderien, Iritakamas, Marcoboelling, Luckey313, Catras, Twillisjr, Hazhk, D Quinn42, Kevin Gorman, Sjmantyl, Space1134, Dream of Nyx, Widr, CostaDax, Tr00rle, Caille91, MerllwBot, Helpful Pixie Bot, LORD CAAZE, Daviddwd, Meerta, Bibcode Bot, Nashhinton, Jeraphine Gryphon, Trunks ishida, Lowercease sigmabot, BG19bot, Krenair, Silentkiller101, Billy from Bath, Meepamoop, James350z, Navhus, Qwerty1214, Sailing to Byzantium, Kangaroopower, Hallows AG, Kaspuhler, Superyoshiw, JohnChrysostom, MusikAnimal, Metricopolus, Exobiologist, Purpleaye, Dan653, Petruss, Bugoyito, Kyurem7, Cadiomals, Olev Vinn, Wld555, BurnerHero2, RakezDurban, MordinSolus0, Awright34, Daniellelynettehardy, MrBill3, Pmhorler, Josepheous, Risingstar12, Doowop62, Gwickwire, Fewny, Omgdinosaurs111, Academictask, Dvq, Edonnelly527, JENNATIFFANY, Achowat, Chip123456, Lashleymullins, Wer900, Dubstepalien, Anbu121,

EricEnfermero, F33mason, Melodychick, BattyBot, Phinizyspalding, Sparsy, Longboardhard, Allylynn9, CrazeGUY1515, Itrollfaceumadbro, EnglishMole0615, Cyberbot II, Bigbottomedgirl, The Illusive Man, ChrisGualtieri, Nick.mon, AshxRoc, Shiltsev, Slordax, EuroCarGT, Gameboy97q, Scoot4life, VoteLobster, Illia Connell, Jmorton2222, Dannyj24, WelshMan1990, Salismasher12, Party102, Qxukhgiels, 123456jp, Stempkik3, Nuf101, Gadgetball, Dexbot, Greggory12, TrollGlaDOS, Anandaraja, Webclient101, Mogism, JimmyTheRacer, Justinwgaines, Lugia2453, 93, Coolbeans443, 069952497a, Wolly123, Reatlas, Erika Leonard, Big boss170, F6Zman, Jubjub01, TishoYanchev, Jamsontoast987, Melonkelon, Eyesnore, Harlem Baker Hughes, Wethar555, Uncledrew, Wikimee34, Doctor-professor-magnificant, Praemonitus, SamoaBot, Purple Door Door, EvergreenFir, Wagnerslove, Worst cooks in America, Adspred, Kentsy, Fatum81, SchroederB, Gruckiller, Ira Bradley, LieutenantLatvia, Andymatches, Legoman 86, Eagle3399, Joppiesaus123, Fatty black man, Spoon in my bellybutton, Aambeienbeffer, NottNott, Jotknktrnfjkrtklg, Johnwatson94, Tjbll, Submarinefy, MagicatthemovieS, Jose.rivera143, XdoomdragonX, Billy Cuppertino, Meteor sandwich yum, Racer Omega, Fixxture, Man of Steel 85, Keltonmadden, SilvestreRiseguy123, Spaghetti1033, Sofia Lucifairy, Chaya5260, Pjmiller1978, IRock123, Lion buisness, Rosario Berganza, Hohohoinyourbutthole, Fafnir1, Monkbot, DK0010, Ware6561, Mickey02, ADHDavid1, MegaGardevoir68, CalebB21, Lagalog, Sofia Koutsouveli, Jack1990tut, Emeryst, GilCarlson, Akanyang, Was an hero, Gurmanmundi, Baconchocolate, Usukwiki, Godzilla1520, Zeus000000, Poiuytrewqvtaatv123321, BicelPhD, AsteriskStarSplat, Amisharajav, Dr Mullen, Deadjoker27, Marky999, Gamebuster19901, Heuh0, TeaLover1996, Joethethird, Rubbish computer, Sam taylor47, Jjmanofdoom, Lilymary123, Anunaki truth, Gweraituh3wa98th9384, PionLax, Nightwingandbats!, Isambard Kingdom, Hades324, Jiggywithe123, DN-boards1, ANOMALIEN, GeneralizationsAreBad, ErdrickLoto, Medinalori, Jerodlycett, KasparBot, Supman23, Wikichipmunk, CAPTAIN RAJU, Mrryanbeogie, Xand123er, Xyndsbjsjdfhbuasdlk.z.kxjvasgvfsadjhf.., Sujai KC, Glawio204, Ehlmao, Prazl001, BluecometFlag, Goofmel, ScottHla2kk3, Leo513, Beatrice is ugly, Julian.Hatcher.1122, Acowmoo, Shebabroward, Prhdbt, ExoeticKnowledge01106, Terrylim81, Yatrollietroll and Anonymous: 2109

- **Astrobiology** *Source:* https://en.wikipedia.org/wiki/Astrobiology?oldid=693098132 *Contributors:* Magnus Manske, Joao, Derek Ross, Brion VIBBER, Mav, Bryan Derksen, The Anome, Berek, Taw, Wayne Hardman, XJaM, Rgamble, Bignose, Edward, Lexor, Gabbe, Gdvorsky, Minesweeper, Docu, Angela, JWSchmidt, Александър, Glenn, Kimiko, Evercat, Astudent, Stone, Doradus, Phoebe, Wiwaxia, Jni, Robbot, Yas~enwiki, Fredrik, Zandperl, Kristof vt, RedWolf, Merovingian, Sverdrup, Flauto Dolce, Rursus, Davodd, Fuelbottle, Joshays, Alan Liefting, Polsmeth, Wolfkeeper, Martijn faassen, Curps, Maverick, Niteowlneils, Eequor, Brockert, Jaan513, Bobblewik, Quadell, Beland, Exigentsky, Karol Langner, APH, Latitude0116, Icairns, Tdent, MakeRocketGoNow, Moxfyre, Thorwald, Ta bu shi da yu, Discospinster, ElTyrant, Rich Farmbrough, Cacycle, Vsmith, Roo72, Mani1, Pavel Vozenilek, Bender235, JustinWick, El C, Huntster, Art LaPella, Danshil, Iridia, John Vandenberg, Dreish, Viriditas, Maurreen, Jumbuck, Alansohn, Willrocksmith, Alex '05, Tony Sidaway, Geraldshields11, Zoohouse, Dan100, Galaxiaad, Angr, Bushytails, Benbest, MatthewJ, Nakos2208~enwiki, Aristotle Pagaltzis, GregorB, Zzyzx11, TheloniousMONK, Palica, Marudubshinki, Mandarax, Ashmoo, Marskell, WBardwin, Teflon Don, Drbogdan, Rjwilmsi, Mayumashu, Koavf, Zbxgscqf, Eyu100, Mike s, DouglasGreen~enwiki, Eubot, Fifthgenerationdeb, MacRusgail, Kevinhksouth, Gareth E. Kegg, Twoeyedhuman, Bgwhite, ColdFeet, YurikBot, Zafiroblue05, Mrboh, Gaus Cornelius, CambridgeBayWeather, Member, NawlinWiki, Nowa, Welsh, Slarson, BOT-Superzerocool, Kortoso, Stefeyboy, Ignitus, Dv82matt, Genjix, Fidelsempre, Lt-wiki-bot, Jules.LT, Reyk, Petri Krohn, Black-Velvet, JoanneB, Geoffrey.landis, Anclation~enwiki, Argo Navis, GrinBot~enwiki, Buldozer, Arcadie, SaveTheWhales, Sardanaphalus, SmackBot, MattieTK, Aflm, Cnguyen, Unyoyega, Vald, Jagged 85, Davewild, Nickst, Jrockley, Beemee, Waynem, Gilliam, JAn Dudik, Chris the speller, Aaadddaaammm, MalafayaBot, DHN-bot~enwiki, Ekrenor, Modest Genius, Jefffire, SashatoBot, Coricus, Scientizzle, Gobonobo, Simongraham, JorisvS, Runningfridgesrule, Loadmaster, Smith609, Andrés D., Smedlorificus, Macellarius, Ossipewsk, Iridescent, Michaelbusch, IvanLanin, Thermochap, Phillip J, CmdrObot, JohnCD, Benwildeboer, Green caterpillar, Mike 7, Vectro, Necessary Evil, Cydebot, Abeg92, Hiprus, VashiDonsk, Tawkerbot4, AndersFeder, Arb, Crum375, Audry2, SuziT9, Wikid77, HappyInGeneral, Headbomb, Tellyaddict, CharlotteWebb, Noclevername, Sbandrews, Hazillow, AntiVandalBot, Luna Santin, Guy Macon, Dr. Submillimeter, Kent Witham, Robert A. Mitchell, JAnDbot, Dmar198, Zenjah, Edwardspec TalkBot, 100110100, Aprhys, MZWilson, Magioladitis, Swpb, Redwoodseed, Stella angela, BatteryIncluded, Nikolaj Christensen, Saxophlute, Tigrisek, Thoukis01, Jopasopa, XRiffRaffx, Tuganax~enwiki, Ben MacDui, CommonsDelinker, Nev1, Tinwelint, AstroHurricane001, All Is One, Nigholith, Ian.thomson, Phayne, DjScrawl, Rominandreu, Cs302b, SoCalSuperEagle, Vladeo, Nonluddite, Remi0o, VolkovBot, Brogdev, Sandcastle84, Rbotti, Victor.champion, TXiKiBoT, Sacramentis, Qxz, JhsBot, Serknap, AxelBrandenburg, Sri.dhyana, RobertFritzius, Disley138, Alemaeonid, AlleborgoBot, Esseh, EmxBot, SieBot, PlanetStar, Graham Beards, KGyST, Mistercraig, Turnspten, Polbot, OKBot, Jeroen888, Kalidasa 777, Martarius, ClueBot, Digitante, Artichoker, Firth m, Niceguyedc, Nanobear~enwiki, Rotational, Reconfirmer, Sun Creator, NuclearWarfare, Ieatbabiesbuhrur, Wikiblastfromthewikipast, Tired time, Thngg, Qwfp, GabrielVelasquez, DumZiBoT, 1wiki08, Sweet as candie, XLinkBot, Teh Rote~enwiki, Avoided, Duckyphysics, Cmungall, WookieeGroomer, Addbot, Non-dropframe, Atethnekos, Bemarko, Wassermensch, CactusWriter, MrOllie, Proxima Centauri, AndersBot, Omnipedian, Tassedethe, Wikbot, Tide rolls, OlEnglish, Zorrobot, Legobot, Luckas-bot, Yobot, Dov Henis, Blameitongravity, Anypodetos, Burkinafaso76, AnomieBOT, Jim1138, JackieBot, AdjustShift, Merube 89, Materialscientist, Hunnjazal, Citation bot, Allen234, Simon Chabrillat, Quebec99, LilHelpa, Xqbot, Ataleh, RibotBOT, Shadowjams, Fotaun, Another disinterested reader, Sanblihac, Paine Ellsworth, Jms903, BoundaryRider, Citation bot 1, Pinethicket, Tom.Reding, Triplestop, Tomcat7, RedBot, Wikigeorgie, Trappist the monk, Cybo122, RjwilmsiBot, Ansaszi, Otutusaus, DASHBot, EmausBot, Aeather96, Immunize, Oracle125, Slightsmile, ZéroBot, Mattersplatter, Thargor Orlando, H3llBot, David J Johnson, Wingman417, L Kensington, Sailsbystars, Ollyoxenfree, Hoibenilord, Ego White Tray, Sharonmil, Michaeltyrmarskell, Gary Dee, Grapple X, ClueBot NG, Horoporo, Catlemur, Satellizer, Telemachus.forward, BIT1982, Keithcowing, Braincricket, SDBruce, BigBark44, Pinguinus, Helpful Pixie Bot, Bibcode Bot, Edward Gordon Gey, Cadiomals, NotWith, Dieter1234, Wer900, Mdann52, ChrisGualtieri, Kuggee, Dexbot, Anderson, Adam2828, GoneMoot00, Shrikarsan, Stilgar27, Charliecat1, Monkbot, PigeonFancyHat, Georgio506, Archiloe, Fried Vegetables, BicelPhD, Isambard Kingdom, DN-boards1, KasparBot, Bullets and Bracelets, Cornelius.mchugh, Enigmaticmechanisms, Xinxinliu and Anonymous: 284

- **Biotic material** *Source:* https://en.wikipedia.org/wiki/Biotic_material?oldid=687601269 *Contributors:* Altenmann, HaeB, Pengo, Rich Farmbrough, Paleorthid, Drbogdan, Chris the speller, Lsjzl, Ventifact, Brewhaha@edmc.net, BatteryIncluded, MartinBot, Sheep2000, Azurelcicle, DRTllbrg, Anxietycello, Erik9bot, Look2See1, Ego White Tray, Mark Arsten, NottNott, Kevt2002, Nahid monavarian and Anonymous: 7

- **Exoplanet** *Source:* https://en.wikipedia.org/wiki/Exoplanet?oldid=694383641 *Contributors:* Paul Drye, Bryan Derksen, Robert Merkel, Szopen, The Anome, Taw, Malcolm Farmer, Css, Danny, XJaM, 0, Little guru, Miguel~enwiki, Roadrunner, SimonP, Dzof, Heron, Zimriel, Hephaestos, Olivier, Bdesham, Patrick, Boud, Michael Hardy, Stormwriter, Llywrch, Lluisanunez, Oliver Pereira, Gmalivuk, DopefishJustin, Fruge~enwiki, Yann, Skysmith, t, Minesweeper, Alfio, Looxix~enwiki, Mdebets, Ahoerstemeier, Muriel Gottrop~enwiki, Caid Raspa, Nanobug, Jeandré du Toit, Evercat, Lancevortex, Mxn, Pizza Puzzle, Nikola Smolenski, The Tom, RickK, Tpbradbury, Nv8200pa, Tempshill, Samsara,

Thue, Nickshanks, Sandman~enwiki, Mignon~enwiki, Raul654, Eugene van der Pijll, Jeffq, Owen, Rossumcapek, Phil Boswell, Gentgeen, Robbot, Fredrik, RedWolf, Stephan Schulz, Nurg, Sverdrup, Rholton, Rursus, Wlievens, UtherSRG, Seth Ilys, Xanzzibar, Filemon, Enochlau, Giftlite, ComaVN, Jyril, Laudaka, Unclebex, CComMack, Dissident, Everyking, Dratman, Curps, Wikibob, Jdavidb, Waltpohl, Maarten van Vliet, Ceejayoz, Eequor, Croxis, SWAdair, Bobblewik, John Abbe, Joseph Dwayne, Pgan002, Andycjp, Antandrus, Q17, OverlordQ, MisfitToys, Quarl, Mcnett, Kaldari, Mzajac, Latitude0116, ScottyBoy900Q, Icairns, Tail, Urhixidur, IcycleMort, Gerrit, BrianWilloughby, Deglr6328, Bluemask, Corti, Rich Farmbrough, LuckyStarr, Guanabot, Florian Cauvin, Cfailde, Vsmith, Jpk, StephanKetz, Dbachmann, MDCore, Bender235, ESkog, Jnestorius, Helldjinn~enwiki, Chewie, RJHall, MisterSheik, El C, DS1953, Kwamikagami, Summer Song, Worldtraveller, Shanes, Art LaPella, RoyBoy, Svdmolen, Causa sui, Gedanken, Sole Soul, Harley peters, Aetherfukz, Meggar, Dreish, Duk, Viriditas, DaveGorman, SpeedyGonsales, La goutte de pluie, VBGFscJUn3, NathanHawking, Pschemp, Ardric47, Krellis, Pearle, Drils, Danski14, Frank101, Youknowandy, Penwhale, Lectonar, Sligocki, Avenue, Idont Havaname, Bart133, Burwellian, Sobolewski, Wtmitchell, Yuckfoo, Cmapm, Haros, Computerjoe, Reaverdrop, BDD, Gene Nygaard, Czolgolz, Kitch, Crispiness, Isfisk, WilliamKF, Angr, Bushytails, Woohookitty, Mu301, Riffsyphon1024, Tripodics, StradivariusTV, Benhocking, Jvsett, BoLingua, DemonKyoto, Pi@k~enwiki, Tabletop, Will.i.am, Eilthireach, Waldir, Zzyzx11, Essjay, Palica, Mattd4u2nv, Rnt20, Yoghurt, Ashmoo, Graham87, Marskell, BenJonson, Don Braffitt, Josh Parris, Drbogdan, Coneslayer, Rjwilmsi, Nightscream, JLM~enwiki, Panoptical, Marasama, Mike s, -SA-, Mike Peel, Brighterorange, Bensin, MarnetteD, Zunix, Fish and karate, Azure8472, Titoxd, RobertG, Nihiltres, Harmil, Themanwithoutapast, RexNL, Jeremygbyrne, Tomer Ish Shalom, Turbinator, SGreen~enwiki, Chobot, Jaraalbe, Bgwhite, Gwernol, YurikBot, Wavelength, Spacepotato, Vuvar1, Hairy Dude, Jimp, Midgley, Freiberg, TheDoober, Witan, Chris Capoccia, SnoopY~enwiki, Epolk, Chaser, Gaius Cornelius, CambridgeBayWeather, Eleassar, NawlinWiki, Mipadi, Bluebird47, Howcheng, Seegoon, Retired username, Banes, Paul.h, RattBoy, Ospalh, Nanouk, Deckiller, Kortoso, Bota47, Perry Middlemiss, Jeremy Visser, Pierpontpaul2351, Dna-webmaster, Yisraelasper, Sandstein, Poppy, 2over0, Chaos syndrome, Blackthorn, Closedmouth, Arthur Rubin, Reyk, CWenger, Hurricane Devon, Geoffrey.landis, Garion96, Ilmari Karonen, Banus, Serendipodous, Narkstraws, Vanka5, A13ean, Remiel, SmackBot, The Dark, AaronM, Zazaban, Royalguard11, Melchoir, Unyoyega, WilyD, Jacek Kendysz, ScaldingHotSoup, Midway, Nickst, Jrockley, BPK2, Delldot, Eskimbot, Vilerage, Orserb7, Alsandro, Shibidee, Yamaguchi先生, Cuddlyopedia, Gwdihw, Skizzik, GwydionM, Rohnadams, Kinhull, Sofsoldier, Saros136, Chris the speller, Bluebot, Kurykh, TimBentley, JonRidinger, CKA3KA, Persian Poet Gal, MK8, Hibbleton, I7s, Xx236, Neo-Jay, DHN-bot~enwiki, DMontes, Modest Genius, Mrwuggs, Doubletruncation, OOODDD, Xiner, Bolivian Unicyclist, Keordina, GrahameS, Stevenmitchell, Maurice45, Aldaron, Kingdon, Bowlhover, Sese~enwiki, Rjp0i, Ohconfucius, Nishkid64, CFLeon, Rory096, Derek farn, L0rents, John, Coricus, General Ization, J 1982, Trotterjt, VirtualDave, Enfolder, JorisvS, Thefirechild, Temple, Mgiganteus1, Ben Jos, Bjankuloski06en~enwiki, Zzzzzzzzzzz, Ckatz, StikEmanon, RandomCritic, Martinp23, Npa213, Optakeover, SandyGeorgia, Mets501, Jstupple7, AdultSwim, Ace Frahm, Artman40, Dr.K., Novangelis, Ryanjunk, DI2000, Dk2852, Iridescent, JMK, Deerisano, Dekaels~enwiki, Richard Nowell, Judgesurreal777, Joseph Solis in Australia, Newone, Twas Now, Dsspiegel, RekishiEJ, MD.astronomer, Raetzsch, Az1568, Bottesini, Tawkerbot2, Spiderboy12, Zampafan, Spacini, CmdrObot, Glanthor Reviol, Shyland, SupaStarGirl, Eric, Epud, ThreeBlindMice, Benwildeboer, DanielRigal, Outriggr (2006-2009), Arnavion, WeggeBot, Meodipt, Singerboi22, Skybon, Ieek~enwiki, Cydebot, Boulderinionian, Kanags, HeseGrande, Gogo Dodo, Wa2ise, Flowerpotman, Duccio, Bryan Seecrets, Kozuch, Sweikart, Robert.Allen, A7x, Gimmetrow, Casliber, Thijs!bot, Saintandrewsfall, Hz12kmblt, HappyInGeneral, Sry85, Etm157, PerfectStorm, Leedeth, Headbomb, Marek69, DmitTrix, Lars Lindberg Christensen, Davidhorman, Cogito ergo sumo, Nick Number, BlytheG, Martin Cash, Escarbot, Dr. Laurance R. Doyle, Mentifisto, Tham153, AntiVandalBot, Teentje, Kevin Nelson, J. Langton, Gdo01, Pixelface, Gökhan, Ingolfson, Leuko, Abyssoft, Stewart Robertson, MER-C, Matthew Fennell, Ribonucleic, Andonic, 100110100, Admiral-Bell, Rothorpe, SteveSims, Magioladitis, WolfmanSF, Murgh, Soulbot, Nyttend, JPG-GR, Bubba hotep, Hekerui, Eamon1916, BatteryIncluded, ArthurWeasley, Torchiest, BilCat, Rupert Nichol, Mlindroo, Talon Artaine, Khalid Mahmood, Johnson Lau, TAA~enwiki, Robin S. Kheider, NatureA16, MartinBot, BetBot~enwiki, Lamboman, WikiManGreen, CalendarWatcher, CommonsDelinker, Sdp1978, Fusion7, Teraflop122, J.delanoy, Nev1, DrKay, Darin-0, AstroHurricane001, SCfan7, Hans Dunkelberg, Maurice Carbonaro, Decaheximal, Acalamari, FrummerThanThou, Mikael Häggström, AppleMacReporter, Quarma, Astre, Plasticup, Nwbeeson, Wesino, Rwessel, Manassehkatz, ParsifalTG, Bofoc Tagar, Andy Farrell, Jimbobob19, Sacredceltic, Jebug, MinawaAsuka, Accordance with terms, Nuke555, Frogfucious, Specter01010, I hate featured article, Paul.Murray, Stupidness, Omgkko, Benrmac129, Tisha hoggs, CardinalDan, Battyboi, Bob2727, Idiomabot, WikiInformer, VolkovBot, ColdCase, RingtailedFox, Maghnus, Fences and windows, Gunnar Guðvarðarson, TXiKiBoT, Vipinhari, DarrynJ, Anna Lincoln, Gekritzl, Henrykus, Domino, Mzmadmike, DoktorDec, Onore Baka Sama, Sardonicone, Vikrant42, Uranometria, AlleborgoBot, Mars2035, Vsst, EmxBot, Glycerinester, Fcady2007, Ponyo, SieBot, PlanetStar, Pallab1234, Love fass, Gerakibot, Ergateesuk, Denislemenoir, Cwkmail, Yintan, Vanished user 82345ijgeke4tg, Hirohisat, Andrew MacInnes, Editore99, Man It's So Loud In Here, Cville roger, Lightmouse, Poindexter Propellerhead, Murlough23, Georgemargaris, Tomemorris, Astrocbt, Xcalibur2, Certayne, Kalidasa 777, Nergaal, Mopskatze, ImageRemovalBot, Nondistinguished, W.E.Ward.III, Twinsday, Ratemonth, ClueBot, Pressforaction, Bobathon71, Awg1010, Zengli, Paul K., Razimantv, McDylan1990, Chech Explorer, Niceguyedc, Blanchardb, Krizanic, Peteruetz, NuclearVacuum, Anoa lynx, Pmronchi, Atomic7732, NuclearWarfare, BSmith821, Dj manton, Foogus, WikiJedits, Thingg, RubenGarciaHernandez, Dana boomer, MelonBot, Supersymetrie, GabrielVelasquez, Gürkan Myczko, Life of Riley, Tokrok24, LostLucidity, Pichpich, Olvegg, ErgoSum88, Little Mountain 5, SilvonenBot, Drewtaylor1978, NellieBly, Jbeans, Addbot, Roentgenium111, DOI bot, Fyrael, Vampdow, CanadianLinuxUser, Richardsonlj, Proxima Centauri, LaaknorBot, Delaszk, Debresser, AnnaFrance, LinkFA-Bot, 84user, Ace45954, Ehrenkater, Friarfrank, DubaiTerminator, Lightbot, Smeagol 17, Teles, Arbitrarily0, Denecktie, Legobot, Yobot, Amirobot, Crispmuncher, Dzied Bulbash, Apollofox, Azcolvin429, Eric-Wester, Tbayboy, Orion11M87, AnomieBOT, Thuvan Dihn, 1exec1, Jim1138, Bubblegumgirl91, Icalanise, Sz-iwbot, Powerzilla, Materialscientist, Citation bot, ArthurBot, LilHelpa, Xqbot, Lucianomendez, Capricorn42, Pj uknown, GetLinkPrimitiveParams, Gap9551, Juamax, Bcz, Ediug, Shadowjams, Raydekk, McDrewNotYou, Fotaun, Shortygetlow, Nagualdesign, FrescoBot, Sanpitch, Charles Edwin Shipp, Pinky34567, Majopius, Gregknicholson, Umawera, Citation bot 1, Rayshi1, Gautier lebon, Pinethicket, Jonesey95, 70virginmouse, Tom.Reding, Lithium cyanide, IndigoRoman, Brian Everlasting, Shadroth, Gmoney484, Istcol, Jenab6, Crusoe8181, Jirka.h23, SkyMachine, Zbayz, Lightlowemon, TobeBot, Trappist the monk, ReflMuffin, Canuck100, Deanmullen09, Gabela2, DARTH SIDIOUS 2, Obankston, Nederlandse Leeuw, RjwilmsiBot, DASHBot, Steve03Mills, EmausBot, John of Reading, Qurq, Dr Aaij, Quantanew, Euphrates1orange, Dewritech, GoingBatty, Oleksiy.golubov, Jmencisom, Tommy2010, Winner 42, Luiscalcada, Bethnim, Solomonfromfinland, Italia2006, ReflectionDivine, ZéroBot, Karthikeyan K200878, Bollyjeff, Chasrob, AvidLearnerReturns, Dmawet, Dondervogel 2, PhreshKiid, H3llBot, AManWithNoPlan, David J Johnson, Tolly4bolly, Resprinter123, Sahimrobot, Donner60, Fanyavizuri, Hypercephalic, EvenGreenerFish, ChuispastonBot, HandsomeFella, ChiZeroOne, Ionvort, Senator2029, Whoop whoop pull up, ClueBot NG, Michaelmas1957, Giggett, Movses-bot, Wdchk, Mharbut, IOPhysics, SenseiAC, Widr, Danm, Helpful Pixie Bot, J.Dong820, Gob Lofa, Bibcode Bot, Highlighteryellow, BG19bot, Pine, Dolerite, Hza a 9, Benefac, Ewigekrieg, Frze, Fischpredigt, Mark Arsten, Q6637p, Cadiomals, Vegan11, Cliff12345, StratosMatt, Electrastorm.redd,

Erwin1517, Minsbot, Wer900, BattyBot, Pendragon5, Stigmatella aurantiaca, E prosser, ChrisGualtieri, Mithoron, DoctorKubla, Tandrum, Temerraria, Dexbot, Dissident93, Nouniquenames, Mogism, Stas1995, Smithy6287, Pottsar, CuriousMind01, Acoma Magic, Richardshao, Lugia2453, Typesometext, Andyhowlett, Adam2828, Talkyinfo565, Reatlas, Rfassbind, Faizan, Greengreengreenred, Nicolas M. Perrault, Exsaol, RichardAlexanderHall, Helloexoworld, Unlikelyuser, Praemonitus, Neo Poz, Serpinium, Redplain, B14709, EJM86, Thevideodrome, Ugog Nizdast, Ginsuloft, Kiwikenton, PunkPositive, Paul2520, Nyanya711477, Astredita, OccultZone, Markutz-enwiki, Ethically Yours, 2bithymes, AppleLover2014, Monkbot, Davidbuddy9, Kathalya jones, ThePunMaster123, Punmaster982, Newfriendly, Fallfish12, Momaxox, Jag1889, Messier8, VortexJasper, Pandapao, YeOldeGentleman, Stormfoogle, ThE-fUtUrE-2014, Tetra quark, Crimsonslide, Isambard Kingdom, DN-boards1, DominicJohnsonStudent, Theo Makhoul, RocketCityMan, KasparBot, JMCatron, The secret guy, Ackoli, Bongeo, Fdfexoex, Chris30lol, Huritisho, Charles06.tang, Yimingguo and Anonymous: 736

- **Copernican principle** *Source:* https://en.wikipedia.org/wiki/Copernican_principle?oldid=679377164 *Contributors:* Bryan Derksen, The Anome,Michael Hardy, Andres, Cimon Avaro, Schneelocke, Joy, Goethean, Forseti, Malyctenar, Matthead, Piotrus, Karol Langner, Sam Hocevar,Rich Farmbrough, Guanabot, Vsmith, Bishonen, RJHall, Pjf, Bobo192, Enirac Sum, Hackwrench, Schaefer, Deacon of Pndapetzim, Reaver-drop, DV8 2XL, Jeff3000, Btyner, MrFix3, Drbogdan, Rjwilmsi, R.e.b., AndyKali, KSchutte, Joshurtree, Enormousdude, Petri Krohn, Finell,SmackBot, Diplomacy Guy, Jagged 85, Kurykh, Tsca.bot, Deiz, Lapaz, JorisvS, Flamboyant-enwiki, Noleander, Baskinmyglory, Chalnoth,CmdrObot, Geremia, Gregbard, Cydebot, Valodzka, Gmusser, Publicola, Headbomb, Peter Gulutzan, Unintelligible, D.H. Uruamme, Escar-bot, JAnDbot, VoABot II, BatteryIncluded, Kheider, J.delanoy, DogcatcherDrew, MickO'Bants, Biglovinb, BotKung, Dickvb4, PaddyLeahy,Nihil novi, METIfan, Mátyás, Hamiltondaniel, Wyattmj, Alexbot, 7&6=thirteen, MilesAgain, RexxS, Richard A. Ryals, SilvonenBot, DOI bot,Proxima Centauri, Aunva6, Legobot, Yobot, AnomieBOT, JackieBot, Materialscientist, Citation bot, ArthurBot, Xqbot, Maygytr, Lithopsian,RibotBOT, Lisybave, Paine Ellsworth, Citation bot 1, Adlerbot, Jonesey95, MondalorBot, Zbayz, Trappist the monk, EmausBot, Solomon-fromfinland, AManWithNoPlan, David J Johnson, Sharktopus, Bibcode Bot, StarryGrandma, Ushau97, ChrisGualtieri, JYBot, Dexbot, Cer-abot-enwiki, CuirassierX, Diamondandrs, Diamondadnrs, Tachyon1010101010, Kogge, Bene71, Xibalban Alchemist, Joe6Pack, Monkbot,Tetra quark, Isambard Kingdom, Rick Ryals and Anonymous: 55

- **Mediocrity principle** *Source:* https://en.wikipedia.org/wiki/Mediocrity_principle?oldid=677129820 *Contributors:* Bryan Derksen, XJaM, Michael Hardy, Skysmith, Ahoerstemeier, Smack, Nickg, Populus, Drernie, Zandperl, Jredmond, Henrygb, Wlievens, Barbara Shack, Cloud200, Quadell, Rich Farmbrough, Aranel, RJHall, Tabletop, Triddle, TotoBaggins, Btyner, RichardWeiss, Drbogdan, Rjwilmsi, Hsrin-ava, Wragge, Lord Patrick, Wjfox2005, Clark Kent, Mesohmbo, Perry Middlemiss, Theplebianking, SMcCandlish, Petri Krohn, ArquiWHAT, Crystallina, SmackBot, Chronodm, Portillo, Stevenmitchell, Dreadstar, Pwjb, Mattpersons, CmdrObot, Vyznev Xnebara, Myasuda, Cydebot, RZ heretic, Fen, Arb, Mbell, Doremitzwr, The Obento Musubi, Mdotley, RadicalPi, Magioladitis, Usien6, BatteryIncluded, Infovarius, Schnloof, Little-How, D.M.N., Earfetish1, VolkovBot, Xresonance, Alan Addison, Lou.weird, Sst557, Bporopat, Spielvogel (renamed), Piperdown, Northfox, Rep07, Paucabot, Nihil novi, METIfan, Thehotelambush, Fahklempt, WikiLaurent, Kalidasa 777, Mr. Granger, General Epitaph, Gabodon, Giralgrathor, Shaantapalooza, DumZiBoT, Gtoffoletto, Addbot, Ainali, Redheylin, FutureDragon, Luckas-bot, Yobot, JustWong, Denispir, Rachmaninoff, Azcolvin429, AnomieBOT, Rubinbot, Galoubet, Sz-iwbot, Haleyga, Xqbot, Sketchmoose, Gigemag76, Gap9551, Luis Felipe Schenone, Erroramong, Paine Ellsworth, Machine Elf 1735, Landmand, MondalorBot, Dinamik-bot, Shiny.Magnum, Diannaa, Mediocreshane, Tesseract2, Einkleinestier, Baruta07, Jaque Hammer, Kuguar03, Chester Markel, Frietjes, Helpful Pixie Bot, JohnChrysostom, Allenmad2234, Danny Sprinkle, Stilgar27, Cltschirhart, Frinthruit, Isambard Kingdom and Anonymous: 78

- **Biochemistry** *Source:* https://en.wikipedia.org/wiki/Biochemistry?oldid=692755530 *Contributors:* AxelBoldt, Tobias Hoevekamp, Magnus Manske, Marj Tiefert, Mav, Bryan Derksen, Tarquin, Andre Engels, Enchanter, Deb, Peterlin-enwiki, AdamRetchless, Heron, Youandme, Someone else, Dwmyers, D. Booyabazooka, Lexor, Card-enwiki, 168..., Ellywa, Ahoerstemeier, Cyp, Mac, Александър, Glenn, °¡°, Rob Hooft, Tobias Conradi, Mxn, Vivin, Lfh, Selket, Steinsky, DJ Clayworth, Paul-L-enwiki, Samsara, Vincent kraeutler, Doug swisher, Donar-reiskoffer, Gentgeen, Robbot, Jotomicron, Kowey, Mayooranathan, Stewartadcock, Blainster, Mendalus-enwiki, Wikibot, Fuelbottle, Quadal-pha, Diberri, Giftlite, Mikez, Netoholic, Curps, Dmb000006, Bensaccount, Andris, Guanaco, Kandar, OldakQuill, Knutux, Antandrus, Be-land, Karol Langner, APH, H Padleckas, Bornslippy, Gseshoyru, Ben Zealley, Kevyn, Bluemask, Ornil, Discospinster, Vsmith, Notinasnaid, Mani1, Bender235, El C, Bobo192, Smalljim, Maurreen, Arcadian, Jag123, Giraffedata, Pschemp, Haham hanuka, Krellis, HasharBot-enwiki, Passw0rd, Jumbuck, Danski14, Alansohn, Etxrge, Arthena, Wouterstomp, Kocio, Alex '05, Ombudsman, ClockworkSoul, Knowledge Seeker, Bsadowski1, Alan, Mattbrundage, Dan East, HenryLi, Ceyockey, Boothy443, Linnea, TigerShark, YannisKollias, JeremyA, Bennetto, JRHorse, TheAlphaWolf, MarcoTolo, Halcatalyst, Palica, Marskell, Chun-hian, Kbdank71, Jclemens, Techneaux, Mayumashu, Alcarreau, Mike Peel, ScottJ, FlaBot, RexNL, Tedder, Chobot, WriterHound, YurikBot, Wavelength, Mushin, RobotE, RussBot, J. M., Splette, Stephenb, Cambridge-BayWeather, Doctorbruno, Wimt, NawlinWiki, Grafen, Orioneight, Gareth Jones, Ragesoss, Gadget850, Bota47, Trainra, Wknight94, Occha-nikov, Djramone, Arthur Rubin, Josh3580, JeramieHicks, LeonardoRob0t, Katieh5584, GrinBot-enwiki, Luk, Itub, SmackBot, Looper5920, Espresso Addict, Hydrogen Iodide, McGeddon, Pgk, KocjoBot-enwiki, AndreasJS, Eskimbot, Kjaergaard, Edgar181, Zephyris, M stone, Gilliam, Cabe6403, Chris the speller, Jethero, Persian Poet Gal, Bduke, Rich.buckman, MalafayaBot, Go for it!, DHN-bot-enwiki, Sbhar-ris, Darth Panda, MyNameIsVlad, Snowmanradio, Rrburke, COMPFUNK2, Chris jf, Flyguy649, Dreadstar, Richard001, Drphilharmonic, DMacks, Lineweaver-enwiki, Sadi Carnot, SashatoBot, Serein (renamed because of SUL), Kuru, J. Finkelstein, Kipala, IronGargoyle, Tarcieri, Mets501, Sasata, Andreworkney, Quaeler, J Di, Twas Now, Chika11, Courcelles, Tawkerbot2, K.murphy, Pi, Firewall62, CmdrObot, Ale jrb, Moreschi, Cydebot, MC10, Rifleman 82, Neuroticopia, Carstensen, Shirulashem, Christian75, Anpetu-We, John Lake, Epbr123, Headbomb, X201, Defeatedfear, Escarbot, I already forgot, Thomaswge, AntiVandalBot, Majorly, Luna Santin, Quintote, TimVickers, Doctor C, Qw-erty Binary, Figma, JAnDbot, MER-C, The Transhumanist, Justacontributor, Jhay116, BenB4, Zeb edee, GoodDamon, Magioladitis, Staib, Pedro, VoABot II, Confiteordeo, Mah159, Bubba hotep, Mikhail Garouznov, Allstarecho, DerHexer, Gludwiczak, Squidonius, DGG, Gwern, Pvosta, MartinBot, ChemNerd, Rettetast, Winkel111, N4nojohn, J.delanoy, TimBuck2, Mikael Häggström, Coppertwig, Chriswiki, Loohe-snuf, The Transhumanist (AWB), Rommandreu, Tanaats, Bob, Cometstyles, Tiggerjay, Potaaatos, 52G, Bonadea, Squids and Chips, Remi00, Deor, VolkovBot, Jeff G., Alexandria, Philip Trueman, TXiKiBoT, Caltechdoc, Amaher, Leafyplant, Ermysted's school, Jackfork, Bran-donrush, UnitedStatesian, Untitled300, Mishlai, Jaqen, Loka1282, RaseaC, Spinningspark, Monty845, HiDrNick, AlleborgoBot, EmxBot, Deafgirl, Nssr 84, SieBot, Logan baum, Calliopejen1, Tiddly Tom, Graham Beards, Caulde, WereSpielChequers, Gerakibot, Keilana, Dra-magirlforjesus, Jesvarela, Hzh, Oxymoron83, Scorpion451, Lightmouse, Pmrich, Superbeecat, Into The Fray, Naturespace, WikipedianMar-lith, Pscott22, Loren.wilton, ClueBot, Avenged Eightfold, PipepBot, The Thing That Should Not Be, Stupid2, Yikrazuul, Arakunem, Dr-mies, Niceguyedc, RebelBodhi, Neverquick, ChandlerMapBot, Promage281, DragonBot, Excirial, Sarakoth, Cbailey7, Sun Creator, 2pac 2007, Jotterbot, KLassetter, Dekisugi, Tbeyett, Bald Zebra, N8mills, Jamesscottbrown, XLinkBot, Fastily, Jytdog, Jovianeye, DPistotallyawesome, Willisis2, Basilicofresco, DOI bot, Tenev, CanadianLinuxUser, MrOllie, Favonian, SamatBot, Numbo3-bot, Tide rolls, Vicki breazeale, Legobot, Luckas-bot, Yobot, Themfromspace, Ptbotgourou, TaBOT-zerem, Salvaje20, Mauler90, Essam Sharaf, Explosivedeath7, Washburnmav, AnomieBOT, Jim1138, Templatehater, Materialscientist, Stevendboy, Qwertyman33, Citation bot, Lil'swallowtail, Frankenpuppy, LilHelpa, Xqbot, Zad68, Timir2, The Banner, Capricorn42, Jburlinson, P99am, Fakih1234, Paulkappelle, BostonBiochem, ThyCantabrigde, RibotBOT, Barefootguy, Doulos Christos, Shadowjams, PiFanatic, PM800, Erik9, StevieNic, A.amitkumar, Gumball456, FrescoBot, Tobby72, Jobseeker-us, Theruchet, Pinethicket, I dream of horses, Abductive, A8UDI, Ntropy25, Alaadabarita, Jauhienij, Gamewizard71, Wotnow, Vanished user kiij3irj4tihns, Kiwakwok, Jeffrd10, Canuckian89, Weewengweng, Riomack, TjBot, Sdneidich, Ash421, Skamecrazy123, Applepie390, EmausBot, H.tjeras, GoingBatty, Babyboy004, Neilmitch02, K6ka, Lucas Thoms, Thecheesykid, AvicBot, JSquish, John Cline, Lalsingh, Jenvargas93, Ohsoserious, Tom the biochemist, Kelly222, Jesanj, L Kensington, Donner60, DASHBotAV, 28bot, ClueBot NG, Rodelapa, Jdcollins13, Frietjes, Widr, EtherROR, Krist Wood, Wertyu739, Curb Chain, Arifur6051, BG19bot, Nighttoy, MusikAnimal, Piguy101, Mark Arsten, Amitashsam, MrBill3, YVSREDDY, Joud98, Fathimaazara, ChrisGualtieri, GoShow, Saltwolf, Fvandrog, JYBot, Webclient101, Marywhi, CaSJer, Frosty, SFK2, David P Minde, Aftabbanoori, Epicgenius, Shafiq.sims, Jamesmcmahon0, AmaryllisGardener, AmericanLemming, Mrs.Stewart, Tony johnsong, Rjdodger, David246, Ugog Nizdast, Rahul and Atasi, Ginsuloft, Lizia7, Nazia Singh, Sirisindhu, DrFGWohler, Crisur, Jasonbourn2, Sixtyn, Ped92man, Shaneandlauren, LaughableStand, Hypervalentanion, Biochemistry&Love, Wiscoeditor, Eatmyjorts, Mintosh Kumar, Lol235689, KasparBot, Han-Jun Cho, Karpywikipedia, Jkbghj and Anonymous: 606

- **Panspermia** *Source:* https://en.wikipedia.org/wiki/Panspermia?oldid=694669039 *Contributors:* Mav, Bryan Derksen, Timo Honkasalo, Taw, Andre Engels, William Avery, SimonP, Shii, Maury Markowitz, Stevertigo, Edward, Alan Peakall, Alfio, CesarB, JWSchmidt, Whkoh, Kimiko, Timwi, RickK, Rednblu, Abscissa, VeryVerily, Populus, Paul-L~enwiki, Omegatron, Samsara, Warofdreams, Proteus, Tonderai, Jeffq, RadicalBender, Northgrove, Chris Roy, Postdlf, Sverdrup, Rholton, Rebrane, Wlievens, David Edgar, Jor, Cyberia23, Per Abrahamsen, Nagelfar, Alan Liefting, Neox, Kbahey, Jyril, Bfinn, Dmb000006, MingMecca, Duncharris, Solipsist, SWAdair, Bobblewik, Pgan002, Alexf, Bcameron54, Piotrus, Melikamp, Csmiller, Taka, Obby~enwiki, Mzajac, Cglassey, Sam, Lacrimosus, Rich Farmbrough, Vsmith, ArnoldReinhold, YUL89YYZ, Dbachmann, BACbKA, RJHall, CanisRufus, RoyBoy, Godfreylouis, Viriditas, Cmdrjameson, Nicke Lilltroll~enwiki, Dejitarob, L33tminion, Hob Gadling, Vicarage, Pschemp, Slipperyweasel, Calebe, Chino, Arthena, JoaoRicardo, Ahruman, Hu, Mnemo, TahitiB~enwiki, Pauli133, Gene Nygaard, Dismas, Stephen, Gmaxwell, Mindmatrix, TotoBaggins, Pictureuploader, DanHobley, Holek, Dbutler1986, Mandarax, Marskell, Drbogdan, Rjwilmsi, Oblivious, Miserlou, SonicSpike, Erkcan, Krash, Dionyseus, FlaBot, Ian Pitchford, John Baez, Papacha, Diza, SteveBaker, Gurubrahma, Chobot, Voodoom, Bgwhite, Wavelength, Acefox, Chris Capoccia, Akamad, Dysmorodrepanis~enwiki, Icelight, Anetode, Wangi, Emdx, DeadEyeArrow, Skepticsteve, Elkman, Pegship, Noosfractal, 2over0, Jules.LT, Chase me ladies, I'm the Cavalry, Abune, SMcCandlish, Reyk, Petri Krohn, Pádraic MacUidhir, Brentt, SmackBot, Judith d, Melchoir, Elfsareus, Kintetsubuffalo, Portillo, Ohnoitsjamie, Hmains, Jushi, Kinhull, Qwasty, Sumthingweird, RDBrown, Thumperward, Nemodomi, Silly rabbit, Hibernian, Zephyr707, WDGraham, Jefffire, Frap, OrphanBot, Nixeagle, Thomqi, Soosed, Wen D House, Tsop, Iamdaniel, Kismetmagic, Wizardman, Bklyce, StN, Virago, BrownHairedGirl, Eaglecros, John, Siddharth srinivasan, John Cumbers, Neodarksaver, ISoron, Extremophile, Skymist, Novangelis, Corykoski, Jason.grossman, FelisSchrödingeris, Greygirlbeast, Courcelles, Brainbark, Centered1, Banedon, Kylu, RagingR2, JFreeman, Daniel J. Leivick, Dancter, Michael C Price, DumbBOT, Iliank, Robertinventor, Arb, UberScienceNerd, Thijs!bot, JAF1970, Mpallen, Headbomb, John254, Second Quantization, Z10x, Iulius, Morgana The Argent, The Hams, NeilEvans, Gnixon, Yellowdesk, Rothorpe, LittleOldMe, Magioladitis, Professor marginalia, Yakushima, Theroadislong, Gabriel Kielland, BatteryIncluded, Joe hill, Thibbs, DerHexer, Lelek, Waninge, Urco, Jim.henderson, Retietast, Ulisse0, AlphaEta, Trusilver, Terrek, Nsande01, Ian.thomson, ABVS1936, Skier Dude, Kukec, RenniePet, Davy p, M-le-mot-dit, EyeRmonkey, Diletante, Idioma-bot, Evolvearth, Soliloquial, Dom Kaos, Philip Trueman, JayEsJay, TXiKiBoT, Mathwhiz 29, Cloudswrest, Stickyhammer, Agmon, AlleborgoBot, Hrafn, SieBot, Coffee, Hugh16, Arbor to SJ, Tiki 92090, Mimihitam, Excutio, LSmok3, Sunrise, Anchor Link Bot, Mnmautner, Martarius, Doyee5, ClueBot, CarolSpears, Foxj, Wanderer57, DrFO.Jr.Tn~enwiki, 1111news, Niceguyedc, Nanobear~enwiki, Sid-Vicious, Sjdunn9, Deselliers, Scog, BSmith821, HRosenberg, Fitzburgh, Unmerklich, Aitias, Apparition11, Crowsnest, Mrmpsy, DumZiBoT, Hotcrocodile, Rror, Ost316, Oogaboogabooga, Fabio6043, BrucePodger, Addbot, Basilicofresco, DOI bot, Potentialten, Ronhjones, Chamberlain2007, Tdchel, MrOllie, Download, Glane23, SamatBot, Scienceislife, Beren, MuZemike, Cannizzaro S, Legobot, Luckas-bot, Yobot, Wikipedian Penguin, Untrue Believer, AnomieBOT, Taylordw, Rubinbot, Citation bot, Donkybotay, Eumolpo, Quebec99, LilHelpa, Xqbot, Silver Spoon Sokpop, Albahna, Luis Felipe Schenone, GrouchoBot, Knightofcydonia49, Omnipaedista, N419BH, Hdrosenberg, Unused0011, A.amitkumar, Nagualdesign, FrescoBot, LucienBOT, Styxpaint, Citation bot 1, Solarflaredigital, Pinethicket, Jonesey95, Tom.Reding, Fumitol, SkyMachine, IVAN3MAN, Trappist the monk, Jonkerz, RjwilmsiBot, Damaavand, Ansaazi, Androstachys, EmausBot, Clive tooth, Emmajanej, Evanh2008, JacobSheehy, Kp grewal, ZéroBot, AManWithNoPlan, David J Johnson, Actsmart, Orange Suede Sofa, Spicemix, Grapple X, ClueBot NG, Njh321, Kikichugirl, Liveintheforests, RocketLauncher2, Old wombat, BrigKlyce, Russellml, Widr, Secret of success, Helpful Pixie Bot, Curb Chain, Gob Lofa, Bibcode Bot, Lilman0509, BG19bot, Mariansavu, Chemistryfan, Walterfarah, Kooky2, BattyBot, Arodr451, MichaelEF71, ChrisGualtieri, Khazar2, Rchouake, Dexbot, SummerWillow, Adam2828, Abepeace, Tomdarkblade, Melonkelon, KingSupernova, AlexeiSharov, Praemonitus, Syd Menon, Npascucci01, Someone not using his real name, HarbingerOfLunch, Monkbot, Denny123123, Formuse, BicelPhD, Mapsfly, HiBlueSky, Cspoleta, ComicsAreJustAllRight, Dsmith125, Alexis Gervais, Jnav7, Dutral and Anonymous: 318

- **Abiogenesis** *Source:* https://en.wikipedia.org/wiki/Abiogenesis?oldid=694849876 *Contributors:* Damian Yerrick, AxelBoldt, Joao, Bryan Derksen, The Anome, Sjc, -- April, Ed Poor, SimonP, Maury Markowitz, AdamRetchless, Zadcat, Mjb, Heron, Someone else, Lexor, Gabbe, Martin BENOIT~enwiki, Bobby D. Bryant, Ixfd64, Cyde, Sannse, Mcarling, Iheoye, Ellywa, Mdebets, Cyp, JWSchmidt, Julesd, Raven in Orbit, Norwikian, Ee5618, Charles Matthews, Timwi, Steinsky, Foodman, Maximus Rex, David Shay, Populus, Omegatron, Samsara, Jackson~enwiki, Raul654, Johnleemk, Finlay McWalter, Skatfman, Twang, Jason Potter, Robbot, Fredrik, Goethean, Altenmann, Nurg, Rursus, Rebrane, Sheridan, Hadal, Wereon, Raeky, Xanzzibar, Xyzzyva, Giftlite, Mshonle~enwiki, Polsmeth, Pretzelpaws, Everyking, Curps, Solipsist, Bobblewik, Pgan002, Andycjp, Keith Edkins, Sonjaaa, Quadell, Beland, Onco p53, Nograpes, Savant1984, JohnArmagh, Deglr6328, Flex, Lacrimosus, Mike Rosoft, Ta bu shi da yu, Rfl, Discospinster, Rich Farmbrough, Vsmith, ArnoldReinhold, Dave souza, Paul August, Bender235, ESkog, Srbauer, RJHall, Mr. Billion, Crunchy Frog, José Gnudista, Lycurgus, Kwamikagami, Liberatus, Sietse Snel, Art LaPella, RoyBoy, Fufthmin, Guettarda, Causa sui, Bobo192, John Vandenberg, Enric Naval, Viriditas, ::Ajvol.., ZayZayEM, I9Q79oL78KiL0QTFHgyc, VBGFscIUn3, Sulai~enwiki, Hob Gadling, A Karley, Orangemarlin, Marwood, DanielVallstrom, Darrelljon, Psychofox, SlimVirgin, Ferrierd, Koeio, InShaneee, Wtmitchell, Velella, Darco, XB-70, Knowledge Seeker, Pauli133, Tainter, BerndH, Bdrasin, Linas, Mindmatrix, Anilocra, LOL, Rocastelo, Schultz.Ryan, Tabletop, Grace Note, Sadettin, GregorB, CharlesC, Wdanwatts, Essjay, Palica, Gerbrant, GSlicer, RichardWeiss, Alienus, V8rik, BD2412, Rkevins, Sjö, Drbogdan, Rjwilmsi, Mayumashu, Nightscream, Koavf, Zbxgscqf, OneWeirdDude,

Bob A. XP1, Crazynas, Mikedelsol, Bfigura, SLi, Duagloth, Margosbot~enwiki, Nihiltres, Alhutch, Geologist~enwiki, Vanished user psd-fiwncf3niurunfiuh234ruhfwdb7, WhyBeNormal, Kaoma Tsujmai, Truthteller, Chobot, DVdm, Bgwhite, Poorsod, YurikBot, Spacepotato, RadioFan2 (usurped), GPS Pilot, The Hokkaido Crow, NawlinWiki, Rick Norwood, DragonHawk, Dysmorodrepanis~enwiki, Uberisaac, Dtrebbien, Seirscius, Zarel, SAE1962, RecSpecz, Apokryltaros, Nick, Kdbuffalo, E rulez, Crasshopper, Kortoso, Stefan Udrea, WAS 4.250, 2over0, Encephalon, Bhumiya, Smoggyrob, Davril2020, Petri Krohn, Fram, DisambigBot, JDspeeder1, NeilN, CIreland, Victor falk, NetRoller 3D, Quadpus, KnightRider~enwiki, SmackBot, Eperotao, PiCo, John Croft, Rtc, TestPilot, David Shear, Lankenau, Bmearns, BiT, Edgar181, Yamaguchi先生, Macintosh User, Gilliam, Portillo, Betacommand, Skizzik, Eloy, Chris the speller, Kaylus, RDBrown, Davep.org, Jprg1966, Thumperward, Silly rabbit, Hibernian, Complexica, Jeff5102, Sewlong, John Hyams, JoelWhy, Jefffire, Viperphantom, Vanished User 0001, Avb, Cfassett, Ines it, Khukri, John D. Croft, Richard001, Archgoon, Smokefoot, Greg.collver, The PIPE, DMacks, Sammy1339, Daniel.Cardenas, Denise from the Cosby Show, Alan G. Archer, Ohconfucius, SashatoBot, Danielrcote, Technocratic, Gloriamarie, At-tys, Atkinson 291, Khazar, John, Writtenonsand, Butko, JoshuaZ, JorisvS, Robert Stevens, Mgiganteus1, Olin, Sectoaux, Fig wright, Ex-tremophile, 041744, Robbins, A. Parrot, Tarcieri, Smith609, Makyen, Stevebritgimp, Tae2z, Mr Stephen, Xiaphias, Larrymep, NJA, No-vangelis, LenW, Dan Gluck, Nehrams2020, Clarityfiend, Twas Now, Lent, The Letter J, George100, Chris55, VinnieCool, DangerousPanda, CRGreathouse, Ale jrb, Memetics, BeenAroundAWhile, Runningonbrains, RoliSoft, ButFli, WeggeBot, Moreschi, Richard Keatinge, Nnp, Myasuda, Ciyean, Abeg92, Peterdjones, Cyhawk, Hughgr, Michael C Price, Doug Weller, DumbBOT, Narayanese, DnimrevO, Ebyabe, Crum375, PKT, Thijs!bot, Barticus88, Ryansca, Pstanton, Mojo Hand, Mungomba, Headbomb, James086, Astrobiologist, Davidhorman, Chandler, Gossamers, AntiVandalBot, Luna Santin, Guy Macon, Dbrodbeck, Gnixon, TimVickers, Cstreet, Smartse, Fluffy654, Danny lost, Princeofexcess, JAnDbot, XyBot, GromXXVII, MER-C, The Transhumanist, Matthew Fennell, Mildly Mad, Andonic, Xeno, Panarjedde, TAnthony, Tstrobaugh, Rothorpe, Kornbelt888, Magioladitis, Carlwev, Sushant gupta, JNW, CattleGirl, Harelx, Trishm, Hubbardaie, Mark PEA, Recurring dreams, Zephyr2k~enwiki, Theroadislong, Cgingold, BatteryIncluded, Allstarecho, Lyonsce, DerHexer, Edward321, Urco, JohanViklund, Mdsats, Robin S, Drm310, Tsinoyboi, Keith D, R'n B, CommonsDelinker, Verdatum, Leyo, Mzaki, Player 03, PhageRules1, Ulisse0, Ifomichev~enwiki, AstroHurricane001, Avkulkarni, Rlsheehan, Hans Dunkelberg, Sidhekin, AmagicalFishy, Dispenser, It Is Me Here, Enuja, McSly, Taroteards, Janet1983, Davy p, RobinGrant, Lbeaumont, Jorfer, Cmichael, KylieTastic, AzureCitizen, IceDragon64, Funandtrvl, Novernae, Jamiejoseph, Speaker to wolves, Philip Trueman, Sub-life, Vipinhari, GcSwRhle, Charlesdrakew, Matthewrossing, Littlealien182, Steven J. Anderson, Awl, AllGloryToTheHypnotoad, Noformation, MacFodder, Mannafredo, Mishlai, Gibson Flying V, Wiki-isawesome, Maxim, Shanata, WinTakeAll, Distinguisher, SheffieldSteel, Wolfrock, Lamro, Synthebot, Zarcoen, Omernar, Northfox, Rep07, Planet-man828, Hrafn, Nachohosking, EGMAG, Tezuel, Macdonald-ross, Gnocchi, Carny, KatieandHandy, Nihil novi, ToePeu.bot, Meldor, Dawn Bard, ConfuciusOrnis, Odd nature, Yintan, 0xFFFF, Abhishikt, Chhandama, Oda Mari, Je-SOCO, Oxymoron83, Lightmouse, He-likophis, RW Marloe, Manifolds~enwiki, Jruderman, RyanParis, Sunrise, Diego Grez-Cañete, Skeptical scientist, StaticGull, Mos bratrud, Tesi1700, Hamiltondaniel, Driftwood87, Kalidasa 777, Marmenta, Lucius Sempronius Turpio, Twinsday, Sfan00 IMG, ClueBot, Tmol42, Fyyer, The Thing That Should Not Be, AstroMark, Sexiestjen4u, Desoto10, Pi zero, Unbuttered Parsnip, Jumacdon, Canopus1, Polyamorph, Timberframe, Tfpsly, Niceguyedc, Baegis, Alexis Brooke M, Rotational, Jandew, Paulcmnt, Excirial, Gustavocarra, Winston365, Vital Forces, Shinkolobwe, Abeo iniuria, Sun Creator, Eznight, Coinmanj, NuclearWarfare, SchreiberBike, Audaciter, BOTarate, Truth is relative, under-standing is limited, Thusled, Thingg, Aitias, AC+79 3888, Johnuniq, Egmontaz, Editor2020, Goodvac, Bentheadvocate, Darkicebot, Cap-tainVideo890, XLinkBot, Roxy the dog, Jytdog, Jovianeye, Rror, Brady, Elfgeek, Ost316, Jungfruchallan, Aloboof123, Opaq87, Aunt En-tropy, Virajelix, Thatguyflint, Janisterzaj, Addbot, Roentgenium111, DOI bot, Landon1980, Swissmeister, Ronhjones, CanadianLinuxUser, Dsmith77, Lindert, Download, Redheylin, Bernstein0275, Camedit, Blade13125, Wildreecleste, Polyp2, LinkFA-Bot, Quietmarc, Partofwhole, U3190, Tide rolls, TL782, Romaioi, Nase, Yobot, StarTroll, Scepticus2, Yngvadottir, The Earwig, Punu, 489thCorsica, Cseppala, Cinch-Bug, Dr.Buttons, AnomieBOT, Brroga, Mike Hayes, Kerfuffler, JWSurf, Trabucogold, Csigabi, Mann jess, Materialscientist, Citation bot, Quebec99, Romandoggie, LilHelpa, FreeRangeFrog, Xqbot, Sventington the Second, Blorblowthno, Wapondaponda, Mmnlaxer, Æ, Nas-nema, Mononomic, Turk oğlan, J JMesserly, Crzer07, 7h3 3L173, DerryTaylor, ProtectionTaggingBot, Gui le Roi, Conquistador, Sophus Bie, Ramssiss, Shadowjams, Metheub, Eugene-elgato, Joaquin008, Biem, FrescoBot, Fiastergeist, Yanima, Hoffmannrungethailand, Riven-tree, BKMBC3, Machine Elf 1735, Trkiehl, Citation bot 1, Redrose64, ANDROBETA, DrilBot, Winterst, Gravityguy, WaveRunner85, Gamocanno, Jonesey95, Helzrule19, Tom.Reding, Deleteduser2015, Hoo man, SpaceFlight89, FormerIP, Tanzania, Jerrywickey, Mikespe-dia, Jandalhandler, Kibi78704, MichaelExe, Fartherred, SkyMachine, IVAN3MAN, Trappist the monk, Silenceisgod, MEPK, Fama Clamosa, Comet Tuttle, Mcfl116, Vrenator, Victorfrogg, Jimmetry, Clarkcj12, Bcoolsdad, Diannaa, 564dude, Jynto, Gregrutz, Myrmidon1, DARTH SIDIOUS 2, Tor1714, Onel5969, RjwilmsiBot, Apotheosa, Hppa, Plommespiser, WildBot, Tesseract2, I belong to Jesus Christ, Emaus-Bot, Jeffhughes22, Immunize, Dominus Vobisdu, Niluop, Dewritech, Ibbn, ResponsibileSQ, Tamtrible, Jmv2009, Pboehnke, Slightsmile, Tommy2010, Kiran Gopi, Maneijeri, Solomonfromfinland, Ofekalef, Kiwi128, H3llBot, Wayne Slam, David J Johnson, Ksarasoli, Korztin, Jesanj, Brandmeister, I. Kensington, Scientific29, Ego White Tray, Tanoan, Renji911, SemanticMantis, Dr. Hipopotamo, JanetteDoe, Sven Manguard, JonRichfield, Ldvhl, Zuky79, Gary Dee, AUN4, ClueBot NG, Don Para, E3cubestore, Afterrock81, Colin Fredericks, Rain-bowwrasse, Jorge 2701, Joefromrandb, Sketchup123, DonaldRichardSands, DS Belgium, Sjmantyl, Asukite, Telpardec, Keenedged, Wiki-wiki180, MerllwBot, Lotterox, Michaeldeans, Helpful Pixie Bot, Elefnose, Cinnaplum, Anentiresleeve, Curb Chain, Bibcode Bot, Mwsegehr, BG19bot, Lebs27, Expewikiwriter, Vevanpelt, Karmstrong909, Knowledge Examiner, Halstedew, Mark Arsten, IraChesterfield, Drewransey, Dkspartan1, Cauhtcoatl, Գարիկ Հայազ, MLearry, Cadiomals, Mthoodhood, Ghostsarememories, Harizotoh9, Blackstar167, HMman, Dontreader, Zedshort, Zetazeros, Dontshootimgay, Benyboy2, BattyBot, Decruft, Sfarney, Hghyux, Marc Tessera, Jinw338, SkepticalRap-tor, David B Stephens, Soulbust, TheJJJunk, Garamond Lethe, Tanookiinashu, Khazar2, Ekren, Nathanielfirst, Elfinanciero222, Cmw255, Cotupnet, Pterodactyloid, RGA1980, Dexbot, Webclient101, Jinx69, Cerabot~enwiki, TippyGoomba, Mbreht, TheTahoeNatrLuvnYaho, Cu-riousMind01, Saehry, Leptus Froggi, 93, TruthOrTruthy, Corinne, Frivolous Consultant, HerbertHuey, FlaviusFerry, Tjmiler, Reatlas, Anas-tronomer, Bret palmer, Faizan, ICameHereToEdit, Surfer43, KnowledgeIncreases07, Analiticus, StewartGrifliths, Nirendeka, DavidLeighEllis, Ronaldo Laranja, Nigellwh, AbioScientistGenesis, SzostakJack, Mj12hoaxwriter, MDPub13, PubMed2015, Andreas Geisler, EunuchRU, Pri-vateMasterHD, SpazAbiogenesis, Leptinresistinadiponectin, NottNott, SuperFreakCell, Anmusna, Stamptrader, Sstur, Suelru, Chaya5260, Inphynite, Kkosman, Baltazorgue, Johngraybosch, Monkbot, BethNaught, Acagastya, Garfield Garfield, Shandek, Signedzzz, Brianbleakley, Pombrand, Ruwdaman, Fried Vegetables, BicelPhD, Yazan atheos, BlueFenixReborn, Strongjam, Sarr Cat, Imradiamyownway, Washington Charter, Chemistryorigin, Michaelo1019, One sanguin, KasparBot, Fernando orrego, Atehoun, Ktns, Paula NK, Joholub123, Shadowblade001, Bik0ser, Dylangenetic and Anonymous: 719

- **Circumstellar habitable zone** *Source:* https://en.wikipedia.org/wiki/Circumstellar_habitable_zone?oldid=692667400 *Contributors:* Michael Hardy, Skysmith, Jeandré du Toit, Mulad, Stone, Furrykef, Omegatron, Nickshanks, Jeffq, Jni, Merovingian, Filemon, Giftlite, Daibhid

C. Pgan002, Beland, Kaldari, B.d.mills, Thorwald, Jayjg, N328KF, Wesha, Gimmick Account, Rich Farmbrough, Tsujigiri~enwiki, Bender235, Cyclopia, Chewie, RJHall, Liberatus, Art LaPella, RoyBoy, Dalf, Dave Fried, Viriditas, Elipongo, Vystrix Nexoth, Gary, Carbon Caryatid, Tabor, Mlm42, Itschris, Pauli133, Alai, Jun-Dai, Woohookitty, Falconer, CWitte, Waldir, Elenes, RichardWeiss, Marskell, Drbogdan, Sjakkalle, Rjwilmsi, Lockley, Rillian, Bruce1ee, Mike s, Thangalin, Bensin, Yug, Pinkville, John Maynard Friedman, YurikBot, Deerslayer, Robert A West, Pigman, Raquel Baranow, Zhatt, Expensivehat, Dmoss, Xompanthy, Samir, Perry Middlemiss, JdwNYC, Zero1328, Arthur Rubin, Hurricanehink, Petri Krohn, Ilmari Karonen, Tzepish, SmackBot, Anarchist42, TBH, Nickst, Cuddlyopedia, Bluebot, Fplay, Gyrobo, Modest Genius, WikiPedant, Rogermw, Trekphiler, Can't sleep, clown will eat me, Wen D House, Kingdon, Metta Bubble, Eynar, Ligulembot, Thor Dockweiler, Ohconfucius, Dmh~enwiki, J. Finkelstein, J 1982, Gobonobo, JorisvS, Tonybaldacci, Ckatz, Mr Stephen, Rock4arolla, Artman40, Clarityfiend, Joseph Solis in Australia, Triumph Sisyphus, CmdrObot, JohnCD, Benwildeboer, Steven Kelly, Cydebot, A876, My Flatley, Arb, Casliber, Headbomb, Noclevername, Escarbot, Bluehen, Kevin Nelson, Wmgries, Lindemulet, JAnDbot, Txomin, Inks.LWC, Rothorpe, Sinnerwiki, Pedro, Murgh, Swpb, BatteryIncluded, Tojo940, Kheider, Kiminatheguardian, CommonsDelinker, Sastrugi1, Hans Dunkelberg, Acalamari, Dadadaddyo, Mikael Häggström, ElectricValkyrie, Ohms law, AzureCitizen, Killeruni, Dorftrottel, Idioma-bot, VolkovBot, Katydidit, Kyle the bot, Fences and windows, TXiKiBoT, Dllahr, Sintaku, MartinKal, Martin451, Telecineguy, AlleborgoBot, Aubri, AusJeb, WereSpielChequers, Sakkura, KGyST, Da Joe, METIfan, Mimihitam, Tomemorris, Timeastor, Martarius, ClueBot, EoGuy, Qhadspeth, Polyamorph, AirdishStraus, Niceguyedc, StigBot, P. S. Burton, Piledhigheranddeeper, Bensci54, Patricius Augustus, Muro Bot, Thingg, Johnuniq, Sparkygravity, GabrielVelasquez, DumZiBoT, Hoobus, Jimmythatdawg, Rreagan007, Addbot, Roentgenium111, DOI bot, Medich1985, Proxima Centauri, LaaknorBot, Glane23, Obsidianspider, LinkFA-Bot, Craigsjones, Qemist, Luckas-bot, Jameboy123, Yobot, Fraggle81, TaBOT-zerem, Azcolvin429, Againme, 4th-otaku, Tbayboy, AnomieBOT, Ciphers, Piano non troppo, Hunnjazal, Citation bot, ArthurBot, Xqbot, Smk65536, Tad Lincoln, WingedSkiCap, Ataleh, Indeedous, Omnipaedista, Bellerophon, Trafford09, Dougofborg, Nlj7b2, лакичи, Paine Ellsworth, Thayts, ClickRick, Citation bot 1, The most interesting man in the world, Manuelzs, Jonesey95, Tom.Reding, Shadroth, IJBall, Trappist the monk, Lotje, Bluefist, Kugellager, Nederlandse Leeuw, RjwilmsiBot, Mr. Anon515, John of Reading, Quru, GoingBatty, Mmeijeri, Solomonfromfinland, ZéroBot, Udvarias, Ὁ οἶστρος, Arbnos, H3llBot, Demomoer, Eniagrom, AManWithNoPlan, Kilopi, Brandmeister, Vanished user fijtji34toksdcknqrjn54yoimascj, Yapanuwan, EvenGreenerFish, WaterCrane, Mike6828, Gary Dee, ClueBot NG, Hallaman3, Violettsureme, Tideflat, Frietjes, Ryan Vesey, RiddledEpitome, Drlectin, Alicegiada, WiseBass, Bibcode Bot, Aepsil0n, BG19bot, Josvebot, GGShinobi, Jonathan2112, Hostager, Zedshort, Weewrfwerw, Aristizle, Thiswatertaken, Chip123456, Wer900, SoylentPurple, BattyBot, Artieboyaa, Khazar2, Plusw, Dexbot, Mogism, G.Kiruthikan, Anderson, Cerabot~enwiki, Jamesx12345, Epicgenius, Dangerous3001, Astredita, Derekdoth, Monkbot, Teddyktchan, Wer902 and Anonymous: 188

- **Planetary habitability** Source: https://en.wikipedia.org/wiki/Planetary_habitability?oldid=694379170 Contributors: AxelBoldt, NathanBeach, Bryan Derksen, Taw, XJaM, Michael Hardy, Bcrowell, Wolfstu, Nikola Smolenski, Vanished user 5zariu3jisj0j4irj, Stone, Denni, Haukurth, Tpbradbury, SEWilco, Thue, Bevo, Fredrik, Stephan Schulz, Modulatum, Johnstone, Mattflaschen, Lupin, Btinn, Everyking, Jacob1207, Curps, Michael Devore, Henry Flower, Foobar, Bobblewik, Kvasir, MisfitToys, Latitude0116, Satori, Neutrality, Adashiel, Aponar Kestrel, RobKohr~enwiki, Alkivar, Yueni, DanielCD, Jiy, Rich Farmbrough, Vsmith, Pie4all88, Florian Blaschke, Mani1, Cyclopia, Chewie, RJHall, Kwamikagami, Art LaPella, Dalf, Eritain, KarlHallowell, Helix84, Mareino, Bob rulz, Steven Watson, Babajobu, Paleorthid, SI, Avenue, Snowolf, Tycho, Amorymeltzer, Cmapm, Reaverdrop, Gene Nygaard, Drbreznjev, Ceyockey, Stemonitis, WilliamKF, Angr, Etacar11, Polyparadigm, Bo, Waldir, Zzyzx11, Wayward, Graham87, Marskell, Bunchofgrapes, Josh Parris, Drbogdan, Rjwilmsi, Mike s, Mred64, Ligulem, Brighterorange, RobertG, Ground Zero, Themanwithoutapast, RexNL, Tomer Ish Shalom, Zotel, King of Hearts, DVdm, Hall Monitor, Bomb319, E Pluribus Anthony, EamonnPKeane, YurikBot, Midgley, Ohwilleke, Briaboru, Bergsten, Gaius Cornelius, Zhatt, Wiki alf, Justin Eiler, Akulaalfa, Wagens, Kortoso, Xino, Wknight94, Noosfractal, BazookaJoe, Bdell555, Poppy, Serendipitous, Jules.LT, Jwissick, Arthur Rubin, Th1rt3en, DynaBlast, HereToHelp, SigmaEpsilon, Kungfuadam, Serendipodous, Clreland, Buldožer, SmackBot, PiCo, TestPilot, Flounderer, Motorcycle~enwiki, Speight, Ifnord, Nickst, Mscuthbert, Imzadi1979, Yamaguchi先生, Skizzik, JMiall, Fetofs, Bluebot, SchrodingersRoot, Mapledell, Colonies Chris, Verrai, Zsinj, Trekphiler, Can't sleep, clown will eat me, Ajaxkroon, Lehiarav, Tamfang, AussieLegend, Alms, Onorem, Wen D House, Bowlhover, Nakon, Exiled from GROGGS, Aelffin, Polonium, Maelnuneb, Jiminy pop, Mrdallaway, BrownHairedGirl, Coricus, TheKeithD, Edwy, JorisvS, PowerCS, RomanSpa, Clone1, Iridescent, Judgesurreal777, Joseph Sohs in Australia, Brainbark, CRGreathouse, CmdrObot, Rambam rashi, Vyznev Xnebara, Ruslik0, Benwildeboer, N2e, Joechao, Myasuda, Cydebot, Mike Christie, Grammargeek, Gogo Dodo, Kozuch, Arb, Thijs!bot, Honeplus, S Marshall, Keraunos, Tjarlds, Headbomb, Escarbot, AntiVandalBot, Paul from Michigan, Kevin Nelson, MECU, Dancingspring, MER-C, Plantsurfer, Simon Burchell, Steveprutz, Jhamilton2087, WolfmanSF, Bennybp, Rich257, Fabricebaro, Ben Ram, BatteryIncluded, Mus Musculus, Jim.henderson, Mathwhiz90601, CommonsDelinker, Fusion7, Leyo, DrKay, Hans Dunkelberg, Acalamari, MikeEagling, AntiSpamBot, LittleHow, Rominandreu, Ohms law, Mikeonatrike, Treisijs, Gemini1980, Jochem Atteveld, Fences and windows, Philip Trueman, TXiKiBoT, Zanardm, Gabhala, Henrykus, MartinKal, AllGloryToTheHypnotoad, BotKung, Jobberone, Northfox, Hellochairesse, PlanetStar, Nite-Sirk, AmrasWolf, Harry~enwiki, Kalidasa 777, SallyForth123, Martarius, Taroaldo, AlptaBot, Niceguyedc, Mightyms, Alexbot, Lartoven, NuclearWarfare, CKCortez, Qwfp, GabrielVelasquez, Mortified penguin94, Beria, Rror, MystBot, Mr. IP, D.M. from Ukraine, Addbot, Kyrkarena, DOI bot, Knight of Truth, OliverTwisted, Proxima Centauri, LaaknorBot, Obsidianspider, Debresser, LinkFA-Bot, Poocat9, 84user, Numbo3-bot, Ehrenkater, DubaiTerminator, Tide rolls, Speaker1978, EugeneZ, Jarble, Luckas-bot, Yobot, Anomalieshunters, Nallimbot, AnomieBOT, Icalanise, Materialscientist, Citation bot, James500, LilHelpa, Xqbot, Ssola, Dkelly1966, SassoBot, Trafford09, SD5, FrescoBot, Citation bot 1, Pinethicket, I dream of horses, Jonesey95, Tom.Reding, Calmer Waters, Posada432, Water.writ, Shadroth, Ki Chjang, Koakhtzvigad, Horst-schlaemma, Trappist the monk, Pbrower2a, DASHBot, EmausBot, WikitanvirBot, Slightsmile, Wikipelli, ZéroBot, SnorksIII, Amir1uph, Nstock, H3llBot, Wingman417, Rimazram, EvenGreenerFish, ChiZeroOne, ClueBot NG, Wikileadspresident, This lousy T-shirt, Colapeninsula, Thejavadrinker, Violettsureme, Tideflat, Stas000D, Russellml, Listsshown, Mightymights, Helpful Pixie Bot, Gob Lofa, Soulcarber, Bibcode Bot, Fischpredigt, Wer900, BattyBot, Dexbot, GoneMoot00, Reatlas, WorldWideJuan, Adi1008, Baconfry, Ginsuloft, Astredita, Monkbot, Davidbuddy9, Esmera en, FACBot, Hampton11235, Pyrotle, RuneMan3, Tmerrillhsfdyewfyqdt and Anonymous: 236

- **Habitability of natural satellites** Source: https://en.wikipedia.org/wiki/Habitability_of_natural_satellites?oldid=684284013 Contributors: Rpyle731, Jacooks, Florian Blaschke, BoLingua, BD2412, Drbogdan, Rjwilmsi, Mike s, Bdell555, SmackBot, Nickst, Trekphiler, Tamfang, J 1982, JorisvS, SMasters, Arb, Magioladitis, BatteryIncluded, Interlaker, Universaladdress, PlanetStar, JL-Bot, Noca2plus, Sun Creator, Sunomi316, Heironymous Rowe, Addbot, Yobot, Gongshow, AnomieBOT, Citation bot, Eugene-elgato, HighFlyingFish, Moonraker, Brucewh, Double sharp, Callanecc, RjwilmsiBot, John of Reading, ZéroBot, Bud Charles, EvenGreenerFish, ClueBot NG, Tideflat, Listsshown, Gob Lofa, Bibcode Bot, Plantdrew, NotWith, Wer900, Qetuth, BattyBot, CrypticLittleNotes, Reatlas, Praemonitus, DavidLeighEllis, Kogge,

Teddy894, Jelle Gouw, Kaystay, Monkbot, Davidbuddy9, DN-boards1 and Anonymous: 22

- **Extremophile** *Source:* https://en.wikipedia.org/wiki/Extremophile?oldid=694407865 *Contributors:* Magnus Manske, Bryan Derksen, Josh Grosse, Rgamble, PierreAbbat, Edward, Patrick, Lexor, Kku, Axlrosen, Skysmith, Ellywa, Mortene, DavidWBrooks, JWSchmidt, Stone, David Latapie, Dragons flight, SEWilco, WormRunner, Rasmus Faber, Jrash, Matthew McVickar, Everyking, JuanitaJP, DragonflySixtyseven, Icairns, Guernica~enwiki, Jmeppley, Julianonions, Pinnerup, Thorwald, CALR, Rich Farmbrough, Guanabot, Vsmith, Kenb215, Chadlupkes, Bender235, Kbh3rd, CanisRufus, Remember, Bobo192, Viriditas, MrTree, Axl, Apoc2400, Katefan0, CaseInPoint, ClockworkSoul, Admiral Valdemar, TShilo12, Stemonitis, JGodman, Ashmoo, Raivein, Drbogdan, Rjwilmsi, Feydey, Naeonneo, Nihiltres, Vossman, Bgwhite, Yurik-Bot, Cathalgarvey, RobotE, Gaius Cornelius, Grafen, Theodolite, Arthur Rubin, Petri Krohn, Fram, Ilmari Karonen, SmackBot, TestPilot, Wschwarz, Clpo13, CMD Beaker, Eskimbot, BiT, Yamaguchi先生, Gilliam, Ohnoitsjamie, Skizzik, Mr Beige, Kurykh, Hibernian, WikiPedant, Valenciano, Richard001, DMacks, Cephal-odd, Vina-iwbot~enwiki, Scharks, Loadmaster, JHunterJ, Bendzh, Dr.K., DJ2000, Natronomonas, AlainD, Yuxinzhu93, CRGreathouse, CmdrObot, Jodawi, Cleanr, AshLin, Cheapestcostavoider, Mato, Gogo Dodo, Eric Martz, Rracecarr, JodyB, Thijs!bot, JAF1970, Epbr123, Runch, Deborahjay, MountainBum83, Mailseth, AntiVandalBot, Saimhe, Danger, Sluzzelin, MortimerCat, Tstrobaugh, Bongwarrior, VoABot II, Nyttend, Rich257, Bubba hotep, Cyktsui, BatteryIncluded, DerHexer, Lmchalwell, Qifeng, Feonaway, Lilac Soul, Effectrode, Century0, Naniwako, OAC, Ajtep, 83d40m, STBotD, Philip Trueman, TXiKiBoT, Oshwah, Antoni Barau, Xarelete, Earthdirt, Aedan cunningham, Ninjataeoshell, Sue Rangell, Jsde, Blpage, SieBot, Jerryobject, Bentogoa, Flyer22 Reborn, Hobartimus, Fratrep, ImageRemovalBot, Touchstone42, ClueBot, Deviator13, PipepBot, The Thing That Should Not Be, Niceguyedc, DragonBot, Shinkolobwe, Estirabot, Thingg, Rror, Avoided, Salam32, Qgil-WMF, RP459, Bioguz, Grayfell, C6541, Gravitophoton, DOI bot, Atethnekos, Fluffernutter, Cst17, LaaknorBot, Bernstein0275, ChenzwBot, Delemon, Legobot, Luckas-bot, Yobot, KamikazeBot, Robert Treat, Campsis, Rubinbot, 1exec1, Hunnjazal, Citation bot, Frankenpuppy, Gifh, LilHelpa, Tzim78, Xqbot, Anna Frodesiak, CeliaRSC, The Wiki ghost, Eugene-elgato, FrescoBot, Akshatrathi294, VI, Doom Order, Citation bot 1, Xirja, Livestronger, Livestrongest, TjBot, Ripchip Bot, Androstachys, Rayman60, Orphan Wiki, WikitanvirBot, Look2See1, Deirovic, ZéroBot, Wintermomi, Miguelmote, Crm1003, ClueBot NG, Horoporo, Russellml, HMSSolent, Bibcode Bot, Skullcandle, Sam48823, Thegreatgrabber, Jonadin93, EuroCarGT, Nugget00, GoneMoot00, Kyletheowl, TheGuy520, Nickosr.p, Iztwoz, Dustin V. S., Muderab11, Ginsuloft, Fewaut001, Officialdrgamer, Monkbot, DeadCyngus, Dr John Martin Wright, WaffleMan19 and Anonymous: 212

- **Biosignature** *Source:* https://en.wikipedia.org/wiki/Biosignature?oldid=687910791 *Contributors:* Ellywa, Julesd, Stone, Phoebe, Thue, Gzornenplatz, IdahoEv, Sparky the Seventh Chaos, TedPavlic, CanisRufus, Drbogdan, Rjwilmsi, Mred64, The ARK, Nihiltres, Gareth E. Kegg, RussBot, Bergsten, Joel7687, SmackBot, Amatulic, Bluebot, Miguel Andrade, Whispering, Arb, RichardVeryard, BatteryIncluded, Adlamajw20041157, Boris.jansen, Niceguyedc, Addbot, Grey Geezer, Atethnekos, Lightbot, Yobot, Citation bot, Mamaberry11, Citation bot 1, Tom.Reding, IVAN3MAN, Trappist the monk, RjwilmsiBot, WikitanvirBot, ZéroBot, Rafiwiki, Helpful Pixie Bot, Bibcode Bot, BattyBot, Divemast, Stamptrader, Monkbot, Tetra quark, CV9933 and Anonymous: 14

- **List of multiplanetary systems** *Source:* https://en.wikipedia.org/wiki/List_of_multiplanetary_systems?oldid=686518507 *Contributors:* Bryan Derksen, The Anome, Chrislintott, Zimriel, Zache, Jeandré du Toit, Zarius, Nickshanks, Phil Boswell, Seth Ilys, Filemon, Giftlite, Jyril, Wikibob, The Singing Badger, Superborsuk, Gunnar Larsson, Csmiller, Bosmon, Icairns, Willhsmit, Trevor MacInnis, Spiffy sperry, Gimmick Account, Bornintheguz, Jkl, Bender235, MasterRegal, RJHall, Kwamikagami, Worldtraveller, -jkb-, Art LaPella, Enric Naval, Tronno, DaveGorman, Alansohn, Melaen, Czolgolz, Mu301, Riffsyphon1024, Benhocking, Waldir, 丁丁, Palica, Rm20, Marskell, Drbogdan, Astronaut, Marasama, Rillian, Mike s, SeanMack, Bensin, FlaBot, Mnatessa, Themanwithoutapast, Bgwhite, Peter Grey, YurikBot, Spacepotato, Wester, Gaius Cornelius, West-I, Canadadunane, Welsh, DarthVader, SAE1962, Seegoon, Wolbo, Bota47, Perry Middlemiss, Smkolins, Emirp, Chesnok, Chaos syndrome, Nae'blis, Hurricane Devon, Geoffrey.landis, Curpsbot-unicodify, SmackBot, Bowzer, Nickst, Eskimbot, Elk Salmon, GwydionM, Kinhull, Mokwella, Hibernian, Trekphiler, Tsca.bot, BrianBird, Speedyprimus, Милан Јелисавчић, Eliyahu S. Chlewbot, OOODDD, WinstonSmith, Aldaron, Wen D House, -Paul-, Shirifan, Enfolder, JorisvS, Bjankuloski06en~enwiki, Zzzzzzzzzzz, Ckatz, Artman40, Newone, Markham, CmdrObot, ZICO, Broc, Teixant, ThreeBlindMice, MarsRover, Xaman, Cydebot, Editor at Large, Robert.Allen, A7x, Casliber, Thijs!bot, Gennytte, BlytheG, Kevin Nelson, Myanw, JAnDbot, MER-C, QuantumEngineer, Rothorpe, Nyttend, Wormcast, Alekjds, BatteryIncluded, Mollwollfumble, NJR ZA, Kheider, Geboy, Flo422, Leyo, Sege1701, Hans Dunkelberg, Katharineamy, Plasticup, Rwessel, Aervanath, Squids and Chips, VolkovBot, Cosmium, Amakthea computer, Aaron Rotenberg, DoktorDec, BotKung, Vikrant42, Stephen J. Brooks, Coronellian~enwiki, PlanetStar, Grundle2600, Raekshea, TitanOne, Harry~enwiki, Georgemargaris, Benoni-iBot~enwiki, OKBot, Anchor Link Bot, Nergaal, JL-Bot, AstroMark, DragonBot, Atomic7732, Thingg, MelonBot, HumphreyW, GabrielVelasquez, LostLucidity, Olvegg, MystBot, Good Olfactory, Addbot, Roentgenium111, DOI bot, Proxima Centauri, LaaknorBot, AndersBot, Ginosbot, LinkFA-Bot, 84user, Ace45954, Zorrobot, Luckas-bot, Yobot, Aldebaran66, HieronymousCrowley, AnomieBOT, ThaddeusB, Icalanise, Crystal whacker, Materialscientist, ArthurBot, Xqbot, Noonehasthisnameithink, Gap9551, Alumnum, Brandon5485, Fobos92, 11cookeaw1, Aaronb121, Nagualdesign, FrescoBot, Thayts, Treklix, Citedvisitnext, RedBot, Ilvon, IJBall, Gabeln2, Nederlandse Leeuw, Thex1, Nate5713, WildBot, EmausBot, WikitanvirBot, Quantanew, Physics16, ArthurRG, ThorX13, Solomonfromfinland, JacobSheehy, Fæ, Shuipzv3, Harbingerdawn, Sailsbystars, EvenGreenerFish, ErokDS, Whoop whoop pull up, ClueBot NG, Mharbut, Danim, Oklahoma3477, BG19bot, Vagobot, Dr.Toonhattan, Q6637p, Wer900, Szczureq, ShellfaceTheStrange, Jawshewah, Astredata, Sdgedfegw, Kepler-777, Huritisho and Anonymous: 167

- **Terrestrial planet** *Source:* https://en.wikipedia.org/wiki/Terrestrial_planet?oldid=679562371 *Contributors:* Bryan Derksen, Zundark, Stephen Gilbert, -- April, Andre Engels, Josh Grosse, JeLuF, Olivier, Boud, Karada, Alfio, Andres, Raven in Orbit, Pizza Puzzle, The Tom, Thue, Bevo, Robbot, RedWolf, Altenmann, Romanm, Lowellian, PedroPVZ, Filemon, Nephelin~enwiki, Giftlite, JimD, Neile, Icairns, Eddpayne, Urhixidur, Danh, Mike Rosoft, DanielCD, Plexust, Rich Farmbrough, Vsmith, Moochocoogle, RJHall, Kwamikagami, Shanes, Mairi, Whosyourjudas, Smalljim, Viriditas, Mpvdm, Alansohn, Idont Havaname, Wtmitchell, Henry W. Schmitt, Blaxthos, Benbest, Sejessey, Tabletop, Waldir, Smartech~enwiki, Christopher Thomas, Phlebas, Paxsimius, Graham87, Marskell, Chun-hian, Kbdank71, Phoenix-forgotten, Drbogdan, Rjwilmsi, Nightscream, Mike s, AndyKali, Chobot, YurikBot, Gaius Cornelius, Wimt, Talklave, Aaron Schulz, Lockesdonkey, Bota47, Dna-webmaster, Wknight94, Chaos syndrome, Pietdesomere, Modify, Argo Navis, Allens, Junglecat, TLSuda, Serendipodous, Phil 1970, Sardanaphalus, SmackBot, Brammers, Jroekley, Provelt, Kintetsubuffalo, DHN-bot~enwiki, Redline, Gyrobo, Tamfang, Mrwuggs, TheK-Man, Nreprm2026, Bob Castle, Evlekis, DDima, SashatoBot, J 1982, Pthag, JorisvS, Bjankuloski06en~enwiki, Ckatz, Frokor, Xiaphias, Novangelis, Danilot, Joseph Solis in Australia, Scooter20, MD:astronomer, Ewulp, Chetvorno, Cwastell, CmdrObot, Drinibot, Ruslik0, Benwildeboer, WeggeBot, Cydebot, Derek Balsam, Acv4b, Codetiger, Paddles, The Lizard Wizard, Sweikart, Thijs!bot, Epbr123, Purple Paint, Basilo, Luna Santin, QuiteUnusual, Tjmayerinsf, 1of3, Joeth, JAnDbot, Tautology, MER-C, Planetary, Something14, Rothorpe, VoABot

II. JamesBWatson, Blackicehorizon, Mlindroo, DerHexer, Edward321, Kheider, NatureA16, MartinBot, Schmloof, Anaxial, J.delanoy, Neolandes, DomBot, Hans Dunkelberg, Weirdoinventor, Ohfosho, Mikael Häggström, Gurchzilla, Hennessey, Patrick, Ginogrz, Pdcook, Izno, Idioma-bot, Justin Forbes, Chinneeb, King Lopez, VolkovBot, TheOtherJesse, Sdsds, TXiKiBoT, Mercurywoodrose, HarryAlffa, Anna Lincoln, Mzmadmike, Wing7990, Synthebot, AjitPD, PlanetStar, Tireless666, Jimbo online, Danieldaglish, Happysailor, Radon210, I'd Buy That for a Dollar, Lightmouse, Nergaal, ClueBot, Justin W Smith, Excirial, Alexbot, Jusdafax, Jacob u47, Matthew R Dunn, Aitias, GrahamDo, Vanished user tj23rpoij4tikkd, Ts41596, HumphreyW, Roxy the dog, WackyBoots, Avoided, Jd027, Good Olfactory, Hoplophile, Addbot, Roentgenium111, Ayrenz, ChenzwBot, Tide rolls, Zorrobot, LuK3, Megaman en m, Legobot, Luckas-bot, 4th-otaku, AnomieBOT, 1exec1, Piano non troppo, Mann jess, The High Fin Sperm Whale, Citation bot, Xqbot, Lucianomendez, Capricorn42, SassoBot, Shadow jams, Fotaun, Thehelpfulbot, FrescoBot, D'ohBot, Citation bot 1, Tom.Reding, Yahia.barie, RedBot, Ltuch, عباد ديرانى ديجاع باد, TobeBot, Copistopplayer, Aniten21, RjwilmsiBot, Lady gaga SJG, Salvio giuliano, EmausBot, Super48paul, IncognitoErgoSum, TuHan-Bot, Traxs7, Wieralee, Gniniv, Leofil2, Tomásdearg92, Sweisel, EvenGreenerFish, GermanJoe, 1er malkin, Terraflorin, 28bot, Cgt, ClueBot NG, Cazorla, Brainericket, Widr, Bibcode Bot, Snaevar-bot, PhnomPencil, ElphiBot, Ewigekrieg, Cadiomals, Vegan11, Smettems, Blackbird5555, Filiosus's Saga, Wer900, Jfkduihv, JYBot, Stas1995, Lugia2453, Reatlas, Terrestrial111, Jepag2012, Harlem Baker Hughes, Baconfry, Bosteeboy, Hughleesonsmith, Astredita, Marikanessa, Monkbot, Davidbuddy9, CamxxCore, TheWhistleGag, Ansh Mathur and Anonymous: 227

- **Planetary system** *Source:* https://en.wikipedia.org/wiki/Planetary_system?oldid=687191267 *Contributors:* Roadrunner, Ahoerstemeier, RotemDan, Nurg, Yosri, Filemon, Jyril, HangingCurve, Mboverload, Foobar, Jackol, Antandrus, Gunnar Larsson, Icairns, Lithorien, RJHall, El_C,Kwamikagami, Worldtraveller, Phoenix Hacker, Jeffmedkeff, Sasquatch, Gary, Wricardoh~enwiki, Bebenko, Drbogdan, Rjwilmsi,Marasama,Mike_s, Vegaswikian, E Pluribus Anthony, Wavelength, Spacepotato, The Merciful, Irishguy, Caiyu, Tony1, Marcelo-Silva, Carlosguitar,Serendipodous, SmackBot, Lotse, Markov, Darth Panda, Jan.Kamenicek, JorisvS, RandomCritic, Zapvet, Civil Engineer III, Courcelles, Dlo-heierekim, KyraVixen, ShelfSkewed, Funnyfarmofdoom, Red Director, A Softer Answer, Alaibot, Wikid77, Honeplus, Headbomb, Cogitoergo_sumo, Reswobslc, Dawnseeker2000, Majorly, D V S, MER-C, LittleOldMe, Guy0307, WolfmanSF, Murgh, Bongwarrior, Kheider,Geboy, MartinBot, BouncingBeatnik, CommonsDelinker, Sdp1978, Leyo, J.delanoy, AltiusBimm, Uncle Dick, Yonidebot, Acalamari, Jeep-day, Plasticup, Johhny jamisondalefordardostien, Johnfos, Blankego, SlateGrey, EewNick, Yksin, BotKung, 1981willy, Finngall, SieBot, Plan-etStar, Rackshea, AngelOfSadness, Bagatelle, LonelyMarble, Nergaal, Martarius, Sfan00_IMG, ClueBot, Wookie501, Polyamorph, Ktr101,LaosLos, PixelBot, Eeekster, X relentless x, Abrech, ChrisHodgesUK, D.M. from Ukraine, Albambot, Addbot, Ronhjones, Proxima Centauri,PFSLAKES1, AndersBot, Anxietycello, Arbitrarily0, Luckas-bot, AnomieBOT, Piano non troppo, Materialscientist, Citation bot, Maxis ftw,LilHelpa, Xqbot, Pigflower74, NFD9001, Ctolsen, Moxy, MeDrewNotYou, Interstellar Man, FrescoBot, DeTru711, Majopius, OgreBot, Cita-tion bot 1, Pinethicket, Theatrefreak09, Tom.Reding, A8 UDI, Trappist the monk, RjwilmsiBot, Ripchip Bot, 2mascottone, Immunize, Mmei-jeri, Vediesciences, ZéroBot, StringTheory11, EvenGreenerFish, Ego White Tray, Terraflorin, Manytexts, ClueBot NG, This lousy T-shirt,Widr, Helpful Pixie Bot, Bibcode Bot, BG19bot, Fischpredigt, Ollieinc, Mclear54, Wer900, BattyBot, Simeondahl, ChrisGualtieri, GoShow,AlexTes, Astredita, Jaiswal Garima, Anasdon, Elenceq, Monkbot, Tigercompanion25, The Last Arietta, Fleivium, Oap deals, KasparBot,Fdfexoex, Huritisho and Anonymous: 107

- **Extragalactic planet** *Source:* https://en.wikipedia.org/wiki/Extragalactic_planet?oldid=683552357 *Contributors:* Czolgolz, Miss Madeline, Drbogdan, Arasaka, StuRat, AndrewWTaylor, SmackBot, J 1982, JorisvS, Colonel Warden, Thijs!bot, Chalence~enwiki, Mdriver1981, Rothorpe, Kheider, Station1, Barath s, PlanetStar, MystBot, Addbot, Luckas-bot, Munin75, H8erade, Matthurricane, DrilBot, Jesse V., EmausBot, Jmencisom, Italia2006, ZéroBot, SSR2000, MillingMachine, Reatlas, Asonofomegasupreme, Mohamed-Ahmed-FG, Monkbot, Tetra quark and Anonymous: 17

- **Drake equation** *Source:* https://en.wikipedia.org/wiki/Drake_equation?oldid=693648347 *Contributors:* The Epopt, Eloquence, Mav, Bryan Derksen, The Anome, Taw, Ed Poor, Shsilver, XJaM, Christian List, Ghakko, Roadrunner, SimonP, Cayzle, AdamRetchless, Olivier, Spiff~enwiki,Edward, Deljr, Gdvorsky, Skysmith, Alfio, Darkwind, Julesd, Bogdangiusca, Susurrus, Mulad, Charles Matthews, Timwi, Jogloran, Judzil-lah, Tpbradbury, Jakenelson, Nv8200pa, Dogface, Dbabbitt, Raul654, JorgeGG, Robbot, Paranoid, Lowellian, Sverdrup, Rursus, Wikibot,Joshays, Mattflaschen, Pengo, Acrider, Giftlite, Graeme Bartlett, Lethe, Bfinn, Mboverload, Peter Ellis, Wmahan, Garrett~enwiki, Academi-cian, Quadell, Antandrus, Beland, Balcer, Jawed, Satori, Elroch, Clemwang, Bhugh, Percy, DanielCD, Mindspillage, Discospinster, RichFarmbrough, Dbachmann, SpookyMulder, Kenb215, Ground, Pmcm, PhilHibbs, Shanes, RoyBoy, Viriditas, Elipongo, HasharBot~enwiki,PCJockey, Jumbuck, Alansohn, Eric Kvaalen, Titanium Dragon, Jm 1234567890, Wtmitchell, BRW, Gene Nygaard, Alai, Vadim Makarov,Wyvern, Dismas, Stephen, Roboshed, Woohookitty, Drseti, GregorB, Eilthreach, Waldir, Zzyzx11, Emerson7, Graham87, BD2412, Drbog-dan, Banditski, Rjwilmsi, Koavf, Eyu100, Rillian, Mike Peel, Bubba73, Bensin, Nandesuka, Eubot, Mathbot, Thegreatloofa, Nihiltres, Norvy,Krackpipe, RexNL, Echo5ive, Ggb667, Chobot, Bgwhite, RussBot, Anonymous editor, Lofty, Zafiroblue05, KSmrq, Ihope127, Jugander,Irishguy, Sir48, Viper5dn, Kortoso, Wknight94, Thnidu, Nikkimaria, Rpvdk, Arthur Rubin, D'Agosta, Reyk, JQF, Petri Krohn, CWenger,Shawne, Geoffrey.landis, Aryah, JJansen, KnightRider~enwiki, A bit iffy, SmackBot, Oub, 1dragon, Davidkevin, KoejoBot~enwiki, Davewild,Eskambot, Kintetsubutfalo, Skizzik, Chris the speller, Someonesdad363616, Kurykh, Thumperward, Ocicat, Sbharris, Dethme0w, Can't sleep,clown will eat me, Fiziker, WinstonSmith, Xyzzyplugh, LouScheffer, Addshore, Alanf777, The-dissonance-reports, Nakon, Ck lostsword, Phydeaux, Foolish Child, 2T, Mets501, Novangelis, MrDolomite, GNB, Michaelbusch, Spark, Jason.grossman, JoeBot, Kartik Agaram, Eye-fragment, Dlohcierekim, Zaphody3k, CmdrObot, IntrigueBlue, Chmee2, Cydebot, Foolfromhell, Mysernnm, Bencope, Rraeecarr, DanielJ, Leivick, Fen, DumbBOT, Robertinventor, JayW, Dragonflare82, Arb, Jed keenan, Ajcee7, Thijs!bot, Wikid77, Headbomb, Marek69,Turkeyphant, Greg L, Natalie Erin, Darekun, Widefox, Obiwankenobi, Kent Witham, .anacondabot, Magioladitis, Ramurf, Connormah, Wolf-manSF, Bongwarrior, VoABot II, Swpb, ZackTheJack, BatteryIncluded, Torchiest, Mike Payne, Gwern, Racepacket, Foraminifera, Falazure,Surcer, Leyo, Tgeairn, Maurice Carbonaro, Athaenara, Balsa10, TomyDuby, Tiberius47, Mikael Häggström, RenniePet, Alevion, LittleHow,NewEnglandYankee, Fountains of Bryn Mawr, EyeRmonkey, KylieTastic, Davidf2281, Taylornm, BernardZ, VolkovBot, Butwhatdouknow,SpaceKangaroo, Barneca, TXiKiBoT, NathanielPoe, Fbs. 13, Ffkling, Cogitoecogito, Paulpv, Falcon8765, JRandomUser, Scottywong, SieBot,LarsHolmberg, BrianFH, Da Joe, METIfan, RJaguar3, Uicyend, Mimihitam, JPBonsain, Verson, Davesp, Fstanchina, Jruderman, Nickorum,Spartan-James, Hamiltondaniel, Naturespace, Mr. Granger, ClueBot, Ropata, Andy Roland, The Thing That Should Not Be, AstroMark,Lunatic83, Leecharleswalker, Pi zero, Ensor76, J8079s, Toad of Steel, ChandlerMapBot, Erebus Morgaine, Estirabot, Scog, Lostraven, BO-Tarate, Thehelpfulone, Tired time, Theking2, DumZiBoT, Rickremember, XLinkBot, Scylentbob, Wertuose, BodhisattvaBot, Vp11_fb9,Pgallert, Whoischristopher, Appleman Crabcakes, Muffin Pants, Trigonomamous, Roentgenium111, DOI bot, Guoguo12, Bte99, Goatstein, Numbo3-bot, Tide rolls, Teles, 维基小 Ben Ben, Legobot, Luckas-bot, Yobot, Sectsisfuns, AnomieBOT, Statie623, 1exec1, Tsuchan, Mintrick,Materialscientist, Citation bot, Kjellmikal, MidnightBlueMan, SouthH, Cedricthecentaur, Pontificalibus, Bsteve7, Jaysmasher, Jeffrey Mall,

XZeroBot, Gap9551, RG72, Aperson1234567, Trafford09, Howwy29, Zman9600, MeDrewNotYou, FrescoBot, Paisiello2, FChurca, Aurelia19, Sae1962, Trdsf, Citation bot 1, The most interesting man in the world, Yitping, Pinethicket, Tom Reding, Hoo man, Mercy11, Sbunny8, A2fwiki, Reach Out to the Truth, Sideways713, RjwilmsiBot, Ienpw III, Venustas 12, TheHunter1337, Tntermini, WikitanvirBot, Super48paul, Racerx11, AlanSiegrist, Megantater, Youncej, Howieluv2, Solomonfromfinland, Fæ, Mh7kJ, Shuipzv3, HylandPaddy, David J Johnson, Wingman417, IllegalKnowledge, A, L Kensington, Donner60, Mungujakisa, Ontyx, Carmichael, Robin Lionheart, Jmperel, Mjbmrbot, Gary Dee, ClueBot NG, Horoporo, SamKater, APL92, Doh5678, Kmchanw, Vault14, Gmansoliver, Helpful Pixie Bot, Bibcode Bot, SidKemp, BG19bot, Fangslayer, Sterling.M.Archer, Tony Tan, Vanischenu, SoylentPurple, BattyBot, Phinizyspalding, Wavadd, Magnus234, ChrisGualtieri, Axentoke, Ducknish, Anderson, Frosty, Graphium, Mayogal, Hillbillyholiday, Leprof 7272, Advanceddeepspacepropeller, Epicgenius, Avirupkundu, Shrikarsan, Thevideodrome, Keplerian, Klloyd3, Jameskirk, Abattoir666, Elenceq, Monkbot, Filedelinkerbot, BangBangClubUK, Neatsfoot, Daveric2k, Kneentem, Ericcarmichael, Baller23456, Tetra quark, Isambard Kingdom, Diiddiidd, BU Rob13, Nateisadick and Anonymous: 473

- **Fermi paradox** *Source:* https://en.wikipedia.org/wiki/Fermi_paradox?oldid=694563541 *Contributors:* AxelBoldt, Derek Ross, Vicki Rosenzweig, Mav, Uriyan, Bryan Derksen, Zundark, The Anome, Tarquin, Gareth Owen, Andre Engels, Shsilver, Darius Bacon, Hai-Etlik, Roadrunner, Shii, Ant, AdamRetchless, Tedernst, Jim McKeeth, Frecklefoot, Paul Barlow, Gdvorsky, Karada, Arpingstone, Minesweeper, Looxix~enwiki, Ihcoyc, Mortene, Stevenj, JWSchmidt, DropDeadGorgias, Mark Foskey, Sir Paul, Cyan, Vzbs34, Cimon Avaro, Tristanb, Ehn, Jengod, Timwi, Dcoetzee, Yggdrasil, Ike9898, David Latapie, Jwrosenzweig, Doradus, Wik, Tpbradbury, Motor, Furrykef, SEWilco, Xaven, Fvw, Pstudier, Pakaran, Pollinator, Lumos3, ChrisO~enwiki, Fredrik, Chris 73, Jotomicron, Sanders muc, Xiaopo, RedWolf, Naddy, Chancemill, Arkuat, Lowellian, Mirv, Babbage, Ukuk~enwiki, Sverdrup, Kencomer, Meelar, Auric, Rhombus, Smb1001, Billranton, Bkell, Dodger~enwiki, Hadal, Wereon, NeoThe1, Anthony, Xanzzibar, Mattflaschen, BradNeuberg, David Gerard, Enochlau, Matt Gies, Giftlite, JamesMLane, Mightieris-thepen, Laudaka, Barbara Shack, Mat-C, Cormac Canales, Pretzelpaws, Tom harrison, Art Carlson, Lupin, Bfinn, Karn, Everyking, Radius.Avsa, Takatoriyama, Matt Crypto, Bobblewik, Wiki Wikardo, Ryanaxp, Wmahan, OldakQuill, Andycjp, Ljhenshall, Zendonut, Sonjaaa, Ke-pion, Quadell, IdahoEv, Fredeondo, Loremaster, Robert Brockway, MisfitToys, Khaosworks, Jossi, Exigentsky, Rdsmith4, OwenBlacker, Michael Rowe, Rlquall, Thineat, Pethan, Sam Hocevar, Neutrality, Sam, Mschlindwein, Danarmak, Lacrimosus, Danh, RobKohr~enwiki, Jason Carreiro, Oskar Sigvardsson, Ta bu shi da yu, Malu5531, Sparky the Seventh Chaos, Eyrian, CALR, Andy Smith, Bornintheguz, Dis-cospinster, Rich Farmbrough, Snap2grid, Pjacobi, Florian Blaschke, ArnoldReinhold, Kostja, HCA, Ivan Bajlo, Ponder, Samboy, Sperling, JPX7, Bumhoolery, SpookyMulder, Corvun, Bender235, ESkog, Wazerface, Tr606~enwiki, Quietly, A purple wikiuser, Kaisershatner, RJHall, Pietzsche, Mr. Billion, El C, Jim127, Lycurgus, Mjk2357, Laurascudder, Worldtraveller, Thickslab, Susvolans, Deanos, Dalf, Leftmosteat, Shoujun, Grick, Sippan, Func, Shenme, Viridius, Dejtarob, 19Q79oL78KiLOQTFHgyc, Rajah, Rje, Shanen, DanB~enwiki, Sebastian Goll, CNash, Jonathunder, JYolkowski, ChristopherWillis, Arthena, Cormaggio, DanielVallstrom, Aduthie, Monk127, Axl, MrBudgens, Velella, BRW, M3tainfo, Suruena, Spiritchaser~enwiki, TenOfAllTrades, Anlala, Alai, Vanished user dfvkjmet9jweflkmdken234, Mullet, Mesee, Chilepine~enwiki, Distantbody, Firsfron, Jeffrey O. Gustafson, Roboshed, Vashti, Linas, Keillan, Mindmatrix, Pmberry, Blair P. Houghton, DoctorWho 42, Nuggetboy, Koshki, Ljfeliu, Norro, M Alan Kazlev, Wayward, Sin-man, RichardWeiss, Ashmoo, Marskell, Johnny Mnemonic, Magister Mathematicae, Galwhaa, Phoenix-forgotten, Drbogdan, Rjwilmsi, Mayumashu, .digamma, Zbxgscqf, War, Jweiss11, Urbane Legend, LoganFive, Mike Peel, Miserlou, Ligulem, LjL, Bubba73, Brighterorange, Cassowary, Ropez, Fish and karate, Exeunt, Strobilomyces, Wragge, FlaBot, Eubot, RobertG, JdforresterBot, Kmorozov, Bubbleboys, Czar, Alexjohne3, Str1977, Diza, Bornhj, DVdm, Bgwhite, Joseph11h, YurikBot, Wavelength, Sean Et Cetera, Paul C. Pratt, Huw Powell, Zafiroblue05, Jengelh, Raquel Baranow, Gaius Cornelius, Miskatonic, DavidR, Ingham, Anomie, SEWilcoBot, Leutha, Jmacaulay, BlackAndy, Thiseye, Shiner~enwiki, Tony1, Xompanthy, Lockesdonkey, Gujamin, Ko-rtoso, Bota47, Stevendahlin, Nick123, Noosfractal, Mütze, Bakkster Man, Thnidu, Reyk, Bondegezou, JQF, Petri Krohn, CWenger, Nae'blis, Peter, Geoffrey.landis, JDspeeder1, Patiwat, The Wookieepedian, Attilios, A bit iffy, JJL, SmackBot, Nahald, Jasonuhl, Unyoyega, Vald, Davewild, RedSpruce, ZerodEgo, AKismet, Edgar181, Eloil, Yamaguchi先生, Cuddlyopedia, Peter Isotalo, Ohnoitsjamie, Evilandi, Skewetoo, Bluebot, Rrohbeck, ElTchanggo, Green meklar, Neurodivergent, Stevage, Jeskeca, Nedlum, Zven, Mikker, Emurphy42, Gyrobo, Scalene, Be-owulf314159, Heapehk, Glloq, OOODDD, Nima Baghaei, Xiner, Rrburke, GRuban, LouScheffer, Andy120290, Aarondude919, Quokkapox, PiMaster3, Bigturtle, Steve Pucci, "alyosha", Richard001, Gregwmay, Mini-Geek, Daveschroeder, Nairebis, Aedx, PhilJ, Clicketyclack, Ten-PoundHammer, Byelf2007, Playanaut, Thesmothete, Kimholder, Harryboyles, Khazar, ML5, Foolish Child, Berek Halfhand, TimWakefield, JorisvS, Minna Sora no Shita, Ben Jos, Mattpersons, Redherring, Ckatz, BoyliciousDarian, Lampman, SQGibbon, Hypnosifl, Vedexent, Doczilla, Davesilvan, TPIRFanSteve, Moretz, Peyre, Dfred, Osame, Celeritas, Vincecate, Hu12, Stephen B Streater, Ossipewsk, Michaelbusch, Alessandro57, Dansiman, Colonel Warden, JoeBot, RokasT~enwiki, Planet-Earth, Courcelles, Loyh, Mostly Zen, Baqu11, Harold f, Eastlaw, Heiseneat, Jpxt2000, Jeremy Banks, CmdrObot, Wafulz, Centered1, Baroquesmguy, Lighthead, Palendrom, AGTMADCAT, Cydebot, O.Harris, Shandydrinker, ANTlearrot, Rraacearr, Michael C Price, Manfroze, Robertinventor, Scarpy, Arb, NotQuiteEXPComplete, Thijs!bot, Mawfive, Fournax, TK421, Headbomb, Marek69, Second Quantization, Fenrisulfr, Db26, Noclevername, Northumbrian, AntiVandalBot, Ma-jorly, Mrshaba, CultureArchitect, Ilsistoday, Bakabaka, Richardhod, P.D., Ingolfson, Serpent's Choice, JAnDbot, Stevedix, Somesuns, Jeff560, Emax0, 100110100, NSR77, Joxernolan, Rothorpe, Geniac, Magioladitis, Murgh, Rhwawn, Dekimasu, Ekrumme, SHCarter, J meandrews, Nyttend, Froid, Akhai, ZackTheJack, Jondeere, BatteryIncluded, Tswsl1989, Emw, DasHermit, Tuviya, Gwern, Otvaltak, Ulkomaalainen, EGI, Threedots dead, CommonsDelinker, N4nojohn, Qwanqwa, Uncle Dick, KrytenKoro, Thaurisil, OttoMäkelä, OingoBoingo2, Friend-lyRiverOtter, Infocat13, OAC, Alphapeta, Plasticup, Cadwaladr, SJP, Mufka, BrettAllen, Equazcion, Ross Fraser, BernardZ, Idioma-bot, Sobrien140, VolkovBot, Amikake3, BoogaLouie, TXiKiBoT, Oshwah, RC Pinchey, GimmeBot, BuickCenturyDriver, Jacob Lundberg, Vin-cent naveen morris, Udufruduhu, Myles325a, Thmazing, Zanardm, Rei-bot, Deep Atlantic Blue, Aymatth2, Macslacker, Pah246, Oxfordwang, SoniaZ, CanOfWorms, Brian Eisley, Lou.weird, Room429, Seb az86556, Waycool27, Itemirus, Sapphic, Rob Pommer, Peter.thelander, Plan-etStar, Paradoctor, Ypps~enwiki, Kashin, METIfan, Wildonrio, Flyer22 Reborn, VideoRanger2525, Mimihitam, Nk.sheridan, Jruderman, Spartan-James, StaticGull, Maxime.Debosschere, Hamiltondaniel, Sirlanz, Albert.a.jackson, Kalidasa 777, DRTllbrg, Finetooth, Romit3, Martarius, Sfan00 IMG, ClueBot, Agaribotti, Reargunner, Deanlaw, Panopticon70, Pi zero, Hungwunfai, Drmies, VQuakr, SuperHamster, Phenylalanine, Homonhilis, BobKawanaka, Sun Creator, Nabukhadnezar, SchreiberBike, Tired time, Nasageek, MelonBot, DumZiBoT, In-ternetMeme, Jeturcotte, Dthomsen8, WalrusLike, Rohsage, Noctibus, WikiDao, Blnewbold, JCDenton2052, Blackspotw, Ianeke, Addbot, Roentgenium111, DOI bot, Maddude11, Yobmod, Silas Stoat, Ronhjones, Fluffernutter, Arthur Hal, MrOllie, MatrixArsenal, Debresser, LinkFA-Bot, PopularOutcast, Tassedethe, Tide rolls, Verbal, Lightbot, OlEnglish, Olsen-Fan, Luckas-bot, Zhitelew, Yobot, Worldbruce, EchetusXe, Reargun, Mdw0, AnomieBOT, Diderot08, Jim1138, Wtachi, Materialscientist, Citation bot, Kjellmikal, Brightgalrs, ArthurBot, Xqbot, Millahnna, Smk65536, Dkelly1966, Gap9551, Pra1998, 3starhunter, GrouchoBot, Bizso, False vacuum, Omnipaedista, MagicalSky-Man, Seligne, Alexanderpopoff, Greengrapes, Locobot, Bigger digger, Vantine84, Amicianthony, Andrewhayes, Twhair, Unbesorgt, Citation bot 2, Mr.rastapopolus, JMilty, Citation bot 1, DrilBot, Pinethicket, Julzes, Jonesey95, Tom.Reding, Geogene, Mithvetr, Medic463, Shadroth, Farmer21, Tlhslobus, Fartherred, Graham france, SkyMachine, IVAN3MAN, Silviu Mihaila, Abpotato, Bernat mussons, DrCrisp, Race911, RjwilmsiBot, Meerwind7, Mrsnuggless, Metaferon, EmausBot, John of Reading, Grrow, Faolin42, Gardaud, 8digits, Lounorte, The Mysterious El Willstro, Winner 42, Mmeijeri, K6ka, Einkleinestier, Solomonfromfinland, Anir1uph, Alpha Quadrant (alt), AarCart, H3llBot, Quondum, Kingaero, SporkBot, Ewa5050, AManWithNoPlan, David J Johnson, Card Zero, Ontyx, EvenGreenerFish, Robin Lionheart, 维基小霸王, Topdownquark, ClueBot NG, Horoporo, Andrei S, David O. Johnson, GodBlessYou2, Dream of Nyx, 78562X, JoshKW, Antiqueight, Helpful Pixie Bot, Strike Eagle, Bibcode Bot, Jeraphine Gryphon, BG19bot, Flax5, Disguised22, M0rphzone, Interchangeable, MusikAnimal, Olev Vinn, Tony Tan, MrBill3, NotWith, Newburyjohn, MisterMorton, Duxwing, Taps333, Wer900, RavelTwig, BattyBot, Phinizyspalding, Pratyya Ghosh, ChrisGualtieri, Tandrum, IsraphelMac, Elric Grey, Dexbot, Bree's Block, Cerabot~enwiki, Acoma Magic, Cainamarques, Frosty, Jamesx12345, Jochen Burghardt, Migratingmynah, Sturmgewehr88, Corinne, Qqminuss, ToFeignClef, Sndeep81, Epicgenius, BreakfastJr, Loganfalco, ProKro, Anrnusna, Isa Thunderfoot, Xenxax, Fixuture, W.carter, Fafnir1, Monkbot, Thibaut120094, Raichu234352, Dguzzo, Unician, Formuse, ChristianJorn, Bar5555, Yeowe, Splićanin, Narky Blert, Andrew Bearne, Gamebuster19901, The Yesterday Trilogy, Stormfoogle, Tetra quark, Isambard Kingdom, Lhachrism, Pulkitmidha, Anarchyte, JJMC89, TheBabbyMammoth, Furious Mythical Beast, Rambunctious Racoon, Athonitscold, Brando2131, Danrhew, Rufsotufs, Alinaresgoenaga and Anonymous: 835

- **Cosmic pluralism** *Source:* https://en.wikipedia.org/wiki/Cosmic_pluralism?oldid=672912485 *Contributors:* Edward, Wetman, Banno, Goethean, Altenmann, Rursus, Mboverload, Karol Langner, Pharos, Uncle G, Marskell, Ian Pitchford, Jimp, RussBot, Chris Capoccia, TheMadBaron, SmackBot, Herostratus, Jagged 85, Chris the speller, Madmedea~enwiki, Nima Baghaei, Polonium, Ultimaga, Gregbard, Cydebot, Arb, PamD, Brusegadi, Uncle Dick, Bearian, Kalidasa 777, ClueBot, Jonas kork, Rockfang, WikHead, MystBot, Addbot, Favonian, Luckas-bot, Yobot, AnomieBOT, Citation bot, Xqbot, Omnipaedista, FrescoBot, Citation bot 1, Rausch, Weedwhacker128, EmausBot, Savh, H3llBot, StaszekLem, TyA, Helpful Pixie Bot, Wbm1058, 2pem, JohnChrysostom, Iryna Harpy, Wer900, Zach Lipsitz, Mbay 2012, Aellithy, Robevans123, Mackbad, Tetra quark and Anonymous: 17

- **Search for extraterrestrial intelligence** *Source:* https://en.wikipedia.org/wiki/Search_for_extraterrestrial_intelligence?oldid=693028439 *Contributors:* Derek Ross, WojPob, Eloquence, Bryan Derksen, Robert Merkel, The Anome, Tarquin, Koyaanis Qatsi, Taw, Andre Engels, Snorre, Shii, Ellmist, Imran, Mbecker, Lorenzarius, Infrogmation, Michael Hardy, Tim Starling, Lexor, Grizzly, Gabbe, Axeloide, Ixfd64, TakuyaMurata, Minesweeper, Kosebamse, Looxix~enwiki, Ellywa, Ahoerstemeier, Urbanus~enwiki, Notheruser, Kingturtle, Kevin Baas, Cimon Avaro, Jiang, Michael T. Richter, Timwi, Wik, Radiojon, E23~enwiki, Furrykef, Omegatron, Wernher, Bevo, Shizhao, Stormie, Pstudier, Carbun-cle, Jni, Robbot, Sdedeo, Noplasma, Lowellian, Sverdrup, Jondel, Hadal, Asparagus, Mattflaschen, Pengo, Davidcannon, Giftlite, DavidCary, Ferkelparade, Marcika, Jacob1207, Gamaliel, Niteowlneils, Xinoph, The zoro, Adam McMaster, SWAdair, Golbez, Joseph Dwayne, PeterEllis, Pgan002, Beland, Robert Brockway, Fehler, Kuralyov, Mschlindwein, Deglr6328, Jimaginator, Urvabara, Discospinster, Rich Farm-brough, Paulo Oliveira, Snap2grid, Vsmith, ArnoldReinhold, Pgabolde, Darren Olivier, Mani1, Night Gyr, ESkog, ZeroOne, NeilTarrant, Wazerface, RJHall, Pietzsche, Chairboy, RoyBoy, Neilrieck, Causa sui, C S, Mtruch, Viriditas, ..Ajvol.., Kjkolb, Minghong, Obradovic Goran, Nsaa, A2Katir, Alansohn, QVanillaQ, Eric Kvaalen, DanielVallstrom, Andrewpmk, Sligocki, Wdfarmer, DreamGuy, Ross Burgess, BRW, Saga City, GL, Suruena, Danielmfonseca, Geraldshields11, Ndteegarden, Johntex, Tom.k, Postrach, Kfitzner, Havermayer, David Haslam, Skyraider, Drseti, Old-copy-editor, Dmol, Eyreland, Eilthireach, Studio34, Wayward, Palica, Graham87, Marskell, BD2412, Darkwand, Drbogdan, Rjwilmsi, Mayumashu, Tim!, Zbxgscqf, Staecker, DrTorstenHenning, Mike Peel, Vegaswikian, Oblivious, Bubba73, TheGWO, Thinkpad, Exeunt, FayssalF, FlaBot, JdforresterBot, Jjhat1, President Rhapsody, RLent, Geimas5~enwiki, Jemecki, Tedder, King of Hearts, Chobot, DVdm, Bgwhite, YurikBot, Wavelength, Borgx, Jimp, Kafziel, Phantomsteve, Eshuy, Fabartus, Zafiroblue05, Hede2000, Lord-Bleen, Kerowren, Hydrargyrum, Gaius Cornelius, Miskatonic, Eleassar, Bruguiea, Dumoren, Irishguy, Rbarreira, Inhighspeed, Raven4x4x, Melly42, Adreamsoul, Wangi, Kurtoso, DeadEyeArrow, Bota47, Cadillac, Stefan Udrea, Noosfractal, Cnmirose, Deeday-UK, Richardcavell, Brianhass, Emijrp, Caleb rosenberg, Closedmouth, Spondoolicks, Reyk, ZekeTheElder, TBadger, 1D4EVER, CWenger, Peter, Codemonkey, Geoffrey.landis, Syndrome~enwiki, GrinBot~enwiki, Brentt, Gabenowicki, A bit iffy, SmackBot, Reedy, KnowledgeOfSelf, CelticJobber, David.Mestel, Unyoyega, Nickst, RedSpruce, Spaceranger137, Leda74, Nil Einne, Portillo, Ohnoitsjamie, Skizzik, Andy M. Wang, Kinhull, Ati3414, Chris the speller, Bluebot, QTCaptain, Thumperward, Zinc5k, Vanessadannenberg, Baa, Wackjum, Brianporter, MaxSem, De-thme0w, Eschbaumer, JonHarder, Nima Baghaei, Stepho-wrs, LouScheffer, TheLimbicOne, John D. Croft, Jellyfisho, Monkeysyodel, MarcusBrute, SashatoBot, Serein (renamed because of SUL), Sanya, Titus III, John, Rigadoun, Faturita, Breno, Wickethewok, Michael miceli, Shat-tered, Loadmaster, Volatileacid, ChrisBianchi, 2T, SandyGeorgia, Wwagner, SETIGuy, BranStark, OnBeyondZebrax, Iridescent, Spebudmak, Superjoe30, Fdp, Tawkerbot2, I5bala, The Letter J, Chris55, MightyWarrior, Tommywommy117, DrStrangedlove, JF Mephisto, CmdrObot, Ale jrb, Amalas, Zarex, Manwithbrisk, R9tgokunks, Ibadibam, N2e, Myasuda, Icek~enwiki, Cydebot, Jetblack101, A876, Gogo Dodo, ST47, Caliga10, Tawkerbot4, Bryan Seecrets, Omicronpersei8, Khamar, Arb, Crum375, Thijs!bot, Epbr123, Wikid77, Pstanton, Dariusz Peczek, Daniel, Doc richard, Glennpicker, Danlibbo, Mojo Hand, Mereda, Headbomb, Marek69, Leon7, Sturm55, Michael A. White, Sbandrews, Dj-cunning, FWL.A.M.F., AntiVandalBot, Stephen Wilson, Leeroy SD, Prolog, Autocracy, Fashionslide, Dane 1981, Chill doubt, Kent Witham, Ingolfson, Caper13, Res2216firestar, JAnDbot, MER-C, Grant Gussie, Emax0, Kirrages, Bongwarrior, VoABot II, Craig Baker, JNW, T0mmy, Think outside the box, Danjoseofluzon, WODUP, BatteryIncluded, Galactic drifter seti, Hamiltonstone, Emw, Peemil, JaGa, Lackey, Mar-tinBot, BetBot~enwiki, Docame, Fethers, J.delanoy, Philcha, Maurice Carbonaro, Nothingofwater, Gzkn, Shawn in Montreal, Katalaveno, Crakkpot, Schinatiger, Nephersir7, Fountains of Bryn Mawr, Bermy88, Ohms law, Gregftzy, Lifeboatpres, Spaceviz, Taylornm, VolkovBot, Derekbd, JohnBlackburne, Michelle Roberts, Jurystar, Station1, Philip Trueman, JayEsJay, TXiKiBoT, MarkusJ, Olly150, Rachaelsulley, Lradrama, Surfereric, Puremage4229, JhsBot, Peterhousehold, Rare4, Wasted Sapience, James McBride, Insanity Incarnate, Thefirstfirefox, Skepper43, Andypp123, Overlord11001001, SieBot, Coffee, Tiddly Tom, WereSpielChequers, KGyST, Caltas, METIfan, Keilana, David Be, Doc Perel, Oxymoron83, AmazeStar, Steven Crossin, Lightmouse, Afernand74, Cyfal, Anchor Link Bot, Hamiltondaniel, Albert.a.jackson, Kalidasa 777, Francvs, Twinsday, WikiBotas, Sfan00 IMG, Tanvir Ahmmed, ClueBot, LAgurl, B1atv, Marwo, AstroMark, VQuakr, Eye.earth, Liam langan, Namazu-tron, Paulcmnt, DragonBot, Tripallokavipasek, PixelBot, Sun Creator, Brews ohare, Ice Cold Beer, Maradona01, MuroBot, Jonverve, DumZiBoT, XLinkBot, Jytdog, Pgallert, LeDiableBrun, NCDane, Mingovia, Addbot, Gravitophoton, M.nelson, Morriswa, Zellfaze, GyroMagician, CanadianLinuxUser, Proxima Centauri, Amy1r4e3, Chzz, Debresser, 5 albert square, Tassedethe, Tide rolls, Teles, अर्शीव महानगर, Luckas-bot, Yobot, OrgasGirl, Ptbotgourou, Fraggle81, Il MusLiM HyBRiD Il, Aldebaran66, Jan Arkesteijn, Archon-Magnus, Againme, Eric-Wester, Backslash Forwardslash, Orion11M87, AnomieBOT, Cantanchorus, Kingpin13, Materialscientist, Wiki isrong, Citation bot, ArthurBot, LilHelpa, FreeRangeFrog, TheAMmollusc, Noonehasthisnameithink, Quazgaa, Capricorn42, Marc9510000, Psycano, Gap9551, Jamesslot3, Swd, RibotBOT, Amaury, Trafford09, FreeKnowledgeCreator, Tangent747, Draytonian, Tiramisoo, Steve

Quinn, Citation bot 1, Pitri2009, Tom.Reding, Yutsi, BlackHades, Trappist the monk, Lb.at.wiki, Marvinandmilo, Michael9422, Dusty777, Ruxda, Obankston, Between My Ken, RjwilmsiBot, TjBot, Slon02, EmausBot, John of Reading, Misterman123, BruceSwanson, KG4SGP, Slightsmile, Billy9999, Akerans, Jonstewartwiki, H3llBot, AManWithNoPlan, David J Johnson, Monsieurjm, Brandmeister, Surajt88, Even-GreenerFish, Carmichael, Ihardlythinkso, Gary Dee, ClueBot NG, Michaelmas1957, Giordanobrunoburning, TruPepitoM, Biophily, Friet-jes, Brainericket, Tr00rle, Gerryharp, Blelbach, Helpful Pixie Bot, The rakish fellow, Theslayer69, Bibcode Bot, Geraldo Perez, Benzband, Altaïr, Harizotoh9, Wer900, BattyBot, ChrisGualtieri, MadGuy7023, Gadgetball, Dexbot, Mogism, Nren4237, Anderson, CuriousMind01, Acoma Magic, Frosty, Ebthi2012, GoneMoot00, Schoolisawsome123, Neonovaz, SomeFreakOnTheInternet, Wamiq, Igor Topilsky, Zscrebr, Sharmin.h, SJ Defender, Matthew2535, Fixature, Monkbot, Mondolkiri1, GreenGazoo, Johndhs, EngineeringIsFun, Elliofino, Tetra quark, Isambard Kingdom, DN-boards1, Supdiop, KasparBot, Gokinetic and Anonymous: 463

- **Wow!signal** *Source:* https://en.wikipedia.org/wiki/Wow!_signal?oldid=692178306 *Contributors:* AxelBoldt, Taw, Christian List, Shii, Nealmcb, JohnOwens, Michael Hardy, ZoeB, Rl, BAxelrod, Charles Matthews, Tpbradbury, Pstudier, Northgrove, Greudin, Postdlf, Bkell, Robinh, Xanzzibar, Everyking, Avsa, DÅugosz, R. fiend, Piotrus, Blazotron, Kaldari, Cubelodyte, Kuralyov, Ukexpat, Wesha, Jpg, Discospinster, Shuff dog, Vsmith, Ross Uber, Dbachmann, JPX7, Nchaimov, Ylai, Bender235, Rubicon, Kwamikagami, Dennis Brown, Alderbourne, JRM, Mdhowe, Mtruch, Reuben, I9Q79oL78KiL0QTFHgyc, Brainy J, Max rspct, Danhash, Alai, Blaxthos, Dismas, Roboshed, OwenX, Xover, Drseti, Pbhj, Robert K S, Commander Keane, Apokrif, Tabletop, Uris, GregorB, 陽炎01, Imperialles, Drbogdan, Rjwilmsi, Mayumashu, Nightscream, Koavf, Mike s, Mike Segal, Mike Peel, ElKevbo, Bubba73, Hauger, Thinkpad, FayssalF, President Rhapsody, Srleffler, GarethE, Kegg, Gwernol, Sceptre, Pleonic, Gaius Cornelius, Eleassar, Dlugosz, Brasswatchman, Voidxor, XXV, Oecono, Perry Middlemiss, TimKMSI, Tuckerresearch, Richardcavell, Gergis, Serendipodous, Cmglee, The Yeti, Borisbaran, KnightRider~enwiki, SmackBot, Hux, Knowl-edgeOfSelf, C.Fred, Kintetsubuffalo, Evanreyes, Septegram, SmartGuy Old, Gilliam, Winterheart, Bluebot, Modest Genius, Madman2001, Pwjb, Solarius, Kendrick7, Lambiam, BHC, John, JorisvS, Cowbert, Filanca, Hvn0413, 2T, Meco, RJNeb2, DabMachine, SETIGuy, Jet-man, Twas Now, Vanisaac, JForget, Banedon, Tim1988, Marc W.Abel, Cydebot, Svend, Gogo Dodo, Caliga10, DumbBOT, Omicronpersei8, Arb, OhioAtty, JohnInDC, Thijs!bot, Headbomb, WilliamH, Gioto, Tangerines, Fashionslide, Spojima, Leevclarke, Davewho2, CosineKitty, Sonicsuns, Ryan4314, Rothorpe, VoABot II, Otterfan, ClovisPt, Tperegrin, DBWikis, Mike Payne, Xtifr, Robin S, MartinBot, Rod57, Stan JKlimas, Osndok, Daniel Rollison, Atheuz, STBotD, Gorba, Equazcion, Hc5duke, Cs302b, JulesVerne, VolkovBot, Safte4~enwiki, TXiKiBoT, Oshwah, Red Act, Miranda, Qxz, Hunterhogan, Bentley4, Mazarin07, Brandonlh15, Steve Smith, Synthebot, Falcon8765, FKmailliW, SieBot, Pallab1234, Sophos II, Rob.bastholm, Yintan, Comu nacho, Mimihitam, AmazeStar, Kontrol Z, R0uge, Chris.Phillips, IdreamofJeanie, Z-angel~enwiki, Tdp1001, Hamiltondaniel, Petzl, Denisarona, Beeblebrox, Elassint, ClueBot, Parkjunwung, Niceguyedc, Dsilverm, Trivialist, Paulemnt, Tezuni, DragonBot, Excirial, M4gnum0n, Sun Creator, Neuronone, Aseals, SchreiberBike, Jonverve, SirKafka, Graham1973, Lx121, Centralhighperson, XLinkBot, Airplaneman, Addbot, DOI bot, JSenek, Download, Robert Zwemmer, Makelelecba, SamatBot, Myth-icalManMoth, Tide rolls, ليده, Zorrobot, Luckas-bot, Yobot, Painstaker, AnomieBOT, DoctorJoeE, AlienDragon, Materialscientist, Em-bram, Xqbot, Felixjin, Grantgw, Swd, RibotBOT, Thehelpfulbot, Rdbrid, Green Cardamom, FrescoBot, Citation bot 1, Tom.Reding, Jaguar, MondalorBot, BlackHades, Meaghan, Mikespedia, Full-date unlinking bot, Shanmugamp7, Tatepac, Elmoro, Batternut, D.tsapkou, Tbhotch, Ineverheardofhim, EmausBot, Racerx11, Slightsmile, Soren84, ZéroBot, David J Johnson, Tiago Penedo, Brandmeister, Maxrossomachin, Report900, Orange Suede Sofa, Mentibot, SemanticMantis, Grapple X, ClueBot NG, Wukai, Astrocog, Jerrychman, Brainericket, Scribalweb, Helpful Pixie Bot, Levdr1lostpassword, Jeraphine Gryphon, BG19bot, Ditto 51, Flax5, TricksterWolf, Piguy101, Gautehuus, Harizotoh9, Mr-Bill3, Glacialfox, TBrandley, Jeremy112233, Pratyya Ghosh, IdiotSnake, YFdyh-bot, 14Debee, Graphium, Tomerha91, GoneMoot00, Yaps8, Reatlas, FallingGravity, Eyesnore, Eltro102, PhantomTech, Marco.bs, The Herald, Asadwarraich, Royalcourtier, Monkbot, Filedelinkerbot, Sofia Koutsouveli, WanderingLost, Hexware, Jerodlycett, Build Until God Shows and Anonymous: 248

- **Communication with extraterrestrial intelligence** *Source:* https://en.wikipedia.org/wiki/Communication_with_extraterrestrial_intelligence?oldid=690550915 *Contributors:* Bryan Derksen, Shii, Dominus, Finlay McWalter, Lowellian, David Gerard, Darrien, Beland, Ylee, Viriditas, Cmdrjameson, Beniz~enwiki, Cavrdg, Kjkolb, Larryv, Vedantm, Brian.mcconnell, Waldir, Drbogdan, Rjwilmsi, Bensin, Mmatessa, MacRus-gail, Alexander Zaitsev, Bgwhite, Wavelength, Wikky Horse, Xihr, Grafen, Equilibrial, Closedmouth, Scoutersig, AndrewWTaylor, SmackBot, Jeppesn, Onebravemonkey, Gilliam, Portillo, Mindeye, George Church, MaxSem, MaxCosta, JonHarder, Seduisant, Yulia Romero, Peyre, Michaelbusch, Halfblue, Prs17, Cydebot, Bellerophon5685, Arb, Headbomb, Ackatsis, Astrobiologist, RBraastad, Gioto, Jj137, Qwerty Bi-nary, LordKael, BranER, Inks.LWC, WolfmanSF, Keith Lynch, J.delanoy, Maurice Carbonaro, Stambolov, Puddytang, METHan, MarkMLl, Denisarona, Twinsday, ClueBot, AvOid3r, Erunafailaro, Rhododendrates, Thingg, DumZiBoT, DrJonre, Morriswa, 84user, OlEnglish, Yobot, AnomieBOT, Xenowiki, Diderot08, Citation bot, Aollongren, Trafford09, Citation bot 1, Rumlin, RjwilmsiBot, Koolmagicguy, David J John-son, Ego White Tray, ClueBot NG, Brainericket, Helpful Pixie Bot, B3tt3rwhothe, Plipsticks, Thegreatgrabber, BattyBot, CharlesRChapman, Praemonitus, JeanLucMargot, Kim9988, Sofia Koutsouveli, Rhadamanthus17, Jerodlycett, Huritisho and Anonymous: 64

- **Nexus for Exoplanet System Science** *Source:* https://en.wikipedia.org/wiki/Nexus_for_Exoplanet_System_Science?oldid=676115118 *Contributors:* Viriditas, Drbogdan, Nikkimaria, Z22, The Anomebot2, BatteryIncluded, Life of Riley, BG19bot, Ceannlann gorm and Anonymous: 1

- **Extraterrestrial liquid water** *Source:* https://en.wikipedia.org/wiki/Extraterrestrial_liquid_water?oldid=694333031 *Contributors:* Bryan Derk-sen, Robbot, Gimmick Account, Florian Blaschke, Bender235, RJHall, Kwamikagami, Art LaPella, Viriditas, Pharos, Pauli133, Kitch, Table-top, Waldir, Drbogdan, Rjwilmsi, Nimur, Ergzay, Dlugosz, Ospalh, E Wing, SmackBot, Nickst, Gilliam, GwydionM, Gyrobo, J 1982, JorisvS, Ckatz, Novangelis, Atakdoug, Joseph Solis in Australia, Robertinventor, Arb, Andyjsmith, Headbomb, Aquilosion, Igodard, Magioladitis, Bat-teryIncluded, CommonsDelinker, Vegasprof, Rominandreu, Larryisgood, Fences and windows, Mercurywoodrose, Quizimodo, Universalad-dress, PlanetStar, Steorra, Yintan, RW Marloe, 03jkeeley, ClueBot, Madcio, NovaDog, Solar-Wind, NuclearVacuum, Dthomsen8, WikHead, Alexius08, Good Olfactory, Addbot, Roentgenium111, 84user, Yobot, HieronymousCrowley, AnomieBOT, Citation bot, LilHelpa, Logos, Jabmsh, FrescoBot, Kafkadeeaf, Tom.Reding, December21st2012Freak, Tim1357, Steve03Mills, John of Reading, Jmencisom, Blm00, De-momoer, Brandmeister, EvenGreenerFish, BlackTarHeron, ChuispastonBot, NTox, Colapeninsula, Catlemur, Bibcode Bot, BG19bot, Tycho Magnetic Anomaly-1, Dexbot, 134340Goat, Astredita, Fench, Filedelinkerbot, Davidbuddy9, Tetra quark and Anonymous: 57

- **Planetary protection** *Source:* https://en.wikipedia.org/wiki/Planetary_protection?oldid=691909696 *Contributors:* The Anome, Julesd, Stone, Icairns, Aidan W, O'Dea, Florian Blaschke, Art LaPella, DanB~enwiki, Ricky81682, Oliphaunt, DanHobley, Drbogdan, Rjwilmsi, Veg-aswikian, 01101001, Bgwhite, Aeusoes1, Dantor, Robert McClenon, SmackBot, Chris the speller, Bluebot, Lucid-dream, Cydebot, DavidMc-Cabe, Robertinventor, Arb, Mpallen, Kent Witham, Ingolfson, Magioladitis, BatteryIncluded, Urco, Jim.henderson, Potatoswatter, Aagtbdfoua,

Celten, WarrenPlatts, Arjayay, Addbot, Gravitophoton, DOI bot, Nohomers48, Legobot, Luckas-bot, Yobot, AnomieBOT, Citation bot, Eumolpo, Steve Quinn, Citation bot 1, Tom.Reding, Fartherred, RjwilmsiBot, Arbnos, DinoSlider, ChiZeroOne, Whoop whoop pull up, Bibcode Bot, BG19bot, Hallows AG, Mogism, Happy-marmotte, Aqua817, Tetra quark and Anonymous: 27

- **Potential cultural impact of extraterrestrial contact** *Source:* https://en.wikipedia.org/wiki/Potential_cultural_impact_of_extraterrestrial_contact?oldid=693854637 *Contributors:* Imc, Everyking, Piotrus, Pie4all88, RJHall, Viriditas, Waldir, Rjwilmsi, Wavelength, Anomie, Raindrift, Srinivasasha, Inhighspeed, Portillo, The Gnome, Jprg1966, Modest Genius, Harryboyles, JorisvS, Mgiganteus1, Mr Stephen, Bobnorwal, Gregbard, Doug Weller, Mbrousseau, Casliber, Wikid77, Hut 8.5, Hamiltonstone, Ljgua124, WereSpielChequers, Stfg, Niceguyedc, Arjayay, Indopug, Gravitophoton, MightySaiyan, Blaylockjam10, David Klompas, Yobot, Againme, AnomieBOT, Bluerasberry, Citation bot, Jonesey95, Tom.Reding, Mjs1991, Trappist the monk, Diannaa, Zidanie5, MyMoloboaccount, RjwilmsiBot, Medeis, David J Johnson, Staszek Lem, JonRichfield, Gary Dee, ClueBot NG, Dru of Id, Helpful Pixie Bot, Bibcode Bot, FutureTrillionaire, ArticlesForCreationBot, Mathew-Townsend, Wer900, BattyBot, Stigmatella aurantiaca, AlexGraal, Khazar2, Jihadcola, Dexbot, Anderson, Cheerioswithmilk, Jamesx12345, Advanceddeepspacepropeller, Fixture, Monkbot, AntHerder and Anonymous: 20

- **Aurelia and Blue Moon** *Source:* https://en.wikipedia.org/wiki/Aurelia_and_Blue_Moon?oldid=665788536 *Contributors:* Bryan Derksen, AlainV, Barbara Shack, IRelayer, Robert Brockway, Latitude0116, Kuralyov, Icairns, Rich Farmbrough, Cyclopia, Art LaPella, Tritium6, DreamGuy, Kay Dekker, Angr, Firsfron, Woohookitty, Zzyzx11, Johnny Mnemonic, Cuchullain, Drbogdan, Ian Pitchford, Chobot, Ytrottier, Hellbus, Ilmaisin, Vicarious, SmackBot, Zazaban, TBH, Davewild, Chris the speller, Bluebot, Colonies Chris, Trekphiler, Tamfang, Moonsword, OrphanBot, Fuhghettaboutit, Kyuss-Apollo, Erimus, Coricus, Soumyasch, Joffeloff, Marhawkman, SubSeven, Joseph Solis in Australia, CmdrObot, Frizaven, Ksbrown, After Midnight, Arb, UberScienceNerd, Dawnseeker2000, Noclevername, E. A. Green, J. Langton, Alphachimpbot, Kaobear, Siddharth Mehrotra, Ben Ram, Frotz, Spellmaster, Saxophlute, R'n'B, CommonsDelinker, J.delanoy, Hans Dunkelberg, Johnny542, VTNC, Nikthestunned, VolkovBot, AllGloryToTheHypnotoad, PlanetStar, TheThingy, ImageRemovalBot, Joanie61, LAX, Bernd Jendrissek, Patricius Augustus, Ngebendi, DumZiBoT, SilvonenBot, Sugmullun, Addbot, Proxima Centauri, Lightbot, Luckas-bot, Legobot II, AnomieBOT, B.Lameira, FrescoBot, Aurelia19, Thorenn, HRoestBot, RedBot, DASHBot, ZéroBot, Westley Turner, Snotbot, Tideflat, ScottSteiner, Dream of Nyx, Chris the Paleontologist, Q6637p, MrBill3, 220 of Borg, Khazar2, Rob J. Elkton, Andyhowlett, Muskie72, Monkbot and Anonymous: 86

- **Kardashev scale** *Source:* https://en.wikipedia.org/wiki/Kardashev_scale?oldid=693890875 *Contributors:* The Epopt, Derek Ross, Bryan Derksen, The Anome, XJaM, SimonP, Shii, Heron, Hephaestos, Leandrod, Bdesham, Michael Hardy, Sannse, Gdvorsky, Karada, Skysmith, Chrishorrocks, Bogdangiusca, Nikai, Evercat, Denny, Charles Matthews, Time, Maximus Rex, Val42, Mignon~enwiki, Pakaran, Lumos3, AlainV, Astronautics~enwiki, Fredrik, Scott McNay, Fifelfoo, Peak, Merovingian, Meelar, Auric, Aetheling, Ruakh, GreatWhiteNortherner, David Gerard, Giftlite, DocWatson42, Laudaka, ShaunMacPherson, Wwoods, Everyking, Anville, Maver1ck, LockeShocke, Gracefool, Eequor, Andycjp, Gdr, Beland, Piotrus, Armaced, Balcer, One Salient Oversight, Cyopardi, Peter bertok, MakeRocketGoNow, Todd Kloos, Absinf, Guppyfinsoup, Alkivar, Freakofnurture, NathanHurst, Rich Farmbrough, Vague Rant, Kenj0418, Liso, Pmsyyz, Florian Blaschke, Azikala, Bender235, ESkog, Ben Standeven, Borofkin, BrokenSegue, Viriditas, Ctrl build, KarlHallowell, Mtreinik, Foant, Somepostman, CyberSkull, Inky, Rwoodsco, Cjthellama, Mac Davis, Stillnotelf, Mad Hatter, BRW, Rhialto, Geraldshields11, Gene Nygaard, K3rb, Alai, Dan East, Kardrak, Wyvern, Lebob (renamed), Dismas, Siafu, Bobrayner, WilliamKF, Woohookitty, Mindmatrix, Madmardigan53, Daniel Case, Oliphaunt, JFG, KevinOKeeffe, SCEhardt, MiG, M412k, Male1979, Pictureuploader, Marudubshinki, Behun, Magister Mathematicae, Jan van Male, Drbogdan, Rjwilmsi, Tim!, Zbxgscqf, Quiddity, Ligulem, Bubba73, FireCrack, FlaBot, Ground Zero, Stoph, JdforresterBot, Jrtayloriv, Goudzovski, Super Jamie, Zotel, Theshibboleth, Jiiling, WouterBot, Chobot, Lord Patrick, Visor, Wjfox2005, YurikBot, Vuvar1, Hairy Dude, Huw Powell, Raccoon Fox, Sasuke Sarutobi, Zelmerszoetrop, JihemD, Ksyrie, Alastair Houghton, Mipadi, Shultz, BertK, FoolsWar, Sandman1142, Andrewqsmith, Ilmaisin, JGoodman, SamuelRiv, Blurble, 2over0, N-Bot, Igglybuff, Nikkimaria, Chase me ladies, I'm the Cavalry, Arthur Rubin, CWenger, Shawnc, Ecnassianer, Geoffrey.landis, Caballero1967, ThunderBird, Nimbex, Mebden, Serendipodous, Xtraeme, SG, KnightRider~enwiki, SmackBot, Kurulanamfok, Senix, Zazaban, Bigbluefish, Vald, Davewild, RedSpruce, GraemeMcRae, Pretendo, Kinhull, Bluebot, Qwasty, TDS, Thumperward, Hibernian, Ranting Martian, Colonies Chris, Emurphy42, Rogermw, Foogod, Nima Baghaei, LouScheffer, GeorgeMoney, Elendil's Heir, Aktron, CanDo, Coolbho3000, John D. Croft, Memiux, AndyBQ, Bdiscoe, Cecil, Kreb Dragonrider, The idiot, Zaphraud, Sosodank, Loodog, Gobonobo, Mataobz, AstroChemist, The Frederick, A. Parrot, Zelaron, Hypnosifl, Intranetusa, JdH, DabMachine, Michaelbusch, Joseph Solis in Australia, Exander, Beno1000, Spongefan, ScottW, CRGreathouse, CmdrObot, N2e, Sahrin, Fordmadoxfraud, Stormwyrm, Malamockq, Cydebot, Mierlo, Vorlon19, Meno25, Alexnye, Teratornis, Rgbatduke, Omicronpersei8, Thirtysilver, Arb, Maziotis, Thijs!bot, Keraunos, Tobz1000, Jofishtrick, Electron9, Tgok, Nick Number, Elert, Paul from Michigan, Spartaz, Canadian-Bacon, Bschott, Ingolfson, HolyT, Barek, Skomorokh, Planetary, Vultur~enwiki, Lord Crayak, Extropian314, Alastair Haines, Just H, SyD!, Lord mrazon, Interrobamf, Falcor84, Johann1870, Osquar F, R n'B, CommonsDelinker, HEL, AltiusBimm, Rrostrom, Adavidb, Hom sepanta, Skier Dude, Murgatroid99, BTaronji, Mikemits42, Ronnmandreu, Jorfer, Mihilz, Ukt-zero, Sarregouset, DASonnenfeld, BernardZ, Izno, VolkovBot, Wavanova, ExarPalantas, RingtailedFox, Udufuduhu, Epsilon8998, Seraphim, JeremyBoggs, AllGloryToTheHypnotoad, Millanead, Lamro, Spinningspark, Seraphita~enwiki, SieBot, YonaBot, WereSpielChequers, Laoris, Yintan, Happysailor, Flyer22 Reborn, Sings-With-Spirits, Minihtam, Vanished user oij8h435jweih3, KathrynLybarger, Ctxppc, NPalmius, Hamiltondaniel, Dabomb87, Velvetron, Martarius, ClueBot, Deanlaw, Retsilla, Mild Bill Hiccup, Frito31382, James Kelvin, Epsilon60198, CohesionBot, Alexbot, Resoru, BobKawanaka, Cantbeatpie, Sun Creator, Kryptonian250, SchreiberBike, JasonAQuest, BOTarate, Dana boomer, Sparkygravity, Brianpeiris, DumZiBoT, BearblokeWiki, Zombie Hunter Smurf, Rickremember, Yunuswesley, Dthomsen8, JCDenton2052, Addbot, Roentgenium111, Imeriki al-Shimoni, DOI bot, Toyokuni3, MamaLuigiYTP, Glashoppah, Megapanphilos, Michaelwuzthere, Debresser, SpBot, Cmissy, FutureDragon, OlEnglish, MuZemike, ScienceApe, Ben Ben, PlankBot, NimblyPimbly, Luckas-bot, Yobot, Fraggle81, Legobot II, CK6569, Aedazan, Celloyd9785, Againme, AnomieBOT, Götz, Nmuselin, Jim1138, Livven, Citation bot, ArthurBot, LilHelpa, Xqbot, SouthH, Smk65536, Ouija2k, Justanothervisitor, Teddks, Crzer07, GrouchoBot, SassoBot, Brutaldeluxe, Wmeg2, Radioaktive, FrescoBot, Andrewhayes, Citation bot 1, Citation bot 4, Winterst, Jonesey95, Tom.Reding, Σ, Jeroen De Dauw, Araxhiel, Trappist the monk, Dinamik-bot, Vrenator, Geras2, Tech12, RjwilmsiBot, Tesseraet2, Piotrek54321, Eantonya, EmausBot, Peaceray, Deoxy99, TeleComNasSprVen, Hhhippo, Classedhaie, Westley Turner, H3llBot, AManWithNoPlan, Wingman417, Olekp, Lampsalot, Terraflorin, Sven Manguard, Whoop whoop pull up, RoboJIM, ClueBot NG, Wukai, Tideflat, Addlertod05, Kevin Gorman, Widr, Helpful Pixie Bot, Bibcode Bot, Sergeant Cribb, BG19bot, Juro2351, Pastaguy12, Peanutbutterrocks, Amisner2k, Akiatu, Harizotoh9, Kydon Shadow, Alicerce21, Darylgolden, Tonusamuel, Togatime, Dexbot, MennoH8472, TheIrishWarden, Berndagon, 069952497a, Alexlyoko13, Reatlas, Condorcraft110, Niketitan, Ekips39, Epicgenius, Curtiss29, Lsmll, Kreindeker, Slickricksweet16, ResearcherQ, Iwantfreebooks, ShephardLR, Nksor, FPSmike, Linuxjava, Hidehicampers,

- **File:201008-2a_PlanetOrbits_16x9-_Transit_timing_of_1-planet_vs_2-planet_systems.ogv** *Source:* https://upload.wikimedia.org/wikipedia/commons/9/96/201008-2a_PlanetOrbits_16x9-_Transit_timing_of_1-planet_vs_2-planet_systems.ogv *License:* Public domain *Contributors:* [http://kepler.nasa.gov/multimedia/animations/scienceconcepts/?ImageID=98 http://kepler.nasa.gov/multimedia/animations/scienceconcepts/?ImageID=98] *Original artist:* NASA Ames Research Center/Kepler Mission

- **File:2014_June_Astrobiology_and_Theology_seminer_01.JPG** *Source:* https://upload.wikimedia.org/wikipedia/commons/e/e3/2014_June_Astrobiology_and_Theology_seminer_01.JPG *License:* CC BY-SA 3.0 *Contributors:* Own work *Original artist:* Geraldshields11

- **File:444226main_exoplanet20100414-a-full.jpg** *Source:* https://upload.wikimedia.org/wikipedia/commons/9/97/444226main_exoplanet2.jpg *License:* Public domain *Contributors:* http://www.nasa.gov/topics/universe/features/exoplanet20100414-a.html *Original artist:* NASA/JPL-Caltech/Palomar Observatory

- {{int:Coll-image-attribution|File:951_Gaspra.jpg|https://upload.wikimedia.org/wikipedia/commons/8/81/951_Gaspra.jpg|Public domain|C from TIFF image from [http://www.solarviews.com/cap/ast/gaspra3.htm Solarviews.com|NASA}}

- **File:ADN_animation.gif** *Source:* https://upload.wikimedia.org/wikipedia/commons/8/81/ADN_animation.gif *License:* Public domain *Contributors:* Own work *Original artist:* brian0918™

- **File:ALH84001_structures.jpg** *Source:* https://upload.wikimedia.org/wikipedia/commons/a/a8/ALH84001_structures.jpg *License:* Public domain *Contributors:* http://web.archive.org/web/2/curator.jsc.nasa.gov/antmet/marsmets/alh84001/ALH84001-EM1.htm *Original artist:* NASA

- **File:A_Swarm_of_Ancient_Stars_-_GPN-2000-000930.jpg** *Source:* https://upload.wikimedia.org/wikipedia/commons/6/6a/A_Swarm_of_Ancient_Stars_-_GPN-2000-000930.jpg *License:* Public domain *Contributors:* Great Images in NASA Description *Original artist:* NASA, The Hubble Heritage Team, STScI, AURA

- **File:Absorption_spectrum_of_liquid_water.png** *Source:* https://upload.wikimedia.org/wikipedia/commons/1/18/Absorption_spectrum_of_liquid_water.png *License:* CC BY-SA 3.0 *Contributors:* I created this work myself. The data curve is based upon various reported values of water absorption in the literature. (E.g. see http://omlc.ogi.edu/spectra/water/abs/index.html for a list of references.) *Original artist:* Kebes (talk)

- **File:Aleksandr_Oparin_and_Andrei_Kursanov_in_enzymology_laboratory_1938.jpg** *Source:* https://upload.wikimedia.org/wikipedia/commons/f/f4/Aleksandr_Oparin_and_Andrei_Kursanov_in_enzymology_laboratory_1938.jpg *License:* Public domain *Contributors:* ? *Original artist:* ?

- **File:Ambox_current_red.svg** *Source:* https://upload.wikimedia.org/wikipedia/commons/9/98/Ambox_current_red.svg *License:* CC0 *Contributors:* self-made, inspired by Gnome globe current event.svg, using Information icon3.svg and Earth clip art.svg *Original artist:* Vipersnake151, penubag, Tkgd2007 (clock)

- **File:Ambox_important.svg** *Source:* https://upload.wikimedia.org/wikipedia/commons/b/b4/Ambox_important.svg *License:* Public domain *Contributors:* Own work, based off of Image:Ambox scales.svg *Original artist:* Dsmurat (talk - contribs)

- **File:Ambox_question.svg** *Source:* https://upload.wikimedia.org/wikipedia/commons/1/1b/Ambox_question.svg *License:* Public domain *Contributors:* Based on Image:Ambox important.svg *Original artist:* Mysid, Dsmurat, penubag

- **File:AminoAcidball.svg** *Source:* https://upload.wikimedia.org/wikipedia/commons/c/ce/AminoAcidball.svg *License:* Public domain *Contributors:* Own work *Original artist:* This vector image was created with Inkscape.

- **File:Amino_acids_1.png** *Source:* https://upload.wikimedia.org/wikipedia/commons/8/82/Amino_acids_1.png *License:* CC-BY-SA-3.0 *Contributors:* ? *Original artist:* ?

- **File:Amylose_3Dprojection.corrected.png** *Source:* https://upload.wikimedia.org/wikipedia/commons/7/71/Amylose_3Dprojection.corree.png *License:* Public domain *Contributors:* Own work *Original artist:* glycoform

- **File:AncientMars.jpg** *Source:* https://upload.wikimedia.org/wikipedia/commons/9/98/AncientMars.jpg *License:* CC BY-SA 3.0 *Contributors:* Own work *Original artist:* Ittiz

- **File:Arecibo_Observatory_Aerial_View.jpg** *Source:* https://upload.wikimedia.org/wikipedia/commons/c/cd/Arecibo_Observatory_Aerial_View.jpg *License:* Public domain *Contributors:* Transferred from en.wikipedia to Commons by Giro720 using CommonsHelper. *Original artist:* H. Schweiker/WIYN and NOAO/AURA/NSF.

  (The original uploader was Quazgaa at English Wikipedia.)

- **File:Arecibo_Observatory_Black_and_White.jpg** *Source:* https://upload.wikimedia.org/wikipedia/en/c/cb/Arecibo_Observatory_Black_and_White.jpg *License:* PD *Contributors:*

  http://history.nasa.gov/SP-419/pxv.htm *Original artist:*

  NASA

- **File:Arecibo_message.svg** *Source:* https://upload.wikimedia.org/wikipedia/commons/5/55/Arecibo_message.svg *License:* CC-BY-SA-3.0 *Contributors:* Own drawing, 2005 *Original artist:* Arne Nordmann (norro)

- **File:Artist'{}s_concept_of_collision_at_HD_172555.jpg** *Source:* https://upload.wikimedia.org/wikipedia/commons/4/4a/Artist%27s_concept_of_collision_at_HD_172555.jpg *License:* Public domain *Contributors:* http://www.nasa.gov/multimedia/imagegallery/image_feature_1454.html *Original artist:* NASA/JPL-Caltech

- **File:Artist_Concept_Planetary_System.jpg** *Source:* https://upload.wikimedia.org/wikipedia/commons/e/e6/Artist_Concept_Planetary_System.jpg *License:* Public domain *Contributors:* ? *Original artist:* The original uploader was 1981willy at English Wikipedia Later versions were up-loaded by WdyD at en.wikipedia.

- **File:Artist's_illustration_of_temperature_inversion_in_exoplanet's_atmosphere.jpg** *Source:* https://upload.wikimedia.org/wikipedia/commons/f/fa/Artist%E2%80%99s_illustration_of_temperature_inversion_in_exoplanet%E2%80%99s_atmosphere.jpg *License:* CC BY 3.0 *Contributors:* http://www.spacetelescope.org/images/opo1525a/ *Original artist:* NASA, ESA, and K. Haynes and A. Mandell (Goddard Space Flight Center)

- **File:Atacama.png** *Source:* https://upload.wikimedia.org/wikipedia/commons/7/7f/Atacama.png *License:* Public domain *Contributors:* ? *Original artist:* ?

- **File:Beagle_2_replica.jpg** *Source:* https://upload.wikimedia.org/wikipedia/commons/a/a8/Beagle_2_replica.jpg *License:* GFDL *Contributors:* Photo by user:geni *Original artist:* user:geni

- **File:Beta-D-Glucose.svg** *Source:* https://upload.wikimedia.org/wikipedia/commons/b/bb/Beta-D-Glucose.svg *License:* Public domain *Contributors:* Own work *Original artist:* Yikrazuul

- **File:Beta_Pictoris.jpg** *Source:* https://upload.wikimedia.org/wikipedia/commons/a/a9/Beta_Pictoris.jpg *License:* CC BY 4.0 *Contributors:* http://www.eso.org/public/images/eso1024a/ *Original artist:* ESO/A.-M. Lagrange

- **File:Blacksmoker_in_Atlantic_Ocean.jpg** *Source:* https://upload.wikimedia.org/wikipedia/commons/6/6f/Blacksmoker_in_Atlantic_Ocean.jpg *License:* Public domain *Contributors:* NOAA Photo Library *Original artist:* P. Rona

- **File:BlueMarble-2001-2002.jpg** *Source:* https://upload.wikimedia.org/wikipedia/commons/1/1c/BlueMarble-2001-2002.jpg *License:* Public domain *Contributors:* ? *Original artist:* ?

- **File:BrownDwarfs_Comparison_01.png** *Source:* https://upload.wikimedia.org/wikipedia/commons/2/23/BrownDwarfs_Comparison_01.png *License:* CC BY 3.0 *Contributors:* First published in "Joergens, Viki, 50 Years of Brown Dwarfs - From Prediction to Discovery to Forefront of Research, Astrophysics and Space Science Library 401, Springer, ISBN 978-3-319-01162-2." *Original artist:* MPIA/V. Joergens

- **File:Brown_dwarf_2M_J044144_and_planet.jpg** *Source:* https://upload.wikimedia.org/wikipedia/commons/d/d1/Brown_dwarf_2M_J044144_and_planet.jpg *License:* Public domain *Contributors:* Original from: HubbleSite.org *Original artist:* NASA

- **File:Buckminsterfullerene-perspective-3D-balls.png** *Source:* https://upload.wikimedia.org/wikipedia/commons/0/0f/Buckminsterfuller.png *License:* Public domain *Contributors:* Own work *Original artist:* Benjah-bmm27

- **File:C_G-K_-_DSC_0421.jpg** *Source:* https://upload.wikimedia.org/wikipedia/commons/0/0e/C_G-K_-_DSC_0421.jpg *License:* CC BY 3.0 *Contributors:* http://www.flickr.com/photos/cgk/1558787110/ *Original artist:* Colby Gutierrez-Kraybill

- **File:Calcidiscus_leptoporus_05.jpg** *Source:* https://upload.wikimedia.org/wikipedia/commons/b/b9/Calcidiscus_leptoporus_05.jpg *License:* CC BY 3.0 *Contributors:* Own work *Original artist:* Hannes Grobe (<a href='//commons.wikimedia.org/wiki/User_talk:Hgrobe' title='User talk:Hgrobe'>talk</a>), Alfred Wegener Institute

- **File:Carbon_Planet.JPG** *Source:* https://upload.wikimedia.org/wikipedia/commons/5/55/Carbon_Planet.JPG *License:* Public domain *Contributors:* Own work *Original artist:* Luyten

- **File:Ceres_Cutaway.jpg** *Source:* https://upload.wikimedia.org/wikipedia/commons/e/e5/Ceres_Cutaway.jpg *License:* Public domain *Contributors:* http://hubblesite.org/newscenter/newsdesk/archive/releases/2005/27/text/ *Original artist:* NASA, ESA, and A. Feild (STScI)

- **File:Champagne_vent_white_smokers.jpg** *Source:* https://upload.wikimedia.org/wikipedia/commons/a/aa/Champagne_vent_white_smokers.jpg *License:* Public domain *Contributors:* http://oceanexplorer.noaa.gov/explorations/04fire/logs/hirez/champagne_vent_hirez.jpg *Original artist:* NOAA

- **File:Circling_Two_Suns.ogv** *Source:* https://upload.wikimedia.org/wikipedia/commons/9/99/Circling_Two_Suns.ogv *License:* Public domain *Contributors:* Goddard Multimedia *Original artist:* NASA/Goddard Space Flight Center

- **File:Color_HD_189733b_vs_solar_system.jpg** *Source:* https://upload.wikimedia.org/wikipedia/commons/b/b1/Color_HD_189733b_vs_solar_system.jpg *License:* Public domain *Contributors:* http://www.nasa.gov/content/nasa-hubble-finds-a-true-blue-planet/#.UyWllYWnzZ5 *Original artist:* A. Feild

- **File:Commons-logo.svg** *Source:* https://upload.wikimedia.org/wikipedia/en/4/4a/Commons-logo.svg *License:* ? *Contributors:* ? *Original artist:* ?

- **File:Crab_Nebula.jpg** *Source:* https://upload.wikimedia.org/wikipedia/commons/0/00/Crab_Nebula.jpg *License:* Public domain *Contributors:* HubbleSite: gallery, release. *Original artist:* NASA, ESA, J. Hester and A. Loll (Arizona State University)

- **File:Crystal_energy.svg** *Source:* https://upload.wikimedia.org/wikipedia/commons/1/14/Crystal_energy.svg *License:* LGPL *Contributors:* Own work conversion of Image:Crystal_128_energy.png *Original artist:* Dhatfield

- **File:DNA_chemical_structure.svg** *Source:* https://upload.wikimedia.org/wikipedia/commons/e/e4/DNA_chemical_structure.svg *License:* CC-BY-SA-3.0 *Contributors:* <a href='//validator.w3.org/' data-x-rel='nofollow'><img alt='W3C' src='https://upload.wikimedia.org/wikipedia/commons/thumb/1/1a/Valid_SVG_1.1_%28green%29.svg/88px-Valid_SVG_1.1_%28green%29.svg.png' width='88' height='30' style='vertical-align: top' srcset='https://upload.wikimedia.org/wikipedia/commons/thumb/1/1a/Valid_SVG_1.1_%28green%29.svg/132px-Valid_SVG_1.1_%28green%29.svg.png 1.5x, https://upload.wikimedia.org/wikipedia/commons/thumb/1/1a/Valid_SVG_1.1_%28green%29.svg/176px-Valid_SVG_1.1_%28green%29.svg.png 2x' data-file-width='91' data-file-height='31' /></a>iThe source code of this SVG is <a data-x-rel='nofollow' class='external text' href='//validator.w3.org/check?uri=https%3A%2F%2Fcommons.wikimedia.org%2Fwiki%2FSpecial%3AFilepath%2FDNA_chemical_structure.svg,&,ss=1#source'>valid</a>. *Original artist:* Madprime (talk· contribs)

- **File:Darwin_restored2.jpg** *Source:* https://upload.wikimedia.org/wikipedia/commons/b/b6/Darwin_restored2.jpg *License:* Public domain *Contributors:* Library of Congress[1] *Original artist:* Elliott & Fry

- **File:David_A._Aguilar'{}s_Red_Dwarf_Stars.jpg** *Source:* https://upload.wikimedia.org/wikipedia/commons/7/72/David_A._Aguilar%27s_Red_Dwarf_Stars.jpg *License:* Public domain *Contributors:* http://www.nasa.gov/images/content/126852main_image_feature_401_ys_full.jpg from http://www.nasa.gov/multimedia/imagegallery/image_feature_401.html *Original artist:* David A. Aguilar (CfA)

- **File:De_Revolutionibus_manuscript_p9b.jpg** *Source:* https://upload.wikimedia.org/wikipedia/commons/e/e8/De_Revolutionibus_manuscript_p9b.jpg *License:* Public domain *Contributors:* www.bj.uj.edu.pl *Original artist:* Nicolas Copernicus

- **File:Dopspec-inline.gif** *Source:* https://upload.wikimedia.org/wikipedia/commons/6/65/Dopspec-inline.gif *License:* Public domain *Contributors:* I created this work entirely by myself. *Original artist:* Reyk

- **File:Dr._Frank_Drake.jpg** *Source:* https://upload.wikimedia.org/wikipedia/commons/6/69/Dr._Frank_Drake.jpg *License:* CC BY 2.0 *Contributors:* Flickr: Dr. Frank Drake *Original artist:* Raphael Perrino

- **File:Dyson_Sphere_Diagram-en.svg** *Source:* https://upload.wikimedia.org/wikipedia/commons/5/5f/Dyson_Sphere_Diagram-en.svg *License:* Public domain *Contributors:* Vector version of en:Image:Dyson_Sphere_Diagram.jpg by ed629 on the English Wikipedia *Original artist:* User:Bibi Saint-Pol

- **File:Dyson_Swarm_-_2.png** *Source:* https://upload.wikimedia.org/wikipedia/commons/b/ba/Dyson_Swarm_-_2.png *License:* CC BY 2.5 *Contributors:* Transferred from en.wikipedia to Commons. *Original artist:* The original uploader was Vedexent at English Wikipedia

- **File:EXPOSE_location_on_the_ISS.jpg** *Source:* https://upload.wikimedia.org/wikipedia/commons/d/da/EXPOSE_location_on_the_ISS.jpg *License:* Public domain *Contributors:* http://spaceflight.nasa.gov/gallery/images/shuttle/sts-124/html/s124e009982.html *Original artist:* NASA

- **File:Earth-moon.jpg** *Source:* https://upload.wikimedia.org/wikipedia/commons/5/5c/Earth-moon.jpg *License:* Public domain *Contributors:* NASA [1] *Original artist:* Apollo 8 crewmember Bill Anders

- **File:Earthlights_dmsp_1994–1995.jpg** *Source:* https://upload.wikimedia.org/wikipedia/commons/e/ea/Earthlights_dmsp_1994%E2%80%931995.jpg *License:* Public domain *Contributors:* http://eoimages.gsfc.nasa.gov/ve//1438/land_lights_16384.tif *Original artist:* Data courtesy Marc Imhoff of NASA GSFC and Christopher Elvidge of NOAA NGDC.

- **File:Earthlike_moon_extrasolar_gas_giant.jpg** *Source:* https://upload.wikimedia.org/wikipedia/commons/7/71/Earthlike_moon_extrasolar_gas_giant.jpg *License:* Public domain *Contributors:* http://planetquest.jpl.nasa.gov/images/NEWextrasolar-medium.jpgNASA] *Original artist:* ?

- **File:Eccentric_Habitable_Zones.jpg** *Source:* https://upload.wikimedia.org/wikipedia/commons/0/00/Eccentric_Habitable_Zones.jpg *License:* Public domain *Contributors:* Eccentric Habitable Zones *Original artist:* NASA/JPL-Caltech

- **File:Edit-clear.svg** *Source:* https://upload.wikimedia.org/wikipedia/en/f/f2/Edit-clear.svg *License:* Public domain *Contributors:* The *Tango! Desktop Project. Original artist:*

  The people from the Tango! project. And according to the meta-data in the file, specifically: "Andreas Nilsson, and Jakub Steiner (although minimally)."

- **File:Enceladus_Cold_Geyser_Model.svg** *Source:* https://upload.wikimedia.org/wikipedia/commons/7/76/Enceladus_Cold_Geyser_Model.svg *License:* CC BY-SA 3.0 *Contributors:* Own work. Derived from <a href=//commons.wikimedia.org/wiki/File:PIA07799.png class='image'><img alt='PIA07799.png' src='https://upload.wikimedia.org/wikipedia/commons/thumb/9/9e/PIA07799.png/25px-PIA07799.png' width='25' height='19'srcset='https://upload.wikimedia.org/wikipedia/commons/thumb/9/9e/PIA07799.png/38px-PIA07799.png 1.5x, https://upload.wikimedia.org/wikipedia/commons/thumb/9/9e/PIA07799.png/50px-PIA07799.png 2x' data-file-width='960' data-file-height='720' /></a>File:PIA07799.png by GPHemsley(en.wp), released under PD-USGov-NASA. The original image was sourced from http://saturn.jpl.nasa.gov/photos/ of theThe National Aeronautics and Space Administration'sJet Propulsion LaboratoryandSpace Science Institute. Coloured usingInkscape.*Original artist:* cflm(talk)

- **File:Enrico_Fermi_1943-49.jpg** *Source:* https://upload.wikimedia.org/wikipedia/commons/d/d4/Enrico_Fermi_1943-49.jpg *License:* Public domain *Contributors:* This media is available in the holdings of the National Archives and Records Administration, cataloged under the ARC Identifier (National Archives Identifier) 558578. *Original artist:* Department of Energy, Office of Public Affairs

- **File:Estimated_extent_of_the_Solar_Systems_habitable_zone.png** *Source:* https://upload.wikimedia.org/wikipedia/commons/7/7b/Estimated_extent_of_the_Solar_Systems_habitable_zone.png *License:* CC BY-SA 3.0 *Contributors:* Own work (Original text: I (EvenGreenerFish(talk)) created this work entirely by myself.) *Original artist:* EvenGreenerFishatEnglish Wikipedia

- **File:Ethanol-3D-balls.png** *Source:* https://upload.wikimedia.org/wikipedia/commons/b/b0/Ethanol-3D-balls.png *License:* Public domain *Contributors:* ? *Original artist:* ?

- **File:Europa-moon.jpg** *Source:* https://upload.wikimedia.org/wikipedia/commons/5/54/Europa-moon.jpg *License:* Public domain *Contributors:* http://photojournal.jpl.nasa.gov/catalog/PIA00502 (TIFF image link) *Original artist:* NASA/JPL/DLR

- **File:EuropaInterior1.jpg** *Source:* https://upload.wikimedia.org/wikipedia/commons/8/88/EuropaInterior1.jpg *License:* Public domain *Contributors:* Transferred from en.wikipedia to Commons. Transfer was stated to be made by User:Mu301.

  (Original text : *NASA's Planetary Photojournal, PIA01669;*

  Original artist: The original uploader was Latitude0116 at English Wikipedia

- **File:Evolsex-dia1a.png** *Source:* https://upload.wikimedia.org/wikipedia/commons/f/fc/Evolsex-dia1a.png *License:* CC-BY-SA-3.0 *Contributors:* ? *Original artist:* ?

- **File:ExoMars_model_at_ILA_2006.jpg** *Source:* https://upload.wikimedia.org/wikipedia/commons/4/46/ExoMars_model_at_ILA_2006.jpg *License:* CC-BY-SA-3.0 *Contributors:* created by Thomas Hagemeyer(User Topper81 on german Wikipedia) *Original artist:* Thomas Hagemeyer(User Topper81 on german Wikipedia)

- **File:Exoplanet_Comparison_Kepler-10_c.png** *Source:* https://upload.wikimedia.org/wikipedia/commons/2/20/Exoplanet_Comparison_c.png *License:* CC BY-SA 3.0 *Contributors:* Own work, incorporating public domain images for reference planets (see below), inspired by Thingg's size comparison *Original artist:* Aldaron, a.k.a. Aldaron

- **File:Kepler-452b_System.jpg** *Source:* https://upload.wikimedia.org/wikipedia/commons/2/23/Kepler-452b_System.jpg *License:* Public do-main *Contributors:* http://www.nasa.gov/press-release/nasa-kepler-mission-discovers-bigger-older-cousin-to-earth *Original artist:* NASA

- **File:Kepler186f-ComparisonGraphic-20140417_improved.jpg** *Source:* https://upload.wikimedia.org/wikipedia/commons/c/e9/0417_improved.jpg *License:* Public domain *Contributors:* This file was derived from Kepler186f-ComparisonGraphic-20140417.jpg: <a href= //commons.wikimedia.org/wiki/File:Kepler186f-ComparisonGraphic-20140417.jpg'class='image'><img alt='Kepler186f-Comparison Graphic-20140417.jpg'src='https://upload.wikimedia.org/wikipedia/commons/thumb/0/02/Kepler186f-ComparisonGraphic-2014041 7.jpg/150px-Kepler186f-ComparisonGraphic-20140417.jpg'width='150'height='84'srcset='https://upload.wikimedia.org/wikipedia/ commons/thumb/0/02/Kepler186f-ComparisonGraphic-20140417.jpg/225px-Kepler186f-ComparisonGraphic-20140417.jpg1.5x, https://upload.wikimedia.org/wikipedia/commons/thumb/0/02/Kepler186f-ComparisonGraphic-20140417.jpg/300px-Kepler186f-ComparisonGraphic-20140417.jpg2x'data-file-width='4534'data-file-height='2550' /></a> *Original artist:* NASA Ames/SETI Institute/

- **File:Keplerspacecraft-FocalPlane-cutout.svg** *Source:* https://upload.wikimedia.org/wikipedia/commons/2/2a/Keplerspacecraft-FocalP svg *License:* Public domain *Contributors:* National Space Agency *Original artist:* Dr. David Koch, Kepler Deputy Principal Investigator

- **File:Liquid_lakes_on_titan.jpg** *Source:* https://upload.wikimedia.org/wikipedia/commons/4/4b/Liquid_lakes_on_titan.jpg *License:* Public domain *Contributors:* http://photojournal.jpl.nasa.gov/catalog/PIA09102 *Original artist:* NASA / JPL-Caltech / USGS

- **File:LombergA1024.jpg** *Source:* https://upload.wikimedia.org/wikipedia/commons/b/be/LombergA1024.jpg *License:* Public domain *Contributors:* http://kepler.nasa.gov/images/LombergA1600-full.jpeg *Original artist:* Painting by Jon Lomberg, Kepler mission diagram added by NASA.

- **File:Los_Alamos_aerial_view.jpeg** *Source:* https://upload.wikimedia.org/wikipedia/commons/4/4d/Los_Alamos_aerial_view.jpeg *License:* Public domain *Contributors:* http://www.lanl.gov/worldview/news/photos/aerials.shtml *Original artist:* Los Alamos National Laboratory

- **File:Lowell_Mars_channels.jpg** *Source:* https://upload.wikimedia.org/wikipedia/commons/f/f3/Lowell_Mars_channels.jpg *License:* Public domain *Contributors:* Яков Перельман - "Далёкие миры", СПб, типография Сойкина (**English transliteration**: Yakov Perelman - "Distant Worlds", St. Petersburg, Soykin printing house), 1914. *Original artist:* Percival Lowell

- **File:Magnetosphere_rendition.jpg** *Source:* https://upload.wikimedia.org/wikipedia/commons/f/f3/Magnetosphere_rendition.jpg *License:* Public domain *Contributors:* http://sec.gsfc.nasa.gov/popscise.jpg *Original artist:* NASA

- **File:Marciano_Genérico.JPG** *Source:* https://upload.wikimedia.org/wikipedia/commons/c/c6/Marciano_Gen%C3%A9rico.JPG *License:* CC0 *Contributors:* Own work *Original artist:* Erechel

- **File:Mars_sunset_PIA00920.jpg** *Source:* https://upload.wikimedia.org/wikipedia/commons/5/50/Mars_sunset_PIA00920.jpg *License:* Public domain *Contributors:* ? *Original artist:* ?

- **File:Masses_of_terrestrial_planets.png** *Source:* https://upload.wikimedia.org/wikipedia/commons/9/93/Masses_of_terrestrial_planets.png *License:* CC BY-SA 3.0 *Contributors:* Transferred from en.wikipedia; transferred to Commons by User:Sir48 using CommonsHelper. *Original artist:* kwami (talk) Original uploader was Kwamikagami at en.wikipedia

- **File:Matrix_sphere.jpg** *Source:* https://upload.wikimedia.org/wikipedia/commons/f/fa/Matrix_sphere.jpg *License:* CC0 *Contributors:* Own work *Original artist:* Medie463

- **File:Methane-2D-stereo.svg** *Source:* https://upload.wikimedia.org/wikipedia/commons/9/92/Methane-2D-stereo.svg *License:* Public domain *Contributors:* Own work *Original artist:* SVG version by Patricia.fidi

- **File:Morgan-Keenan_spectral_classification.png** *Source:* https://upload.wikimedia.org/wikipedia/commons/8/8b/Morgan-Keenan_spect classification.png *License:* CC-BY-SA-3.0 *Contributors:* ? *Original artist:* ?

- **File:Msl20110519_PIA14156-full.jpg** *Source:* https://upload.wikimedia.org/wikipedia/commons/a/a9/Mars_Science_Laboratory_Curiosity_rover.jpg *License:* Public domain *Contributors:* http://marsprogram.jpl.nasa.gov/msl/multimedia/images/?ImageID=3504 *Original artist:* NASA

- **File:Myoglobin.png** *Source:* https://upload.wikimedia.org/wikipedia/commons/6/60/Myoglobin.png *License:* Public domain *Contributors:* self made based on PDB entry *Original artist:* →<a href='//commons.wikimedia.org/wiki/User:AzaToth' title='User:AzaToth'>Az</a>,><a href='//commons.wikimedia. org/wiki/User:AzaToth' title='User:AzaToth'>Tot</a>

- **File:NASA-KeplerSpaceTelescope-ArtistConcept-20141027.jpg** *Source:* https://upload.wikimedia.org/wikipedia/commons/9/91/NASA jpg *License:* Public domain *Contributors:* http://www.nasa.gov/sites/default/files/transits2_on_starfield_12x7-med.jpg *Original artist:* NASA Ames/ W Stenzel

- **File:NASA-SETI-Sensitivity.jpg** *Source:* https://upload.wikimedia.org/wikipedia/commons/0/0e/NASA-SETI-Sensitivity.jpg *License:* Public domain *Contributors:* http://history.nasa.gov/CP-2156/ch5.5.htm *Original artist:* John H. Wolfe, Robert E. Edelson, John Billingham, R. Bruce Crow, Samuel Gulkis, Edward T. Olsen, Bernard M. Oliver, Allen M. Peterson, Charles L. Seeger, Jill C. Tarter.

- **File:NGC_7331_zoomed.jpg** *Source:* https://upload.wikimedia.org/wikipedia/commons/d/d8/NGC_7331_zoomed.jpg *License:* Public domain *Contributors:* ? *Original artist:* ?

- **File:NH-Pluto-SputnikPlanum-HillaryMontes-NorgayMontes-20150714.jpg** *Source:* https://upload.wikimedia.org/wikipedia/commons/ a/ad/NH-Pluto-SputnikPlanum-HillaryMontes-NorgayMontes-20150714.jpg *License:* Public domain *Contributors:* http://www.nasa.gov/sites/ default/files/thumbnails/image/nh_04_mckinnon_02b.jpg *Original artist:* NASA/JHUAPL/SwRI

- **File:NewKeplerPlanetCandidates-20150723.jpg** *Source:* https://upload.wikimedia.org/wikipedia/commons/5/5a/NewKeplerPlanetCandi jpg *License:* Public domain *Contributors:* http://www.nasa.gov/sites/default/files/thumbnails/image/fig10-new_kepler_planet_cand.jpg *Original artist:* NASA Ames/W. Stenzel

- **File:Nitrous-oxide-3D-balls.png** *Source:* https://upload.wikimedia.org/wikipedia/commons/9/93/Nitrous-oxide-3D-balls.png *License:* Public domain *Contributors:* Own work *Original artist:* Ben Mills

- **File:Nuvola_apps_kcmsystem.svg** *Source:* https://upload.wikimedia.org/wikipedia/commons/7/7a/Nuvola_apps_kcmsystem.svg *License:* L. *Contributors:* Own work based on Image:Nuvola apps kcmsystem.png by Alphax originally from [1] *Original artist:* MesserWoland

- **File:OGLE-2005-BLG-390Lb_planet.jpg** *Source:* https://upload.wikimedia.org/wikipedia/commons/6/6b/OGLE-2005-BLG-390Lb_planet. jpg *License:* Public domain *Contributors:* http://www.nasa.gov/topics/universe/features/exoplanetHouseOfHorrors.html *Original artist:* NASA

- **File:Office-book.svg** *Source:* https://upload.wikimedia.org/wikipedia/commons/a/a8/Office-book.svg *License:* Public domain *Contributors:* This and myself. *Original artist:* Chris Down/Tango project

- **File:Open_Access_logo_PLoS_transparent.svg** *Source:* https://upload.wikimedia.org/wikipedia/commons/7/77/Open_Access_logo_PLoS_transparent.svg *License:* CC0 *Contributors:* http://www.plos.org/ *Original artist:* art designer at PLoS, modified by Wikipedia users Nina, Beao, and JakobVoss

- **File:Operation_Upshot-Knothole_-_Badger_001.jpg** *Source:* https://upload.wikimedia.org/wikipedia/commons/7/79/Operation_Upshot--Badger_001.jpg *License:* Public domain *Contributors:* This image is available from the National Nuclear Security Administration Nevada Site Office Photo Library under number XX-34. *Original artist:* Federal Government of the United States

- **File:Orbit3.gif** *Source:* https://upload.wikimedia.org/wikipedia/commons/5/59/Orbit3.gif *License:* Public domain *Contributors:* Own work *Original artist:* User:Zhatt

- **File:Orbits_of_some_Kepler_Planetary_Systems.jpg** *Source:* https://upload.wikimedia.org/wikipedia/en/7/7e/Orbits_of_some_Kepler_Planetary_Systems.jpg *License:* PD *Contributors:*
  http://www.jpl.nasa.gov/spaceimages/details.php?id=PIA15264 *Original artist:*
  NASA

- **File:Outer_Space_Treaty-SVG.svg** *Source:* https://upload.wikimedia.org/wikipedia/commons/1/1b/Outer_Space_Treaty-SVG.svg *License:* Public domain *Contributors:*

- Map: File:BlankMap-World-Microstates.svg *Original artist:* WillemBK (talk)

- **File:PIA01130_Interior_of_Europa.jpg** *Source:* https://upload.wikimedia.org/wikipedia/commons/7/7b/PIA01130_Interior_of_Europa.jpg *License:* Public domain *Contributors:* http://photojournal.jpl.nasa.gov/catalog/PIA01130 *Original artist:* unknown author of the NASA

- **File:PIA17934-MartianSlope-SeasonalDarkFlows-20140210.jpg** *Source:* https://upload.wikimedia.org/wikipedia/commons/e/ee/PIA17 jpg *License:* Public domain *Contributors:* http://photojournal.jpl.nasa.gov/jpeg/PIA17934.jpg *Original artist:* NASA/JPL-Caltech/UA/JHU-APL

- **File:PIA18410-TitanSunsetStudies-CassiniSpacecraft-20140527.jpg** *Source:* https://upload.wikimedia.org/wikipedia/commons/d/d3/P jpg *License:* Public domain *Contributors:* http://photojournal.jpl.nasa.gov/jpeg/PIA18410.jpg *Original artist:* NASA/JPL-Caltech

- **File:PIA19088-MarsCuriosityRover-MethaneSource-20141216.png** *Source:* https://upload.wikimedia.org/wikipedia/commons/6/69/P png *License:* Public domain *Contributors:* http://mars.jpl.nasa.gov/msl/images/methane-source-mars-rover-curiosity-pia19088-full.jpg *Original artist:* NASA/JPL-Caltech

- **File:PIA19656-SaturnMoon-Enceladus-Ocean-ArtConcept-20150915.jpg** *Source:* https://upload.wikimedia.org/wikipedia/commons/7/70/PIA19656-SaturnMoon-Enceladus-Ocean-ArtConcept-20150915.jpg *License:* Public domain *Contributors:* http://photojournal.jpl.nasa.gov/figures/PIA19656_fig1.jpg *Original artist:* NASA/JPL-Caltech

- **File:PIA19827-Kepler-SmallPlanets-HabitableZone-20150723.jpg** *Source:* https://upload.wikimedia.org/wikipedia/commons/f/fe/PIA jpg *License:* Public domain *Contributors:* http://photojournal.jpl.nasa.gov/jpeg/PIA19827.jpg *Original artist:* NASA/Ames/JPL-Caltech

- **File:Panspermie.svg** *Source:* https://upload.wikimedia.org/wikipedia/commons/4/48/Panspermie.svg *License:* CC BY-SA 3.0 *Contributors:* Own work; For the proto-bacteria I used an adapted version of File:Bacteria-.svg by JrPol and for the DNA I used an adapted version of File:DNA chemical structure.svg by Madprime. Earth from File:Earth Flag.svg by Himasaram *Original artist:* Silver Spoon Sokpop

- **File:Parkes.arp.750pix.jpg** *Source:* https://upload.wikimedia.org/wikipedia/commons/4/4b/Parkes.arp.750pix.jpg *License:* Copyrighted free use *Contributors:* ? *Original artist:* ?

- **File:People_icon.svg** *Source:* https://upload.wikimedia.org/wikipedia/commons/3/37/People_icon.svg *License:* CC0 *Contributors:* OpenClipart *Original artist:* OpenClipart

- **File:Phospholipids_aqueous_solution_structures.svg** *Source:* https://upload.wikimedia.org/wikipedia/commons/c/c6/Phospholipids_solution_structures.svg *License:* Public domain *Contributors:* Own work *Original artist:* Mariana Ruiz Villarreal ,LadyofHats

- **File:Phylogenic_Tree-en.svg** *Source:* https://upload.wikimedia.org/wikipedia/commons/5/58/Phylogenic_Tree-en.svg *License:* CC BY-SA 3.0 *Contributors:* This file was derived fromPhylogenic Tree.jpg: <a href='//commons.wikimedia.org/wiki/File:Phylogenic_Tree.jpg' class='image'><img alt='Phylogenic Tree.jpg' src='https://upload.wikimedia.org/wikipedia/commons/thumb/d/de/Phylogenic_Tree.jpg/50px-Phylogenic_Tree.jpg' width='50' height='33' srcset='https://upload.wikimedia.org/wikipedia/commons/thumb/d/de/Phylogenic_Tree.jpg/75px-Phylogenic_Tree.jpg 1.5x, https://upload.wikimedia.org/wikipedia/commons/thumb/d/de/Phylogenic_Tree.jpg/100px-Phylogenic_Tree.jpg 2x' data-file-width='1069' data-file-height='713' /></a> *Original artist:* Phylogenic_Tree.jpg: John D. Croft

- **File:Pikaia_gracilens_B.jpg** *Source:* https://upload.wikimedia.org/wikipedia/commons/e/e8/Pikaia_gracilens_B.jpg *License:* CC BY-SA 4.0 *Contributors:* Own work *Original artist:* Apokryltaros

- **File:Pioneer10-plaque_tilt.jpg** *Source:* https://upload.wikimedia.org/wikipedia/commons/f/f5/Pioneer10-plaque_tilt.jpg *License:* Public domain *Contributors:* Ames Pioneer 10 *Original artist:* Designed by Carl Sagan and Frank Drake. Artwork prepared by Linda Salzman Sagan. Photograph by NASA Ames Resarch Center (NASA-ARC)

- **File:Plagiomnium_affine_laminazellen.jpeg** *Source:* https://upload.wikimedia.org/wikipedia/commons/4/49/Plagiomnium_affine_lamin jpeg *License:* CC-BY-SA-3.0 *Contributors:* photographed by myself *Original artist:* Kristian Peters -- Fabelfroh

- **File:Planet_Discovery_Neighbourhood_in_Milky_Way_Galaxy.jpeg** *Source:* https://upload.wikimedia.org/wikipedia/commons/0/05/Planet_Discovery_Neighbourhood_in_Milky_Way_Galaxy.jpeg *License:* Public domain *Contributors:* http://planetquest.jpl.nasa.gov/atlas/images/galaxy-graphic.jpg *Original artist:* NASAJet Propulsion Laboratory

- **File:Planet_sizes.svg** *Source:* https://upload.wikimedia.org/wikipedia/commons/f/f6/Planet_sizes.svg *License:* Public domain *Contributors:*
- Planetsizes.jpg *Original artist:* Planetsizes.jpg: Marc Kuchner/NASA GSFC
- **File:PlanetaryHabitability.ogg** *Source:* https://upload.wikimedia.org/wikipedia/commons/8/82/PlanetaryHabitability.ogg *License:* CC-BY-SA-3.0 *Contributors:*
- Derivative of Planetary_habitability *Original artist:* **Speaker:** Mrdallaway
  Authors of the article
- **File:Planets2013.jpg** *Source:* https://upload.wikimedia.org/wikipedia/commons/a/a9/Planets2013.jpg *License:* CC BY-SA 3.0 *Contributors:* Planets2008.jpg *Original artist:* WP
- **File:Planets_everywhere_(artist's_impression).jpg** *Source:* https://upload.wikimedia.org/wikipedia/commons/9/9b/Planets_everywhere_%28artist%E2%80%99s_impression%29.jpg *License:* CC BY 4.0 *Contributors:* http://www.eso.org/public/images/eso1204a/ *Original artist:* ESO/M. Kornmesser
- **File:Polycyclic_Aromatic_Hydrocarbons.png** *Source:* https://upload.wikimedia.org/wikipedia/commons/e/e0/Polycyclic_Aromatic_Hyd png *License:* Public domain *Contributors:* Own work by uploader, Accelrys DS Visualizer *Original artist:* Inductiveload
- **File:Portal-puzzle.svg** *Source:* https://upload.wikimedia.org/wikipedia/en/f/fd/Portal-puzzle.svg *License:* Public domain *Contributors:* ? *Original artist:* ?
- **File:Porto_Covo_February_2009-2.jpg** *Source:* https://upload.wikimedia.org/wikipedia/commons/9/93/Porto_Covo_February_2009-2.jpg *License:* CC BY-SA 3.0 *Contributors:* Own work *Original artist:* Alvesgaspar
- **File:Pythagorean.svg** *Source:* https://upload.wikimedia.org/wikipedia/commons/d/d2/Pythagorean.svg *License:* CC-BY-SA-3.0 *Contributors:* Transwikied from en:. Originally created by en:User:Michael Hardy, then scaled, with colour and labels being added by en:User:Wapcaplet, transformed in svg format by fr:Utilisateur:Steff, changed colors and font by de:Leo2004 *Original artist:* en:User:Wapcaplet
- **File:Question_book-new.svg** *Source:* https://upload.wikimedia.org/wikipedia/en/9/99/Question_book-new.svg *License:* Cc-by-sa-3.0 *Contributors:*
  Created from scratch in Adobe Illustrator. Based on Image:Question book.png created by User:Equazcion *Original artist:*
  Tkgd2007
- **File:RocketSunIcon.svg** *Source:* https://upload.wikimedia.org/wikipedia/commons/d/d6/RocketSunIcon.svg *License:* Copyrighted free use *Contributors:* Self made, based on File:Spaceship and the Sun.jpg *Original artist:* Me
- **File:SETI@Home_Logo.svg** *Source:* https://upload.wikimedia.org/wikipedia/en/0/04/SETI%40Home_Logo.svg *License:* Fair use *Contributors:*
  The logo may be obtained from Seti@home.
  Original artist: ?
- **File:SETI@home_Multi-Beam_screensaver.png** *Source:* https://upload.wikimedia.org/wikipedia/commons/b/b4/SETI%40home_Multi-screensaver.png *License:* LGPL *Contributors:* self make *Original artist:* Namazu-tron
- **File:STS-46_EURECA_deployment.jpg** *Source:* https://upload.wikimedia.org/wikipedia/commons/b/bf/STS-46_EURECA_deployment.jpg *License:* Public domain *Contributors:* NASA http://images.jsc.nasa.gov/luceneweb/caption.jsp?photoId=STS046-08-010 *Original artist:* NASA
- **File:Sagan_Viking.jpg** *Source:* https://upload.wikimedia.org/wikipedia/commons/e/e8/Sagan_Viking.jpg *License:* Public domain *Contributors:* http://solarsystem.nasa.gov/multimedia/display.cfm?IM_ID=244 *Original artist:* JPL
- **File:Schematic_relationship_between_biochemistry,_genetics_and_molecular_biology.svg** *Source:* https://upload.wikimedia.org/wikipedia/commons/2/25/Schematic_relationship_between_biochemistry%2C_genetics_and_molecular_biology.svg *License:* CC BY 2.5 *Contributors:* No machine-readable source provided. Own work assumed (based on copyright claims). *Original artist:* No machine-readable author provided. OldakQuill assumed (based on copyright claims).
- **File:Size_of_Kepler_Planet_Candidates.jpg** *Source:* https://upload.wikimedia.org/wikipedia/commons/c/cc/Size_of_Kepler_Planet_Candidates.jpg *License:* Public domain *Contributors:* http://www.nasa.gov/content/nasa-kepler-results-usher-in-a-new-era-of-astronomy *Original artist:* NASA
- **File:Size_planets_comparison.jpg** *Source:* https://upload.wikimedia.org/wikipedia/commons/3/3c/Size_planets_comparison.jpg *License:* CC BY-SA 3.0 *Contributors:* Own work *Original artist:* Lsmpascal
- **File:Sound-icon.svg** *Source:* https://upload.wikimedia.org/wikipedia/commons/4/47/Sound-icon.svg *License:* LGPL *Contributors:* Derivative work from Silsor's versio *Original artist:* Crystal SVG icon set
- **File:Spinoloricus.png** *Source:* https://upload.wikimedia.org/wikipedia/commons/b/b4/Spinoloricus.png *License:* CC BY 2.0 *Contributors:* Danovaro R., Dell Anno A., Pusceddu A., Gambi C., Heiner I. & Kristensen R. M. (2010). "The first metazoa living in permanently anoxic conditions". *BMC Biology* **8**: 30. doi:10.1186/1741-7007-8-30. Imported in 300dpi from http://www.biomedcentral.com/content/pdf/1741-7007-8-30.pdf Figure 1c, retouched. *Original artist:* Roberto Danovaro, Antonio Dell Anno, Antonio Pusceddu, Cristina Gambi, Iben Heiner & Reinhardt Mobjerg Kristensen
- **File:Spinvelocity-vs-mass-BetaPicb-and-solarsystem.jpg** *Source:* https://upload.wikimedia.org/wikipedia/commons/a/ad/Spinvelocity-jpg *License:* CC BY 4.0 *Contributors:* http://www.eso.org/public/images/eso1414b/ *Original artist:* ESO/I. Snellen (Leiden University)
- **File:Stromatolites.jpg** *Source:* https://upload.wikimedia.org/wikipedia/commons/e/e0/Stromatolites.jpg *License:* Public domain *Contributors:* National Park Service - http://www.nature.nps.gov/geology/cfprojects/photodb/Photo_Detail.cfm?PhotoID=204 *Original artist:* P. Carrara, NPS

- **File:Wikinews-logo.svg** *Source:* https://upload.wikimedia.org/wikipedia/commons/2/24/Wikinews-logo.svg *License:* CC BY-SA 3.0 *Contributors:* This is a cropped version of Image:Wikinews-logo-en.png. *Original artist:* Vectorized by Simon 01:05, 2 August 2006 (UTC) Updated by Time3000 17 April 2007 to use official Wikinews colours and appear correctly on dark backgrounds. Originally uploaded by Simon.

- **File:Wikiquote-logo.svg** *Source:* https://upload.wikimedia.org/wikipedia/commons/f/fa/Wikiquote-logo.svg *License:* Public domain *Contributors:* ? *Original artist:* ?

- **File:Wikisource-logo.svg** *Source:* https://upload.wikimedia.org/wikipedia/commons/4/4c/Wikisource-logo.svg *License:* CC BY-SA 3.0 *Contributors:* Rei-artur *Original artist:* Nicholas Moreau

- **File:Wikiversity-logo-Snorky.svg** *Source:* https://upload.wikimedia.org/wikipedia/commons/1/1b/Wikiversity-logo-en.svg *License:* CC BY-SA 3.0 *Contributors:* Own work *Original artist:* Snorky

- **File:Wikiversity-logo.svg** *Source:* https://upload.wikimedia.org/wikipedia/commons/9/91/Wikiversity-logo.svg *License:* CC BY-SA 3.0 *Contributors:* Snorky (optimized and cleaned up by verdy_p) *Original artist:* Snorky (optimized and cleaned up by verdy_p)

- **File:Wiktionary-logo-en.svg** *Source:* https://upload.wikimedia.org/wikipedia/commons/f/f8/Wiktionary-logo-en.svg *License:* Public domain *Contributors:* Vector version of Image:Wiktionary-logo-en.png. *Original artist:* Vectorized by Fvasconcellos (talk · contribs), based on original logo tossed together by Brion Vibber

- **File:Wow_signal.jpg** *Source:* https://upload.wikimedia.org/wikipedia/commons/d/d3/Wow_signal.jpg *License:* Public domain *Contributors:* http://www.bigear.org/Wow30th/wow30th.htm *Original artist:* Credit: The Ohio State University Radio Observatory and the North American AstroPhysical Observatory (NAAPO).

- **File:Wow_signal_location.jpg** *Source:* https://upload.wikimedia.org/wikipedia/commons/6/69/Wow_signal_location.jpg *License:* CC BY-SA 3.0 *Contributors:* Transferred from en.wikipedia
Original artist: Benjamin Crowell. Original uploader was Fashionslide at en.wikipedia

- **File:Wow_signal_profile.svg** *Source:* https://upload.wikimedia.org/wikipedia/commons/1/15/Wow_signal_profile.svg *License:* CC BY-SA 3.0 *Contributors:* Own work *Original artist:* Maxrossomachin